Lecture Notes in Networks and Systems

Volume 316

The series "Lecture Notes in Networks and Systems" publishes the latest developments in Networks and Systems—quickly, informally and with high quality. Original research reported in proceedings and post-proceedings represents the core of LNNS.

Volumes published in LNNS embrace all aspects and subfields of, as well as new challenges in, Networks and Systems.

The series contains proceedings and edited volumes in systems and networks, spanning the areas of Cyber-Physical Systems, Autonomous Systems, Sensor Networks, Control Systems, Energy Systems, Automotive Systems, Biological Systems, Vehicular Networking and Connected Vehicles, Aerospace Systems, Automation, Manufacturing, Smart Grids, Nonlinear Systems, Power Systems, Robotics, Social Systems, Economic Systems and other. Of particular value to both the contributors and the readership are the short publication timeframe and the world-wide distribution and exposure which enable both a wide and rapid dissemination of research output.

The series covers the theory, applications, and perspectives on the state of the art and future developments relevant to systems and networks, decision making, control, complex processes and related areas, as embedded in the fields of interdisciplinary and applied sciences, engineering, computer science, physics, economics, social, and life sciences, as well as the paradigms and methodologies behind them.

Indexed by SCOPUS, INSPEC, WTI Frankfurt eG, zbMATH, SCImago.

All books published in the series are submitted for consideration in Web of Science.

More information about this series at http://www.springer.com/series/15179

Naida Ademović · Edin Mujčić ·
Zlatan Akšamija · Jasmin Kevrić ·
Samir Avdaković · Ismar Volić
Editors

Advanced Technologies, Systems, and Applications VI

Proceedings of the International Symposium on Innovative and Interdisciplinary Applications of Advanced Technologies (IAT) 2021

Springer

Editors
Naida Ademović
Faculty of Civil Engineering
University of Sarajevo
Sarajevo, Bosnia and Herzegovina

Zlatan Akšamija
University of Massachusetts Amherst
Hadley, MA, USA

Samir Avdaković
Faculty of Electrical Engineering
University of Sarajevo
Sarajevo, Bosnia and Herzegovina

Edin Mujčić
Faculty of Technical Engineering
University of Bihać
Bihać, Bosnia and Herzegovina

Jasmin Kevrić
International Burch University
Ilidža, Bosnia and Herzegovina

Ismar Volić
Department of Mathematics
Wellesley College
Wellesley Hills, MA, USA

ISSN 2367-3370 ISSN 2367-3389 (electronic)
Lecture Notes in Networks and Systems
ISBN 978-3-030-90054-0 ISBN 978-3-030-90055-7 (eBook)
https://doi.org/10.1007/978-3-030-90055-7

This Springer imprint is published by the registered company Springer Nature Switzerland AG
The registered company address is: Gewerbestrasse 11, 6330 Cham, Switzerland

Preface

This book contains papers focusing on the implementation of new and innovative technologies which were presented at the 12th annual days of BHAAAS in Bosnia and Herzegovina—International Symposium on Innovative and Interdisciplinary Applications of Advanced Technologies (IAT) 2021 in Mostar, Bosnia and Herzegovina, during June 24–27, 2021. The symposium covered an extensive variety of disciplines and innovative technologies in information and communication technologies, advanced electrical power systems, computer science, mechanical engineering, civil engineering, geodesy and geoinformatics, computer modeling, and simulation for engineering applications, trends, and innovation in geographic information systems and sustainable urban development: designing smart, inclusive, and resilient cities.

Days of BHAAAS are a unique conference with a long tradition where prominent experts in a variety of special sessions present their research and new developments. The International Symposium on Innovative and Interdisciplinary Applications of Advanced Technologies was an international forum that allowed these experts to share their work with scientists, engineers, and students. The conference provided an opportunity for researchers, academics, scientists, and engineers to present their research, exchange ideas, and network for scientific and industrial collaborations. This volume contains papers that were peer-reviewed by at least two expert reviewers. In total, sixty-three papers were accepted for publication across all sessions.

The first chapter covers the information and communication technologies—ICT 2021 where the innovative and interdisciplinary applications of information and communication technologies were presented. The papers in the symposium described the development and application of the latest technologies in various areas of digital workplace transformation, the application of the fuzzy system, the application of the IoT system, exploring customer's behavior through data analysis, and simulation of communication systems. Seven peer-reviewed papers were presented. The co-chairs of the session were Aljo Mujčić and Edin Mujčić.

The second chapter is devoted to the advanced electrical power systems session. This was an international forum for scientists, engineers, and students that provided them an opportunity to present their latest research and development results in all areas of power systems. Five peer-reviewed papers were presented. The session was moderated by co-chairs Mirza Šarić, Maja Muftić Dedović, and Tarik Hubana.

The computer science symposium is presented in **the third chapter**. This session was primarily dedicated to the modern developments and innovative techniques in computer science. Topics covered were mostly related to natural language processing, text mining, video games, health care, and COVID-19. Research studies from four different continents were presented (Europe, Australia, Asia, and Africa), giving the symposium a truly international character. Moreover, almost all of the presented papers involved the contributions of Master's and Ph.D. candidates, giving them an exceptional opportunity to meet, network, and share their research ideas and results with world-class programmers, students, and researchers. Nine peer-reviewed papers were presented, and the session was moderated by co-chairs Zerina Mašetić, Jasmin Kevrić, and Dželila Mehanović.

The fourth chapter is devoted to the mechanical engineering session. The authors were final year students or Ph.D. candidates together with their mentors-professors from the Faculty of the Mechanical Engineering Department of the University of Sarajevo, Technical University in Bihać, as well as researchers from the industry. Thermal energy, hydropower, production engineering, and the field of motor vehicles are professional and scientific fields from which the results of the research projects were presented. The session emphasized the transfer of knowledge and the potential of cooperation between universities and industries. The results of the research presented at this session provided clear guidelines to companies about collaborating with the faculty at the mechanical engineering departments at universities in Bosnia and Herzegovina. Seven peer-reviewed papers were presented in this session. The symposium was co-chaired by Izet Smajević, Hajrudin Džafo, and Anes Kazagić.

The fifth chapter covers different areas of civil engineering. The papers dealt with up-to-date themes, such as the high concentration of small harmful airborne particles encountered in the transportation sector. One of the aspects that were tackled during the session was the quality of the construction project management process as well as factors affecting the decision-making process at construction sites in Bosnia and Herzegovina. The importance of the energy-efficient building was also discussed at the session. In addition, interesting research regarding the seismic vulnerability of buildings in Bosnia and Herzegovina was presented, in hope of raising awareness of the important research in earthquake engineering. This was especially timely in light of the devastation of structures in the Zagreb and Petrinja earthquakes. Papers dealing with sanitary landfills and waste disposal centers as well as assessment of regional analyses methods for spatial interpolation of flood quantiles in the basins of Bosnia and Herzegovina and Serbia were presented as well. Nine peer-reviewed papers were presented. The session was moderated by co-chairs Naida Ademović, Adnan Ibrahimbegović, and Mirza Pozder.

Chapter six is devoted to the symposium on geodesy and geo-information which was organized for the first time. This meeting provided the opportunity for presentations of state-of-the-art analyses of a broad spectrum of geo-spatial data. Thanks to tremendous advances in instrumentation and computational abilities in recent years, geodesy and geoinformatics became an even more exciting scientific field. Technological advances over recent years enabled geodesy and geoinformatics to develop under a wide spectrum of interactions, including theory, science, engineering, technology, observation, and practice-oriented services. This session primarily showcased various kinds of investigations and results in advanced methods of solving geodetic problems, estimating parameters, contributions illustrating the use of new methods of satellite positioning on real data in various branches of geodetic sciences, such as the geodetic reference frames, photogrammetry, and laser scanning, metrology, and geo-information system applications.

Five invited speakers, all eminent European and regional scientists, delivered lectures on the state-of-the-art topics using geodetic observation satellite data: contributions of geodesy to monitoring natural hazards and global change; atmosphere monitoring; monitoring the earthquake in Zagreb, Croatia; applications of geo-information systems; 3D scanning for repairing the damage on buildings after the earthquake; and challenges and benefits of the academia-business cooperation for excellence in geoinformatics. Seven peer-reviewed papers were presented, and the session was moderated by Medžida Mulić and Džana Halilović.

Chapter seven covers computer modeling and simulation for engineering applications. Many engineering problems are nowadays solved very efficiently and with good accuracy using computer models and simulations. These methods are in many cases preferred to traditional, experimental ones, due to their repeatability, flexibility, and immense data processing speed. Thanks to continuous and significant progress in the development of computer technologies, numerical and other mathematical methods, and software design, very complex problems regarding geometry, underlying physical and chemical processes, as well as data structures and amount of data, are successfully solved. The papers included in this session presented and discussed experiences and research results for several instances of modeling, computing, and simulation relevant to modern technologies and engineering applications. Eight peer-reviewed papers were presented, while the session was moderated by co-chairs Adnan Mujezinović and Muris Torlak.

The session on trends and innovation in geographic information systems is covered in **chapter eight**. This symposium aimed to bring together scientists, researchers, and research scholars to exchange and share their experiences and results in geo-spatial data and knowledge, observation, representation, processing, visualization, sharing, and managing. The papers focused both on information communication and technologies (ICT) as well as management information systems and knowledge-based systems. The session also provided an interdisciplinary platform for researchers, practitioners, and educators to present and discuss the most recent innovations, trends, and concerns as well as practical challenges, as well as the solutions adopted in the fields of geographical information systems. Five peer-reviewed papers were presented. The session was moderated by co-chairs Mirza Ponjavić, Jasminka Hasić Telalović, and Almir Karabegović.

The ninth chapter is devoted to the symposium on sustainable urban development: designing smart, inclusive, and resilient cities which was organized for the first time, creating an opportunity for a multidisciplinary approach in planning for an urban future and knowledge exchange between experts in different fields. With 70% of the global population projected to live in cities by the year 2050, human action continuously threatening to create irreversible damage to the planet and acute shocks like the current pandemic, it is imperative to put the focus on smart, equitable, inclusive, resilient, and sustainable development approaches. The topics presented at the symposium varied from solutions to sustainability challenges in informal settlements of the Philippines and Bosnia and Herzegovina, lessons from vernacular architecture, healthy urban environments to renewable energy solutions through photovoltaic power plants along national corridors. Six peer-reviewed papers were presented, while the session was moderated by co-chairs Tijana Tufek-Memišević, Maja Muftić Dedović, and Ajla Merzić.

Contents

Advanced Electrical Power Systems

Computer Science

Information and Communication Technologies

MATLAB Simulation of IEEE 802.11p Technology on High User's Mobility

Abderrahim Mountaciri(✉), El Mostafa Makroum, and My Abdelkader Youssefi

Laboratoired'Ingénierie, de Management Industriel Et d'Innovation (LIMII), Faculté Des Sciences Et Techniques (FST), Hassan First Université of Settat, Hassan, Morocco

Abstract. In this article proposed IEEE 802.11p Physical layer (PHY). A MATLAB simulation is performed to analyze the baseband processing of the transceiver. Orthogonal Frequency Division Multiplexing (OFDM) is applied in this project according to the IEEE 802.11p standard, which allows data transmission rates from 3 to 27 Mbps. Separate modulation schemes, bit phase shift modulation (BPSK), quadrate phase shift modulation (QPSK), and quadrature amplitude modulation (QAM), are used for different data rates. These schemes are combined with time interleaving and a convolutional error correction code. A guard interval is inserted at the start of the transmitted symbol to reduce the effect of intersymbol interference (ISI). This article studies the PHY physical layer of the IEEE 802.11p vehicular communication standard. An IEEE.802.11p PHY model, with many associated phenomena, is implemented in the V2V vehicle-to-vehicle, and the vehicle-to-vehicle ad hoc network (VANET) provides convenient coordination between moving vehicles. A moving vehicle could move at a very high speed, producing a Doppler effect that damages OFDM symbols and also causes intercarrier interference (ICI). This article has discussed VANET technology versus 802.11a technology, as they have many differences when it comes to user mobility. The Doppler effect resulting from the mobility of the user with a high speed of 25 to 400 km/h has been studied as the main parameter, the estimation of the channel based on the lms algorithm has been proposed in order to improve the performance of the physical chain.

Keywords: VANET · OFDM · 802.11p · Doppler effect · Power spectral density

1 Introduction

Vehicle to vehicle and vehicle to infrastructure communications aim to reduce road accidents, automatic fare collection, route information, working lights, etc. The Vehicle-to-Vehicle Communication Consortium (C2C-CC) is a non-profit industrial organization supporting the creation of a European standard for future communicating vehicles covering all brands. The 802.11p standard is designed for use in vehicle-2-vehicleand vehicle-2-infrastructure communications. This standard is based on the same structures as that of the 802.11a standard; The IEEE 802.11p standard typically uses 10 MHz bandwidth channels in the 5.9 GHz band (5.850–5.925 GHz). This is half the bandwidth, or double the transmission time for a specific data symbol, as used in 802.11a.

N. Ademović et al. (Eds.): IAT 2021, LNNS 316, pp. 3–14, 2022.
https://doi.org/10.1007/978-3-030-90055-7_1

This allows the receiver to better manage the characteristics of the radio channel in vehicular communication environments, for example, signal echoes reflected from other cars or homes.

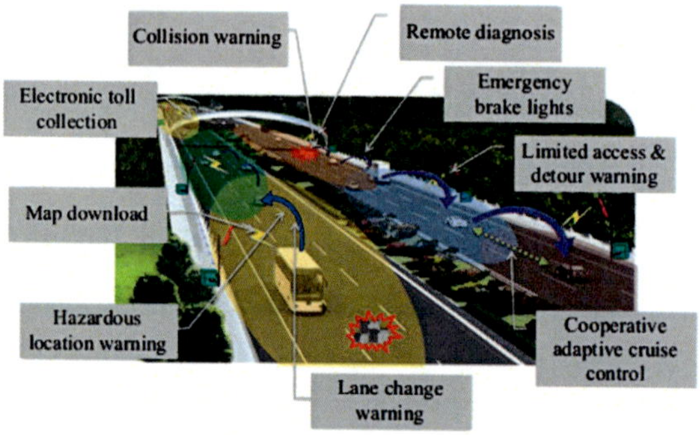

Fig. 1. The intelligent transport system (ITS)

The intelligent transport system (ITS) [1] given in Fig. 1 is developed for communication between vehicles and operates in the 802.11p standard at a frequency of 5.9 GHz, as opposed to the 802.11a standard, which operates at 5 GHz This document is organized as follows: In Sect. 2, the physical layers (PHY) of 802.11p and 802.11a are described and compared. The fading in the v2v channel and also the Doppler Effect is described. The simulation results are given in Sect. 3, which show BER (Bit Error Ratio) performance versus Eb/N0 (energy spectral density ratio per bit. on noise power) for different two standard IEEE 802.11 a and IEEE 802.11p it presents the results of simulation and discusses the results, Sect. 4 conclude the article.

2 Phy IEEE 802.11 a and IEEE 802.11 p

The RF LAN (Radiofrequency local area network) system is initially planned for 5.15–5.25, 5.25–5.35, and 5.725–5.825 GHz unlicensed; in the National Information Structure (U-NII) bands, the OFDM (orthogonalfrequency-divisionmultiplexing) system provides a wireless LAN with useful data rates of 6, 9, 12, 18,24, 36, 48 and 54 Mbit/s. The support for data transmission and reception rates of 6, 12, and 24 Mbit/s are mandatory. The system uses 52 subcarriers which are modulated by binary or quadrature-phase modulation PSK/QPSK or quadrature amplitude modulation QAM64. Forward error correction coding (convolutional coding) is used with a coding rate of 1/2, 2/3, or 3/4. Figure 2 shows the format of the PPDU (procedure protocol data unit), where the preamble of the OFDM PLCP, the OFDM PLCP header, the PSDU, the tail bits, and the range bits are shown. The PLCP (physical layer convergence procedure) header contains the following fields: LENGTH, RATE, a reserved bit, an even parity bit, and the SERVICE field. In terms of modulation, LENGTH, RATE, reserved bit and parity bit (with 6 tail bits "zero" in the appendix) form a single, separate OFDM symbol, called SIGNAL, which is transmitted with the most robust combination of BPSK modulation and a coding rate of

R = 1/2. The SERVICE field of the PLCP header and the PSDU (with 6 "zero" tail bits and padding bits), called DATA, are transmitted at the data rate described in the RATE field and can constitute several OFDM symbols. The tail bits in the SIGNAL symbol allow the decoding of the RATE and LENGTH fields immediately after receiving the tail bits. The RATE and LENGTH fields are required to decode the DATA part of the packet. IEEE 802.11p is a standard based on 802.11–2007 [2] PHY is based on OFDM and is analogous to 802.11a. to analyze the physical layer of 802.11p, with the most important parameters making the comparison with IEEE 802. 11 a, developed for communication between vehicles and operates within the 802.11p standard at a frequency of 5, 9 GHz, the bandwidth of 802.11p is 10 MHz, which is half the bandwidth of 802.11a. The distribution of any signal is accompanied by attenuation, which depends on the distance from the point of transmission and the frequency of the signal. The received signal can vary due to reflections from moving vehicles, many copies of the signal arrive at the receiving antenna with different levels and delays, in order to avoid the negative influence of signal interference. The received signal may vary due to reflections from moving vehicles and many copies of the signal arrive atthe receiving antenna with different levels and delays, in order to avoid the negative influence of the interference. OFDM is used, its realization in the transmitter using the inverse Fourier transform (IFFT), consider the first N-carriers multiplexed on the signal ofthe temporal frequency representation.The 802.11p standard uses the IFFT with a window of 64 frequency (N) subcarriers. Since the channel has a bandwidth of 10 MHz (Bw), each orthogonal frequency subcarrier has a spacing of Bw/N = 0.15625 MHz (Δf). Among the 64 orthogonal subcarriers, 56 are used: 52 - data tones, 4 pilottones. The cyclic prefix (CP) is an important part of OFDM. The CP is the last part of the signal, which is added at the beginning. A fragment of the desired signal is transmitted to the CP; it guarantees the orthogonality of the subcarriers in the received signal. The CP (cyclic prefix) is redundant information that reduces the transmission speedbut protects against inter-symbol interference. The duration of the CP depends on the decoder and the maximum receiver. The longer the delay, the longer the CP duration should be. The IEEE 802.11p [3]. standard offers data transmission rates of 3, 4.5, 6, 9, 12, 18, 24, and 27 Mbps; the frequency range used is 5,850 to 5,925 GHz, which is divided into 7 channels of 10 MHz. The standard is based on communications. The transmission bandwidth is divided into several narrow sub-channels, which are transmitted in parallel. Ideally, each sub-channel is narrow enough that the fading it experiences is flat.

2.1 Flow and Delay Limit

To achieve maximum flow and minimum delay for a real WAVE system where several vehicles are struggling for wireless support [2]. The lowest data rate of 3 Mbps is used. The physical layer convergence procedure protocol data unit (PPDU) of the IEEE 802.11 PHY consists of a Layer Convergence Procedure (PLCP) preamble, PLCP header,tail bits, and fill bits of the physical layer service data unit (PSDU), such as illustrated in Fig. 2.

The PSDU format includes a MAC header, Frame body, and Frame Check Sequence (FCS) field. Values the duration of Tpm, Tsg, and Tda, as well as other keys parameters, are summarized in Table 1.Because the broadcast does not require an ACK frame, the

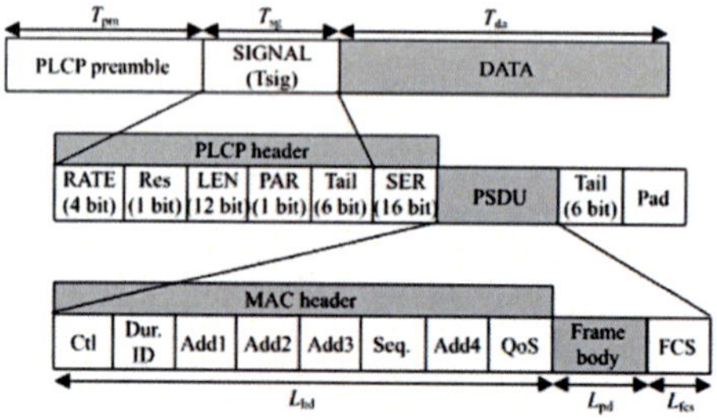

Fig. 2. PPDU frame t ieee802.11p OFDM PHY

maximum throughput in [3] can be further simplified *Lpd is the data payload with* [0 – 2304] × 8 *bits*, ***Tdata*** is the data transmission duration for given in (1)

$$T_{data} = \text{Tpm} + \text{Tsg} + \text{Tsym}\left\lceil \frac{16 + \text{lhd} + \text{Lpdd} + \text{Lfcs} + 6}{\text{N}_{\text{DPBS}}} \right\rceil \tag{1}$$

lhd is the data MAC Layer header with a size of 328 bits lfcs is the frame check sequence (FCS) with a size of 4 × 8 bits, ***NDPBS*** is the data bit per OFDM symbol given n Table 1.The packet delay is defined as the time duration of packet transmission and its successful reception. The minimum delay is given by $M_D = \text{Tdata} + \text{Tprop} + \text{TDIFS} +$ **CWavg** the maximum throughput limit (MTL) and minimum delay limit (MDL) are shown The MTL and MDL are obtained by assuming that the NDBPS are infinite such that the time duration for data transmission is zero. The maximum throughput of IEEE 802.11p is approximately 18 Mbps for the 27-Mbps PHY mode with a payload of 2,000 bytes, whereas the upper limit is around 80 Mbps for a payload of 2,000 bytes, whereas The Maximum Throughput Limit (MTL) and Minimum Delay Limit (MDL) are obtained by assuming that the NDBPSis infinite so that the duration of data transmission is zero. The maximum data rate for IEEE 802.11p is approximately 18 Mbps for the 27-Mbps PHY mode with a 2000-byte payload, while the upper limit is approximately 80 Mbps for an infinite data rate. The minimum delay is about 0.8 ms for 27-Mbps PHY mode with QAM; the data rates that the 802.11a standard allows are 6, 9, 12, 18, 24, 36, 48, and 54 Mbps. An OFDM symbol is sent to each 4 μs, of which 0.8 μs is the cyclic prefix.

As 250 000 symbols are sent every second, and one symbol uses 48 data carriers, a bit rate of 6 Mb/s is obtained with BPSK modulation and a half rate convolutional code (48*1/2 *250000 = 6 Mb/s). To determine the differences between the 802.11a and 802.11p standards, this document uses the BER (bit error rate) as a comparison parameter. The channel used is an AWGN and fading rayleigh channel; these channels can be used for the 802.11a and 802.11p standards because they have the same structure. The characteristics of the channels are taken from the literature The impulse channel response decreases exponentially with delay. Characterized by the Rayleigh a Typical office environment for NLOS condition with 50 ns RMS delay spreading a Gaussian channel – channel with additive white Gaussian noise (AWGN).

Table 1. PHY modes and NDBPS of IEEE 802.11p

	Data rate (Mbit/s)	Modulation	Coderate	*NDBPS*
1	3	BPSK	1/2	24
2	4.5	BPSK	3/4	36
3	6	QPSK	1/2	48
4	9	QPSK	3/4	72
5	12	16QAM	1/2	96
6	18	16QAM	3/4	144
7	24	64QAM	2/3	192
8	27	64QAM	3/4	216

3 Resultats and Discussion

3.1 Fading in v2v Channel

The V2V channel, in which both transmitter and receiver can be mobile, is a relatively newer area of study. Wireless channels can be modeled either deterministically, or statistically [4, 5]. statistical models are the most widely used because they do not attempt to provide an exact estimate of the small size of a scale chain of fading characteristics at points in space at a given time; instead, they try to faithfully minimize the variation of these effects channels. small-scale discoloration, which most often occurs due to destructive interference from several replicas of the transmitted signal arriving at the receiver with different delays due to several reflections from several beams from buildings and other installations as shown in Fig. 3. As an example autoroute to 6-lane highways with speeds ranging from 25 to more than 40 m/s and different traffic densities. Two-lane rural roads with speeds up to 30 m/s. There are few buildings and the vehicle density is low. One and two-lane suburban roads streets. Traffic density is generally low and speeds are limited to about 15 m/s or less on urban streets. They are characterized by wider streets and higher traffic density than suburban streets. In order to describe the frequency and temporal behavior of the channel. For information-carrying signal transmitted by the channel having *BT* bandwidth. the channel is said to have frequency selective fading if Bc is less than *BT*, it is said to be non-frequency selective or to uniform fading, if bc is greater than bt the consistency band is defined.

Maximum of the delay propagation delay, respectively. *rmax* is defined to be less than the upper-value propagation with limited delay $rlim = TGI = 1.6\,\mu s$ The coherence time of the channel event is linked to the vehicle speed v frequency, c is the speed of light and v is the speed of the vehicle. A slowly changing channel has a small Doppler frequency or, equivalently, a large coherence time. If the coherence time Tc is small, compared to the measured time interval Tm of the transmitted signal, the channel is said to be fast fading. Otherwise, the channel is said to be slowly fading. the channel parameters include *BT*, Tm, and τlim. $BT = 10$ MHz; the measured signal interval Tm = packet time = 840 μs; each transmitted packet contains 100 orthogonal frequency division

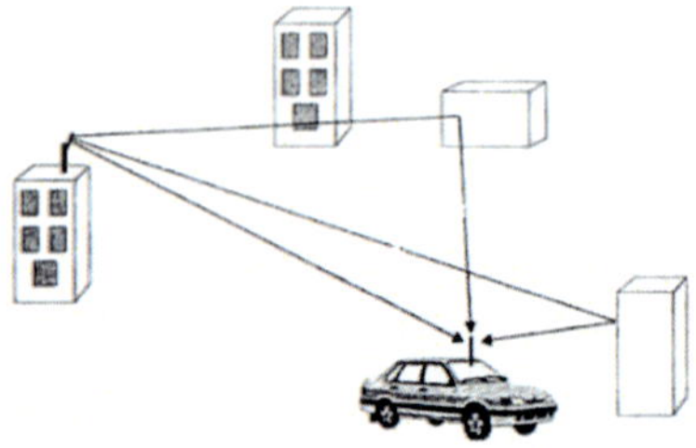

Fig. 3. Illustrationofthemulti-path physicalenvironment

multiplexing (OFDM) symbols; the maximum delay spreading time τmax is set to be less than the guard interval (GI) TGI of the OFDM symbol, i.e. the limit of the maximum delay spreading time τlim = TGI = 1.6 μs table 4 abstract four types of multipath fading channels.The simulation of the model follows the structure of the transmitter and the receiver of the standards 802.11a and 802.11p in Fig. 4 The transmitter has been implemented with the steps: convolutional encoding with an encoding rate of R = 1/2, BPSK modulation mapping, IFFT implementation for 64 points, add CP. The effects of fading channel and white Gaussian noise are added. The receiver includes the following steps: CP removal, FFT implementation for 64 points, BPSK modulation demapping, and error code decoding using Viterbi's hard decision algorithm. To determine the differences between the 802.11a and 802.11p standards, [6]. This article calculates the BER ratio as a function of Eb/N0 for a channel with fading of type AWGN and RAYLEIGH1 despite the increase in momentum, Doppler shift, and difficult channel characteristics. the increase in momentum, Doppler shift, and difficult channel characteristics. The typical propagation of the effective delay for a LOS signal component can be up to 320 ns, seen in Table 2 this can be more or less interpreted as (root mean square of) the time difference between the LOS component and the last non-component.

Table 2. Condition of fading channel in DSRC

	Flatfading	Frequency selectivefading
Slow fading	$0 \leq rRMS \leq 20$ ns	20 ns $\leq rRMS \leq$ 320 ns
	$0 \leq fd \leq$ 595,2 Hz	$0 \leq fd \leq$ 595,2 Hz
	$0 \leq v \leq$ 109 km/h	$0 \leq v \leq$ 109 km/h
Fastfading	$0 \leq rRMS \leq 20$ ns	20 ns $\leq rRMS \leq$ 320 ns
	595,2 $\leq fd \leq$ 595,2 Hz	595,2 $\leq fd \leq$ 2185,2 Hz
	109 km/h $\leq v \leq$ 400 km/h	109 km/h $\leq v \leq$ 400 km/h

The guard time is 1.6 s, therefore the guard time is sufficient to eliminate ISI. However, according, If we assume a maximum speed of 400 km/h, the maximum Doppler propagation will be 2185.2 kHz and this indicates a channel coherence time of 228.8 s.

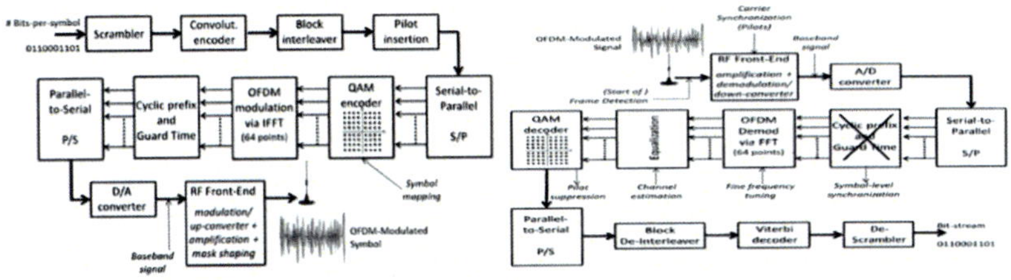

Fig. 4. IEEE 802.11 p transmitter and receiver

A transmitted frame should not be "on-air" longer than this period, or the drive symbols at the start of a frame will not be long enough and the signal may become too distorted. The size of the frame which it depends on the bit rate, but it is only 686 bits for the lowest data rate (3 Mbit/s 228.8 μs). Since the probability of bit error in a AWGN (Additive White*Gaussian noise*) channel of a communication system depends only on the SNR per bit or Eb/N0, the energy per bit is doubled in 802.11p. Symbol time in 802.11p is doubled and data rate is divided by If the data rate is halved and the transmit power remains the same, a symbol contains twice the energy it makes in 802.11a, given in Eq. (2) which means energy doubled by symbol and energy doubled per bit:

$$P_s = \frac{1}{2}Q(\sqrt{\frac{E_b}{N_0}}); \tag{2}$$

$\frac{E_b}{N_0}$ is the effective signal-to-noise ratio (SNR) per bit where Eb is the energy per bit, N0 is the noise density in W/Hz Q (x) is the function Q, which is the distribution function of a Gaussian distribution i.e. the integral which calculates the area under the right tail). Figure 5 gives the differences between the 802.11a and 802.11p standards The type of modulation used in this simulation is binary phase shift modulation (BPSK). Although BPSK has a lower transmission rate, it is often used for security or high mobility applications due to its lower BER property By comparing the BER as a function of the SNR we can notice that the IEEE 802.11 p standard is better than the IEEE 802.11 a standard for a fixed SNR the p standard presents a low value of the BER in comparison with a standard due to its better reliability.

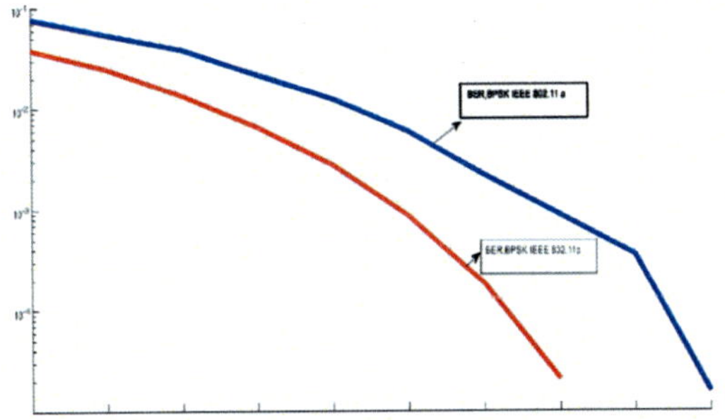

Fig. 5. BER curves BPSKforIEEE802.11a and IEEE 802.11p

3.2 Doppler Shift in IEEE802.11p

Doppler shift When a source vehicle and a receiver vehicle move relative to each other, the frequency of the received signal will not be the same as the source. [7] When they get closer to each other, the frequency of the received signal is higher than the source, and it decreases when they get closer as seen in [8]. The Doppler effect has a great effect on vehicle networks due to the VANET (Vehicular ad hoc networks) mobility Doppler shift When a source vehicle and a receiver vehicle move relative to each other, the frequency of the received signal will not be the same as the source. When they get closer to each other, the frequency of the received signal is higher than the source, and it decreases when they get closer as. The Doppler effect has a great effect on vehicle networks due to the VANET mobility the frequency Doppler shift can be calculated $\Delta f = \pm f_c {}^{V}\!/_{C} \cos \beta_c$; Δf where is the deviation of the source vehicle frequency at the receiver. The frequency of the source is $\boldsymbol{f_c}$, $\boldsymbol{v}$ is the speed difference between the source and transmitter vehicles, and c is the light velocity. In vehicular networks,the range of $fc = 5.9$ GHZ for simulation we choose v ϵ [25, 400 km/h] which gives $\Delta f \epsilon [136, 2185{,}2\ Hz]$ for Doppler shift, OFDM symbol is very sensitive to the Doppler Shift. An OFDM signal can be represented by $v(t) = \sum_{i=1}^{N_s-1} A_i \cos(\omega_i t + \varphi_i) A_i,\ \omega_i,\ \varphi_i$ be represented as respective values of the amplitude, angular frequency, and phase of subcarrier used from the OFDM signal. If Ns is the number of subcarriers used in OFDM, then each subcarrier must be orthogonal to one another. This can be obtained if $f_i = \omega_i / 2\pi$ which is the integer multiplication of ${}^{1}\!/_{2T}$, with T is the period of the data, and f1 is the frequency range with a value of $Rs = {}^{1}\!/_{T}$. If we continued Eq. (3) with processing at up converter side to produce 5,9 GHz frequency, becomes,

$$Vi = Ai\ \cos(\omega it + \varphi i)e - (j2\pi f\ \text{ct})\ \text{if he value}\ Ai\ \cos(\omega it + \varphi i) = Si\ \text{then}\ \boldsymbol{Vi = Sie} - (j2\pi \boldsymbol{f}\,\text{c t}) \tag{3}$$

When the value of $e - (j2\pi fc)$ and τi represents the carrier frequency and delay to the component of the received signal, respectively. If the vehicle moves at a constant velocity v, and each vehicle moves straight with β = 0 then the value of the transmitted signal is affected by the value of ${}^{V}\!/_{C} \cos \beta$ so the Eq. (3) become $\boldsymbol{Vi = Sie - (j2\pi fc)\ \tau ie(j2\pi \Delta f)t}$ with $(j2\pi \Delta ft)$ is a value from phase-changing at the transmitted signal. If the value of t = ¼ Δf, then the equation becomes $Vi = je - (j2\pi fc)\ \tau i$ Assuming the period is ¼ of the Doppler shift $0 \leq \Delta f \leq \pm 2.185\ \boldsymbol{kHz}$, the MATLAB simulation model see APPENDIX 1 is calculated for a variation in vehicle speed of 25.75,125 175 225 and 400 km/h, the SNR varies from 1 to 40 dB Fig. 6 gives the result of parameters such as defined by the IEEE 802.11p standard, the channel used is a combination of a Rayleigh and an AWGN. For the BER 10e−3 value if the mobile speed is 30 km/h. no communication can be performed when the SNR is less than 10 dB; for communication to be effective at SNR = 16 dB, the maximum relative speed must be less than or equal to 175 km/h. Figure 6 describes the impact of the Doppler effect in the IEEE 802.11 p standard due to the high mobility of the channel, we observe that the BER decreases if the speed increases which deters the transmission, a Matlab program describing the operation of the IEEE 802.11a PHY layer gives in APPENDIX 2 encoding process consists of many steps, the transmitted bits are generated with the data source component. These data are modulated using phase-shift modulation (BPSK or QPSK) or amplitude modulation (16-QAM or 64 QAM) the transmitted signal rate which can vary by 3 Mb/s (with BPSK and 1/2 code rate) at 27 Mb/s (with 64-QAM and 3/4 code rate. In order to compare the modulation, M-ary, Fig. 7 and Fig. 8, respectively give the BER the QAM 4 and the QAM 64, Figs. 9 and 10 respectively give the constellations which are the distance between points I and Q modulation system for M = 4 and M = 64 4-QAM has been found to offer the best performance (lowest BER) compared to 16-QAM and 64-QAM. The results of the comparison also show that a modulation scheme with a lower constellation value has better BER (bit error rate) performance due to the higher bit rate we note the influence of the Rayleigh type channel [9]. We conclude the error performance of the 4QAM-OFDM, 8QAM-OFDM and 16QAM-OFDM systems on the AWGN channel and the Rayleigh fading channel. It is observed from the simulation results that the signal power increases the error rate decreases in both AWGN and Rayleigh fading channels but the error rate increases as the value of the modulation scheme the error data can be reduced as well as transmission easily influenced by noise. the error data can be reduced as well as transmission easily influenced by noise. 64 QAM is much better than 4 QAM when BER decreases SNR increases because transmission is easily influenced by noise. 64 QAM is much better than 4 QAM when BER decreases SNR increases because transmission is easily influenced by noise. 64 QAM is much better than 4 QAM when BER decreases SNR increases because the signal is stronger than noise. 64QAM modulation requires because the signal is stronger than noise the error data can be reduced as well as transmission is easily influenced by noise. 64 QAM is much better than 4 QAM when BER decreases SNR increasesbecause the signal is stronger than noise. 64QAM modulation requires because the signal is stronger than noise. 64QAM modulation requires 4QAM We also validate the communication performance of OFDM systems proposed by the MATLAB simulation is the result of experimentation with the BER (Bit Error Rate) test with different SNR (Signal Noise Ratios).

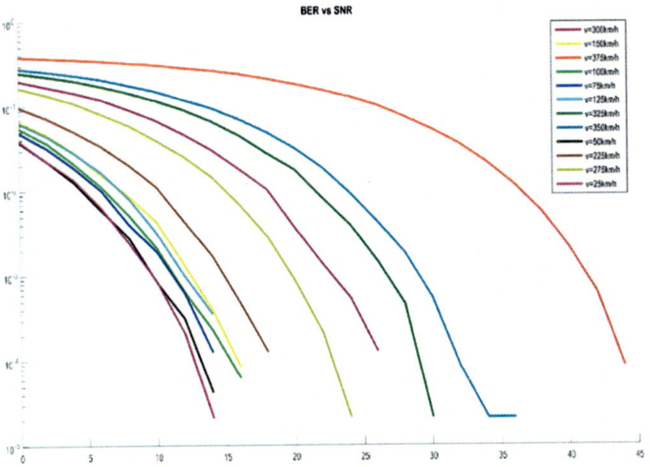

Fig. 6. BER BPSK. V ϵ [25,400 km/h]

It may be observed that the system has better noise-canceling capability, BER decreases rapidly with increasing SNR. Also, in the same state of SNR, 16QAM has a lower BER than BPSK.

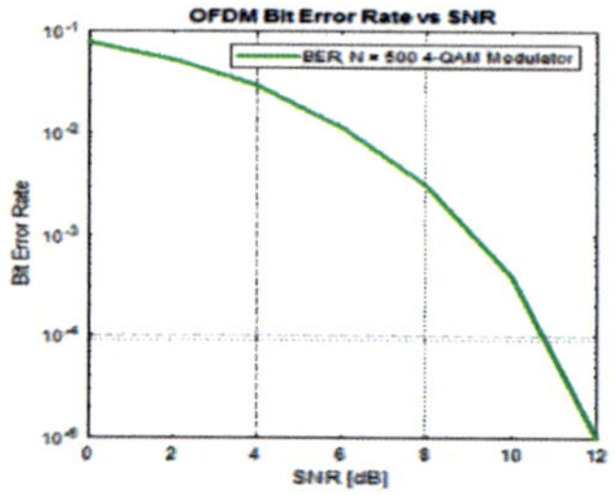

Fig. 7. BER Vs SNR 4QAM

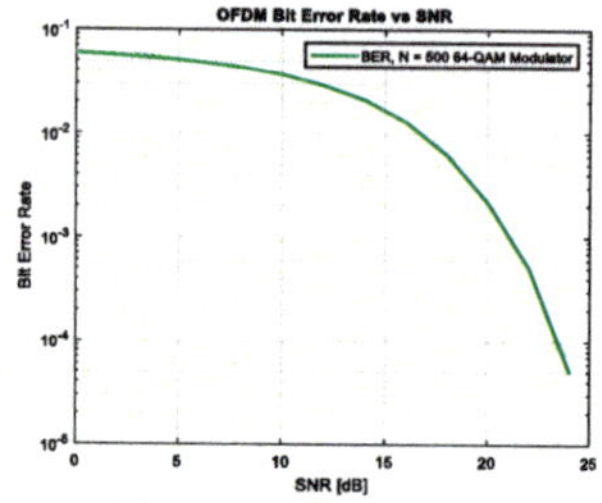

Fig. 8. BER Vs SNR 64QAM

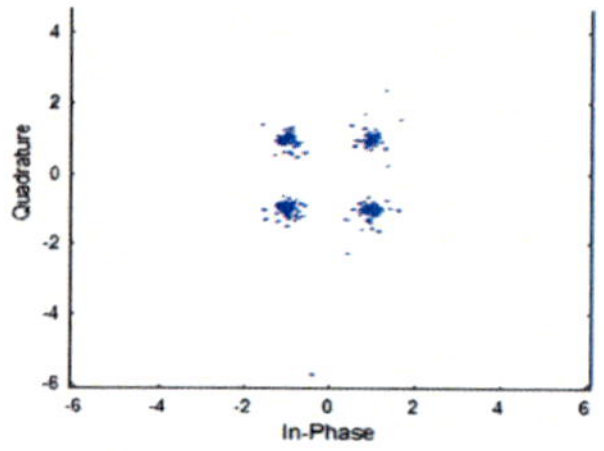

Fig. 9. Received constellation 4 QAM

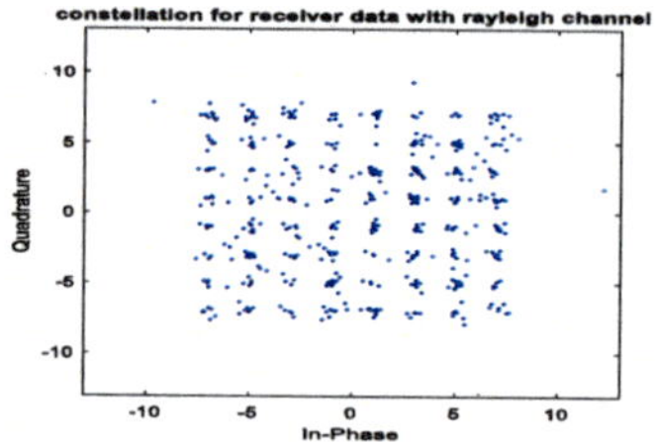

Fig. 10. Received constellation 64 QAM

It shows that the M-ary modulation system has higher bandwidth utilization than binary modulation but it has fewer advantages in noise-canceling capability. Figure 11 a) is the time domain waveform of OFDM. Figure 11 b) represents the power spectrum density of the modulated OFDM signal which contains 52 subcarriers [10], it shows that the data of the OBUs are correctly modulated on the relevant subcarriers.

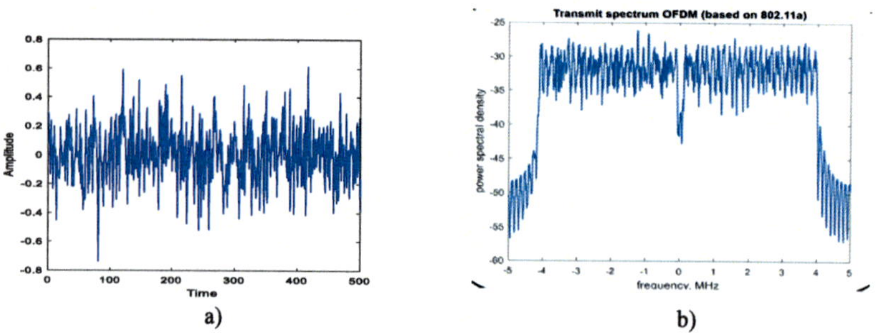

Fig. 11. a) Time domain waveform of OFDM in BPSK b) represents the power spectrum density

4 Conclusion

In this work, we focused on the IEEE 802.11 standard which has been adapted for inter-vehicular communication aiming at providing allow latency and high reliability. A near realistic IEEE.802.11p PHY model for vehicular communication is developed taking account of DOPPLER SHIFT phenomena associated. The BER performance of OFDM and M-QAM modulation schemes under AWGN and Rayleigh channel has been investigated. The 4-QAM, 16-QAM, and 64-QAM modulation schemes under Rayleigh fading channels have been simulated and the results of these three modulation schemes have been compared. It was found that 4-QAM gives the best performance (the lowest BER) compared to 8-QAM and 16-QAM. The comparison results also show that a modulation scheme with a lower constellation value has better BER performance because of having a higher bit rate. The performance of QAM and QPSK modulation schemes has also been compared. It was found that the QPSK gives lower BER compared to the QAM modulation scheme. It is also observed that the lower the order of modulation techniques is, the better system performance is obtained without the consideration of spectral efficiency. Finally, the performances of QPSK-based OFDM, 16-QAMbased OFDM, and 64-QAM-based OFDM in the Rayleigh channel have also been compared. It was found that the QPSK-based OFDM always gives the lowest BER compared to the 16-QAM-based OFDM and 64-QAM-based OFDM in order to increase the throughput, performance of IEEE 802 a with BPSK modulation and IEEE802.11 p with BPSK and also Doppler effect it concludes that IEEE 802.11 p offer the lower BER.The technology (MIMO) implemented in the V2V technology is used for higher rate bit and lower BER will be the subject of my future work,

References

1. Quality, Reliability, Security and Robustness in Heterogeneous Networks: 12th International Conference, QShine 2016, Seoul, Korea, July 7–8, 2016, Proceedings ... Included format: EPUB, **PDF**; ebooks can be used on all reading devices ... International Workshop on 5G Communication Architecture and accessed on 27 June 2021
2. Wang, Y., Duan, X., Tian, D., Lu, G., Yu, H.: Throughput and Delay Limits of 802.11 p and its Influence on Highway Capacity. Procedia Soc. Behav. Sci. **96**, 2096–2104 (2013)
3. Xiao, Y., Rosdahl, J.: Throughput and delay limits of IEEE 802.11. IEEE Commun. Lett. **6**(8), 355–357 (2002)
4. Vivek, N., Srikanth, S.V., Saurabh, P., Vamsi, T.P., Raju, K.: On field performance analysis of IEEE 802.11 p and WAVE protocol stack for V2V & V2I communication. In: International Conference on Information Communication and Embedded Systems (ICICES2014), pp. 1–6. IEEE, February 2014
5. Ho, K.Y., Kang, P.C., Hsu, C.H., Lin, C.H.: Implementation of WAVE/DSRC devices for vehicular communications. In: 2010 International Symposium on Computer, Communication, Control and Automation (3CA), Vol. 2, pp. 522–525. IEEE, May 2010
6. Lin, W.Y., Li, M.W., Lan, K.C., Hsu, C.H.: A comparison of 802.11a and 802.11p for V-to-I communication: A measurement study. In: Zhang, X., Qiao, D. (eds.) Quality, Reliability, Security and Robustness in Heterogeneous Networks, pp. 559–570. Springer Berlin Heidelberg, Berlin, Heidelberg (2012). https://doi.org/10.1007/978-3-642-29222-4_39

7. Xie, L.F., Ho, I.W.H., Liew, S.C., Lu, L., Lau, F.C.: Mitigating Doppler effects on physical-layer network coding in VANET. In: 2015 IEEE 26th Annual International Symposium on Personal, Indoor, and Mobile Radio Communications (PIMRC), pp. 121–126. IEEE, August 2015
8. Pamungkas, W., Suryani, T.: Doppler effect in VANET technology on high user's mobility. In: 2018 International Conference on Information and Communications Technology (ICOIACT), pp. 899–904. IEEE (2018)
9. Chyne, P., Dhilip Kumar, V., Kandar, D.: Network on wheels: leveraging vehicular communication to newer heights for intelligent transportation system. The International Journal of Electrical Engineering & Education. 0020720919891071 (2019)
10. Abdelgader, A.M., Lenan, W. The physical layer of the IEEE 802.11 p WAVE communication standard: the specifications and challenges. In Proceedings of the world congress on engineering and computer science, Vol. 2, pp. 22–24, October 2014

Exploring Customers' Behavior – Analysing Customer Data, Customer Segmentation and Predicting Customers' Behavior on Black Friday

Amela Vatreš(✉) and Zerina Mašetić

Faculty of Engineering and Natural Sciences, International Burch University, Sarajevo, Bosnia and Herzegovina
{amela.vatres,zerina.masetic}@ibu.edu.ba

Abstract. Due to massive expansion of trading businesses, especially e-commerces, it is necessary for companies to compete for their status among customers. As most of the customers value customer experience and claim that customer experience is the most important parameter when it comes to their decision making, in this work we focus on exploring those claims using "Black Friday" dataset. The aim of this study is to perform customer segmentation and suggest a strategy based on customers' analysis and division, and to propose an optimal model for predicting customers' behavior. Data mining and machine learning techniques were used to perform analysis, clustering, and model building. At first, data cleaning and preprocessing was done, followed by an extensive analysis of data. Then, K-means clustering algorithm was used to perform segmentation. Finally, three different models were built with a goal of predicting customers' purchase behavior (in this case purchase value). After comparing the results of the models, it was concluded that Random Forest Regressor performed the best. In this paper we discuss results that we got, compare them to the previous studies and draw conclusions that might be useful for marketers and decision makers, as they can be the base for their new strategies. Potential improvements and upgrades were suggested at the end of this paper.

Keywords: Customer segmentation · Analysis of customers behavior · Buying behavior · Machine learning · Data mining

1 Introduction

As trade is one of the main sources of earnings necessary for human existence, it experiences constant expansion. The popularity of this industry, especially online businesses, has led to increasing the competition in the market, so that businesses ought to compete for their status.

According to Gartner [1], 81% of marketers expect to compete mostly (or entirely) based on the customer experience. Therefore, as they concluded, most of the budget invested should be directed towards improving CX, for the efforts such as defining

N. Ademović et al. (Eds.): IAT 2021, LNNS 316, pp. 15–26, 2022.
https://doi.org/10.1007/978-3-030-90055-7_2

personas, customer segmentation or voice of the customer. Another statistic provided by Arm Treasure Data and Forbes Insight [2] reveals that 74% of consumers are at least somewhat likely to buy based on the experience alone. The importance of CX continues to grow, and to meet customers' ever-growing expectations, businesses must focus on identifying critical points with a focus on designing an optimal customer experience. To do so, customer segmentation is necessary.

Customer segmentation is a way to split customers into groups based on certain characteristics that those customers share [3]. While business-to-business marketers mostly segment customers according to the industry or products; direct-to-consumer brands focus on collecting data more specific to the customers. A great contribution to grouping customers was made by introducing ML and data mining techniques into the customer segmentation process, which were also used in this study.

This paper is based on the research conducted with an aim of providing better insight into customer segmentation and creating a machine learning model that can be used for buying prediction purposes. Firstly, exploratory data analysis was performed on a dataset, which was followed by building prediction models. The approach used in this paper for segmentation purposes is dividing the market based on demographic characteristics, like age, gender, marital status, occupation – commonly known as demographic segmentation [4]. By the end of the paper, the approach switches towards behavioral aspects, as we then predict spending and purchasing habits. The main contributions of this study are an extensive demographic data exploration with summarized conclusions, customer segmentation model with proposed strategies to improve performance, and optimized prediction model for predicting customers' behavior.

2 Literature Review

Customer segmentation and predicting customers' behavior is a popular research topic, which is reasonable due to the popularity of this industry. In this section we introduce previous research conducted on this topic.

The research conducted by Ceren Iyim in 2020 [5] had a goal of enabling companies to customize the relationship with their customers. That was achieved through customer segmentation, performed using an unsupervised ML algorithm with Python. The dataset used contains data collected from the e-commerce website within 6 months. The study was conducted in phases, and the approach used for customer grouping was a behavioural aspect, with focus on the average return rate and total spending per customer. Some of the results of this research show that the group of most favorable customers is a group of customers that spend high amounts of money, having an average return rate as 0. However, the percentage of these customers is the lowest. The strategy suggested was to keep working on preserving these customers, while moving other categories towards higher average spend and lower return rates.

An interesting study on this topic was provided by Jeffri Sany in 2019 [6] who investigated Starbucks customers segmentation using unsupervised ML. The data used for this research was taken from Udacity and it mimics customer behavior on the Starbucks rewards app. After data preparation, cleaning, preprocessing and feature extraction, EDA was performed. Some of the interesting findings were that customers were mostly males, the average income of females was higher than that of males, but the unknown gender made the highest number of transactions. Afterwards, K-means clustering was applied, so that a researcher was able to identify different groups of customers and propose a new business strategy.

A great dynamic and data driven framework for predicting whether a customer is going to make a purchase was established by Andres Martinez, Claudia Schmuck, Sergiy Preverzyev and Clemens Pirker [7]. The main contribution of this research is the proposition of new customer relevant features that are updated every month based on the number of previous purchases and values of those. Researchers used a gradient tree boosting method with a dataset containing more than 10000 customers, which provided an accuracy of 89% for predicting next month purchases.

Baris Karaman in his research [8] predicted the next purchase day using Python. After performing data wrangling, feature extraction, model selection, multi-classification and hyperparameter tuning, he tried using different models. The best performing one was Naive Bayes with an accuracy of ~64%. To see the flexibility of this model using different datasets, he performed cross validation. Therefore, we were provided with a stable ML model that can be used generically for predicting the next purchase day. Karaman continued this research using the same dataset and procedures, but with a goal of predicting sales [9]. For his second research a linear regression and LSTM models were created, and he was able to predict the next six months' sales numbers. The third Karaman's study [10] was strictly focused on customer segmentation and creating clusters based on different characteristics. He applied clustering algorithms based on recency, frequency, revenue, and an overall score. As a result of this research, we were provided with a scale of customer groups according to their contribution to business.

The previous research show what different approaches can be followed for exploring customers' behavior. The contribution of this research in comparison to the previously conducted ones is the focus on strategy-improvement, as we believe that methodologies proposed for this case study, can be applied to the other businesses.

3 Materials and Methods

To achieve all the goals that were previously set, the research included many steps. The whole workflow can be divided into preparing data, exploratory data analysis, clustering/segmentation and building models. The main steps and the flow in which the research was performed are presented in Fig. 1.

3.1 Dataset Description

The data used in this research was collected from a store containing both electronics and clothes product categories. All the sales were made on Black Friday. This dataset was downloaded from Kaggle [11], and contains information mostly about customers, including age, gender, city category, occupation category, etc. and product category that they purchased.

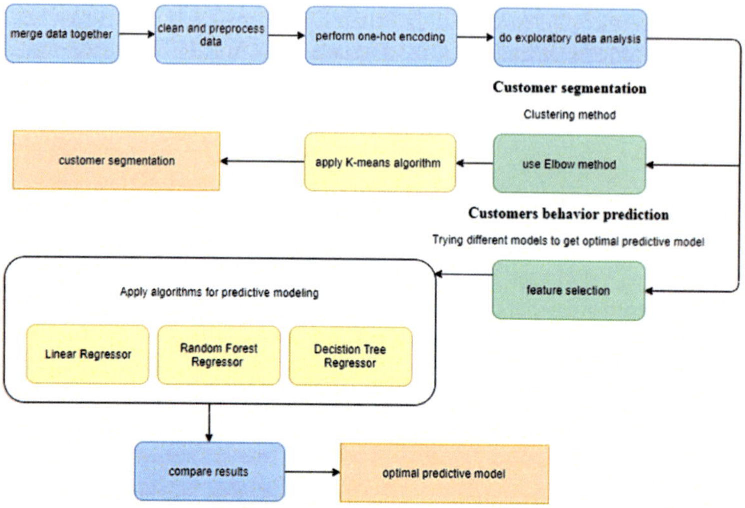

Fig. 1. Flow chart with key steps of the research

The original dataset was divided into training and test data. When concatenated, a total number of records is 783667. The total number of features is 12. The number of customers whose purchase was included into the dataset is 5891, and they were able to choose among 3677 products of different categories, including electronics and clothing items.

Among all customers, 75% of them were males, as can be seen in Fig. 2. If age groups were taken into consideration, the highest number of customers was in the age group of 26–35 years, in total 40% of all customers. Table 1 represents the distribution of customers per age groups.

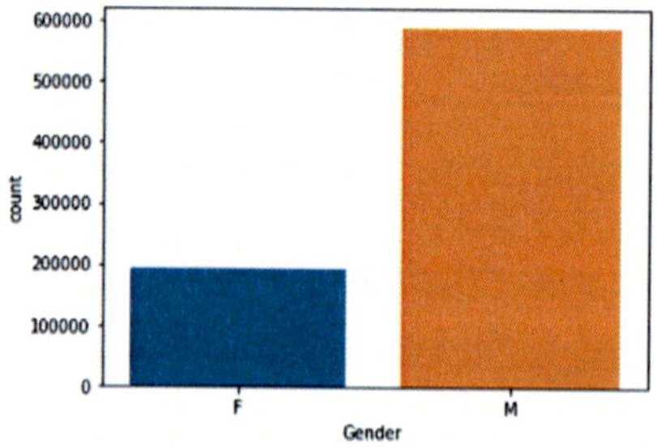

Fig. 2. Customer distribution per gender

Table 1. Distribution of customers per age groups

Age group	0–17	18–25	26–35	36–45	46–50	51–55	55+
Percentage of customers	2.72%	18.11%	40%	20%	8.32%	7%	3.90%

An interesting metric is the marital status of customers. Around 59% of customers were not married at the time they made the purchase. Among 21 different occupation groups, the most popular one was group 4, followed by groups 0 and 7.

When it comes to the location of the customers, all the customers were located within 3 different city categories. The most popular city group is group B, followed by groups C and A. Most of the customers, in total 276425 of them have been staying in the current city for at least a year.

Two product categories that were presented in the dataset are clothes and electronics categories. The results of analysis show that the purchase value was almost doubled for electronics. However, the purchase count for clothes is much higher than that of electronics. Therefore, we can assume that this occurrence might be due to the pricing of electronics items in comparison to clothes.

3.1.1 Data Cleaning and Preprocessing

Data cleaning is the process of preparing data for analysis by removing or modifying data that is incorrect, incomplete, irrelevant, duplicated or improperly formatted [12]. Similarly, data preprocessing was applied, which can be defined as a data mining technique whose main purpose is transforming raw data into a format which will be understandable by machine learning model [13].

The "Black Friday" dataset consisted of two separate files, which were firstly merged to get relevant data analysis results. Concatenation was done cumulatively. No duplicates were present. Third product has been removed as the percentage of missing values was above 60%, and missing values in the second product category were imputed using mean values. Afterwards, data types were checked and casted.

Pandas, NumPy, SciKit-learn, Matplotlib and Seaborn are some of the libraries used for the purpose of data cleaning, preprocessing and analysis.

3.1.2 Exploratory Data Analysis

Exploratory data analysis refers to the process of performing initial investigations on data to discover patterns, spot anomalies, test hypotheses and check assumptions with the help of summary statistics and graphical representations [14]. EDA is mostly used to see what data can reveal beyond the formal modeling and provides a better understanding of dataset variables and relationships between them [15].

In this specific study, an in-depth data analysis was performed. Each of the attributes was observed separately and in combination with others, resulting in the multiple patterns found at the end of this phase.

3.2 Customer Segmentation using Clustering

There are 4 different ways to perform customer segmentation: demographic, geographic, psychological, and behavioral [16]. To get the best results of segmentation, specific clustering model is applied, depending on the data available.

Clustering is the task of dividing the population or data points into several groups such that data points in the same groups are more similar to other data points in the same group and dissimilar to the points in other groups. Shortly it is a collection of objects based on similarity and dissimilarity between them [17]. There are different types of clustering: connectivity model, centroid model, and density model [18]. In this research, centroid model was used, in which the similarity is based on the proximity of a data point to the centroid of the clusters.

3.2.1 Application of Clustering in this Study

The initial step in this phase included transforming data. The initial data was categorical data – variables that contain label values rather than numeric values [19]. For K-Means clustering, we had to convert it into a model understandable format.

Among other clustering algorithms, K-means clustering was selected for this research, as it is one of the simplest and most popular unsupervised clustering algorithms, whose objective is grouping similar data points together and discovering underlying patterns [20]. In this case, a demographic way of clustering was chosen, that being the most suitable one for our dataset.

To decide on optimal cluster number for segmentation, different methods can be used. The elbow method is the most significant for this research. The elbow method is used to help choose the optimum 'k' by fitting the model with a range of values of 'k' (where k represents the number of clusters) [21].

As presented in Fig. 3, an optimal number of clusters is between 2 and 5. After trying different combinations, it was concluded that 3 clusters performed the best.

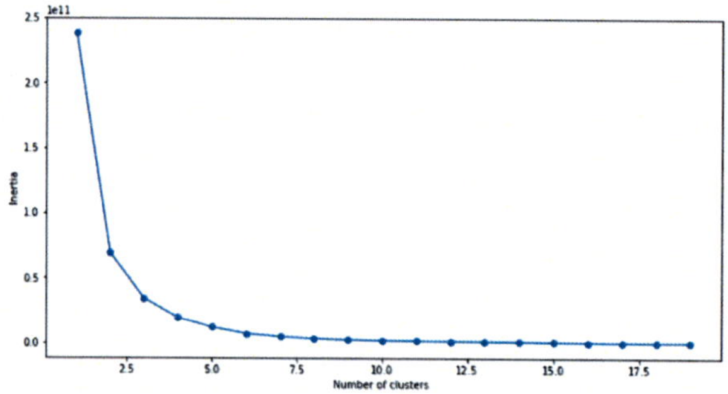

Fig. 3. The application of the elbow method

3.3 Prediction of Customers' Purchase Value Using ML Models

Predictive modeling is the general concept of building a model that can make predictions. Mostly, that kind of a model includes an ML algorithm that learns properties from a given dataset to make predictions [22].

Before applying models, data had to be converted into numerical data through one hot encoding. As a result, we were provided with 22 different attributes. Worth mentioning, feature scaling was applied as well. Feature scaling is used for reducing the number of input values by selecting features which contribute the most to the output – prediction

[23]. The last pre-model-building step was dividing data into training and testing data in the ratio 70%-30%. Finally, linear regressor, random forest regressor and decision tree regressor were applied. The purpose of building three different models was to find the best performing for predicting purchase value.

4 Results

In this section, we present results of our study. Results are divided into three groups: the fundamental conclusions from analyzing and visualizing the data, customer segmentation results and model building results.

Performance evaluation of built models was compared through 2 different metrics: mean absolute error and mean squared error. Mean absolute error is the average of all absolute errors [24]. Mean squared error is the average of the square of the difference between the observed and predicted values [25].

4.1 Data Analysis Findings

As previously mentioned, most of the customers were males of age 26–35. When we consider both age and gender together, we can see that males prevail in all the age groups. This can be seen in Fig. 4.

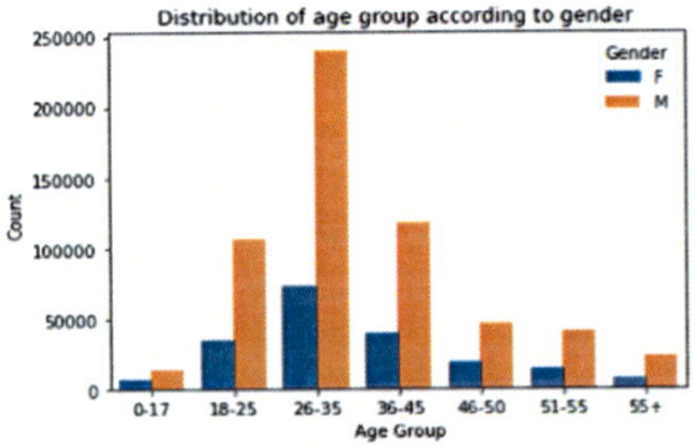

Fig. 4. Customer gender according to age group

Among all male customers, a higher number of them were not married. Situation with female customers is similar. If we observe marital status per age groups, conclusion would be that in most of the age divisions customers were not married. However, there are a couple of exceptions, for age groups of 46–50, 51–55 and 55 + years. Therefore, we can conclude that most of the married customers belong to the older demographic, while younger customers were not married.

An important conclusion from EDA could be delivered by observing purchase distribution with respect to different properties. An example is presented in Fig. 5, where the contribution of male gender towards purchase is almost 3 times that of a female gender. However, mean purchase value for males is not much higher in comparison to the mean purchase for females. In short, males tend to order more, while both genders spend similar amounts of money on their orders.

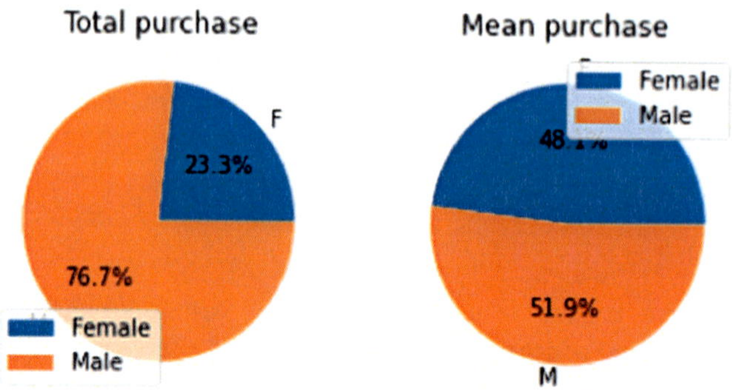

Fig. 5. Purchase distribution with respect to gender

Overall, the maximum purchase amount was around $23 961, while the minimum purchase amount was $12. The average purchase amount was $9264. Occupation categories 0, 4 and 7 benefit the organization the most, by having the highest purchase value. Similarly, the contribution of City category B is the highest.

4.2 Divisions of Black Friday Customers

As previously described, clustering was performed with a goal of segmenting the market, using K-Means algorithm. Different clusters were created based on the attributes, and they are presented below.

The results of clustering when only age and gender are taken into consideration can be seen in Fig. 6. There, we can see 3 clusters, mostly horizontal oriented, that can be defined as following:

- Mostly young males who are not willing to purchase high values,
- Mostly young males who are willing to purchase middle values,
- Customers who are willing to purchase high values.

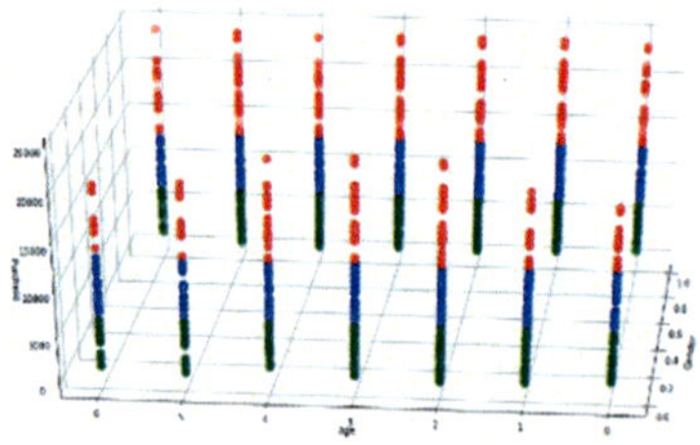

Fig. 6. Clusters with respect to gender & age

The results of clustering when only age and occupations are taken into consideration is presented in Fig. 7. Unlike the previous case, the cross section is not that sharp, and clusters can be defined as following:

- Mostly customers with an occupation category 0–10 who are not willing to spend high amounts,
- Mostly younger customers with and occupation categories 10–20 who are willing to spend middle amounts,
- Mostly younger customers with an occupation category 0–12 who are willing to spend high amounts.

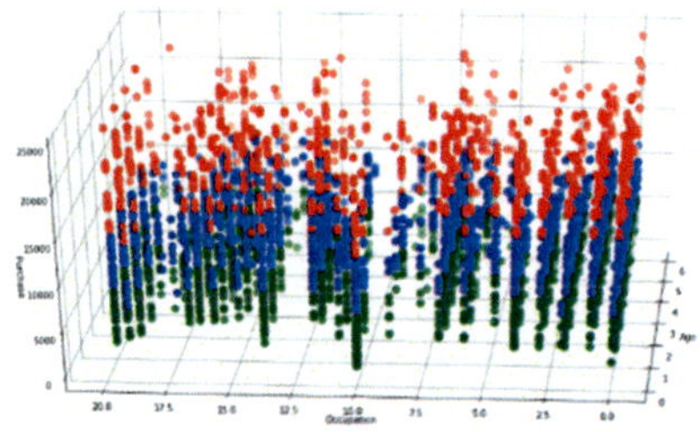

Fig. 7. Clusters with respect to occupation & age

An interesting approach to clustering is using product-related attributes. In Fig. 8, we can see clustering based on product categories – clothes and electronics. There, we can see that most of the red points are those of Product category 1 being equal to 0–2 and Product category 2 being spread across.

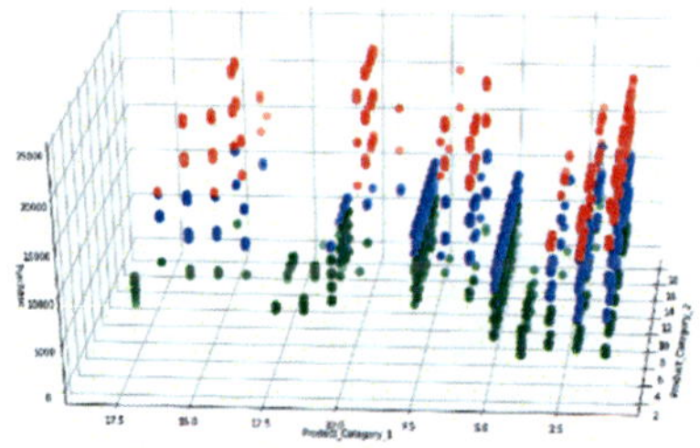

Fig. 8. Clusters with respect to product categories

4.3 Purchase Value Prediction

Results of training data with different models are represented through performance metrics: mcan absolute error and mean squared error.

Data was trained using 3 different models: linear regressor, random forest regressor and decision tree regressor. For easier interpretation, it is important to say that MAE and MSE are presented in dollars. Table 2 represents the comparison of the results that we got.

Table 2. Comparison of metrics for different models

Metric/Model	linear regressor	Random forest regressor	Decision tree regressor
MAE	~43. 61	~43.04	~49. 95
MSE	~2567.70	~2526. 90	~3571. 82

5 Discussion

Based on the Fig. 6, it can be said that the red cluster (above described as customers who are willing to spend high amounts of money) is the most beneficial to the company. It mostly contains young customers with an occupation category 0–12. Therefore, the company's strategy could be to target customers with occupation categories 0–10 (especially 0,7,12) who are younger than 46.

In addition to that, from the Fig. 7 it could be concluded that clothes category is not as beneficial as electronics category. In other words, electronic products bring higher revenue to the company than clothes. However, there are a couple of products within clothes category that are more beneficial than the other ones – and are labeled as products 0, 1 and 2. A new approach of the company could be to put focus on electronics and on those specific products, by advertising them and making them more accessible to the customers. However, since the number od clothes orders was more than double that of electronics, clothes is also a significant product category for this company.

From the exploratory data analysis results we concluded that males tend to purchase more, especially the ones that are not married. Customers whose occupation was within categories 0, 4 or 7 are the most beneficial to the organization. All these conclusions can be used as a base for targeting customers in the future. These are not only important for targeting when it comes to advertisements, but also for adjusting products. After multiple attempts to create an optimal predictive model, we have finally found the right approach to predicting purchase value. Table 2 shows that the random forest regressor performed the best, with MAE being around $43.04. Linear regression performed slightly worse, while decision tree regressor was the worst performing model in this case study. Therefore, in future studies of this kind, the best solution would be using random forest regressor.

6 Conclusion

In this study we investigate customers' behavior during Black Friday, with a goal of understanding patterns to be able to segment customers and predict their future behavior. The dataset that was used in the experimental part contains demographic information about the customer and product category that they chose.

After data cleaning and preprocessing, the first goal was to perform exploratory analysis. Most important findings from EDA show that young/middle aged males tend to purchase the most; the purchase value is higher for unmarried customers; males tend to purchase more in total value – but the mean purchase value is almost the same for both genders; occupation groups 0, 4 and 7 benefit purchase value the most; customers that live in city category B contribute the most to the e-commerce.

Exploratory analysis was followed by performing customer segmentation. Different factors were used to create clusters. Most importantly, based on clusters, new marketing strategies were proposed. Among others, it was suggested to focus on targeting customers with occupation categories 0–10, especially those who are younger than 46. Also, products that could be advertised and should drive the most attention are electronics products, as they bring the most value.

Finally, three different models were offered, and the best performing model was Random Forest Regressor.

Future improvements would include optimizing built models to increase their performance and looking into different perspectives when it comes to customer's buying decisions. Namely, this question can also be observed from the perspective of products and product characteristics. Similarly to what we did in this research, it is possible to perform product-based segmentation and round out this topic.

To sum up, this topic is interesting to be investigated. Experiments of this kind can be useful to the organization before acting and decision making. We believe that in the future most of the businesses will start to build their strategies based on research like this.

References

1. Pemberton, C.: Key Findings From the Gartner Customer Experience Survey. Available at: https://www.gartner.com/en/marketing/insights/articles/key-findings-from-the-gartner-customer-experience-survey (2018). Accessed 25 Jan 2021
2. Levine, B.: CX is as important as product or price, says new report. Available at: https://www.clickz.com/cx-is-as-important-as-product-or-price-says-new-report/260845 (2020). Accessed 25 January 2021
3. LeBlanc, D.: Creating actionable customer segmentation models. Available at: https://looker.com/blog/creating-actionable-customer-segmentation-models (2019). Accessed 20 February 2021
4. Mialki, S.: What is Demographic Segmentation with 5 Examples. Available at: https://instapage.com/blog/demographic-segmentation (2020). Accessed 20 February 2021
5. Iyim, C.: Customer Segmentation with Machine Learning. Available at: https://towardsdatascience.com/customer-segmentation-with-machine-learning-a0ac8c3d4d84 (2020). Accessed 20 February 2021
6. Sandy, J.: Investigating Starbucks Customers Segmentation using Unsupervised Machine Learning. Available at: https://medium.com/@jeffrisandy/investigating-starbucks-customers-segmentation-using-unsupervised-machine-learning-10b2ac0cfd3b (2019). Accessed 20 February 2021
7. Martinez, A., Schmuck, C., Pereverzyev, S.: A Machine Learning Framework for Customer Purchase Prediction in the Non-Contractual Setting. Available at: https://www.researchgate.net/publication/324973822_A_Machine_Learning_Framework_for_Customer_Purchase_Prediction_in_the_Non-Contractual_Setting (2018). Accessed 11 November 2020
8. Karaman, B.: Predicting Next Purchase Day. Available at: https://towardsdatascience.com/predicting-next-purchase-day-15fae5548027 (2021). Accessed 20 February 2021
9. Karaman, B.: Predicting Sales. Available at: https://towardsdatascience.com/predicting-sales-611cb5a252de (2019). Accessed 20 February 2021

10. Karaman, B.: Customer Segmentation. Available at https://towardsdatascience.com/data-driven-growth-with-python-part-2-customer-segmentation-5c019d150444 (2019). Accessed 20 February 2021
11. Black Friday. Available at: https://www.kaggle.com/sdolezel/black-friday (2018) Accessed 16 January 2020
12. What is Data Cleaning. Available at: https://www.sisense.com/glossary/data-cleaning (n.d). Accessed 04 February 2021
13. Sharma M.: Data Preprocessing: 6 Necessary Steps for Data Scientists. Available at: https://hackernoon.com/what-steps-should-one-take-while-doing-data-preprocessing-502c993e1caa (2020) Accessed 28 January 2021
14. Patil P.: What is Exploratory Data Analysis?. Available at: https://towardsdatascience.com/exploratory-data-analysis-8fc1cb20fd15 (2018) Accessed 21 February 2021
15. IBM Cloud Education: Exploratory Data Analysis. Available at: https://www.ibm.com/cloud/learn/exploratory-data-analysis (2020). Accessed 20 February 2021
16. Kelly A.: Customer Segmntation with K Means Clustering. Available at: https://towardsdatascience.com/customer-segmentation-with-kmeans-e499f4ebbd3d (2020) Accessed 20 February 2021
17. Priy S.: Clustering in Machine Learning. Available at: https://www.geeksforgeeks.org/clustering-in-machine-learning (2020). Accessed. 21 February 2021
18. Kumar A., Khanteymoori A.: Clustering in Machine Learning. Available at: https://training.galaxyproject.org/training-material/topics/statistics/tutorials/clustering_machinelearning/tutorial.html (2020). Accessed 20 February 2021
19. Brownlee, J.: Why One-Hot Encode Data in Machine Learning. Available at: https://machinelearningmastery.com/why-one-hot-encode-data-in-machine-learning (2017). Accessed 04 February 2021
20. Usai, D.: A Gentle Introduction to Customer Segmentation – Using K-Means Clustering to Understand Marketing Response. Available at: https://diegousai.io/2019/05/a-gentle-introduction-to-customer-segmentation/ (2019). Accessed 20 February 2021
21. Franklin, S.J.: Elbow method of K-means clustering using Python. Available at: https://medium.com/analytics-vidhya/elbow-method-of-k-means-clustering-algorithm-a0c916adc540. (2019). Accessed. 18 February 2021
22. Raschka, S., Predictive modeling, supervised machine learning and pattern classification. Available at: https://sebastianraschka.com/Articles/2014_intro_supervised_learning.html (2014). Accessed 20 February 2021
23. Brownlee, J.: 82019) How to Choose a Feature Selection Method For Machine Learning. Available at: https://machinelearningmastery.com/feature-selection-with-real-and-categorical-data/. Accessed 23 February 2020
24. Absolute Error & Mean Absolute Error. Available at: https://www.statisticshowto.com/absolute-error/ (n.d.) Accessed 10 March 2021
25. Calculating Mean Squared Error in Python. Available at: https://www.educative.io/edpresso/calculating-mean-squared-error-in-python (n.d.) Accessed 10 March 2021

Dental Clinic in the e-Health Cloud: A Prototype of the SaaS Platform

Ahmet Durmić, Ajna Fetić, Dženan Bejdić, Mejra Dizdarević, Harun Bečić, Mujo Hodžić, and Sanid Muhić(✉)

Polytechnic Faculty, University of Zenica, Zenica, Bosnia and Herzegovina
{ahmet.durmic2018,ajna.fetic2018,dzenan.bejdic2018, mejra.dizdarevic2018,harun.becic2018,mujo.hodzic}@size.ba

Abstract. With development of technology came significant changes in every aspect of people's lives. Primarily, in today's society the main focus is on web and mobile applications which have been designed in a way that makes our lives easier. These applications provide fast and easy access for their users and easy problem and task solving. With this in mind, emerged the need for Dental cloud application which would be used by private dental practices and their patients. The authors' proposal on this problem is the application cloud architecture which would provide easy and simple management of staff, invoices, materials and equipment used in an appointment as well as patient management. One of many functionalities that comes with is detailed insight in history of dental illnesses of every patient that has been registered on the system through electronic patient records directory module. In this paper, authors' goal is to show in great detail advantages and disadvantages as well as functionalities of solution proposal. Throughout the history human kind yearned for the better and have always tried their best to develop some tool or mechanism that would make their lives and work easier, so we tried to follow our primary instinct and continue our tendencies by creating this application.

Keywords: Cloud · e-Health · Saas · UML · Software architecture · C# · .NET · S.O.L.I.D · DRY · KISS · YAGNI · Visual studio · Adobe Xd

1 Introduction

In today's society Information and Communication Technologies (ICT) are no longer a privilege accessible only to developed countries, but rather a greater good. Implementation of ICT in numerous aspects of human life can give different advantages and can be implemented in various ways such as applications for "multimedia information and news services, electronic transactions with legally binding effect, cooperative workflow management in distributed organizations to information management for multimodal platform, and e-applications" [4]. Due to countless ways ICT can be implemented, it is a great way to improve not only the quality of life but also the economy. They improve the economy by viewing users' data as "an economic product—technically manageable and affordable." [4] Our goal was to create an e-application which can be used for any sphere of medicine, thus our application is an e-health application. Medicine has a lot of

N. Ademović et al. (Eds.): IAT 2021, LNNS 316, pp. 27–34, 2022.
https://doi.org/10.1007/978-3-030-90055-7_3

branches, hence we decided to demonstrate our idea on a less popular branch of dental medicine (dentistry).

Integration of ICT into medicine through e-health is widely spread in developed countries as well as developing countries. With development of 5G and 6G and wider usage of WI-FI, almost every citizen is connected to the Internet. "ICTs enable online communication about medical issues and diagnosis of complicated diseases by linking medical practitioners who are separated geographically.

They have the potential to change the delivery of healthcare services and patient care, as well as the management of health-care systems. According to Eysenbach (2001), e-health is an emerging field in the intersection of medical informatics, public health and business, with referral and information delivery enhanced through the Internet and related technologies. In a broader sense, the term characterises not only a technical development, but also a state-of-mind, a way of thinking, an attitude, and a commitment for networked, global thinking to improve healthcare locally, regionally and worldwide by using information and communication technologies" [5]. Hence, ICT are viewed as a bridge between urban and rural communities, because they improve access and provide better health services to unprivileged people who live in less developed parts of a country.

The prototype in this paper is adapted to the conditions of poor development of e-health systems in the Balkans and includes a prototype and an example of the first steps of private practices, all in order to achieve a level of development that will enable integration with future national e-health development programs.

2 Industry 4.0 and SaaS e-Health Services in the Technologically Advanced Countries of the World

The world is entering the era of the Industrial Revolution 4.0, an era that is a period of new development when several technologies including physical technology, digital technology and biological technology have each achieved unprecedented developments in their respective fields. These three technologies are the main or most fundamental technology drivers for the Industrial Revolution 4.0. This technology is very closely related to the world of health. The advances that have been achieved in the Industrial Revolution 4.0 era are reducing costs, increasing the ease and efficiency in technology and genetic sequencing, activation and editing [1].

From the strictly technical point of view, this so-called fourth industrial revolution is based mainly on the concept of Cyber-Physical Systems or CPS (integration of computing, communication, and control), and heavily relies on three groups of technologiesb(pillars): the Internet of Things (IoT) paradigm (characterized by the pervasive presence of a variety of uniquely addressable cooperating objects such as mobile phones, sensors, and actuators); Cloud and Fog Computing (providing virtually unlimited computing, storage, and communication resources as utilities, i.e. on-demand and pay-per-use); Big Data Analytics (to extract value from challenging amounts of data). [2].

The advanced technologies related to AR and VR have a great effect on the healthcare industry with their adoption in virtual training of surgeons in 3D operating room simulations for difficult surgeries and as phobia buster in mental health treatment as well as for chronic pain management. Also, VR plays a major role in eye movement desensitization and reprocessing (EMDR) therapy to enable reframing of traumatic memories through certain eye movements. [3].

Besides the technologies involved, there are in fact also development plans including aspects of enterprise management and work organization, regulatory frameworks, and dissemination and training.

The development of e-health is strictly guided and regulated through national and state programs and the involvement of the state is crucial. On the other hand, in countries that have not yet adopted regulatory plans for Industry 4.0, private practices (private health companies) can take the first steps in changing the way patients and external associates work and engage in order to prepare for integration with national programs on the one hand as and workers and patients for a new era of medical services.

3 Prototype and Technical Solutions

A common problem for people with any kind of health issues around the globe is reaching and contacting their health care providers (family doctor, dentist, nurse, technician…), booking their appointments due to a big influx of patients, and of course spending hours in the waiting room in front of the doctor's office. It is also difficult to keep track and have a clear insight into your own medical record and the history of your diagnosis because of the excessive paperwork. As previously mentioned, we have decided to resolve these problems on a dental medicine (dentistry) example.

Dental industry is growing rapidly and thus you can see more and more privatized dental practises opening day by day. Every dental office, privatized or not, has a large scaled set of patients and that implies a problem for both the medical staff and the patients. Patients need an easier way to book their appointments, stay connected with their doctors, have their dental record and keep track of their payments to the dental office. The dental office on the other hand needs an easier way to keep track of their patients, their diagnosis, payments, debts and many more.

Another issue that occurs for the dental office is their business and communicating with their external associates, mainly with their suppliers. An inevitable solution for resolving these problems and making life easier for patients and business easier for medical staff will be demonstrated in the following.

Given the described problem that the dental health system has been facing for many years, below we will present you with a conceptual solution that will not only solve the described problem, but also raise the level of the dental health system. The application should cover all business aspects of private dental practices as well as their external associates. Users of the application will have the opportunity to access it both through the mobile application and through their favorite web browser. The aim of the application is to create a unique eCard of the patient, where it is expected to enter personal data and history of dental diseases in order to gain better insight into the health status of the patient. The user will be provided with the possibility of online appointments, evaluation of professional staff and the possibility of card payment is included. In addition to the already mentioned possibilities, the user of the service will have reminders on their profile for their scheduled appointments, where there will also be the possibility of canceling the desired appointment no later than 24 h before the scheduled appointment. The dentist, who is also a user of the application, will have a view of the schedule on a weekly basis and details of each appointment. They will be able to assign responsibilities to assistants and have an overview of their responsibilities. They will also have the ability to make diagnoses as well as a list of employees.

The implementation of the solution is planned with the help of S.O.L.I.D, DRY, KISS and YAGNI design principles that allow scalability and robustness of the application, and independent testing of each module. Flexibility when migrating a database is also ensured by using repository pattern, and by using CQRS principle we are able to achieve the most optimal system performance (Figs. 1, 2, 3, 4 and 5).

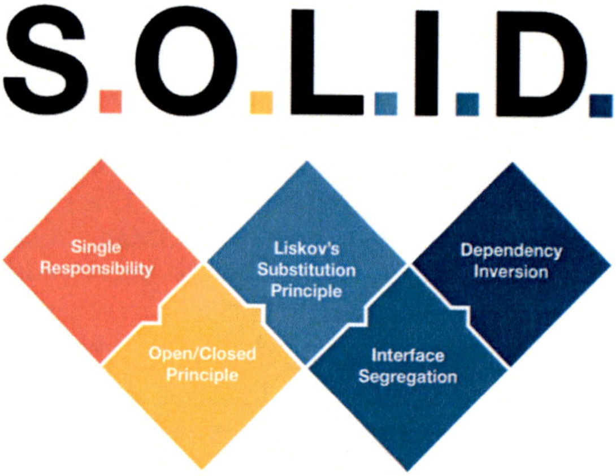

Fig. 1. SOLID principles [6]

Technologies, frameworks and programming languages for the implementation of the proposed Sass Platforms include: C #,.NET 5, SQL, JavaScript, HTML, CSS and UI/UX components, as well as tools: Adobe xD, Visual Paradigm, Visual Studio, MS SQL Server.

The software development methodology must include agile methods, as well as the use of project management and collaboration tools.

4 Solution Architecture

In this paper, the architecture of the system is observed, from the aspect of functional modules, as well as from the aspect of production infrastructure and server-client components of the system.

During the analysis and design, our team identified 3 main functional modules, which are:

eCard Module

The primary goal of the eCard module, which will be available at the national level, is both to more efficiently organize health records and to reduce the time patients spend in waiting rooms. It is necessary to provide a unique data repository for patients with a comprehensive view of their dental health. The patient has a detailed insight into the history of dental diseases, the price list and the services it provides dental clinic. The medical staff of the dental practice also has access to eCard module. eCard module also offers an online appointment scheduling feature. When scheduling an examination, the patient is required to fill out a form in which specifies the date and time of the desired inspection.

eWorker Module

The main goal of the eWorker module is to better organize work responsibilities within a dental practice. The users of the eWorker module are employees of the dental practice, each of whom has a unique username and password. Employees include the owner of the dental practice (manager), dentists, medical assistants and administrative staff.

eMaterial Module

The eMaterial module is intended for the owner of a dental clinic. The whole process of procurement of consumables requires certain procedures before the cooperation between the owner of the practice and the supplier of equipment. The task of the owner of the dental office is to know exactly what material he needs and in what quantities, and to know where he can look for these funds. If the owner is a dentist, he will do it himself, and if he is only the owner of the office, he will consult with the main employed dentist regarding the details. The owner of the dental practice is expected to know in advance what material he needs to order.

Connection Between Modules

The modules are interconnected in such a way that the e-Worker module defines the rights and roles of workers within the application as a user, which includes formal rights to enter and edit data as well as access to individual functionalities of the module. For example, a worker who has the role of a medical technician cannot edit the procurement of materials, but he has all the rights to enter and edit patient data.

The e-Maerial module updates the available work and consumables within the clinic and introduces order in their physical consumption and use. Restrictions, which could arise as a result of the non-existence of some material in stock, for the purpose of performing a medical request, would prevent the creation of appointments through the reservation system in the e-Card module, with notification of lack of material in storage. Therefore, the reservation of a medical procedure is directly related to the reservation of materials in the warehouse.

The elaboration of the infrastructure modules and server-client components is presented in the diagram below. The main components of the server infrastructure are the Application Web server with RuntimeEnviroment for.Net Core and the database server.

The basis on the client side is a web browser, which can be on a PC or a portable device, considering the responsiveness of the design.

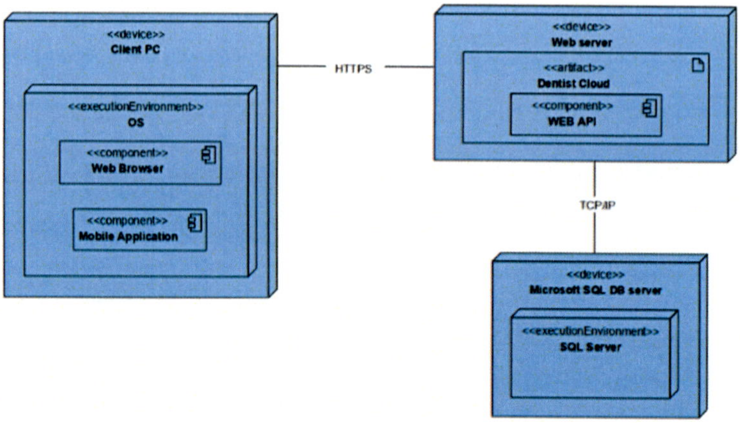

Fig. 2. Platform architecture

5 UI Prototype

A prototype GUI interface was made through several iterations of development and testing. The final proposal is shown in the figures below.

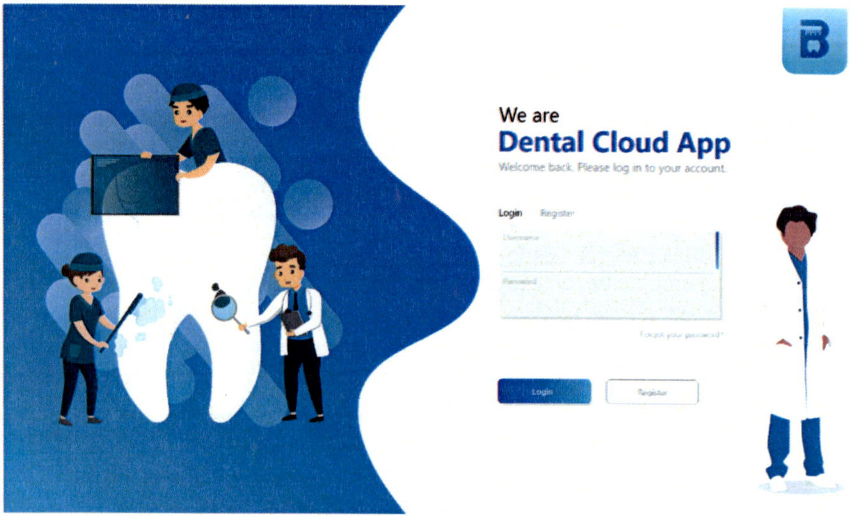

Fig. 3. Login screen

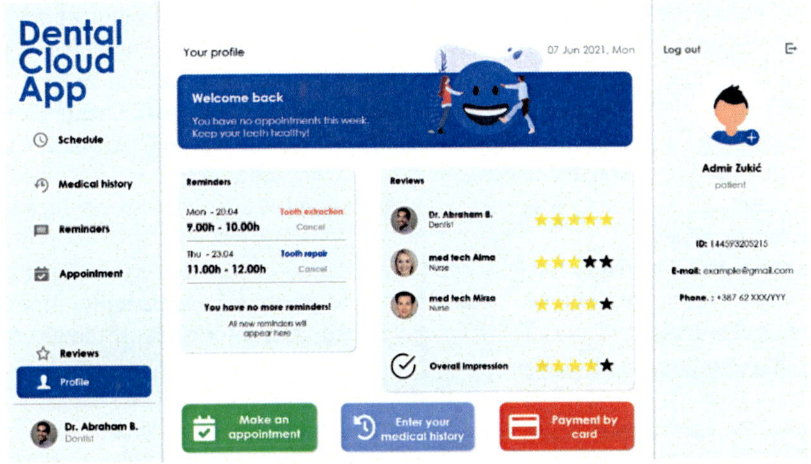

Fig. 4. User profile

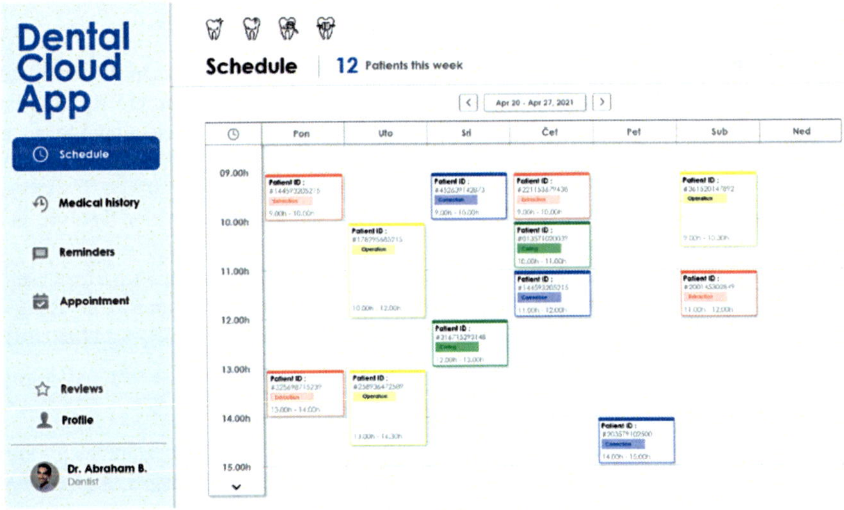

Fig. 5. Schedule

6 Conclusion

By far, the problem, its solution and practical realization are presented and elaborated in detail. Dental Cloud App Prototype is a tool designed for dental professionals and patients and many benefits, which are offered, are extremely important for users, both dental staff and patients. The Dental Cloud application prototype software is a tool that allows dentists to organize patient data and administrative documentation.

It is specially designed for private dental practices in the conditions of poor development of e-health systems in the Balkans area and their preparation for future integration with the upcoming national Industry 4.0 programs.

Patients, as users of this application, are offered a number of benefits that are extremely important, especially now that the pandemic has become part of everyday life. With this application, the patient can schedule his appointment and the scheduled appointment can be changed or canceled. Notifications of appointments are forwarded to patients by email.

In the world we live in today, a good organisation of time is very important and therefore, by promoting innovative strategies, and exploiting increasingly affordable wireless technologies and inexpensive computer hardware, anyone can use the advantage of digital opportunities and can join the information age.

Future research and development will extend to the development and use of mobile applications. Special attention will be paid to the integration with tertiary services in terms of direct connection with various devices for additional tests in order to accurately diagnose the disease and trauma (X-rays, laboratory tests…).

References

1. Purnomo, W.A., Prima, W., Yusran, Efendi, R., Suryadimal: Analysis and design of web-based health service information systems (E-Health), in the industrial revolution era 4.0. Journal of Physics: Conference Series, Volume 1764, The 1st Paris Van Java International Seminar on Computer, Science, Engineering, and Technology (PVJ_ISComSET), 15–16 July 2020, Tasikmalaya, Indonesia (2020)
2. Aceto, G., oPersico, V., Pescapé, A.: Industry 4.0 and health: internet of things, big data, and cloud computing for healthcare 4.0. J. Ind. Inf. Integr. **18**, 100129 (June 2020)
3. Le, C.V., Le, D.-N., Nguyen, N.G., Tromp, J.G.: Emerging Technologies for Health and Medicine: Virtual Reality, Augmented Reality, Artificial Intelligence, Internet of Things, Robotics, Industry 4.0, John Wiley & Sons; Salem, ISBN: 978–1–119–50981–3,1119509815 (2018)
4. Pagani, M.: Encyclopedia of Multimedia Technology and Networking, Second Edition (3 Volumes). Bocconi University, Italy (2009)
5. Ruxwana, N., Herselman, M., Conradie, D.: ICT Applications as E-health solutions in rural healthcare in the eastern cape province of South Africa, Health Inf. Manag. J. **39**(1), ISSN 1833–3583 (PRINT) ISSN 1833–3575 (ONLINE) (2010)
6. BGL Tech: SOLID design principles explained [Online]. Available at: https://medium.com/bgl-tech/what-are-the-solid-design-principles-c61feff33685 (2019). Accessed 15 May 2021

Application of UML in Designing Beekeeping Production and Sales, Using Visual Paradigm and Other IT Tools

Merjem Bajramović[1], Mujo Hodžić[1], Nermin Goran[2(✉)], Sanid Muhić[1], and Aljo Mujčić[3]

[1] Faculty of Polytechnic, University of Zenica, Zenica, Bosnia and Herzegovina
{merjem.bajramovic,mujo.hodzic,sanid.muhic}@size.ba

[2] Faculty of Transport and Communication, Faculty of Polytechnic, University of Sarajevo, University of Zenica, Zenica, Bosnia and Herzegovina
nermin.goran@size.ba

[3] Faculty of Electrical Engineering, University of Tuzla, Tuzla, Bosnia and Herzegovina
aljo.mujcic@untz.ba

Abstract. The application of IC technologies in all branches of the economy and society is becoming more common, including agriculture. Beekeeping, as a part of agriculture, strives for maximum yield per each hive, therefore it requires the different software solutions. This paper describes a software solution that allows recording and monitoring of all hive yields with different control procedures over a period of time. These solutions can be used simultaneously to promote and sell bee products.

Keywords: UML · IK · Visual paradigm · PHP · MySql · PHPRunner enterprise · Inkscape · WampServer

1 Introduction

The idea for application and cloud-oriented beekeeping appeared due to the shortcomings and complexity of beekeeping in the traditional way (the use of physical hive diaries and the maintenance of different records) [1]. Traditional methods of record conditions in the hive often lead to inaccuracies in data management of individual bee colonies (hives), which often results in their loss, due to bad decisions during prevention [2].

On the other hand, analysis using data in paper format becomes inefficient and unnecessarily time-consuming. In addition, there is a weak interaction between beekeepers, which is limited to only occasional lectures within beekeeping societies. It was concluded that it is necessary to create an efficient management system, using modern IC technologies. In that way, it would contribute to the maintenance and increase of bee productivity in this area [5]. The application of this system would be possible at the level of individual beekeepers and beekeeping societies, as well as at the level of cantonal, federal and state ministries of economy. We used UML (Unifield Module Language) language to design this applicative solution along with the Visual Paradigm software tool.

N. Ademović et al. (Eds.): IAT 2021, LNNS 316, pp. 35–43, 2022.
https://doi.org/10.1007/978-3-030-90055-7_4

In addition to them, PHP, MySql, PHPRunner Enterprise, Inkscape and WampServer were used for design and implementation of this solution [3, 4].

2 System Specifications

The segment of the cloud related to the beekeeping has to provide the following functionalities:

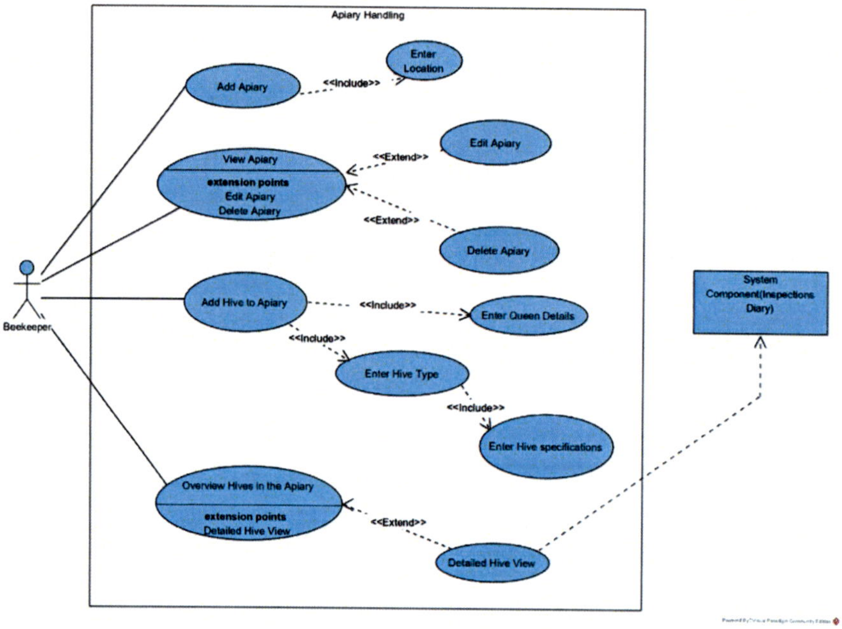

Fig. 1. UML diagram for guiding beehives (Use Case)

• Beekeeping diary - includes the ability to add, edit and remove one or more apiaries and add and edit individual hives for each apiary. For added hives, the entry of inspection date should be provided for each hive in the calendar and add records of important observations and treatments performed on the bee colony. After entering the data for the apiary, beekeepers should be allowed to enter certain details about the inspection of one or more hives that they own. In addition to details about inspections, beekeepers must have access to a special section for entering records and details about the means and medicines used in the treament of certain hives - the treatment period. The application should, in order to facilitate the work of beekeepers, also enable the entry of details on the nutrition of bee colonies, which includes recording the quantity and type of food used. Beekeeping log modules should include elements depicted in Fig. 1:

- Apiary management
- Hive management

Diary of observations during hive inspection, recording of feeding and treatment of hives.

• The inventory management module should allow beekeepers to enter the required number of tools, resources and materials from different manufacturers and other data needed for the purpose of re-procurement. In addition to entering the quantity balance, the minimum allowed quantities for each item should be defined. This information will initiate the sending of a notice for the purpose of executing the order for the new procurement.

• Product sales is a section intended for the sale of bee products. This allows beekeepers to have a list of the products they offer with the addition of product descriptions. In addition to online shopping, customers have access to information on other locations where they can purchase products. Each product presented in the sales section should be evaluated by an authorized laboratory. An additional product ranking section is available to customers who can leave product reviews and ratings based on their personal experiences. Figure 2 depicts UML diagram of all necessary classes.

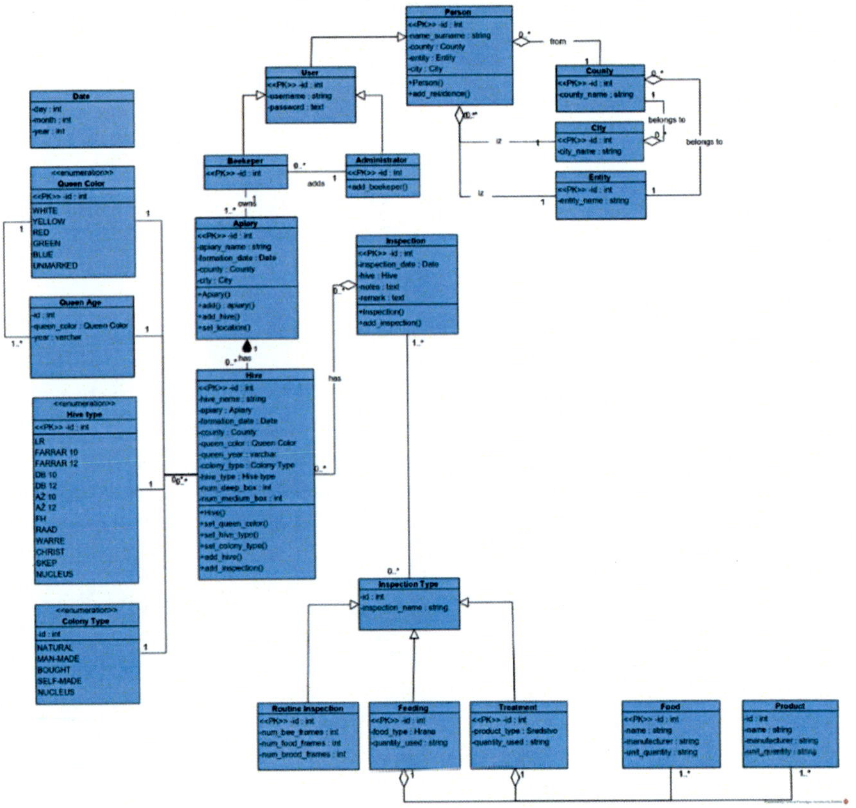

Fig. 2. UML class diagram for the diary

3 Prototype of UI

Prototype of UI has modules for:

Administrators

- Beekeepers (beekeepers registered in a beekeeping society)
- Users (customers) - only a part of the software intended for the purchase of products is available to users.

The layout of the prototype of user interface is presented in Fig. 3.

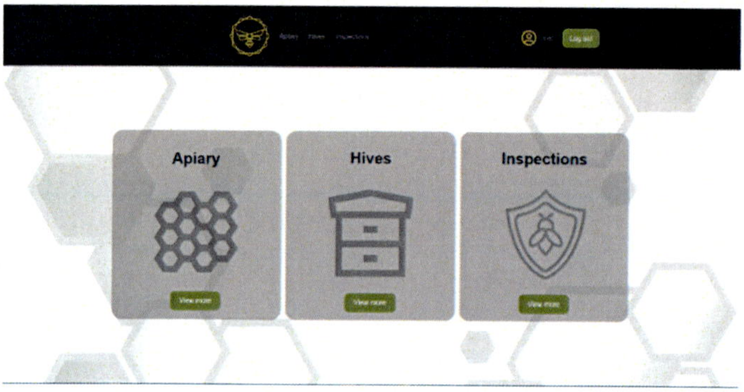

Fig. 3. Prototype UI with a logo

3.1 User Requirements

3.1.1 Administrator

The administrator as part of the beekeeping system can be the president of the beekeeping society. All beekeepers, before being able to use the system, must register as members of a beekeeping society. The administrator assigns a username and a temporary password to the registered members, which each user can change after the first login to the system. The administrator has the ability to add components to the system modules (for example: in the hive section he/she can add module and the system provides a dropdown menu in which the specific hive type is selected. The administrator can enter additional elements such as creating a new hive type,…). In addition to the above, the administrator has the obligation and the ability to add a final quality score for all presented and certified products.

3.1.2 Beekeepers

Beekeepers are users of application which is intended for honey producers. When they get credits, obtained by the administrator, beekeepers can automatically login with all abilities offered by the part of the system specified exclusively for them.

In addition to access to the user part intended for honey producers, all of them are allowed to use the part which is intended for the purchase of products.

3.1.3 Users

The role assigned to users is intended for potential customers. Users have access only to the part of the system intended for sale to whom they access via a web browser. By accessing the system, users can select the category of honey products they want to search (categories are divided in the same way as the categories of the entire cloud, i.e., products of the beekeeping industry, livestock,...), or search all products without first selecting a category. Purchase of products is not possible without prior registration. Users define the username and password that are associated with their e-mail address. Registration is not possible without prior confirmation (verification) of the e-mail address. In addition to purchasing products, users are allowed to leave reviews (rankings) of purchased products. In this way, the average customer rating is formed for the product. Login UI is presented in Fig. 4.

Fig. 4. Login access to the application

4 Functional Requirements

4.1 Beekeeping

The apiary management section allows the entry of one or more apiaries. The process of entering the apiary specifies the name of the apiary and its location. A list of all available beehives with options for removing or modifying them is provided as shown in Fig. 5.

In this site, modification relates to the change of the state in the apiary and its location. In the list of available apiaries, we have some information about available apiary, the location and the number of beehives/companies (related to the section on hive management).

4.2 Management of Beehives/Society of Bees

At the start there must be at least one apiary in order to the hives to be brought in. After we have an apiary, we can add at least one beehive. In order to add the hive, in the menu we need to pre-selected an apiary. After selecting the bee colony, we can notice all

Fig. 5. Prototip UI (general interface)

avaliable beehives which are located in the menu. The hive management system provides the ability to add, edit, give detailed insight as well as a global view of all hives in the selected apiary with the ability to remove individual hives.

4.3 Entering Hives

The entry of a hive or a society of bees means the entry of all characteristic data for that society, as follows:

- Name of the bee company (the user is allowed to specify a special name or auto default name),
- Selection of the type of the bee queen and the year when the queen was born (age). The system provides a selection of predefined colors (white-2001/2006/2011/2016…, yellow-2002/2007/2012/2017…, red-2003/2008/2013/2018…, green-2004/2009/2014/2019…, blue-2005/2010/2015/2020… or unmarked). According to the color and year selection, the user selects the exact year or user can leave it to the system to automatically generate the color/year.
- Hive type specification. The last requirement for successfully adding a hive to an apiary is the hive type specification. The system provides the user a menu from which the desired type of hive is selected. The hive is added on the principle of selecting the hive type. Hive types available in the menu are: Langstrit Rutova (LR), Farrarova (Farrar 10, Farrar 12), Dadan Blatova (DB 10, DB 12), Anton Zindariceva (AZ 10, AZ 12), Francois Huberova (FH), Raadova Raad), Warre (Warre), Christova (Christ), Pletara (Pletara), Nukleus. After selecting the hive type, the user enters the number of extensions and/or half-extensions which the hive is composed of, which completes the addition of the hive. The hive is stored in the apiary. The beekeeper can add new hives as desired.

4.4 View the Hive List

Insight into the records-list of hives is available after adding a new hive or, after selecting the beehives in the main menu. The hive record contains all the hives in that apiary with the name, hive type and the date of the last inspection. For each hive in the records, a more detailed insight into the information about that hive is available, which is related to the inspection log module.

4.5 Review Log

The inspection log is a module that becomes available after entering the first hive into the apiary. Thus, the user does not have the option of entering any notations related to the inspection while there are no hives in the apiary. The user selects the apiary to be inspected and then the hive addition module. In that way, you get an insight into the record list of all hives in that apiary and the possibility of adding an overview for each of them (Fig. 6).

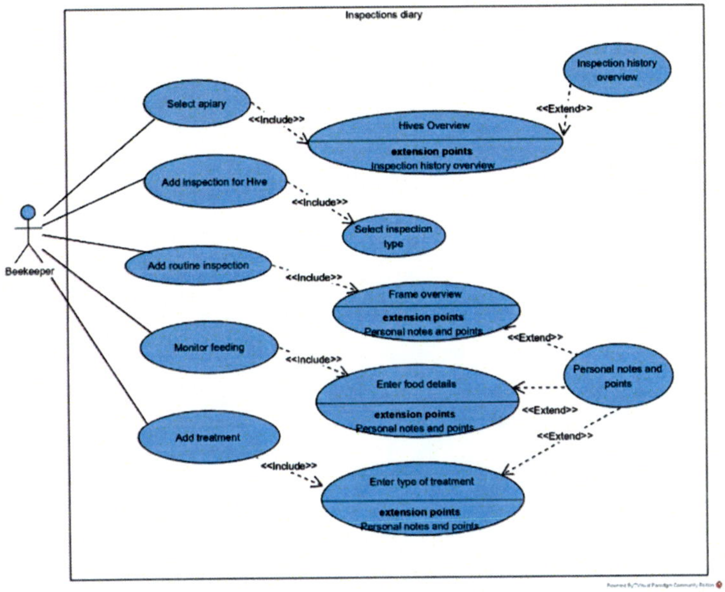

Fig. 6. UML diagram for keeping a review log

The user can add one or more views for each hive and save them all at once or each of them individually.

4.6 Add a Preview

For each preview, the user enters the wanted date. Before entering the observation, the user selects the type of review to be performed. Three types of examinations are defined; routine examination, feeding and treatment. The user selects the type of examination which is performed, and according to the choice, a diary is kept.

4.7 Routine Examination

The option of routine inspection allows the user to enter the number of frames with food (honey), litter… In addition to entering the number of frames, according to the inspection, the user can add their own observations (for example, whether a nut was found or similar).

4.8 Feeding

The option that allows you to enter data on the feeding of the hive allows the user to enter the type and the amount of used food. It is also possible for the user to enter their own observations and remarks related to food or feeding time.

4.9 Treatments

The option that allows you to enter data on the treatment performed on the hive allows the user to enter the name and amount of the medicine which is used. Logging modules and the Treatments module provide the user with the possibility to enter own additional observations.

4.10 Save Reviews and View Reviews

As previously described, the user can add multiple views to one or more hives at the same time. After completing the inspection, the user has the opportunity to inspect individual hive in the apiary.

4.11 Inventory Management

The inventory management module allows the user to monitor the quantitative status of beekeeping tools, hive bodies, frames, preparations and used food. When entering the contents of the inventory, the user enters the name of the product, the quantity in stock and the minimum allowed quantity. The contact of the manufacturer from whom we purchase the specified inventory item is also entered. In case of reduction of quantities at the barn below the minimum stated quantity, the user receives a notification for the procurement of missing products. When entering the product next to the name, the user is required to select one of the categories to which the product belongs to: bodies, frames, additional tools, food and preparations. After adding the products, the user has an available list of all added products in the inventory. In addition, for better visibility, the user has the possibility of insight into the inventory by index (bodies, frame, additional tools, food and preparations) where only products from the selected category are given to him for review.

4.12 Product Sales

The product sales module requires prior entry of the product for sale. This module provides the option to add your own products. When adding a product, the following is stated: product name, composition, quantity in stock, product quantity per unit and unit price. Own products are added to a separate indexed table that separates the user's products from the rest of the inventory and thus allows the connection to the sales module. After adding the product, the user gets an insight into all the products to be offered. When entering the product, the user has the opportunity to submit a request for testing the same by accredited laboratories for quality control.

5 Conclusion

This solution, which refers to cloud beekeeping, can be applied to other branches of agriculture. If it is cattle breeding, then instead of bees we can use cattle with their peculiarities. Instead of adding beehives, in this solution (cloud livestock), a farm can be added. Instead of adding bee hives, cattle pens would be added.

In addition, the idea and most of the diagrams could be applied to fruit growing in the cloud. Some of the objects do not exist in fruit growing (frame) but could be replaced with some other objects which can be related to fruit growing.

We suppose that all diagrams with minor modifications can be adapted in other branches of agriculture and this we plan to develop in future works.

References

1. Karagiannis, D., Burzynski, P., Miron, E.T.: The "IMKER" Case Study- Practice with the Bee-Up tool. Universitat Wien, Wiena (2017)
2. FAO-Food and Agriculture Organization of the United Nations: Good beekeeping practices: Practical manual on how to identify and control the main diseases of the honeybee (Apis mellifera). Rome (2020). https://doi.org/10.4060/ca9182en
3. Rumbaugh, J., Jacobson, I., Booch, G.: The Unified Modeling Language Reference Manual. Addison Wesley Longman Inc (1999)
4. Greenspan, J., Bulger, B.: MySQL/PHP Database Applications. M&T Books, Foster city, Californija (2001)
5. Vapa-Tankosic, J., Miler-Jerkovic, V., Jeremic, D., Stanojevic, S.: Investment in Research and Development and New Technological Adoption for the Sustainable Beekeeping Sector. https://www.mdpi.com/journal/sustainability (2020)

Digital Workplace Transformation: Small Scale Literature Review and Study of Managers in Bosnia and Herzegovina

Mićić Ljubiša[1,2](✉)

[1] Faculty of Economics Banja Luka, University of Banja Luka, Banja Luka, Bosnia and Herzegovina
ljubisa.micic@ef.unibl.org
[2] Faculty of Economics Subotica, University of Novi Sad, Novi Sad, Serbia

Abstract. Modern workplace requires not just technological adaptation but also organizational and process changes in order to achieve functional and operational transformation of traditional workplace into digital future workplace. Different sociological and technological changes have influenced that traditional ways of communication, cooperation and collaboration are achieved with use of specific technological tools and organizational adaptation. Paper provides literature review on the subject of digital transformation correlation to technological, and organizational conditions. Search and review include search of key research databases including Web of Sciences, Scopus, Springer Link and Google scholar. Additional, paper provides short quality field research with 44 managers in small and medium companies in Bosnia and Herzegovina and their perception of importance and correlation of above mentioned factors and digital workplace transformation. Based on their professional perception, experience and recommendation, the paper also provides a proposal of framework for successful digital workplace transformation which could be a base for further research and use by academics and digitalization professionals.

Keywords: Digital workplace · Digital transformation · Technology change · Management

1 Introduction

In the last 10 years, digital transformation has manifested itself as a need in all spheres of business. Recently, major developments have attracted immense attention in the world of work. All over the world debates about challenges and opportunities that current trends seem to bring, are taking place among policymakers, academics, scholars and the others. Opinions are divided about what the changes are bringing to the future of work. Technology represents a one of the key megatrends in the upcoming years that influence of work in the world and the society. According to Coetzee [1] in twenty-first century "organizations of the Industry 4.0 era are rapidly becoming smart networked workspaces as a result of the exponential revolution brought about by technological advancement".

N. Ademović et al. (Eds.): IAT 2021, LNNS 316, pp. 44–55, 2022.
https://doi.org/10.1007/978-3-030-90055-7_5

According to a study conducted by Nokia, the average American smartphone user checks their phone every six and a half minutes [2]. This represent increasing use of technology, everything is changed in the way that people live their lives. Marina Dimova, as The Governance Chief Technical Specialist, UNDP Bosnia and Herzegovina, stated in her interview that the region SEE (South East Europe) have made significant breakthroughs in the e-government space by using new technologies to increase the efficiency and effectiveness of their service delivery [3]. UNDP (United Nations Development Programme) in Bosnia and Herzegovina and the British Embassy in Sarajevo held a digital transformation conference in partnership with PwC (Central and Eastern Europe Private Business Survey). Conclusion of the conference was that policymakers need to put in place a whole- of-government vision for digital transformation. Next, institutions need to improve their digital skills and embrace innovation to create progressive tools for creating policy and service delivery tools. Since the digital transformation can only benefit comprehensive, people-centered e-government, all citizens of Bosnia and Herzegovina will be in a better position. What is also an important conclusion of the mentioned conference is certainly that the enthusiasm of innovative ideas is a powerful driver of change. Based on a report of Central and Eastern Europe Private Business Survey 2019 [4] that many private businesses in CEE (Central and Eastern Europe) were in just the initial stages of digitalization. Digitalization as a way of improving processes saw the most leaders (about 68%), only 35% of them seeing it as a way to develop end-to-end digital solutions that ultimately meet customers' needs. Digital workplace is a fairly broad term, and it practically means that it is a virtual replacement for physical office space. It allows employees to successfully complete their work responsibilities through an appropriate platform, software or device. Based on Report "The digital workplace: Think, share, do: Transform your employee experience" [5] the digital workplace framework includes the next four components:

1) Use: collaborate, communicate, connect; A productive business relationships within and beyond natural work groups and to enable knowledge sharing across the organization
2) Technology: the digital toolbox; the key is to adopt the right tools for your employees to do their jobs.
3) Control: governance, risk and compliance; Information flow and use must also comply with your organization's policies and industry regulations.
4) Business drivers: measurable business value; the direction of your organization should guide the direction of your digital workplace.

2 Basic Terminology in Literature

Authors Vairinhos, Matos and Edvinsson [6] of the first book chapter in the book "Knowledge, People, and Digital Transformation Approaches for a Sustainable Future" analyzed main issues of Digital Transformation (DT) and relate this concept with National Intangible Capital (NIC), having in mind people's attitudes and reactions to this transformation process. In this study was used data from the EU (year 2017), where it presented in a holistic way the Europeans countries attitudes of Digital Transformation (DT) and specifically in relation to Robotics and Artificial Intelligence (AI). Conclusion from this study in general way, that there was not point out strong opposition to DT. However, in some advanced countries informed people manifest negative reactions and distrust to AI and Robotics.

Digital transformation is the process that cannot be stopped. The digital workplace according to [7] "On the one hand, provides a sample opportunity for learning and growth through up skilling and innovative and smarter ways of working and collaborating. On the other hand, employer surveys show that people are working longer hours and asking employers to provide solutions for well-being, productivity and stress reduction, all of which could point to symptoms of trying to survive the bombardment of constant change and fast pace of digital work"… "While managers are learning how to capitalize on the power of smart digital technology in order to improve business performance, they also need to take heed of the mitigating potential negative impacts on employee well-being, thriving and productivity". Organizations need to reinvent themselves and focus on becoming a self-driven, agile and data-driven learning organization, in other words SADL organization.

Definition of SADL organization is reflected through four critical shifts that need to take place: [8].

- Shifting towards a Self-Driven culture;
- Adopting an Agile Philosophy and Mindset;
- Embracing Data-Driven Insight;
- Cultivating a Learning
- Culture.

Colbert, Gerard and Gerard [9] analyzed in their study transformation of work, work practices, and workplaces. The results showed that digital competencies of the workplace and the ways in which technology is used in the workplace will continue to develop and change. This will provide an increasing level of organizational effectiveness.

In digital workplace transformation, it is important that each puzzle is put together. In that sense, the people who participate in the changes are also important. Author Salgueiro [10] analyzed the relation between technology and the Human Mind. In his study it stands out that technology is a projection of human reason in the world. While the animals analyses the world around them, they only see and act according to what is given to them, but man creates artefacts, the visible effect of the synthetic nature of human thought. …"The scientific-technological development model is as follows: Human mind Creation of artefacts Incorporation of the new artefact into the mind Science Human mind Creation of new artefacts." Many researchers regarding the digital transformation have asked the main question: Can the connection between people and technology be regulated and predicted? Various scenarios of this relationship are set, where people will be subordinated in this process. "…Digital transformation is no longer an issue, everything is digitized, we are beyond digitization…" [6].

As the notion of the digital workplace has expanded, it is reflected through multiple technological developments through the digital transformation of many areas of life. In his study Gartner [11] states the digital workplace "enables new, more effective ways of working; raises employee engagement and agility; and exploits consumer- oriented styles and technologies". Based on four main research topics on the digital workplace that are present in academic-practitioner literature: 1) Collaboration, 2) Compliance, 3) Mobility, and 4) Stress and overload, Köffer [12] in his study categorizes practical implications on the digital workplace into 15 concepts. In total, the systematic review identified 79 relevant publications, whose practical implications on the digital workplace could be categorized into 15 concepts. Conclusion was that the many concepts seem to be relevant for the digital workplace as a whole beyond the specific research context they were investigated in. Authors Wiliams and Schubert [13] identified digital workplace designs through an analysis of 13 Enterprise Collaboration Systems (ECS) using organization. This study represents six distinctive workplace designs. These designs are divided into two groups: 1) people-focused designs (three designs) and 2) process-focused designs (the other three designs). Conclusions were that the different organizations require different specific requirements, which may relate to the specificity of the organization's business, historical path or the nature of the project implementation.

3 Research Methodology

Literature Review Methodology

Systematic literature review was conducted following methodology developed by Barbara Kitchenham. First phase includes research of specific databases connected to subject research and filtration of papers, books and journal editions according to connectivity with subject research. Literature review process consisted from several steps: primary scanning of databases results, filtration of the results based on title, filtration of the results based on abstract/table of content and final selection based on content of the paper/book (Fig. 1).

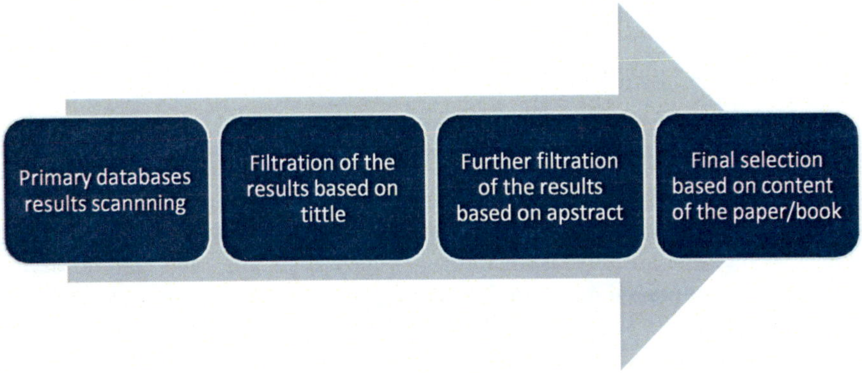

Fig. 1. Literature review process. Source: Authors' contribution

Keywords for the database search have been "digital workplace". Databases included in research are SpringerLink (EISZ), Scopus (Elsevier), Web of Science – WoS + ESCI backfiles 2005–2014 (Thomson Reuters) and Google Scholar (Google) as additional supportive research for different reports and research addition. Time limitation was 2016–2021 (some databases allow titles and abstracts of the papers that will be published in 2021). First output of the bases for the identified keywords are: Scopus 1113, WoS 757, SpingerLink 12803 and supportive, but not fully used Google Scholar with 84800 results. Database search was conducted in the period February - September 2020 and we used only publications published in English language.

Based on year, Schopus had a "peak" of paper with these keywords published in 2019 (319 papers) followed by 2018 (252 papers), 2020 (201 paper), 2017 (198 paper) and 2016 (142 papers). Additionally, it included one document that will be published in 2021.

Based on year, WoS had also a "peak" of number of published papers with these keywords in 2019 (204 papers) followed by 2018 (200 papers), 2020 (124 paper), 2017 (122 paper) and 2016 (107 papers).

Number of published papers in WoS correlates with the number in the Scopus database, which is a clear signal that research is present as a subject in those highly important databases and it is of high interest to researchers in this field.

Having in mind that SpingerLink is a publisher more focused on publishing monographs, books and conference proceedings and that this database search had a significantly high number of outputs, we used a filter "relevance" and took only the first page with the most relevant publication in this field. However, the total number of more than 12000 publications in this time framework, including several in 2020, confirmed that this research subject is up-to-date and relevant for the scientific and business community, which confirmed our motivation to do a following research addition to the literature review phase.

Next step included identification of possible copies of papers in different databases in order not to use the same paper in overall statistics. After filtering and finalization of the papers, books and other publications, we limited our review to those papers/books/publications relevant for this specific paper and which correlates the most with filed quality research we planned to implement.

Trough process of the literature review, we identified possible research question and tried to examine them through qualitative research using Delphi technique in order to provide professionals a useful framework for digital workplace development.

Field Quality Research Methodology

Field research we implemented was a quality research based on Delphi technique, which is used when research is connected to consensus views of experts [12]. Technique includes quality and quantity data gathering, which is complex and requires several phases, in the most cases three or until experts agree on a specific issue, at least 70% or until one of the experts would not change his view in any circumstances (Fig. 2).

In this phase, we selected more than 40 experts (44) in small and medium companies from Bosnia and Herzegovina from different industries but which have some kind of need for digital transformation of the workplace, especially having in mind COVID-19 pandemic in 2020. SMEs are defined by the local legal framework in Bosnia and Herzegovina based on size, profit and operation and local framework is used as a ground for SMEs definition.

All three phases in the mentioned Delphi technique are conducted with the same 44 experts. After each of the phases and gotten answers, we compared them and discussed them with other research participants in order to achieve 70% of the agreement on framework development. Each participant had a chance to adapt new proposals, change the answer and possibly stay with his previous answers which would, in case other participants agree on a final framework, eliminate his proposals and opinion.

Participants were from different industries: ICT sector, technical services and crafts, education, transport, consulting and beauty industry. All of them have been identified as professionals with experience in digital workplace transformation.

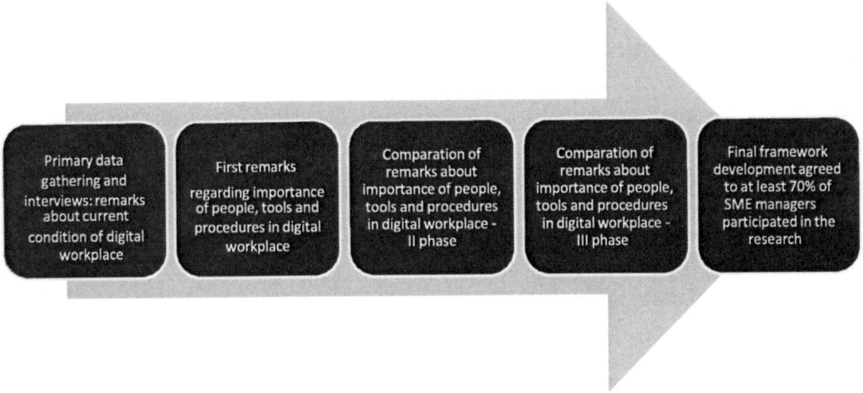

Fig. 2. Field research process. Source: Authors' contribution

Data gathering included a first phase – semi-structured interview that included open questions as well as limited answer questions, including Likert scale, checklists and check tables but also open questions. All answers have been written and after it content analyzed and transformed to usable format. We used MS Excel as a tool for insertion and evaluation of gathered data. Second phase was introduced after answers gathering and their comparison. Participants have been asked for possible adaptation after they saw answers and proposals from other participants. After second and third step adaptation, we successfully succeeded in getting 90% agreement of the developed framework (40 of 44 participants) which we propose in research results.

4 Research Results

In the first phase of interviews, we gathered data about opinions, remarks and experience regarding the current condition of digital workplace in SMEs in Bosnia and Herzegovina. In the continuation, we will provide descriptive statistics of gotten answers in this phase.

Majority of the managers have answered that their employees are familiar with the concept of the digital workplace. More than 90% of participant confirmed employees are familiar with the concept fully or partially (Fig. 3).

Majority of the managers also confirmed that their employees are supporters of digital workplace transformation (more than 2/3 are supporters). However, more than 30% of

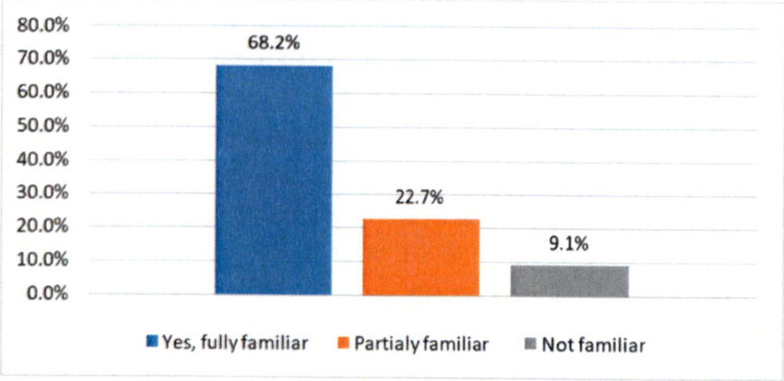

Fig. 3. Employees are familiar with the concept of digital workplace (N = 44). Source: Authors' contribution

managers confirmed that there is latent resistance to new technologies and transformation of the workplace to digital (Fig. 4).

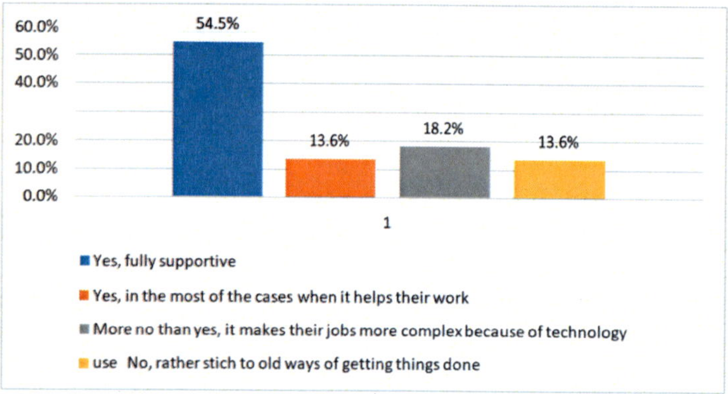

Fig. 4. Employees are supporters of digital workplace transformation (N = 44). Source: Authors' contribution.

Based on the previous answer, managers also confirmed that people are key factors to digital transformation but also, some of the managers, identified other key factors of successful digital transformation (Fig. 5).

When being asked to compare three key factors of successful digital workplace transformation, managers identified tools as the least important, focusing on people and procedures as more important ones. In interviews, most of them agreed that even the best and up-to-date tools are not any kind of confirmation that digital workplace transformation will be successful. From another point of view, several of them agreed that even with the basic tools, successful digital workplace transformation can be achieved so therefore, they advised managers to focus on people, procedures and processes rather than on purchasing expensive tools with no effect to success of transformation (Fig. 6).

More than half of managers agreed that we are following the latest trends in digitization and that, even if we have some difficulties, we are not far behind the modern

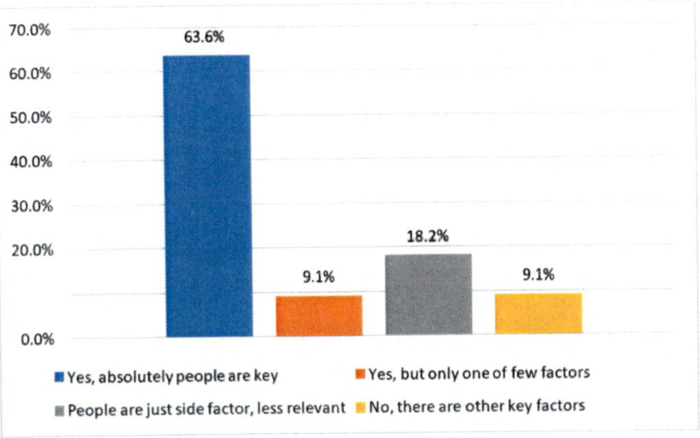

Fig. 5. Managers identified people – employees as a "key factor" of successful digital workplace transformation (N = 44). Source: Authors' contribution.

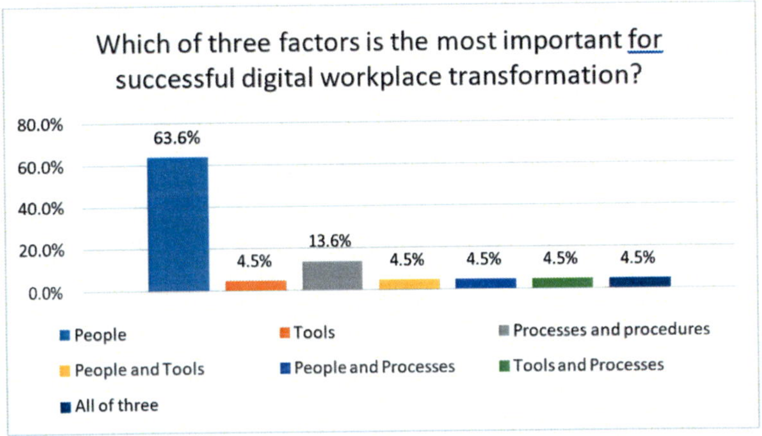

Fig. 6. Tools are the least important in process of successful digital workplace transformation (N = 44) Source: Authors' contribution

way of doing businesses. Less than 1/5 of participants, approximately 18% stated that we are too below average level of digitalization or going in the wrong direction (Fig. 7). This evaluation was promising and confirmed that even with public misjudgment we are far below the modern digital way of doing business, this confirmed that we are still following trends and trying to make our business more digital. Additionally, this data is important having in mind that managers were coming not only from the ICT sector, which is usually modern and trying to follow trends, but also from other usually consider "non-technological" sectors.

Following current condition assessment by managers regarding the level of the digital transformation, we moved to more quality research and framework development. After three phases of evaluation by the interviews and possible adaptations of the managers' answers, we identified a framework for successful workplace transformation to digital

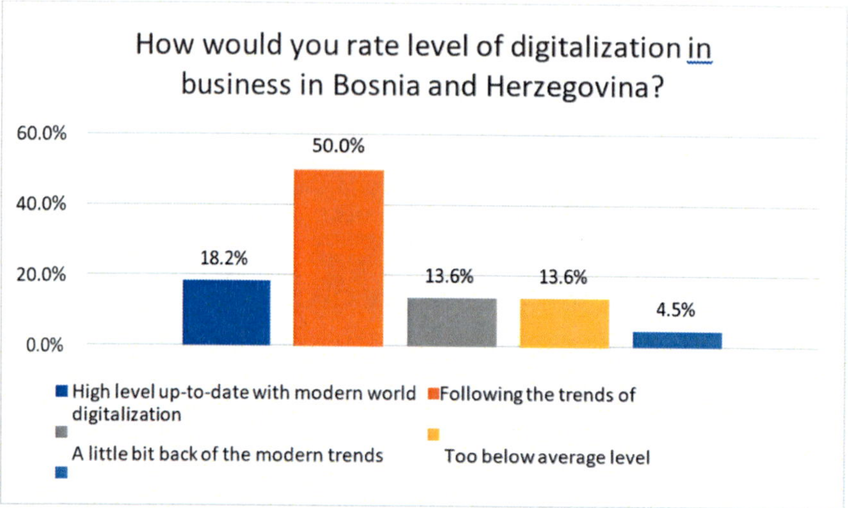

Fig. 7. Following a lead to digital transformation in business in Bosnia and Herzegovina (N = 44). Source: Authors' contribution

one, combining three factors identified as key factors for successful transformation process. Proposed framework included recommendations based on local and regional obstacles and limitations, including complexity and local characteristics of doing business in the limited territory of Bosnia and Herzegovina in several represented industries.

Framework is presented and explained in further text and finally it is the result of more than 90% agreement of participatory managers (40 of 44 agreed to this final framework and stated that this is acceptable one for successful digital transformation of the workplace).

Steps in the framework for successful digital transformation of the workplace are divided in three phases: evaluation of technology needs and current condition for specific SME, identification of key processes and operational challenges at workplace and development of specific transformation model (Fig. 8). Evaluation model would include analysis of enterprise size and industry where it operates, identification of clients and partners profiles and evaluation of need for technology for every day operation in the company. Second phase would include key process challenges such as identification of time consumers, typical mistakes and cost influenced by it. Additionally, it would include interview and self-evaluation of the employees in technical skills and knowledge (Fig. 8).

Proposed framework components for successful digital transformation would include different levels of transformation process: strategy, culture, attitudes and skills but also tools and processes that are familiar and welcomed by employees (Fig. 9).

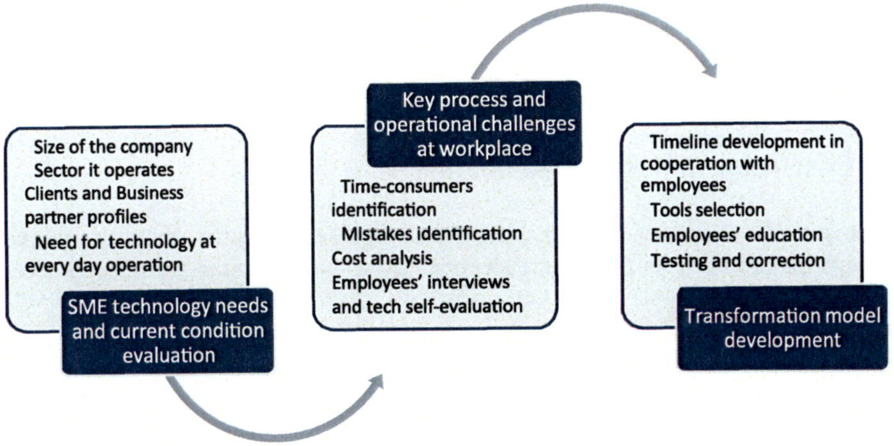

Fig. 8. Proposed framework steps for successful digital transformation of the workplace in SMEs in Bosnia and Herzegovina, considering regional limitation and obstacles. Source: Authors' contribution

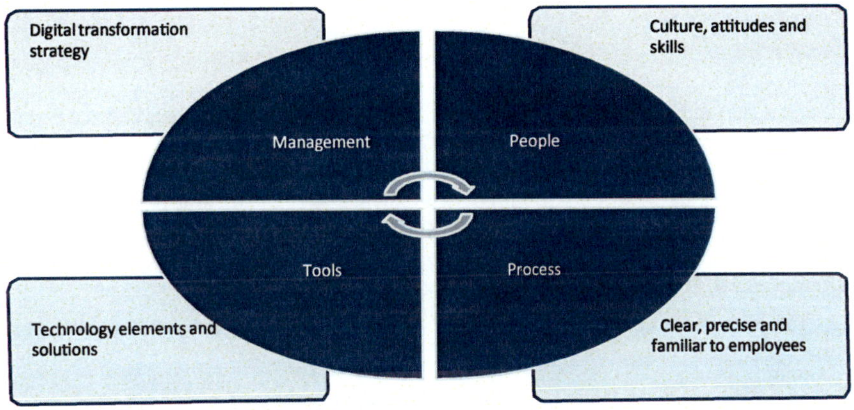

Fig. 9. Proposed framework components for successful digital transformation of the workplace in SMEs in Bosnia and Herzegovina, considering regional limitations and obstacles. Source: Authors' contribution

5 Research Limitations

There are several limitations for this research, which can be a ground for future papers in this field. First, the size of the sample is something that can be improved in further research. Although this was a quality research based on Delphi technique, bigger sample size could only improve quality of the results. Furthermore, broader regions and SMEs from more sectors could just benefit from the quality of the research and could help identify possible limitations that are not obvious in this geographical region and these sectors. Additionally, literature review could be improved with more databases and papers included in the literature review, but having in mind time limitation and complexity of the subject it can be only a ground of more complex and longer research in the future.

6 Conclusion and Recommendation

Paper provides evaluation based on quality technique of key success factors to digital workplace transformation. People are identified as the most important factor but process regulation as well as tools are also considered as important. Based on three phase evaluation, there is a framework provided as a ground for success in digital transformation which could be a starting point for professionals but also researchers who could improve the framework and identify other key factors in digital workplace transformation.

Recommendation for further research would include further increase of sample size, diversification of the industries from which SMEs come from and possible comparison between SMEs and bigger companies in reference to success of digital workplace transformation. Additionally, the framework could be improved with additional factors that could play an important role in specific business sectors and business segments.

References

1. Coetzee, M. (ed.): Thriving in Digital Workspaces, Emerging Issues for Research and Practice. Springer Nature Switzerland AG, Pretoria, Gauteng, South Africa (2019)
2. Spencer, B.: Mobile users can't leave their phone alone for six minutes and check it up to 150 times a day. Daily Mail. http://www.dailymail.co.uk/news/article-2276752/Mobile-users-leave-phone-minutes-check-150-times-day.html (2013)
3. Dimova, M.: Digital transformation in Central and Eastern Europe. (E.Koenig, Interviewer) Retrieved 9 9, 2020. from http://www.pwc.com (2020)
4. Central and Eastern Europe Private Business Survey 2019: Time to act: moving from good to great in times of uncertainty and digital transformation. ww.pwc.com (2019)
5. Deloitte: The digital workplace: Think, share, do: Transform your employee experience. Belgium. Retrieved 9 1, 2020. from http://www.deloitte.com/ (2014)
6. Brucker-Kley, E., Keller, T.: Beyond digitalization: "My Boss Is Artificial." In: Matos, F., Vairinhos, V., Salavisa, I., Edvinsson, L., Massaro, M. (eds.) Knowledge, People, and Digital Transformation. CMS, pp. 37–54. Springer, Cham (2020). https://doi.org/10.1007/978-3-030-40390-4_4
7. Bersin, J.: Talent, technology, and HR predictions for 2019. Deloitte Consulting LLP: Bersin. Retrieved 9 2, 2020, from Deloitte Consulting LLP: Bersin: https://joshbersin.com/hr-tech-disruptions-for-2019/ (2019)
8. Kohl, K., Swartz, J.: Building industry 4.0 talent. In: Coetzee, M.. (ed.) Thriving in Digital Workspaces, pp. 15–39. Springer, Cham (2019). https://doi.org/10.1007/978-3-030-24463-7_2
9. Colbert, A., Gerard, N., Gerard, G.: The digital workforce and the workplace of the future. Acad. Manag. J. **3**(59), 731–739 (2016)
10. Salgueiro, I.: Knowledge and technology: manas a technological animal. In: Matos, F., Vairinhos, V., Salavisa, I., Edvinsson, L., andamp, Massaro, M.: Knowledge People, and Digital Transformation: Approaches for a Sustainable Future, pp. 25–37. Springer Nature, Switzerland AG (2020)

11. Gartner.: Digital Workplace. Retrieved from Gartner IT Glossary: http://www.gartner.com/it-glossary/digital-workplace (2015)
12. Köffer, S.: Designing the digital workplace of the future – what scholars recommend to practitioners. Thirty Sixth International Conference on Information Systems, Fort Worth 2015. ICIS, Muenster, Germany (2015)
13. Williams, S.P., Schubert, P.: Designs for the digital workplace. Procedia Comput. Sci. **138**, 478–485 (2018). https://doi.org/10.1016/j.procs.2018.10.066

Analysis of Blood and Urine Findings Using the Fuzzy System

Aldina Sušić, Una Drakulić(✉), and Edin Mujčić

Faculty of Technical Engineering, University of Bihać, Bihać, Bosnia and Herzegovina

Abstract. Laboratory analysis of medical findings is one of the basic methods for diagnosing the disease. Based on the findings, the existence of one or more diseases in the organism can be approximately determined. In this paper, an analysis of blood and urine findings was performed, where using fuzzy logic, an application was created that based on the input parameters of the system as an output gives feedback on the health status of the patient. The implemented application consists of several separate fuzzy systems created for analysis of blood findings, lipid status, urine and general liver function, where based on actual data the disease is diagnosed and a recommendation containing the optimal choice of medical treatment is given. Considering the input parameters of the fuzzy system differ by gender, during the initial startup of the application, the menu is offered and it has option of selecting the analysis of blood and urine findings separately for men and women.

Keywords: Analysis of blood and urine findings · Diagnosis of the disease · Fuzzy logic · Lipid status · General liver function

1 Introduction

The fuzzy logic, which was founded in 1995 by Lofti Zadeh, is highly suitable for developing knowledge-based systems for many fields in medicine. Fuzzy logic in field of medicine is mainly used for tasks such as the interpretation of sets of medical findings, diagnosis of diseases, selection of medical treatments, etc. [1]. In fuzzy set theory, there is a degree of belonging of an element to a set, while in classical set theory, an element belongs or does not belong to a set [2]. Despite all the methods in field of medicine, making a correct diagnosis is still considered an art. The most common laboratory tests include blood and urine tests. The significance of laboratory findings is that deviations from the reference values indicate possible diseases of individual organs. Blood analysis is important for the analysis of blood picture, lipid status, assessment of liver function, and urine analysis is great importance for assesing the findings of urinary system [3].

The practical implementation of the application was achived by creating the following fuzzy system:

1. Fuzzy system for analysis of blood findings
2. Fuzzy system for analysis of lipid status based on blood findings

N. Ademović et al. (Eds.): IAT 2021, LNNS 316, pp. 56–82, 2022.
https://doi.org/10.1007/978-3-030-90055-7_6

3. Fuzzy system for analysis of urine findings
4. Fuzzy system for analysis of general liver function

In addition to the created fuzzy systems, graphical interfaces have been created that the user uses when testing and analyzing the results. The proposed system allows the user to choose one of the four types of testing, and as feedback returns the user a conclusion about the patient's health. The success of the implemented system was examined on a number of actual blood and urine findings for men and women. Using the created application is simple, and therefore accessible to all users. The basic functionalities of the created fuzzy system, ie application, for the analysis of blood and urine findings are:

1. Testing of all individual systems
2. General health assessment that includes advice and recommendations
3. List of all tested components and entered numerical values
4. Textual description of the general state of health of the patient based of the analysis of blood and urine findings

The goal of the system is for user to have an insight into health in an acceptable way based on the performed blood and urine analyses, whereby the users are doctors, laboratory technicians and patients who enter data from blood and urine findings into the application. The most important part of the medical profession is making precise decisions.

In this process doctors rely on gained knowledge and experience. However, it seems necessary for them to have the ability to think logically, to use reasoning, to infer, to precisely and clearly express their thoughts and justify the assertions made [4–6]. This paper proofs the importance of fuzzy logic in the field of medicine.

2 Creating Fuzzy System for Analysis of Blood and Urine Findings

In all created fuzzy systems, all input variables are created using trapezoidal membership functions, and output variables using triangular memberships functions.

2.1 Fuzzy Blood Analysis System

When analyzing blood findings, it is very important whether the user is a male or a female. In both cases, the concentrations of erythrocytes, leukocytes, hemoglobin and hematocrit are measured. For male or female patients, the different values are the referenced values of erythrocytes, hemoglobin and hematocrit.

Erythrocytes (red blood cells or RBCs) are anucleate, biconcave cells, filled with hemoglobin, that transport oxygen and carbon dioxide between the lungs and tissues. Hemoglobin is the protein molecule in red blood cells that carries oxygen from the lungs to the body's tissues and returns carbon dioxide from the tissues back to the lungs. Hemoglobin is made up of four protein molecules (globulin chains) that are connected together. White blood cells leukocytes are the cells of the immune system that are involved in protecting the body against both infectious disease and foreign invaders.

The hematocrit is the volume percentage of red blood cells in blood, measured as part of a blood test. The measurement depends on the number and size of red blood cells. It is normally 40.7%–50.3% for male and 36.1%–44.3% for female. It is a part of a patient's complete blood count results, along with hemoglobin concentration, white blood cell count and platelet count [7, 8].

The first step in creating a fuzzy system is to create a block diagram on which the input and output variables are set. The block diagram is shown in Fig. 1.

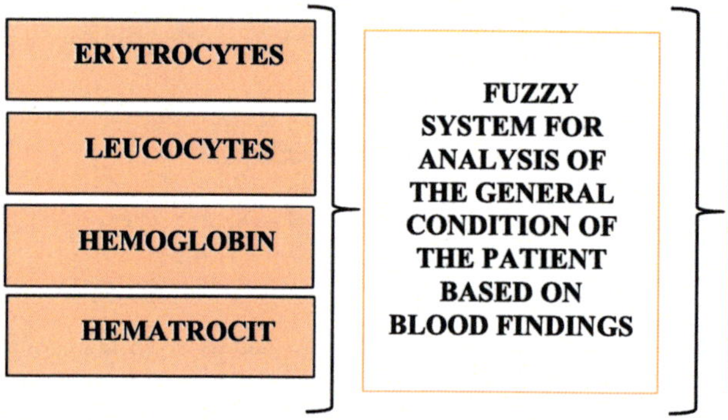

Fig. 1. Block diagram of a fuzzy system for analysis of blood findings

Based on the block diagram in Fig. 1 we can see there are four inputs and one output from the fuzzy system for analysis of blood findings. In Table 1 are defined the states for the input and output variables from Fig. 1. The output variable is defined as the assessment of the patient's general condition.

Table 1. The input and output states of variables

Input	Eritrocytes	Leukocytes	Hemoglobin	Hematocrit
State	Reduced			
	Optimal			
	Increased			
Output	Assessment of the patient's general condition			
State	Reduced			
	Moderately reduced			
	Optimal			
	Moderately increased			
	Increased			

In Table 2 are shown expressions for obtaining the values for inputs and for the output of the created fuzzy system.

Table 2. The expressions for obtaining the values for input variables

Reduced	Optimal	Increased
Electrocytes		
$\mu(X; 1.2, 2.8, 4.116, 4.8)$ $0, X < 1.2$ $\frac{X-1.2}{1.6}, 1.2 \leq X \leq 2.8$ $1, 1.28 \leq X \leq 4.116$ $\frac{4.8-X}{0.687}, 4.116 < X < 4.8$ $0, X > 4.8$	$\mu(X; 4.34, 4.6, 5.39, 5.72)$ $0, X < 4.34$ $\frac{X-4.34}{0.26}, 4.34 \leq X \leq 4.6$ $1, 4.6 \leq X \leq 5.397$ $\frac{5.72-X}{0.323}, 5.39 < X < 5.72$ $0, X > 5.72$	$\mu(X; 5.21, 5.94, 7.21, 8.81)$ $0, X < 5.21$ $\frac{X-5.21}{0.73}, 5.21 \leq X \leq 5.94$ $1, 5.94 \leq X \leq 7.21$ $\frac{8.81-X}{0.6}, 7.21 < X < 8.81$ $0, X > 8.81$
Leucocytes		
$\mu(X; -2.05, 1.5, 3.39, 5)$ $0, X < -2.05$ $\frac{X+2.05}{3.6}, -2.05 \leq X \leq 1.5$ $1, 1.5 \leq X \leq 3.39$ $\frac{5-X}{1.617}, 3.39 < X < 5$ $0, X > 5$	$\mu(X; 3.4, 5, 8, 9.7)$ $0, X < 3.4$ $\frac{X-3.4}{1.6}, 3.4 \leq X \leq 5$ $1, 1.5 \leq X \leq 8$ $\frac{9.7-X}{1.7}, 8 < X < 9.7$ $0, X > 9.7$	$\mu(X; 8, 9.6, 11.45, 15.05)$ $0, X < 8$ $\frac{X-8}{1.6}, 1.6 \leq X \leq 9.6$ $1, 9.6 \leq X \leq 11.45$ $\frac{15.05-X}{3.6}, 11.45 < X < 15.05$ $0, X > 15.05$
Hemoglobin		
$\mu(X; 46, 94, 136, 154)$ $0, X < 46$ $\frac{X-46}{48}, 46 \leq X \leq 94$ $1, 94 \leq X \leq 136$ $\frac{154-X}{18}, 136 < X < 154$ $0, X > 154$	$\mu(X; 140, 154, 166, 180)$ $0, X < -2.05$ $\frac{X-140}{14}, 140 \leq X \leq 154$ $1, 154 \leq X \leq 166$ $\frac{180-X}{14}, 166 < X < 180$ $0, X > 180$	$\mu(X; 166, 184.3, 266, 274)$ $0, X < 166$ $\frac{X-166}{18.3}, 166 \leq X \leq 184.3$ $1, 184.3 \leq X \leq 266$ $\frac{274-X}{48}, 266 < X < 274$ $0, X > 274$
Hematocrit		
$\mu(X; -0.45, -0.05, 0.36, 0.48)$ $0, X < -0.45$ $\frac{X+045}{0.4}, -0.45 \leq X \leq -0.05$ $1, -0.05 \leq X \leq 0.36$ $\frac{0.48-X}{0.08}, 0.366 < X < 0.48$ $0, X > 0.48$	$\mu(X; 0.41, 0.44, 0.49, 0.53)$ $0, X < 0.41$ $\frac{X-0.41}{0.031}, 0.41 \leq X \leq 0.44$ $1, 0.44 \leq X \leq 0.49$ $\frac{53-X}{0.0313}, 0.49 < X < 0.53$ $0, X > 0.53$	$\mu(X; 0.5, 0.56, 1.05, 1, 45)$ $0, X < 0.5$ $\frac{X-0.5}{0.0662}, 0.5 \leq X \leq 0.56$ $1, 56 \leq X \leq 1.05$ $\frac{1.45-X}{0.4}, 1.05 < X < 1.45$ $0, X > 1.45$

In Table 3 are shown expressions for obtaining the values for the output of the created fuzzy system.

Table 3. The expressions for obtaining the values for output variable

Reduced	Moderately reduced
$\mu(X; -0.228, -0.00787, 0.1113)$ $0, X < -0.228$ $\frac{X+0.228}{0.22013}, -0.228 \le X \le -0.00787$ $\frac{0.1113-X}{0.11917}, -0.00787 < X < 0.1113$ $0, X > 0.1113$	$\mu(X; 0.0648, 0.181, 0.28)$ $0, X < 0.0648$ $\frac{X-0.0648}{0.1162}, 0.0648 \le X \le 0.181$ $\frac{0.28-X}{0.099}, 0.181 < X < 0.28$ $0, X > 0.28$
Optimal	**Moderately**
$\mu(X; 0.0648, 0.181, 0.28)$ $0, X < 0.237$ $\frac{X-0.237}{0.1217}, 0.237 \le X \le 0.3587$ $\frac{0.48-X}{0.1213}, 0.3587 < X < 0.48$ $0, X > 0.48$	$\mu(X; 0.434, 0.552, 0.6667)$ $0, X < 0.434$ $\frac{X-0.434}{0.118}, 0.434 \le X \le 0.552$ $\frac{0.6667-X}{0.1147}, 0.552 < X <$ 0.6667 $0, X > 0.6667$
Increased	
$\mu(X; 0.6173, 0.7297, 0.8403)$ $0, X < 0.6173$ $\frac{X-0.6173}{0.1124}, 0.6173 \le X \le 0.7297$ $\frac{0.8403-X}{0.0106}, 0.7297 < X < 0.8403$ $0, X > 0.8403$	

In the next chapter is define the creation of the fuzzy system for lipid status analysis based on blood findings.

2.2 Fuzzy System for Lipid Status Analysis Based on Blood Findings

Unlike the fuzzy blood analysis system for where the component values are different for male and female patients, the lipid status analysis system uses one fuzzy system and one graphical interface for both genders. The components which are testing are cholesterol, HDL cholesterol, LDL cholesterol and triglycerides.

In the Fig. 2 is shown block diagram for fuzzy system for analysis of lipid status of the patient on the basis of blood findings.

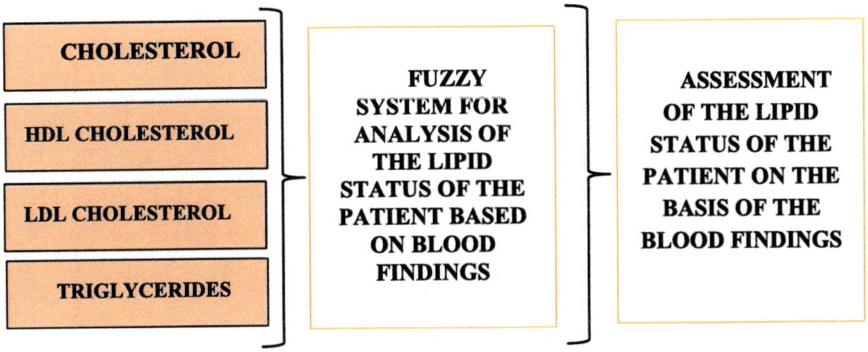

Fig. 2. Block diagram of a fuzzy system for analysis of lipid status of the patient on the basis of blood findings

In Table 4 are shown input and the output states of variables.

Table 4. The input and output states of variables

Input	Cholesterol	HDL cholesterol	LDL cholesterol	Triglycerides
State	Reduced			
	Optimal			
	Increased			
Output	Assesment of the patient's general condition			
State	Reduced			
	Moderately reduced			
	Optimal			
	Moderately increased			
	Increased			

In Table 5 are shown expressions for obtaining the values for inputs and for the output of the created fuzzy system.

Table 5. The expressions for obtaining the values for input variables

Reduced	Optimal	Increased
Cholesterol		
$\mu(X; -2.15, 0.65, 2.8, 3.8)$ $0, X < -2.15$ $\frac{X+2.15}{2.8}, -2.15 \leq X \leq 0.65$ $1, 0.65 \leq X \leq 2.8$ $\frac{3.8-X}{1}, 2.8 < X < 3.8$ $0, X > 3.8$	$\mu(X; 3.1, 3.8, 4.8, 5.5)$ $0, X < 3.1$ $\frac{X-3.1}{0.7}, 3.1 \leq X \leq 3.8$ $1, 3.8 \leq X \leq 4.8$ $\frac{5.5-X}{0.7}, 4.8 < X < 5.5$ $0, X > 5.5$	$\mu(X; 4.8, 5.824, 8.35, 11.2)$ $0, X < 4.8$ $\frac{X-4.8}{1.024}, 4.8 \leq X \leq 5.824$ $1, 5.824 \leq X \leq 8.35$ $\frac{11.2-X}{2.85}, 8.35 < X < 11.2$ $0, X > 11.2$
HDL cholesterol		
$\mu(X; -1.125, 0.125, 0.35, 1)$ $0, X < -1.125$ $\frac{X+1.125}{1}, -1.125 \leq X \leq 0.125$ $1, 0.125 \leq X \leq 0.35$ $\frac{1-X}{0.65}, 0.35 < X < 1$ $0, X > 1$	$\mu(X; 0.46, 0.82, 1.835, 2.28)$ $0, X < 0.46$ $\frac{X-0.46}{0.36}, 0.46 \leq X \leq 0.82$ $1, 0.82 \leq X \leq 1.835$ $\frac{2.28-X}{0.445}, 1.835 < X < 2.28$ $0, X > 2.28$	$\mu(X; 1.75, 2.37, 2.63, 3.62)$ $0, X < 1.75$ $\frac{X-1.75}{0.6}, 1.75 \leq X \leq 2.37$ $1, 2.37 \leq X \leq 2.63$ $\frac{3.62-X}{1}, 2.63 < X < 3.63$ $0, X > 3.63$
LDL cholesterol		
$\mu(X; -2.7, -0.3, 1.42, 2.3)$ $0, X < -2.7$ $\frac{X+2.7}{2.4}, -2.7 \leq X \leq -0.3$ $1, -0.3 \leq X \leq 1.42$ $\frac{2.3-X}{0.879}, 1.42 < X < 2.3$ $0, X > 2.3$	$\mu(X; 1.55, 2.2, 3.8, 4.53)$ $0, X < 1.55$ $\frac{X-1.55}{0.65}, 1.55 \leq X \leq 2.2$ $1, 2.2 \leq X \leq 3.8$ $\frac{4.53-X}{0.73}, 3.8 < X < 4.53$ $0, X > 4.53$	$\mu(X; 3.8, 4.58, 6.3, 8.7)$ $0, X < 3.8$ $\frac{X-3.8}{0.779}, 3.8 \leq X \leq 4.58$ $1, 4.58 \leq X \leq 6.3$ $\frac{8.7-X}{2.4}, 6.3 < X < 8.7$ $0, X > 8.7$
Triglycerides		
$\mu(X; -1.35, -0.15, 0.42, 0.78)$ $0, X < -1.35$ $\frac{X+1.35}{1.2}, -1.35 \leq X \leq -0.15$ $1, -0.15 \leq X \leq 0.42$ $\frac{0.782-X}{0.3653}, 042 < X < 0.78$ $0, X > 0.78$	$\mu(X; 0.46, 0.77, 1.94, 2.28)$ $0, X < 0.46$ $\frac{X-0.46}{0.314}, 0.46 \leq X \leq 0.77$ $1, 0.77 \leq X \leq 1.94$ $\frac{2.28-X}{0.332}, 1.94 < X < 2.28$ $0, X > 2.28$	$\mu(X; 1.92, 2.53, 3.15, 4.35)$ $0, X < 1.92$ $\frac{X-1.92}{0.433}, 1.92 \leq X \leq 2.35$ $1, 2.35 \leq X \leq 3.15$ $\frac{4.35-X}{1.2}, 3.15 < X < 4.35$ $0, X > 4.35$

In Table 6 are shown expressions for obtaining the values for output and for the of the created fuzzy system.

In the next chapter is define the creation of the fuzzy system for analysis of urinary system findings.

Table 6. The expressions for obtaining the values for output variable

Reduced	Moderately reduced
$\mu(X; 0.05, 0.15, 0.25)$	$\mu(X; 0.18, 0.28, 0.41)$
$0, X < 0.05$	$0, X < 0.18$
$\frac{X-0.05}{0.1}, 0.05 \le X \le 0.15$	$\frac{X-0.18}{0.11}, 0.18 \le X \le 0.29$
$\frac{0.25-X}{0.15}, 0.15 < X < 0.25$	$\frac{0.41-X}{0.12}, 0.29 < X < 0.41$
$0, X > 0.25$	$0, X > 0.41$
Optimal	**Moderately**
$\mu(X; 0.33, 0.44, 0.54)$	$\mu(X; 0.48, 0.59, 0.67)$
$0, X < 0.33$	$0, X < 0.48$
$\frac{X-0.33}{0.1096}, 0.33 \le X \le 0.44$	$\frac{X-0.48}{0.099}, 0.48 \le X \le 0.57$
$\frac{0.54-X}{0.0974}, 0.44 < X < 0.54$	$\frac{0.67-X}{0.098}, 0.57 < X < 0.67$
$0, X > 0.54$	$0, X > 0.67$
Increased	
$\mu(X; 0.61, 0.7, 0.8)$	
$0, X < 0.61$	
$\frac{X-0.61}{0.09}, 0.61 \le X \le 0.7$	
$\frac{0.8-X}{0.098}, 0.7 < X < 0.8$	
$0, X > 0.8$	

2.3 Fuzzy System for Analysis of Urinary System Findings

The most important organs of the urinary system are the kidneys, ureters, bladder and urethra. As with the fuzzy blood test analysis system, the gender of the patient is important. Figure 3 shows a block diagram based on which the imput and output variables in the fuzzy system are created.

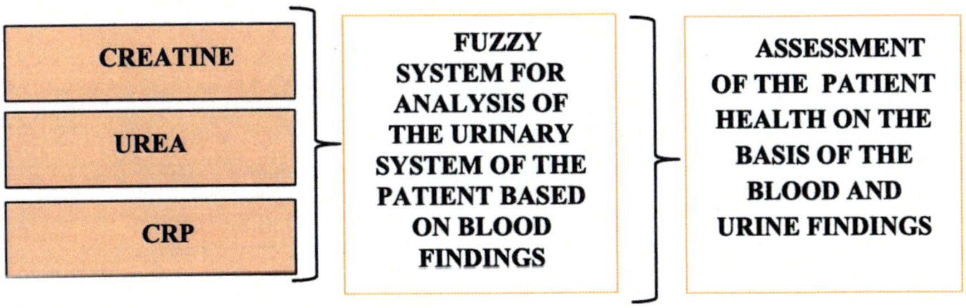

Fig. 3. Block diagram of a fuzzy system for analysis of lipid status of the patient on the basis of blood findings

In Table 7 are shown input and the output states of variables.

Table 7. The input and output states of variables

Input	Creatine	Urea	CRP	Triglycerides
State	Reduced			
	Optimal			
	Increased			
Output	Assesment of the patient's general condition			
State	Reduced			
	Moderately reduced			
	Optimal			
	Moderately increased			
	Increased			

Table 8. The expressions for obtaining the values for input variables

Reduced	Optimal	Increased
Creatine		
$\mu(X; 11.8, 45.8, 77.1, 88.3)$ $0, X < 11.8$ $\frac{X-11.8}{34}, 11.8 \leq X \leq 45.8$ $1, 45.8 \leq X \leq 77.1$ $\frac{88.3-X}{11.2}, 77.1 < X < 88.3$ $0, X > 88.3$	$\mu(X; 80, 88.25, 107, 115)$ $0, X < 80$ $\frac{X-80}{8.25}, 80 \leq X \leq 88.25$ $1, 88.25 \leq X \leq 107$ $\frac{115-X}{8}, 107 < X < 115$ $0, X > 115$	$\mu(X; 108, 119, 139, 173)$ $0, X < 108$ $\frac{X-108}{11.4} 108 \leq X \leq 119$ $1, 119 \leq X \leq 173$ $\frac{173-X}{34}, 139 < X < 173$ $0, X > 173$
Urea		
$\mu(X; -6.75, -0.75, 2.08, 3.2)$ $0, X < -6.75$ $\frac{X+6.75}{6}, -6.75 \leq X \leq -0.75$ $1, 0.7 \leq X \leq 2.08$ $\frac{3.2-X}{1.117}, 2.08 < X < 3.2$ $0, X > 3.2$	$\mu(X; 2.5, 3.5, 7.3, 8.5)$ $0, X < 2.5$ $\frac{X-2.5}{1}, 2.5 \leq X \leq 3.5$ $1, 3.5 \leq X \leq 7.3$ $\frac{8.5-X}{1.2}, 7.3 < X < 8.5$ $0, X > 8.5$	$\mu(X; 7.8, 8.7, 15.8, 21.8)$ $0, X < 7.8$ $\frac{X-7.8}{0.9}, 7.8 \leq X \leq 8.7$ $1, 8.7 \leq X \leq 15.8$ $\frac{21.8-X}{6}, 15.8 < X < 21.8$ $0, X > 21.8$

(*continued*)

Table 8. (*continued*)

Reduced	Optimal	Increased
CRP		
$\mu(X; -7.4, -2.6, -0.4, 1)$ $0, X < -7.4$ $\frac{X+7.4}{4.8}, -7.4 \leq X \leq -2.6$ $1, -2.6 \leq X \leq -0.4$ $\frac{1-X}{1.429}, -0.4 < X < 1$ $0, X > 1$	$\mu(X; 0, 1.5, 3.5, 5)$ $0, X < 0$ $\frac{X-0}{1.5}, 0 \leq X \leq 1.5$ $1, 15 \leq X \leq 3.5$ $\frac{5-X}{1.5}, 3.5 < X < 5$ $0, X > 5$	$\mu(X; 4, 5.4, 10.6, 15.4)$ $0, X < 4$ $\frac{X-4}{1.4}, 4 \leq X \leq 5.4$ $1, 5.4 \leq X \leq 10.6$ $\frac{15.4-X}{4.8}, 10.6 < X < 15.4$ $0, X > 15.4$

In Table 8 are shown expressions for obtaining the values for inputs and for the output of the created fuzzy system (Table 9).

Table 9. The expressions for obtaining the values for output variable

Reduced	Moderately reduced
$\mu(X; -0.33, -0.17, -0.008)$ $0, X < -0.33$ $\frac{X+0.33}{0.161}, -0.33 \leq X \leq -0.17$ $\frac{-0.0082-X}{0.166}, -0.17 < X < -0.008$ $0, X > -0.008$	$\mu(X; -0.07, 0.104, 0.25)$ $0, X < -0.07$ $\frac{X+0.07}{0.17}, -0.07 \leq X \leq 0.104$ $\frac{0.25-X}{0.146}, -0.07 < X < 0.104$ $0, X > 0.25$
Optimal	**Moderately**
$\mu(X; 0.2, 0.35, 0.51)$ $0, X < 0.2$ $\frac{X-0.2}{0.155}, 0.2 \leq X \leq 0.355$ $\frac{0.51-X}{0.1548}, 0.35 < X < 0.51$ $0, X > 0.51$	$\mu(X; 0.43, 0.6, 0.7)$ $0, X < 0.43$ $\frac{X-0.43}{0.16}, 0.43 \leq X \leq 0.6$ $\frac{0.7-X}{0.1784}, 0.6 < X < 0.7$ $0, X > 0.7$
Increased	
$\mu(X; 0.7, 0.8, 1.605)$ $0, X < 0.7$ $\frac{X-7}{0.184}, 0.7 \leq X \leq 0.8$ $\frac{1.605-X}{0.09}, 0.446 < X < 0.54$ $0, X > 0.54$	

In the next chapter is define the creation of the fuzzy system for analysis of general liver function based on blood findings.

2.4 Fuzzy System for Analysis of General Liver Function Based on Blood Findings

The liver is the largest gland in our body that is exclusively important for the metabolism of fats, carbonhydrates and proteins. The Fuzzy system for analyzing general liver function is the same for both genders.

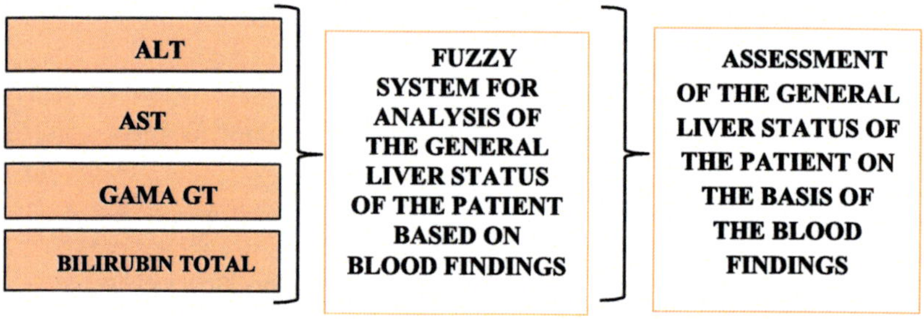

Fig. 4. Block diagram of a fuzzy system for analysis of lipid status of the patient on the basis of blood findings

In the Fig. 4 the block diagram of fuzzy system for analysis of lipid status of the patient on the basis of blood findings is shown. The general liver function, ALT, AST, GGT and Bilirubin total, whose concentrations are measured in the blood, are tested (Tables 10, 11 and 12).

Table 10. The input and output states of variables

Input	ALT	AST	Gama GT	Bilirubin total
State	Reduced			
	Optimal			
	Increased			
Output	Assesment of the patient's general condition			
State	Reduced			
	Moderately reduced			
	Optimal			
	Moderately increased			
	Increased			

Table 11. The expressions for obtaining the values for input variables

Reduced	Optimal	Increased
ALT		
$\mu(X; -27, -3, 9, 18)$ $0, X < -27$ $\frac{X+27}{-24}, -27 \leq X \leq -3$ $1, -3 \leq X \leq 9$ $\frac{18-X}{9}, 9 < X < 18$ $0, X > 18$	$\mu(X; 10, 20, 38, 48)$ $0, X < 10$ $\frac{X-10}{10}, 10 \leq X \leq 20$ $1, 20 \leq X \leq 38$ $\frac{18-X}{9}, 38 < X < 48$ $0, X > 48$	$\mu(X; 40, 49, 63, 87)$ $0, X < 40$ $\frac{X-40}{9}, 40 \leq X \leq 49$ $1, 49 \leq X \leq 63$ $\frac{87-X}{24}, 63 < X < 87$ $0, X > 87$
AST		
$\mu(X; -22.5, -2.5, 6, 15)$ $0, X < -22.5$ $\frac{X+22.5}{20}, -22.5 \leq X \leq -2.5$ $1, -2.5 \leq X \leq 6$ $\frac{6-X}{8.5}, 6 < X < 15$ $0, X > 15$	$\mu(X; 8, 15, 31, 38)$ $0, X < 8$ $\frac{X-8}{7}, 8 \leq X \leq 15$ $1, 15 \leq X \leq 31$ $\frac{38-X}{7}, 31 < X < 37$ $0, X > 38$	$\mu(X; 31, 40, 52.5, 72.5)$ $0, X < 31$ $\frac{X-31}{9}, 31 \leq X \leq 40$ $1, 40 \leq X \leq 52.5$ $\frac{72.5-X}{20}, 52.5 < X < 72.5$ $0, X > 72.5$
Gama GT		
$\mu(X; -19.8, 2.25, 9, 20)$ $0, X < -19.8$ $\frac{X+19.8}{22.05}, -19.8 \leq X \leq 94$ $1, 2.25 \leq X \leq 9$ $\frac{20-X}{11}, 9 < X < 20$ $0, X > 20$	$\mu(X; 11, 20, 46, 55)$ $0, X < 11$ $\frac{X-11}{9}, 11 \leq X \leq 20$ $1, 20 \leq X \leq 46$ $\frac{55-X}{9}, 31 < X < 55$ $0, X > 55$	$\mu(X; 46, 57.25, 62.75, 84.75)$ $0, X < 46$ $\frac{X-46}{11.25}, 46 \leq X \leq 57.25$ $1, 57.25 \leq X \leq 62.75$ $\frac{84.75-X}{22}, 62.75 < X < 84.75$ $0, X > 84.75$
Bilirubin total		
$\mu(X; 17, 21, 26.25, 36.25)$ $0, X < 17$ $\frac{X-17}{4}, 17 \leq X \leq 21$ $1, 21 \leq X \leq 26.25$ $\frac{36.25-X}{10}, 26.25 < X < 36.25$ $0, X > 36.25$	$\mu(X; -11.2, -1.25, 2, 7)$ $0, X < -11.25$ $\frac{X+11.25}{10}, -11.25 \leq X \leq -1.25$ $1, -1.25 \leq X \leq 2$ $\frac{7-X}{5}, 2 < X < 7$ $0, X > 7$	$\mu(X; 3, 8, 15, 20)$ $0, X < 3$ $\frac{X-3}{5}, 3 \leq X \leq 8$ $1, 8 \leq X \leq 15$ $\frac{20-X}{5}, 15 < X < 20$ $0, X > 20$

Table 12. The expressions for obtaining the values for output variable

Reduced	Moderately reduced
$\mu(X; -0.1, 0.03, 0, 18)$ $0, X < -0.1$ $\frac{X+0.1}{0.13}, -0.1 \leq X \leq 0.03$ $\frac{0.18-X}{0.15}, 0.03 < X < 0.18$ $0, X > 0.18$	$\mu(X; 0.13, 0.25, 0.43)$ $0, X < 0.13$ $\frac{X-0.13}{0.15}, 0.13 \leq X \leq 0.28$ $\frac{0.43-X}{0.15}, 0.28 < X < 0.43$ $0, X > 0.43$
Optimal	**Moderately**
$\mu(X; 0.35, 0.5, 0.65)$ $0, X < 0.35$ $\frac{X-0.35}{0.15}, 0.35 \leq X \leq 0.5$ $\frac{0.65-X}{0.15}, 0.5 < X < 0.65$ $0, X > 0.65$	$\mu(X; 0.57, 0.72, 0.87)$ $0, X < 0.57$ $\frac{X-0.57}{0.15}, 0.57 \leq X \leq 0.72$ $\frac{0.87-X}{0.15}, 0.72 < X < 0.87$ $0, X > 0.87$
Increased	
$\mu(X; 0.82, 0.97, 1.13)$ $0, X < 0.82$ $\frac{X-0.82}{0.154}, 0.82 \leq X \leq 0.97$ $\frac{0.13-X}{0.155}, 0.97 < X < 1.13$ $0, X > 1.13$	

In the next chapter is define the creation of the fuzzy system for analysis of general liver function based on blood findings.

3 Defining Rules and a Graphical Interface

3.1 Defining Rules and a Graphical Interface for the Fuzzy System to Analyze Blood Findings

The rules are executed at the same time, so their schedule is not important. If then rules represent the rules on the basis of which the fuzzy system performs certain calculations and conclusions. The maximum number of rules is determined by the number of input quantities and the number of linguistic values. In this case, the number of rules is not determined that way. The rules are formed on the basis of the knowledge of experts in the field of medicine. The rules are executed at the same time, so their schedule is not important. The If-Then rules represent the rules on the basis of which the fuzzy system performs certain calculations and conclusions. The rules are formed on the basis of the knowledge of experts in the field of medicine. Most often, the concentrations of erythrocytes, leukocytes, hemoglobin and hematocrit are in the same or approximately the same conditions. Most often, the concentrations of erythrocytes, leukocytes, hemoglobin and hematocrit are in the same or approximately the same conditions.

Most often, the concentrations of erythrocytes, leukocytes, hemoglobin and hematocrit are in the same or approximately the same conditions. Common to all input variables of the fuzzy blood test analysis system is that reduced concentrations of any of them indicate anemia. Optimal values of blood count parameters are when the concentrations of all are within optimal limits. Moderately reduced and moderately elevated conditions are concentrations that are on the border with optimal values. In the Fig. 5 is shown the graphical interface for the the fuzzy system to analyze blood findings.

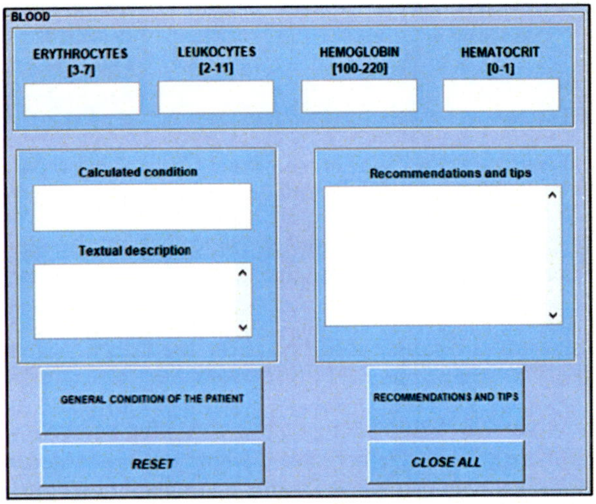

Fig. 5. The graphical interface

The general goal of the graphical interface is to achieve easy, productive and pleasant work with the computer, and a quality graphical interface is based on quality content organization, efficient use of elements, and effective interaction.

Reference values are different for male and female, so it was necessary to create two fuzzy systems and two graphical interfaces. The user is offered both on the home page of the application, and will choose the one.

3.2 Defining Rules and a Graphical Interface for the Fuzzy System to Analysis of Lipid Status Based on Blood Findings

The Fuzzy system for lipid status analysis based on blood findings consists of 95 rules. They are formed based on the state of all inputs and outputs. In this system, the input variables are not interrelated, however, large deviations within the limits are very rare (Figs. 6, 7 and 8).

A significant feature is that HDL cholesterol can be elevated to the highest limit, without indicating compromised lipid status. Unlike HDL cholesterol, LDL can be lowered to the lowest limits and is suitable for the body. Elevated triglyceride concentrations were followed by elevated LDL cholesterol concentrations and decreased HDL cholesterol concentrations.

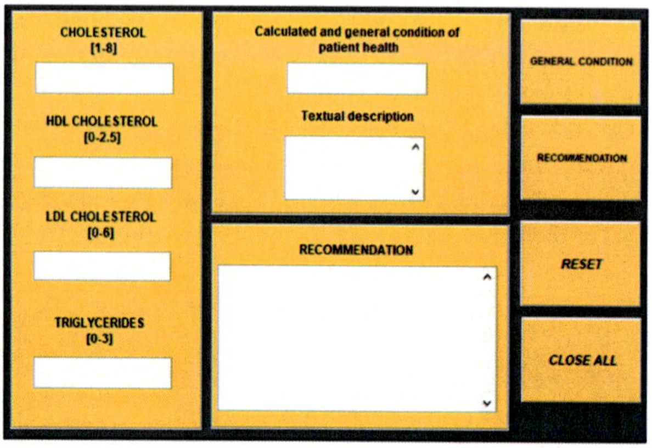

Fig. 6. Graphical interface used to analyze the patient's lipid status (M/F)

Since the limits of input values for lipid status testing are the same for male and female, one graphical interface was created that is valid for both gender.

3.3 Defining Rules and a Graphical Interface for the Fuzzy System to Analysis Urine Findings

It is important to emphasize that the components creatinine and urea are interrelated. Their concentrations are always together lowered, optimal, or elevated. It is most desirable for the organism that both components be in optimal limits, and that they strive for a moderately reduced state.

Fuzzy system for analysis of urine findings consists of 45 rules that are formed based on the knowledge of experts in the field of medicine.

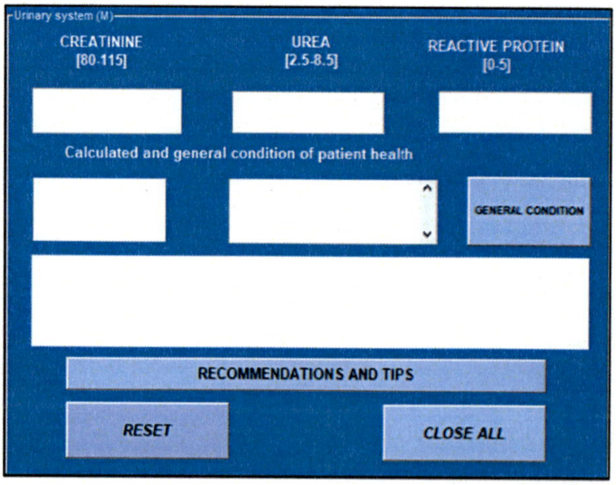

Fig. 7. Graphical interface used to analyze urine findings (M/F)

3.4 Defining Rules and a Graphical Interface for the Fuzzy System to Analysis of General Liver Function Based on Blood Findings

A significant feature of the input variables of this system is that not all reduced conditions are clinically significant.

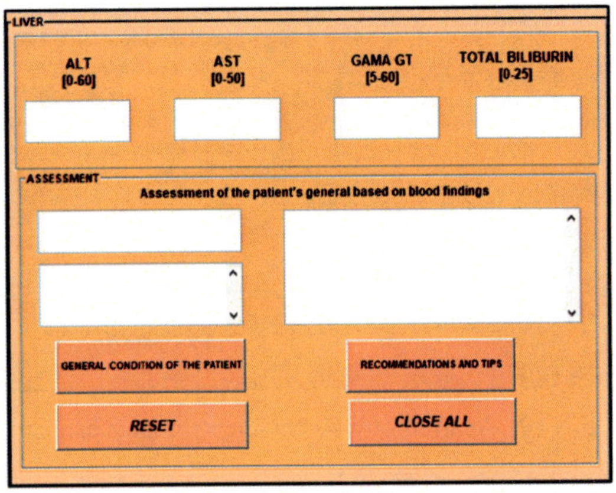

Fig. 8. Graphical interface used to analyze general liver function

4 Experimental Work Analysis for Analysis of Blood and Urine Findings Using a Fuzzy System

When the system starts, the main graphical window is shown to the user and is shown in Fig. 9.

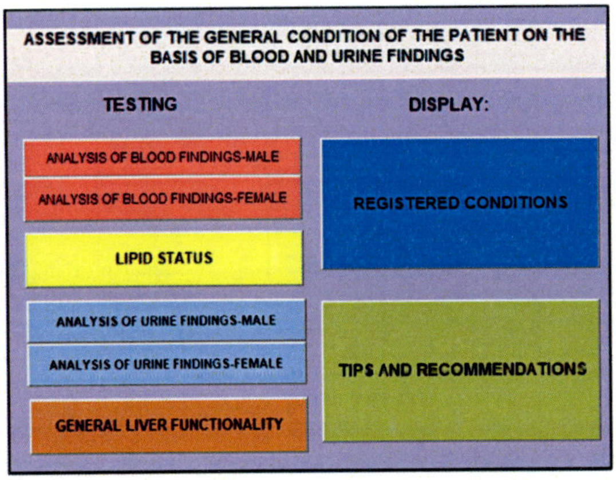

Fig. 9. Graphical interface for the assessment of the general condition of the patient on the basis of blood and urine findings

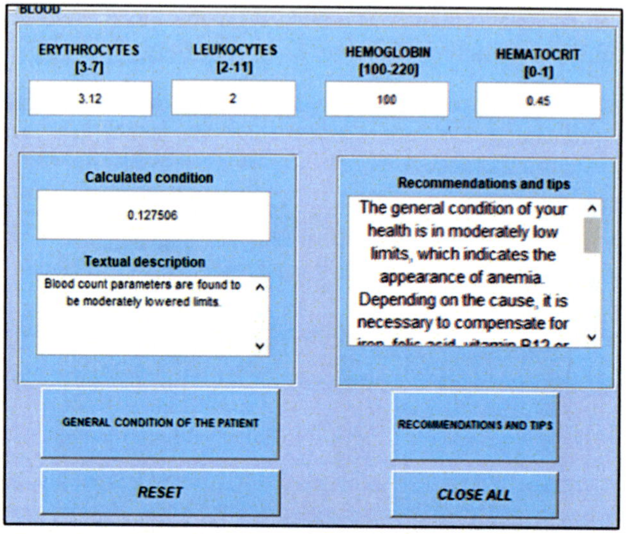

Fig. 10. Analysis of blood findings for male patient

Because fuzzy blood and urine analysis systems are different for male and female patients, the following section presents two examples of analysis of findings using the fuzzy system, and using a graphical interface. The first example relates to the analysis of blood and urine findings for male patients, and the second to analysis of blood and urine findings for female patients. The first test is performed based on the real blood findings for male patients, as shown in Fig. 10.

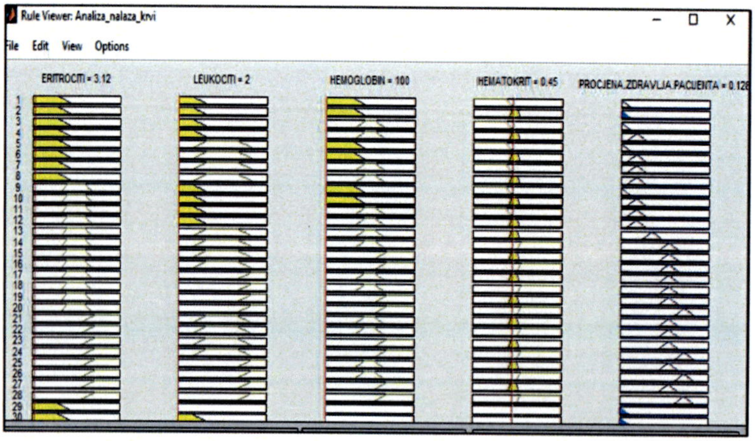

Fig. 11. Testing a fuzzy system to analyze blood findings for men patient

The Fig. 11 shows the analysis of blood findings in the fuzzy system. The same input values were entered as in the previous figure, and the output value, ie the patient's health assessment, is the same as in the graphical interface, which means that the system is working properly. The second system tested is the system for analysis of lipid status based on blood findings (Figs. 12, 13, 14 and 15).

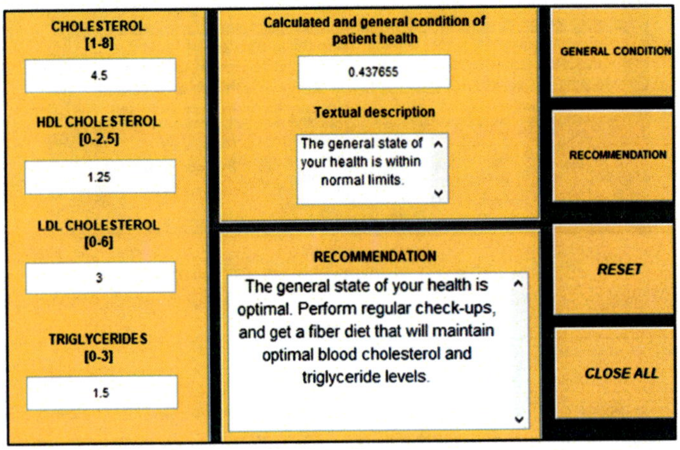

Fig. 12. Analysis of the patient's lipid status based on blood findings

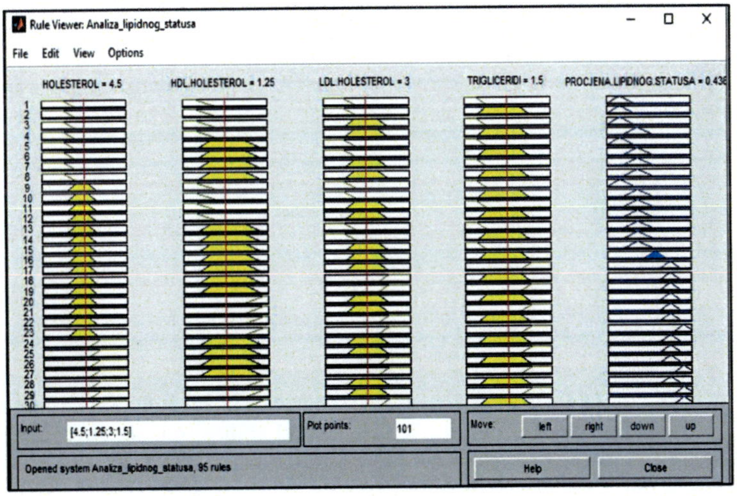

Fig. 13. Testing of fuzzy system for analysis of lipid status based on blood findings

The third fuzzy system tested relates to the analysis of urine findings. The analysis of urine findings differs for male and female patient, and in this case the analysis of findings for male was chosen. In urine findings, creatinine and urea concentrations always move in the same states, and in this case it is a moderately reduced state.

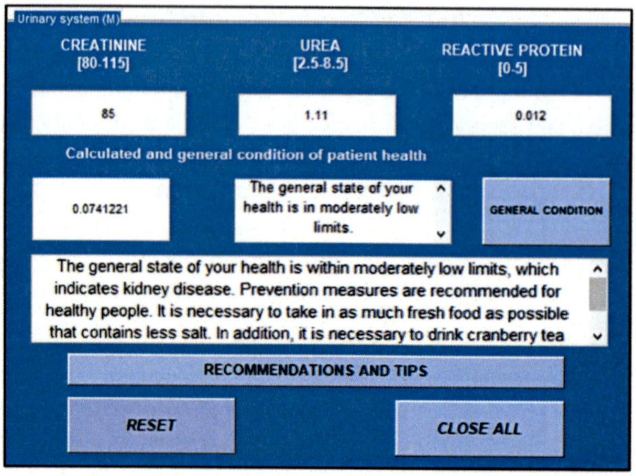

Fig. 14. Analysis of urine findings for men patient

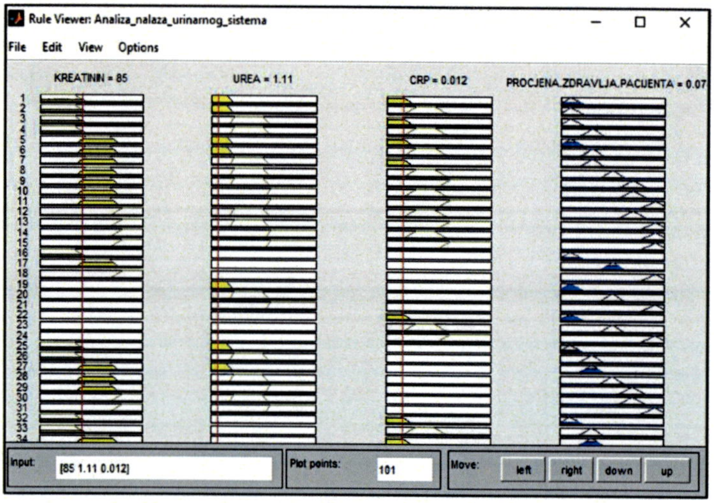

Fig. 15. Testing a fuzzy system to analyze urine findings for men patient

Testing of the fuzzy system for the analysis of general liver function based on blood findings is shown in Fig. 16. The entered numerical values indicate the optimal state (Figs. 17 and 18).

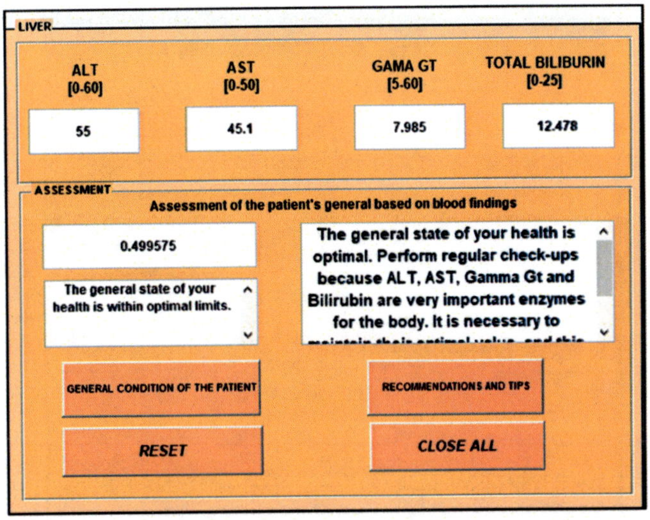

Fig. 16. Anaysis of general liver function based on blood findings

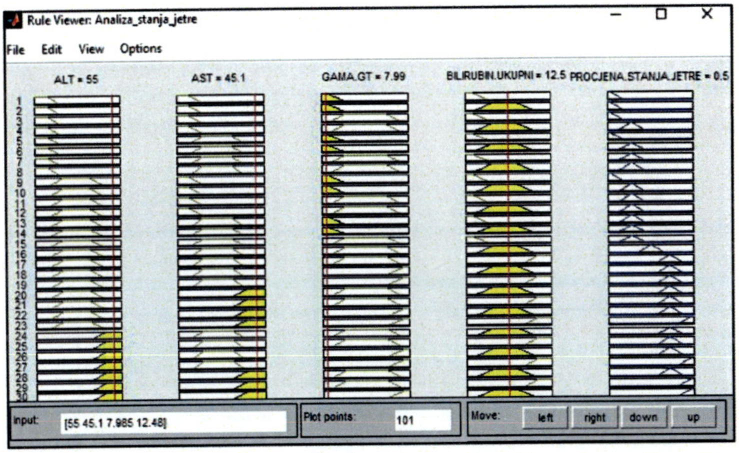

Fig. 17. Testing a fuzzy system to analyze general liver function based on blood findings

The user can easily conclude whether his general state of health is within the reference limits because they are next to the list of entered values. If the user has tested only certain systems and not all, empty fields will remain on this list next to variables that have not been tested.

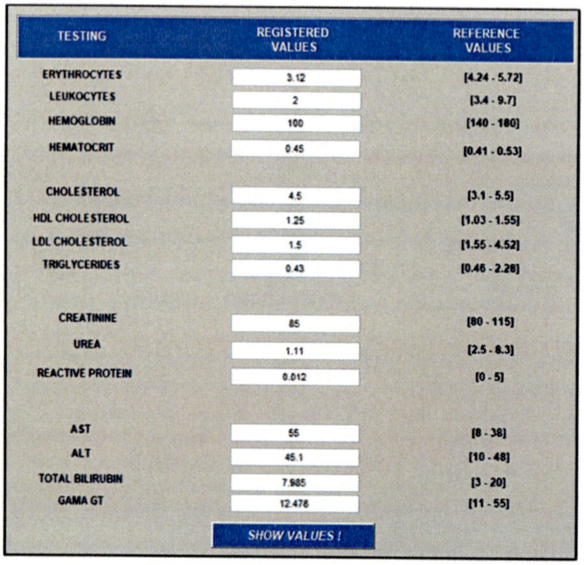

Fig. 18. Display of all entered values

The Fig. 19 shows the patient health description, tips and recommendations (Figs. 20 and 21).

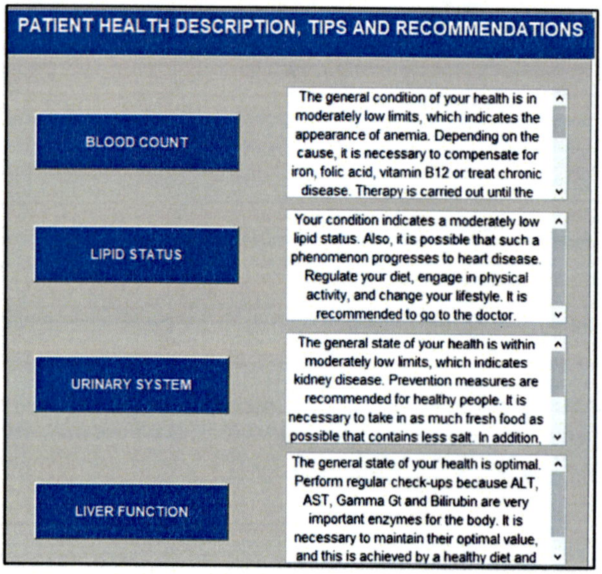

Fig. 19. Patient health descriptions, tips and recommendations for all analyzed findings

After a several number of tests of the actual findings using this realized system, it is concluded that the system works completely correctly, and that it provides correct analyzes of the findings. The following example shows the testing of all systems in female patient.

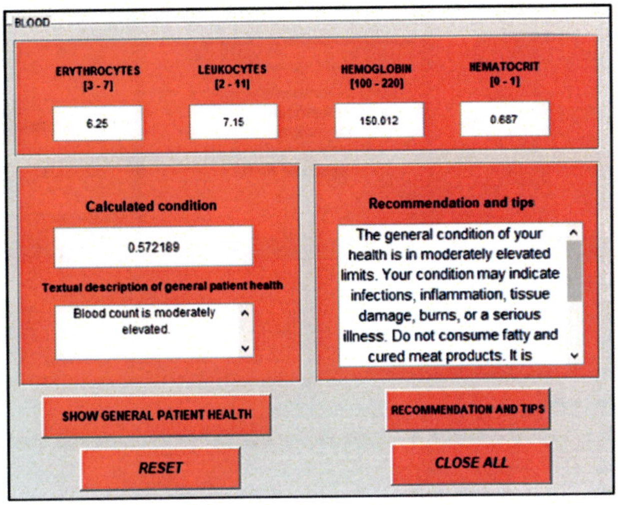

Fig. 20. Analysis of blood findings in women

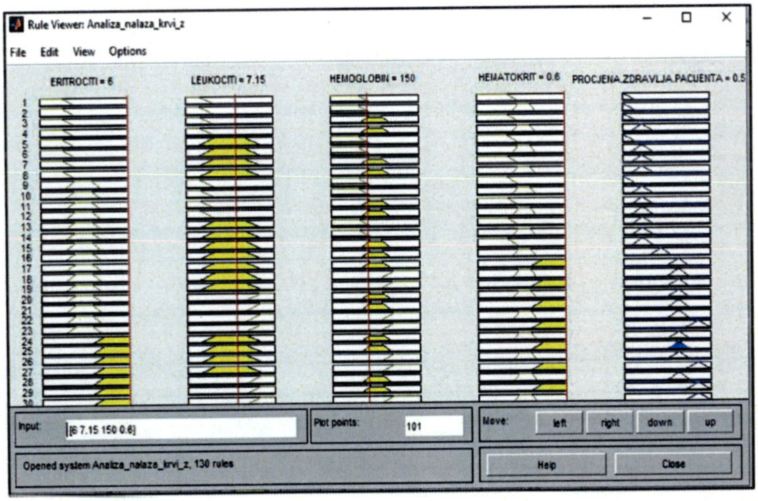

Fig. 21. Testing of fuzzy system for analysis of blood findings for female patient

The Fig. 22 shows the lipid status analysis based on blood findings for female patients (Fig. 23).

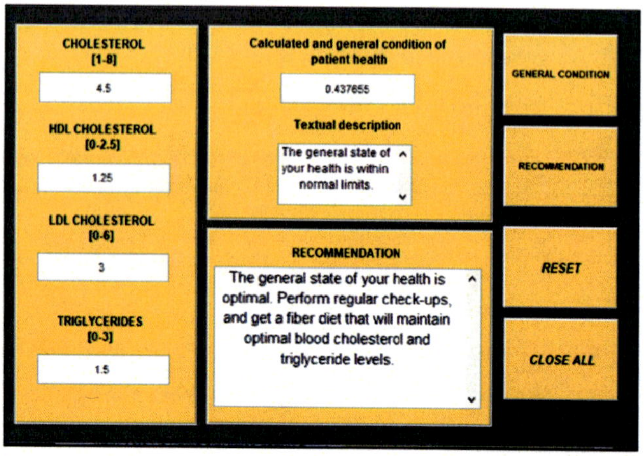

Fig. 22. Lipid status analysis based on blood findings

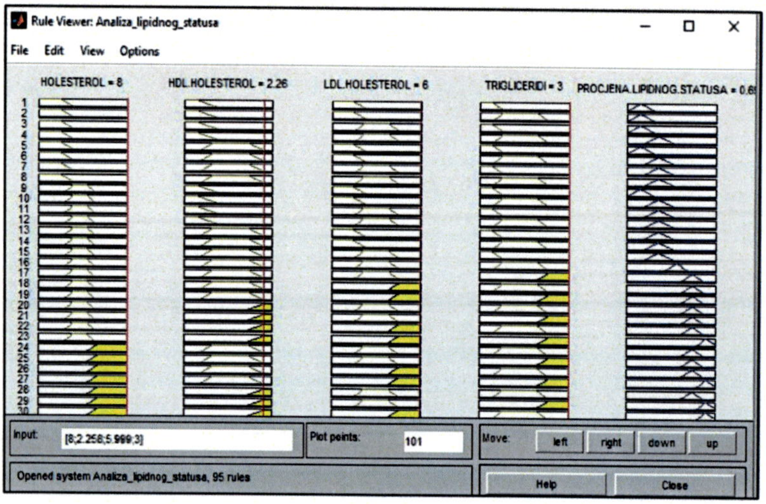

Fig. 23. Testing of fuzzy system for analysis of lipid status based on blood findings

The next test system is the urinary system, which is analyzed by analyzing urine findings. The Fig. 24 shows the analysis of urine findings in women.

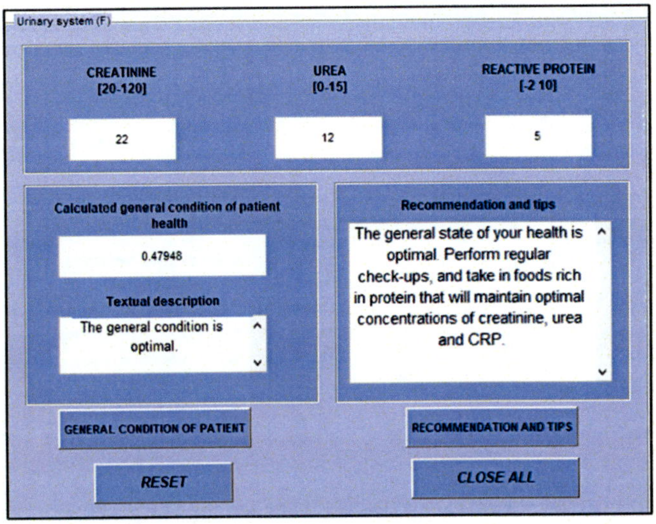

Fig. 24. Analysis of urine findings for female patient

Figure 25 shows an analysis of urine findings in women patient in the fuzzy system. The data is the same as the data from the graphical interface, as well as the solution, which means that the system is connected correctly.

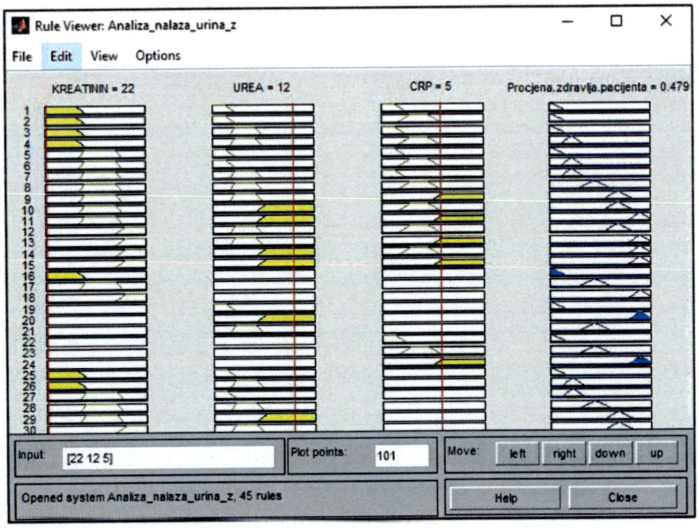

Fig. 25. Tesing of fuzzy system for analysis of urine findings for women patient

The Fig. 26 shows the general liver testing for female patient (Fig. 27).

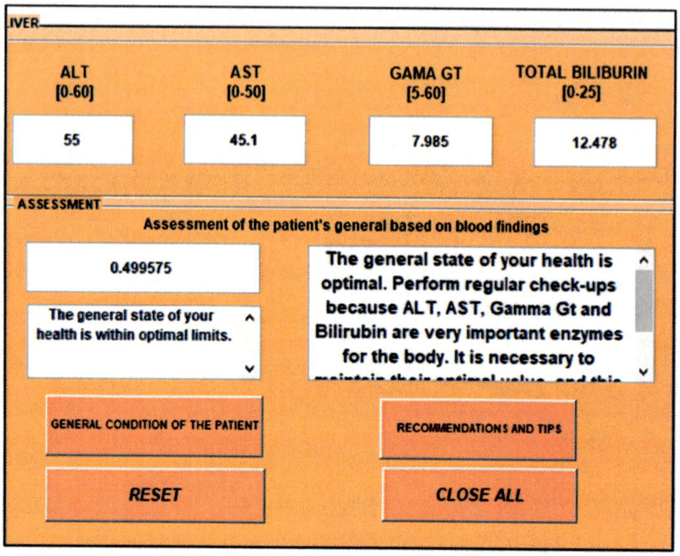

Fig. 26. General liver testing for female patient

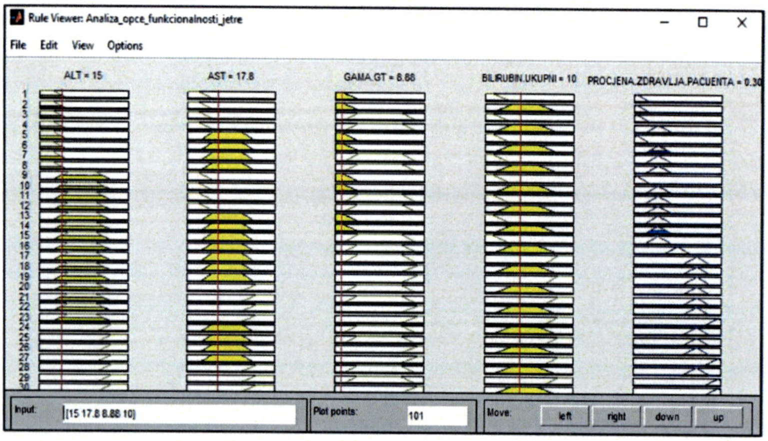

Fig. 27. Fuzzy system testing with default combination on inputs for liver condition analysis for female patient

The Fig. 28 shows all entered data used in the analysis of blood and urine findings.

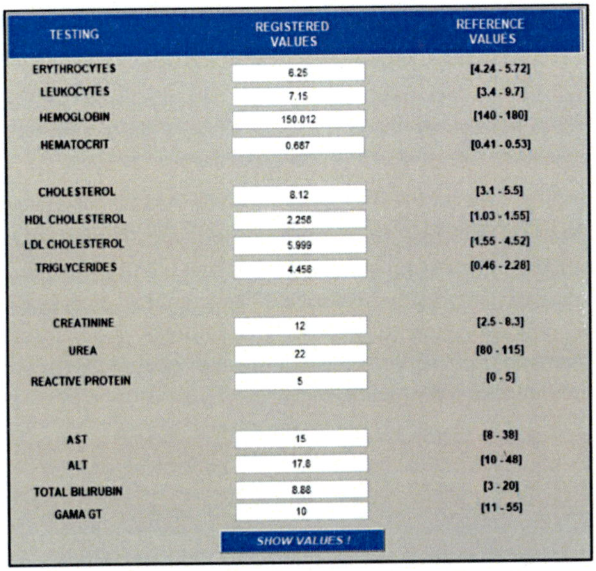

Fig. 28. Display of all tested data for female patient

The Fig. 29 shows textual descriptions of health, tips, and recommendations related to the analyzes shown in the previous figure.

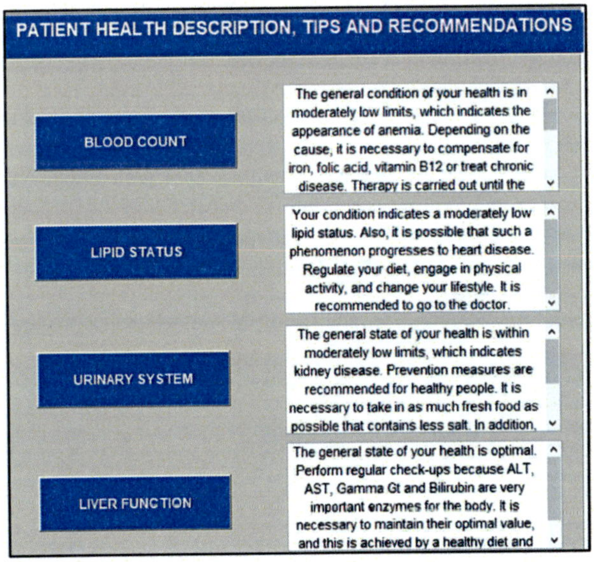

Fig. 29. Display of patient health descriptions, tips and recomendations for all tested systems for women patient

Based on the reference values found in Fig. 28 and the descriptions shown in Fig. 29, it is observed that the system provides accurate information on the general state of health of the user.

5 Conclusion

Based on a large number of testing results, and various tested findings, it has been proven that the system is working correctly. It is concluded that the implemented system is very easy to use which is acceptable among users. The graphical interfaces are very clear which also allows easy analysis of blood and urine findings. When testing each system, it is recommended to load tips and recommendations because in this way the user can get a very clear insight into their health.

References

1. Phuong, N.H., Kreinovich, V.: Fuzzy logic and its applications in medicine. Int. J. Med. Inf. **62**(2–3), 165–173 (2001)
2. Allahverdi, N.: Design of fuzzy expert systems and its applications in some medical areas. Int. J. Appl. Math. Electron. Comput. **2**(1), 1 (2014)
3. Mujčić, E.: MATLAB programski jezik za matematičare i tehničke proračune, Univerzitet u Bihaću, ISBN: 987-9958-533-05-1 (2015)
4. Abrishami, Z., Tabatabaee, H.: Design of a fuzzy expert system and a multi-layer neural network system for diagnosis of hypertension Bull. Environ. Pharmacol. Life Sci. **4** (2015)
5. Toroman, A., Mujčić, E., Drakulić, U.: Fuzzy systems and their application in the study enrollment process. In: 12th International Scientific Conference on Production Engineering, Devolopment and Modernization of Production, RIM 2019, Sarajevo (2019)
6. Guzman, J.C., Melin, P., Prado-Arechiga, G.: Design of a fuzzy system for classification of blood pressure load. In book: Computational Intelligence and Mathematics for Tackling Complex Problems, January (2020)
7. Zahra, S.B., Hussain, T., Atta, A., Saleem Khan, M.: Human blood pressure and body temp analysis using fuzzy logic control system. Int. J. Comput. Sci. Network Secur. **18**(12) (2017)
8. Bhoi, B.K., Satpaty, S., Pradhan, M.C.: A smart fuzzy based system design for future prediction of medical condition. In: 2nd International Conference on Emergent Trends in Computing and Communication, May (2015)
9. Koller, D.Y.: Sampling methods urine/blood analysis. Am. J. Respir. Crit Care Med. **162** (2000)

The Smart Security Alarm System for Vehicles

Sandra Pajazetović, Una Drakulić(✉), and Edin Mujčić

Faculty of Techical Engineering Bihac, University of Bihac, Bihac, Bosnia and Herzegovina

Abstract. The need for a security alarm system for the vehicles begin at the end of the 19th century when increased vehicles production began. The first security systems for vehicles were much simpler than security alarm systems for vehicles today. The truth is that the vehicle thieves are more resourceful so the simple and known security alarm systems for the vehicles does not have a purpose anymore. The development of technology has modernized and perfected the security alarm systems for vehicles. The security alarm systems for vehicles now enables two-way communication between the vehicle owner and the security alarm system inside the vehicle. However, despite progress in this field, vehicles theft is still one of the most common. In the past year, approximately 775 cases of theft of expensive vehicles have been documented in Bosnia and Herzegovina. Worldwide, according to the Insurance Information Institute, over a million vehicles are stolen annually in the USA, corresponding to vehicle theft every 23 s. In today's time there is a need to develop a high technology quality security alarm system for a vehicles that will concur with the requirements of today's technology in terms of performance and functionality in the conditions of vehicle theft. In this paper is a proposed the smart security alarm system for vehicle that can fulfill all the requirements. The proposed system is realized using GSM/GPRS technology for two-way communication for system-user, and the GPS technology for locating the vehicle.

Keywords: Smart security alarm system for the vehicle · GSM · GPS/GPRS · SMS · Google Map · Remote control

1 Introduction

With the increase in the standard of living, the number of bought vehicles has rapidly increased in the past decade. As the number of vehicles purchased increases, the theft rate also increases [1]. The rate of increase in vehicle theft in this part of the world has reached an alarming rate [2]. The simplest definition of the vehicle security alarm system is that is a device built in a vehicle to make a sound if someone tries to break in the vehicle [3]. Through the years, security alarm systems for vehicles have evolved in terms of functionality and performance [4].

The first documented case of vehicle theft occurred in 1913 [3]. From that time until today, vehicle theft has become quite a common thing. Along with the development of different types of vehicles and the emergence of the need to protect them, security alarm systems for vehicles have been developed. The emergence of the need for security alarm systems for vehicles started at the end of the nineteenth century when vehicle production began.

N. Ademović et al. (Eds.): IAT 2021, LNNS 316, pp. 83–95, 2022.
https://doi.org/10.1007/978-3-030-90055-7_7

The first security alarm system for vehicle was invented 1913 by prisoner in Denver. It was adjusted manually and when someone tried to start the engine also start the sound alarm. After few years, Victor Helman 1954 patented his ‘Automatic burglar alarm’ with an sound signal which can only be turned off by the vehicle owner. Also, Joseph Yourtz invented a vehicle security system based on a motion sensor. Next known security alarm system is from 1974 when Chrysler Corporation patented a more sophisticated security system that contains multiple sensors placed in different places inside the vehicle. After activating one of the sensors, the warning lights and sound turn on. It also had the possibility of automatic doors locking [5].

The first security systems for vehicle were far from these today, however, the development of technology has enabled their progress. The security systems for vehicle have been developed to such an extent that two-way communication between the vehicle owner and system is enabled. In the past few years, it was not even possible to imagine that the security system inside the vehicle would inform the vehicle owner that it had been stolen, or that the vehicle owner would be able to stop the vehicle from a distance [6, 7].

The security alarm systems for vehicle have been very popular in recent times and widely used in most vehicles owned by companies, business organizations, and even by some private individuals who can afford the cost.

In today’s modern time the security alarm system for vehicles are made of the latest technology and have much more ability and functionality then those described. There are different types of the security alarm system for vehicles based on several criteria [8]. According to the method of activating the system can be divided on active and passive. The security alarm system, by the time of installation in the vehicle, can be divided in factory and subsequently installed. Also, according to the type of communication between the vehicle owner and the vehicle security system can be divided in one-way and two-way, etc. [8].

Today, most used and efficient security alarm system for vehicles are Compustar CS6900-AS, Crimestopper SP-502, Viper VSS5000, Python PS5000, Avital 5303L, Compustar CS700-AS, Viper 5906V.

Viper is a brand that has been in the car alarm industry for years. This company produces good quality products that vehicle owners benefit from. This system has a keyless entry feature with four buttons in the remote. If needed extra protection to vehicle, you can get additional features such as a remote starter and GPS tracking. The Viper 350 has a FailSafe starter kill, a panic alarm and anti-carjacking feature, a siren, and a parking light alarm response. The status LED has a bright blue color that warns anyone who tries to steal vehicle. This system has shock sensors that can be adjusted to reduce the instances of a false alarm. Avital also offers a keyless entry system that is very convenient when you are in a hurry. You can open or lock your car even from a distance. The LED status has a bright red color that warn-off thieves. The alarm system can also be hooked to your vehicle’s own siren to make a louder sound when someone tries to break into vehicle. It also has a parking light flash and a failsafe starter kill feature which prevents thieves to start the vehicle without the remote. The Compustar CS7900 system allows user to unlock and lock vehicle without a delay. It will work even if the owner is 1000 m away from the vehicle. What’s even more special about this system is that the owner can start vehicle remotely [9].

In this paper the proposed security alarm system for vehicle is realized using the GSM/GPRS and GPS technology. GPS is known as the Global Positioning System used to trace the location of the vehicle. A GPS framework computes its position by accurately timing the signal sent by GPS satellites high over the Earth. GPS Receiver gets the location information from satellites. It consists of internal RTC backup and can be directly connected to the USART of the microcontroller. The current date, time, longitude, altitude, speed, and travel direction/heading among other data are provided by the module and can be used in many applications including navigation, fleet management, tracking system, mapping, and robotics [4].

GSM is known as Global System for Mobile Communication used for communication purposes. GSM module is basically used here for receiving and sending confirmation SMS [5].

This paper describes the design of the vehicle security system that is specified in terms of system functionality in case of attempted or succeeded in vehicle theft. The proposed system has a built-in GSM/GPRS module, which in case an unauthorized person enters the vehicle, notifies the vehicle owner via SMS. The GPS receiver within the system recalculates the position of the vehicle and the system forwards the obtained coordinates to the vehicle owner, also via SMS. The system continuously providing the vehicle owner with updated coordinates until the owner turns off the system willingly using the remote control device.

In the next part of the paper are mentioned some of the first vehicle security systems and some of today's most used ones.

2 The Design and Implementation of the Smart Security Alarm System for Vehicle

In this part of the paper is described the design of the smart security system for vehicles using GPS and GSM/GPRS technology. The used microcontroller is an industrial ATMega328 microcontroller which has characteristics that can fullfill the need of the proposed system. To enable GPS positioning of the vehicle, we chose the Ublox NEO-6M GPS module. The Ublox NEO-6M GPS module is relatively small, has only four pins marked VCC, RX, TX and GND and easily connects to the microcontroller system via serial communication. To detect unauthorized persons in the vehicle, we used PIR (Passive InfraRed Sensor), which is widely used today in security alarm systems [4]. We have selected the SIM900 GSM/GPRS module for sending and receiving SMS messages from the security alarm system for vehicle to the vehicle owner. The SIM900 GSM/GPRS module is easy to connect to the ATMega328 microcontroller system using serial communication via the UART protocol. This module is supporting serial transfer rates ranging from 1200 bps to 115200 bps and frequency bands of 850/900/1800/1900 MHz. This makes this SIM900 GSM/GPRS module compatible with all GSM networks around the world, which would mean that our security alarm system for vehicle vehicle is usable in all areas of the world. To remotely turning the security alarm system for vehicle on and off we use an industrial remote control device that has an operating frequency of 433 MHz. We choose the industrial remote control device shown in the Fig. 1. This remote control device comes with a transmitter and a receiver. The sensitivity of the receiver in this remote control device is ≥105 dBm. Due to lower consumption, we redesigned the receiver in remote control device. We have replace the relay because energy consumption is high with an optocoupler and a MOSFET transistor.

Fig. 1. The remote control device

In order to enable the vehicle owner to act on the vehicles drive system by using his mobile device, we have designed a subsystem for remote stopping the vehicle. We connected a power amplifier and a relay to the output pin of the ATMega328 microcontroller, which can act on the drive part of the vehicle and stop the further movement. The relay can be connected by, for example, on the fuel pump supply and enable remote stopping the vehicle.

After selecting all required components for the smart security alarm system for vehicle we made a block diagram shown in Fig. 2.

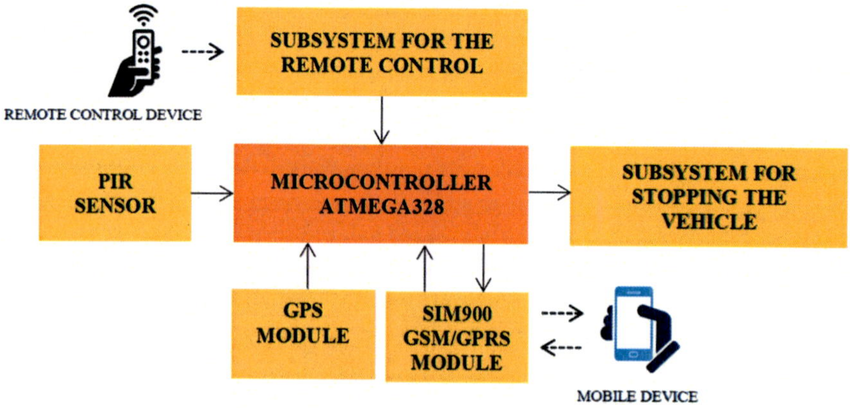

Fig. 2. The block diagram of the smart security alarm system for vehicle

Based on the block diagram in the Fig. 2 and selected components, we made a diagram for hardware part of the smart security alarm system for vehicle. This is shown in Fig. 3.

From Fig. 3 we can see that the SIM900 GSM/GPRS module and the GPS module are powered with relay because of the lower power consumption.

We have significantly reduced the energy consumption by not turning on GPS and SIM900 GSM/GPRS modules immediately when turning the proposed smart security alarm system for vehicle. These two modules are turn on when the pir sensor is active, i.e. when the unauthorized person is in the vehicle.

The power supply of the proposed smart security alarm system for vehicle can be separately only for the system or can use the power supply located in the vehicle (battery installed in the vehicle). Since the smart security alarm system for vehicle operates at 5 V and the battery provides 12 V, it is necessary to reduce this voltage to 5 V. To reduce voltage for 12 V to 5 V we use integrated voltage stabilizers 7805.

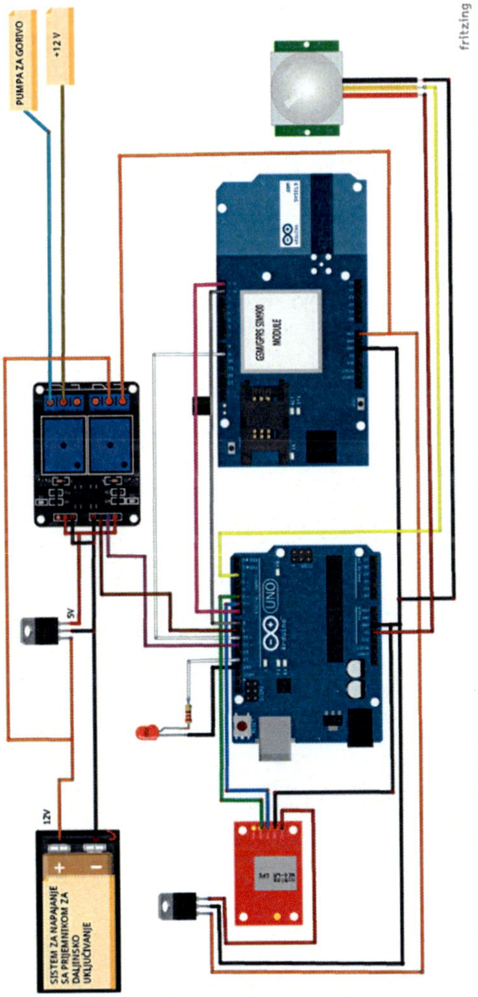

Fig. 3. The hardware part of the smart security alarm system for vehicle

After connecting all the components we place them in housing that we design and make from plexiglass and wood. This is shown in Fig. 4.

Fig. 4. The appearance of the designed smart security alarm system for vehicle in the housing

Figure 5 shows the final appearance of the security alarm system for vehicle.

Fig. 5. The final appearance of the smart security alarm system for vehicle

After finish the hardware part of the smart security alarm system for vehicle, it was necessary to program the ATMEGA328 microcontroller.

The flow diagram of the smart security alarm system is shown in Fig. 6.

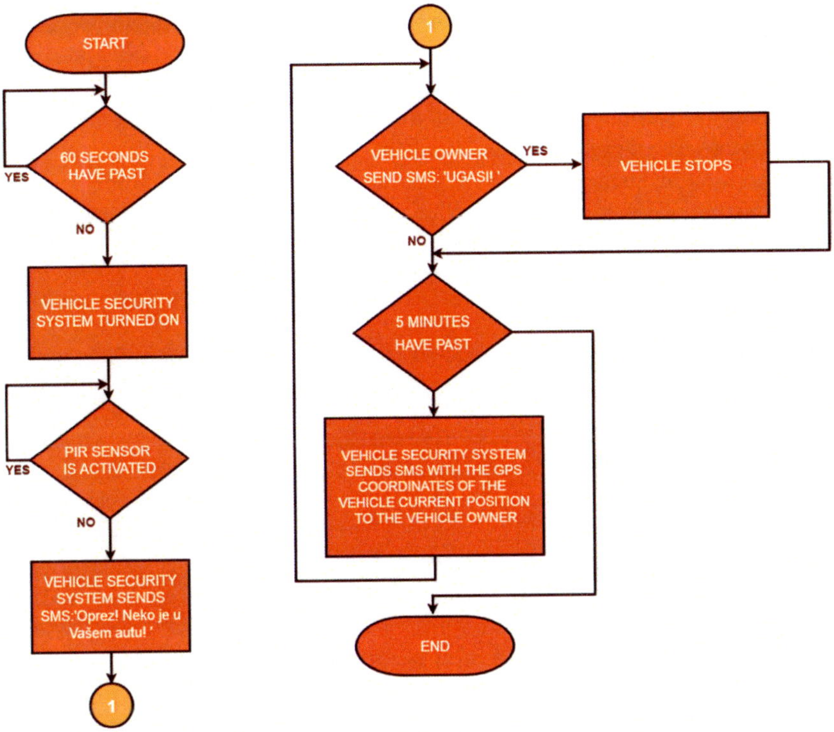

Fig. 6. The block diagram of the vehicle security alarm system

The programming of the smart security alarm system is based on the desired principle of operation:

The vehicle owner activates the smart security alarm system with a remote control device after exiting the vehicle. After 60 s, the smart security alarm system turns on and pir sensor is ready to detect any entry into the vehicle. The reason for the subsequent activation of the pir sensor is due to the possible return of the vehicle owner to the vehicle in case he forgot something. If anybody entry in vehicle, the PIR sensor will detect and that information send to the ATMEGA328 microcontroller. Based on this information, the ATMEGA328 microcontroller will turn on a SIM900 GSM/GPRS and a GPS Ublox NEO-6M module. After GPS and SIM900 GSM/GPRS modules turn on, the ATMEGA328 microcontroller sends the first SMS message to the vehicle owner using the SIM900 GPS/GPRS. We decide that first SMS is only a warning and does not contain information about the current position of the vehicle. If the vehicle owner does not turn off the smart security alarm system with remote control device in less then 5 min, the smart security alarm system sends second SMS message. The second SMS content is the current vehicle position. If the vehicle owner still does not turn off the smart security alarm system using remote control device, the current vehicle position will be continuously sending as SMS to the vehicle owner. This process is repeated every 5 min until the vehicle owner turns off the smart security alarm system with remote control device.

When designing the smart security alarm system, we also enabled two-way communication between the vehicle owner and the designed smart security alarm system. After activating the smart security alarm system, the vehicle owner can at any time send an SMS message with content: 'UGASI!' to the smart security alarm system in order to stop further movement of the vehicle.

After connecting all the components, programming the ATMega328 microcontroller the measurement of electricity consumption was performed. The current was measured for 9 different states in which the proposed security alarm system can be.

The results of this measurement are shown in Table 1.

Table 1. The measurement of energy consumption for the vehicle security alarm system

State	Measured value of current (mA)		The description of measured state
	1st	2nd	
1	10	10	The electronic circuit is not turned on via the remote control, only receiver consumes power
2	46	46	The electronic circuit is turned on via the remote control, the sensor is not activated
3	60.30	60.4	The sensor is activated
4	46	46	The sensor is not activated
5	60	60	The sensor is activated
6	270	270	The GPRS GSM and GPS modules are turned on
7	305	303	The system sends SMS
8	330	330	The system receive SMS and activate the relay
9	330	330	All components are turned on, the system sends SMS

From Table 1 it can be seen that after repeated measurements we get almost the same values. The proposed security alarm system mostly operates in state 2, ie when the security alarm system is turned on but not activated. The energy consumption in this state, as can be seen from Table 1, is 46 mA, ie if the battery voltage is 12 V, the power consumption is 552 mW.

The proposed security alarm system can be connected to the vehicle battery or have a separate power supply. If we connect the security alarm system to a battery with a capacity of 75 Ah, the security alarm system can work constantly for 1630, 43 h, or approximately 68 days.

3 The Experimental Work Analysis of the Security Alarm System for Vehicle

In the first step of the experimental work analysis, we performed an experimental work analysis of the remote control device.

For experimental work analysis of the remote control device, we have measured the distance from which the remote control device can turn on or turn off the smart security alarm system. The results of this analysis are shown in Table 2.

Table 2. The measurement of the distance for the use of remote control device for turn on or off the vehicle security alarm system

No.	Measured distance in meters	The security alarm system for vehicle is turn on/off
1	20	Turn on
2	30	Turn on
3	40	Turn on
4	50	Turn on
5	60	Turn on
6	70	Turn off
7	63	Turn off
8	63	Turn off
9	63	Turn on

Based on the experimental work analysis, we can conclude that the smart security alarm system can be turned on or turned off from a distance of less than 60 m. We tested the remote control device 9 times and obtained similar results shown in Table 2. Figure 7 shows how the experimental work analysis of the remote control device was performed.

In the second step of the experimental work analysis, we performed an experimental work analysis of the smart security alarm system for vehicle. For this experimental work analysis of the proposed security alarm system, we performed the simulation of the theft of our own vehicle. The first step of the simulation of vehicle theft is the vehicle owner leaves and locks the vehicle. Next, after one minute the security alarm system is turned on.

Now, the security alarm system is turned on but not active.

After a while, we simulate vehicle theft. The vehicle owner as a thief enters the vehicle, turn it on, and start driving the vehicle. When the 'thief' enters the vehicle, the pir sensor detects him and the security alarm system is activated. After one minute the SMS from the security alarm system is sent to the vehicle owner with content: 'Oprez! Neko je u Vašem vozilu!'. This first SMS is considered as a warning. This is shown in Fig. 8a).

Fig. 7. The experimental work analysis of the remote control device

The 'thief' continues to drive the vehicle not knowing that the security alarm system in the vehicle report the vehicle owner. After 5 min, after the last received SMS, the second SMS is received. The second SMS consists of the coordinates of the current vehicle position. This SMS is in form of a web link that, when clicked, opens the Google Maps application and shows the current position of the vehicle. This is shown in Fig. 8b).

By clicking the received web link for the Google Maps application, we can see that the coordinates for the current vehicle location are the same as the real current vehicle position. The 'thief' continues to drive the vehicle. If the vehicle owner doesn't respond with the SMS to the security alarm system, the vehicle owner receives the third SMS. The third SMS is the same as the second SMS, with updated coordinates for the current vehicle position. If we check the received coordinates for the current vehicle position we can conclude the real current position of the vehicle is the same. This is shown in Fig. 8c).

Figure 9a) shows the current location of the vehicle, as well as a satellite view of the same location in Fig. 9b).

This process will repeat every five minutes, i.e. the vehicle owner keeps receiving the SMS with the coordinates of the current vehicle position in form of a web link to the Google Maps application. In the third step of the experimental work analysis, we performed an experimental work analysis of stopping the smart security alarm system

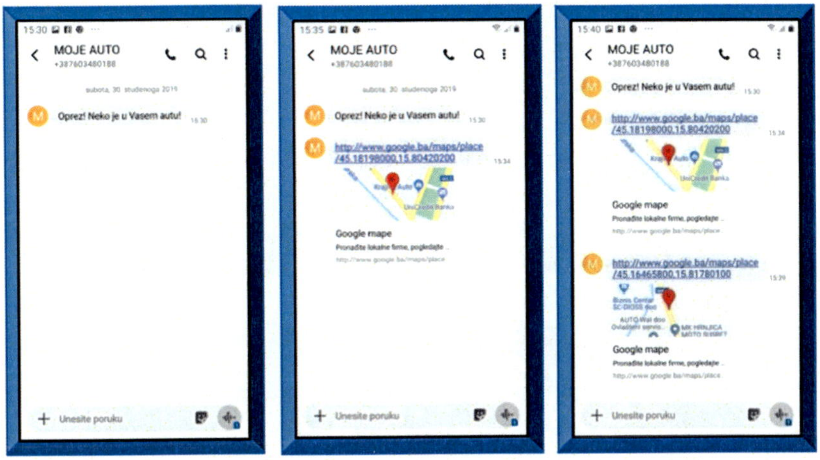

a) b) c)

Fig. 8. Testing the proposed system: a) The first received SMS with content "Oprez! Neko je u Vasem autu!", b) The second received SMS with Google Maps link, c) The tird received SMS with Google Maps link

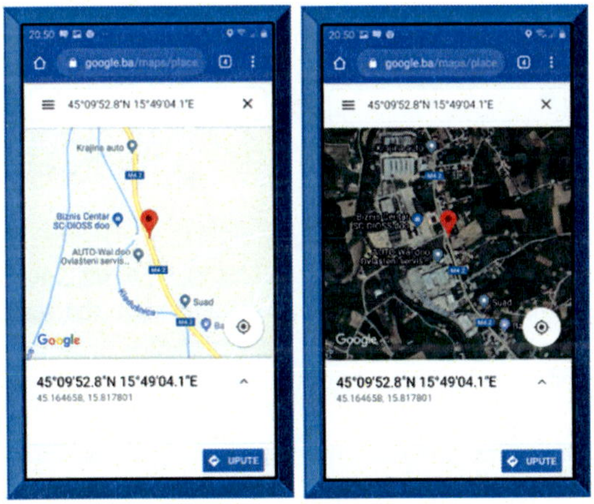

a) b)

Fig. 9. The proposed system: a) SMS with opened Google Maps link, b) The satellite view of the same location

for vehicle. The vehicle owner sends an SMS with content: 'UGASI!' to the security alarm system. The security alarm system recognizes received SMS as a code word and turns off the relays. We use the light bulb which simulates any of the important drive parts of the vehicle. When the security alarm system turns off the relay it means that the light bulb stops lighting, i.e. if the light bulb is, for example, a fuel pump the vehicle stops driving. Figure 10 is shown that the SMS with content: 'UGASI!' is recognized by the security system and the light bulb is turned off.

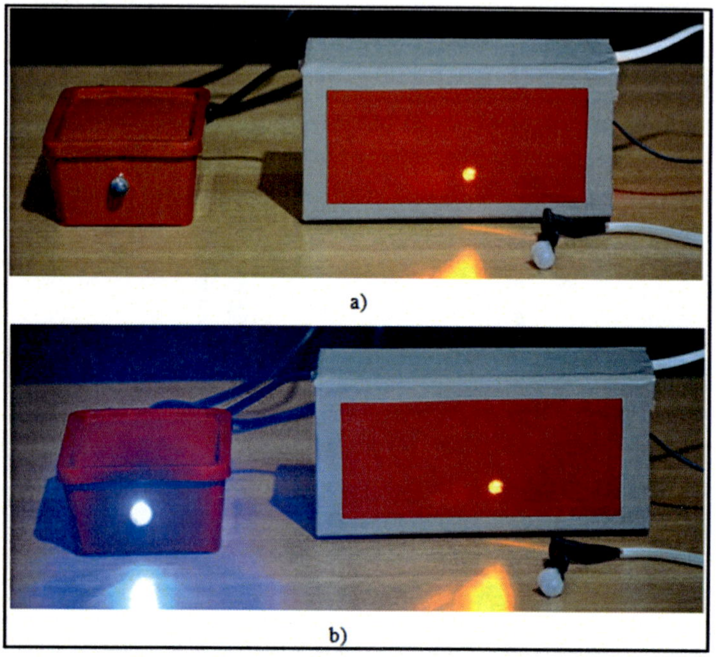

Fig. 10. Test of stopping the car with the remote control device: a) before SMS "UGASI!" is sent, b) after SMS "UGASI!" is sent

After performing an experimental work analysis of the security alarm system, we can conclude that:

- the remote control device works properly in less or 60 m distance from the vehicle,
- the security alarm system turns on after one minute when the vehicle owner exit and locks the vehicle,
- the security alarm system is turn on but not active until the pir sensor detect someone in the vehicle,
- the security alarm system is activated when the pir sensor detects someone in the vehicle
- the security alarm system sends the first warning SMS in the first minute from the intrusion detection
- the security alarm system continuously sends SMS to the vehicle owner every five minutes with updated coordinates for the current vehicle position in form of a web link that, when clicked, opens the Google Maps application and shows the current position of the vehicle
- the security alarm system recognizes the received SMS with content: 'UGASI!', from the vehicle owner and turn off the relays
- the security alarm system stops further driving after turning the relays off

We repeated this experimental work analysis of the security alarm system several times and obtained similar results.

4 Conclusion

In this paper, a smart security alarm system for vehicles is proposed and experimental work analysis is performed. The proposed smart security alarm system for vehicles is designed based on the desired input characteristics of the security alarm system. The smart security alarm system for vehicles has been optimized for less energy consumption. Remote control of the smart security alarm system for vehicles is enabled. Activation of the smart security alarm system is performed using PIR sensor that detects any intrusion in the vehicle to avoid "accidental" activation. Using the proposed smart security alarm system is possible to "quietly" inform the vehicle owner about the theft.

There is enabled two-way communication between the vehicle owner and the smart security alarm system. The vehicle owner can send SMS to the smart security alarm system and stop further driving the vehicle.

The design of the smart security alarm system is small in size and can be placed anywhere inside the vehicle.

The experimental work analysis of the smart security alarm system is performed several times and we obtained similar results. Based on experimental work analysis we can conclude that the proposed smart vehicle security system is working properly.

References

1. Mathankumar, M., Shanmugasundaram, S., Thirumoorthi, P., Rajkanna, U.: Develo-pment of smart car security system using multi sensors. Int. J. Pure Appl. Math. **117**(22 2201), 19–23 (2017)
2. Alli, K.S., Ijeh-Ogboi, C., Gbadamosi, S.L.: Design and construction of a remotely controlled vehicle anti-theft system via gsm network. Int. J. Educ. Res. **3**(5) (2015)
3. Agarwal, R., Boominathan, P.: Vehicle security system using iot application. Int. Res. J. Eng. Technol. e-ISSN: 2395-0056, p-ISSN: 2395-0072, **05**(04) (2018)
4. Mujčić, E., Drakulić, U., Merisa, Š.: Advertising LED system using PIC16F4550 microcontroller and LED lighting. In: International Symposium on Innovative and Interdisciplinary Applications of Advanced Technologies – IAT, Neum (2016)
5. Šabić, Z., Drakulić, U., Mujčić, E.: The smart greenhouse system based on the the mobile network and iot. In: Avdaković, S., Volić, I., Mujčić, A., Uzunović, T., Mujezinović, A. (eds.) Advanced Technologies, Systems, and Applications V: Papers Selected by the Technical Sciences Division of the Bosnian-Herzegovinian American Academy of Arts and Sciences 2020, pp. 285–298. Springer International Publishing, Cham (2021). https://doi.org/10.1007/978-3-030-54765-3_19
6. Yasashree, A., Lakkar, P.: Smart automobile security system. Int. J. Eng. Res. Technol. ISSN: 2278-0181, **8**(04) (2019)
7. Singh, P., Sethi, T., Balabantaray, B.K., Biswal, B.B.: Advanced vehicle security system. In: International Conference on Innovations in Information, 20 March 2015. https://doi.org/10.1109/ICIIES.2015.7193276, India, IEEE 13 Aug (2015)
8. Singh, P., Sethi, T., Biswal, B.B., Pattanayak, S.K.: A smart anti-theft system for vehicle security. Int. J. Mater. Mech. Manuf. **3**(4), 249–254 (2015)
9. Shah, R.J., Gharge, A.P.: GSM based car security system. Int. J. Eng. Innovative Technol. **2**(4), ISSN: 2277-3754 (2012)

Advanced Electrical Power Systems

Correlation of Day-Ahead Electric Energy Market Price with Renewable Energy Sources Generation and Load Forecast

Amer Aščerić[1,2](✉), Miloš Pantoš[1], and Marko Čepin[1]

[1] Faculty of Electrical Engineering, University of Ljubljana, Ljubljana, Slovenia
amer.asceric@epbih.ba

[2] JP Elektroprivreda BiH d.d.-Sarajevo, Sarajevo, Bosnia and Herzegovina

Abstract. An increase in the share of renewable energy sources has an impact on generation variability and generation forecast error, which is also reflected in the formation of the day-ahead electric energy market price. This paper analyzes the impact of variable renewable energy sources and load forecast on the day-ahead German electric energy market. A correlation between the market price, renewable energy sources generation and load forecasts was analyzed. Results show that there is a significant correlation between the forecast of renewable energy sources generation as well as the load on day-ahead electric energy market price. However, the biggest impact on day-ahead price has a residual load, which represents the difference between the load and the renewable generation.

Keywords: Renewable energy sources · Electric load · Day-ahead market price · Renewables generation forecast · Correlation

1 Introduction

Energy generation based on renewable energy sources (RES) has become a strategic commitment of the world to enable a sustainable society [1]. Integration of RES into power energy system is related to benefits like reduction of energy losses, decentralization of the power system and contribution to a cleaner environment by reduction of amounts of pollutants in the atmosphere [2].

Energy policies around the world stimulate an increase in renewable energy production [3], so the installed capacity of RES increases notably, worldwide. This largest contribution of new RES includes new installations producing electric energy primarily from solar and wind energy. The latter accounted for 84% of all new facilities installed in 2018, finally pushing the overall share of hydropower to just under 50% of the world's renewable installed capacity [4]. Technological development and decrease in production costs are factors that contribute notably to the fast spread of electric energy production from solar and wind energy in addition to the subsidies, which in the first place pushed these technologies forward years ago.

Although, the introduction of RES connected to power distribution system has its benefits, due to the stochastic nature of RES, it also introduces various issues to the

N. Ademović et al. (Eds.): IAT 2021, LNNS 316, pp. 99–109, 2022.
https://doi.org/10.1007/978-3-030-90055-7_8

power distribution system, such as voltage deviation, system stability, line congestion, reactive power compensation, and other power quality issues [5].

The challenges for today's electric energy market are the variability of generation and prices, the error in the generation forecast, negative prices, etc., which are a result of the growing share of RES [6]. An increase in the share of RES affects various aspects of the electric energy market, primarily the economic aspect as well as the shape of the supply curve [7].

The introduction of RES has an effect on the merit order curve. Typical merit order curve sorts ascending different types of power plants based on their marginal costs. Base and medium load generators are located on the left-hand side of the merit order curve with high fixed costs and low variable costs. Peak load generators like gas plants are located on the right-hand side of the merit order curve with lower fixed costs and higher variable costs. Therefore, peak load plants receive revenue from electric energy sales only during high demand or scarcity situations. Due to low variable cost, increasing the share of RES shifts the merit order curve to the right-hand side, which leads to less utilization of peak power plants. The effect of increasing the share of RES on merit order is graphically presented in Fig. 1 [8]. Due to determining market price as the intersection of supply and demand, the average market electric energy price will be lower, which causes less revenue for peak power plants. This effect is discouraging for future investments in peak power plants and thermal power plants with high variable costs (coal, gas, oil, carbon emission coupons, etc.).

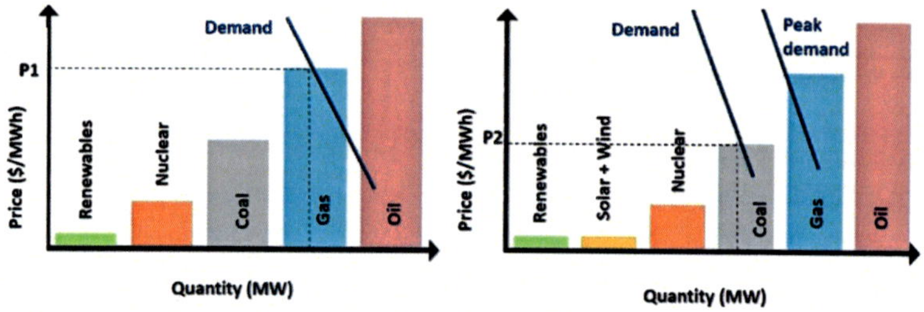

Fig. 1. Impact of RES on merit order curve

On the other hand, generation from renewables is intermittent and not reliable enough. Then, the market will be faced with a resource adequacy problem, which means the generation adequacy in the market will not be high enough to meet the peak demand. A well-functioning energy market should be able to provide cost recovery for all types of generators to ensure an adequate level of resource adequacy [9].

In previous years electric energy market is facing an increase in the share of RES in the generation portfolio and this trend is expected to continue in the future with the highest growth estimated for wind and solar PV technologies [10]. In this paper, the term "renewable energy source" refers to the wind (onshore and offshore) and photovoltaics (PV) RES.

RES variability is a challenge for the power system operators, as it requires additional actions to balance the power system. Power system with a large share of variable RES needs sufficient flexsibility to ensure that generation and demand are adjusted.

Intermittent and uncertain characteristic of RES generation implies difficult planning of generation from RES. Precise RES generation forecast provides the basis for optimized power system operation under market conditions, so minimum difference between forecast and realized RES generation is required. Greater accuracy in the forecast of RES generation is one of the preconditions for further increase in the share of RES.

In order to determine price at day-ahead market auction, it is essential to provide a day-ahead renewable generation forecast. System operators provide a forecast of renewable generation for the next day with an hour or eventually 15-min resolution. The difference between real and forecasted generation is traded in the intra-day market. Day-ahead forecast of RES generation and load is important for the day-ahead market, because trading in the day-ahead market is based on day-ahead forecast values of generation and load.

Germany is one of the most progressive countries in RES installed capacity. In addition, Germany has a liquid electric energy market, so it is convenient to use its example to analyze the impact of RES on various market parameters. For the purpose of this paper following data is used:

- Day-ahead PV and wind (onshore and offshore) generation forecast [11]
- Actual PV and wind generation [11]
- Day-ahead load forecast [11]
- Actual load [11]
- Day-ahead market electric energy price[11]
- Day-ahead transfer capacities [11]

The analysis is performed on data from the ENTSO-E database for the period from 1.1.2019 to 27.6.2020 with hourly resolution. This paper does not propose a forecast method for day-ahead values of RES generation, load and market electric energy price. Forecast values are obtained from the ENTSO-E database.

The objective of this paper is to consider the impact of day-ahead forecast values of RES generation and load on day-ahead electric energy market price. Section 2 presents the analysis of variability and forecast error of RES. The third section presents correlation analysis of day-ahead RES generation and load forecast on day-ahead electric energy market price and a discussion of the obtained results. The last section presents the conclusion of this work.

2 Introduction

2.1 RES Generation Variability

Large variations in RES generation in markets with a large share of RES can cause several issues in power system operation and planning [12]. The installed capacity of wind and PV generation in Germany in 2018 amounted to 58.84 GW and 45.18 GW, respectively. Taking into account the capacity factor of 23% for wind generation and 13% for PV generation [13], the average hourly generation from variable RES would equal 19.41 GWh.

The mean value of hourly RES (wind and PV) generation in Germany for the considered period is 20.1 GWh with a standard deviation of 11.6 GWh. A histogram of hourly RES generation is presented in Fig. 2. Maximum, minimum, mean and standard deviation values of hourly PV and wind generation, as well as load, are given in Table 1.

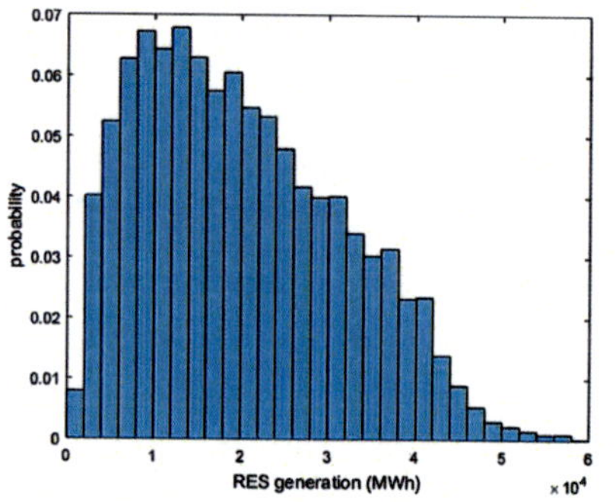

Fig. 2. Histogram of hourly RES generation

Table 1. Descriptive statistics of hourly values of RES and load

Parameter	Max (GW)	Min (GW)	Mean (GW)	St.dev. (GW)
PV	33.2	0.0	5.1	7.7
Wind	46.2	0.4	15.0	10.6
PV + wind	59.5	0.5	20.1	11.6
Load	77.2	29.6	55.7	10.0

Hourly variability of each parameter presents hourly gradients of that parameter. The maximum and minimum hourly RES generation gradients are +11.34/−10.05 GW. The ratio of the maximum and minimum RES generation gradient to the average hourly RES generation is +56.42% and −50%, respectively. The average hourly gradient of RES generation is approximately 1.72 GW and the ratio of the average RES generation gradient to the average hourly RES generation is 8.56%.

Figure 3 and Fig. 4 show histogram of RES generation hourly gradient for the considered period in MWh and as a percentage of RES generation for a specific hour. The ratio of RES generation hourly gradient and RES generation for a specific hour have a mean value of 10.84%.

These findings prove that the variability of RES generation in the German electric energy market is high. By increasing the share of variable RES, the market will consequently face a higher level of variability of the generation profile.

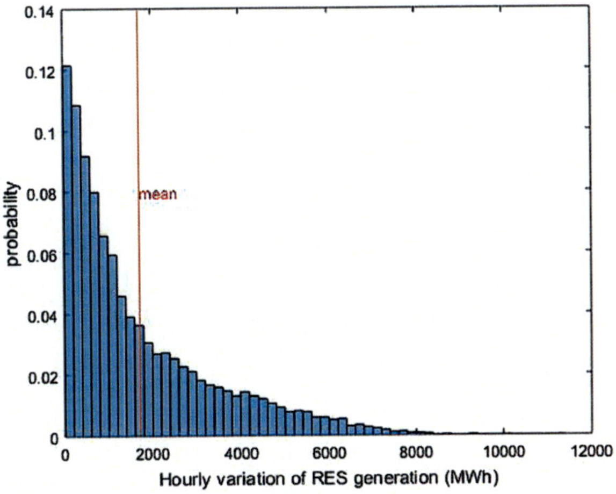

Fig. 3. Histogram of RES generation hourly gradient

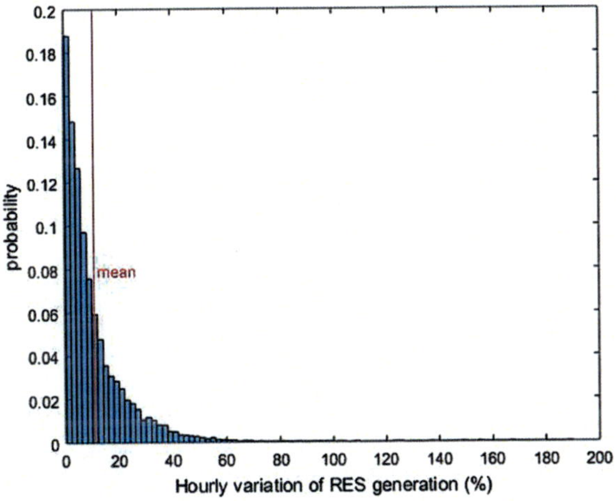

Fig. 4. Histogram of RES generation hourly gradient (%)

2.2 RES Generation Forecast Error

There are several ways for calculating forecast error. In this analysis, forecast error is calculated by subtracting realized RES generation from day-ahead forecast values for each hour of considered period. Forecast values are obtained from the ENTSO-E database for the period from 1.1.2019 to 27.6.2020 with hourly resolution.

The histogram of RES day-ahead generation forecast error is shown in Fig. 5. RES generation forecast error for the considered period was in the range of −11.79 GWh to 11.26 GWh with a mean absolute percentage error of 9.15%.

Figure 6 presents renewable generation forecast error as a percentage of the hourly load, which has a normal distribution with a mean value of 0.15% and a standard deviation of 3.4%. Renewable generation forecast error as a percentage of hourly load is in the range of −25.59% to 21.22%.

Renewable generation forecast error as a percentage of the hourly load for the German electric energy day-ahead market for years 2012 and 2013 had a mean value of 0.3% [12]. Compared to results for the German electric energy day-ahead market for years 2012 and 2013 [12], a mean value of renewable generation forecast error has a smaller value for 50%. This can be explained by a larger share of renewables, so the forecast generation error of individual generation units is compensated by other units. As expected, range of forecast error is larger than in years 2012 and 2013 because RES installed capacity and thus RES generation has significantly increased.

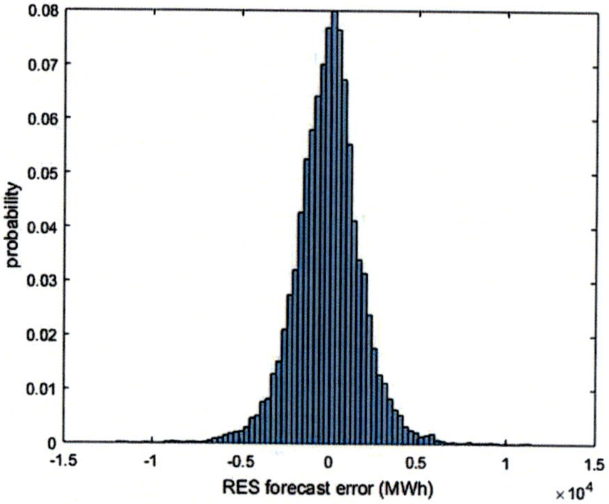

Fig. 5. Day ahead RES generation forecast error

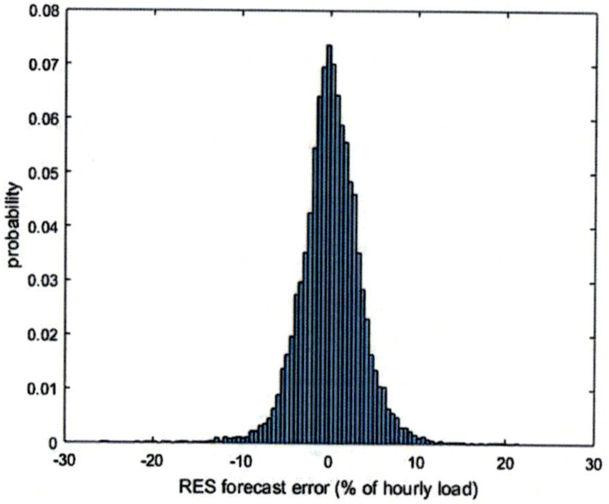

Fig. 6. Day ahead RES generation forecast error

3 RES Impact on the Day-Ahead Market Price

In this section impact of RES on day-ahead market price is analyzed based on hourly day-ahead forecast values of RES generation, load and market price data.

In Germany, day-ahead electric energy price for the considered period was in the range from −90.00 €/MWh to 121.46 €/MWh, with the mean value of 32.99 €/MWh and standard deviation of 16.81 €/MWh. The histogram of the hourly day-ahead price for the considered period is shown in Fig. 7.

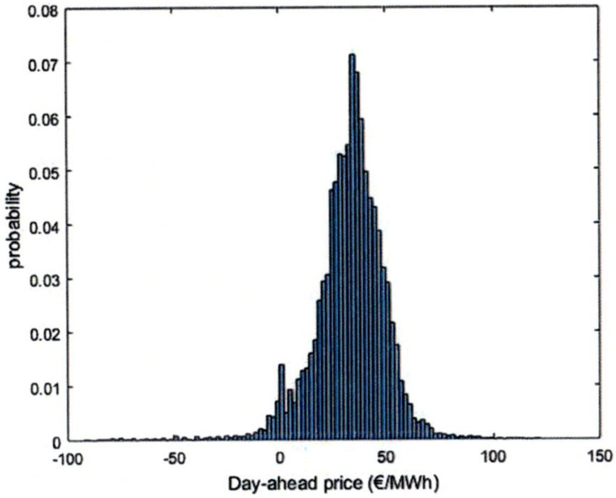

Fig. 7. Histogram of day ahead price

In order to examine the impact of RES on the day-ahead market price, it is necessary to analyze the correlation between hourly values of day-ahead RES generation forecast and hourly day-ahead market price. A scatter plot of hourly day-ahead RES generation forecast versus hourly day-ahead market price is shown in Fig. 8. Results show that by increasing renewables infeed to the market the day-ahead price decreases.

Negative day-ahead electric energy price is recorded for 3.23% of hours during the considered period. Previously, negative electric energy price was usually related to the holiday season because of low consumption. But with the increase share of RES, due to their nature, periods of higher generation than consumption are happening more often, so negative price is distributed through the year.

The period of peak price was 23.01.–25.01.2019. When load had maximum values for the considered period and renewable generation was extremely low.

A scatter plot of hourly day-ahead load forecast versus hourly day-ahead market price is shown in Fig. 9. As can be seen in Fig. 9, an increase in consumption predominantly implies an increase in price, because in periods of high demand peak power plants with high variable costs are engaged.

Figure 8 indicates that an increase in generation from RES has the effect of reducing electric energy prices, while Fig. 9 suggests that higher consumption has the effect of increasing prices. However, it is interesting to observe the impact of the difference

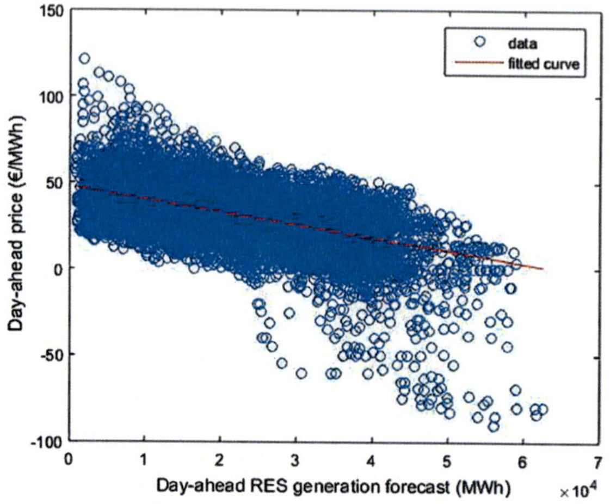

Fig. 8. RES generation and day ahead market price

between the hourly day-ahead forecast of load and RES generation, i.e. residual load. The residual load is supplied from non-renewable sources, which are characterized by higher variable cost than RES. Figure 10 presents a scatter plot of hourly day-ahead residual load forecast versus hourly day-ahead market price. As can be seen, there is a clear correlation between residual load and day-ahead price.

Table 2 provides an overview of the correlation coefficients between hourly day-ahead price and hourly day-ahead forecast of RES generation, load and residual load.

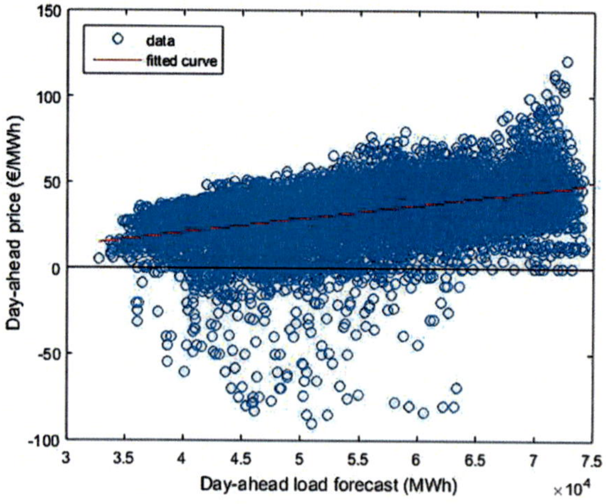

Fig. 9. Load forecast and day ahead market price

The correlation coefficient between day-ahead forecasted load and day-ahead price shows that a larger load significantly affects an increase in day-ahead price.

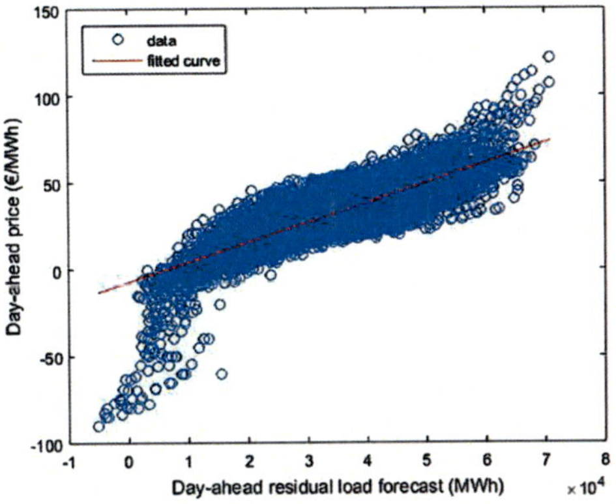

Fig. 10. Residual load and day ahead market price

Table 2. Correlation coefficients

Corr. coef.	RES generation forecast	Load forecast	Residual load forecast
Price	−0.5077	0.4437	0.8245

Presented results in Table 2 suggest that day-ahead price is more affected by the day-ahead forecast of RES generation than load. The negative sign of the correlation coefficient implies that an increase in RES generation brings to decrease in the day-ahead price.

Results show that day-ahead electric energy market price is dominantly correlated to residual load. Residual load presents load that cannot be covered with RES generation, so non-renewable sources from the right-hand side of the merit order curve must be engaged. Low residual load occurs in periods of low demand, high RES generation, or a combination of both. On the other side, high residual load occurs in periods of high demand, low RES generation, or a combination of both. Higher residual load implies the need for a larger amount of generation from non-renewable sources. The operation cost of non-renewable sources is much higher than renewable ones, so electric energy price increases.

Hourly variations of hourly day-ahead RES generation forecast versus hourly variations day-ahead price are shown in Fig. 11, where it can be seen that in periods of low RES generation variability there is a big spread of price variation. Increase in hourly variation of RES generation leads to an increase in price variation. This is expected due to additional cost for balancing. The correlation coefficient between hourly variations of day-ahead RES generation forecast and hourly variations of day-ahead price is 0.14. The reason for the low correlation coefficient lies in the fact that price variability depends on several factors such as hourly electric energy load, load variations, available generation from different generators and the share of RES in the generation mix. Therefore, price variability must be analyzed by considering all related factors.

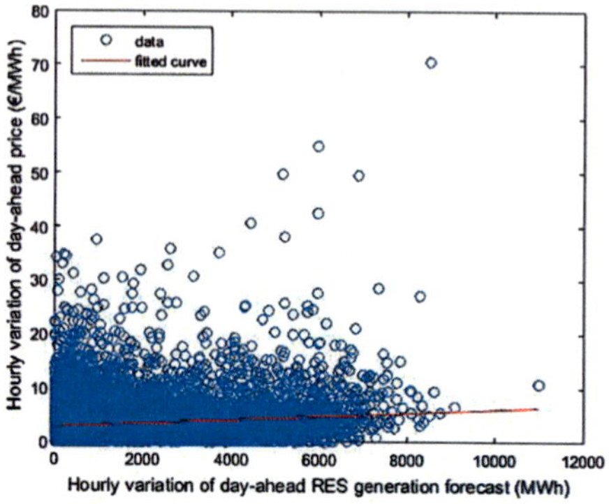

Fig. 11. Hourly variation of RES generation and day ahead price

Interstate day-ahead transfer capacities correlation with RES day-ahead generation and the day-ahead price are expressed in terms of correlation coefficients and presented in Table 3. Results show that there is no significant correlation between the day-ahead generation of German RES with interstate transfer capacities limit between Germany (DE), Austria (AT), Slovenia (SLO), Italy (IT) and (Hungary). Although correlation coefficients between interstate transfer capacities and day-ahead price are slightly larger than for RES generation, they do not show a significant correlation.

Table 3. Descriptive statistics of hourly values of RES and load

Corr. coef.	DE > AT	AT > SLO	AT > IT	AT > HU
RES generation	0.0350	0.0843	0.0721	0.0528
Price	−0.0296	−0.1792	0.0818	−0.1226

4 Conclusion

Increasing RES generation brings new challenges to the electric energy market. For the considered period, RES generation was 36.1% of consumption with average hourly RES generation variability of 1.72 GW. Hourly variation of the load is 1.88 GW, which shows that load has higher variation than RES generation. However, that is not a big problem because the load can be very well predicted compared to RES generation.

Increase in share of RES contribute to reduction of RES forecast generation error, because other units compensate the forecast generation error of individual generation unit.

Due to low variable costs, RES shifts the merit order curve to right-hand side and thus reduce electric energy price. In the periods when RES generation is higher than demand, electric energy market price is negative. With increasing the share of RES, periods with negative price are becoming more frequent and are distributed trough year. Price peaks are recorded in periods of high consumption and extremely low RES generation.

Results show that the correlation between price and residual load is much higher than the correlation between price and RES generation. Increasing the share of RES in generation portfolio contributes to reducing residual load, i.e. less demand is supplied from conventional non-renewable sources. Therefore, generation from non-renewable sources decreases. However, power system still needs conventional energy sources in order to balance RES generation.

Opposite to expected, an increase in variation of RES generation does not imply an increase in price volatility. The reason for this is that price does not depend only on RES generation, but also on other factors, such as hourly electric energy load, load variations, available generation from different generators and the share of RES in the generation mix.

Interstate transfer capacities do not show a noticeable correlation to day-ahead price or day-ahead RES generation.

References

1. He, Y., Xu, Y., Pang, Y., Tian, H., Wu, R.: A regulatory policy to promote renewable energy consumption in China: review and future evolutionary path. Renew. Energy **89**, 695–705 (2016). https://doi.org/10.1016/j.renene.2015.12.047
2. Bhattacharya, M., Paramati, S.R., Ozturk, I., Bhattacharya, S.: The effect of renewable energy consumption on economic growth: evidence from top 38 countries. Appl. Energy **162**, 733–741 (2016). https://doi.org/10.1016/j.apenergy.2015.10.104
3. Comission, E.: Best practices on renewable energy self-consumption. J. Chem. Inf. Model. **53**(9), 1689–1699 (2015). https://doi.org/10.1017/CBO9781107415324.004
4. IRENA: Renewable energy capacity highlights (2019)
5. Venkateswaran, V.B., Saini, D.K., Sharma, M.: Approaches for optimal planning of the energy storage units in distribution network and their impacts on system resiliency — a review. CSEE J. Power Energy Syst. **6**, 816–833 (2020). https://doi.org/10.17775/cseejpes.2019.01280
6. Morales, J.M., Conejo, A.J., Madsen, H., Pinson, P., Zugno, M.: Integrating Renewables in Electricity Markets. Springer US, Boston, MA (2014)
7. Sioshansi, F.P.: Evolution of Global Electricity Markets: New Paradigms New Challenges New Approaches. Academic Press (2013)
8. Aghaie, H.: Resource Adequacy and Optimal Investment in Energy-only Markets, pp. 1–16 (2015)
9. Aghaie, H., Haas, R.: Efficient energy only markets. In: International Conference on the European Energy Market EEM, vol. 2015-Augus, pp. 8–12 (2015). https://doi.org/10.1109/EEM.2015.7216736
10. Gielen, D., Boshell, F., Saygin, D., Bazilian, M.D., Wagner, N., Gorini, R.: The role of renewable energy in the global energy transformation. Energy Strategy Rev. **24**, 38–50 (2019). https://doi.org/10.1016/j.esr.2019.01.006
11. ENTSO-E: ENTSO-E Transparency Platform. Available at: https://transparency.entsoe.eu/dashboard/show
12. Aghaie, H.: Statistical Analysis of the German Electricity Market in Presence of Renewables (2017)
13. Joint Research Centre: Energy Technology Reference Indicator Projections for 2010–2015, vol. 57, no. 3 (2014)

Smart Meter Based Non-intrusive Load Disaggregation and Load Monitoring

Tarik Hubana(✉) and Elma Begić

P.E. Elektroprivreda BiH d.d., Sarajevo, Bosnia and Herzegovina
t.hubana@epbih.ba

Abstract. With the advancement in metering technology, a large field of possibilities has opened. The presence of smart meters that from year to year have better metering capabilities is rising, and they are common for both commercial and residential consumers. By using these single-point measurements of current and voltage with the appropriate load disaggregation techniques, a feasible solution that separates and classifies the loads at that metering point could be achieved. On the other hand, the increasing interest in load management and energy efficiency drives the need for non-intrusive load monitoring (NILM) techniques with a minimum invasion of privacy. The research presents the NILM technique based on the smart meter voltage and current measurements. The developed algorithm in MATLAB simulation software is tested with the laboratory test case with the smart meter measurements and four different appliances, and the results demonstrate the satisfactory accuracy and applicability of the technique. This research is contributing to the existing body of knowledge in terms of developing and testing the non-intrusive load monitoring technique that is seen as a natural upgrade to the existing smart meter functionalities.

Keywords: Monitoring · Load · Smart metering · Signal processing · Energy management · Energy efficiency

1 Introduction

The energy market is going through a historic change in the past decades. This is particularly expressed at the distribution level, where the emergence of new technologies changed the traditional operation of the distribution networks. In the first place, it is regarded to the advances in smart metering, operation, control, bidirectional power flows, renewables, electric vehicles, and energy storage devices. The rising awareness about climate change and energy efficiency and the more challenging requirements of the energy regulative is a strong driving force of the integration of the smart technologies in the distribution system and at the end consumers. Motivated by these factors, and with the increasing amount of data, different applications emerged to improve both operation and efficiency of the distribution systems. With the available data, machine learning methodologies found their place in this area as well, resulting in a number of applications for intelligent forecasts [1], operation [2], protection [3–5], and finally metering

N. Ademović et al. (Eds.): IAT 2021, LNNS 316, pp. 110–116, 2022.
https://doi.org/10.1007/978-3-030-90055-7_9

[6, 7]. By collecting the data with smart meters, new methodologies can be utilized, such as non-intrusive load monitoring (NILM) and load disaggregation.

Therefore, significant works have been reported in the past 2–3 decades on NILM using disaggregation. These can be broadly classified under the following categories: the techniques using basic power analysis, steady-state analysis, and transient analysis of power supply data [8]. Also, different conventional and artificial intelligence techniques are used for feature extraction and classification purpose [8]. The NILM has several important advantages [9]. Some of these advantages include low cost of installation and maintenance due to fewer components to install, maintain and remove, greater reliability, and smaller space requirements [9]. The ease of installation will allow more appliances to be monitored in more homes, providing broader data and in many cases more accurate data [9]. In addition, the NILM system only depends on the sophistication of the algorithms and hence can decompose the total load [9]. Furthermore, it can be used by customers who would not allow the intrusion of power utility into their premises [9].

Having that said, the methodology that will be presented in this paper is based on the data collected by the smart meter and afterwards used to non-intrusively disaggregate and monitor the total load. The paper is organized as follows: firstly, the theoretical background of smart metering, advanced metering infrastructure and non-intrusive load monitoring is given, after which the methodology, research results, and discussion is presented. Finally, the conclusions and the future research directions are given.

2 Theoretical Background

2.1 Smart Meters and Advanced Metering Infrastructure

The smart meter is an advanced energy meter that obtains information from the end consumers' load devices and measures the energy consumption of the consumers and then provides added information to the utility company and the distribution system operator for better monitoring and billing [10]. With a smart meter, electrical data such as voltage, current, and power are measured, and real-time energy consumption information is recorded. Smart meter supports bidirectional communications between the meter and the central system [10].

In that sense, the advanced metering infrastructure (AMI) presents an integrated system of smart meters, communications networks, and data management systems that enables two-way communication between utilities and customers [6]. The system provides several important functions that were not previously possible or had to be performed manually, such as the ability to automatically and remotely measure electricity use, connect and disconnect service, detect tampering, identify, and isolate outages, and monitor different parameters [6].

The main driving force for the implementation of smart meter installation is the goal of achieving increased energy efficiency [11]. The smart metering system is expected to provide more information to the consumers and utilities, where better-informed customers will use less energy [11]. In addition, a need for improved billing accuracy is also a driver for smart metering [11]. The need to reduce losses due to fraud has also been a strong driver for companies to introduce smart metering [11]. The advancement of smart grids and incorporation of renewable energy sources and distributed generation is also an important part of the implementation of smart metering [9].

2.2 Non-intrusive Load Disaggregation and Load Monitoring

The NILM is a process of analyzing changes in the consumers' voltage and current and deducing what appliances are used, as well as their individual energy consumption [9]. It is called non-intrusive because it does not require intruding into the house or consumer premises when measuring the power consumption of different appliances [9]. NILM is considered a low-cost alternative to intrusive monitoring techniques [9]. G. W Hart [12] introduced the concept of NILM in 1991 with his research at MIT supported by the Electric Power Research Institute (EPRI). His work was based on a change of levels in steady-state reactive and real power consumption.

However, the main idea behind the NILM system is to determine the energy consumption of individual appliances in a building based on the analysis of the aggregated data measured from a single measuring point outside the building [13]. The total aggregated measurement of the household energy consumption of the individual appliance from the main metering point can be estimated according to the following equation [13, 14]:

$$p_t = \sum_{i=1}^{n} p_i(t) + e_t$$

where;

p_t is the aggregated household energy consumption, $p_i(t)$ is the individual appliance energy consumption at a given period and e_t the measurement error and line loss.

Generally, domestic appliances can be classified into the following categories based on the operational state or working principle of the appliances: two-state appliances, multi-state appliances, continuously-varying power appliances, and permanent consumer appliances [13, 15–17].

3 Methodology

To successfully detect the changes in the power consumption, the proposed algorithm is using the current and voltage measurements from the smart meter to obtain the power consumption over time. The smart meter is connected via a standard infrared probe to the laptop with the installed smart metering support software. In this manner, the data collected by the smart meter is transferred to the laptop and prepared for the next step. This data is further used to detect the abrupt changes in the power signal, which denotes that one or more appliances started or stopped working.

Firstly, the power signal is analysed with the gradient function that is used to detect the abrupt changes in the signal. Finally, the detected changes are categorized and paired with the appropriate appliance to start or stop the energy consumption algorithm, by using the user-guided classification and pairing. The algorithm is developed by using the MATLAB software environment. The flowchart of the load-monitoring algorithm is presented in Fig. 1.

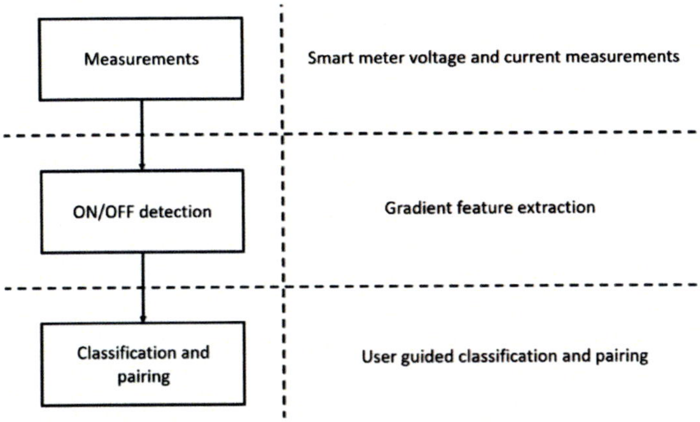

Fig. 1. Non-intrusive load disaggregation algorithm

4 Results and Discussion

To test the algorithm, an experimental setup consisted of several appliances is used. Appliances are connected to the 230 V low voltage supply over a smart meter that is used to measure the total power consumption of the connected devices. The rated power consumption of each connected appliance is shown in Table 1.

Table 1. Rated power consumption of the appliances in the test system

Appliance	Rated power [W]	Rated voltage [V]
Single phase motor	350	230
CFL light bulb	28	230
Single phase motor	150	230
Incandescent light bulb with dimmer	75	230

For testing purposes, various ON/OFF switching sequences of the appliances in different order are conducted. A 30-s time frame of the testing procedure will be used to demonstrate the algorithm operation. The measured voltages and currents are used to calculate the power consumption over time that is shown in Fig. 2 with the sampling frequency defined by the meter (in this case 200 ms).

With the calculated power consumption, the features that characterize the load switching can be extracted. The signal is afterwards processed with the gradient filter, and as a result, the switching states of the appliances can be determined, as shown in Fig. 3.

By monitoring these states and the changes in the power consumption waveform the switching moments and values can be determined. After completing this step, user guidance is necessary to confirm the appliances that were switched on or off during

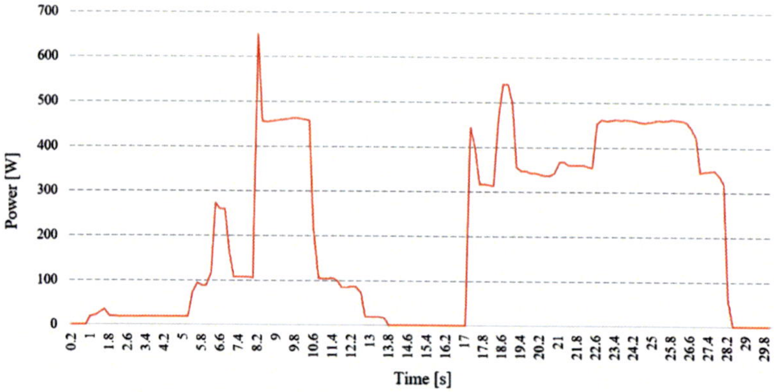

Fig. 2. Power consumption in the 30 s time frame during the test

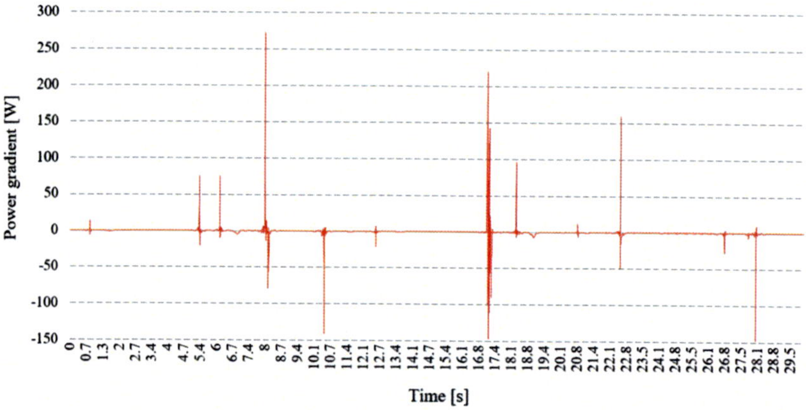

Fig. 3. Power gradient in the 30 s time frame during the test

the testing phase. Having this step done, the non-intrusive load monitoring algorithm is ready to be used. New devices that are connected to the system will be detected and again the user confirmation will be required. During the testing phase, small load changes (CFL light bulb) could not be detected in all scenarios, therefore resulting in a detection accuracy of 87.5%. Events that represent the on or off switching states are shown in Fig. 4.

For the experimental setup used in this research, the data from the smart meter is transferred via an infrared probe to the local laptop, however, in the future application, the meter can be integrated into the advanced metering infrastructure with the non-intrusive load monitoring software installed on the remote computer or cloud. In this manner, the application of the non-intrusive load monitoring could be offered as an additional service by the distribution system operator to the end consumers to improve their energy efficiency.

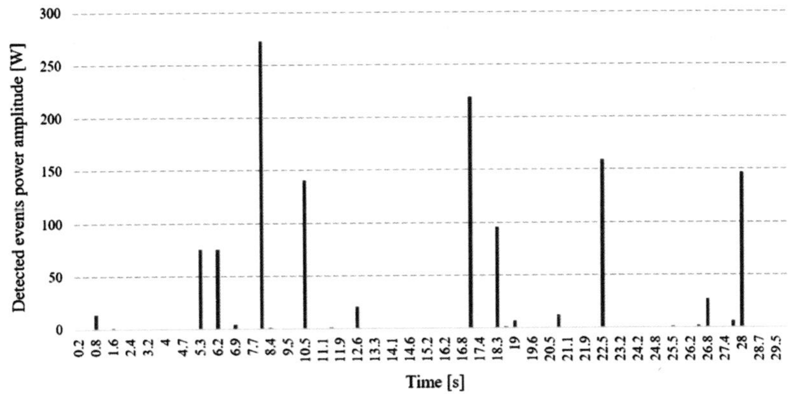

Fig. 4. Detected switching events in the 30 s time frame during the test

5 Conclusion

In this paper, the non-intrusive load disaggregation and load monitoring method is presented. With the constant advancements in the smart metering area and with the rising awareness about energy efficiency and climate impacts, this topic is widely researched in the past decades. The results presented in this paper demonstrated the applicability and accuracy of the proposed algorithm. With the real measurements obtained from the experimental setup, the algorithm accuracy is tested, resulting in a total accuracy of 87.5%.

Even though the results showed satisfactory accuracy, there are significant improvements that could be part of the future research directions. The increased number of appliances and the appliance waveforms, especially with the intermittent operation should be examined. In the light of the machine learning algorithms that emerged as great tools for classification, utilization of these algorithms could also improve the accuracy.

This research is a part of ongoing larger research with the ultimate goal to achieve completely unsupervised non-intrusive load disaggregation and monitoring, by using the existing metering equipment and data collection methods. Having that said, it is believed that this research and will contribute to the existing body of knowledge by designing and testing the simple yet effective load disaggregation methodology.

References

1. Aladesanmi, E.J., Folly, K.A.: Overview of non-intrusive load monitoring and identification techniques. IFAC-PapersOnLine **48**(30), 415–420 (2015)
2. Barai, G.R., Krishnan, S., Venkatesh, B.: Smart Metering and Functionalities of Smart Meters in Smart Grid – A Review. IEEE, London (2015)
3. Baranski, M., Jürgen Voss, J.: Non-Intrusive Appliances Load Monitoring Based on an Optical Sensor. IEEE, Bologna (2003)
4. Hart, G.W.: Nonintrusive appliance load monitoring. Proc. IEEE **80**(12), 1870–1891 (1992)
5. Hubana, T.: Artificial Intelligence based Station Protection Concept for Medium Voltage Microgrids. IEEE, East Sarajevo, Bosnia and Herzegovina (2020)

6. Hubana, T., Saric, M., Avdakovic, S.: High-impedance fault identification and classification using a discrete wavelet transform and artificial neural networks. Elektrotehniški Vestnik **85**(3), 109–114 (2018)
7. Hubana, T., Šarić, M., Avdaković, S.: New approach for fault identification and classification in microgrids. In: Avdaković, S., Mujčić, A., Mujezinović, A., Uzunović, T., Volić, I. (eds.) IAT 2019. LNNS, vol. 83, pp. 27–39. Springer, Cham (2020). https://doi.org/10.1007/978-3-030-24986-1_3
8. Hubana, T., Šemić, E., Laković, N.: Machine Learning Based Electrical Load Forecasting Using Decision Tree Algorithms. In: Avdaković, S., Volić, I., Mujčić, A., Uzunović, T., Mujezinović, A. (eds.) IAT 2020. LNNS, vol. 142, pp. 107–116. Springer, Cham (2021). https://doi.org/10.1007/978-3-030-54765-3_6
9. Karadza-Komic, J., Hubana, T., Lakovic, N.: Voltage unbalance analysis in low voltage distribution networks using smart meters. In: CIGRE – Bosnia and Herzegovina Committee. Neum, Bosnia and Herzegovina (2019)
10. Lakovic, N., Hubana, T., Karadza-Komic, J.: Estimation of missing smart meter data using artificial neural networks. In: CIGRE – Bosnia and Herzegovina Committee. Neum, Bosnia and Herzegovina (2019)
11. Norford, L.K., Leeb, S.B.: Non-intrusive electrical load monitoring in commercial buildings based on steady-state and transient load-detection algorithms. ELSEVIER Energy Build. **24**(1), 51–64 (1996)
12. Rathore, V., Jain, K.S.: Non-Intrusive Load Monitoring and Load Disaggregation using Transient Data Analysis. IEEE, Jabalpur, India (2018)
13. Semic, E., Hubana, T., Saric, M.: Distributed Generation Allocation in Low Voltage Distribution Network Using Artificial Neural Network. IEEE, Novi Sad, Serbia (2019)
14. Wang, Z., Zheng, G.: Residential appliances identification and monitoring by a non-intrusive method. IEEE Trans. Smart Grid **3**(1), 80–92 (2012)
15. Zeifman, M., Roth, K.: Non-Intrusive load Monitoring: review and outlook. IEEE Trans. Consum. Electron. **57**(1), 76–84 (2011)
16. Zheng, J., Gao, D.W., Lin, L.: Smart Meters in Smart Grid: An Overview. IEEE, Denver, USA (2013)
17. Zoha, A., Gluhak, A., Imran, M., Rajasegarar, S.: Non-intrusive load monitoring approaches for disaggregated energy sensing: a survey. Open Access Sens. 16838–16866 (2021)

Comparison of Different Maximum Power Point Tracking Algorithms

Aida Škamo(✉), Mirza Šarić, and Lejla Vuić

International Burch University, Sarajevo, Bosnia and Herzegovina

Abstract. Use of renewable energy sources is constantly increasing throughout the world. Considerable attention is dedicated to solar energy systems which are interesting due to possibility of direct conversion of solar energy to electrical energy with use of solar cells. However, it is also becoming increasingly important to improve the efficiency of the systems during design stages. This paper presents comparison, simulation, and analysis of standalone photo-voltaic systems in terms of different maximum power point algorithms. First, various algorithms are proposed, and their advantages and disadvantages are investigated. Furthermore, a simulation model is developed that can work as generalized model of many commercially made PV modules. All proposed maximum power point (MPPV) algorithms are implemented and simulated. Finally, comparison of the results based on predetermined criteria is given along with a conclusion drawn from the presented results. This paper presents further contributions in this field as it presents evidence on importance of MPPV algorithms, their quantitative comparison and practical design.

Keywords: DC/DC converters · Irradiance · MATLAB/SIMULINK · Photovoltaic · PV parameters · Simulation · Temperature

1 Introduction

Increased pollution and climate changes around the globe have scientists competing to find a way to make clean energy more accessible in everyday use. Solar energy as a source of renewable and clean energy is particularly interesting because it provides a possibility for direct conversion of light (solar) energy to electrical using solar cells (modules, arrays) and it requires little maintenance whilst producing no noise [1, 2]. Solar energy conversion can not, however, be implemented without any negative implications whatsoever. Power system regulation, market, and operation requrements continues to place increasing preassure in terms of PV solar systems efficiency, reliability, and performance. One of the possible ways to increase the efficiency is implementation of the appropritate MPPT algorithm. This paper makes contribution in this field as it develops, analyses, and compares various MPPT algorithms which might be used in real applications. Standalone system, described further in this paper, operates not only using the photovoltaic (PV) module, but in a connection with a DC/DC converter which is controlled by a micro-controller [3]. This switching using micro-controller is a particular interest of this paper.

N. Ademović et al. (Eds.): IAT 2021, LNNS 316, pp. 117–132, 2022.
https://doi.org/10.1007/978-3-030-90055-7_10

2 Literature Overview

Mathematical modeling as well as circuit modeling of PV cell helps to analyze its performance with no need for installing the system on site and analyzing it in on-site conditions. They can be simulated using different program packages with negligible degrees of inaccuracy which enables in depths analysis during design stage. Multiple different models of the PV array have been presented and published. Some are simple-circuit one-diode models which do not consider parallel or serious resistance that are the result of junction impurities, while others consider multiple diodes and the effect of those impurities [4–9]. These mentioned models are different on many aspects and each bring a different level of complexity. Whether it is in number of diodes, finite or infinite shunt resistance or it is a changing diode ideality factor [10]. Key aspect in the PV module comparison is its efficiency. Efficiency of the PV array (PV plant) is affected by three factors: efficiency of the PV panel, efficiency of the used DC/DC converter and efficiency of the maximum power point tracking algorithm – MPPT. Influencing the efficiency of solar panel is hard, since it depends on the technology readily available for its production and is mostly in the range determined by the manufacturer. Nowadays it is up to 17.2% [11]. Efficiency of the DC/DC converter is 95% to 98%, leaving very little place for improvement [12]. Thirdly, into account is taken efficiency of the MPPT algoritm. Improving efficiency of MPPT with new control algorithms is easier, less expensive and can be done in plants which are already in use. These algorithms are necessary due to non-linear characteristic of a PV module. This non-linearity results in one point where the power produced by the PV module is maximum. This point depends on several factors, but mainly on the ambient temperature and irradiation levels. Both mentioned conditions are subject to change during day as well as the season of the year. Irradiation can change very rapidly when there is shading due to atmospheric conditions such as clouds. Fundamental issue addressed by the various MPPT algorithms is automatic assessment of output current or output voltage for which the PV array produces maximum power for any given set of conditions [13, 14]. MPPT algorithms not only enable maximum power delivered to the load, but also increase the operating lifetime of a solar cell, since it is known that sudden changes and oscillations can damage the cell.

Different MPPT techniques can be categorized as online or offline methods, depending on the fact that the algorithm relies heavily on the solar cell model, or if it in fact does not specifically rely on solar cell behaviour. Third type are so called hybrid methods which are a combination of the other two mentioned above. Offline methods – which do rely on the solar cell behaviour and model use physical measures of the PV panel as control signals. Those methods are fractional open circuit voltage (FOC) and fractional short circuit current (FSC) methods. Open circuit voltage method uses approximately linear relationship between open circuit voltage and voltage at the maximum power point [15, 16]. Similarly, fractional short circuit current method suggests that, short circuit current I_{SC} is approximately linearly related to current at the maximum power point [17, 18]. In online methods, instantaneous values of the PV output voltage and PV output current are used to generate the control signal. Control signal is applied in this case to the DC/DC converter. Some of the methods addressed in the paper are hill climbing, perturb and observe (P&0), incremental conductance (IncCond) algorithm as well as optimized approaches to the before mentioned and hybrid methods which consider both online and offline methods [19–24]. Figure 1 how the V-P characteristic of solar cell behaves under different conditions mentioned above.

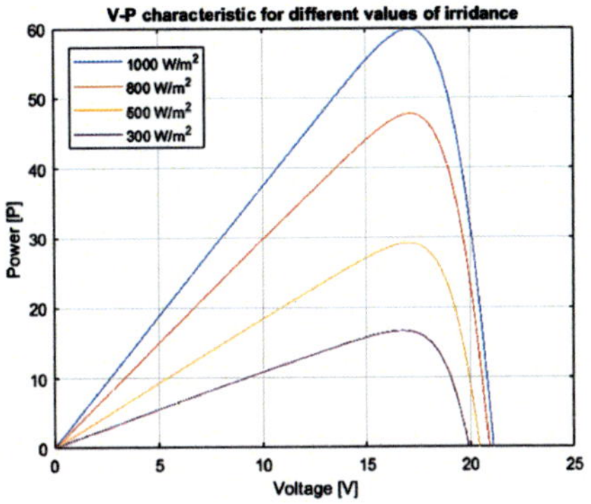

Fig. 1. V-P characteristic of solar cell under varying irridance conditions

3 Mathematical Model

PV system usually consists of solar array (module, cell), DC-DC converter, controller, and a battery. DC-DC converters can be of several different types: boost, buck, or buck-boost. The choice of the converter depends on voltage levels in question. Here, we are interested in the solar cell itself. There are numerous models of solar cell, but single-diode model and two-diode models are two best known ones. Figure 2 shows PV cell equivalent circuit using single diode model of PV cell:

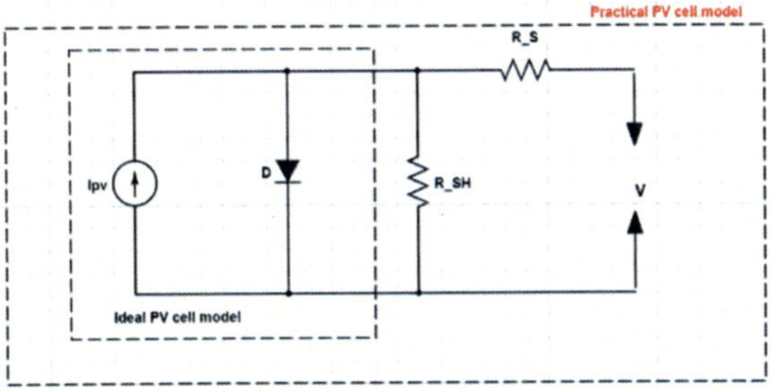

Fig. 2. PV cell equivalent circuit

Equation (1) describes voltage-current characteristic of a solar cell:

$$I = I_{PV} - I_0\left(e^{\frac{q(V-IR_S)}{akT}} - 1\right) - \frac{V - IR_S}{R_{SH}} \quad (1)$$

Where,
I: Solar cell current
I_{PV}: Light generated current
I_0: Diode saturation current
q: Electron charge, $q = 1.60217646 \cdot 10^{-19} C$
K: Boltzmann Constant $k = 1.3806503 \cdot 10^{-23} \frac{J}{K}$
T: Cell temperature in Kelvin (K)
V: Solar cell output voltage
R_S: Solar cell series resistance
R_{SH}: Solar cell shunt resistance
a: Diode ideality factor

Ideally R_S is 0 and R_{SH} is infinite. However, considering that solar cells come in series, parallel or some series-parallel combination and simplification that implies R_S is 0 a R_{SH} is infinite, Eq. (1) can be rewritten as (2):

$$I \approx n_p I_{PV} - n_p I_0 \left(e^{\frac{q(V - IR_S)}{akTn_s}} - 1 \right) \tag{2}$$

I_{PV} – light generated current depends mainly on the solar irradiance and cell temperature and is described by Eq. (3):

$$I_{PV} = I_{PV,STC} + K_i \Delta T \frac{G}{G_{STC}} \tag{3}$$

Where,
$I_{PV,STC}$: short circuit current at standard test conditions (STC)
K_i: temperature coefficient of cell's short circuit current
G: Solar irridiance of a cell
$$G_{STC}$: Solar irridiance at standard test conditions ($1000 \frac{W}{m^2}$)
$\Delta T = T - T_{STC}$ – referent temperature of a cell, T_{STC} – temperature at standard test conditions 25 °C.

Diode saturation current is described using Eq. (4):

$$I_0 = I_{0(STC)} \cdot \left(\frac{T}{T_{STC}} \right)^3 \cdot e^{\frac{qE_g}{ak}\left[\frac{1}{T} - \frac{1}{T_{STC}}\right]} \tag{4}$$

Here E_g is a band gap energy of a semiconductor, and $I_{0(STC)}$ is calculated as depicted by Eq. (5):

$$I_{0(STC)} = \frac{I_{SC}}{e^{\frac{qV_{OC}}{akT_{STC}}} - 1} \tag{5}$$

3.1 Open Circuit Voltage, Short Circuit Current and Maximum Power Point

There are three specially interesting points on the V-I characteristic of a solar cell – open circuit voltage point, short circuit current point and point of maximum power. In first two points power produced by the solar cell equals 0. They can both be derived from the Eq. (1) when the simplifications are taken into consideration. Open circuit voltage is calculated as:

$$V_{OC} \approx \frac{akT}{q} \ln\left(\frac{I_{PV}}{I_0} - 1\right) \tag{6}$$

Short circuit current is current when V = 0 and is therefore approximated as:

$$I_{SC} \approx I_{PV} \tag{7}$$

Figure 3 shows all three significant points depicted on one graph:

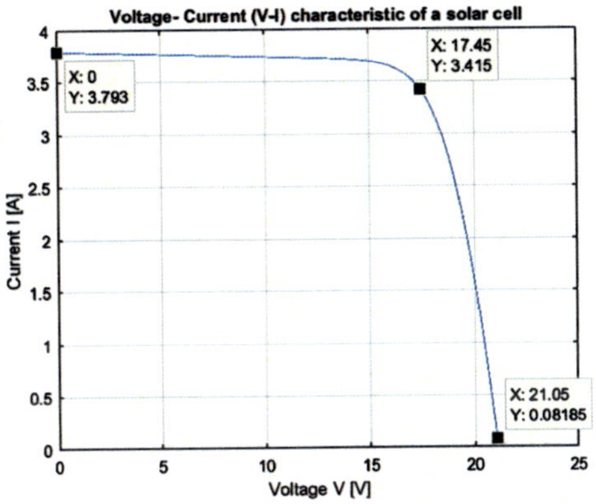

Fig. 3. V-I characteristics of a solar cell with three significant points marked

4 Development of the Proposed Algorithms

MPPT algorithm is an essential part of PV system since it ensures that the system operates at a point of maximum power production. Multiple algorithms have been developed that ensure this behaviour of the system in steady state. Naturally, these algorithms differ from one another in terms of number of used sensors, efficiency, speed of convergence – how fast the algorithm tracks the maximum power point, ease of implementation, cost and many others that go into fine details of algorithm specification. In this section, the details of algorithms implemented in this paper are presented.

4.1 Perturb & Observe

Perturb & Observe (P&O) is one of the online methods used frequently. It is also one of the simplest methods used. This method operates on the principle of periodic perturbations to the reference voltage. This perturbation can represent increase or decrease in the panel's voltage. Output power is then compared to output power of previous perturbation. Since reference voltage after the perturbation changes, power does too. If the power increases operating point is moved in that direction, however if the power decreases (compared to the previous perturbation power) algorithm needs to move in the opposite direction [25]. Process is repeated until maximum power point (MPP) is reached. In the steady state operating point then oscillates around the MPP. These oscillations result in the loss of PV power. Oscillations can be reduced by reducing the perturbation size, but the perturbation size decrease leads to increase of the convergence time – algorithm needs more time to find the MPP. Another downside to the P&O algorithm is the fact that it preforms poorly during rapid changes of the atmospheric conditions, since the the algorithm will deviate from the point until condition change slows down. Algorithm diagram can be seen in the Fig. 4.

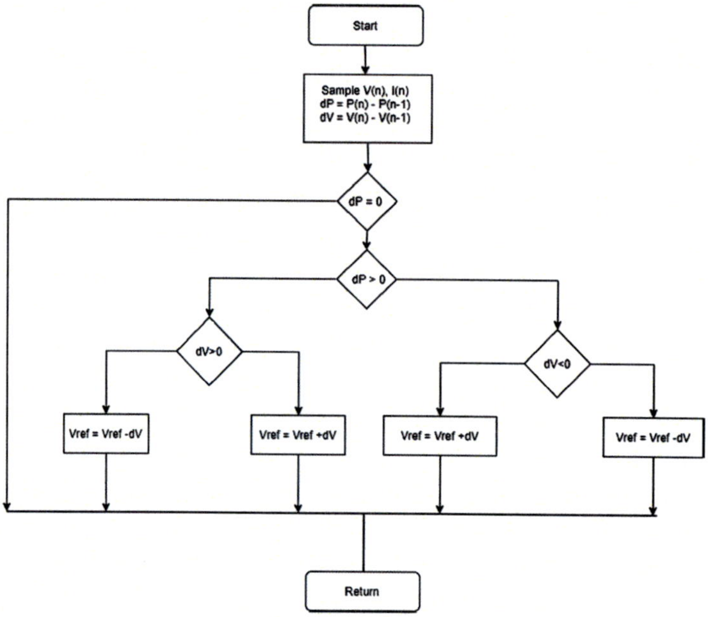

Fig. 4. Diagram of perturb and observe algorithm

4.2 Incremental Conductance

Incremental conductance (IncCond) is another online method that examines the slope of the P-V curve (power vs. voltage curve). It relies on the fact that slope of the P-V curve is zero at MPP, positive for the voltage smaller than the MPP voltage and negative for values of voltage greater than voltage at MPP:

- $\Delta V/\Delta P = 0 \quad \Delta I/\Delta P = 0$ *at MPP*
- $\Delta V/\Delta P > 0 \quad \Delta I/\Delta P < 0$ *on the left*
- $\Delta V/\Delta P < 0 \quad \Delta I/\Delta P > 0$ *on the right*

Change is determined by comparing power and voltage increments between two consecutive samples, which means that the change in the MPP voltage. Like the P&O

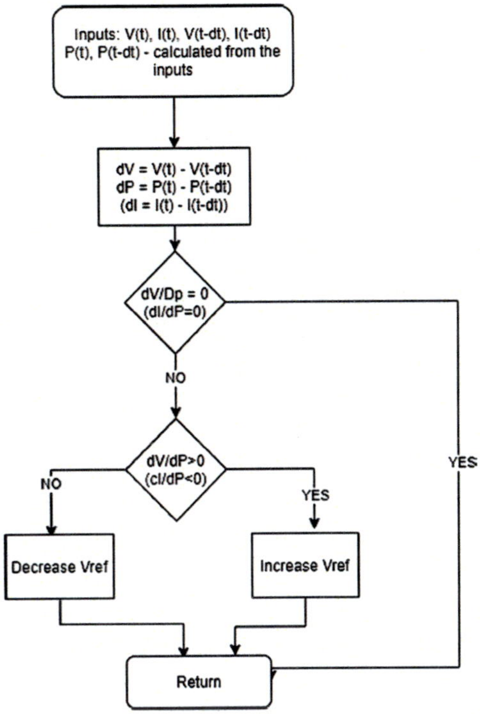

Fig. 5. Diagram of incremental conductance algorithm

algorithm, speed of the convergence depends on the size of the increment. Basic IncCond algorithm requires measurements of PV voltage and current to perform the MPP tracking. Downsides of the IncCond algorithm are like those of P&O algorithm, where a certain trade-off is made depending whether operating point needs to be more accurate, or the stress is put on how fast the MPP is found. Several improved algorithms have been proposed where first operating point is brought close to the MPP, then IncCond algorithm is used to exactly track the MPP. Algorithm diagram can be seen in the Fig. 5.

4.3 Hill Climbing Algorithms

Several works and papers treat both IncComd and P&O algorithms as a type of Hill Climbing algorithm. The biggest advantage of this method is its simplicity, but the disadvantages come from rapidly changing atmospheric conditions. Hill climbing algorithm uses the ratio of change in power and change in duty cycle of the converter used to determine the MPP. When the condition ${}^{dP}/_{dD}$ is met algorithm has found the MPP.

4.4 Fractional Open Circuit Voltage

Fractional Open Circuit Voltage (FOC) method is an offline method that uses the approximate linear relationship between the MPP voltage and the open circuit voltage V_{OC}.

Open circuit voltage method is the simplest MPPT method also know as the constant voltage method. The open circuit voltage depends on the properties of the solar cell used so this poses some difficulties in the implementation. Shortly the method can be described using following simple Eq. (8):

$$V_{MPP} = k \cdot V_{OC} \tag{8}$$

It has been determined that the factor k varies between 0.71 and 0.8. Commonly used value is 0.76 so algorithm can be sometimes found under name 76% algorithm. Disadvantage of the open circuit method is the fact that the constant k is non-flexible and stays same regardless of the changing conditions. Selecting correct value of k requires calibration of the solar panel (series of experimental measures of the used modules in a plant). Several enhancements have been proposed where the value of the constant k can change depending on the relative proximity to the maximum power point. Algorithm diagram can be viewed in Fig. 6:

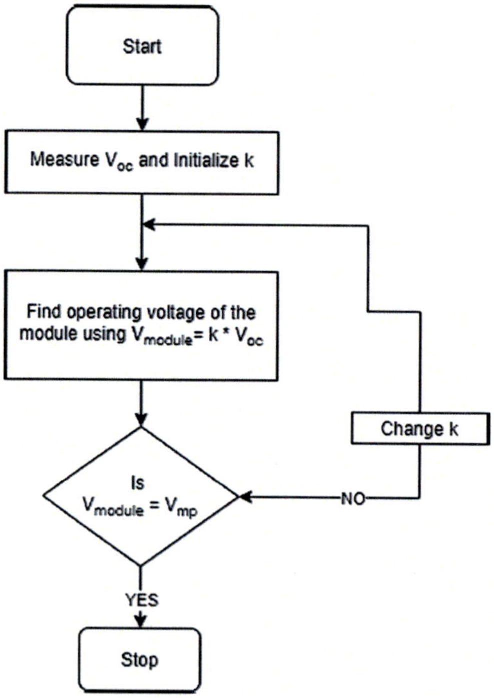

Fig. 6. Diagram of open circuit voltage method

4.5 Fractional Short Circuit Current

Short circuit current method (FSC) is one of the offline methods. It is a very simple method that, just like the open circuit voltage method does not result in the true MPP being found. It is based on the approximately linear relation of the short circuit current I_{SC} and current measurement at the MPP I_{MPP}.

$$I_{MPP} = k \cdot I_{SC} \quad (9)$$

Value of the factor k is different for different PV modules and needs to be determined before the PV module installations. Factor k is always <1 and varies between 0.78 and 0.92. Commonly used value is k = 0.9 since it has been proven that when PV array output current is approximately 90% of the short circuit current it operates at its MPP [28]. If the short circuit current needs to be measured when the PV module has been installed, additional switch needs to be installed which shorts the PV array and I_{SC} is measured.

4.6 Fuzzy Logic Controller

Fuzzy logic controllers can operate with imprecise inputs and they do not require accurate mathematical model. Inputs to the fuzzy system are usually values E and ΔE which can be calculated as following:

$$E(n) = \frac{P(k) - P(k-1)}{V(k) - V(k-1)} \quad (10)$$

$$\Delta E = E(k) - E(k-1) \quad (11)$$

As can be seen from the equations above, values of power and voltage are used to calculate E and ΔE. In the fuzzy system these values are expressed using linguistic terms. Usually between 5 and 7 membership functions are used with 7 fuzzy levels: NB (Negative Big), NM (Negative Medium), NS (Negative Small), ZE (Zero), PS (Positive Small), PM (Positive Medium) and PB (Positive Big). Triangular membership function can be used for inputs and outputs and the output of the controller, which is a duty cycle D is read from the rule base. Advantages of this method are fast convergence and minimal oscillations around the operating point [29].

4.7 Hybrid Methods

Hybrid methods employ the online and offline methods. It is proposed that initially offline method is used to determine approximate maximum power. After the approximate value has been determined several different online methods can be used to find the exact value of the MPP.

5 Results and Discussion

In Fig. 7 model of PV module depicted as a constant dc source. Model has 3 inputs irradiation, temperature and voltage which is the solar cell output voltage reading.

Figure 8 represents MATLAB/Simulink model of a Buck converter.

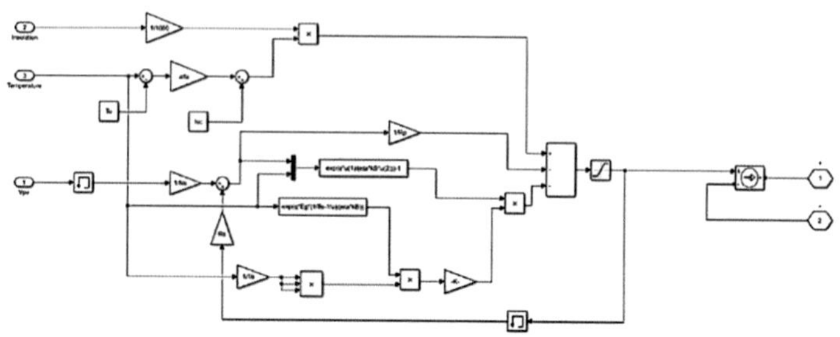

Fig. 7. MATLAB/Simulink simulation model of a PV cell

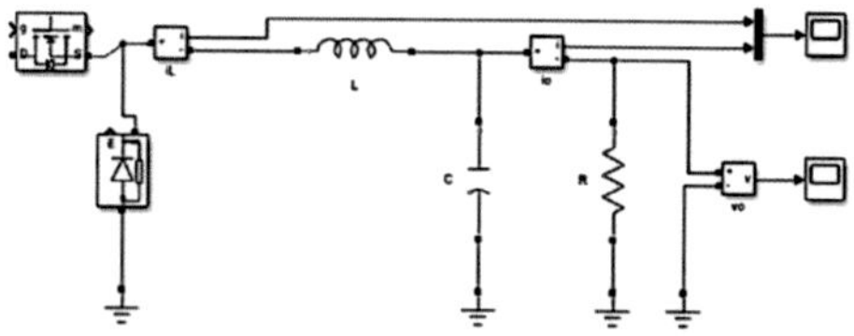

Fig. 8. MATLAB/Simulink simulation model of a Buck converter

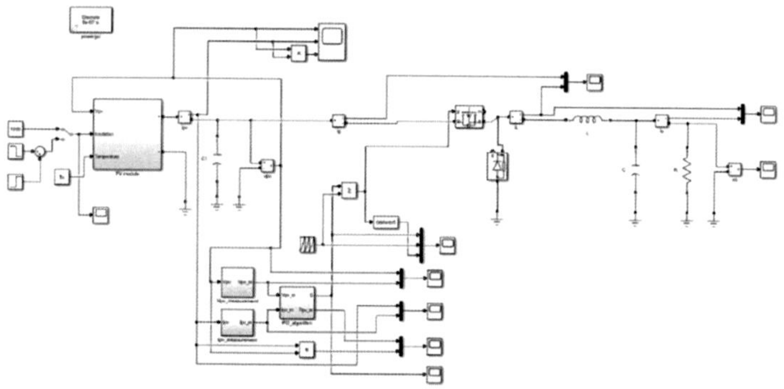

Fig. 9. MATLAB/Simulink model of a PV cell, Buck converter and P&O algorithm implementation

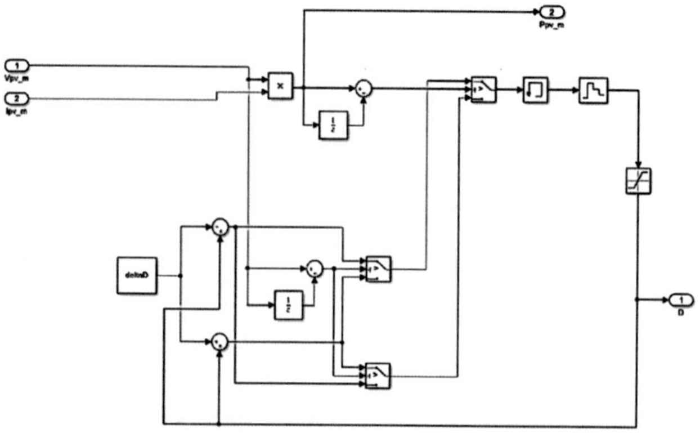

Fig. 10. MATLAB/Simulink model of a Preturb & Observe algorithm implementation

Figure 9 shows full model of the system observed, with P&O algorithm implemented. Same model, with the changed algorithm is used throughout the paper and for comparison based on different previously established criteria. Furthermore, Fig. 10 shows model implementation for P&O algorithm. Fractional Open Circuit Voltage, Fractional Short Circuit Current and Incremental Conductance algorithms have been implemented using Matlab Function block. Additionally, MPPT algorithm using fuzzy logic has been implemented using Fuzzy Logic Controller in Matlab/SIMULINK environment. Figure 11 shows implementation of Fuzzy Logic Controller.

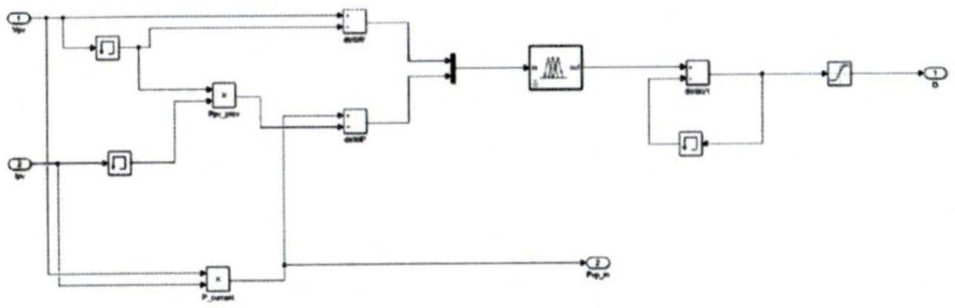

Fig. 11. MATLAB/Simulink model fuzzy controller

6 Simulation Results

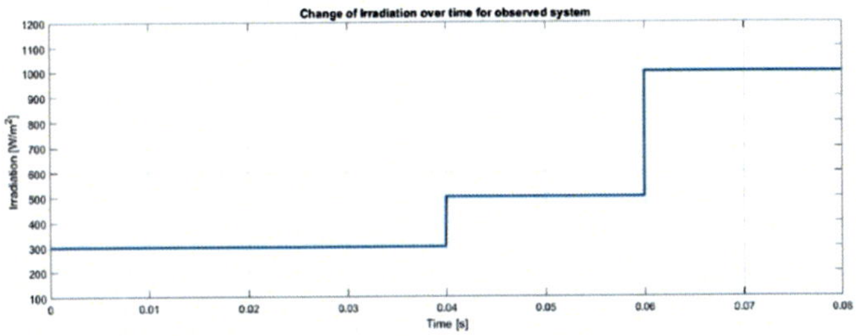

Fig. 12. Change in irradiation over time for system under observation

Proposed criteria for comparison consist of speed of convergence, ease of implementation, number of components used, cost and ability to implement in different environments under changing conditions. Emphasis is placed on changing irradiation conditions which are depicted on Fig. 12.

To compare the implemented algorithms, PV cell power response – power as function of time is displayed. Following can be seen on Fig. 13, Fig. 14, Fig. 15, Fig. 16 and Fig. 17. For the selected algorithms to be comparable, as it was stated before, same model of the solar cell is used, only the algorithm implementation was changed. Observing Fig. 9 – MATLAB block marked as PV module, as well as the inputs to the system, in terms of input temperature and input variable irradiance, remain same throughout the simulations as well as the parameters of the selected converter, while the algorithm implementation changes – the way we calculate duty cycle, as the input to the converter changes in the simulation. Output power under changing irradiation conditions for each

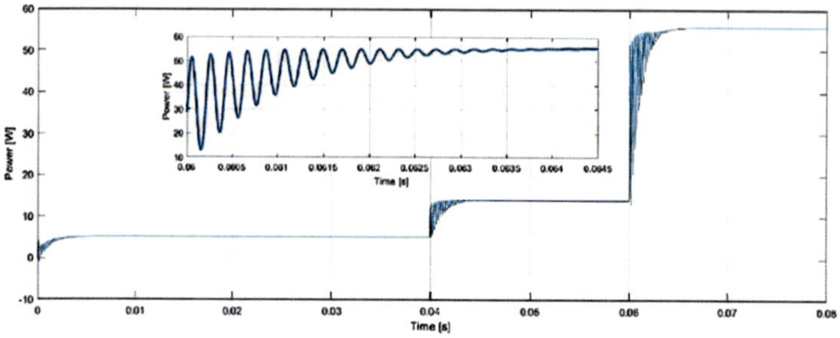

Fig. 13. Output power of a solar cell when P&O algorithm is implemented

implemented algorithm does not vary a lot once the system reached the steady state. However, observable from Figs. 13, 14, 15, 16 and 17 is that the convergence speed – time needed for the system to reach the maximum power point does.

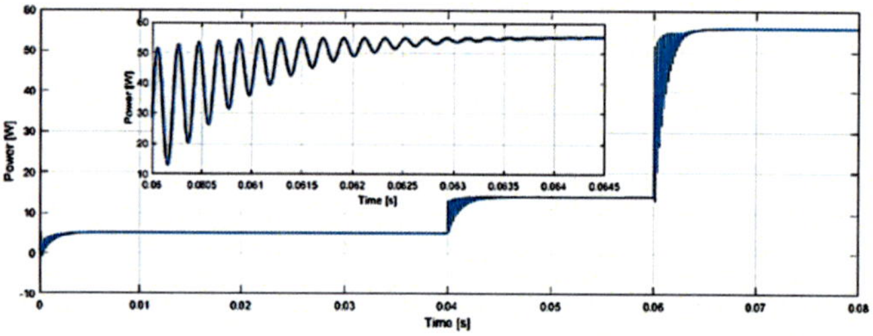

Fig. 14. Output power of a solar cell when IncCond algorithm is implemented

The algorithms in question are very different, differing in terms of their dependency on the panel (some methods are online, while others are offline), some require periodic tuning such as FSC and FOC methods, since due to wear and tear of the solar panel, values that are given in panel documentation relating to the values of factors k_1 and k_2 are susceptible to change. One of the criteria is the implementation complexity – even though very subjective, it was included in the comparison.

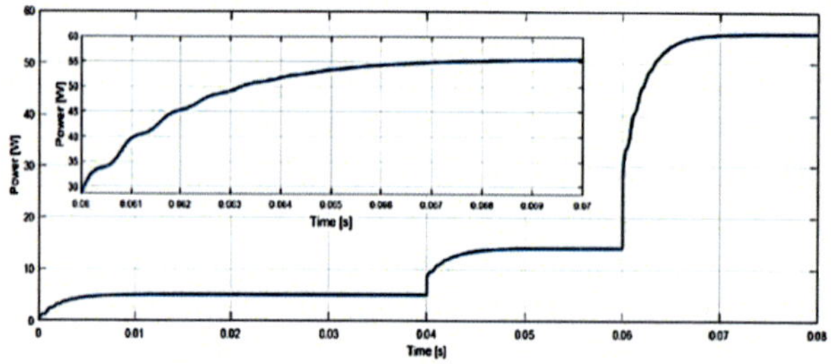

Fig. 15. Output power of a solar cell when FOC algorithm is implemented

Some of the algorithms under observation require more measuring devices. It is important to note that current measurement is usually avoided, and it is turned to measuring the voltage instead, so the methods that require current measurement are usually looked down upon. High number of measuring devices increases the overall price of solar panel installation as well as the size of the system. Speed of convergence of P&O and InCond algorithm is variable, and while integration complexity for P&O is very low, InCond complexity can be considered medium because division by 0 could be encountered and must be dealt with. Two offline methods have medium speed of convergence and very low implementation complexity but a significant flaw – often the calculated MPP value is not the real MPP value of that system.

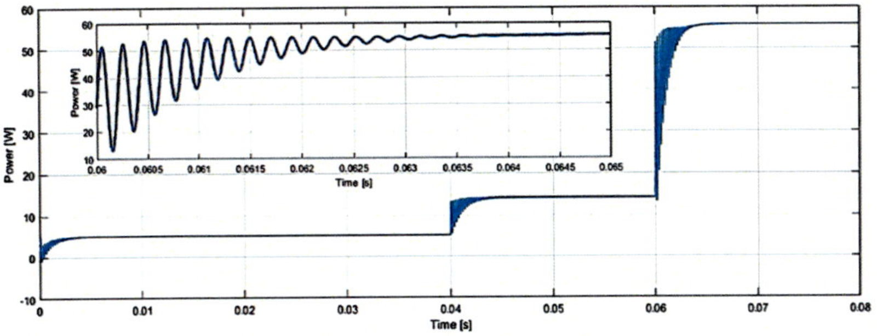

Fig. 16. Output power of a solar cell when FSC algorithm is implemented

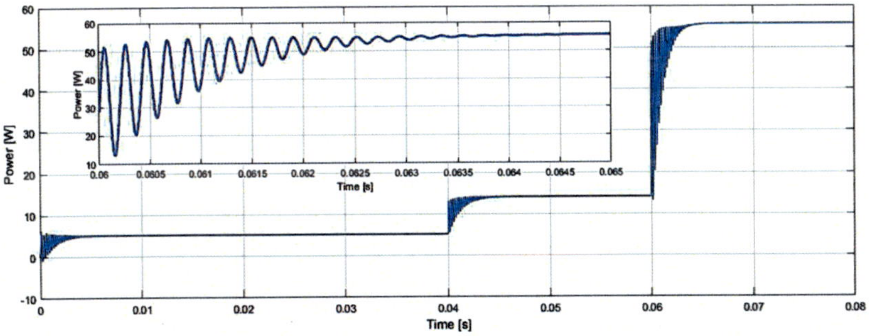

Fig. 17. Output power of a solar cell when Fuzzy controller is implemented

From the Figs. 13, 14, 15, 16 and 17, following is observable: the fastest algorithm is Fuzzy controller with speed of convergence of 4 ms (however, it requires in this case both voltage and current sensing and has medium implementation complexity). FOC and FSC need 10 ms and 5 ms respectively and can find real MPP value for irradiation levels of STC, otherwise the MPP value is not the real one. Speed of convergence of InCond algorithm in this case is 4.5 ms like P&O that has speed of convergence of 5 ms. Both algorithms require both current and voltage sensing. Perturb and Observe algorithm is thought to have low efficiency if it is implemented with fixed perturbation size (which is the case in this paper). FOC and FSC methods have even lower efficiency. Fuzzy controller and Incremental Conductance methods stand out there with what is considered high efficiency of 93.11%. All above findings can be seen summarized in Table 1.

Results presented in this paper confirm the existing knowledge on the importance of the PV solar system efficiency improvements and the role of MPPT implementation [30, 31]. It also confirms that it is not possible to propose a unique MPPT solution. It is therefore required to customize design for every application. This paper by no means exhausts the possible consideration of MPPT algorithm design comparison and implementation. Future research directions include coupling some of the offline methods presented in the work with improved online algorithms, specifically using a variable perturbation size for P&O online method.

Table 1. Speed of convergence and efficiency of the MPPT algorithms under observation

Algorithm	Speed of convergence	Efficiency
Preturb & observe	5 ms	92.64%
Incremental conductance	4.5 ms	93.11%
Fractional open circuit voltage	10 ms	92.607%
Fractional short circuit current	5 ms	91.43%
Fuzzy controller	4 ms	93.11%

7 Conclusion

In this paper several different MPPT, that are classified into two different categories, online and offline, have been implemented. Each presented method has its pros that are greatly dependent on the implementation – and with wide variety of implementations of solar panels, starting with solar farms, electrical vehicles to use of solar panels in space and space research it is easy finding the use of each algorithm that would justify both its pros and cons nowadays. It is impossible to design universal algorithm since every implementation seeks a different outlook – implementing algorithm for use in space satellites and stations does not consider price of installation, but rather puts most of the emphasis on the efficiency and trustworthiness of the applied algorithm. Solar vehicles value speed of convergence as the most important characteristic. As it is presented in this paper, only limited number of algorithms have been observed with satisfactory results, however one deficiency of presented paper is application of hybrid methods that have previously been proven to display some better results in terms of convergence speed tests. In this paper the results show that best performing algorithm in terms of speed of convergence is implementation that contains Fuzzy Controller with speed of convergence of 4 ms. Next best performing algorithm is IncCond algorithm with speed of convergence of 4.5 ms and efficiency of 93.11%. Just as was expected, worst performing algorithms are FSC and FOC with considerably smaller speed of convergence of 6 ms and 8 ms.

Future research considerations involve coupling some of the offline methods presented in the work with improved online algorithms, specifically using a variable perturbation size for P&O online method.

References

1. Hlaili, M., Hfaiedh, M.: Comparison of different mppt algorithms with a proposed one using a power estimator for grid connected PV systems. Int. J. Photoenergy **2016**, 1–10 (2016). https://doi.org/10.1155/2016/1728398
2. Wai, R.-J., Wang, W.-H., Lin, C.-Y.: High-performance stand-alone photovoltaic generation system. Ind. Electron. IEEE Trans. **55**, 240–250 (2008). https://doi.org/10.1109/TIE.2007.896049
3. Sanz Morales, D.: Maximum Power Point Tracking Algorithms for Photovoltaic Applications, Thesis submitted for examination for the degree of Master of Science in Technology. Faculty of Electronics, Communications and Automation, Aalto University, 2010, Espoo 14.12.2010 (2010)
4. Leban, K., Ritchie, E.: Selecting the Accurate Solar Panel Simulation Model. In: Proceedings of the Nordic Workshop on Power and Industrial Electronics (NORPIE/2008). Aalborg University, Pontoppidanstræde 101, Aalborg, Denmark (2008)
5. Sera, D., Teodoresc, R., Rodriguez, P.: PV panel model based on datasheet values. In: Industrial Electronics, June 2007. ISIE 2007 (2007)
6. Elshatter, T.F., Elhagry, M.T., Abou-Elzahab, E.M., Elkousy, A.A.T.: Fuzzy modeling of photovoltaic panel equivalent circuit. In: Proceedings of the Conf. Record 28th IEEE Photovoltaic Spec. Conf., pp. 1656–1659 (2000)
7. Balzani, M., Reatti, A.: Neural network-based model of a PV array for the optimum performance of PV system. In: Proceedings of the Ph.D. Res. Microelectron. Electron., vol. 2 (2005)
8. Ghute, K., Kale, V.: A simplified matlab based simulation of PV module with effect of temperature and irradiation. Int. J. Eng. Tech. Res. **2**, 2014 (2014)
9. Tan, Y.T., Kirschen, D.S., Jenkins, N.: A model of PV generation suitable for stability analysis. IEEE Trans. Energy Convers. **19**, 2004 (2004)
10. El-Ahmar, M.H., El-Sayed, A.-H.M., Hemeida, A.M.: Mathematical Modeling of Photovoltaic Module and Evalute The Effect of Varoius Paramenters on its Performance, Eighteenth International Middle East Power Systems Conference (MEPCON) (2016)
11. Wasfi, M.: Solar energy and photovoltaic systems. Cyber J. Multidiscip. J. Sci. Technol. J. Sel. Areas Renewable Sustainable Energy (2011)
12. SMA Solar Technology AG – Energy That Changes: SMA product catalogue, 2020. Accessed 10 Nov 2020
13. Salas, V., Olías, E., Barrado, A., Lazaro, A.: Review of the maximum power point tracking algorithms for stand-alone photovoltaic systems. Sol. Energy Mater. Sol. Cells **90**, 1555–1578 (2006). https://doi.org/10.1016/j.solmat.2005.10.023
14. Reisi, A.R., Moradi, M.H., Jamas, S.: Classification and comparison of maximum power point tracking techniques for photovoltaic system: a review. Renewable Sustainable Energy Rev. **19** (2013)
15. Schoeman, J.J., Wyk, J.D.V.: A simplified maximal power controller for terrestrial photovoltaic panel arrays. In: IEEE Power Electronics Specialists Conference, pp. 361–367. Cambridge, MA, USA (1982). https://doi.org/10.1109/PESC.1982.7072429
16. Masoum, M.A.S., Dehbonei, H., Fuchs, E.F.: Theoretical and experimental analyses of photovoltaic systems with voltage and current-based maximum power-point tracking. IEEE Trans. Energy Convers. **17**, 2002 (2002)

17. Hart, G.W., Branz, H.M., Cox, C.H.: Experimental tests of openloop maximum-power-point tracking techniques. Solar Cells **13**, 185–195 (1984)
18. Noguchi, T., Togashi, S., Nakamoto, R.: Short-current pulse based adaptive maximum-power-point tracking for photovoltaic power generation system. In: Proceedings of the 2000 IEEE Int. Symp. Ind. Electron., pp. 157–162 (2000)
19. Brambilla, A., Gambarara, M., Garutti, A., Ronchi, F.: New approach to photovoltaic arrays maximum power point tracking. In: Proceedings of the 30th Annu. IEEE Power Electron. Spec. Conf. (1999)
20. Kim, T.Y., Ahn, H.G., Park, S.K., Lee, Y.K.: A novel maximum power point tracking control for photovoltaic power system under rapidly changing solar radiation. In: IEEE International Symposium on Industrial Electronics, vol. 2, pp.1011–1014 (2001). https://doi.org/10.1109/ISIE.2001.931613
21. Sera, D., Kerekes, T., Teodorescu, R., Blaabjerg, F.: Improved MPPT Algorithms for Rapidly Changing Environmental Conditions, pp. 1614–1619 (2006). 10.1109 EPEPEMC.2006.4778635
22. Femia, N., Petrone, G., Spagnuolo, G., Vitelli, M.: Optimization of perturb and observe maximum power point tracking method. IEEE Trans. Power Electron. **20**(4), 963–973 (2005)
23. Xiao, W., Dunford, W.G.: A modified adaptive hill climbing MPPT method for photovoltaic power systems. In: 35th Annual IEEE Power Electronics Specialists Conference, vol. 3, pp. 1957–1963 (2004). https://doi.org/10.1109/PESC.2004.1355417
24. Jain, S., Agarwal, V.: Comparison of the performance of maximum power point tracking schemes applied to single-stage grid-connected photovoltaic systems. Electr. Power Appl. IET **1**(5), 753–762 (2007)
25. Eltamaly, A., Rezk, H.: A comprehensive comparison of different MPPT techniques for photovoltaic systems. Sol. Energy **112**, 1–11 (2015). https://doi.org/10.1016/j.solener.2014.11.010
26. Sholapur, S., Mohan, K., Narsimhegowda, T.: Boost converter topology for pv system with perturb and observe MPPT algorithm. IOSR J. Electr. Electron. Eng. **9**, 50–56 (2014). https://doi.org/10.9790/1676-09425056
27. Ch, S.B., Kumari, J., Kullayappa, T.: Design and analysis of open circuit voltage based maximum power point tracking for photovoltaic system. Int. J. Adv. Sci. Technol. **2**, 51–60 (2011)
28. Kumari, J., Ch, S.B., Yugandhar, J.: Design and investigation of short circuit current based maximum power point tracking for photovoltaic system. Sci. Acad. Publisher Int. J. Res. Rev. Electr. Comput. Eng. **1**, 2046–5149 (2011)
29. Robles, C., Taborda, J., Rodriguez, O.: Fuzzy logic based MPPT controller for a PV system. Energies. **10**(12), 2036 (2017). https://doi.org/10.3390/en10122036
30. Arpadzic, A., Saric, M., Hivziefendic, J., Avdakovic, S.: Power system efficiency improvement using solar PV systems. In: International Symposium on Industrial Electronics and Applications, INDEL 2020 – Proceedings (2020)
31. Bandic, L., Hivziefendic, J., Saric, M., Tesanovic, M.: Voltage regulation of PV system with MPPT and battery stor-age in microgrid. In: Proceedings of 2020 International Conference on Smart Systems and Technologies, SST 2020, pp. 161–166, 9264071 (2020)

Secondary Arc Extinguishing During a Single-Phase Fault at the Long High Voltage Overhead Lines - Case Study of a Real Power Network

Belmin Memišević[1](✉), Mirza Šarić[2], and Jasna Hivziefendić[2]

[1] Department of Substation Automation Systems and SCADA, Energoinvest d.d.-Sarajevo, Sarajevo, Bosnia and Herzegovina
belmin.memisevic@ibu.edu.ba

[2] Faculty of Engineering and Natural Sciences, International Burch University, Sarajevo, Bosnia and Herzegovina

Abstract. A power system design is determined by transient conditions, rather than by steady state behavior, hence the significant attention is payed to the analysis of transients in long overhead lines (OHLs). There is an attempt to reduce the secondary arc current magnitude and the extinction time that is created by the stray and ground capacitance in these long lines. The usual concept in solving these problems is by using line shunt reactors (SR) that are grounded by neutral grounding reactors (NGRs). This paper analyzes the performance of proposed SRs and NGRs with an example of a real high voltage transmission power system operating in Tanzania, investigates the overvoltages during the single-phase to ground fault, secondary arc current magnitude and the extinction time. The analyzed quantities are important for further power system protection settings parameters, since the successful single-phase autoreclosing (SPAR) strictly depends on the arc extinction time. It is intended to reduce this time in order to enable fast SPAR to prevent the system instability while ensuring that the equipment will withstand the caused stresses. It is shown that the implementing such reactors and NGRs would be beneficial in terms of reducing the secondary arc current, and extinction time. The caused overvoltages does not negatively affect the other equipments' performance in terms of stress. Besides, such configuration will help the power system during the normal operation, not only during the fault, since the inductive load, such as the shunt reactors, have other benefits in long OHLs. Moreover, this paper provides a strong base for the further analysis of similar phenomena, that can be translated to other models, i.e. network conditions and transients with the SPAR implementation, shunt reactors switching transients, surge arresters performance etc. All the analysis have been performed by applying PSSE and EMTP-RV professional software tools.

Keywords: Shunt reactors · Secondary arc · Neutral grounding reactors · Transients

N. Ademović et al. (Eds.): IAT 2021, LNNS 316, pp. 133–151, 2022.
https://doi.org/10.1007/978-3-030-90055-7_11

1 Introduction

The power system is designed to withstand worst possible stresses to which it may be subjected. These stresses occur during transients. Consequently, power system design should be determined by transient conditions, rather than by steady state behavior. The size of transmission line towers, the clearances for transmission line conductors, insulation of windings in power apparatus, rating of circuit breakers, loading capability of equipment - all of these specifications are dictated by considerations of power system transients.

In this research, a real case study of 400 kV power network from Tanzania was modelled in the EMTP-RV software, where the single-phase to ground short-circuit faults were simulated in order the test, and show whether the proposed SRs and NGRs are sufficient and well-sized in order to enable successful and fast secondary arc extinguishing.

The main function of the proper shunt reactor placing is to enable secondary arc extinguishing when singe-phase to earth fault occurs at 400 kV overhead lines. The reactors are placed in the new 400/220/33 kV substation (SS) Lemugur that is to be built in Tanzania and throughout the network as well.

The function of shunt reactors in transmission networks is also, to consume the excessive reactive power generated by overhead lines under low-load conditions, and thereby stabilize the system voltage.

Shunt reactors which are grounded through neutral reactors are unique to extra high voltage (EHV) systems. Such schemes are applied to achieve better single pole fault clearing of the faults and assist in successful reclosing.

Application of automatic single-phase reclosing makes it possible to increase system stability even for extremely high voltage transmission lines. To enable successful fast reclosing, NGR is normally used when transmission line is compensated with shunt reactors [1].

In case of SPS on a double circuit EHV line, the secondary arc is maintained by the inter-phase capacitive coupling between faulted phases and energized phases of the same circuit, and also by the mutual coupling of the other healthy circuit.

The magnitude of the secondary arc current and the recovery voltage are the most important factors, which determine whether or not the secondary arc will be self-extinguishing. Use of properly rated NGR at the neutral point of the shunt reactor ensures secondary arc extinction and successful SPS [1].

The parameters of NGR are initially determined based on the steady state analysis and is previously defined by the other studies for this case study, and this analysis will check whether they are of optimum size.

The methodology and analysis can serve as a base for the further analysis of similar phenomena, that can be translated to other models, networks etc. The paper will provide clear guidelines for modelling and the investigation of such phenomena. Since the transients are integral part of many power system analysis, it is important to know how to properly define and model each element and use adequate parameters. Finally, the proper observations to the simulation results are needed in order to draw relevant conclusions.

In the next section the brief literature review is presented, followed by the network parameters, after which the methodology, results and discussion will be presented.

2 Literature Review

The guidelines for proper NGR and shunt reactors sizing is presented in [2]. It is explained that the shunt reactors are provided on long EHV transmission lines to limit overvoltage during line energization and load rejection. NGR at the neutral point of the SR on long EHV lines are adopted for ensuring the secondary arc extinction and successful single pole switching.

The auhtors of [3] also presented the modelling and analysis guidelines for SR and NGR.

In [4] a model of three-phase variable shunt reactor (VSR) for simulation of switching transients in EMTP-RV software is described. The calculation provided shows that the grading capacitors equalize potential distribution across the breaking chambers, which reduces the probability of restrike inside the circuit breaker.

In [5] a number of details regarding HV SRs and their protection and control schemes is described in order to help the end user to properly select and apply multifunctional numerical relays for HV SR protection and control.

Studies and simulations indicated that the controlled switching method can shift and delay system recovery voltage peak and prevent reignition. In fact, this shifting/delaying provides the necessary time for circuit breaker contacts to separate completely for applying the system recovery voltage across them.

The authors of [6] presented the characteristics of a real case SRs, grounded via NGR and their effect on the power network and the overvoltages that they generate. Controlling methods are proposed to help in understanding this phenomenon.

The reactor parameters influence the generated overvoltages, and consequently the applied electric field, in a very complicated manner. The detailed analysis, introducing some general tendencies that illustrate the significance of parameter variations is presented in [7]. The control availability of these parameters in practice in order to improve the transient response of the reactor was also analyzed.

The unavoidable high frequency transient recovery voltage (TRV) existing in circuit breakers due to inductive switching can be depressed to a reasonable and safely level by equipping suitable arrestor proposed in the [8].

A Variable dead time auto-reclosing scheme for adaptive SPAR has been proposed in [9]. The proposed algorithm analyzes distinct characteristics of the voltage waveforms and accurately classifies the faults by its nature and provide a precise secondary arc extinction time. Its validity is demonstrated with simulation studies in relation to the Korean 765 kV system using MODELS in EMTP. The developed algorithm has shown to be accurate and effective under various fault locations. The realization of the variable dead time control algorithm is simple and it only requires information of voltage transients at the ends of line.

When the circuit breaker poles correspondent to the faulted phase open at both ends, the faulted phase conductor connects to ground by the impedance of the secondary arc. If the secondary arc is extinguished this phase conductor will be floating. That leads to the observation of a power frequency voltage in this conductor, due to the coupling with the other energized phases. Hence, the measurement of this power frequency voltage in the faulted phase conductor may indicate the extinction of the secondary arc as introduced in [10]. This is beneficial in determining whether to block the single-phase auto reclosing shot to avoid the closing of the circuit breakers poles to prevent a subsequent opening.

In [11] the three-phase reactor model obtained by technical manufacturer data and implemented in EMTP-ATPDRAW program permitted to reproduce the overvoltage phenomena and the self-maintaining secondary arc contributions after single fault phase and during single auto-reclose cycle in an appropriate way.

The authors of [12] presented the transients due to switching of 400 kV SR, where different modelling methods are presented.

In [13], authors paid special attention on the effects of three-phase reactors on secondary arc current. The possibility to neglect this current or influence on it using an additional neutral reactor was analyzed. The necessity and influence of factors such as shunt and neutral reactors, capacitive and inductive coupling on the extinction of secondary arc was described and investigated in order to probe the importance for SPAR.

A method to use a single-phase reactor in the neutral of the SR, when transmission line is compensated with SR for fast reclosing is used in [14]. In this research work, traditional method for determining neutral reactor parameters and arc extinction time based on steady-state model was studied.

The presence of one circuit in a double circuit line influences the magnitude of capacitive coupling in addition to inter-phase coupling. Magnitude of secondary arc current also depends on the line length, power flow, source strength etc. This phenomenon has been clearly demonstrated in [15] implementing different methods on a real case study, using EMTP software.

The power system stability can be significantly affected by the connection of new generators and this phenomenon needs to be carefully considered during the connection planning process. The influence of generators on the power system transient stability and practical stability improvement solution is presented [16]. It is important to analyze this phenomen in order to understand the system performance during and after to fault in order to set SPAR and prevent the instabilities.

An investigation of a wind farm transient stability under symmetrical three phase fault conditions, backed-up by a theoretical background of induction generator, simulation model and dynamic response analysis is discussed in [17] with a real case-study network. It introduced the base for the understanding the system performance depending of the fault duration and clearing time.

With the proper combination of different technologies used for supplying the consumers, the longer outages can be allowed, which is discussed in [18]. The conclusion drawn is that proper hybrid power system (HPS) configuration and implementation can positively affect environment, reduce power supplying costs and secure better exploitation and utilization of natural renewable energy sources. In this way, possible solution for distant places enabling autonomous power supply for consumers, which will completely rely on local sources in forms of wind and solar power, combined with electricity storage in battery is proposed. By the implementation of the similar system, the system could endure longer dead-time and possibly need for fast auto reclosing would be decreased.

Many studies have dealt with the impact of different methods, both classical to more technologically evolved methods to suppress the secondary arc extinguishing time [2, 4, 7, 9]. It is highly relevant topic, especially since the most HV transmission lines are long. For every new OHL and substation to be built, these investigations and analysis shall be performed. Many scientific approaches are also available to support these analysis [10, 11, 13, 14, 16] and emphasize its significance. Some researchers propose the interdisciplinary approach in the control of SR switching and auto-reclosing in order to improve the system reliability [9].

Furthermore, this work can be improved by applying the proposed methodologies and technologies to the other similar problems studies such as SPAR settings, switching overvoltages, sending end, receiving end overvoltages with the inductive load studies etc.

3 Analyzed HV Transmission Network and Problem Formulation

A case-study of a part of a real HV transmission network in Tanzania is presented. It includes the new SS 400/220/33 kV that is to be built and will serve as an interconnection between Kenya and Tanzania. A single-line diagram is shown in Fig. 1.

It consists of two main parallel overhead 400 kV lines Singida-Lemugur, and Isinya-Lemugur and other 220 kV lines. The subject SS of this paper is SS Lemugur.

In this case-study the network is modelled in the Electromagnetic Transients Program (EMTP-RV).

In the Fig. 1, it is shown that the analyzed substation is supplied by the 4 AC sources with the proper impedances. These values are obtained from the Electric utilities from Kenya and Tanzania from their PSSE models, when the three-phase and single-phase short circuits were simulated. They are presented as Thevenin equivalents as per IEC 60909-0:2016, Short-circuit currents in three-phase a.c. systems.

OHLs parameters are also obtained from the same utilities and as such, they reflect the actual data. The Singida-Lemugur 400 kV OHL is double-circuit OHL with the total length of 300 km. Isinya-Lemugur is a double circuit OHL as well, with the total length of 210 km.

Each of these OHLs is modelled with the following data: operating voltage, double three phase circuits, conductor and shield wire data, x and y coordinates for each conductor and shield wire, vertical height at midspan of the conductor above ground line length, span, ground return resistivity, tower data – tower geometry, surge impedance and footing resistance. Additionally, each of these 400 kV OHLs are modelled with the transposition included, hence the OHLs are segmented based on the length of each transposition at the certain OHL.

Since the OHLs in this network are long (300, 210 km, etc.), the shunt reactors need to be used. When referring to the single-line diagram, each of the 400 kV OHLs has proper shunt reactors that are grounded via NGR. Only the shunt reactor at Babati 220 kV substation is busbar reactor which is directly grounded. The following Table 1. shows the power of each SR and ground impedance for each NGR, and the transposition lengths are presented in the Table 2.

3.1 Problem Formulation

This paper aims to present some aspects regarding the relevant transients in power systems with the long OHLs, by simulating single-phase to ground fault at different locations, and the means to mitigate them. It will present the classical means i.e. shunt reactors and NGRS. The main purpose of this research is to show that proposed shunt reactor dimensioning and positioning is properly developed. The illustrative cases are presented in order to evaluate approximately the maximum arc duration that may be expected. It is important to know the worst dead time that must be allowed for complete arc extinction, to prevent the arc restriking when voltage is reapplied.

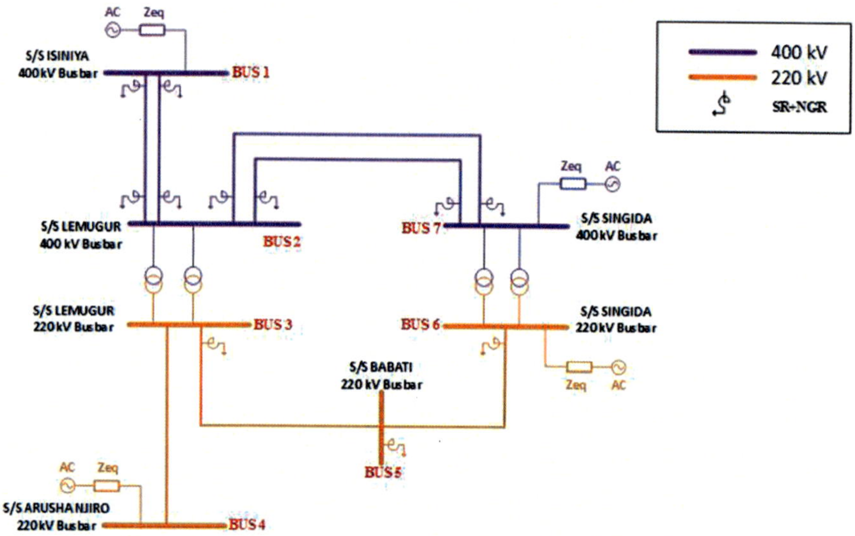

Fig. 1. Single-line diagram of the analyzed network

The paper will also present one simulated case, when there are no SRs and NGRs employed in the system for the comparative purposes in order to illustratively show the main goal and purpose of installing the SRs and NGRs on the long OHLs.

It is also intended to provide the modelling and the analysis guidelines for the investigation of such phenomena in other networks. Finally, this paper presents a method for the measurement of the secondary arc current on a double-circuit transmission line employing single phase switching (SPS). There is a lot of literature studying the concerned transients, yet authors have found it difficult to investigate it and the methods used differs from one author to the other. Hence, the straightforward guidelines and analysis are presented in this paper, combining the different methods and guidelines presented in other papers.

Table 1. Shunt reactors used in the analyzed network

Location	Qn (MVAr)	Method of grounding	If grounded via impedance XE (Ohm)
OHL Singida-Lemugur 1&2 at Singida end	70	Via NGR	710
	70	Via NGR	710
OHL Singida-Lemugur 1&2 at Lemugur end	60	Via NGR	850
	60	Via NGR	850
OHL Isinya-Lemugur 1&2 at Isinya end	40	Via NGR	4058.70
	40	Via NGR	4058.70
OHL Isinya-Lemugur 1&2 at Lemugur end	40	Via NGR	800
	40	Via NGR	800
OHL Singida-Babati at Singida end	18.15	Via NGR	800
OHL Lemugur-Babati at Lemugur end	18.15	Via NGR	800
Busbar reactor at Babati busbar	10	Direct grounding	–

Table 2. Transposition segments for double-circuit 400 kV OH Lines

OHL	No of segments	1st in km	2nd in km	3rd in km	4th in km	5th in km	6th in km	7th in km	Total in km
Lemugur-Singida 1&2 400 kV	7	25	50	50	50	50	50	25	300
Lemugur-Isinya 1&2 400 kV	4	34.5	70	70	34.5	–	–	–	209

4 Methodology

Single phase-to-ground fault (at phase A) was simulated along 400 kV transmission lines. This fault may be eventually eliminated by disconnecting only the faulted phase, while allowing the energy flow on remaining phases. Circuit breakers on both sides of the faulted transmission line are tripped (open). Secondary arc is generated. The simulated switching operation sequence was as follows: fault occurs at 100 ms, circuit breakers trip at 180 ms. After the single-phase to ground fault occurs at the defined point of the OHL circuit (points defined in Table 3) at 100 ms of the total length of the simulation, the software will simulate firstly the primary arc, which will be present in the system up to the point of 180 ms which is the time when all line (phase A) related circuit breakers open, hence isolating the fault. The substation configuration is breaker-and-a-half per bay, so the two circuit breakers open in the SS Lemugur for the faulted line (circuit) only faulted phase, and the third one on the other end of the faulted OHL circuit.

The values of primary arc can be observed from the signal that is available after the simulation is completed. After the fault is cleared at 180 ms, then the secondary arc is induced. Model of the secondary arc is discussed in the following subsection. The magnitude and the extinction time of secondary arc can also be observed after the simulation is completed. After some time, based on the system parameters and configuration, the secondary arc shall extinguish. These extinction times will play a significant role in the further studies for analysis of the system performance with the SPAR.

Table 3. Methodology scenarios for the analysis of the Single-phase to earth fault at 400 kV overhead lines

Scenario	Shunt reactors ON	OHL 400 kV	Location of the fault
S0	No	Lemugur-Singida 1	Near Lemugur
S1-1	Yes	Lemugur-Singida 1	Near Lemugur
S1-2	Yes	Lemugur-Singida 1	25 km from Lemugur
S1-3	Yes	Lemugur-Singida 1	75 km from Lemugur
S1-4	Yes	Lemugur-Singida 1	125 km from Lemugur
S1-5	Yes	Lemugur-Singida 1	175 km from Lemugur
S1-6	Yes	Lemugur-Singida 1	225 km from Lemugur
S1-7	Yes	Lemugur-Singida 1	275 km from Lemugur
S1-8	Yes	Lemugur-Singida 1	300 km from Lemugur (near Singida)
S2-1	Yes	Lemugur-Isinya 1	Near Lemugur
S2-2	Yes	Lemugur-Isinya 1	34.5 km from Lemugur
S2-3	Yes	Lemugur-Isinya 1	104.5 km from Lemugur
S2-4	Yes	Lemugur-Isinya 1	174.5 km from Lemugur
S2-5	Yes	Lemugur-Isinya 1	209 km from Lemugur (near Isinya)

The single-phase to ground fault was simulated at the four circuits of the two 400 kV OHLs (2 parallel circuits for each). As it can be seen from the table, scenarios only treat the single-phase fault at circuit 1 for each of the OHLs. The simulations were performed for both of the circuits for each of the OHL, but for the easy interpretation only results of the circuit 1 will be presented. As it is already mentioned, these circuits are in parallel with the same line parameters, the results obtained for one OHL are similar for both circuits.

The results obtained from these simulations are going to presented by these scenario names in the results and the discussion section. Other power system elements ant their modelling is discussed in the previous section.

These particular points for the simulation of the single-phase to earth fault are chosen merely for the demonstration purposes, and based on other researches, the slightly differences in observed quantities occur after approximately 50 km. That is why the faults were simulated there. Since the OHLs in the EMTP software are modelled in segments due to transposition, that segments are used as a fault location throughout the OHLs. In that way, the fault could be simulated every 25–50 km for Singida and 34.5–70 for Isinya OHLs.

4.1 Secondary Arc Model

The electric arc is the basic element in the process of interrupting currents in electric energy transmission systems.

For simulation of the single-phase fault, the simplified arc model is usually applied. The fault is represented by time varying resistance model during primary arc period and for secondary arc until self-extinguish occurs. The permanent arc extinction will occur when the voltage impressed across the discharge path lower than the arc re-ignition voltage.

The arc model which was implemented in EMTP, according to [16] have been used for these simulations and analysis.

4.2 Shunt Reactors and NGR Model

The reactors and NGRs are modelled by three phase self-inductances, mutual inductances between phases, equivalent capacitances (stray and insulation capacitance), winding resistance and insulation resistance. Self-inductance, equivalent capacitance and winding resistance are obtained from the routine and factory acceptance tests and as such are used in the modelling. Insulation resistance is also measured by dielectric tests during the FATs. In the Fig. 2 the equivalent scheme of a shunt reactor with NER is presented.

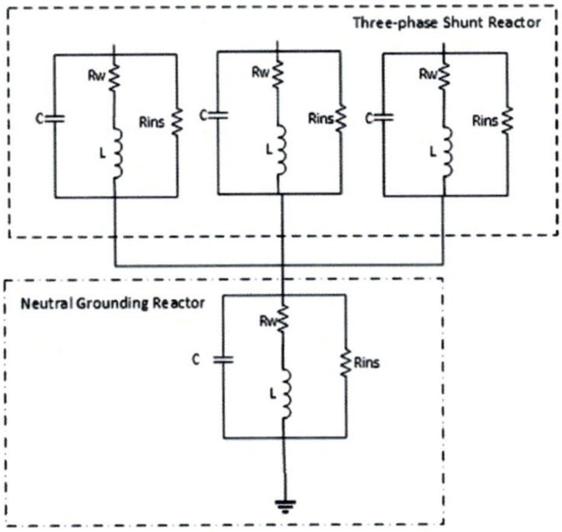

Fig. 2. The equivalent single-line scheme of a shunt reactor with NER

5 Results and Discussion

The following quantites were analyzed and measured during the simulations:

- Secondary arc current (magnitude)
- Arc votage (Recovery voltage)
- Voltage at sending bus (Singida for S1-x scenarios and Isinya for S2-x scenarios)
- Voltage at receiving bus in Lemugur substation for each scenario

In the following Fig. 3, the result of S0 scenario is presented. Only the secondary arc response was presented, since it is the most relevant quantity, analyzed in this paper. It can be clearly observed that, compared to the S1-1 scenario results, the secondary arc has greater magnitude, and longer duration when the shunt reactors are not connected to the system, in fact during this particular simulated fault, the secondary arc does not extinguish during the whole simulation time, i.e. 1 s.

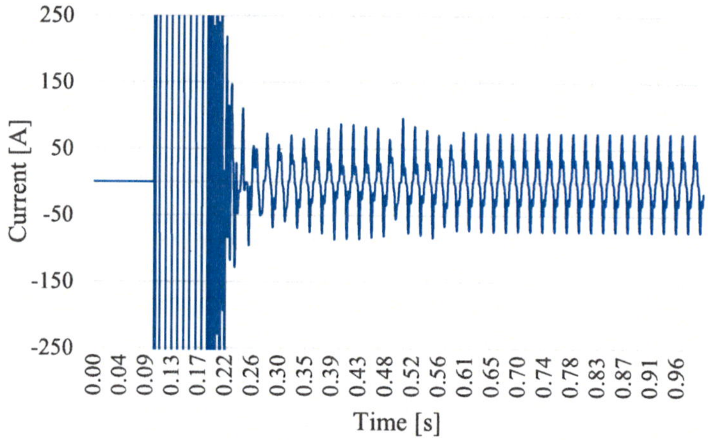

Fig. 3. Secondary arc magnitude for S0 scenario – shunts reactors off

In the following Fig. 4, the results for S1-1 scenario are presented.

In the Fig. 4a, the secondary arc current is presented. As seen, the secondary arc is extinguished before 310 ms. If 180 ms, which is the time at which circuit breakers open the faulted phase (phase A), is subtracted from 310 ms, the value of 130 ms is obtained, which means that the arc extinction time is less than 130 ms for this scenario. This time will play an important role in the time setting for single-phase auto reclosing. Since the SPAR cannot happen until the arc is extinguished, this is the important factor than needs to be measured. It is also important to reduce the dead time, because the long dead time can cause the system instability.

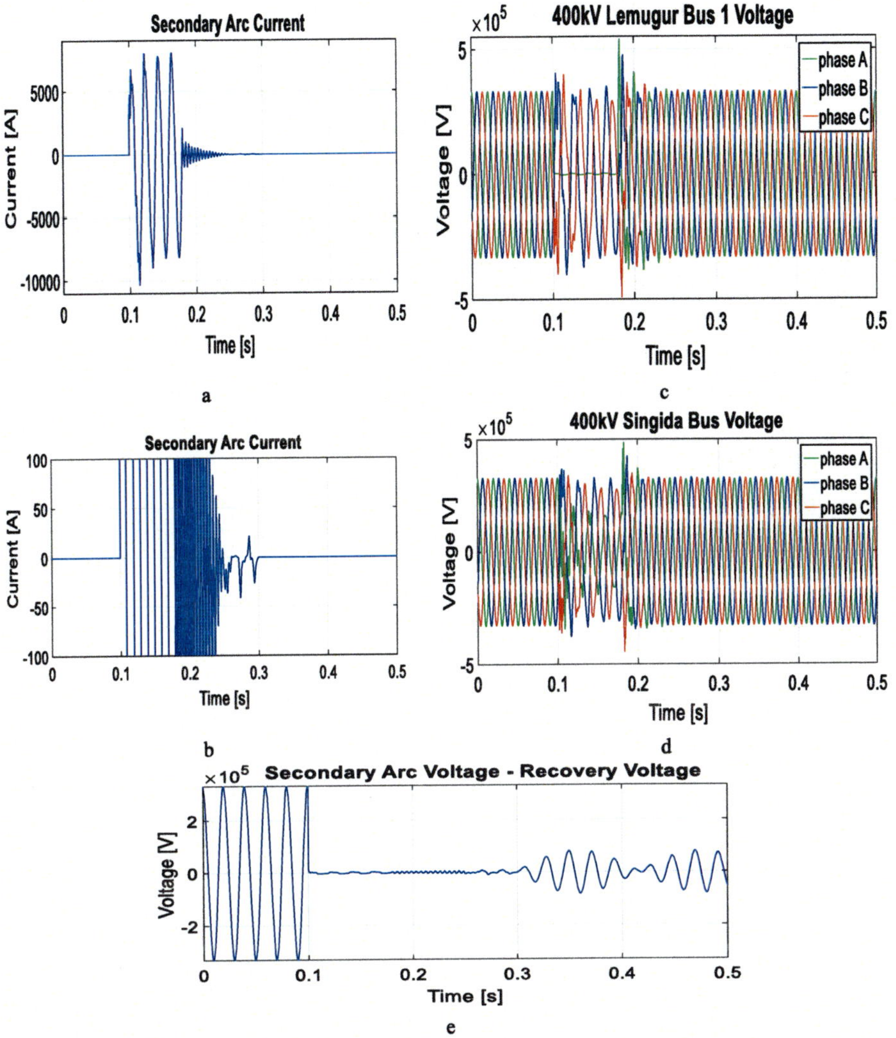

Fig. 4. Results and measured quantities for S1-1 scenario a) Secondary arc current, b) Secondary arc current (zoomed in), c) Voltage at Lemugur 400 kV Bus, d) Voltage at Singida 400 kV Bus and e) Secondary arc Voltage

Another quantity that is analyzed is arc voltage (recovery voltage) – voltage that induces on the secondary arc path after the arc is extinguished at the system frequency. The recovery voltage will further enable successful auto reclosing. This can be observed from the Fig. 4e.

The sending and receiving buses, Fig. 4c and Fig. 4d, respectively, show certain overvoltages at the time when the fault is cleared. This is important to observe in order to analyze the strain of materials and elements in the system. In order to reduce the

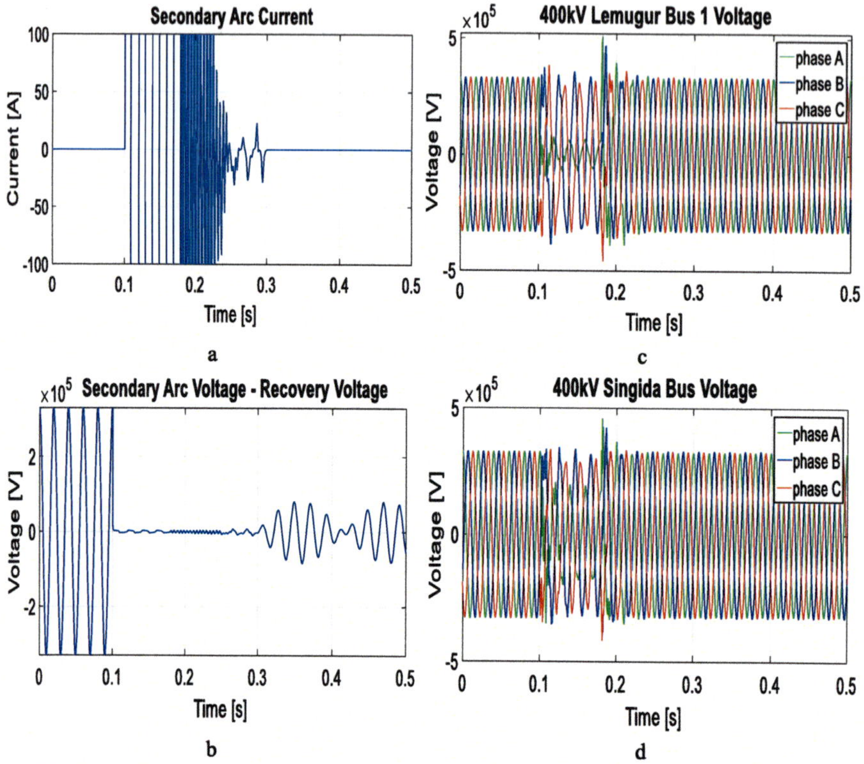

Fig. 5. Results and measured quantities for S1-2 scenario a) Secondary arc current, b) Secondary arc Voltage, c) Voltage at Lemugur 400 kV Bus, d)Voltage at Singida 400 kV Bus

overvoltages, the surge arresters are used. They are especially of great importance at the shunt reactor terminals, and are used to reduce the switching overvoltages.

This is going to be a subject of a different analysis.

In the Fig. 5, the results for S1-2 scenario are presented.

In the Fig. 5a, the secondary arc current is presented. As seen, the secondary arc is also extinguished before 350 ms. Again, the secondary arc extinction time is less than 170 ms. The recovery voltage is induced after the secondary arc is extinguished. It can be observed from the voltage at Singida end that the phase A voltage is of greater magnitude that at the receiving end of the OHL. It is the proper representation since the fault for this scenario is located 25 km from Lemugur. It is also important to observe that the other phases (phase B and C) adapt very fast and maintain within the normal operating voltage levels. The overvoltages at both ends pass within 50 ms after the fault is cleared.

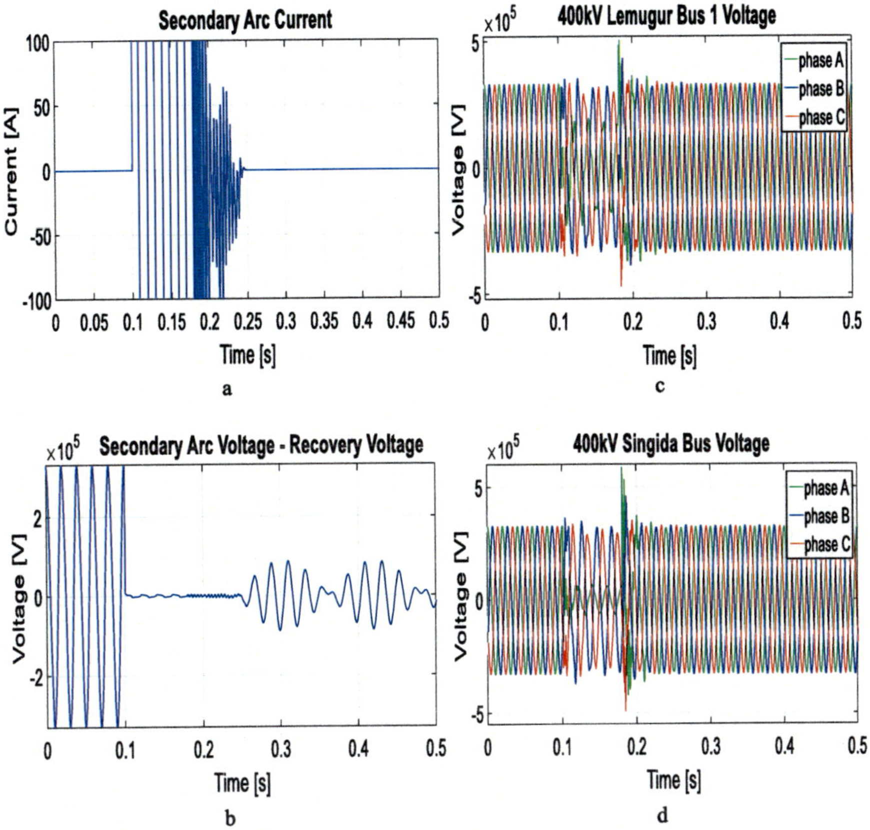

Fig. 6. Results and measured quantities for S1-7 scenario a) Secondary arc current, b) Secondary arc Voltage, c) Voltage at Lemugur 400 kV Bus, d) Voltage at Singida 400 kV Bus

The results for the scenarios S1-3, S1-4, S1-5 and S1-6 are very similar to the previously presented results, as could be analyzed from the Table 4, in which the summary of the all-simulated scenarios are presented. The results start to change in scenario S1-7 and S1-8, hence the detailed results for these two scenarios are presented separately in Fig. 6 and Fig. 7, respectively.

As the fault is nearer to Singida, it can be observed that the secondary arc extinguishing time is less than of the scenarios when the fault is near Lemugur. The shunt reactors are of higher power in Singida than in Lemugur, as well as the grounding impedance of the NGRs. This can be observed from Sect. 3.

With the fault near Singida, the voltage of phase A, is closer to zero than the voltage at the receiving end. It is similar as the case when the fault is near Singida. Even though there is an AC source at the Singida bus, it is still near zero, because of the fault location.

In the Fig. 8 the results for the S2-1 scenario are presented - fault at OHL Isinya Lemugur 1.

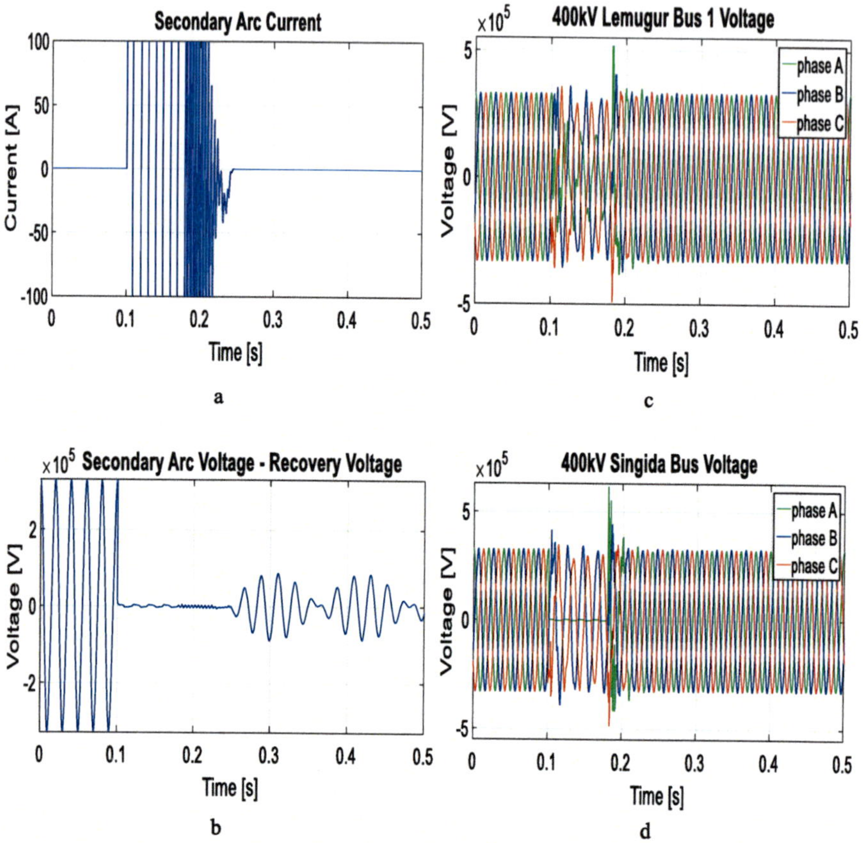

Fig. 7. Results and measured quantities for S1-8 scenario a) Secondary arc current, b) Secondary arc Voltage, c) Voltage at Lemugur 400 kV Bus, d) Voltage at Singida 400 kV Bus

With the fault near Lemugur at the Isinya 1 OHL, it can be observed that the arc extinguish time is very short, extinguishes before 280 ms. This line is compensated with the shunt reactors of 40 MVAr from both sides. The NGR impedances for shunt reactors at Lemugur side is 800 Ω, while of the NGRs at Isinya end 4058 Ω. This Isinya OHL line is shorter than Singida, with total length of 209 km, which is the reason why the arc extinguishes faster. The same observation as in the previous scenarios can be made for the sending and receiving end overvoltages.

The results for the scenarios S2-2, S2-3, S2-4 and S2-5 are very similar to each other in terms of the responses, magnitudes; with the difference in the secondary arc extinguishing time, as presented in the Table 4. The detailed results, are hence presented only for the scenario S2-5 in order to make the interpretation easier in Fig. 9.

As it was expected, on the shorter line Isinya-Lemugur, secondary arc is extinguished very fast, so there is no time for it to reach "steady state" and get the shape that can be observed on S1-x scenarios secondary arc. Hence the recovery voltage appears faster compared to the Singida-Lemugur line.

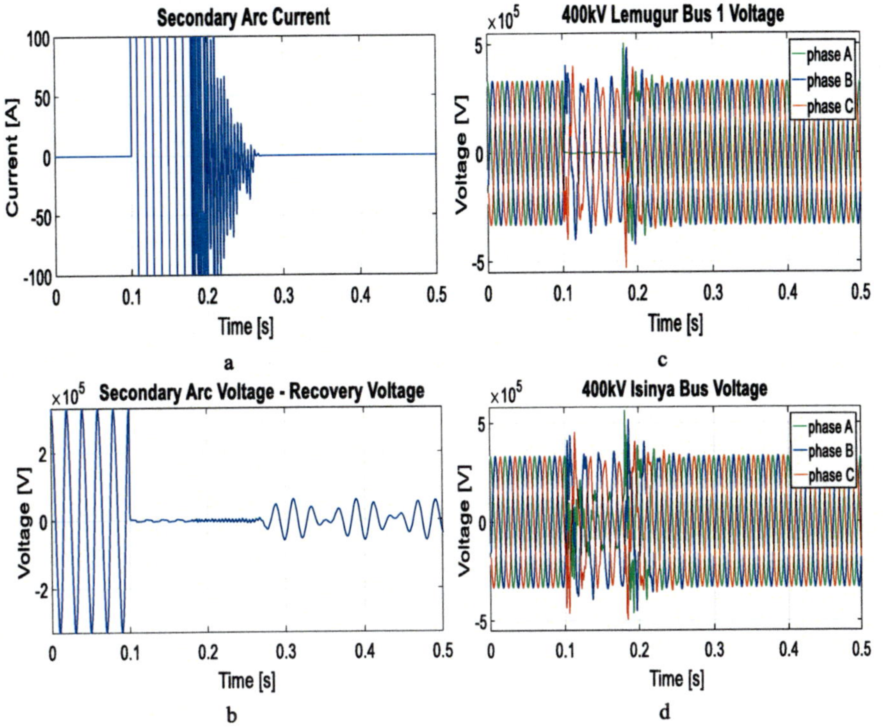

Fig. 8. Results and measured quantities for S2-1 scenario a) Secondary arc current, b) Secondary arc Voltage, c) Voltage at Lemugur 400 kV Bus, d) Voltage at Isinya 400 kV Bus

After analyzing the simulation results, several conclusions can be drawn: for each of the scenarios, the secondary arc is extinguished. Recovery voltage is induced across the path of secondary arc at system frequency. Overvoltages at sending and receiving ends are within the allowed limits. The NGRs, used to ground shunt reactors are properly dimensioned, which enables successful and fast arc extinguishing.

The Table 4 shows the results summary. It is easy to determine the longest time needed for the arc extinguishing, hence the proper dead time for auto reclosing can be determined.

As it can be observed from Table 4, secondary arc current on OHL Singida nor Isinya does not restrike.

In most of simulated faults estimated secondary arc current magnitude in is around 30–45 A. Highest magnitude is estimated to be 42 A when the fault occurs in the middle of OHL Singida. The secondary arc duration is around 125 ms with the faults near Lemugur, and around 70 ms near Singida. The longest duration of secondary arc is with the fault in the middle of Singida OHL, 175 km from Lemugur.

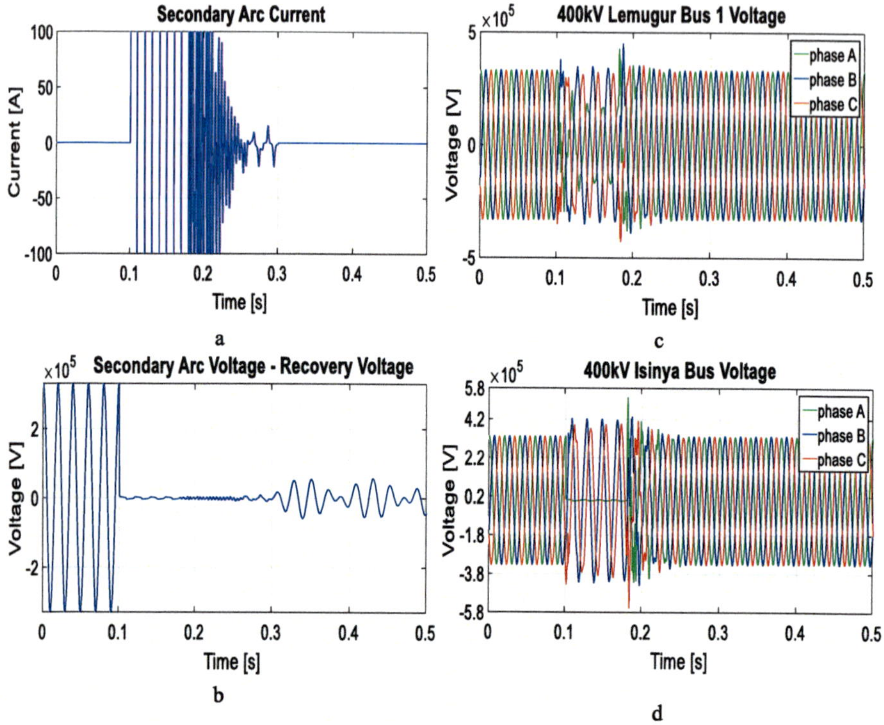

Fig. 9. Results and measured quantities for S2-5 scenario a) Secondary arc current, b) Secondary arc Voltage, c) Voltage at Lemugur 400 kV Bus, d) Voltage at Isinya 400 kV Bus

The change of the arc current magnitude over the distance from SS Lemugur for all simulated faults, Singida and Isinya can be observed from the Fig. 10, while the secondary arc extinction time is shown in the Fig. 11.

This model could be used as a template with the guidelines for modelling and analysis of very fast transients in the power system. The case study presented in this research is modelled with all real data collected from the relevant sources and as such provides a strong basis for further research not only for this particular network, but with slight modifications for others as well. It can also be improved by adding more OHL sections in order to enable simulating more scenarios and faults throughout the network. The several more aspects of the system transients can be anaylzed such as system conditions with the SPAR, inductive load switching etc., but it was not the the scope of this work. Further studies will propose methodology and analyze this as well.

Table 4. Results summary for each scenario of the Single-phase to earth fault at 400 kV overhead lines

Scenario	Fault location (From Lemugur) in km	Secondary Arc current peak magnitude [A]	Secondary Arc extinction time -duration [ms]	Recovery voltage peak value [kV]	Overvoltage at sending end - Singida/Isinya	Overvoltage at receiving end - Lemugur
S0	0 - near Lemugur	82	800	–	457	579
S1-1	0 - near Lemugur	41	0.1215	80.85	487.19	539.37
S1-2	25	28	0.1211	81.59	459.79	508.74
S1-3	75	38	0.1207	83.98	430.67	449.03
S1-4	125	42	0.1211	84.39	439.12	457.54
S1-5	175	43	0.1227	83.98	482	455.85
S1-6	225	38	0.1227	83.57	510.86	455.88
S1-7	275	<10	0.0687	87.47	586.77	505.56
S1-8	300 - near Singida	<10	0.0668	87.93	616.34	520.25
S2-1	0 – near Lemugur	<10	0.0915	61.49	563.96	531.20
S2-2	34.5	28	0.1015	57.73	566.05	495.06
S2-3	104.5	23	0.1212	57.94	538.03	439.55
S2-4	174.5	22	0.1212	58.17	544.62	441.34
S2-5	209 – near Isinya	26	0.1217	57.81	558.79	450.36

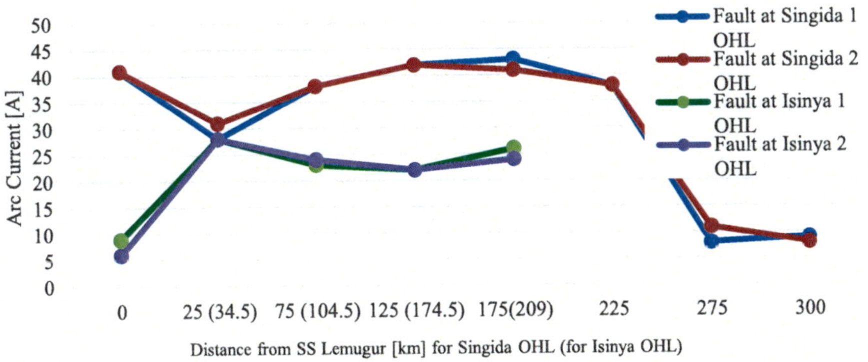

Fig. 10. Change of the secondary arc current magnitude over the distance of the fault

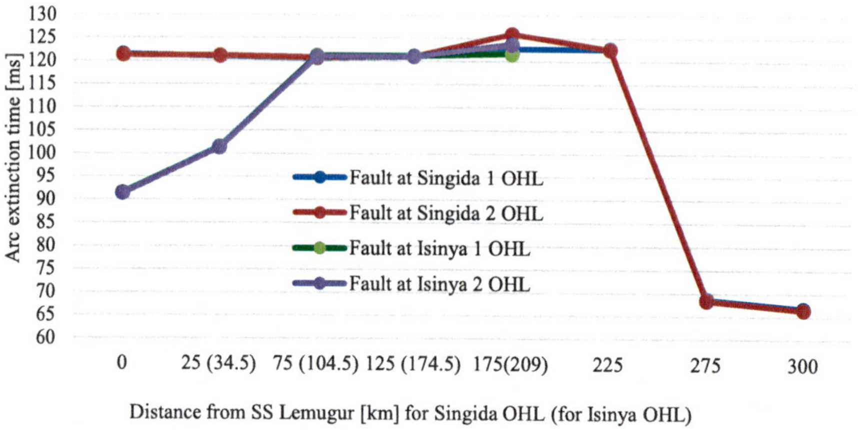

Fig. 11. Change of the secondary arc extinction time over the distance of the fault

6 Conclusions

The impact of shunt reactors and NGRs is investigated and analyzed in this research. A part of the real HV transmission network was modelled. A single-phase to ground faults were simulated on the 400 kV overhead lines in the proposed network. A secondary arc was modelled as well, in order to determine whether the proposed shunt reactors powers and NGR sizing are adequate for the analyzed system. This study shows that the NGRs used in this network are sufficient to enable the fast secondary arc extinguishing.

The illustrative cases that are presented helped in evaluating the maximum arc duration that may be expected in the network and with the obtained results, further improvements can be achieved. Knowing the worst-case scenario, the dead time that must be defined to provide enough time for complete arc extinction. This will make it easier to set the protection in the power system in order to prevent the arc restriking when voltage is reapplied and enable the successful auto-reclosing and improve system stability.

By analyzing the sending end and receiving end overvoltages, it can be observed that all equipment in the network can withstand these transients in times of a fault and switching.

This paper provided the modelling and the analysis guidelines for the investigation of the subject phenomena in other networks. Finally, it presented a method for the measurement of the secondary arc current on a double-circuit transmission line which is both illustrative and indicative and as such can be used in wider analysis.

Further research can improve this analysis by adding several more analysis, i.e. analysis of the transients due to inductive load switching i.e. shunt reactors switching, single-phase auto-reclosing study etc.

References

1. Atmuri S., Thallam R.: Nuetral reactors on shunt compensated EHV lines. In: Proceedings of the 1994 IEEE Power Engineering Society Transmission and Distribution Conference (1994)

2. Bathini, V., Kiran, P.: Guidelines for selection of neutral reactors rating for shunt compensated EHV transmission lines (2012)
3. Povh, D., Valcker, O., Wutzmann, R.: Modelling and analysis guidelines for very fast transients. IEEE Trans. Power Deliv. (1996)
4. Župan, A., Filipović-Grčić, B., Uglešić, I.: Modelling of variable shunt reactor in transmission power system for simulation of switching transients. In: MIPRO, Opatija, Croatia (2017)
5. Gajić, Z., Hillström, B., Mekić, F.: HV shunt reactor secret for protection engineers. In: 30th Western Protective Relaying Conference, Washington (2003)
6. Barghandan, A., Sedaghat, B.: Simulation of switching overvoltages of 400 kV shunt reactor. Transform. Mag. **2**(1) (2015)
7. Haziah, A.H.: Transients in reactors for power system compensation, School of Engineering Cardiff University, Doctoral thesis (2012)
8. Hsiao, H.-C., Cheng, C., Fan, H.-L.: Shunt reactor switching characteristics and maintenance planning in 161 kV system. J. Electromagn. Anal. Appl. **1**, 214–219 (2009)
9. Ahn, S.-P., Kim, C.-H., Aggarwal, R.K., Johns, A.T.: An alternative approach to adaptive single pole auto-reclosing in high voltage transmissionsystems based on variable dead time control. IEEE Trans Power Deliv. **16**(4) (2001)
10. Miguel, P.M.: A methodology to prevent failure in single pole reclosing operations. Int. J. Emerg. Electr. Power Syst. (2017)
11. Gutierrez, G.A.: Control of unsymmetrical overvoltages during open ole condition on overhead transmission lines compensated at one end. In: International Conference on Power Systems Transients (IPST2013), Canada (2013)
12. Uglešić, I., Hutter, S., Krepela, M., Filipović-Grčić, B., Jakl, F.: Transients due to switching of 400 kV shunt reactor. In: IPST - International Conference on Power Systems Transients (2001)
13. Ramold, M., Idarraga, G., Jager, J.: Transient shunt reactor dimensioning for bulk power transmission systems during normal and faulty network conditions. In: International Conference on Power System Technology (2006)
14. Zadeh, M.R.D., Sanaye-Pasand, M., Kadivar, A.: Investigation of neutral reactor performance in reducing secondary arc current. IEEE Trans. Power Deliv. **23**(4) (2008)
15. Nayak, R.N., Sehgal, Y.K., Sen, S., Gupta, M.: Optimization of neutral grounding reactor parameters - an analysis for a double circuit EHV line. In: IEEE Power India Conference (2006)
16. Šarić, M.: Generator dynamic response analysis and improvement following distribution network disturbance. Indones. J. Electr. Eng. Comput. Sci. **7**(2) (2017)
17. Šarić, M., Penava, I.: Transient stability of induction generators in wind. In: 14th International Conference on Environment and Electrical Engineering (2014)
18. Mašić, F., Memišević, B., Bosović, A., Merzić, A., Musić, M.: Implementation of microgrid on location Rostovo with installation of sustainable hybrid power system (case study of a real medium-voltage network). In: Avdaković, S. (eds.) IAT 2018. LNNS, vol. 59, pp. 224–242. Springer, Cham (2019). https://doi.org/10.1007/978-3-030-02574-8_18
19. Prikler, L., Kizilcay, M., Bán, G., Handl, P.: Modeling secondary arc based on identification of arc parameters from staged fault test records. Int. J. Electr. Power Energy Syst. (2003)

Analysis and Control of DG Influence on Voltage Profile in Real Low-Voltage Distribution Network Using Variable Power Factor

Ahmed Balihodžić(✉), Mirza Šarić, and Jasna Hivziefendić

International Burch University, Sarajevo, Bosnia and Herzegovina

Abstract. Photovoltaic PV solar systems continue to be one of the most promising and widely used renewable energy source. PV systems are increasing used as distributed generation technology connected to low voltage (LV) network. This paper investigates the influence of PV connection on voltage profile in LV network. This analysis includes scenarios with voltage control based using variable power factor. The analysis is performed on model of realistic LV network, fed from 10 (20)/0.4 kV distribution transformer currently operated in Bosnia and Herzegovina. This paper also discusses the issues of voltage stability and voltage regulation in distribution network. The main objective of this work is to investigate the possibility of practical voltage control in distribution network. Results indicate importance of appropriate operation strategy for PV systems connected on the LV network. The results confirmed that DG can have both positive and negative effect on the grids depending on control strategy. Methodology is based on load flow analysis which is performed using DIgSILENT Power Factory software. In conclusion, this paper contributed to practical power system operation and management as it provides additional case analysis which confirm the importance of appropriate power system planning and operation strategies.

Keywords: Distributed generation · Power factor · DIgSILENT PowerFactory · Power factor control

1 Introduction

The power supply quality is a very important aspect of the power system operation, planning and development. In modern power systems, power quality is not only a problem that concerns technical aspect. In recent years, also financial aspect draws more attention since a certain amount of power is inherently lost in power system due quality issues, which amounts to significant financial losses [1]. Technology development, especially improvement in the area of power electronics applications, enabled many technical advantages and financial benefits. At the same time, however, it created new challenges in terms of investigations, deployment and applications of new technologies in power systems [2]. Development of distributed generation (DG) technologies, investigation of their influence on power system, combined with the investigation of new emerging problems is a very challenging task for engineers and scientists. One such problem is voltage

N. Ademović et al. (Eds.): IAT 2021, LNNS 316, pp. 152–163, 2022.
https://doi.org/10.1007/978-3-030-90055-7_12

stability and its improvement in the case of a power system operational challenges. Considering the inherent nature of electrical load and their need to operate under safe and reliable conditions, it is important to deliver sufficient amount of power which complaint with predefined quality standards. The power supply quality and voltage stability is a very important aspect especially to power system customers, operators and regulators.

The use of renewable energy sources, in particular PV solar systems continues to increase throughout the world. In particular, due to different policies and practical reasons, there has been an increase in low voltage (LV) connection of PV systems. Even if this topic is well referenced, practical problems continue to puzzle Distribution Network Operators (DNO). Traditionally, most studies on the influence of the DG grid connection has been reserved to higher voltage levels and LV network were omitted from investigations due to their localized influence. However, the amount of renewable energy sources connected to network is increasing sharply and which accumulated challenges faced by DNO. In particular, voltage variations due to intermittent nature of renewable sources continues to be number one in LV network for DNO. Further, DNO normally request DG units to either be capable of operating within a specific range of power factor (e.g., 0.95 inductive/capacitive) or to operate at a fixed power factor (e.g., unity) for large DG units. This requirement is based on the characteristics of the network and the penetration of DG, and aims to keep voltage profiles within limits and minimize interference with the voltage regulation provided by on load tap changer transformers (OLTCs) or line drop compensators (LDS). For smaller DG units, this requirement usually proves to be problematic due to reactive power absorption limitations. According to the [3] it can be achieved and explained by implementing the dual mode scheme which means the combination of two operation modes of a generator:

1. Constant power factor control when conditions are normal and measured voltage is within the statutory upper and lower limits, the generator will operate in constant power factor control mode that will in the same time cause on load tap change transformer to work normally without constant disruptions. In this mode the ratio of active to reactive power is kept constant with the reactive power following the variation of active
2. Voltage control when conditions are abnormal and measured voltage is not within the statutory upper and lower limits, the generator will adopt the voltage control mode and act as PV bus, that is, by varying the excitation of the automatic voltage regulator. On this mode, the automatic voltage controller is activated to vary excitation and move the operating point within the bus voltage limits

This paper is a part of the ongoing investigation and description of the realistic operational challenges faced by the DNO with the increasing penetration levels of solar PV systems in LV network. In particular, this paper presents and analysis symmetrical voltage components and their behavior in case of single-phase customer connection and connection of photovoltaics. DIgSILENT Power Factory software is used to model real low-voltage distribution network supplied by distribution transformer 10/0.4 kV with installed power of 0.16 MVA. The program is also used to calculate three-phase power flow using Newton-Raphson method. This work is important for the following reasons:

1. Provides additional evidence of PV influence in practical power systems
2. It designs practical voltage control schemes for LV connected PV systems
3. Investigates the common operational issues faced by DNO on a daily base and show the importance of appropriate distributed generation (DG) systems

2 Literature Review

Paper [4] states that using dual mode control P/Q should be constant. Any fluctuation in P bring about proportional variation of voltage. If Q can be compensated for the voltage variation generated by P by adjusting in the opposite direction, then the voltage variation can be maintained within statutory limits. For voltage rise situation, a more leading power factor is required. This control is also known as automatic voltage/power factor control since it is implemented by means of combining the advantages of power factor control and automatic voltage regulator. However, according to [5], usage certain voltage control including power factor (for example VAR/PF control) can decrease system's ability to recover from low voltage conditions. For large machine applications the VAR/PF controller needs to be equipped with additional slow integrated VAR function. Constant power factor controller operating in the lagging mode can lead to the minimization in the reactive power drawn from the substation at the expense of increased customer level voltage deviations, while the power factor operating in leading mode can minimize the customer level voltage deviations at the expense of increasing the reactive power drawn from the substation [6]. Traditional transmission and distribution systems can work close to stability limits due to growing demand for electricity [7]. Operation of systems close to stability limits can expose system security to risks [8]. In [9], transformers regulated voltage at low-voltage side by changing taps to keep voltage inside limits. Paper showed that variations have been high enough to prevent transformers to keep voltage inside limits leading to need for reactive control. It also concluded that fast reactive devices (STATCOM) can ensure continual reactive power change. In [10], the method that can suppress voltage spikes and voltage fluctuations is based on PV power factor while keeping voltage stability through phase-modifying equipment.

In [11] it was demonstrated that connection of PV systems have positive effects on the environment, but if their penetration level is above certain values it can cause voltage fluctuations, overloading of some network elements (such as transformers), deviation in frequency etc. Results in same paper showed that renewable energy sources and PV can have positive impact on power system parameters in case of appropriate planning, allocation and renewable energy source operation. In [12], PV generator with the maximum power point tracking (MPPT) control and battery energy storage systems in the stand-alone mode of operation is used for voltage regulation for a microgrid. Analysis showed that PV and battery energy storage systems have good performance where system has an appropriate reaction to the environmental and technical variations.

Based on the short review of the existing developments in this area, it can be concluded that operational issues faced by DNO in the LV networks with high share of DG continues to be an important practical topic.

3 Problem Definition

Voltage stability is the ability of power system to maintain voltage values in range that is technically acceptable in normal working regime (stationary) as well as after the disturbance. Power system becomes instable when disturbance (for example large increase in load demand or some other action that changes working regime) causes progressive and uncontrollable voltage drop. Main reason for this is that at the moment power system does not have resources to provide additional reactive power when it is needed. Voltage stability criterion comes down to that there should be positive increase in voltage module at every node for every positive increase in reactive power in the same node. In case that it is not the case for just one node, system is instable.

Voltage stability analysis is simplified by means of indicators based on small signal properties of system, derived from linearized models such as Voltage Sensitivity Factor that is defined as a function of reactive power injection at the node, $\Delta U/\Delta E$ indicator and Qg/Q1 criterion. Together with voltage stability analysis, problems related to voltage profile analysis are one of the most important and elaborated questions when it comes to the analysis including connection of DG. In this work, the aspect of voltage variations in context of power quality according to EN-50160 standard is analyzed by means of connecting the distributed generator (PVs of different capacity) to the low-voltage network. Parameters considered by this standard include various aspect of voltage behavior including harmonics in voltage, unbalance, transient overvoltage, overvoltage, long interruptions, short interruptions, voltage sags, flicker, rapid voltage variations, voltage variation and frequency. This standard does not consider abnormal operating conditions [13].

According to the Regulatory Commission for Electricity in the Federation of Bosnia and Herzegovina, whose jurisdictions is to supervise and regulate the relations between power generation, distribution and electricity customers including power traders, maximum allowed voltage deviations from steady-state nominal values in distribution systems are, [14]:

$\Delta um = \pm 10\%$ for the medium voltage (MV) grid.
$\Delta um = +5\%$ and -10% for the low voltage (LV) grid.

This means that upper voltage limit for this analysis is 1.05 p.u. and 0.90 p.u. Power factor, as measure of the phase difference between voltage and current in an AC power system, is set to be determined based on network parameters and operational requirements since National Electricity Rules in Bosnia and Herzegovina do not provide information about power factor requirements. In Bosnia and Herzegovina, DG cannot bill the reactive energy delivered to the network. However, DGs are regularly billed for the excess reactive power absorbed [13].

Voltage profiles mainly is affected by the amount of reactive power and DG can be used in order to control voltage development by either absorbing or delivering reactive power and it can contribute to voltage control by implementing the pf operation scheme [13]. The amount of reactive power produced by DG must approximately match the amount of reactive power that is being consumed. Power factor can be:

Leading power factor in the system (due to capacitive loads) causes the voltage to rise.

Lagging power factor (due to inductive loads) will cause the voltage to fall.

If reactive power is either under or over supplied, the voltage on the network may rise or fall to a point where generators must switch off for protection. That will cause less generation and further problems [15].

4 Results and Discussion and Future Work

In this section results are presented and discussed, followed by the future research direction. First of all, model of realistic power system is presented in details. Then various operational scenarios are defined and considered followed by results discussion and future research directions.

4.1 Test Network Modelling

The network consists of 33 loads and 56 buses. Data that was provided by meter readings that were collected every 15 min (15-min interval), showed that peak consumption of P = 124.272 kW. Total power of loads was 30.13 kW. Network that is modelled is shown in Fig. 1.

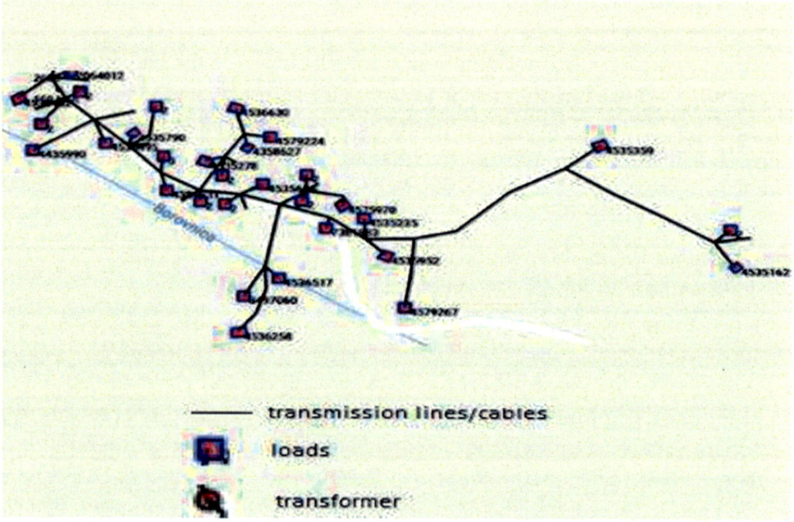

Fig. 1. Scheme of the analyzed network

Topology of network modelled in DIgSILENT PowerFactory is shown in Fig. 2.

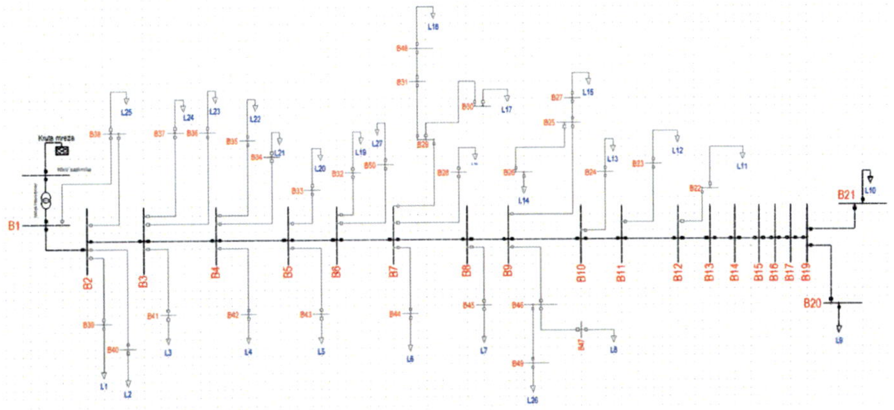

Fig. 2. Topology of the network in DIgSILENT PowerFactory

4.2 Base Case Scenario

First case that is analyzed shows voltage profile along main feeder (from bus B2 to bus B20) for unity power factor without any generations. Unity power factor means that ratio of active an apparent power is equal to 1 and that no reactive power is either absorbed nor delivered by the generator nor loads. This analysis is performed in order to determine the base case scenario. DG is connected at bus B20 at the end of main feeder. In Fig. 3, the voltages along the main feeder from bus B2 to bus B20 are shown. Based on the results shown in Fig. 3, it can be concluded that voltages remain within voltage tolerance limits between 0.9 p.u, and 1.05 p.u.

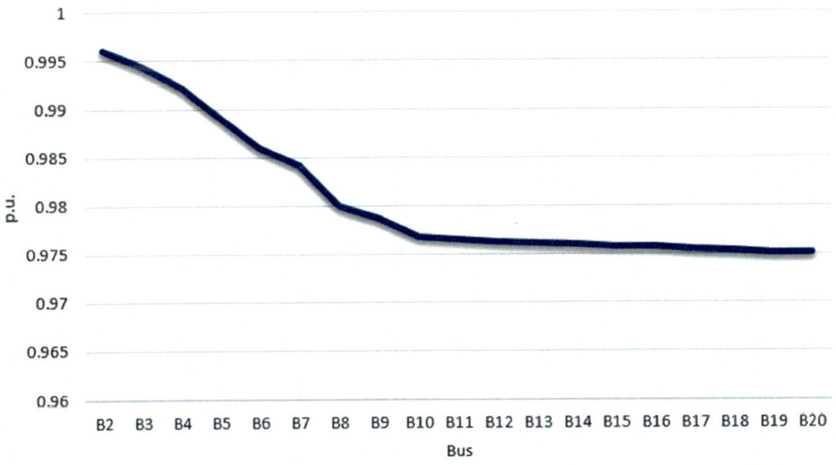

Fig. 3. Voltages along main feeder without DG connected

4.3 Unity Power Factor Operation of DG

Effect of DG connection of various capacity at the end of main feeder is shown in Fig. 4, which depicts problems that may occur if DG of larger capacity is used. For cases when DG capacity is 10 kW, 20 kW or 30 kW voltages at all buses are inside voltage tolerance limits. In case that DG has capacity of 40 kW or greater, voltage at certain buses rises to more than 10% compared to case when DG is not connected at all. Table 1 shows the lowest amount of DG capacity that will cause voltage rise above upper limit. All the results were performed with unity DG power factor of solar PV systems which is in general their default operational state in LV network. Table 2 reveals more details on the voltage violations and their values in p.u. for DG capacity that is even greater than smallest DG capacity that will cause voltage rise above upper limit. Based on the presented results, unity power factor operation is a plausible operational state in terms of voltage control. After the tap position change option is exhausted, it is proposed to analyze voltage control in LV network using variable power factor operation of the PV solar generator. Results of leading power factor operation and their influence on voltage levels is presented in following sections.

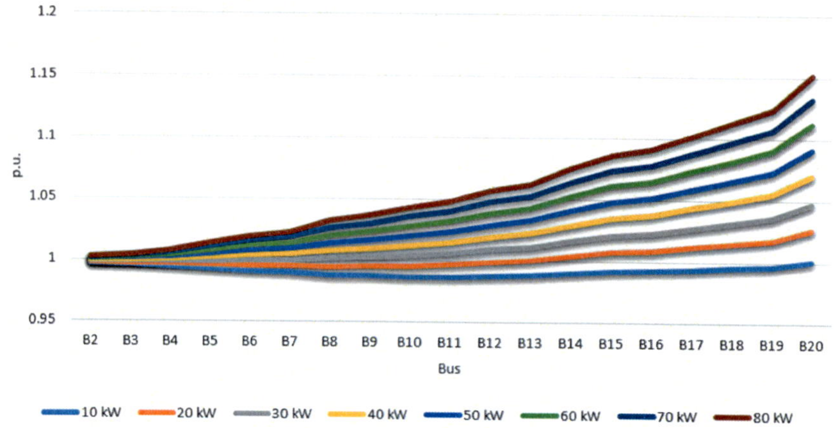

Fig. 4. Voltages along main feeder for unity power factor when DG is connected

4.4 Leading Power Factor Operation vs Voltage Tolerance Limits

In this section, the values of leading power factor operation that causes to voltage excursion reduction and return within tolerance limits for different values of DG are be presented. This analysis is partially based on the data presented Table 1 and Table 2. When DG of 32 kW is connected, the leading power factor that will cause voltage reduction at B20 back to voltage tolerance limits is 0.99 which requires absorption of 4.56 kVAr of reactive power. For largest capacity of DG the values of power factor and reactive power absorbed required for all bus voltages to be inside voltage tolerance limits are shown in Table 3.

Table 1. DG capacity that will cause voltage rise above upper voltage limit

DG capacity	Voltage violation	Voltage value (p.u.)
30 kW	No voltage violation at any bus	–
31 kW	No voltage violation at any bus	–
32 kW	Voltage violation at bus B20	1.0520
33 kW	Voltage violation at bus B20	1.0542
34 kW	Voltage violation at bus B20	1.0565
35 kW	Voltage violation at bus B20	1.0587
36 kW	Voltage violation at bus B20	1.0609
37 kW	Voltage violation at bus B20	1.0632
38 kW	Voltage violation at buses B19 and B20	B19: 1.0510 B20: 1.0654
39 kW	Voltage violation at buses B19 and B20	B19: 1.0529 B20: 1.0676

DG of larger capacity with leading power factor which caused the absorption or reactive power causes all voltages to be inside voltage tolerance limits are shown in Table 4. Figure 5 depicts the amount of reactive power absorbed that will lead to decrease in voltage magnitude, forcing the operation of DG with power factors that are different from unity.

From Fig. 4 it can be seen that DG capacity up to 30 kW will not cause rise in voltage magnitudes along main feeder at any bus. However, Fig. 4 and Table 1 and Table 2 lead to information that DG capacity of 32 kW is the smallest one that will cause first problems at the last bus (B20). By increasing the DG capacity, the problems at other buses appear, mostly at the end of the feeder. Voltage values at the beginning of the feeder are not affected for DG values from 10 kW to 80 kW. The effect of DG capacity of 80 kW or more is not analyzed in this paper. That led to conclusion that system does not need the generator to generate reactive power, but to absorb. Table 3 and Table 4 and Fig. 5 and Fig. 6 shows power factor vs Q absorbed diagram. Power factor is leading and moves away from unity with absorption of reactive power. The requirements for reactive power absorbed increase as size of DG increases. The calculated values of reactive power absorption vs voltage magnitude can be used for practical voltage control purposes.

Table 2. DG of larger capacity that will cause voltage rise above upper voltage limit

DG capacity	Voltage violation	Voltage value (p.u.)
40 kW	Voltage violation at bus B19 and B20	B19: 1.0547 B20: 1.0698
50 kW	Voltage violation at bus B16, B17, B18, B19 and B20	B16: 1.0516 B17: 1.0587 B18: 1.0658 B19: 1.0728 B20: 1.0912
60 kW	Voltage violation at bus B14, B15, B16, B17, B18, B19 and B20	B14: 1.0527 B15: 1.0609 B16: 1.0652 B17: 1.0736 B18: 1.0819 B19: 1.0903 B20: 1.1120
70 kW	Voltage violation at bus B13, B14, B15, B16, B17, B18, B19 and B20	B13: 1.0527 B14: 1.0640 B15: 1.0735 B16: 1.0784 B17: 1.0880 B18: 1.0976 B19: 1.1073 B20: 1.1321
80 kW	Voltage violation at bus B12, B13, B14, B15, B16, B17, B18, B19 and B20	B12: 1.0567 B13: 1.0622 B14: 1.0750 B15: 1.0856 B16: 1.0912 B17: 1.1020 B18: 1.1129 B19: 1.1237 B20: 1.1517

4.5 Future Work

This paper paves the path for numerous future research directions, such as installation of smart transformers in the LV networks and their use in voltage control applications. Further, various power system load could be added such as electrical vehicles and battery storage systems. Investigations of voltage profile behavior according to different standards (such as EN-50160 or other) in network with both electrical vehicles (charging stations) and DG connected can be done, especially when it comes to voltage unbalance when both electrical vehicles and solar power plants as possible DG are connected.

Table 3. DG capacity with pf and Q absorbed leading to all voltages to be inside limits

DG capacity	Power factor (leading)	Q (kVAr)
32 kW	0.99	4.56
33 kW	0.97	8.27
34 kW	0.94	12.34
35 kW	0.91	15.95
36 kW	0.88	19.43
37 kW	0.85	22.93
38 kW	0.82	26.52
39 kW	0.79	30.27

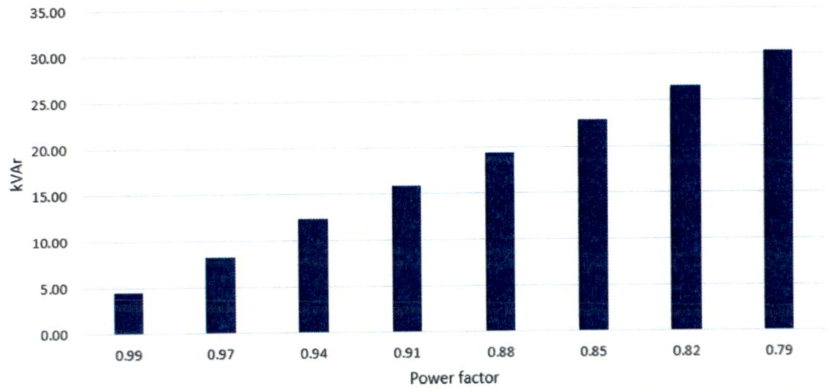

Fig. 5. Power factor vs Q absorbed for DG with capacity from 32 kW to 39 kW following the data from Table 3

Table 4. DG of larger capacity with pf and Q absorbed leading to all voltages to be inside limits

DG capacity	Power factor (leading)	Q (kVAr)
40 kW	0.77	33.15
50 kW	0.62	63.27
60 kW	0.58	84.27
70 kW	0.56	103.56
80 kW	0.55	121.48

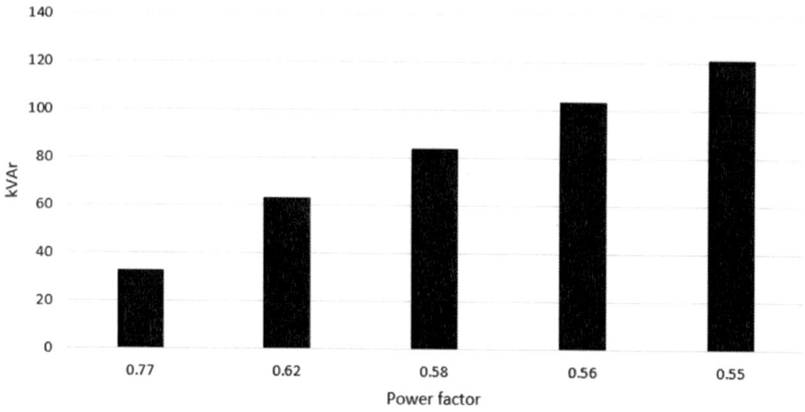

Fig. 6. Power factor vs Q absorbed for DG with capacity from 40 kW to 80 kW following the data from Table 4

5 Conclusion

In this paper, analysis of various DG connection at the end of main feeder of real LV distribution network and possibility of power factor voltage control is performed. Analysis showed that DG capacity up to 32 kW will not cause any rise of fall in voltage values. However, smallest DG capacity leading to voltage rise at only one bus (B20) is 32 kW with unity power factor. With increase in DG capacity, voltage at other buses at the end of main feeder will also rise. All voltage values that violate voltage tolerance limits are above upper voltage limit. In the analysis, the highest rise in percentage is around 15% (for DG of 80 kW at the end of main feeder). That rise lead to conclusion that voltage should be decreased meaning power factor should be leading. The objective is to keep power factor as close as possible to unity power factor. Power factor values will decrease as reactive power Q absorbed rises leading to power factor to be 0.55 for highest DG capacity analyzed (80 kW). For DG capacity from 32 kW to 35 kW, power factor is higher than 0.9. In addition, connection of DG at the end of main feeder will lead to voltage rise. However, regardless of value the DG generator capacity that operate with a leading power factor can be subject of unstable voltage regulation and increased heating. This paper demonstrated the importance of appropriate DG planning and operation in LV network. It highlights the severity and importance of everyday operational challenges faced by DSO in realistic LV networks with high share of renewable energy.

References

1. Ali, A.-J.F., Zayer, W.H., Shukir, S.S.: Power quality problems and improvement techniques. J. Eng. Appl. Sci.**14**, 5888–5898 (2019)
2. Bečirović, V., Hasanić, M., Nikolić, B., Hanjalić, S.: Naponske prilike u jednostrano napajanoj distributivnoj mreži sa distribuisanim generatorom
3. Tengku Hashim, T.J., Mohamed, A., Shareef, H.: A review on voltage control methods for active distribution networks. Przeglad Elektrotechniczny (2012)

4. Vovos, N., Kiprakis, A.E., Wallace, A.R., Harrison, G.P.: Centralised and distributed voltage control: impact on distributed generation penetration. IEEE Trans. Power Syst. (2007)
5. Eberly, T.W., Schaefer, R.C.: Voltage versus VAr/power-factor regulation on synchronous generators. IEEE Trans. Ind. Appl. (2002)
6. Agrawal, A.: Performance of PV Generation Feedback Controllers: Power Factor versus Volt-VAR Control Strategies. Virginia Polytechnic Institute and State University, Blacksburg (2015)
7. Ettehadi, M., Ghasemi, H., Vaez-Zadeh, S.: Voltage stability-based DG placement in distribution networks. IEEE Trans, Power Deliv. **28**(1) (2012)
8. Al Abri, R.S., El-Saadany, E.F., Atwa, Y.M.: Optimal placement and sizing method to improve the voltage stability margin in a distribution system using distributed generation. IEEE Trans. Power Syst. **28**(1) (2013)
9. Hubana, T., Begić, E., Šarić, M.: Voltage control in distribution network using variable power factor operated distributed generation. In: IEEE SYPC Sarajevo 2016, Sarajevo (2016)
10. Aida, S., Ito, T., Iwamoto, S., Kitada, R., Ariyoshi, N.: Voltage control using PV power factor, static capacitor, and transformer tap for large PV penetration. Electr. Eng. Jpn. **196**, 21–31 (2016)
11. Arpadžić, A., Šarić, M., Hivziefendić, J., Avdaković, S.: Power system efficiency improvement using solar PV systems
12. Badnić, L., Hivziefendić, J., Šarić, M., Tešanović, M.: Voltage regulation of PV system with MPPT and battery storage in microgrid
13. Šarić, M., Hivziefendić, J., Bandić, L.: Analysis and control of DG influence on voltage profile in distribution network. In: Hadžikadić, M., Avdaković, S. (eds.) IAT 2017. LNNS, vol. 28, pp. 30–43. Springer, Cham (2018). https://doi.org/10.1007/978-3-319-71321-2_3
14. ferk.ba. http://www.ferk.ba/. Accessed 8 Feb 2021
15. G. T. Papers. www.gses.com.au. https://www.gses.com.au/wp-content/uploads/2016/03/GSES_powerfactor-110316.pdf. Accessed 8 Feb 2021

Computer Science

Traffic Flow Influence on the Bitumen Distortion Through the Back Propagation Algorithm Based on the Real on Road Formed Dataset Parameters

Sabahudin Vrtagić[1,2](✉), Edis Softić[3], Aditya Gmanjunath[1], Mirza Ponjavić[2], Željko Stević[4], Marko Subotić[4], and Jasmin Kevrić[2]

[1] College of Engineering and Technology, American University of the Middle East, Block 6 Street 250, 54200 Egaila, Kuwait
sabahudin.vrtagic@stu.ibu.edu.ba

[2] Faculty of Engineering and Natural Sciences, International Burch University, Francuske revolucije bb, 71210 Ilidža, Sarajevo, Bosnia and Herzegovina

[3] Technical Faculty, University of Bihac, Irfana Ljubijankića bb, 77000 Bihać, Bosnia and Herzegovina

[4] Faculty of Transport and Traffic Engineering, University of East Sarajevo, Vojvode Mišića 52, 74 000 Doboj, Bosnia and Herzegovina

Abstract. Due to more radical escalation in essentials for goods exchange an increase in the traffic is observed which is proportional to the roadway breakdown. Research conducted is to measure the traffic and erosion proportions. Digital data collection is in the form of a video file. From audiovisual file foundations information concerning the number of vehicles, class of vehicles, speed, traveling direction, and time is collected with the help of a linear support vector machine model (Linear SVM). Additionally, reference asphalt erosion measurement is done by friction coefficient of asphalt with "pendulum method", asphalt texture with "sand patch method" and finally the irregularities or deformations in the road structure with "Lath and nail method". All the following measurements are the output parameters for modeling. Finally, Ministry of Transport data and data collected from video and field analysis are combined and are forming the final dataset. A prepared dataset is pre-processed for a machine learning technique that is used to predict erosion coefficients. Results established on the prepared back-propagation algorithm model simulation are showing that heavy/light vehicles in combination with the low flow rate intersections are one of the key factors which increase road erosion. The following model result indicates that the intersection asphalt lifecycle cannot be considered to be the same and that it depends on the traffic flow. This is proved by the developed multi-layer perceptron (MLP) model on the nearby intersection results and by a simulation of smart intersection where results indicate that the smart intersections will reduce parameters by 50%.

Keywords: Modeling erosion coefficients · Back propagation · Lath and nail method · Sand patch method · Pendulum method (SRT)

N. Ademović et al. (Eds.): IAT 2021, LNNS 316, pp. 167–186, 2022.
https://doi.org/10.1007/978-3-030-90055-7_13

1 Introduction

Since the increase in the human population and a more drastic increase in the needs for goods exchange the roadways are becoming more occupied with the vehicles. In addition to the weather conditions the traffic flow plays an important role in road erosion. Research conducted is to find the impact relationship of the traffic flow and to suggest some possible solution for better optimization and a longer life for roadways. Road surface texture is the main component that defines asphalts tire contact or rolling friction which is one of the key factors for road safety. From a general viewpoint, improvement of surface and its abundancy, shape and frequency are significant concerns, these requires contemplations of a wide range of factors contributing to impacts [1].

The computed roadway responses with the usage of firsthand study of roadway derivation revels that an instantaneous asphalt response is needed, and is the key for a long-term asphalt response. The Monte Carlo models settled in National Cooperative Highway Research Program Project 1–37A were engaged to forecast asphalt out-turn. The revision has presented that the properties of amplified tire stress and navigating axle load on a characteristic asphalt assembly for 2 million truck passes. The Monte Carlo replication technique and models used in this work conveyed valued understanding for the progress of prospect elastic pavement design and examination techniques [2]. The tire force standards are used to inspect the road erosion physical appearance of a simple tandem-axle for a automobile model with different speed range and road roughness disorders. It determined that for the projected measures, the imaginary lifespan of highway asphalt surfaces that are prone to lethargy failure and may be bargain considerably by the dynamic module of wheel forces. The erosion done to around five per cent of the road surface area during the passage of a hypothetical model vehicle at standard main road speeds may be improved by as plentiful as four times [3]. Additionally, as an important point the connections between grinding obstruction and macro texture got from various strategies were examined with assessed surface profundities anticipated from laser profiler, laser surface scanner, and CT meter were practically identical to the MTD got from sand fix tests [4].

Scope of presented research is based on simulated scenarios or under the controlled environment as can be concluded from paragraphs before. The aim and achievement of the current research is to use available on the road traffic flow data to model out dependency of road erosion with the traffic flow. This is achieved by pre-processing of existing traffic data with traffic video data sample. In this way, the needed data is extracted. The following data was then used with multi-layer perceptron (MLP) to obtain a model for erosion coefficients.

The presented study is systematized as follows: Introduction and literature review delivers a momentary picture of the correlated research in this area. The detailed procedure used in this study to obtain erosion coefficients is given in methodology section. The outcomes of research findings using the presented research models and their discussion are presented in result section. Finally, the last section provides concluding considerations with specific values and directions for further research.

2 Literature Review

The process of learning and development is constantly evolving as new methods and instruments are introduced in all engineering fields. One way of asphalt erosion measurements for defined period can be performed by methods for civil engineering field, namely by Pendulum method, Sand patch method, Lath and nail method. The pendulum test valuations attained are symbolized to as either the Slip Resistance Value (SRV) or Pendulum Test Value (PTV). To represent the different road friction coefficient conditions the rubber 'Slider' which imitates the tire is interchangeable using rubber of different rubber hardness and resilience [5].

Road surface roughness is measured with sand patch method. The following technique is appropriate for bituminous surface sequence and concrete roadway surface with roughness depth greater than 0.25 mm. [6]. Main conclusion as a criterion is then that texture depth (sand-patch) must be higher than 1.5 mm and shall still be higher than 1.0 mm after one year of operation [7]. A 4-m-long lath is used to measure the pavement's straightness. There are also 2-m-long laths that are used to measure the depth of water in tracks. As a rule, they are rectangular in section, made of solid wood or light metal, without any possibility of guessing. The measuring nail is 300 mm long metal and 25 to 30 mm wide. When measuring, it is especially important that the lath has at least two supports [8]. In addition to summarize a new approach termed as imaging spectroscopy is given and the model performed by hyper-spectral image to forecast and record the abrasion standards of each motorway pixel in the section. Merging the typical method with imaging spectroscopy may perhaps be responsible for the required enlargement of the accessible data to deliver result architecture with a complete picture of the roads' status. This modus operandi boundaries originate mostly in compositional differences between roadways, and the prerequisite for the application of various calibrations between scenes [9]. A summary of some asphalt erosion measurement methods can be described shortly like that the British pendulum, dynamic grating, outpouring test, sand test, and round surface meter are moderate tests that need lane closure during their estimation because of that the procedure is time-consuming for data collection [10, 11]. The capacity for constant road erosion estimation is a bit hard task for most estimation strategies. It is positive to utilize techniques with low activity and gear costs, strategies, for example, the bolted wheel test and variable slip test have high essential and working expenses [12]. Very interesting study of tire/road friction and texture properties claims that in the tire/road slip initial state to the peak is mainly dependent by tire properties and that the part after mainly by road surface properties [1, 13].

Observations form civil engineering data, combined with digital data with the application of machine learning modeling approaches for possible solutions on road safety are a key point for a purposed research beside asphalt erosion measurements on road vehicle or entity detection which will form the dataset. Plentiful methods are associated to entity divulgation like Edge matching, also Gray scale matching and Gradient matching and most of methods is based on the graphic details by distinguishing shape, motion, color and texture of the objects [14]. Some of the real factors and causes why the SVM was a selection of can be perceived from the accompanying outcomes where the examinations of two assorted gathering techniques with K-Nearest-Neighbor (KNN) and Support-Vector-Machine (SVM) is practical. Where a principle favorable position of SVM grouping is that SVM has prevalent execution on data-sets also being different highlights, in any event, when there are lacking of free information tests for the preparation cycle. In any case, a few faults of SVM groupings are in its asset concentrated computational prerequisites, which are seen both during preparing stage and testing stage. Also, execution is dependent upon kernel work boundaries being utilized. The general precision is high and for the SVM classifier the rate was 92%. Unlike KNN, SVM characterization has proved to be more capable at automobile grouping. This has given us the reason to say it is better in large executions [15]. In general models are custom on assortments of vehicle pictures assessed and so clear Eigen spaces are procured. Structures that are molded with Eigen spaces for the purpose of Support Vector Machine (SVM) become a dependable classifier for automobile identification. The results are unmistakable to be strongly exact, also in expediency and exactness for automobile classification. Critical resources of Eigenvectors are invariance in heading during a change, which can be utilized as a portrayal set of the entire huge dataset. Those components are entitled Eigen-faces in face acknowledgment use [15] and Eigen-vehicles in motor vehicle acknowledgment use [9]. Examination has exhibited that a little arrangement of the eigenvectors is proper to frame up the entire picture trademark. In the field of execution, given structure gives a significantly higher rate of classification (around 94%) compared to the other alternatives namely, 55% for Line Segment Detector (LSD) and 70% for Artificial Neural Network (ANN). This approves the estimation of upgraded classifier of automobile pictures throughout evening time traffic conditions [16]. A different method for low light time traffic scenes is with the speculation that the photos of a section under changeable enlightenment conditions lie in an angled cone outlined in the image space. Likewise, varieties due to varieties in light fixation can be void by standardizing images. Considering these bits of knowledge, recommended strategy joins paired classifications using discriminant hyper planes in the normalized image space. It is demonstrated that the issue brings about a mix of parallel classifications utilizing hyper planes in the (1-D) dimensional standardized picture space. It is tentatively affirmed that the utilization of hyper planes for standardized picture space is compelling, and that SVMs are more qualified than Fisher's straight discriminant FLDs for getting discriminant hyper planes [17].

Strategy altering different highlights of vehicle indication at the two epochs of sunshine and of the evening, four symbols containing under, upright edge, evenness, and rear light are joined for automobile revelation [18, 19]. Particle filter starting testing, proliferation, perception, sign combination and assessment conclusively produces the vehicle dispersion. With recently portrayed technique mixes, the proposed system can adequately recognize and follow vehicles. Also, it has a ground-breaking and strong method of working under different illumination circumstances. In the evening time, rear light glow is the main property for automobile location [18, 19]. Likewise, a significant level after methodology with BSAS grouping technique is chosen to manage the situation when the space is under bending for a brief period [20]. Elective technique is the utilization of Gabor channels which give a creation to overseeing levels of invariance to fixation because of worldwide brightening that are discernment surmounted, and have exercising judgment in direction. These are arrangement and scale tunable edge and line tags. Automobiles comprehend strong edges and blueprints that may well take wide scope of shapes in the arrangement and scales, in this way, the insights of these substances (e.g., mean, standard deviation and sticks) end up being capable for vehicle recognition. Partner the exhibition of NNs with SVMs, SVMs outmatched NNs utilizing whichever head segment examination (PCA) or Gabor elements. In one explicit case, the aftereffect of SVM classifier had grasped about 9% more noteworthy exactness than the NN with PCA substances and 11% more prominent precision in simultaneous utilization of Gabor elements. Kindred of SVM sizes, the normal numeral of support vectors by Gabor technique elements was 200, which is 1441 not as much as utilizing PCA substances. These capitals that SVM established automobile recognition utilizing Gabor substances are quick [21].

True to form, the most troublesome conditions introduced the low light object detection. As practically all systems and recommendations are managing vehicle moving including some of them actualized from a host vehicle moving, a significant agreement with respect to execution of various channels and diverse sort of vector characteristics is noted. As the information for an introduced study incorporates day and night video chronicles, picture coordinating is utilized to recognize day and night spans.

Additionally, some regression, machine learning and statistical methods are overviewed and used as the next important point for the following research. Like a model founded on the multi-layer perceptron (MLP) that sequences using the back-propagation algorithm with square error as the loss function. The ensuing model adjusts the squared-loss using LBFGS or stochastic gradient descent [22]. Limited-memory BFGS (L-BFGS or LM-BFGS) is an optimization set of rules in the kinfolk of quasi-Newton methods that methodologies the Broyden–Fletcher–Goldfarb–Shanno algorithm (BFGS) using certain degree of volume of workstation memory [22].

3 Methodology

The first step was to collect data and form the dataset to study the effects of traffic loads on asphalt deformations with developed back propagation model which is trained by combining collected data (video data- linear support vector machine, road erosion measurements) and samples of statistical data.

In this research, python, OpenCV, and machine learning methods are implemented for data collection and analyses. OpenCV and python with some external python libraries are used as software tools for in frame objects detection, object tracking and to export the collected data as comma separated value file. Experimental data collection is done with civil engineering methods. Method is providing the following measurements:

1. Road friction coefficient
2. Road surface texture

The steps for the following research can be summarized as:

a) Video data processing
b) Civil engineering reference measurements
c) Video data filtering and combining with statistical data

d) Input/output dataset modeling
e) Implementation of machine learning and regression
f) Comparison of obtained model results

The scope of part "a" is to extract information related to the vehicle's motion. Information extracted is the number of vehicles their speed, type, time spent on the intersection, and information on the vehicles going to a complete stop. Mentioned stage is done with the help of Linear SVM and HOG [23]. The object's features are extricated by first normalizing the substance's factors and afterward processing the principal request inclinations of the edge. In the following stage, building encoding that stays powerful to little changes while being touchy to neighborhood outline content is settled. At last, the casing is standardized again and the HOG descriptors are gathered. This turns into the substances vector with which will prepare the Linear SVM classifier.

The aftereffect of SVM and different strategies for regulated learning is information input subordinate, thus pre-preparing of train information is a significant advance since these substances removed from this information are fit onto the support vector machine (SVM) mode [23], this is exemplified by scaling all measurements as [−1, 1]. Having unstructured information, the utilization of RBF piece is decent, anyway rechecking preparing occurrences is significant just when the informational collection is lopsided. In the event that the information isn't directly divisible, the radial basis function (RBF) or different pieces are received and both expense and gamma utilizing "Matrix Search" calculation for SVM is optimized. A HOG dataset will comprise boundaries from positive and negative pictures. The positive pictures are the Region of interest (ROI) with vehicles separated from the source videos. On the other hand, the negative pictures are all together that are not vehicles seen from any bearing that is street, structures, and so on. These pictures are additionally resized to (64, 64) pixels. The element Histogram of Oriented angles vector of each picture is acquired in more than 9 directions, with 8 pixels for every cell and 2 cells for each square alongside the histogram containers of the slopes, and the spatial picture is produced from a resized picture of (16, 16) [23]. The component vector comprises the histogram of oriented gradients HOG descriptor, the histogram of the arranged inclinations, and the spatial picture vector. Standard scaling the element vector would eliminate focus intend to 0 and scale the difference to 1. A variety of support vector machine SVM called a Linear SVM has been prepared to recognize the objects of interest and those that are definitely not. The recently created highlight vector was parted into 70% and 30% as train and test set individually, with the revealed precision score of 98–100% on the test set.

Next step is applying sliding window idea across the district of interest with a classifier to check if the vehicle can be identified. Note that the size of the vehicle will change contingent upon how far the vehicle is concerning the camera. Warmth map is added for all vehicles distinguished in outline for which forecast is positive or 1 to keep just 1 window rather than different windows for one vehicle. Likewise, some vehicles will be dismissed based on determined edge esteems (values for pixels of 120 and beneath) [23].

The part "b" scope is related to the civil engineering methods implemented. The following civil engineering methods are used to collect different types of data on road erosion. Methods used with their brief explanation are "pendulum method" which measuring the friction coefficient of asphalt, "sand patch method" which measures the asphalt texture and "Lath and nail method" which measures irregularities or deformations in asphalt structure.

Video data filtering refers to removing all detections that do not correspond to the vehicles in the dataset. One of the most important steps that are able to filter most of the error is based on speed. Namely, if the speed into the intersection is zero a conclusion is that a different type of object is detected which is not of interest. Examples for this are water accumulations. This step also includes modeling of the video data with statistical data. With the help of programs like excel and python, inputs and outputs are computed in the part "d". The following data sets inputs and outputs are used to generate a model capable to predict road erosion. The following inputs and output sets are generated based on a combination of statistical data and data obtained from videos established on the total number of days. The following input parameters were selected based on their impact on the erosion parameters. The stage "e" denotes and is related to this part. The final part "f" relates to the multiple tests on generated models and a comparison of obtained model results. Finally, after selecting the best model an application is created so that real data tests can be simulated. The following application is useful as a simulation tool that can help with road safety improvements. In the following paper only one model is presented.

A numerical model and detailing for SVM order with a basic portrayal can be depicted with the accompanying conditions and the following equations:

SV classification:

$$\min_{f,\xi i}\|f\|_k^2 + C\sum\nolimits_{i=1}^{I} \xi i\, y_i f(x_i) \geq 1 - \xi i \qquad \text{for all } i\ \xi i \geq 0 \tag{1}$$

SVM classification, Dual formulation:

$$\min_{a_i}\sum\nolimits_{i=1}^{I} a_i - \frac{1}{2}\sum\nolimits_{i=1}^{I}\sum\nolimits_{j=1}^{I} a_i a_j y_i y_j K(x_i, x_j) 0 \leq a_i \leq C \qquad \text{for all } i \sum\nolimits_{i=1}^{I} a_i y_j = 0 \tag{2}$$

Where ξi is called loose variables and it represents the error based on the (xi, yi). As the preparation of SVM turns out to be very perplexing with the quantity of preparing focuses is expanded, some valuable strategies for quick SVM preparing have been suggested [24–26].

As the generated data has oscillation or deviations for the different inputs normalization between 0 and 1 was the key solution for the model accuracy improvement. The following was done by applying StandardScaler python function which is based on the standard deviation mathematical equation. Standard scores (z) are calculated as:

$$z = \frac{x - \mu}{\sigma} \tag{3}$$

In the given equation μ is the mean average and σ is the standard deviation with respect to the mean value.

Additionally, a variety of Microsoft Excel 2016 statistical function is used to speed up data modeling and to simplify final dataset generation.

The following research and data sampling is performed on the Bosnia and Herzegovina roadway namely on the intersection of the road M223 and R363 near Tuzla city as can be seen from Fig. 1. The following intersections two lanes are observed and used in the conduced study. Mentioned two lanes are of interest as shown in Fig. 1 where it can be seen that intersection entrance form Tuzla city side is the region of interest.

The following section will give detail on the data results of the multiple stages used in the research.

One of the first and the key point's results are visualized in Fig. 2 which shows on road vehicle detection in a specified region of interest. Observed shape of histogram generated with oriented gradients broadly known as (HOG) as well as support vector machine (SVM) are used to collect the data on vehicle detection in the region of interest. Data set of HOG is comprised of boundaries obtained as positive pictures; additionally negative pictures are used as well.

Fig. 1. Roadway intersection and data measurements representation

Positives set is described as Region of interest (ROI) where vehicles have been removed from the source recordings likewise the negative pictures are generally those that are not vehicles seen from any course like street, structures, and so forth. So the pictures containing a vehicle comprise the positive train information with a sum of 2954 examples, at the same time pictures seeing the streets including similar static items that are not defined as vehicles establish what is called negative preparing dataset with 2860 examples. These pictures are further resized, a new format or size that is used is 64 by 64 pixels. Further Histogram element with oriented inclinations vector obtained from each picture has got more than 9 directions, distributed are 8 pixels for every cell, also 2 cells for every square alongside the histogram canisters of the slopes and so a spatial picture is produced from picture the size of 16 by 16 pixels. Element vector is now comprised of HOG descriptor as well as a histogram of the situated angles and the spatial picture vector. With given details a variety of SVM that is more broadly known as a Linear SVM has been prepared to recognize objects that are designated as of interest, additionally with high degree of accuracy those that are most certainly not. The Newly created vector was added into 70% and 30% as a train and test sets respectively. Detailed precision score is 98–100% on the examination set [26].

Fig. 2. Linear-SVM vehicle detection and classification

The following video data is than combined with statistical data from traffic department and on road erosion measurements data to form dataset as:

Number of LW vehicle - The number that defines the count of light vehicles (LW) for a specific interval;
Number of HW vehicle - The number that defines the count of heavy vehicles (HW) for a specific interval;
Time period (h) - Time counted from 27.10.2015–12.06.2020 to obtain specified intervals;
Time on the intersection/junction (h) for HW - Calculated with respect to the average value found from video data and multiplied with the number of (HW) vehicles;
Time on the intersection/junction (h) for LW - Calculated with respect to the average value found from video data and multiplied with the number of (LW) vehicles;
Number of HW going to a complete stop - Calculated with respect to the percentage amount;
Road erosion (lath) - An assumption on a percentage basis is used to generate so machine learning can be implemented;
Road erosion (send) - An assumption on a percentage basis is used to generate so machine learning can be implemented;
Road erosion (SRT) - An assumption on a percentage basis is used to generate so machine learning can be implemented.

The final dataset is created on the daily basis for two road intersection lanes. Because the lanes are next to each other weather and temperature are not considered in this study. Total number of data forming the final data set is composed of 3382 days for both lanes. An example of the dataset based on the daily percentage traffic flow generated for the left lane can be seen in Table 1.

Multilayer Perceptron (MLP) which is used in this study is supervised learning which learns from input and output correlations through the training process. With the given set of input variables and a target or output value, MLP can develop or learn a non-linear dependence approximation for regression. In the Fig. 3 a simple structure with one hidden layer for MLP is visualized.

Figure 3 on the left side presents a set of input variables represented with x that are actually variables presented in Table 1. The target value or Y(x) are values based on the on-road measurement types which are namely “lath”, “sand” and “SRT”. The process behind finding the x and y dependency is achieved by the hidden layers which transmutes the values from the former layer with a weighted linear summation given as w1x1+w2x2… where “m” is the integer of proportions for input x and w is weight. The

Table 1. Dataset sample overview (HW-heavy vehicles, LW-light vehicles)

Day	Number of LW vehicle	Number of HW vehicle	Time period (h)	Time on junction (h) HW	Time on junction (h) LW	Number of HW going to a full stop	Road erosion (lath)
1	2977	225	24	0.8725725	9.74092262	51	0.02069781
2	6235	424	48	1.6443144	20.4012941	96	0.04139562
3	9423	625	72	2.4238125	30.83262138	142	0.06209343
4	12784	839	96	3.2537259	41.83001504	190	0.08279124
5	16018	996	120	3.8625876	52.41185708	225	0.10348905
6	18627	1069	144	4.1456889	60.94866162	242	0.12418686
7	21828	1251	168	4.8515031	71.42252568	283	0.14488467
8	24944	1437	192	5.5728297	81.6182646	325	0.165582

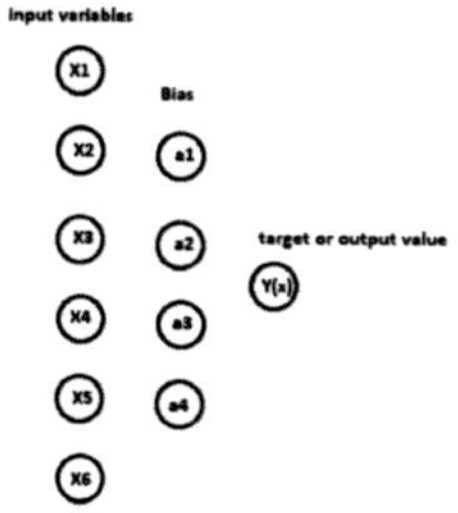

Fig. 3. MLP with one hidden layer

learning process in the perceptron MLP is achieved by changing association loads or weights after each bit of information is processed, in light of the measure of mistake in the yield contrasted with the normal or real outcome. The error is calculated by calculating the difference between the target and predicted values which are then used to weights recalibrations. After which the different activation functions can be used to achieve best accuracy.

4 Results

In the next pages, road erosion and its connectivity to the traffic flow is presented. Models are generated and trained for each of the road erosion measurement individually. The first one presented will be for the "lath" measurements. Graphical model results or prediction vs measured statistical value for "lath" dataset set can be observed in the Fig. 4.

As the test includes 30% of the dataset which is too large to be clearly visualized, the calculation of the overall test set average error difference is 4.88% which with the R = 0.9947763661294762 and (Mean squared error) MSE = 0.00042021130941892996

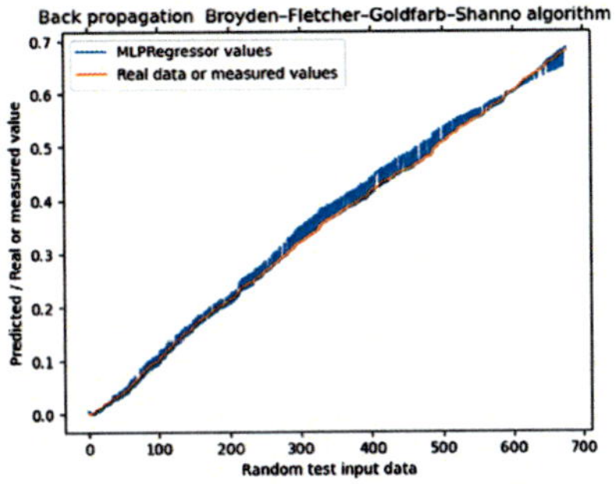

Fig. 4. MLP "lath" predictions vs measured statistical value

computed from learning process acceptable result for the model. The model improvements depend on the spectrum and size of the input data set which will then if the size is increased be dependent on the computational power. The R value indicates and gives the alteration of the correlation in the reliant variable that is foreseeable with respect to the independent variables. The next step is to show the other two models and their accuracy. So, the second model presented is for the "sand" measurements with graphical results presented in Fig. 5.

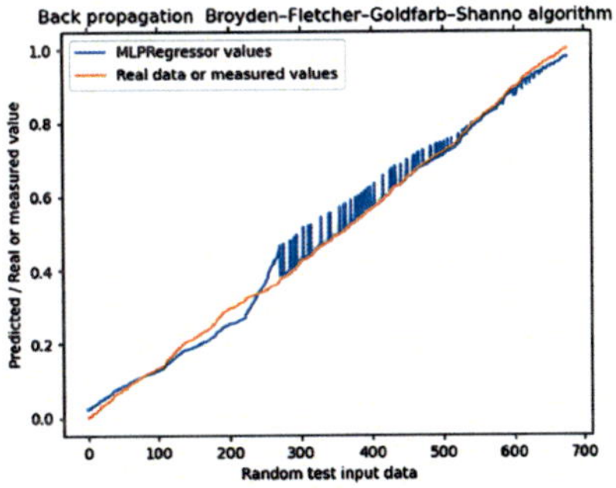

Fig. 5. MLP "sand" predictions vs measured statistical value

The calculation of the overall test set average error difference for "sand" statistical real value calculated as 8.31%, with the given model parameters R = 0.9921404210496664 and MSE = 0.0006601426345023692 calculated during the learning process shows that the model's correlation can be used as a solution. So now the final piece of the puzzle is to check prediction and accuracy for an "SRT" model. The "SRT" model is presented in Fig. 6.

The calculation of the overall test set average error difference for "SRT" statistical real value calculated as 4.04%, with the given model parameters R = 0.9993539863230343 and MSE = 5.298541489983264e−05 calculated during the learning process shows that the model's correlation can be used as a solution.

The Figure 6, 7, 8, 9, 10 and 11 are presenting prediction on the test samples, the following test samples are linearized. All the MLP predictions are with the random set

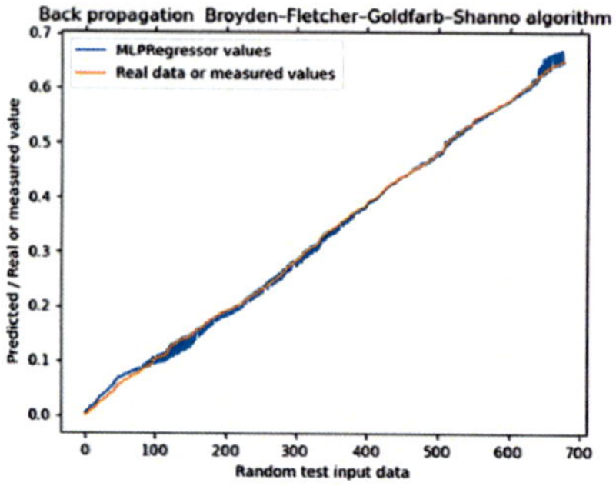

Fig. 6. MLP "SRT" predictions vs measured statistical value

data, with the split of the dataset given as follows: 70% for training and 30% for testing. The MLP model is based on the multi-layer perceptron (MLP) that trains using the back-propagation algorithm with square error as the loss function. The following model optimizes the stochastic gradient descent.

But to more vibrant a tabular form of data with completely new inputs and the MLP prediction are shown in Table 2, Table 3 and Table 4 respectively. As it can be concluded erosion modeling is performed separately for each of the civil engineering sampling method.

Table 2. MLP "lath" predictions with data normalization on a new dataset

Number of LW vehicle	Number of HW vehicle	Time period (h)	Time on junction (h) HW	Time on junction (h) LW	Number of HW going to a full stop	Road erosion (lath)	Predicted value
11551382	1144688	40584	57.50618947	9787.591466	10063.63262	16	16
11551382	1144688	40584	100	19000	20000	30	30

As it can be seen in Table 3 MLP (lath) prediction are having maximum accuracy as predicted and measured values are identical. The next step is to check out the result for MLP "sand" and "SRT".

From Table 4 it can be easily calculated out that the error difference for measured and predicted values is 0.2% for the first and 0.13% for the second value. All the following results are indicating that the following model accuracy and precision is high. From the above tables, it can be concluded that the model input-output correlation was very high which is seen from real vs predicted values in the tables.

5 Discussion

The data set used for modeling and learning generates 3382 samples for both lanes and in combination with multi-layer perceptron (MLP) that trains using the back propagation

Table 3. MLP "sand" predictions with data normalization on a new dataset

Number of LW vehicle	Number of HW vehicle	Time period (h)	Time on junction (h) HW	Time on junction (h) LW	Number of HW going to a full stop	Road erosion (sand)	Predicted value
3515789	482009	23808	1771.460196	12324.8796	108838	76.53459712	76.18699
6076189	869029	40200	3193.82062	21300.5666	196227	60.378478	60.73705

Table 4. MLP "SRT" predictions with data normalization on a new dataset

Number of LW vehicle	Number of HW vehicle	Time period (h)	Time on junction (h) HW	Time on junction (h) LW	Number (complite stop)	Road erosion (SRT)	Predicted value
6137288	877722	40584	3225.768786	21514.75407	198190	35.99999587	36.07260557
4565288	220681	33816	855.8229861	14937.89625	49830	52.50561117	52.43648827

algorithm with square error as the loss function was the key point for the given results. MLP model results are visible in Tables and Figure above. Normalization is one of the key factors as the data range of inputs varies a lot. This is why the modeling is performed with normalized data samples to the range from 0 to 1. The following MLP model result for R and RMS are satisfying but the additional test was performed with new input data.

So the next step is to combine given models into a single one so an overall model precision could be executed. The following is performed by creating an "exe" executable that is used to simulate scenarios of new data for a completely new intersection or a road. The following application depends on a model of both normalizing and predictions and those models must be provided in the same folder with the application.

The following application helps in an understanding of connectivity between traffic flow and erosion. So the first test will be run from existing data used for modeling for the left and right lane, only final values are used. The following data are given in Table 5.

Table 5. Left and right lane final data sets

	Number of LW vehicle	Number of HW vehicle	Time period (h)	Time on junction (h) HW	Time on junction (h) LW	Number of HW going to a full stop	Road erosion (lath)	Road erosion (send)	Road erosion (SRT)
Left	5415785	268656	40584	1041.874834	17720.77347	60663	34.99999671	90.00000094	42.99999183
Right	6137288	877722	40584	3225.768786	21514.75407	198190	59.99999436	60.00000376	35.99999587

The results for the following final data values are given in Fig. 7 and Fig. 8.

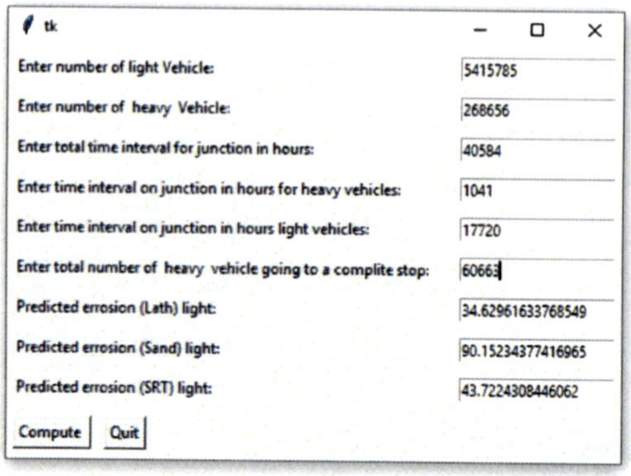

Fig. 7. MLP model prediction based on the left lane data

The Fig. 8 is a print screen of "exe" executable model which shows results for left lane. The next step is to run the same model for the right lane which is shown in Fig. 9.

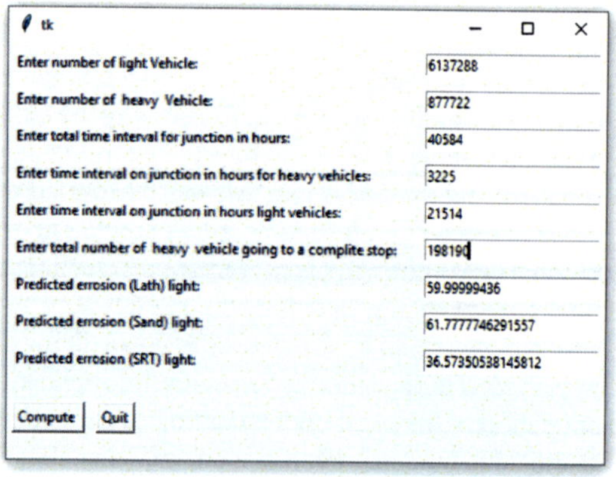

Fig. 8. MLP model prediction based on the right lane data

If the error difference for measured and predicted values from Fig. 7 and Fig. 8 is calculated it can be easily seen that the maximum is 2.8% which is related to the right lane "Sand" measurement.

The real test is performed on the measurements from the different intersection summarized in Table 6.

The first two input parameters in Table 5 are taken from statistical traffic data and the time interval is the same as for the first intersection. The next three parameters in Table 6 are generated by random of the first two parameters because the measurements are done 75 m from the intersection. As the observation point of the first intersection was in the range of 25 m and the observation for the second one is at 75 m the values reading must be much smaller with respect to data generated for first intersection.

Table 6. Dataset generated from the new nearby intersection (with the measurements taken 75 m from the intersection)

Number of LW vehicle	Number of HW vehicle	Time period (h)	Time on junction (h) HW	Time on junction (h) LW	Number of HW going to a full stop	Road erosion (lath)	Road erosion (send)	Road erosion (SRT)
11551382	1144688	40584	4206	5049	40000	16	60	40

The model results summarized in Table 7 are a given as statistical errors calculated on the real range erosion which confirms and opens different possibilities for continuation on the following research.

Table 7. Real data scale errors

	Mean squared error	Root mean squared error	Mean absolute error	Relative absolute error (RAE)	Root relative squared error (RRSE)
Sand	1.575	1.255	0.781	0.082	0.111
Lath	0.304	0.551	0.385	0.052	0.064
SRT	0.273	0.523	0.367	0.036	0.045

In the Table 7 different type of error calculations is shown. Those error values are indicating the MLP model overall accuracy for different civil engineering erosion measurements.

In the next Fig. 9 the prediction based on intersection two are given.

In the following results, the observation of the "Sand" standards error can be seen. But as "Sand" method is very sensitive and the other two standards prediction is in a range a conclusion is that the following model can be used but only for limited purposes as more new intersections data is needed to generate a general model. The "Sand" method sensitivity means that if a measurement is taken some distance before the point of view the new data readings can be observed.

As a conclusion, it is obvious that heavy vehicles do have an impact on all standards like "Lath", "Sand" sand "SRT" in the given research. But it is not linked directly to the number of vehicles as can be seen if the intersection one and two are compared. It is clear that roadway without intersection can last longer as can be seen from the model. So the final conclusion is that retention and braking of vehicles have a major effect on the roadways. But as the following model is generated on a small data sample from only one intersection it cannot be generalized without including more data. Just to point out one more time so that there is no confusion the number of vehicles is an important parameter as the data calculations for the vehicle parameters like retention and braking are computed based on it.

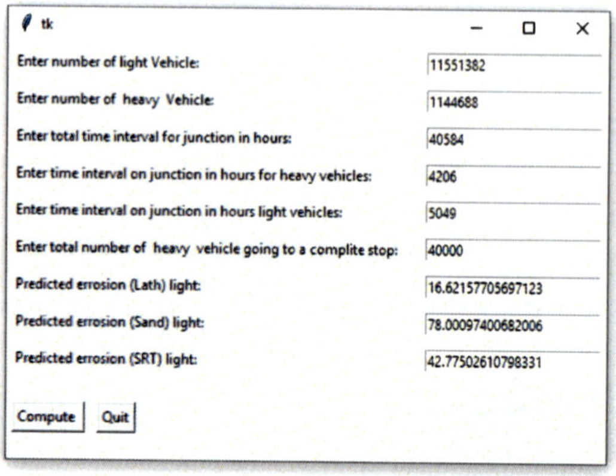

Fig. 9. MLP model prediction based on the new intersection data

An incomplete answers for lessening and improve street conditions lies in the extent of AI. Specifically, as finished up from research power over the traffic flow can radically change the street conditions. So, a significant perspective is to accomplish control of the traffic flow-ability. Various arrangements are as of now out there. By picking the ideal arrangement a decrease in the information factors introduced in this exploration will be created. To be specific, a model is that because of the keen traffic control some of the vehicle setting off to a stand-still will be decreased. So by producing better traffic flow-ability a decrease in vehicle time stretches on the convergence will be watched.

In Fig. 10 and Fig. 11 a mimicked effect of the referenced brilliant framework results is introduced. In Fig. 10 the information scope from the contemplated convergence is utilized. Specifically, to be more explicit it is the correct path last information boundaries are utilized. As of now referenced before the keen framework will improve traffic flow-ability, reenactment in which a savvy framework lessens boundaries by 50% is introduced in Fig. 11.

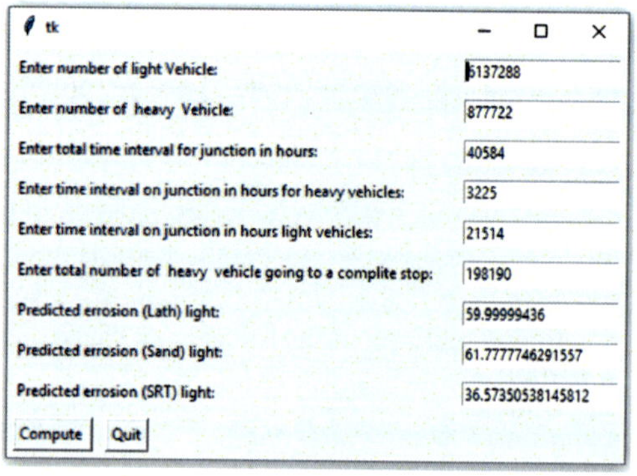

Fig. 10. Road erosion prediction without smart traffic system control

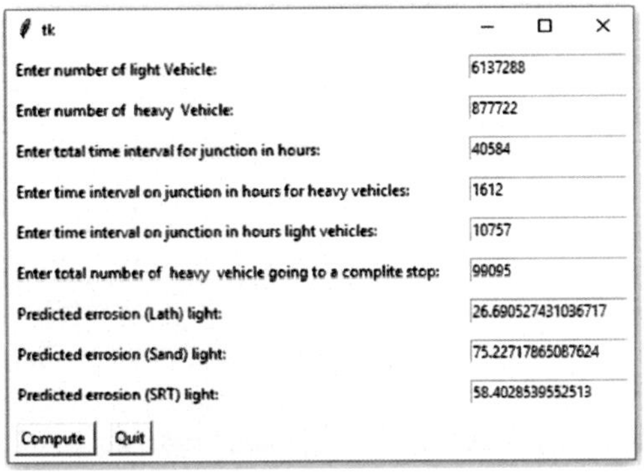

Fig. 11. Road erosion prediction with smart traffic system control

So the final conclusion is that there is a huge impact generated by vehicles on road erosion, but at the same time, it can be seen that number of vehicles is not the greatest factor. Namely, in the last two figures, it can be seen that even if the number of vehicles is kept the same but other parameters like time interval on the intersection and the number of vehicles going to a complete stop are reduced, a decrease in the erosion decay is observed.

But again as the time on the intersection parameter and the other parameters are statistically computed from the number of vehicles and are proportional to it, it can be stated that heavy/light vehicles in combination with the low flow rates intersections are the key factor affecting the road erosion.

As most similar research is performed under the controlled environment a big step for furthering and expanding of presented modeling is opened. It can be seen that research model (MLP) testing on intersection two overall research results are matching which proves and finds correlation behind traffic flow impact on the road erosion. With more data and with different bitumen types a more robust model can be driven. The key benefit is that for the following research a controlled environment is not the key as for the researches before which were limiting research and increasing the cost.

6 Conclusion

The extent research is to collect data identified with the vehicle's movement with Linear SVM model. Data separated is the number of vehicles their speed, type, time spent on the intersection, and data on the vehicles setting off to a stand-still. Research identified and is based on the tree structural scheming techniques to gather various kinds of information on asphalt disintegration. Techniques utilized with their short clarification are "pendulum technique" which estimating the grating coefficient of black-top, "sand technique" which gauges the black-top surface and "Lath technique" which estimates abnormalities or disfigurements in black-top structure. The accompanying informational indexes data sources and yields are utilized to produce a model competent to anticipate road disintegration. The accompanying information boundaries were chosen dependent on their effect on the disintegration boundaries. The final pint is to run the numerous tests on produced models to test its accuracy. Finally, subsequent to choosing the best model an application is made so genuine information tests can be reenacted. The accompanying application is valuable as a reproduction instrument that can help with street safety upgrades. The model benefits, limitations, and the overall impact are given in the paragraphs that follow.

The following model can be a very reliable tool for road erosion control for maximum safety. As mentioned in the result and discussion chapter it is obvious that heavy vehicles do have an impact on all standards like "Lath", "Sand" sand "SRT" in the given research. But it is not linked directly to the number of vehicles as can be seen if the intersection one and two are compared. It is clear that roadway without intersection can last longer. Final conclusion is that retention and braking of vehicles have a major effect on the roadways, but as mentioned before there is a limitation as the following model is based on data from one intersection. According to the presented MLP model, it can be seen that with the increase in the traffic load an increase in the retention and braking is generated but at the same, it is concluded that traffic fluidity is the key factor. Finally, a conclusion based on the left and right lane is that the different numbers of heavy vehicles will generate different retention and braking data for the left and right lane where it can be seen that erosion for the right lane is greater, this is so because there is a greater traffic load and with smaller flow-ability on the right lane.

As the man-made brainpower and processing power is improving, different designing fields will discover the utilization and another examination fields will arise. Models such as self-driving vehicles and savvy street intersections will build information assortment; thus, innovative methods for traffic safety and traffic stream will be presented. As a forthcoming work, a vehicle fuel utilization and climate impact could be acquired for an improvement in street development arranging from video information of street intersections and traffic roundabouts. A partial solution to reduce and improve roadways conditions lies in the scope of AI and image processing. Namely as concluded from research control over the traffic flow can drastically change the road conditions. Important aspect is to achieve control of the traffic flow-ability. Different solutions are already out there. So, by choosing the optimal solution a reduction in the input variables presented in this research will be generated. Namely, an example is that due to the smart traffic control a number of the vehicle going to a complete stop will be reduced. By generating better traffic flowability a reduction in vehicle time intervals on the intersection will be observed.

As the presented research considered one type of bitumen intersections and limited amount of data, it can be used as a guideline for a research on the roundabout, the difference in erosions generated at intersections and roundabouts can be studied. By studying the following differences an overall impact on the erosion can be studied and related to the traffic flow.

References

1. Sandberg, U.: Influence of road surface texture on traffic characteristics related to environment, economy and safety: a state-of-the-art study regarding measures and measuring methods. Statens väg-och transportforskningsinstitut, VTI notat 53A-1997 (1998)

2. Wang, F., Machemehl, R.B., Popova, E.: Toward Monte Carlo simulation-based mechanistic-empirical prediction of asphalt pavement performance. J. Transp. Eng. **136**(7), 678–688 (2010)
3. Cebon, D.: Theoretical road damage due to dynamic tyre forces of heavy vehicles Part 2: simulated damage caused by a tandem-axle vehicle. Proc. Inst. Mech. Eng. C J. Mech. Eng. Sci. **202**(2), 109–117 (1988)
4. Sezen, H., Fisco, N.: Evaluation and comparison of surface macrotexture and friction measurement methods. J. Civ. Eng. Manag. **19**(3), 387–399 (2013)
5. Wallman, C.G., Åström, H.: Friction measurement methods and the correlation between road friction and traffic safety: a literature review (2001)
6. Benson, C., Twigg, J., Rossetto, T.: Tools for mainstreaming disaster risk reduction: guidance notes for development organisations. ProVention Consortium (2007)
7. Button, J.W., Fernando, E.G., Middleton, D.R.: Synthesis of pavement issues related to high-speed (2004)
8. Softic, E.: Odžavanje puteva i saobraćajnica. Tehnički fakultet, Bihać (2018)
9. Carmon, N., Ben-Dor, E.: Mapping asphaltic roads' skid resistance using imaging spectroscopy. Remote Sens. **10**(3), 430 (2018)
10. Zimmer, R.A., Choubane, B., Holzschuher, C.R.: Friction testing method for open-grated steel bridge decks. Transp. Res. Rec. **1860**(1), 137–143 (2003)
11. Ergun, M., Iyinam, S., Iyinam, A.F.: Prediction of road surface friction coefficient using only macro-and microtexture measurements. J. Transp. Eng. **131**(4), 311–319 (2005)
12. Kotek, P., Kováč, M.: Comparison of valuation of skid resistance of pavements by two device with standard methods. Procedia Eng. **111**, 436–443 (2015)
13. Smith, D.E.: A Primer on Modern Pavement Surface Friction Measurement, Concept highway maintenance vehicle final report: phase one and two, Center for Transportation Research and Education, Iowa State University Research Park, Ontario (1997)
14. Gupta, B., Chaube, A., Negi, A., Goel, U.: Study on object detection using open CV-Python. Int. J. Comput. Appl. **162**(8), 17–21 (2017)
15. Kim, J, Kim, B.S., Savarese, S.: Comparing image classification methods: K-nearest-neighbor and support-vector-machines. In: Proceedings of the 6th WSEAS international conference on Computer Engineering and Applications, and Proceedings of the 2012 American conference on Applied Mathematics, vol. 1001, pp. 48109–2122 (2012)
16. Thi, T.H., Robert, K., Lu, S., Zhang, J.: Vehicle classification at nighttime using eigenspaces and support vector machine. In: 2008 Congress on Image and Signal Processing, vol. 2, pp. 422–426. IEEE, May 2008
17. Okabe, T., Sato, Y.: Object recognition based on photometric alignment using RANSAC. In: 2003 IEEE Computer Society Conference on Computer Vision and Pattern Recognition, 2003. Proceedings, vol. 1, pp. I-I. IEEE, June 2003
18. Chern, M.Y., Hou, P.C.: The lane recognition and vehicle detection at night for a camera-assisted car on highway. In: 2003 IEEE International Conference on Robotics and Automation (Cat. No. 03CH37422), vol. 2, pp. 2110–2115. IEEE, September 2003
19. Wang, C.C., Huang, S.S., Fu, L.C.: Driver assistance system for lane detection and vehicle recognition with night vision. In: 2005 IEEE/RSJ International Conference on Intelligent Robots and Systems, pp. 3530–3535. IEEE, August 2005
20. Chan, Y.M., Huang, S.S., Fu, L.C., Hsiao, P.Y., Lo, M.F.: Vehicle detection and tracking under various lighting conditions using a particle filter. IET Intell. Transp. Syst. **6**(1), 1–8 (2012)
21. Sun, Z., Bebis, G., Miller, R.: On-road vehicle detection using Gabor filters and support vector machines. In: 2002 14th International Conference on Digital Signal Processing Proceedings. DSP 2002 (Cat. No. 02TH8628), vol. 2, pp. 1019–1022. IEEE, July 2002
22. Pedregosa, F., et al.: Scikit-learn: machine learning in Python. J. Mach. Learn. Res. **12**, 2825–2830 (2011)

23. Vrtagić, S., et al.: Video data extraction and processing for investigation of vehicles' impact on the asphalt deformation through the prism of computational algorithms. Traitement du Signal **37**(6) (2020)
24. Smola, A.J., Schölkopf, B.: A tutorial on support vector regression. Stat. Comput. **14**(3), 199–222 (2004)
25. Cristianini, N., Shawe-Taylor, J.: An Introduction to Support Vector Machines and Other Kernel-Based Learning Methods. Cambridge University Press (2000)
26. Vapnik, V., Vapnik, V.: Statistical learning theory Wiley. N. Y. **1**(624), 2 (1998)

Modern ABI Platforms for Healthcare Data Processing

Hala Shaari[1] (✉), Nermina Durmić[1], and Nuredin Ahmed[2]

[1] Department of Information Technologies, Faculty of Engineering and Natural Sciences, International BURCH University, Sarajevo, Bosnia and Herzegovina
hala.shaari@stu.ibu.edu.ba

[2] Control Department -Technical Computer College, National Board for Technical and Vocational Education, Tripoli, Libya

Abstract. The cost of healthcare has risen significantly and demand for high quality care grows rapidly in our society. At the same time, hundreds of tables are gathered daily with patient-related data. Therefore, the modern analytics and business intelligence (ABI) platforms play a crucial role in this endeavor to improve performance while reducing costs in healthcare organizations. Utilizing the appropriate modern analytics and business intelligence (ABI) platform is a challenging task due to the large number of tech vendors providing the BI and analytics applications landscape. The goals of this study are to identify and compare the leader ABI platforms for healthcare organizations. Firstly, to assist managers with better platform selection to improve the quality of services in the healthcare domain. Secondly, to expand IT specialists' knowledge with deep technical details. These goals have been achieved by technically analyzing the leading six modern analytics and business intelligence (ABI) platforms in the healthcare domain, based on the suggested criterion to measure their applicability, general characteristics, and standard requirements.

Keywords: Modern ABI · Healthcare organization · Business intelligence · ETL tools · Business analytics

1 Introduction

Today's specialists have access to vast quantities of data in the healthcare world in the form of personnel documents, electronic patient records, clinical results, imaging procedures, diagnostics, prescription medications, and mobile medical health. For administrators and healthcare practitioners, it is also a task to manage and evaluate the data to better grasp it and use it to make well-informed decisions. Applications for data analytics, also known as business analytics or analysis software, need to turn this rapidly grown data into usable knowledge to enhance the decision-making process. The science of research, using data to make decisions, is analytics [1]. Analytics include the use of data, analysis, and modeling to resolve issues or to explore further possibilities.

N. Ademović et al. (Eds.): IAT 2021, LNNS 316, pp. 187–202, 2022.
https://doi.org/10.1007/978-3-030-90055-7_14

Generally, data analysis has come under the purview of a wider definition, called business intelligence, or BI. In 1989, business intelligence was described as "the concepts and methods which improve business decision-making by using fact-based support systems" [2]. BI is a conceptual framework for supporting decisions that incorporates system design, analytical methods, databases, data storage facilities, and applications software [1]. It is one of the disciplines of decision support systems, suggesting a range of decision-making developments designed to gather, integrate, delay, analyze, and view information on association exercises and procedures in order to assist in an increasingly informed and therefore better decision [2]. The decision-making process can be described in four phases: intelligence, design, choice, and implementation [3, 4]. The Intelligence Stage is the observation of fact and the acquisition of information by data analysis. The design process consists primarily of the creation of decision-making requirements and alternatives for decision-making. During the choice phase, the decision-makers analyze the different options for decision-making and propose actions that best meet the requirements for decision-making. Finally, in the implementation phase, decision-makers should create an implementation plan, secure the necessary resources, and execute the plan. However, the evaluations of decisions need to be reconsidered as they assess the implications of the recommendations [4]. In fact, there are various types of decision-making activities that involve different types of analytics, primarily these types are four types: descriptive, diagnostic, predictive, and prescriptive [5].

BI is an established term applicable in many areas, including healthcare. Transformation initiatives in healthcare allow decision-makers to use the knowledge to consider all aspects of the success of an organization. Healthcare should be seen as a decision-making process from an unbiased perspective. A nurse or a doctor gathers data about a patient, examines an MRI/CT image, then assesses the condition, and decides on the next move towards healing. In addition, a Manager of the hospital medical unit gathers data on the number of hospital patients, the number of beds available, the occupancy of the last year in the unit and the predicted seasonal influenza trends to predict personnel needs and to make certain management decisions. A radiologist also has access to a digital image (e.g., x-ray, magnetic resonance imaging (MRI), ultrasound, and computed tomography (CT)) to detect and record the occurrence of a disease.

Healthcare organizations must take action to measure, monitor, and report the quality, efficiency, and value of care. To reach this goal, an appropriate modern analytics and business intelligence (ABI) tool must be used by these organizations to improve quality, cost-effectiveness, patient outcomes, and safety performance. Modern analytics and business intelligence (ABI) platforms enable to extract knowledge from a vast volume of data that enables managerial decisions to be made not just based on intuition, but also on facts. Adapted to the needs of the healthcare organization, this knowledge extraction may be accomplished by choosing the appropriate tool. Many of these tools are relatively expensive, while others may be too difficult to use. However, the most critical aspect of starting to define business requirements is the selection of the right ABI platform.

To assist these healthcare organizations, this study's main goals are to identify and compare the leader ABI platform for healthcare organization by encouraging the analysis work of managers and decision-makers who intend to deploy the Business Intelligence (BI) platform in their health system and are unaware of the most suitable tool to their business process. Moreover, the goal is to expand IT specialist's knowledge with deep technical details to choose an appropriate ABI platform. To achieve this goal, a comprehensive analysis of the literature has been conducted using scientific and business data sources. The most used BI platforms have been identified through this phase. Thereafter, the top six platforms used in healthcare domains (BORD, Looker, Microsoft Power BI, MicroStrategy, Oracle, and Qlik) were selected to be compared.

In addition to this introduction, this paper includes four sections. The following background section discusses the value of decision-making in healthcare, the main components of healthcare analytics, and the need to improve the efficiency of healthcare organizations. The third section provides an approach to the concept of business intelligence in which the ETL component and a brief overview of popular ETL tools are presented. The fourth section concerns the implementation of new analytics and business intelligence (ABI) platforms in the healthcare sector. Finally, the paper ends with a discussion and conclusion section which highlights current achievements of ABI platforms, and the limitation of the study and future research direction section.

2 Background

This section points out essential aspects of the healthcare system and the need for data analytics and business intelligence in healthcare, describing the value of decision-making of healthcare components and the need to increase the efficiency of healthcare organizations.

Analytics allows one to make decisions by either explaining what has happened in the past, anticipating what could happen in the future, or even suggesting what course of action can be taken to reach a certain goal. Healthcare BI tools and Patient Centric Data Analytics together offer suppliers of health organizations more influential pieces of information for understanding enhancement, cost reduction, and making better decisions [6]. One of the early Healthcare BI definitions given by Madsen [7]: "The integration of data from clinical systems, financial systems, and other disparate data sources into a data warehouse that requires a set of validated data to address the concepts of clinical quality, effectiveness of care, and value for business usage". It also introduced the five tenets of Healthcare BI in accordance with their concept of BI by considering the main aspects of efficiency, effectiveness, and value.

Throughout healthcare, data analysis aims to improve the quality and well-being of patient care, while at the same time reducing costs. Patient care is a human-driven effort; thus, healthcare analytics need stakeholder feedback to determine what is useful and what is required. For better impact, the performance that healthcare analytics provide must be used by managers, quality management teams, and other decision-makers. There are several levels and components in an analytics system between the initial input and the resulting output that make evidence-based decision making possible. Forrester Research, Inc., identified the "business intelligence (BI) stack" [8]. It offers a very comprehensive overview of the components required to construct a BI system for any business enterprise. Nevertheless, Trevor Strome has updated a stack optimized for healthcare research, which focuses on finding market issues and creating knowledge [9]. The fundamental layers of this performance and the quality enhancement analytics program are: data, analytics, business context, quality and performance management, and presentation.

The business context layer is the basis of an analytical framework and it reflects the quality and efficiency priorities and goals of healthcare organizations. Each organization should have its own set of priorities and targets due to specific circumstances, demographics, and other factors. The business's priorities and objectives, and the tactics that healthcare organizations employ to achieve them, drive demands at every other level. The analytics stack data layer represents the consistency, management, integration, and storage of the data and the related infrastructure. The analytics layer consists of the tools and techniques used by analytics teams to generate information and actionable insight-driven decision-making. The quality and performance layer aligns analytics with the processes that need to be improved, the metrics on which processes and results are measured, and the performance goals the healthcare organizations want. Finally, the presentation layer handles the form in which the decision-makers obtain insights and information.

The key elements of healthcare improvement are improving safety, quality, and value. Quality must be quantified to enable measurement, monitoring, analysis, and most critically, decision-making and action. For that end, managers need to set metrics [10] and indicators [11]. Metrics are quantitative instruments used to assess the efficiency or performance of healthcare on a specific scale [9]. The indicator lets managers define the actual output status and how far it is from the target. Metrics alone though, are not enough; we need to connect the metric to the objective so that we can determine if a certain desired goal has been achieved. Metrics related to such targets are referred to as indicators; indicators are measures of progress or accomplishment [12]. An indicator can also be used to measure the level of service and the efficacy of the hospital. Indicators may be assembled on a computer using various types of analysis methods, such as mathematical tables, diagrams, colors or numbers. Such indicators are displayed in an easy-to-use and easy-to-understand manner, called a dashboard.

The healthcare sector includes many types of functional units, ranging from large facilities such as hospitals to medical practices. The majority of healthcare facilities themselves contain several divisions, units, systems, resources, and administrative roles within a single facility. As a result of this diversity of facilities and resources, it is important to take into account the various methods that a healthcare organization must conduct in order to determine quality and how those quality assessments are to be implemented. High-performance healthcare organizations are highly unlikely to achieve high quality in all facets of their services. Therefore, quality and success in healthcare must also be viewed from a variety of perspectives; one way is to see healthcare from the viewpoint of structure, process, and outcomes [13]. These three components (structure, process, and outcomes) form a continuum for quality assessment. Yet, "process measures must be linked to outcomes if they are to be effective measures of quality." [14]. The outcomes of patients rely on the processes and workflows undertaken, how effective and efficient they are, and how well they are administered. The structure of a healthcare organization would determine the capacity of the business to create, execute, and manage the most reliable and effective processes and workflows. All these types of assessments within an organization are necessary for a full efficient quality evaluation. Some effective healthcare organizations, if not most, use an established improvement or management methodology. Many of these approaches are currently used in healthcare, in particular Lean, Six Sigma, the Plan-Do-Study-Act (PDSA), total quality management, constraint management, etc. While these structures differ in their concepts, techniques and procedures, they are similar in that they provide a systemic approach to improve productivity and performance within a complex organization.

3 BI Architecture and Technologies

The key objective of BI is to allow interactive access to data and models, to facilitate data manipulation, and to provide administrators, analysts and clinicians with the ability to carry out the right analysis to fulfill their needs. The BI architecture has four major components: data warehouse, business analytics, business performance management (BPM), and user interface. Data warehouse is a type of database containing data sources, such as patient medical records. The second main component is data analytics, which is a set of tools, techniques, and processes used to manage, retrieve, and interpret data contained in data storage. Business performance management (BPM) is another main component, which includes the tools used to monitor, measure, analyze, and manage business performance and commercial success. Finally, the user interface that provides a comprehensive graphical summary of organizational performance indicators, patterns, and exceptions.

The full BI perspective does not just expand the features through standard reporting, but also with two components needed for any effective BI solution: a data warehouse to store data for analysis, and ETL (Extraction-Transformation-Loading) tools and processes to make data usable and pre-fed into the data warehouse. Data warehouses are a shared archive of existing and historical data from one or more separate sources. Extraction-Transformation-Loading (ETL) is the core process of data warehouses. ETL is a process used to retrieve data from multiple sources, clean it up, and translate it into the desired state and load it into a target database.

3.1 ETL Process

ETL solutions provide a basis for the standardization and incorporation of heterogeneous data in a local data store. It can be a useful tool in creating and maintaining reliable and effective databases and data warehouses [15]. In addition, ETL solutions enable the fast, straightforward, and reproducible loading/reloading of datasets, and can thus be used in systems requiring continuous data updates [16]. ETL essentially consists of the following process:

Extraction: The main aim of the extract process is to retrieve all the data from the source system with as little resources as possible. These sources may have different data formats (different internal representations) which make the extraction a difficult process. The decision on how to extract your data will be influenced by several factors, such as the source structure, business needs, and the expected volume of data to be collected. Depending on these factors, you must determine how to logically and physically extract the data. There are three types of logical extraction: complete extract, gradual extract, and update notice, whereas there are two types of physical extraction: online and offline extraction [17].

Transformation: Transformation is the act of transferring data from one system to another. The transformation process ensures data integrity and data cleaning before it is loaded into the data warehouse. Transformation may be combined with an extraction or loading process depending on when it is performed. Data transformations are also the most complex and the most expensive aspect of the ETL process in terms of processing time. They range from basic data transformations to highly complex techniques for data scrubbing [17].

Loading: The final ETL process step is to load data into the dimensional structures of a data mart or data warehouse accessible by end-users and application systems. The loading process consists of both: tables for dimensions and tables of facts [18], that are used for different analytical purposes.

3.2 ETL Tools

There are various ETL resources available on the market to promote the selection of an organization predominantly based on its needs and specifications. This variety starts from powerhouses at the enterprise level to simple suites for the open source integration. These tools have matured over time and are now able to do more than just extracting, transforming, and loading data. Enhancements include capability in conjunction with data profiling, high-quality data management, monitoring and cleaning, real-time and on-demand statistical integration into provider-oriented framework and metadata access [19]. In addition, these ETL tools can now be customized according to an enterprise data warehouse's functional necessity.

In today's market, a variety of commercial/open source ETLs, from corporate level powerhouses to basic open-source integration suites, are used for data integration and transformation. Several are commercial, licensed tools and a few are free, open source tools. Choosing the proper ETL tool is an essential part in achieving the objectives. ETL tools may be classified into different categories, depending on their features [20] such as: ease of use, visual flow, operational resilience, advanced data profiling and cleansing, performance, etc. Another way to categorize these tools is a list provided by [21] which is organized into four categories based on business needs. These categories are: 1) Current batch ETL tools: the majority of ETL tools worldwide is based on batch processing such as, Microsoft SSIS, IBM InfoSphere DataStage, and Oracle Data Integrator. 2) Cloud based ETL tools: more and more cloud based ETL platforms have recently begun to appear. Many of them retained the same basic batch model, while others began offering real-time support and/or intelligent schema detection, such as: Alooma and Stitch Data. 3) Open source ETL tools: some of these tools have been proposed as a new layer of management for planned workflows and batch processes such as Apache Airflow, Apache Kafka, and Apache NiFi. 4) Real-time ETL tools: The transition to real-time demand has brought about a radical shift in architecture: from a batch-based approach to a distributed message queues and stream processing model. Popular Real-time ETL tools include Alooma and Confluent.

4 Adoption of BI Solutions in the Healthcare Sector

4.1 Business Model of Modern Healthcare Organizations

A business model is essentially a tale of how and why businesses excel or fail [22]. Business models play a critical role within organizations as they help owners and investors build and create value through activities. Although many early studies have not been focused on the healthcare domain, there have been some. Authors in [23, 24] particularly discussed "value-based healthcare" which suggests that healthcare organizations should be paid for the value that they generate and not the activity that they perform. Also, authors in [23] clearly investigated to what extent health care has to be rearranged into "Strategic Business Units". Additionally, [25] presented "Integrated Practice Units" to recognize how healthcare should function in comparison with other industries.

One of the widely accepted formulations for business models stems from Christensen's work [26]. Basically, it proposed a benefit formula that defines the processes used, then selects and implements the required resources and then iteration. Another commonly discussed method used to explain the essential factors of the business model in generating and collecting value by companies is the "Canvas Business Model" [27, 28]. It is a methodology for the development of a business model of nine structural blocks such as, key partners, key resources, key activities, value proposition, etc. [27]. The model's benefit is a simple segmentation of all the main parameters representing the rationale of the organizational activities and aiming to make a profit. Furthermore, this model has tool support, wide practical implementations, and it is used in a significant number of productive companies and corporations.

Nonetheless, the ecosystem perspective is a growing theme in technology-oriented business model research that emphasizes the complexity of the emerging technology's market environment [29, 30]. In [29] four main areas of importance for digital ecosystem business models have been identified: drivers, nodes, exchanges, and extraction dynamics. However, the core concepts of such creative thought for the creation of modern business models are the emerging existence of goods that can anticipate consumer demands, as well as the anticipated reduction in "one-and-done" product assumptions, and, ultimately, a larger area for product/service customization [29].

The success of the business model is informed by the willingness of the company to adjust its business strategies to environmental developments in terms of consumers, technology, competition, and strategic management [31]. According to experts, the concept of managing a modern medical organization is focused on digital procedures assisted by modern IT technology developed to enhance existing processes and implement new patient care capabilities [32]. In addition, implementing management solutions for a healthcare organization is one of the focus fields for major organizations providing IT solutions. The next section introduces the current management level trend in healthcare organizations.

4.2 Current Trends in Healthcare Organizations

Healthcare organization's managers must understand the development strategy of the company and have a well-developed business model, which allows modern technological solutions to be introduced. Effective management models improve consumer satisfaction and benefits and help build a sustainable cost environment when selecting the best choice to simplify and digitize the operations of a healthcare organization.

Healthcare management systems are influenced by recent medical concepts, on the one hand, and the technology that enables medical concepts to be realized, on the other hand. One significant concept is the 4P medicine concept which consists of four components: predictive medicine, preventative medicine, participatory medicine, and personalized medicine [33, 34]. P4 is the product of two convergences: medical systems and digital revolution (IoT, Big Data, Blockchain). The digital revolution has contributed to P4 medicine in three ways: the ability to manage big data sets, social and business networking, and the development of digital personal devices [35]. Another important concept is that value-based medicine requires special medical practice, offering the best level of evidence-based medicine analysis along with the assessment of care effectiveness by the quality of life of the individual receiving the service [36]. Such new medical management systems require major improvements in the organization's operations using digital technologies. In addition, medical organization strategies, business models, business process structures, IT infrastructure, services architecture, and data architecture must be built in order to make timely effective decisions. The medical trends considered can only be implemented with success on the basis of modern digital technologies [35]. Health 4.0's conceptual health (based on Business 4.0) intention is to ensure the introduction of value-based and customized medicine concepts through the massive use of emerging IT technologies such as big data, machine learning, cloud computing, the internet of things, mobile networks, and blockchain technology. However, the efficiency of a medical organization must be measured using appropriate tools. Currently, BI systems are the main tools used to monitor the complexities of business performance measures.

4.3 Selected ABI Healthcare Platforms

Today, a large number of tech vendors span the BI and analytics applications landscape. Some of these technology suppliers are well-known computing firms such as IBM, Microsoft, and Oracle, some of which are important contributors to statistics such as SAS, and others are smaller and more specialist providers such as Tableau and Qlik. The most common used commercial and freeware (open source) BI platforms for the healthcare sector, according to [37–40], are listed below with some functionality details:

1. BOARD: It is an "end-to-end decision-making platform" that offers a comprehensive set of features within a unified, user-friendly environment. BOARD released version 11 of its software in May 2019, based on a re-engineered in-memory calculation engine that replaces its long-standing multidimensional online analytical processing (MOLAP) solution. In addition, it introduces powerful cloud analytics and self-service data modeling. BOARD offers outstanding planning tools for healthcare organizations. These tools can help to forecast potential achievements, schedule needs, and financial patterns. Healthcare professionals can make forecast adjustments and observe the possible effect of these changes on the anticipated conditions [41].

2. Looker: It is a modern analytics and business intelligence (ABI) dashboard functionality application which reports using an agile, unified data model and a database architecture designed for different cloud databases. Looker's product improvements in 2019 include integration with Slack, a revised dashboard interface and content organization structure, as well as automated-model generation capabilities which turn SQL scripts into Looker data models. Looker has also strengthened its development tools and launched a new developer portal, an API sandbox and a marketplace. Specially for healthcare, Domo offers real-time dashboards to help track some of the most sophisticated analyses. Some of the metrics you can track in these dashboards include quality of care, cost per outcome, and efficiency of new treatments. Some of the analyses that Looker also focuses on include comparing reimbursement for different specialties, analyzing average user spending for different locations, and analyzing supplier spending [42].
3. MS Power BI: In Power BI, Microsoft provides data planning, visual data exploration, interactive dashboards, and augmented analytics. It is available as a SaaS option which runs in Azure cloud or as an on-premises option in Power BI Report Server. Power BI Pro cloud service has outstripped other rivals in terms of functionality [37]. It has outstripped others by having creative and enhanced analytics capabilities and automated ML. Inside Power BI, AI-powered services such as text, sentiment, and image analyses are available and rely on Azure's capabilities. MS Power BI supports the healthcare domain with Agile Analytics Healthcare dashboards which are based on Microsoft Power BI and the latest cloud technology. It allows clinics, hospitals, doctors, departments, and managers to easily and confidently take data-driven business decisions while saving significant time and resources [43].
4. MicroStrategy: MicroStrategy provides one of the most robust ABI systems, meeting the analytical and reporting criteria for both Mode 1 and Mode 2. Mobile, cloud, embedded, and identity analytics products complementary supply the core analytical product family for data access, data visualization, and advanced analytics. MicroStrategy applications provide hospitals a deep insight into all aspects of their operations, enabling efficiency and productivity improvements that lead to improved profitability while also delivering a better patient experience. The powerful analytical tools and mobile capabilities of MicroStrategy enable healthcare providers to employ new care models that leverage deeper insights into medical histories and risk profiles of patients and provide more interactive care through mobile devices [44].
5. Oracle Analytics Cloud (OAC): Oracle Analytics Cloud (OAC) offers an integrated design experience with interactive analyzes, reports, and dashboards. It provides an end-to-end cloud approach, including storage, data processing, analytics applications with cloud data centers in nearly all regions of the world. It also continues to develop its groundbreaking augmented analysis and integration of NLPs with OAC. To optimize results, Oracle Healthcare cloud technologies serve the diverse needs of the healthcare payer and provider organizations and their patients. Oracle empowers healthcare organizations with the most extensive business solutions, as well as integrated cloud applications for ERP, finance, HCM, supply chain, and EPM [45].
6. Qlik Sense: Qlik is known as one of the leading ABI solutions in today's market. Qlik Sense works on the powerful Qlik Associative Engine, which has powered Qlik services over the last 20 years. The engine enables users to combine data and

explore information with no query-based tool limitations. The cognitive engine of Qlik applies AI/ML features to the product and interacts with the Associative Engine to provide context-aware recommendations and augmentation of analysis to improve interpretation. Qlik helps healthcare organizations to evaluate clinical, financial, and operational data through visual analysis. This uncovers insights that contribute to better treatment, lower costs, and higher value for patients. Qlik provides numerous healthcare-specific applications to track critical performance metrics [46].

4.4 Applying the Adopted Criteria on the Selected ABI Platforms

A generic criterion based on Gartner's and G2 Crowd Grid criteria was developed to compare the selected ABI platforms. Gartner is a consulting company that publishes its "Magic Quadrant for Analytics and Business Intelligence Platforms" annually. This Magic Quadrant concentrates on products that follow Gartner's latest analytical and BI platform criteria. These criteria are grouped into five cases of usage and fifteen critical analytical and BI platform capabilities. In addition, the fifteen critical capabilities are sub-categorized into five main capabilities: Infrastructure, Data Management, Analysis and Content Creation, Sharing of Findings and Overall Platform [37]. Whereas G2 is one of the biggest innovation markets in the world where companies can find, review, and handle the technologies they need to meet their full potential. The G2 Crowd Grid Report is also released annually by G2, which compares the leading BI and Analytics software resources and technologies based on adoption, ROI, and consumer feedback [38]. In this study, only the data management criteria provided by Gartner, as well as G2 criteria are applied to the selected ABI platforms to explore how well these tools support these critical capabilities. These criteria are categorized into:

- Metadata management: Tools that allow a shared semantic and metadata model to be leveraged by users. These tools can provide a robust and cohesive way for administrators to find, collect, archive, reuse, and publish metadata objects (for example, dimensions, hierarchies, measurements, performance metrics, key performance indicators, etc.).
- Basic ETL functions: A system functionality for displaying, integrating, translating, and loading data into a self-contained engine, with data indexing ability, data load management, and refreshing scheduling features.
- Cloud-based: It is essential that ABI platforms are able to integrate with cloud-based systems. Cloud-based systems have many advantages including seamless updates and allowing users to easily collaborate across departments.
- Data preparation: User-driven data combinations of different sources and analytical models as a "Drag and drop" option, such as user-defined measurements, sets, and groups.
- Advanced capabilities: These include semantic auto discovery enabled by machine learning, intelligent joins, intelligent profiling, generation of hierarchy, data lineage, and data blending on various data sources.
- Scalability and data model complexity: The ability to which high data volumes, complex data models, performance optimization, and large user deployments are managed by the in-memory engine or in-database architecture.

- Self-service: It helps users develop their own data analysis model without the involvement of IT or developers, allowing the analyst to have a better knowledge of their company using ABI without having any technical skills.
- Archive data: It is essential that ABI tools backup data in the event that it needs to be restored. The data can also be easily stockpiled for future analysis. This is helpful for users that work with large amounts of data, as it can be easy to lose track of important data.
- Record integration data: ABI tools must record integration data for every transaction. In the event that an audit needs to take place, it is easier for users to retrieve the relevant data. Additionally, it helps to identify when an issue was introduced to a system.

Modern ABI platforms are recognized to be efficient features that support a full analytical workflow, from data preparation to visual exploration and insight generation, with a focus on self-service and augmentation. The criteria for incorporating the ABI platforms in healthcare organizations are primarily to provide knowledge to support the decision-making process at a strategic level, with some operational ramifications. To promote the analysis work of managers and decision makers who are preparing to deploy Business Intelligence (BI) platforms in their organization and are not aware of the most appropriate tool for their business operations, Table 1 connects the criteria alluded to the specifications of business intelligence platforms in the healthcare context with the previously selected ABI platforms. Additionally, it expands IT specialists' knowledge with deep technical details to choose an appropriate ABI platform.

Table 1. Modern ABI platforms comparative table

BI vendor	BOARD international	Looker	Microsoft	MicroStrategy	Oracle	Qlik
Platform	BOARD 11	Looker 7	MS Power BI	MicroStrategy Platform Analytic	Oracle Analytics Cloud (OAC)	Sense
Criteria						
Metadata Management	Unified metadata in BOARD Engine	i_looker database	Common Data Model (CDM)	Metadata repository	Essbase Classic User Interface	In-memory data model (Master Library)
Basic ETL Functions	In Data warehouse + BOARD DUAL Engine (Data Reader component)	In Data warehouse	Dataflows + Power Query	In Data warehouse	Data Visualization Cloud Service (DVCS)	Qlik Sense Engine (QIX)
Cloud-Based	BOARD server (Engine)	Browser-based experience with Looker APIs	Power BI online service + cloud-based Azure service	Intelligence Server	Data Visualization Cloud Service (DVCS)	Qlik Sense Proxy + Qlik Sense Schedule (QSS)

(continued)

Table 1. (*continued*)

BI vendor	BOARD international	Looker	Microsoft	MicroStrategy	Oracle	Qlik
Data Preparation	BOARD DUAL Engine	LookML Model	Simple interface/navigation	Intelligent Cubes	OAC's self-service data prep	Qlik Sense Engine (QIX)
Advanced Capabilities (ML + NLP)	Board Enterprise Analytics Modeling (BEAM)	LookML Model	Azure Data Lake Storage Gen2	HyperIntelligence (HyperCard)	Oracle Day by Day + OBIEE	Qlik Sense Engine (QIX) + Qlik's + Cognitive engine
Scalability and Data Model Complexity	In-memory clustering technology (HBMP)	Looker Persistent Derived Tables (PDTs)	Dataflows + Model-driven calculation engine	Intelligence Server (Clustering Multiple MicroStrategy Servers) + in-memory cube technology	Executes Apache Spark workloads to process data at scale	Qlik Associative Big Data Index engine (ABDI Engine)
Self-Service	Self-Service Analysis (PinBOARD)		Power Query		Data Visualization + Smart SaaS Integration	Qlik Sense Engine (QIX)
Archive Data	Unified Metadata/Data Federation	i__looker database	Historical Data Preservation	Metadata Repository	Oracle Cloud Infrastructure Object Storage + Catalog Archive	Qlik Sense Repository (QSR)
Record integration data	HBMP Dual Engine + Stored Cubes + Virtual Cubes	Action hub server + i__looker database	Azure Backups integration	Natively integrating the connectors to its metadata engine + REST APIs to open that metadata engine via its semantic graph	Oracle Analytics Cloud Integration	Qlik Replicate

5 Discussion and Conclusion

The goal of this study was to analyze six leading modern analytics and business intelligence (ABI) platforms in the healthcare domain to assist managers and decision makers, who intend to implement a Business Intelligence (BI) platform in their healthcare organization and are unaware which tool is the most fitting tool for their business process. Many modern analytics tools are released each year, often offering similar features and functionality. The deployment of modern analytics and business intelligence (ABI) systems within healthcare organizations is seen as daunting and difficult as they go beyond simple software and hardware implementations and are more complex to install and run. Successful use of the modern ABI is more than just implementing and using technology and resolving data quality problems, but also the technological implementation challenges. There are also managerial and organizational challenges. Modern ABI needs to be viewed holistically and strategically within organizations and needs to be invested in and assisted by managers.

Today, modern analytics and business intelligence (ABI) platforms differ significantly from traditional BI platforms by not needing substantial intervention by IT experts to identify data models or store data in traditional data warehouses, but rather by relying on self-service and agility. A healthcare organization would then be able to use its own independent data warehouse(s) to adapt new BI platforms such as BOARD, Looker, and MicroStrategy platforms to its business process.

In addition, the emerging use of data lakes and logical data warehouses combines with the capabilities of modern BI platforms which can ingest data from these less modeled data sources. In this context, most modern BI platforms have their own self-contained in-memory engine to ensure fast performance and to support rapid prototyping, such as BOARD, Qlik Sense, and MS Power BI. Others however can optionally use existing modeled data sources such as Looker.

Currently, AI healthcare services fall into two groups. The first group uses machine learning methods to evaluate organized data in order to infer the probability of disease outcomes for instance. The second group uses NLP to process unstructured data in order to provide new insights that can relate to structured data. As a result of the use of these applications to promote high-quality and collaborative professional decision-making in the healthcare sector, doctors and radiologists are able to make more informed decisions, benefit patients, and increase health outcomes. This has led to a new wave of change to conventional BI, evolving as augmented analytics, with insights into exponentially large volumes of data produced by machine learning techniques. Natural language processing (NLP) is also used in augmented analytics as a means of querying data and generating scenarios to explain graphics for instance. All selected modern ABI platforms are augmented analytics in different ways and degrees. According to [37], modern analytics and business intelligence acquisitions, data analysis and machine learning systems, and embedded analytics will be dominated by augmented analytics soon. Nevertheless, 50% of analytical queries are either generated by search, natural language processing or speech, or are generated automatically. Augmented analytics technology, on the other side, will be a commonplace by 2022, but only 10% of analysts will harness its full potential [47].

Finally, modern ABI saves time for specialists and doctors who are immersed in many persistent sources of information, which is a huge task for them to concentrate on. Modern ABI allows time-lashed specialists and doctors to track information efficiently on a continuous basis and thereby empowers them to devote extra time for better patient consideration. Healthcare ABI platforms offer health organizations more influential knowledge allowing them to enhance awareness, cost advancement, and outcomes.

6 Limitations and Future Research Direction

Researchers have proposed many ways to categorize modern ABI platforms over the last ten years, among the various styles of modern ABI platforms. However, not all of the methods in the classification can be implicitly linked to a specific individual. According to [48], modern BI systems can be classified as follows: 1) capable of data transformation, analysis, and report generation, 2) capable of binding in key data from various perspectives into real insights and forecasts for effective business results, and 3) capable of binding in key data from various perspectives into real insights and forecasts for successful business performance, 4) on the basis of skills that aid in policy development, strategy, and integration capabilities, and 5) on the basis of considerations such as real-time data collection, cost-cutting, time-saving, business process simplification, and efficiency.

The benefits of the methodology chosen for this study, which was built on the leaders Gartner & Software Advice and G2 Crowd Grid, include a dedicated ranking system focused on filters such as sector, business size, and region, as well as referring throughout to composite measures of quality based on a survey of vendors' reference customers, and the main product performance audit conducted by a specialized product performance audit team. However, the main limitation of this study is the difficulty of applying the selected criteria on the selected modern ABI platforms due to limited technical specifications provided by the vendors of the chosen modern ABI platforms.

Modern ABI platforms can be ranked based on the market, ease of use, difficulty, features, pricing models, and other factors for potential analysis. It also has an impact on a variety of decision-making metrics, and can be studied by prospective researchers using relevant hypotheses and models. It is also worth noting that the modern ABI platforms discussed and studied in this paper are rapidly changing, and features that are not available now may become available in the near future and should be addressed in further studies.

References

1. Delen, D.: Business Intelligence and Analytics: Systems for Decision Support. Pearson, London (2014)
2. Lawrence, K., Klimberg, R.: Contemporary Perspectives in Data Mining Volume 3. IAP-Information Age Publishing (2017)
3. Simon, H.A.: Administrative Behavior-A Study of Decision-Making, New York (1997)
4. Lim, C.P., Jain, L.C.: Advances in intelligent decision making. In: Jain, L.C., Lim, C.P. (eds.) Handbook on Decision Making. Intelligent Systems Reference Library, vol 4, pp. 3–28. Springer, Heidelberg (2010). https://doi.org/10.1007/978-3-642-13639-9_1
5. R. RoseBusiness Technologies: Descriptive Diagnostic Predictive Prescriptive Analytics, April 2013
6. A. S. T. Saviance Technologies Whitepaper: How Data Analytics can help in Decision Making in Healthcare (2014)
7. Madsen, L.: Healthcare Business Intelligence: a Guide to Empowering Successful Data Reporting and Analytics. Wiley, Hoboken (2012)
8. Evelson, B.: It's Time To Reinvent Your BI Strategy, October 2007
9. Strome, T.L.: Healthcare Analytics for Quality and Performance Improvement. Wiley, Hoboken (2013)
10. Stevanovic, S., Pervan, B.: A GPS phase-locked loop performance metric based on the phase discriminator output. Sensors **18**(1), 296 (2018)

11. Olson, K.D.: In reply physician burnout a leading indicator of health performance and head-down mentality in medical education I and II. Mayo Clin. Proc. **93**(4), 545–547 (2018)
12. C. for D. C. CDC: Developing Evaluation Indicators, April 2018
13. Donabedian, A.: Exploratings in quality assessment and monitoring definition of quality and approaches to its assessment. Ann. Arbor., 79–83 (1980) https://ci.nii.ac.jp/naid/10010151458/en/
14. Kelly, D.L.: Applying Quality Management in Healthcare: A Process for Improvement. Health Administration Press (2003)
15. Castellanos, M.G., Dayal, U., Simitsis, A., Wilkinson, W.K.: Quality-driven ETL design optimization. Google Patents (2014)
16. Astriani, W., Trisminingsih, R.: Extraction, transformation, and loading (ETL) module for hotspot spatial data warehouse using Geokettle. Procedia Environ. Sci. **33**, 626–634 (2016)
17. O. Oracle Corporation: Oracle9i Data Warehousing Guide Release 2 (9.2) (2002)
18. Woodall, P., et al.: A framework for detecting unnecessary industrial data in ETL processes. In: 2014 12th IEEE International Conference on Industrial Informatics (INDIN), pp. 472–476 (2014)
19. Sen, A., Sinha, A.P.: A comparison of data warehousing methodologies. Commun. ACM **48**(3), 79–84 (2005)
20. Badiuzzaman Biplob, M., Sheraji, G.A., Khan, S.I.: Comparison of different extraction transformation and loading tools for data warehousing. In: 2018 International Conference on Innovations in Science, Engineering and Technology (ICISET), pp. 262–267 (2018)
21. A. Alooma Team: ETL Tools: A Modern List, May 2018
22. Magretta, J.: Why Business Models Matter. Harvard Business School, Boston, MA (2002)
23. Porter, M.E., Teisberg, E.O.: Redefining Health Care: Creating Value-Based Competition on Results. Harvard Business Press (2006)
24. Lee, T.H., Kaiser, L.S.: Turning value-based health care into a real business model. N. Engl. J. Med. (2016)
25. Porter, M.E., Lee, T.H. : What 21st century health care should learn from 20th century business. NEJM Catal. **4**(5) (2018)
26. Hwang, J., Christensen, C.M.: Disruptive innovation in health care delivery: a framework for business-model innovation. Health Aff. **27**(5), 1329–1335 (2008)
27. Osterwalder, A.: The business model ontology a proposition in a design science approach. Université de Lausanne, Faculté des hautes études commerciales (2004)
28. Osterwalder, A., Pigneur, Y.: Business Model Generation: A Handbook for Visionaries, Game Changers, and Challengers. Wiley, Hoboken (2010)
29. Westerlund, M., Leminen, S., Rajahonka, M.: Designing business models for the Internet of Things. Technol. Innov. Manag. Rev. **4**, 5–14 (2014). https://doi.org/10.22215/timreview/807
30. Iivari, M.M., Ahokangas, P., Komi, M., Tihinen, M., Valtanen, K.: Toward ecosystemic business models in the context of industrial internet. J. Bus. Model. **4**(2) (2016)
31. Voelpel, S.C., Leibold, M., Tekie, E.B.: The wheel of business model reinvention: how to reshape your business model to leapfrog competitors. J. Change Manag. **4**(3), 259–276 (2004)
32. Tian, S., Yang, W., Le Grange, J.M., Wang, P., Huang, W., Ye, Z.: Smart healthcare: making medical care more intelligent. Glob. Heal. J. **3**(3), 62–65 (2019)
33. Hood, L.: A personal journey of discovery: developing technology and changing biology. Annu. Rev. Anal. Chem. **1**, 1–43 (2008)
34. Hood, L., Flores, M.: A personal view on systems medicine and the emergence of proactive P4 medicine: predictive, preventive, personalized and participatory. N. Biotechnol. **29**(6), 613–624 (2012)
35. Hood, L.: Systems biology and p4 medicine: past, present, and future. Rambam Maimonides Med. J. **4**(2) (2013)

36. Baumhauer, J.F., Bozic, K.J.: Value-based healthcare: patient-reported outcomes in clinical decision making. Clin. Orthop. Relat. Res. **474**(6), 1375–1378 (2016)
37. Howson, A.C., Richardson, J., Sallam, R., Kronz, A.: Magic quadrant for analytics and business intelligence platforms. Gart. Inc., February 2019
38. G. R. G2 Crowd Grid Hub: G2 Crowd Grid Scoring Methodology (2020)
39. 2020 Best in KLAS: Healthcare Business Intelligence & Analytics (2020)
40. Conrad, A. : Healthcare BI Software: Which is the Best Solution? (2020)
41. B. I. Board Inc.: BOARD International for Healthcare (2020)
42. I. Looker Data Sciences: Looker Data Sciences for Healthcare (2020)
43. Microsoft: Microsoft Power BI for Healthcare (2020)
44. M. Incorporated MicroStrategy: MicroStrategy for Healthcare (2020)
45. O. Oracle Corporation: Oracle for Healthcare (2020)
46. Q. International Qilk: Qilk for Healthcare Analytics (2020)
47. Richardson, J., Sallam, R., Schlegel, K., Kronz, A., Sun, J.: Magic quadrant for analytics and business intelligence platforms. Gart. Inc., February 2020
48. Tripathi, A., Bagga, T., Scholor, P.: Leading Business Intelligence (BI) Solutions and Market Trends Study of Consumer Behavior towards Real Estate Sector Post Demonetization View project International Conference on Innovative Computing and Communication (ICICC 2020) Leading Business Intelligence (BI) Solutions and Market Trends. papers.ssrn.com. https://doi.org/10.2139/ssrn.3568414

Covid-19 Twitter Data Analysis Using Natural Language Processing

Dželila Mehanović(✉), Zerina Mašetić, and Amela Vatreš

Faculty of Engineering and Natural Sciences, International Burch University, Sarajevo, Bosnia and Herzegovina
dzelila.mehanovic@ibu.edu.ba

Abstract. The world and everyday life are completely changed due to novel Severe Acute Respiratory Syndrome Coronavirus 2 (SARS-CoV-2) which caused COVID-19 disease. New decisions and rules are set from day to day. Reactions and attitudes of people are often shared on social media, especially using Twitter which makes Twitter a good source of data for many researches. In this work, we analyse a dataset of tweets that have at least 1000 retweets to perform sentiment analysis. Python libraries were used for analysis which includes language detection and translation, preprocessing, tokenization and stemming. Moreover, sentiment analysis and classification is performed using Naive Bayes. It is found that most of the tweets have positive sentiment and achieved accuracy is 60%. Classifier is able to predict positive and neutral tweets with high rate, while it has lower accuracy rate for negative sentiment.

Keywords: Twitter · Covid-19 · Natural language processing · Sentiment analysis · Classification · Naive Bayes

1 Introduction

The first confirmed cases of Covid-19 appeared in December 2019 in China, but it spread all over the world very fast. There are more than 23 million confirmed cases in the world [1]. The virus caused huge problems and made changes in the people's lives. The way how education, healthcare and finance systems operate, the way how people work and interact were affected and changed completely to adjust to the new situation [2].

People have different opinions regarding the virus. There are those who do not believe that the virus exists, and those who believe that it exists and is dangerous to the humans. Also, people tend to share their opinions and feelings over social networks. Twitter is social media that allows people to share their thoughts in form of tweets which can be long up to 280 characters. These tweets are used as a source of data for many researchers since it provides insight in the state of mind of people.

According to the statistics there are 500 million tweets every day which means that around 6000 tweets are shared in one second. Also, there are 145 million active users that visit and exchange tweets every day [3]. Number of tweets and users make twitter a

N. Ademović et al. (Eds.): IAT 2021, LNNS 316, pp. 203–212, 2022.
https://doi.org/10.1007/978-3-030-90055-7_15

great data source for many different research topics [4]. Twitter API is available for use to retrieve and manipulate twitter data [5].

Natural language processing is used for a long time to manipulate natural language data and provide outputs useful to build different applications that make life easier. Sentiment analysis is one of the natural language processing methods that is used to decide whether data is positive, negative or neutral. Sentiment analysis can help to follow opinions, needs and problems of people. Also, it helps to identify most common topics discussed in public.

In this work, we apply natural language processing techniques to analyse twitter data, perform sentiment analysis and implement classification of tweets. The rest of the work is organized as follows: in section two we provide literature review, that is we give overview of related works and description of natural language processing, text analysis, classification, and twitter data. In section three we describe methodologies and steps that are performed in the experiment. Section four contains results of the experiment described in section three. Finally, with section five we conclude our work.

2 Literature Review

In this section we present details about works that are related to our topic. After that natural language processing, data analysis, classification and classification methods commonly used with natural processing and twitter data are described.

2.1 Related Works

Authors in [6] public sentiment about covid-19 using twitter data. They tried a two classifiers Naive Bayes which achieved accuracy 91% and logistic regression whose performance was 74%. [7] performed analysis of twitter data with hashtags coronavirus and covid-19 from 9th April to 15th April 2020. After analysis they found that more than 50% were neutral and a large portion of tweets were objective. But also, they found that public sentiment varies from day to day. [8] searched for twitter data with terms: "corona", "2019-nCov," and "COVID-19". Tweets are analyzed using word frequencies of unigrams and bigrams and using latent Dirichlet allocation tweets are categorized in four topics. Ten of the topics had positive and two of them had negative sentiment. The highest mean of likes was 15.4 for tweets related to economic loss and lowest was 3.94 for tweets about travel bans and warnings. [9] analyzed twitter data from March 20 to April 19, 2020 using sentiment analysis 48.2% tweets are classified as positive, 20.7% as neutral and 31.1% as negative. Also, five topics are identified as the most common health care environment, emotional support, business economy, social change, and psychological stress. After quarantine started, [10] processed twitter data from the Philippines using natural language processing and sentiment analysis to determine effects of quarantine to lifestyle. It is determined that most users have negative sentiments and the most problems are related to the food supply and support from the government. They identified that negative sentiment rises over time and it is expected to continue.

2.2 Natural Language Processing

Term natural language denotes language that is used in standard communication between people. Natural language processing (NLP) refers to the manipulation of natural languages performed by computers [11]. It includes various text analysis such as word frequency analysis, sentiment analysis, language translation or giving response to some request.

NLP became widely used in the last two decades, so there are many real word applications that utilize natural language processing such as smart assistants like Siri, Cortana and Alexa, autocorrect and autocomplete functionalities, email classification, chatbots [12, 13]. SignAll application [14] uses natural language processing to convert language to text, which enables deaf people to communicate with those who do not know sign language. Well known Google Translate application [14, 15] uses natural language processing to translate text.

2.3 Text Analysis

Text analysis is a process of text processing to extract useful information [16]. There are different types of methods that can be used in text analysis from basic methods such as word frequency, collocation, concordance to those more advanced such as text classification which includes sentiment analysis, keyword extraction and topic detection.

Word frequency analysis is used to find words that are mostly used in the given context or conversation. Collocation reveals words that are commonly used together with. In the collocation, bigrams, trigrams or quadrans are identified. Concordance is used to find context of the words in the text or set of words [1].

Classification is used to assign labels to the text. Sentiment analysis is used to detect emotions in a text such as positive, negative or neutral. Topic or category detection is another type of classification used in NLP. Text is classified as one of categories, for example news, sport, entertainment or music. Keyword extraction refers to extraction of the most important words from the text. It can be used to summarize text, create indexes on data that is going to be searched or to create word clouds to provide visual representation of the text [1].

2.4 Classification

Classification is a technique used to assign the correct label to the given input [1]. Label is selected from the list of predefined labels. Email filtering, text topic and word context detection are examples of classification tasks.

There are many machine learning classification algorithms that can be applied to perform classification tasks. Some of these classification algorithms are Naive Bayes, Decision Trees, Artificial Neural Network, K-Nearest Neighbor. Decision trees are used to generate tree charts which contain assigned labels to input based on input features. Naive Bayes is a classification algorithm in which each feature contributes to output independently [1]. It is called a probabilistic classifier since it selects the class that has the highest probability to be the correct class.

There are three datasets used in the process of classification: training, test and dev-test dataset. Training set is used to train the classifier, the test set is used to evaluate classifier performance and the dev-test set is used to adjust model features.

Accuracy is used to measure the performance of a classifier and represents the ratio of correctly classified items over total number of items [17].

3 Methodology

3.1 Data

We used a tweeter dataset downloaded from [18]. This dataset contains 48751 tweets related to the covid-19.

The following are several examples of tweets from dataset:

- It looks like the Democrats, along with the left-wing media, are going to try to use the Coronavirus to weaponize it against the 2020 Election Process!
- Every single person should be participating in Coronavirus response by taking the same simple precautions every day.
- While we cannot stop the spread of #coronavirus, we can SLOW it, decreasing the risk of overwhelming our hospitals.
- Another existing med shows great promise killing the virus!
- A few minutes ago, I just found out that my aunt who was a #healthcareworker in #NewYork has died from #COVID19. She contracted the virus from one of her patients. We had hoped she would recover. I am heartbroken and devastated.

Total number of generated tweets in the period between January and May is 628,809,016 which presents a large data set. For purposes of this study, we used a smaller version of the dataset. This version contains 43751 tweets that have at least 1000 retweets. These tweets are collected from January 22 to April 23. It is important to mention that these tweets are considered as important due to the number of retweets, but those do not have to be the most important tweets related to the covid-19. Table 1 and Fig. 1 shows distribution of tweets by period of time.

Table 1. Distribution of tweets in time

Date period	# of tweets
January 22–March 11	5531
March 12–March 21	15000
March 21–April 7	15000
April 7–April 13	4110
April 13–April 23	2945

Dataset contains multiple features such as user, date, tweet, permalink, retweet count and likes count. For our research we used tweet data to perform sentiment analysis and classification.

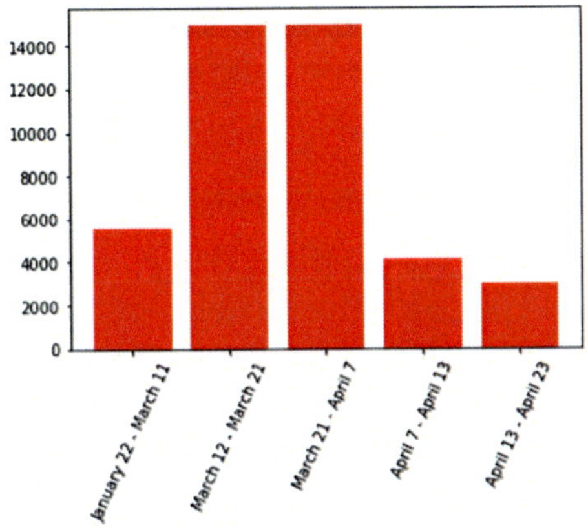

Fig. 1. Distribution of tweets in time

3.2 Language Detection and Translation

Data set contains tweets in 43 different languages such as English, French, Spanish, Portuguese and others. To detect the language of tweets we used python library langdetect [19] text given as argument. It supports 55 languages and returns the ISO 639-1 Code.

After we detected language in which tweets are written, we used another python library to translate all tweets to the English language. Googletrans library [20] implements Google Translate API and it is used to translate text from one language to another.

3.3 Data Preprocessing

Collected tweets contained mixed data that included urls, special characters and tags. To proceed with analysis, it was necessary to perform data preprocessing. Data preprocessing can include several stages such as data cleaning, integration, transformation, reduction and discretization [21]. To perform data preprocessing we mostly used python library re [22] for regex. HTML tags, punctuation marks, numbers, hashtags and urls are removed from tweet data. Also, all letters are transformed into lowercase letters. These steps done in preprocessing made creating a bag of words easier.

3.4 Word Tokenization

Word tokenization is a process of separating text or sentences into words [23]. Words have to be separated and identified in order to be used with natural language processing. To tokenize words we used the word_tokenize method from the nltk tool available in python [24].

3.5 Stemming and Stopwords

Words that have the same base are used in different formats within text, that is with different suffixes. Stemming is a process that is used to return words to their base format

[25]. Before we applied stemming, more precisely Porter Stemmer [26] to our dataset, we firstly removed all stopwords. To remove stop words we used corpus of stopwords from python nltk [27].

3.6 Sentiment Analysis

As described before, sentiment analysis is used to detect public sentiment about some topic. Here we want to find sentiment related to tweets about covid-19. Our dataset does not contain labeled data, so before classification we had to detect sentiment of each tweet.

Two packages were used to detect sentiment of tweets. Those are nltk sentiment analyzer [28] and textblob package [29]. By review of results we found that textblob provided more accurate sentiment analysis. After this phase, tweets are labeled as positive, neutral or negative.

3.7 Classification

To classify tweets, we used Naive Bayes classifier. Firstly, we build a list of tuples which contain tweet and sentiment value. Secondly, we build a list of all unique words which are checked to be English words. Using these words and a list of tweets we build three datasets of features: those with only positive, only neutral and only negative tweets. Finally, the dataset is divided into training and testing sets with ratio 75:25.

4 Results

In this section we present results of the experiment described in the previous section.

We managed to detect 43 different languages in which tweets were written. Table 2 presents the distribution of tweets per language.

After detection of language, all tweets are translated into English language. Average tweet length and number of words are counted using translated values. Average tweet length was 253 characters and average number of words was 51. Furthermore, we created word clouds using English values. Word cloud is used to present most frequently used words. Several word clouds are created, below we provide two of them. Figure 2 and Fig. 3 show two-word clouds. It is noticeable that coronavirus, covid19, China, Trump and cases are words that appear in both of them and seem to be most frequently used.

Sentiment analysis was used to label tweets as positive, neutral or negative. Distribution of labels is presented in Table 3 and Fig. 4.

From the sentiment analysis, we found that most tweets are positive, and the number of negative tweets is two times lower than the number of positive tweets. Moreover, we analysed sentiment distribution of tweets per language. Table 4 presents a number of positive, negative and neutral tweets for 4 languages for which we have more than 1000 tweets. It can be noticed that for all of them the number of positive tweets is highest.

For classification, we build list of tweet-sentiment value pairs that comes in the following form:

$$[(\text{tweet. sentiment}),\ (\text{tweet2},\ \text{sentiment}), \ldots]$$

Table 2. Tweets distribution per language

Language code	# of tweets	Language code	# of tweets	Language code	# of tweets
en	26331	Ta	33	sw	3
es	9080	nl	17	hr	3
pt	3196	ur	16	sv	3
fr	1744	lt	12	mk	2
th	933	no	11	pl	2
id	689	et	7	fi	2
ca	372	af	7	zh-cn	2
ja	340	ro	5	hu	1
it	243	da	5	bn	1
Hi	225	so	5	el	1
de	119	vi	5	zh-tw	1

Fig. 2. Word cloud 1

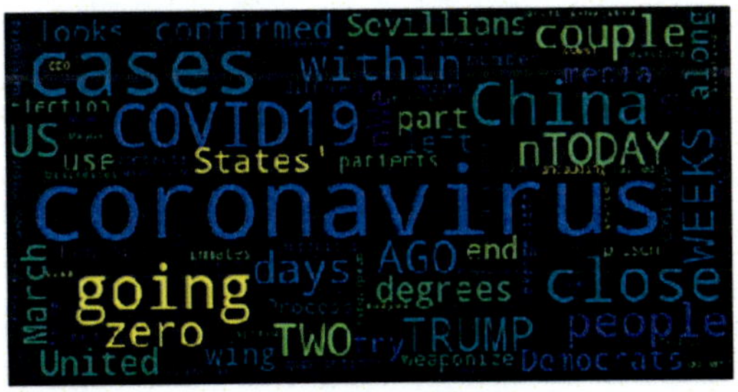

Fig. 3. Word cloud 2

Table 3. Sentiment distribution

Sentiment	# of tweets
Positive	20324
Neutral	12582
Negative	10845

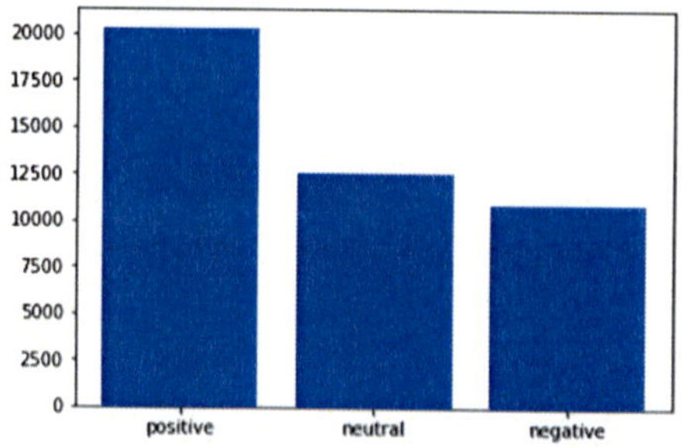

Fig. 4. Sentiment distribution

Table 4. Tweets distribution per language

Language	# of positive tweet	# of neutral tweets	# of negative tweets
en	12732	6793	6806
es	3986	2882	2212
pt	1328	1132	736
fr	679	669	396

Next step was to build a list of all unique words. For this task, we used words that are given as output from the stemming phase. Finally, datasets for positive, negative and neutral features are obtained. All three datasets are combined in one which is further divided into training and test sets.

Classifier is built using a training set and evaluated using a test set. But, when we applied the entire training dataset, we experienced the problem with the memory. Because of that, classification is applied on a smaller dataset with randomly selected instances. To evaluate the performance, we used accuracy which was 60% where the classifier seemed to work better with positive and neutral tweets.

5 Conclusion

In this work we present analysis of tweets related to the covid-19. Dataset includes tweets that have at least 1000 retweets. We performed word frequency analysis and detected the language of tweets. In the language detection phase, we found that tweets are written in 43 different languages, and the python library is used to translate all tweets to English language to make further steps easier.

Tweets did not contain sentiment labels, so it was necessary to perform sentiment analysis to label those. Using two libraries, sentiment of tweets are discovered and a more reliable analyzer TextBlob is selected. It is found that most analysed tweets are positive and there are two times more positive than negative tweets.

Before sentiment analysis, tweets are preprocessed and word tokenization is performed. Moreover, we removed stop words and applied Porter Stemmer to perform stemming. Finally, we created datasets with features for positive, negative and neutral tweets. These datasets are combined to build a training and test set with a ratio 75%–25%.

Classification is performed using Naive Bayes classification algorithm and achieved accuracy is 60%. It is found that positive and neutral tweets are detected with higher accuracy rates, while accuracy for negative sentiment was lower.

This kind of analysis using natural language processing is useful to track public opinions, problems and reactions to decisions that arc madc and affect groups of people. To expand research, we could use a dataset with geographical location to have more specific insight to sentiment based on location. Also, a classifier should be improved to detect negative sentiment more precisely.

References

1. Coronavirus Update (Live): 23,838,486 Cases and 817,636 Deaths from COVID- 19 Virus Pandemic - Worldometer (no date). https://www.worldometers.info/coronavirus/?utm_campaign=homeAdUOA?Si. Accessed 25 Aug 2020
2. Badnjević, A., et al.: Risks of emergency use authorizations for medical products during outbreak situations: a COVID-19 case study. Biomed. Eng. Online **19**(1), 75 (2020)
3. Kim, A.E., et al.: Methodological considerations in analyzing Twitter data. J. Natl. Cancer Inst. Monogr. **2013**(47), 140–146 (2013)
4. Getting Started with the Twitter API (no date). https://developer.twitter.com/en/docs/twitter-api/getting-started/guide. Accessed 25 Aug 2020
5. How many tweets about Covid-19 and Coronavirus? 508 MM tweets so far (2020). https://www.tweetbinder.com/blog/covid-19-coronavirus-twitter/. Accessed 24 Aug 2020
6. Samuel, J., et al.: (no date) COVID-19 public sentiment insights and machine learning for tweets classification. https://doi.org/10.31234/osf.io/sw2dn
7. Manguri, K.H., Ramadhan, R.N., Mohammed Amin, P.R.: Twitter sentiment analysis on worldwide COVID-19 outbreaks. Kurdistan J. Appl. Res., 54–65 (2020). https://doi.org/10.24017/covid.8
8. Abd-Alrazaq, A., et al.: Top concerns of tweeters during the COVID-19 Pandemic: infoveillance study. J. Med. Internet Res. **22**(4), e19016 (2020)
9. Hung, M., et al.: Social network analysis of COVID-19 sentiments: application of artificial intelligence. J. Med. Internet Res. **22**(8), e22590 (2020)
10. Sentiment analysis of filipinos and effects of extreme community quarantine due to coronavirus (COVID-19) pandemic. J. Crit. Rev. (2020). https://doi.org/10.31838/jcr.07.07.15
11. Bird, S., Klein, E., Loper, E.: Natural Language Processing with Python: Analyzing Text with the Natural Language Toolkit. O'Reilly Media, Inc., Sebastopol (2009)
12. 8 common examples of natural language processing and their impact on communication (no date). https://www.tableau.com/learn/articles/natural-language-processing-examples. Accessed 25 Augt 2020
13. Kharkovyna, O.: Natural Language Processing (NLP): Top 10 Applications to Know, Towards Data Science (2019). https://towardsdatascience.com/natural-language-processing-nlp-top-10-applications-to-know-b2c80bd428cb. Accessed 25 Aug 2020
14. Coldewey, D.: SignAll is slowly but surely building a sign language translation platform, TechCrunch (2018). http://techcrunch.com/2018/02/14/signall-is-slowly-but-surely-building-a-sign-language-translation-platform/. Accessed 25 Aug 2020
15. Google Translate (no date). https://translate.google.com/. Accessed 25 Aug 2020
16. Text Analysis (no date). https://monkeylearn.com/text-analysis. Accessed 25 Aug 2020

17. van Halteren, H., Zavrel, J., Daelemans, W.: Improving accuracy in word class tagging through the combination of machine learning systems. Comput. Linguist., 199–229 (2001). https://doi.org/10.1162/089120101750300508
18. 60 Incredible and Interesting Twitter Stats and Statistics (no date). https://www.brandwatch.com/blog/twitter-stats-and-statistics/. Accessed 25 Aug 2020
19. langdetect (no date). https://pypi.org/project/langdetect/. Accessed 24 Aug 2020
20. googletrans (no date). https://pypi.org/project/googletrans/. Accessed 24 Aug 2020
21. Ramezani, M., Fatemizadeh, E.: Comparison of supervised classification methods with various data preprocessing procedures for activation detection in fMRI data. Comput. Neurosci., 75–83 (2010). https://doi.org/10.1007/978-0-387-88630-5_5
22. re — Regular expression operations — Python 3.8.5 documentation (no date). https://docs.python.org/3/library/re.html. Accessed 24 Aug 2020
23. Python - Word Tokenization (no date). https://www.tutorialspoint.com/python_data_science/python_word_tokeniza-tion.htm#:~:text=Word%20tokenization%20is%20the%20process,for%20a%20particular%20sentiment%20etc. Accessed 25 Aug 2020
24. Code Faster with Line-of-Code Completions, Cloudless Processing (no date). https://www.kite.com Accessed 25 Aug 2020
25. Stemming and lemmatization (no date). https://nlp.stanford.edu/IR-book/html/htmledition/stemming-and-lemmatization-1.html. Accessed 25 Aug 2020
26. Rahmatulloh, A., et al.: Comparison between the stemmer porter effect and Nazief-Adriani on the performance of winnowing algorithms for measuring plagiarism. Int. J. Adv. Sci. Eng. Inf. Technol., 1124 (2019). https://doi.org/10.18517/ijaseit.9.4.8844
27. 2 Accessing Text Corpora and Lexical Resources (no date). https://www.nltk.org/book/ch02.html. Accessed 25 Aug 2020
28. Python Examples of nltk.sentiment.vader.SentimentIntensityAnalyzer (no date). https://www.programcreek.com/python/example/100005/nltk.sentiment.vader.SentimentIntensityAnalyzer. Accessed 25 Aug 2020
29. API Reference — TextBlob 0.16.0 documentation (no date). https://textblob.readthedocs.io/en/dev/api_reference.html. Accessed 25 Aug 2020

Comparison of Semantic Role Labeling Tools on Previously Unseen Hand Annotated Texts

Daniel Vasić(✉), Tomislav Volarić, Emil Brajković, Hrvoje Ljubić, and Robert Rozić

Faculty of Natural Science and Education, University of Mostar, Matice Hrvatske b.b, 88 000 Mostar, Bosnia and Herzegovina
daniel.vasic@fpmoz.sum.ba

Abstract. In this article state of the art semantic role labeling (SRL) tools have been compared on published lexical resources and custom hand annotated previously unseen texts. Semantic role labeling is one of the first steps into understanding the meaning of the sentence. The process of identifying semantic roles includes several steps from which the main are predicate identification, argument identification and argument classification. All the approaches compared in this article suggest supervised machine learning algorithms on lexical resources. The comparison of the tools for semantic role labeling are tested against hand annotated corpora independent of FrameNet or PropBank corpus. The precision, recall tests were made on PropBank data as well. Because the results of classification can be biased, the objective of this research was to comprise all the mentioned tools with hand annotated previously unseen data. The results have shown that mate-tools has achieved best results on independent hand annotated data with the best micro average F-score of 78% on custom hand annotated dataset. Results from this article were used to evaluate which tool can be used for knowledge extraction tasks and question generation in communication module of intelligent tutoring system. The overview of all freely available tools, mostly from classifier implementation point of view is given in this article.

Keywords: Semantic role labeling · Mate-tool · Senna msrsplat · Comparison propbank · Framenet

1 Introduction

In natural language, semantic role is a relationship that syntactic constituent has with a predicate in sentence. Automatic processing of semantic roles is a relatively new subject of research in natural language processing where different methods are applied to specially structured linguistic data also known as lexical resources. Certain natural language processing tasks such as assigning semantic roles can be seen as assigning labels to words. This can be achieved in two different steps. In the first step the system is trained on already labeled words (with semantic roles in this task). In second step system automatically assigns label to previously unseen words.

N. Ademović et al. (Eds.): IAT 2021, LNNS 316, pp. 213–220, 2022.
https://doi.org/10.1007/978-3-030-90055-7_16

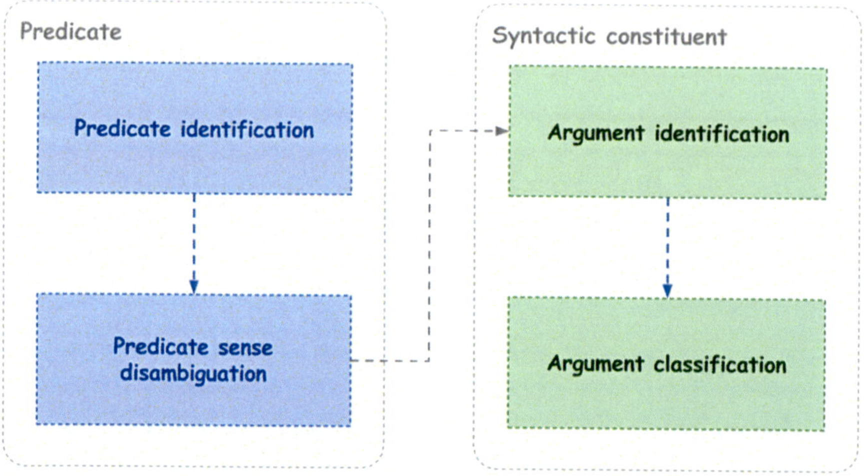

Fig. 1. Predicate and syntactic constituent classification in semantic role labeling task

Typically, semantic role labeling (SRL) consists of four different steps: predicate identification, predicate sense disambiguation, argument identification and argument classification as shown in Fig. 1. In each step, statistical techniques are used to determine all predicates in given sentence, after that for every identified predicate its syntactic constituents are identified and classified into one of the semantic role labels.

In this article we propose evaluation methodology for modern SRL tools. Most modern methods for SRL arose from competition in semantic parsing. These competitions are organized as part of the CoNLL (Conference on Natural Language Learning) conferences, which deal with the issue of natural language processing and are organized every year. In 2005 (Carreras and Márquez 2005) and 2009 (Hajič et al. 2009) competitions in semantic text parsing were organized where representative samples of tree banks were delivered at which the authors evaluated their approaches and proposed their own models.

The main problem of evaluation of different tools for automatic semantic parsing is how to evaluate the performance of the system. The evaluation is usually preformed on the same test data supplied in mentioned CoNLL shared task. The test data for evaluation is very similar to train data because treebanks are constructed from newspaper texts. For this reason, is often hard to develop a method that can easily transfer to new domains, since the meaning representations may vary greatly between different domains. We propose evaluation on out of domain data from different resources such as scientific articles, schoolbooks and other to properly evaluate SRL tool for usage in real world scenario. The same evaluation we done by authors in Hartmann et al. (2017), Li et al. (2020) where authors develop custom out-of-domain corpora developed purposely for out-of-domain evaluation.

2 Lexical Resources

The semantic role labeling tools are usually the last step in syntactic natural language processing pipeline and it's the first step in semantic processing. Semantic predicates represent semantic meaning of events and participants in a sentence. For example, CAUSE and MOTION are used to show the meaning of a sentence. Predicates are words in sentence that are used to "encode" this meaning. So, the verb THROW can bc that verb which implies meaning such as CAUSE and MOTION. So the verb THROW implies that there is something that is moving and that someone caused that motion. Something and someone are called semantic arguments to the verb. The verb in a sentence can also be in passive voice which causes more problems in argument identification. This problem is also called alternation problem. So, for the problem of semantic role labeling we can't provide basic sets of rules that find semantic arguments of verbs. Therefore, resources such as VerbNet (Schuler 2005), PropBank (Kingsbury and Palmer 2002) and FrameNet (Baker et al. 1998) where built.

FrameNet is based on a theory called frame semantics which states that the meanings of most words in a sentence can best be understood on a basis of a semantic frame, a description of an event type, relation and the participants in an event. The FrameNet corpora includes more than 17000 manually annotated sentences that provides unique training set for semantic role labeling. Other similar corpora annotated with verbal propositions and their arguments is PropBank.

PropBank corpora has been extremely influential in natural language processing research. It differs from FrameNet in several ways. PropBank does not annotate events or states of affairs described by nouns. FrameNet is oriented on semantic frames where the descriptions are generalized across similar verbs as well as nouns and other words. The first SRL system used FrameNet as a training set in the creation of model for prediction of semantic roles. Many tools for automatic semantic role identification were developed and trained on these corpora.

VerbNet is organized into verb classes extending Levin classes through refinement and addition of subclasses to achieve syntactic and semantic coherence among members of a class. Each verb class represents some meaning of that verb in some context. Verbnet is largest online verb lexicon and its most commonly used tool for verb sense disambiguation (Kawahara and Palmer 2014).

PropBank is lexical resource which aims to provide widespread hand-labeled corpus for semantic processing thereby allowing the construction of a robust system for natural language processing. PropBank corpus defines a set of semantic roles for certain verb, the semantic arguments are annotated within the text of the original Penn Treebank 2 corpus. Each verb role is numbered. While PropBank annotations are close to syntactic level, FrameNet annotations are semantically motivated. PropBank project annotates the whole Penn TreeBank corpus with semantic roles while in FrameNet rarely consecutive sentences are annotated.

Providing this level of semantic labeling is very important for applications that require the extraction of information, question answer systems and machine translation. During last decade complex rule-based systems are replaced with

3 Automatic Semantic Language Annotation

One of the fundamental approaches for automatic SRL proposed by Gildea and Jurafsky (2002) suggests supervised machine learning algorithms on lexical resources. Lexical resources that are used in SRL tasks are corpora like FrameNet, PropBank and NomBank (Gerber, Chai and Meyers 2009). Statistical methods are applied on these lexical resources, to build a predictive model that automatically determines predicate sense and semantic roles of the syntactic constituents in text. Such systems are developed using supervised learning algorithms where the classifier learns to identify predicate, its meaning, connected syntactic constituents also known as arguments and semantic role labels for every argument based on features from training data. Training data is usually corpus of texts that contain explicit label describing which semantic role is assigned to which segment of text for specific a predicate. Features used for training the classifiers are based Predicate Syntactic constituent Predicate identification Predicate sense disambiguation Argument identification Argument classification on the fact that there is a relation between syntactic information and semantic form, so the syntactic information can be used as an indicator of the semantic roles inside the text.

Today there are many tools that perform SRL automatically, the important difference is on which lexical resource are the tools trained on. Some of the tools for semantic role labeling that are trained on FrameNet are SEMAFOR (Das et al. 2012), Shalmaneser (Erk and Pado 2006), SwiRL (Surdeanu and Turmo 2005), the tools that are trained on PropBank and in some cases NomBank are Mate-tools (Björkelund et al. 2009), SENNA (Collobert et al. 2011), MSRSplat (Quirk et al. 2012), EasySRL (Lewis et al. 2014), PathLSTM (Roth and Lapata 2016). In this article only tools that are trained on PropBank are taken into account. Property of the PropBank annotation is that predicates exhibiting the same diathesis alternations share similarly-labeled arguments provided that the verbs belong to the same Levin verb class (Giuglea and Moschitti 2004). This makes the process of training fast and accurate SRL system easier. For this reason, most of state-of-the-art systems that achieve best accuracy are trained on PropBank corpora. There are also other approaches (Vasic and Vasic 2020) (Vasić et al. 2021) for SRL of morphologically rich languages, such tools where not included in this evaluation but this can be done in future research.

3.1 Performance of Existing Tools by Standard CoNLL Benchmark Tests

Tools like SwiRL and SENNA show best results in terms of memory and CPU usage. But in tasks such as knowledge extraction the labelling performance is key in building a good system. Most of the tools comprised achieve near state-of-the-art performance. The best by CoNLL tests as shown in Table 1 is PathLSTM that builds on top of mate-tool with 87,90% F-score and out of domain 76,60% F-score. Mate-tool follows and achieves out of the domain F-score of 74.38%. All the other tools are compared on different CoNLL tasks so the scores cannot be compared. Also, the MSRSplat SRL tool is not evaluated on any CoNLL task.

Problem in evaluating MSR Splat is that the tool uses its own tokenizer, producing its own trees and cannot be properly aligned with CoNLL format data. In this article the comparison of the tools is presented. All the tools for semantic role labeling are

Table 1. Results of the evaluated systems based on reported CoNLL evaluation

Name	CoNLL-task	F-score (domain)	F-score (Out-of-domain)
Mate-tool	CoNLL-2009	85,63%	74,38%
SENNA	CoNLL-2005	79,49%	N/A
MSRSplat	N/A	N/A	N/A
EasySRL	CoNLL-2008	83,50%	N/A
PathLSTM	CoNLL-2009	87,90%	76,60%

tested against hand annotated corpora independent of FrameNet or PropBank. Three hundred sentences form article about computer are hand annotated by human experts, this corpus is used as a gold standard for tool comparison, from these twenty sentences are hand picked randomly and used in evaluation of these tools. We state that the number of sentences is not relevant as opposed to the difference of the sentences. The results showed that the results on twenty sentences are more or less the same when testing to the whole dataset. The CoNLL dataset for testing includes 5 462 sentences for testing of the approach.

4 Evaluation on Hand Annotated Texts

The sentences are annotated with semantic constituents using state of the art semantic role labeling tools. The precision of the tools is measured and analyzed. Also, the tools are automatically tested against PropBank corpora. The comparison of the different tools for semantic role labeling has been done using the confusion matrices for every of the tools. The results of classification can be biased because all the tools used PropBank as corpus for training. So, the objective of this research was to comprise all the mentioned tools with hand annotated previously unseen data. Subset of twenty sentences from three hundred hand annotated sentences was used to compute per class F-score to give detailed comparison of SRL tools. The results for mate-tools are shown in Table 2. The mate-tools results included results for also nominal predicate and its arguments identification. When the problem of classification involves the search for the positive class samples and are very rare compare to the negative classes the precision and recall approach is used.

This method for evaluation is more useful in needle-in-haystack type problems where the positive class is more interesting than negative class. This is why used the PR to evaluate our classifiers.

Micro average of the results is used to determine what is the precision of the approach on individual level. Micro average is used when the size of the dataset may vary, so it is a logical choice for SRL task. Macro average which is used to find average of micro averages for every individual class. Other tools performed not as good as mate-tools that is clearly visible in the PR analysis. The tables have 15 classes which occur depending on the sentences (the more sentences its likely to have more semantic arguments). The

Table 2. Results of the evaluated systems based on hand labeled previously unseen text

Tool name	Micro average F1-score (%)	Macro average F1-score (%)	Weighted average F1-score (%)
Mate-Tools	78.00	62.00	83.00
MSRSplat	73.00	68.00	75.00
SENNA	68.00	47.00	73.00
EasySRL	40.00	73.00	78.00
PathLSTM	71.00	82.00	86.00

arguments are A0-A5 and AM the arguments are the modifier arguments. Although the macro average of the MSR Splat tool is bigger than mate-tools, micro average is worse. PathLSTM showed the best results on weighted and macro scale but it was worse in micro tests by our evaluation methodology, the reason for that is that PathLSTM model is trained only on development training set of official CoNLL dataset for English, for better performance it must be trained on whole CoNLL dataset.

5 Conclusion

In this article the most popular and freely available tools today for semantic role labeling are comprised. For comparison, human annotated texts were used, subset of three hundred hand annotated arbitrary Wikipedia articles where used. Data in PropBank contains sentences that are continuous, as opposed to handpicked sentences used in evaluation of classifiers, so the classifiers are tested in various scenarios. The results have shown that Mate-tools has achieved best results in PR tests on PropBank and independent hand annotated data. One of the advantages of Mate-tool is the integration of NomBank for nominal predicates identification. Mate-tool SRL is most accurate tool for semantic role labeling for this dataset.

Tools used in this article are freely available open-source projects. The integration of these tools was done in Python using some Java components. MSR Splat API and precomputed binaries for Senna were used to run the classifiers. As far as performances in terms of CPU usage, Senna showed fastest response (due to the implementation details), and mate-tools showed worst performance results. Main advantages remain its precision and its support for nominal predicates. In this article, it was shown that mate-tool API (application programming interface) is the best freely tool for semantic role labeling based on PropBank. This article was used to evaluate which tool can be used (and which will be most precise) for further semantic processing tasks, such as knowledge extraction and question generation in communication module of intelligent tutoring system, also for evaluation of answers written on natural language (Brajkovic and Vasic 2017).

References

Baker, C.F., et al.: The Berkeley FrameNet project. In: Proceedings of the 36th Annual Meeting on Association for Computational Linguistics-. Morristown, NJ, USA, p. 86. Association for Computational Linguistics (1998). https://doi.org/10.3115/980845.980860

Björkelund, A., Hafdell, L., Nugues, P.: Multilingual semantic role labeling. In: Proceedings of the Thirteenth Conference on Computational Natural Language Learning (CoNLL 2009): Shared Task, June, pp. 43–48 (2009). https://doi.org/10.3115/1596409.1596416

Brajkovic, E., Vasic, D.: Tree and word embedding based sentence similarity for evaluation of good answers in intelligent tutoring system. In: 2017 25th International Conference on Software, Telecommunications and Computer Networks, SoftCOM 2017 (2017). https://doi.org/10.23919/SOFTCOM.2017.8115592

Carreras, X., Márquez, L.: Introduction to the CoNLL-2005 shared task: semantic role labeling. In: CoNLL 2005 - Proceedings of the Ninth Conference on Computational Natural Language Learning, pp. 152–164 (2005)

Collobert, R., et al.: Natural language processing (almost) from scratch. J. Mach. Learn. Res. **12**, 2493–2537 (2011)

Das, D., Martins, A.F.T., Smith, N.A.: An exact dual decomposition algorithm for shallow semantic parsing with constraints. In: Proceedings of the First Joint Conference on Lexical and Computational Semantics - Volume 1: Proceedings of the Main Conference and the Shared Task, and Volume 2: Proceedings of the Sixth International Workshop on Semantic Evaluation, pp. 209–217. Association for Computational Linguistics (2012)

Erk, K., Pado, S.: Shalmaneser - a flexible toolbox for semantic role assignment (2006)

Gerber, M., Chai, J.Y., Meyers, A.: The role of implicit argumentation in nominal SRL (2009)

Gildea, D., Jurafsky, D.: Automatic labeling of semantic roles. Comput. Ling. **28**(3), 245–288 (2002). https://doi.org/10.1162/089120102760275983

Giuglea, A.-M., Moschitti, A.: Knowledge Discovering using FrameNet, VerbNet and PropBank (2004)

Hajič, J., et al.: The CoNLL-2009 shared task: Syntactic and semantic dependencies in multiple languages. In: CoNLL 2009: Shared Task - Proceedings of the Thirteenth Conference on Computational Natural Language Learning, CoNLL: Shared Task, pp. 1–18 (2009)

Hartmann, S., et al.: Out-of-domain FrameNet Semantic Role Labeling. The Association for Computational Linguistics (2017)

Schuler, K.K.: VerbNet: a broad-coverage, comprehensive verb lexicon. University of Pennsylvania (2005)

Kawahara, D., Palmer, M.: Single classifier approach for verb sense disambiguation based on generalized features. In: Proceedings of the Ninth International Conference on Language Resources and Evaluation (LREC 2014) (2014)

Kingsbury, P., Palmer, M.: From TreeBank to PropBank. In: Proceedings of the Third International Conference on Language Resources and Evaluation (LREC 2002) (2002)

Lewis, M., Lewis, M., Steedman, M.: A* CCG parsing with a supertag-factored model. In: Proceedings of EMNLP (2014)

Li, Z., et al.: Domain adaptation for semantic parsing. In: Bessiere, C. (ed.) Proceedings of the Twenty-Ninth International Joint Conference on Artificial Intelligence, IJCAI 2020. International Joint Conferences on Artificial Intelligence Organization, pp. 3723–3729 (2020)

Quirk, C., et al.: MSR SPLAT, a language analysis toolkit. In: NAACL-HLT, pp. 21–24 (2012)

Roth, M., Lapata, M.: Neural semantic role labeling with dependency path embeddings, pp. 1192–1202 (2016). https://doi.org/10.3115/v1/P15-1109

Surdeanu, M., Turmo, J.: Semantic role labeling using complete syntactic analysis. In: Proceedings of the Ninth Conference on Computational Natural Language Learning, pp. 221–224, pp. 221–224. Association for Computational Linguistics (2005)

Vasic, D., Vasic, M.K.: Syntax-aware neural semantic role labeling for morphologically rich languages. In: 2020 28th International Conference on Software, Telecommunications and Computer Networks, SoftCOM 2020. Institute of Electrical and Electronics Engineers Inc. (2020). https://doi.org/10.23919/SoftCOM50211.2020.9238179

Vasić, D., Žitko, B., Ljubić, H.: Automatic question generation using semantic role labeling for morphologically rich languages. Technical Gazette (2021, prepress). https://doi.org/10.17559/TV-20200402175619

Comparison of Feature Extraction Methods for Heart Disease Classification

Anas Fares(✉) and Zerina Mašetić

Department of Information Technologies, Faculty of Engineering and Natural Sciences, International Burch University, Ilidža, Bosnia and Herzegovina

Abstract. Cardiovascular disease is the most common cause of deaths in the world, being 1 in every 4 deaths from the total deaths. Electrocardiogram (ECG) is essential in monitoring and diagnosing a patient's heart status as it plays a significant role when the features being extracted. This work aims to develop an accurate feature extraction model of ECG signals for early heart disease detection. Moreover, the comparison of various feature extraction methods will be given and discussed. ECG UCI Cardiac Arrhythmia database was used. Reverse biorthogonal spline wavelet filters were used to detect P, Q and R points. Five classification models using the support vector machine (SVM) polynomial, SVM Gaussian, decision tree, SVM ensemble trees, and Naïve Bayes were developed. The developed models were optimized using Bayesian optimization. Our results suggested that the discrete wavelet transform (DWT) and SVM classifier are applicable in the heart disease diagnosis, which aids the physicians in early and accurate detection.

Keywords: ECG · Classification · DWT · SVM · Machine learning · Feature extraction · Heart disease

1 Introduction

There are many different heart conditions and problems which are collectively called heart disease. Over the last 20 years, heart disease has been the leading cause of death at the global level [1]. Cardiovascular disease (CVD) is a common heart disease, it is the name for the group of disorders of heart and blood vessels, CVDs are the number 1 cause of death globally: more people die annually from CVDs than from any other cause. An estimated 17.9 million people died from CVDs in 2016, representing 31% of all global deaths. Of these deaths, 85% are due to heart attack and stroke. Most cardiovascular diseases can be prevented by addressing behavioral risk factors such as tobacco use, unhealthy diet and obesity, physical inactivity and harmful use of alcohol using population-wide strategies [2]. Diagnose can include blood tests, chest x-ray, electrocardiogram (ECG), echocardiogram, CT scan and MRI scan [3].

Electrocardiogram (ECG) is a graphical representation of electric activity that occurs in the heart. These electrical activities are waves that include heart rate, rhythm and morphology results ECG is currently a bioelectric signal that is easily accessible, which provides significant information regarding cardiac abnormalities. Numerous new feature

N. Ademović et al. (Eds.): IAT 2021, LNNS 316, pp. 221–236, 2022.
https://doi.org/10.1007/978-3-030-90055-7_17

extraction techniques have been established to calculate the current state of cardiac activity through analysis of rhythms and distortions found in the electrocardiogram [4]. Many techniques have been used previously to detect these features; these techniques were based on time-domain methods. Time-domain methods were not always accurate to extract all the features from the ECG signals [5]. Therefore, the graphical frequency representation of the signals is required to detect possible heart diseases. ECG is essential in monitoring and diagnosing a patient's heart status as it plays a significant role when the features being extracted. In the framework of biomedical signals, ECG occupies a central location [6]. Due mainly to its versatility and the vast number of details it offers, it deserves a high prevalence of ECG signals. Moreover, based on the appearance and location of the ECG signal components, it is possible to monitor the activity of the heart as well as identify different heart diseases and disorders, in the continuation of arrhythmia. Until now, the diagnosis of disturbances has been carried out mainly by cardiologists based on their expertise and experience, but the increasing need for continuous surveillance of the patient's condition makes space for automatic devices to diagnose arrhythmias or other cardiac irregularities. By irregular heartbeat, it does not perform its function satisfactorily which damages the organism and also, the heart is endangering itself and can develop more serious damage to the heart muscle [7].

According to the Centre for disease control and prevention [8], one person dies every 36 s in the United States from cardiovascular disease. About 655,000 Americans die from heart disease each year which means 1 in every 4 deaths from the total deaths. Figure 1 shows the interactive atlas of heart disease and stroke in the USA.

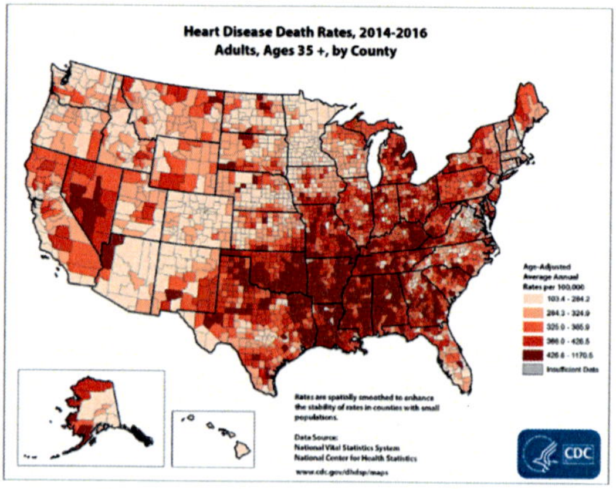

Fig. 1. Heart disease death rates [8]

ECG signals are a recording of the various waveforms that signify and specify the heart's characteristics. Extracting these characteristics has been a challenging challenge because of the noise around the heart [9]. Nowadays many methods help in showing heart status and diagnosing heart diseases such as ECG, CT scan, MRI and several more [10]. Therefore, it is extremely important to detect possible heart diseases fast and accurately, therefore, the development of accurate and fact feature extraction models is necessary.

This work aims to develop an accurate feature extraction model of ECG signals for early heart disease detection. Moreover, the comparison of various feature extraction methods will be given and discussed.

The main contributions of this work are:

1. The developed feature extraction model for the ECG signal extraction with an accuracy of more than 90%.
2. The comparison of feature extraction methods for ECG signals, with main advantages and disadvantages.
3. Selected the features of the ECG signal that have the highest impact on heart disease detection.

At the end of this paper, the results are expected to detect the best feature extraction method from ECG signals to diagnose heart disease.

The rest of the paper is organized as follows: Sect. 2 reviews the related literature, Sect. 3 presents the material and methods used, feature extraction and classification algorithms, Sect. 4 explains the results and discusses the obtained results. Finally, Sect. 5 concludes the paper.

2 Literature Review

Heart disease detection is one of the most important tasks for physicians. The whole process of heart disease detection could be improved by including an automated system that aids in easier and faster disease detection.

Several research in the field of heartbeats classification using different feature extraction and machine learning techniques have been done. Many researchers have applied different pre-processing techniques, various feature extraction techniques and classifiers.

Zhao and Zhang [11] used wavelet transform (WT) and Support Vector Machine (SVM) methods to extract features from ECG signals. The paper showed a new approach that is used in feature extraction for reliable heart rhythm recognition. The proposed system consists of three components which include data pre-processing, feature extraction and classification of ECG signals. To achieve the feature vector of ECG data, two separate feature extraction methods are implemented together. To obtain the transition and features of each ECG section, the wavelet transform procedure is used. Autoregressive modelling (AR) is also applied to get hold of the temporal structures of ECG waveforms. Finally, the SVM and Gaussian kernel are used to classify different ECG heart rhythms. The accuracy of this method reached an overall 99.68%.

Tamil et al. [12] proposed a feature extracting method using Discrete Wavelet Transform (DWT). This method was used to extract relevant information from ECG input data to perform classification tasks. Data acquisition, pre-processing beat detection, feature extraction and classification were the steps the author took to achieve this method. This approach is designed to resolve the issue of ECG signals that are not stationary. By translation and dilation operations, it was taken from a single generating function called the mother wavelet. Using this approach in the extraction of features can result in the highest frequency resolution in all frequency ranges and have stable ECG waveform characteristics.

A new approach to feature extraction developed by Jager [13] is the Kahunen Lo'eve Transform (KLT), which is an attractive and efficient approach to the method of feature extraction and shape representation. If the probability densities of the population of pattern vectors of a problem domain are not specified, it has the solution. The issue with this approach is that it is too sensitive to ECG signal noise patterns.

Following Ranjith et al. [14], the WT is obtained using the quadratic spline wavelet, which used WT to diagnose myocardial ischemia. Both correspond to T wave and P wave identification. Comparatively higher sensitivity and nominal positive predictivity value were demonstrated by their methods. It can quickly be generalized to detect other ECG signal anomalies. But this technique also has the drawback that the necessary measurements are larger than those required by other approaches.

Dallali et al. [15] used DWT, the RR interval was extracted, and the RR interval was transformed using the Z score. To distinguish ECG beats, they used FCM. They accomplished 99.05% accuracy. RR interval and R point position by using DWT are extracted functions. Using FCM, classification was done and final classification using 3-layer MLPNN was done. They reached 99.99% accuracy. ECG classification was done using MLPNN. The obtained Mean Squared Error (MSE) was 0.00621. The RR interval was extracted by Khorrami and Moavenian [16] using DWT. This paper contrasted the efficiency of MLPNN and SVM.

Patra et al. [17] has extracted the R peak and RR interval from the MIT-BIH arrhythmia database annotation file manually. Using Zero mean, characteristics are normalized and using FCM decreased. three-layered FFNN is used as a classifier with a backpropagation algorithm.

Using Db4 discrete wavelet transform, Naima and Timemy [18] have extracted R location and RR interval. A back-propagation algorithm-trained FFNN is used. NN's sensitivity, specificity and accuracy are 90%, 90%, 95%, and 95%, respectively. Khazaee [19] calculated the NxtRR, prevRR, and ratRR were calculated from the RR interval and normalized with a mean of zero and standard deviation of unity. The ECG classification uses the PSO-SVM as a classifier. The RR interval is determined from the MITBIH arrhythmia database recordings. In this article, MLPNN and the SVM classifier were compared. The findings show that MLPNN is excellent for performance checking, while SVM demonstrates good performance in training.

Vishwa et al. [20] have used DWT and normalized by Zero mean to extract RR interval and R peak, by using a bandpass filter, the noise was removed. The authors used two databases: MIT-BIH arrhythmias database and Normal sinus rhythm database. They used FFNN with error backpropagation algorithm as the classifier. Features extracted were R peak, RRt interval (current RR interval at t time), RRt + 1 interval (next RR interval at t + 1 time), QRSh (QRS complex height), and QRS width. The Pan Tompkins algorithm is used for extracting the R peak and QRS complex. For noise reduction, the low pass linear phase filter is used and the median filter for baseline correction is used. 96.251% and 99.104% were the sensitivity and specificity achieved using the PSO-RBFNN classifier.

3 Materials and Methods

The doctor begins the analysis of the electrocardiogram by analyzing the technical quality of the image. This is followed by an analysis of heart rhythm, determination of frequency and review of characteristic quantities.

The analysis of the technical inspection of the image quality is a check of the correctness of the camera itself. Along with the signal itself, parasitic currents can occur, which distort this signal and thus do not provide the required accuracy of the image. If the camera is properly calibrated at the beginning and end of the recording, it appears at the beginning and end of the recording and represents a deviation of the current discharge of 1 mV [21].

The process of visual analysis by a doctor is long. Unfortunately, experts who specialize in reading electrocardiograms can often misinterpret the signal, but for machine learning, their diagnosis is taken as a reference. Understanding the ECG has undergone tremendous progress since the 19th century and Einthoven; however, cardiovascular disease is still one of the biggest causes of mortality. Currently, many devices are in use that helps doctors with proper diagnosis, and this paper should contribute to reducing the classification error by using machine learning techniques [22].

This paper shows the possibility of normal and successful collaboration between man and machine by using the parallel already described in human action and memory. The ECG signal arrives as an input to the machine and is captured using an electrocardiograph or some other measuring device and transformed into a form appropriate for further processing by going through different filters. The first step is signal acquisition. In other words, it is the step of converting a signal from analogue to digital form. Signal recording assumes analysis of the recorded signal, i.e., knowledge of features such as voltage, frequency range, etc. These parameters largely depend on the patient, his age, weight, height, etc. The voltage amplitude of the ECG signal is 5 mV in the worst case. It is necessary to choose a suitable gain that will amplify this signal to make all changes noticeable. This gain is usually taken from 500 to 1000. In the worst case, the frequency can be as high as 150 Hz. The recording speed is generally 25 mm/s, however speeds such as 12.5 mm/s are used at slower frequencies and faster 50 mm/s [3].

3.1 Data Collection and Analysis

The UCI Cardiac Arrhythmia Database was used in this research. The database consists of 452 samples (patients) with 279 attributes, of which 206 are linear (values belong to the set of real numbers) and 73 are nominal (values from the set {0, 1}). The first four attributes give quantitative anamnesis values (age, sex, height and weight), while the rest of the attributes describe recorded electrocardiograms (QRS complex duration, PR interval, QT interval, T wave height, QRS height, frequency, QRS area, and so on) derivatives: average width of Q wave, R wave, R 'wave, the existence of split R, wave and many other measured features mentioned in the first chapter).

The selected database contains for each sample and the class diagnosed by the doctor. These are 16 classes, of which the first class represents the physiological electrocardiogram, while the other classes represent one of the arrhythmias given, as shown in Table 1.

Table 1. The review of the given classes in the UCI cardiac arrhythmia database

Class number	Type of arrhythmia	No. of samples
2	Coronary artery disease	44
3	Anterior myocardial infarction	15
4	Inferior myocardial infarction	15
5	Sinus tachycardia	13
6	Sinus bradycardia	25
7	Ventricular extrasystoles	3
8	Supraventricular extrasystoles	2
9	Block of the left branch of his bundle	9
10	Block of the right branch of his bundle	50
11	First-degree atrioventricular block	0
12	Second-degree atrioventricular block	0
13	Third-degree atrioventricular block	0
14	Hypertrophy of the left ventricle	4
15	Atrial fibrillation and flutter	5
16	Other arrhythmias	22

The database contains missing values. They are replaced by a constant mean value of the attribute for the diagnosed class. The average value is calculated by Eq. (1), where n is the number of samples for the diagnosed class and yi the value of the attributes for the individual sample.

$$\bar{y} = \frac{1}{n}\sum\nolimits_{i-1}^{n} yi \tag{1}$$

In addition, by inspecting the values of the attributes, it was noticed that the values match the theoretical ones. A simple example is sinus tachycardia has the highest frequency and sinus bradycardia the lowest, as shown in Table 2.

Table 2. Sinus rhythm arrhythmia

Physiological ECG	Sinus tachycardia	Sinus bradycardia
73,2 rpm	113,1846 rpm	53,92 rpm

Large mean values of the QRS complex width belong to the samples of the block class of the left and the class of the right branch of His bundle as shown in Table 3.

These arrhythmias are also characterized by the existence of an R 'tooth in special leads, first for lead V5 and second for lead V1, as shown in Table 4.

Table 3. Bundle branch in an ECG with a widened QRS complex

Physiological ECG	His bundle branch left block	Bundle branch right block
84,2776 ms	154,4444 ms	98,5 ms

Table 4. Bundle branch block cardiac Arrhythmias based on ECG

Physiological ECG	His bundle branch left block	His bundle branch right block
0,0033 mV	0,6444 mV	0,01 mV
0,0298 mV	1,928 mV	0 mV

To obtain the best possible results, it is necessary to reduce all samples by attributes to a standard form, which is that they are centered on zero and that their variance is of the same order. If the attribute has an order of magnitude variance greater than the others, it may dominate the training process and the classifier does not pay enough attention to the other attributes.

To center all samples of a given attribute on zero, it is necessary to determine the mean value of each attribute and subtract it from the value of that attribute for each sample, and then scale (divide) with the standard deviation calculated as the root of the variance:

$$z_i = \frac{y_i - \overline{y}}{\sigma_y} \tag{2}$$

$$\left(\sigma_y\right)^2 = \frac{1}{n-1}\sum\nolimits_{i-1}^{n}\left(y_i - \overline{y}\right)^2 \tag{3}$$

Upon completion of the analysis, the database was divided into a ratio of 80% for training and 20% for testing in such a way that each of the sixteen classes was represented in the process of training and testing.

Then, an input matrix in which the individual inputs of an individual patient are set by vector type is prepared. Each vector has as many columns as there is attributes. The matrix is created in the dimensions [361 × 279] and it represents the input to the classifier. The column vector of dimensions [361 × 1] represents the desired outputs based on which classifier will be trained.

3.2 Feature Extraction Methods

One of the most crucial aspects is to prepare the details in a way that helps the algorithm to recognize the critical properties more quickly. To pay attention to different parts of the signal, the doctor's eye is used. This portion is referred to as the process of extracting functions. The classification of ECG signals often requires nonlinear mechanisms that extract properties and attributes from the signal to make the later classification as acceptable as possible [12]. There are many methods of extracting properties, among which can be:

- Wavelet transformation.
- Support vector machine.
- Discrete wavelet transformation.
- Artificial neural networks.
- Cross-correlation.
- Matched filters.
- A combination of the above, etc.

The problem of the classification itself and this thesis is the right side of Fig. 2. It is noticeable that the block diagram splits into two parts. The first is training where, based on the input processed signal, the system is trained according to an algorithm and the trained set is stored in a database [12].

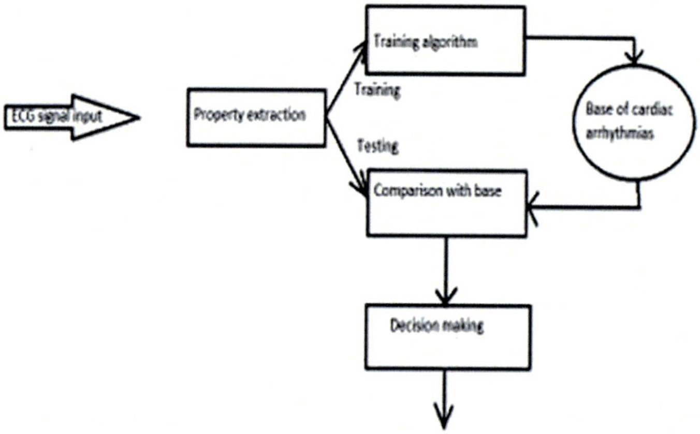

Fig. 2. Classification of the ECG signal

The second part happens after the training is over and is called testing. Based on the processed input signal and based on the trained set, a comparison is made, and a decision is made, i.e., a diagnosis is established.

Diagnosis of cardiac arrhythmias as a classification problem is not a simple problem, therefore there is a need for the application of machine learning methods and techniques. Some of the methods and techniques that have been used over the years are syntactic, probabilistic, graph search methods, rule-based systems, and fuzzy systems [16].

To extract features of the ECG, the wavelet transformation - reverse biorthogonal spline wavelet filters were used to detect P, Q and R points. Machine learning techniques were used also to develop classification models. Five classification models using the SVM polynomial, SVM Gaussian, decision tree, SVM ensemble trees, and Naïve Bayes were developed. Wavelet transformation and Machine learning shows promising results for developing an accurate feature extraction model of ECG signals for early heart disease detection [23, 24].

3.3 Classification Methods

Supervised learning is a type of learning that uses the knowledge of an external supervisor [25]. For classification, as the subject of this paper, it is necessary to present the inputs and desired outputs in front of the system. The extent to which the actual outputs match the desired one is determined by the supervisor. This measure is usually given by the Least Mean Square Error, so the goal is to train the system in a way that this criterion is minimal. Different approaches are encountered in the literature, among which non-parametric classification tools have given excellent results because they do not strictly require specific relationships between variables and have great theoretical and practical significance.

Diagnosis of cardiac arrhythmias as a classification problem is not a simple problem, therefore there is a need for the application of machine learning methods and techniques. Some of the methods and techniques that have been used over the years are syntactic, probabilistic, graph search methods, rule-based systems, and fuzzy systems [22].

There are several developed classification methods used today. Some have been developed using statistics, some are methods of artificial intelligence, etc. In our work, five classification models were used: the support vector machine (SVM) polynomial, SVM Gaussian, decision tree, SVM ensemble trees, and Naïve Bayes.

Each decision tree of the ensemble is trained in an expanded feature space. This approach would encourage the current Ensemble Approaches to leverage the expressive power of the kernels while each member of the Ensemble uses the kernel features along with the original features. Our philosophy of using expanded space features is close to the approach suggested by Rwebangira [26]. In other words, decision trees learn simpler decision boundary (linear decision boundary) in kernel features than the decision tree that learns nonlinear decision boundary. Decision trees trained on the expanded feature space are more likely to be more effective than decision trees trained on the initial features. Decision trees make a list of features at each node. This property is useful for datasets whose original features are complemented by kernel features, as kernel features can only be chosen at higher tree levels if such features are stronger than or identical to the original features for classification [27].

Decision trees made by algorithms such as ID3 (Iterative dichotomiser 3) and C4.5 are widely used for heartbeat classification. ID3 is a mathematical algorithm and is thus less vulnerable to training data errors. However, it is not appropriate for our application because the decision tree that it produces has discreet attributes. Instead, we use the C4.5 algorithm, which makes the use of numeric attributes. The C4.5 algorithm builds the decision tree by evaluating the uncertainty of the training data features and choosing the most stable ones [28].

Support Vector Machines is a set of related supervised learning methods, applicable to classification and regression problems. SVM creates hyperplanes that separate the transformed input space, maximizing the distance between classes. The parameters of the hyperplane equation are calculated as an optimization problem through quadratic programming [28]. The SVM algorithm is the most used learning algorithm for the kernel. The kernel approach allows the SVM algorithm to locate a hyperplane in the kernel space by mapping data from the function space to higher dimensional kernel space and leading to nonlinear separation in the kernel space. Kernel representations have an additional solution to maximize the computing efficiency of linear learning machines. The kernel function and its parameters must be selected to construct the SVM classifier. To date, no theoretical or empiric experiments have determined the superiority of one kernel over another. In our experiment, the SVM classifier is educated based on the Gaussian kernel, which in previous studies proved to be the most suitable kernel for the ECG signal classification, in the subspace features of different dimensions [29].

The Naïve Bayes Net kernel can also be used for numerical attributes as opposed to the Naïve Bayesian classifier [30]. The Naive Bayes Classifier is defined as an elementary probabilistic classifier based on the implementation of the Bayes Theorem along a series of conditional independence assumptions [26]. Naive Bayes algorithm is a classification technique based on Bayes' Theorem with an assumption of independence among predictors. In simple terms, a Naive Bayes classifier assumes that the presence of a particular feature in a class is unrelated to the presence of any other feature. Naive Bayes model is easy to build and particularly useful for very large data sets. Along with simplicity, Naive Bayes is known to outperform even highly sophisticated classification methods [31].

The methods defined in this way work as individual classifiers. In contrast, ensemble methods represent a combination of several basic classifiers. The goal of using these methods lies in improving the performance of the base classifier. Classifier sets are proposed to improve predictive accuracy by combining multiple classifiers, and outputs of individual classifiers are analyzed to make the final prediction [32].

4 Results and Discussions

4.1 Results

Bayesian optimization extraction was performed and results from each model plotted the minimum observed value of the objective function for each model over every iteration of the hyperparameter optimization. The objective function value corresponds to the misclassification rate measured by five-fold cross-validation using the training data set. The plot compares the performance of each model.

The performance of the developed models based on ROC curve was shown in Fig. 3. The figure shows that the SVM polynomial and SVM Gaussian have performed the highest performance compared to other classification methods that were used based on the true positive and false-positive rates.

A confusion matrix associated with a classifier shows the predicted and actual classification. Figure 4, 7, and 9 show the confusion matrices whose entries have the number of correct negative predictions, the number of incorrect positive predictions, the number of incorrect negative predictions, and the number of correct positive predictions. The prediction accuracy of the classification models was computed in Table 5, 6, and 7 based on values in the confusion's matrices.

The confusion matrices in Fig. 4, 7 and 9 show the true positive and true negative for predicted classes, the first 30 iterations, and the second 30 iterations were performed, respectively, have possessed the highest results which belong to SVM polynomial and SVM Gaussian classifiers.

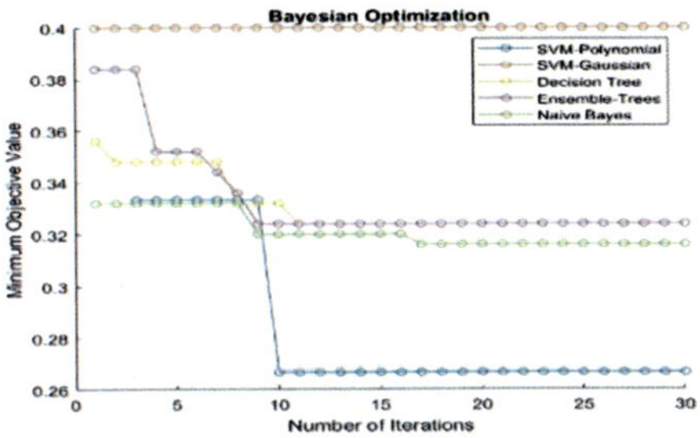

Fig. 3. Performances comparisons of each model

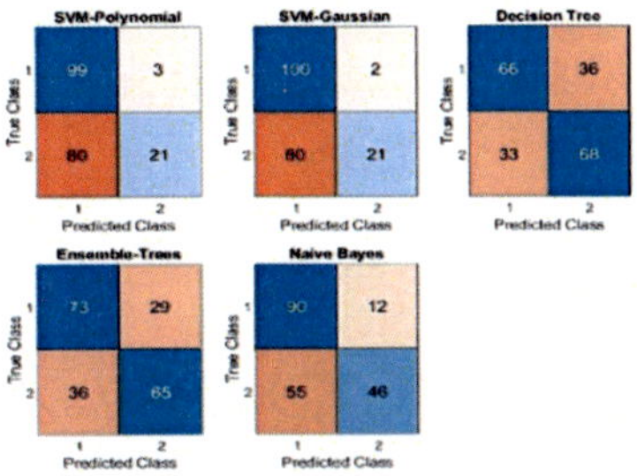

Fig. 4. The confusion matrix with predicted classes

Sensitivity, specificity, and accuracy are statistical measures of the performance of a binary classification test that are widely used in medicine: Sensitivity (True Positive rate) measures the proportion of positives that are correctly identified (i.e., the proportion of those who have some condition (affected) who are correctly identified as having the condition). Specificity (True Negative rate) measures the proportion of negatives that are correctly identified (i.e., the proportion of those who do not have the condition (unaffected) who are correctly identified as not having the condition). A test method is said to be accurate when it measures what it is supposed to measure. This means it can measure the true amount or concentration of a substance in a sample [33].

The best performance of the predicted model was the SVM Gaussian classifier with 82.64%, 97.56%, and 88.67% for sensitivity, specificity, and accuracy, respectively (Table 5). The best performance of the trained models for the first 30 iterations was to the SVM Gaussian classifier with 81.15%, 97.56%, and 87.19% for sensitivity, specificity, and accuracy, respectively (Table 6). The best performance of the trained models for second 30 iterations was to the SVM Gaussian classifier with 87.93%, 100%, and 93.10% for sensitivity, specificity, and accuracy, respectively (Table 7).

Table 5. Performance of the predicted model

Classification model	Sensitivity %	Specificity %	Accuracy %
SVM polynomial	82.50	96.38	88.17
SVM Gaussian	82.64	97.56	88.67
Decision tree	49.25	47.82	48.760
SVM ensemble trees	71.57	55.38	53.69
Naïve Bayes	58.06	82.08	65.31

The Receiver Operating Characteristic (ROC) curve (Fig. 5) is a performance curve for classifier output. Inspection of the classifier performance was performed by plotting a ROC curve for each classifier. Perfcurve function was used to obtain the X and Y coordinates of the ROC curve and the area under the curve (AUC) value for the computed

X and Y. The column order is the same as the category Class 1 and 2 in order of the response variable in the training data set.

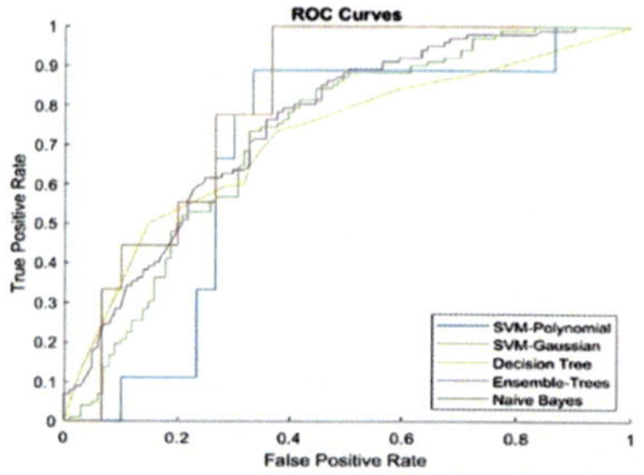

Fig. 5. ROC curve for each classifier

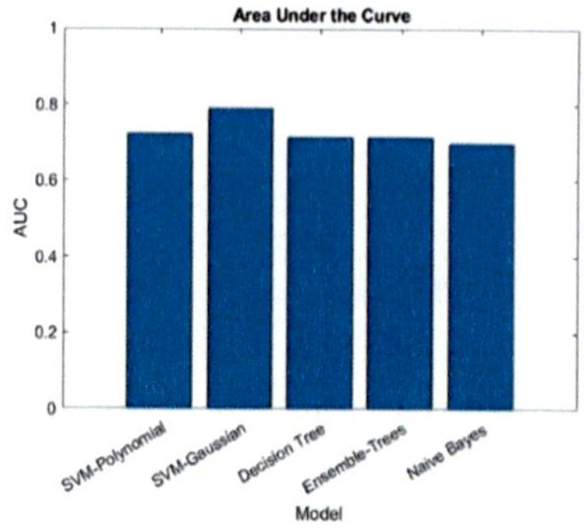

Fig. 6. Area under the curve

The performance of the developed models based on the ROC curve is shown in Fig. 5. The figure shows that the SVM polynomial and SVM Gaussian have achieved the highest performance compared to other classification methods that were used, based on the number of iterations and minimum objective value.

ROC curve shows the true positive rate versus the false positive rate for different thresholds of the classifier output. AUC values were plotted using a bar graph (Fig. 6). For a perfect classifier is true positive rate always 1 regardless of the thresholds. As a result, based on the confusion matrix and the AUC bar graph, we can compare the performance of classifiers.

Bayesian optimization on all models for 30 more iterations was run, using the resume function. Figure 7 shows the minimum observed values of the objective function for each iteration of Bayesian optimization. The first 30 iterations correspond to the first round of Bayesian optimization (Fig. 7). The next 30 iterations correspond to the results of the resume function (Fig. 9). Resuming optimization is useful because the loss continues to reduce further after the first 30 iterations. The ROC curves after the optimization is applied is shown in the Fig. 8.

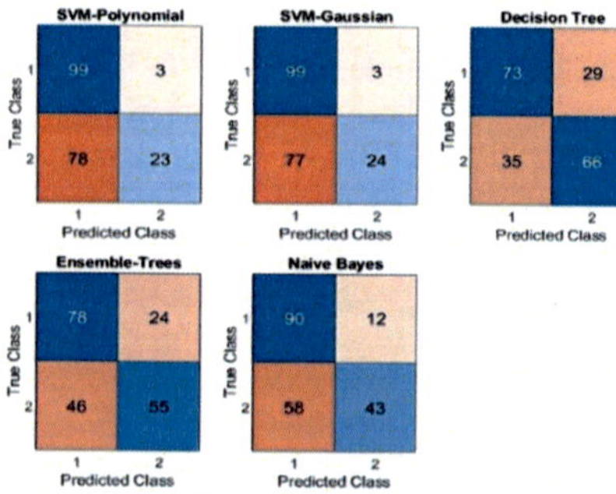

Fig. 7. Confusion matrix for the first 30 iterations

Table 6. Performance for second 30 iterations

Model	Sensitivity	Specificity	Accuracy
SVM polynomial	81.15	96.30	87.19
SVM Gaussian	80.48	96.25	86.70
Decision tree	52.51	54.69	52.94
SVM ensemble trees	58.65	66.67	61.46
Naïve Bayes	67.66	82.85	72.90

Table 7. Performance for second 30 iterations

Model	Sensitivity	Specificity	Accuracy
SVM polynomial	87.93	100.00	93.10
SVM Gaussian	80.49	96.25	86.70
Decision tree	56.93	63.64	59.11
SVM ensemble trees	59.40	67.14	62.07
Naïve Bayes	54.61	58.82	56.15

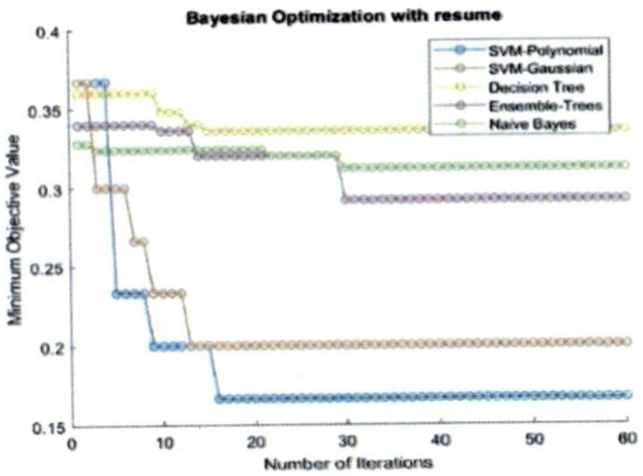

Fig. 8. Bayesian optimization with resume

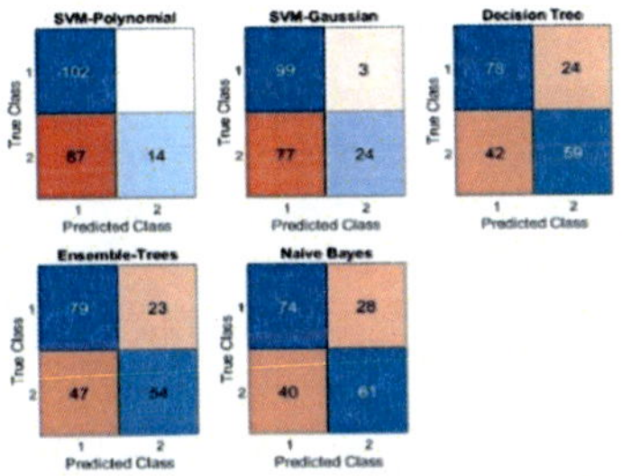

Fig. 9. Confusion matrix for the second 30 iterations

4.2 Discussions

This work aims to develop an accurate feature extraction model of ECG signals for early heart disease detection. Normal ECG signal, duration 6 s from the UCI Cardiac Arrhythmia database was selected. The Infinite Impulse Response (IIR) filter Butterworth order six was used for filtering the ECG signal in the time domain. Wavelet Transformation - reverse biorthogonal spline wavelet filters were used to detect P, Q and R points. The ECG dataset was divided into training and testing groups (80% for training and 20% for testing). A constant mean value was computed using Eq. 1 for removing the missing values from the ECG signal. To obtain the best possible results, it is necessary to reduce all samples by attributes to a standard form, all the ECG values were centered on zero and that their variance by using Eq. 2 and Eq. 3, respectively.

Five classification models using the SVM polynomial, SVM Gaussian, decision tree, SVM ensemble trees, and Naïve Bayes were developed using the training dataset. Five-fold cross-validation was used in that development. Then, the developed models were optimized using Bayesian optimization.

To measure the performance of the developed models, sensitivity, specificity, accuracy, ROC, AUC was computed as shown in Fig. 3, 4, 5, 6, 7, 8, 9. The best performance of the predicted model was the SVM Gaussian classifier with 82.64%, 97.56%, and 88.67% for sensitivity, specificity, and accuracy, respectively (Table 5). The best performance of the trained models for the first 30 iterations was to the SVM Gaussian classifier with 81.15%, 97.56%, and 87.19% for sensitivity, specificity, and accuracy, respectively (Table 6). The best performance of the trained models for second 30 iterations was to the SVM Gaussian classifier with 87.93%, 100%, and 93.10% for sensitivity, specificity, and accuracy, respectively (Table 7).

Also, our result shows the Wavelet Transformation-reverse biorthogonal spline wavelet filters is a good method for extracting features from ECG signal to detect heart disease with an accuracy of more than 90% and this finding is consistent with the findings of Dallali et al. [15], Naima and Timemy [18] and Vishwa et al. [20].

The DWT was used in different studies to detect cardiac ailments and it showed good accuracy [28]. Our analysis suggested that DWT and SVM classifier can be combined to develop a good diagnosing model for cardiac ailments applications.

Venkatesan et al. [28] indicated that the performance of the SVM classifier is better than other machine learning-based classifiers. Our analysis suggested that DWT and SVM classifier can be combined to develop a good diagnosing model for cardiac ailments applications.

We can conclude that the SVM Gaussian classifier has the highest potential in the detection of heartbeats, based on Arrhythmia analysis using ECG signal.

5 Conclusions

This paper mentions a couple of feature extraction methods that are used in detecting heart disease using ECG signals. Each method has its advantages and disadvantages; therefore, it is extremely important to find out and use the most accurate and reliable method especially in health care. R peak and QRS detection are necessary to indicate specific data which is useful for ECG classification. Comparison is made to find out the most accurate method in diagnosing heart disease. This paper has limitations, comparing the performance is depending on the feature extractions mentioned, and in the future, these features can be increased. The benefits of this project are helping healthcare in making decisions about what are the best ways to detect heart disease, in which it can decrease heart disease in the future. It can also help in decreasing the amount of youth death.

References

1. World Health Organization: WHO reveals leading causes of death and disability worldwide: 2000-2019, WHO (2020). https://www.who.int/news/item/09-12-2020-who-reveals-leading-causes-of-death-and-disability-worldwide-2000-2019. Accessed 16 Feb 2021
2. World Health Organization: Cardiovascular diseases (CVDs), WHO (2017). https://www.who.int/news-room/fact-sheets/detail/cardiovascular-diseases-(cvds). Accessed 16 Feb 2021
3. Vecht, R., Gatzoulis, M.A., Peters, N.: ECG Diagnosis in Clinical Practice. Springer, London (2009). https://doi.org/10.1007/978-1-84800-312-5
4. Vaneghi, F.M., Oladazimi, M., Shiman, F., Kordi, A., Safari, M.J., Ibrahim, F.: A comparative approach to ECG feature extraction methods. In: 2012 Third International Conference on Intelligent Systems Modelling and Simulation, pp. 252–256. IEEE, February 2012
5. Shaffer, F., McCraty, R., Zerr, C.L.: A healthy heart is not a metronome: an integrative review of the heart's anatomy and heart rate variability. Front. Psychol. **5**, 1040 (2014)
6. Pourbabaee, B., Howe-Patterson, M., Reiher, E., Benard, F.: Deep convolutional neural network for ECG-based human identification. In: CMBES Proceedings, vol. 41 (2018)
7. Gacek, A., Pedrycz, W. (eds.): ECG Signal Processing, Classification and Interpretation: A Comprehensive Framework of Computational Intelligence. Springer, London (2011). https://doi.org/10.1007/978-0-85729-868-3
8. CDC: Heart Disease Statistics and Maps, CDC (center for diseases control and Preventation) (2020). https://www.cdc.gov/heartdisease/facts.htm
9. Chauhan, S., Vig, L.: Anomaly detection in ECG time signals via deep long short-term memory networks. In: 2015 IEEE International Conference on Data Science and Advanced Analytics (DSAA), pp. 1–7. IEEE (2015)
10. Mayo Clinic: Heart disease, Mayo Clinic (2020). https://www.mayoclinic.org/diseases-conditions/heart-disease/diagnosis-treatment/drc-20353124
11. Zhao, Q., Zhang, L.: ECG feature extraction and classification using wavelet transform and support vector machines. In: 2005 International Conference on Neural Networks and Brain, pp. 1089–1092. IEEE (2005)
12. Tamil, E.B.M., et al.: A review on feature extraction & classification techniques for biosignal processing (Part I: Electrocardiogram). In: Abu Osman, N.A., Ibrahim, F., Wan Abas, W.A.B., Abdul Rahman, H.S., Ting, H.N. (eds.) 4th Kuala Lumpur International Conference on Biomedical Engineering 2008, IFMBE Proceedings, vol. 21, pp. 107–112. Springer, Heidelberg (2008). https://doi.org/10.1007/978-3-540-69139-6_31
13. Jager, F.: Feature extraction and shape representation of ambulatory electrocardiogram using the Karhunen-Lòeve transform. Electrotech. Rev. **69**(2), 83–89 (2002)
14. Ranjith, P., Baby, P.C., Joseph, P.: ECG analysis using wavelet transform: application to myocardial ischemia detection. ITBM-RBM **24**(1), 44–47 (2003)
15. Dallali, A., Kachouri, A., Samet, M.: Classification of cardiac arrhythmia using WT, HRV, and fuzzy c-means clustering. Signal Process. Int. J. (SPJI) **5**(3), 101–109 (2011)
16. Khorrami, H., Moavenian, M.: A comparative study of DWT, CWT and DCT transformations in ECG arrhythmias classification. Expert Syst. Appl. **37**(8), 5751–5757 (2010)
17. Patra, D., Das, M.K., Pradhan, S.: Integration of FCM, PCA and neural networks for classification of ECG arrhythmias. IAENG Int. J. Comput. Sci. **36**(3) (2009)
18. Naima, F.A., Timemy, A.A.: Neural network based classification of myocardial infarction: a comparative study of Wavelet and Fourier transforms. In: Pattern Recognition. BoD–Books on Demand, pp. 337–351 (2009)
19. Khazaee, A.: Heart beat classification using particle swarm optimization. Int. J. Intell. Syst. Appl. **5**(6), 25 (2013)

20. Vishwa, A., et al.: Clasification of arrhythmic ECG data using machine learning techniques. IJIMAI **1**(4), 67–70 (2011)
21. Tikkanen, J.: Early repolarization in the inferolateral leads of the electrocardiogram. Acta Universitatis Ouluensis. Oulu: Juvenes Print (2013)
22. Sattar, Y., Chhabra, L.: Electrocardiogram. StatPearls [Internet]. StatPearls Publishing (2020)
23. Tao, R., et al.: Magnetocardiography-based ischemic heart disease detection and localization using machine learning methods. IEEE Trans. Biomed. Eng. **66**(6), 1658–1667 (2018)
24. Sahoo, S., et al.: Machine learning approach to detect cardiac arrhythmias in ECG signals: a survey. IRBM **41**, 185–194 (2020)
25. Zhang, L., et al.: From machine learning to deep learning: progress in machine intelligence for rational drug discovery. Drug Disc. Today **22**(11), 1680–1685 (2017)
26. Rwebangira, M.R.: Techniques for Exploiting Unlabeled Data. Carnegie-mellon Univ Pittsburgh PA Dept of Computer Science (2008)
27. Ahmad, A.: Decision tree ensembles based on kernel features. Appl. Intell. **41**(3), 855–869 (2014). https://doi.org/10.1007/s10489-014-0575-4
28. Venkatesan, C., Karthigaikumar, P., Varatharajan, R.: A novel LMS algorithm for ECG signal preprocessing and KNN classifier based abnormality detection. Multimedia Tools Appl. **77**(8), 10365–10374 (2018). https://doi.org/10.1007/s11042-018-5762-6
29. Karpagachelvi, S., Arthanari, M., Sivakumar, M.: Classification of ECG signals using extreme learning machine. Comput. Inf. Sci. **4**(1), 42 (2011)
30. Khanna, D., Sharma, A.: Kernel-based naive bayes classifier for medical predictions. In: Bhateja, V., Coello Coello, C.A., Satapathy, S.C., Pattnaik, P.K. (eds.) Intelligent Engineering Informatics. AISC, vol. 695, pp. 91–101. Springer, Singapore (2018). https://doi.org/10.1007/978-981-10-7566-7_10
31. Gu, J., Lu, S.: An effective intrusion detection approach using SVM with naïve Bayes feature embedding. Comput. Secur. **103**, 102158 (2021)
32. Cao, Z., et al.: The lncLocator: a subcellular localization predictor for long non-coding RNAs based on a stacked ensemble classifier. Bioinformatics **34**(13), 2185–2194 (2018)
33. Zhu, W., Zeng, N., Wang, N.: Sensitivity, specificity, accuracy, associated confidence interval and ROC analysis with practical SAS implementations. In: NESUG Proceedings: Health Care and Life Sciences, Baltimore, Maryland, vol. 19, p. 67 (2010)

Influences of Puzzle Videogames on Logical Reasoning

Rijad Hubana(✉)

Faculty of Mechanical Engineering, Computing and Electrical Engineering, University of Mostar, Mostar, Bosnia and Herzegovina
rijad.hubana@fsre.sum.ba

Abstract. From the very beginnings of the video game industry, one genre of videogames has stood out – puzzle games. Puzzle games are based around using different skills, mostly methods involving logical tasks, for solving problems presented on those videogames. The puzzle genre is polarizing when it comes to the player base since it has a different concept and gives the players a different sense of accomplishment. Furthermore, puzzle games have a different effect/influence on the human brain than the other genres of videogames. This research analyzes the influence of puzzle videogames on logical reasoning, using the analytical hierarchy process for evaluating and ranking the analyzed videogames. Using a set of six well-known puzzle videogames, results indicate a significant influence on logical reasoning as well as a set of other analyzed parameters. This paper contributes to the already established knowledge base, by analyzing and comparing the influence of different puzzle games, which improve the logical reasoning by their own positive influences and presents a very important factor in the early mental development of young users.

Keywords: Puzzle · Videogames · Logical reasoning · Brain

1 Introduction

Videogames have come a long way since their inception, which is best seen by comparing videogames of today with their forerunners, i.e., PONG with the videogames of today [1]. The videogame industry has always been dependent on the computer industry, and the limitations in designing a videogame were always set by the computer equipment capabilities of that time. Since the year of 1985, the number of household computers in the UK has been in a constant increase, which represented a "green light" for the expansion and integration of the videogame industry to the computers [2]. One of the first videogames that would define the genre of puzzle games was Flag Capture, released in 1978 [3]. Despite the unfavorable reviews, this game had a great influence on one of the most popular puzzle videogames of all time, Minesweeper [4]. Videogames can act as a "teacher", depending on the purpose of the game. The most interesting results show that playing videogames can change cognitive abilities and some functionalities of the brain. Studies have demonstrated that the videogame playing experience is directly

N. Ademović et al. (Eds.): IAT 2021, LNNS 316, pp. 237–247, 2022.
https://doi.org/10.1007/978-3-030-90055-7_18

connected to better cognitive abilities, especially in the fields of visual attention, short-term memory, reaction time and working memory [5]. Since the playing of puzzle games involves different parts of the brain, it is assumed that they improve the abilities of thinking on the user's part, therefore some research has already been done related to this subject. The research from the Nanyang Technological University shows that the users which played the puzzle game Cut the Rope for one hour daily, had improved executive functions, which are crucial for decision making and managing under sudden changes [6]. The abilities that were tested included the following: task switching, the speed of accommodating to new tasks instead of relying on the same strategy, and at last, the level of focus while removing distractions and undesirable answers. After 20 h of playing Cut the Rope, the test subjects had a 33%, 30% and 60% improvement in task switching, accommodating to new tasks, and the level of focus, respectively [7]. The research conducted in 2014, in which was centered around the effects of Portal 2 and Lumosity on cognitive and non-cognitive abilities, also shows positive results on the abilities mentioned. Namely, the results of the research indicate that playing Portal 2 has a noticeable positive effect on spatial skills, as well as the level of persistence on the player's behalf [8].

The continuation of this paper will contain the theoretical background that should enable better understanding of the methodology and the research that is conducted in the chapter number three. Results of the conducted research are presented in chapter number four, whilst the fifth chapter contains the conclusions of this paper.

2 Theoretical Background

2.1 Puzzle Videogames

Most commonly, puzzle videogames have an appointed goal, which is achieved through thinking, planning, analyzing the surrounding, strategizing, and thought-out decision making. By solving the presented tasks in puzzle games, players get a sense of accomplishment which pushes them forward. This aspect of puzzle videogames makes this genre more appealing, and more unique than other genres of videogames. The tasks in puzzle videogames are often demanding, they pose as a challenge, and present a peculiar test of mental capabilities and dexterity in different environments for the player. Some of the skills tested in puzzle videogames include deduction, pattern recognition, decision making under pressure, memory capabilities, creativity, spatial and temporal abilities etc. The evolution of puzzle videogames is best noticed through the increase of complexity and level of the challenges that are presented in the videogames. The games that are considered as classics, such as Tetris and Minesweeper, have a very simple premise [4, 9]. On the contrary, puzzle videogames of today are much more complex and demand a different set of skills than the aforementioned ones. The perfect examples of these videogames would be the Portal franchise, The Talos Principle and The Witness [10–12]. Each of these games has a unique learning curve and most often demands thinking outside of the box as an additional concept of challenge.

2.2 Logical Reasoning as an Activity of the Brain

Logical reasoning represents an array of processes that are going on in the brain, most often in the frontal lobe [13]. The frontal lobe controls many skills that are directly influenced by puzzle videogames, i.e., problem solving, concentration, planning, decision making etc. The part of the frontal lobe that oversees these tasks is the prefrontal cortex. Aside from the aforementioned functionalities, the prefrontal cortex is also in charge of other processes, such as intuition, pattern recognition, short-term memory, spatial awareness, and decision making based on environmental variables and available information [14]. The part of the brain that oversees long-term memory, organizing and cognitive sequencing is the temporal lobe [15]. These functionalities are directly connected to the tasks of the prefrontal cortex and the frontal lobe, by which the parts of the brain supplement each other.

2.3 Analytical Hierarchy Process

In practice, a problem of multicriterial decision-making is commonly encountered, for which different methods could be of use. One of such methods is the analytical hierarchy process (AHP). AHP is one of the most useful mathematical tools for the analysis of complex decisions with multiple alternatives based on multiple criteria [16].

The first step in solving the problem is defining the problem and setting a clearly established goal. Firstly, dividing a complex problem into a hierarchy of easily understandable problems is an imperative, so that those problems could be individually analyzed. After forming a structured and clearer hierarchy, the next step is scoring different elements by comparing them in pairs, according to their overall impact on the element that is above them in the formed hierarchy. Since the essence of AHP is using human estimation, and not only the basic information that is connected to a certain problem, during comparison, alongside using specific information about elements, decision makers can use their own personal opinion about the significance of specific elements [16]. The target is to convert descriptive scores into numeric values that can be further processed and compared. Every element is assigned a weight value (priority value). By assigning weight values, the comparison of elements that are seemingly different and incompatible is enabled. The last step includes calculating numerical priorities for each of the set alternatives.

The AHP methodology uses matrices for ranking and comparing alternatives and criteria. Standard numerical values from the Saaty's scale are used inside the matrices. It should be mentioned here that Thomas L. Saaty invented and developed the AHP method [17, 18]. The Saaty's scale contains nine different numeric scores, which are used for differentiating the intensity of relationships between certain elements of the hierarchical structure in their pair-wise comparisons.

The AHP method uses the axiom of reciprocity, which means if alternative no. 1 has one of the values from the Saaty's scale in comparison with the alternative no. 2, then the alternative no. 2 has the reciprocal value when compared to the alternative no. 1.

$$a_{ij} \cdot a_{ji} = 1 \tag{1}$$

The same applies whilst comparing criteria [16]. Calculating the weight of the criteria is based on the pair-wise comparison, which is used to form a decision matrix. After forming the mentioned matrix, a normalized pair-wise comparison is conducted, in which every element of the matrix is divided by the sum of elements of their affiliated column:

$$\overline{a_{ij}} = \frac{a_{ij}}{\sum_{k=1}^{m} a_{kj}} \tag{2}$$

where m represents the number of criteria.

The consistency check in the AHP method is possible, and it is based on the consistency index (CI) and the consistency ratio (CR), both of which are calculated by the following formulas:

$$CI = \frac{\lambda_{max} - n}{n - 1} \tag{3}$$

$$CR = \frac{CI}{RI} \tag{4}$$

where n represents the number of alternatives, λ_{max} represents the maximum eigenvalue of the matrix, and RI represents the random index. The values of RI are obtained based on the number of alternatives n. The maximum eigenvalue of the matrix is equal to the series n:

$$\lambda_{max} = n \tag{5}$$

An ideal consistency would be accomplished if $CI = 0$, but smaller inconsistencies are tolerable so that $0 < CR < 0.1$ [19]. If the consistency ratio is in bounds of $0 < CR < 0.1$, the repeating of the procedure is no longer needed, and it is considered that it is sufficiently correct. If the consistency ratio was larger than 0.1, the decision matrix would be inconsistent, and the investigation of the reasons that caused the unacceptably high inconsistency would be needed, so that the causes may be removed by partially repeating the pair-wise comparison. After the repeated comparison, if the consistency calculated would be higher than 0.1 (10%), the procedure would have to be discarded completely, and performed from the beginning [16].

3 Methodology

3.1 Questionnaire About Puzzle Videogames

This part of the paper describes the questionnaire that is specifically designed and used for collecting data about the scores of individual criteria for each of the analyzed videogames. Each question contains a scale from 1–10 for scoring the criteria. The questionnaire was made by using Google Forms [20], and distributed online to the subjects that played the analyzed games. For each of the analyzed videogames, the following questions were asked:

1. Rate the importance of reaction time on a scale of 1–10.
2. Rate the importance of pattern recognition on a scale of 1–10.

3. Rate the importance of short-term memory (fast logical thinking i.e. multiple calculations in a short span of time) on a scale of 1–10.
4. Rate the importance of spatial abilities on a scale of 1–10.
5. Rate the importance of task-solving under pressure on a scale of 1–10.
6. Rate the importance of sequencing (i.e. fast movement-related thinking) on a scale of 1–10.
7. Rate your enjoyment level on a scale of 1–10.
8. Rate the premise complexity (extensiveness, difficulty, convolutedness) on a scale of 1–10.
9. Rate the graphics on a scale of 1–10.

3.2 Calculation Using the Method of Analytical Hierarchy Process

This paper uses AHP methodology for analyzing the influences of six different videogames on logical reasoning. For conducting such an analysis, according to the theory of AHP methodology, which is presented in the previous chapter, the definition of criteria used for ranking videogames is essential. The analysis was conducted via AHP Online System [21]. The criteria selected for this analysis are the following:

1. Reaction time
2. Pattern recognition
3. Short-term memory
4. Spatial abilities
5. Task-solving under pressure
6. Cognitive sequencing
7. Entertainment level
8. Premise complexity
9. Graphics
10. Metacritic score.

After assigning coefficients during the process of pair comparison by the Saaty's scale, the obtained results are presented in the Table 1, and the decision matrix is presented in the Table 2.

Since ten different criteria are used, 45 pair comparisons were conducted, and the consistency factor CR with a value of 0.069 was obtained, which presents a valid and satisfying result. For the purposes of the analysis, the following videogames were used:

- Minesweeper
- Tetris
- Cut the Rope
- Portal 2
- The Talos Principle
- The Witness.

After scoring individual games by every parameter according to the playing experience, and by normalizing the values, the obtained results are shown in the next chapter.

Table 1. Weight coefficients of the criteria based on the pair-wise comparison analysis

Criteria	Priority (%)
Reaction time	2,5
Pattern recognition	13,0
Short-term memory	4,0
Spatial abilities	13,3
Task-solving under pressure	16,9
Cognitive sequencing	2,4
Entertainment level	29,4
Premise complexity	8,7
Graphics	1,6
Metacritic score	8,2

Table 2. The decision matrix showing relationships between criteria based on the pairwise comparison

	1	2	3	4	5	6	7	8	9	10
1	1	0.25	0.33	0.17	0.20	1.00	0.12	0.33	2.00	0.33
2	4.00	1	6.00	2.00	1.00	4.00	0.20	1.00	7.00	2.00
3	3.00	0.17	1	0.20	0.14	3.00	0.14	0.33	5.00	0.25
4	6.00	0.50	5.00	1	0.33	6.00	0.33	3.00	6.00	3.00
5	5.00	1.00	7.00	3.00	1	4.00	1.00	1.00	6.00	2.00
6	1.00	0.25	0.33	0.17	0.25	1	0.11	0.25	2.00	0.25
7	8.00	5.00	7.00	3.00	1.00	9.00	1	6.00	9.00	4.00
8	3.00	1.00	3.00	0.33	1.00	4.00	0.17	1	6.00	1.00
9	0.50	0.14	0.20	0.17	0.17	0.50	0.11	0.17	1	0.12
10	3.00	0.50	4.00	0.33	0.50	4.00	0.25	1.00	8.00	1

4 Results and Discussion

A convenient sample of 22 questionnaire results are used for calculating average scores for every given criterion for each of the analyzed videogames. The reason behind such a relatively small sample size is because not a lot of people know all the videogames used in this research. As a result, the questionnaire results were collected and processed, and the normalized average scores for every videogame are presented in the Tables 3, 4 and 5. The criteria is scored as "better" if its normalized score is higher. The criteria "Metacritic score" is the score taken for every game from the site metacritic.com. That score was static and independent from the questionnaire results [22].

Table 3. Normalized values of scores for each of the analyzed videogames

Criteria/Game	Reaction time	Pattern recognition	Short-term memory	Spatial abilities
Minesweeper	0.23	0.79	0.52	0.44
Tetris	0.86	0.63	0.70	0.78
Cut the Rope	0.78	0.68	0.60	0.77
Portal 2	0.84	0.86	0.82	0.92
The Talos Principle	0.84	0.84	0.80	0.89
The Witness	0.50	0.90	0.68	0.78

Table 4. Normalized values of scores for each of the analyzed videogames

Criteria/Game	Task-solving under pressure	Cognitive sequencing	Entertainment level
Minesweeper	0.37	0.42	0.44
Tetris	0.68	0.83	0.65
Cut the Rope	0.63	0.86	0.77
Portal 2	0.78	0.88	0.87
The Talos Principle	0.52	0.71	0.80
The Witness	0.68	0.70	0.52

Table 5. Normalized values of scores for each of the analyzed videogames

Criteria/Game	Premise complexity	Graphics	Metacritic score
Minesweeper	0.56	0.22	0.83
Tetris	0.48	0.31	0.68
Cut the Rope	0.60	0.65	0.93
Portal 2	0.77	0.78	0.95
The Talos Principle	0.84	0.81	0.85
The Witness	0.92	0.72	0.87

With all the results represented in Tables 3, 4 and 5, all the requirements for ranking the videogames are ready (Fig. 1).

The results from the table of every criteria for each game by priority, made by using the weight coefficients presented in Table 1 and normalized scores presented in Tables 3, 4 and 5, is presented in Tables 6, 7 and 8, as well as visually represented on the Fig. 2.

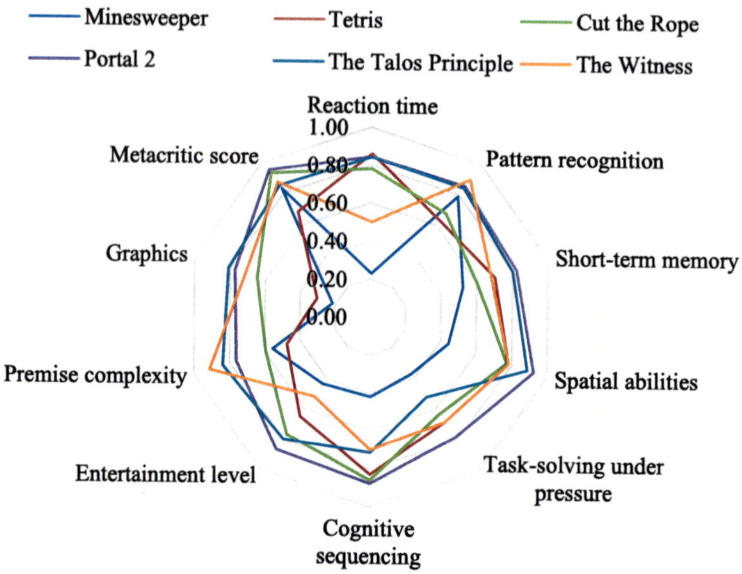

Fig. 1. Visual representation of the normalized values of scores for each of the analyzed videogames

Table 6. Parameters by priority for each of the analyzed videogames

Criteria/Game	Reaction time	Pattern recognition	Short-term memory	Spatial abilities
Minesweeper	0.006	0.103	0.021	0.059
Tetris	0.022	0.082	0.028	0.104
Cut the Rope	0.020	0.088	0.024	0.102
Portal 2	0.021	0.111	0.033	0.123
The Talos Principle	0.021	0.110	0.032	0.118
The Witness	0.013	0.117	0.027	0.104

Table 7. Parameters by priority for each of the analyzed videogames

Criteria/Game	Task-solving under pressure	Cognitive sequencing	Entertainment level
Minesweeper	0.063	0.010	0.129
Tetris	0.115	0.020	0.191
Cut the Rope	0.106	0.021	0.226
Portal 2	0.131	0.021	0.255
The Talos Principle	0.087	0.017	0.235
The Witness	0.115	0.017	0.153

Table 8. Parameters by priority for each of the analyzed videogames

Criteria/Game	Premise complexity	Graphics	Metacritic score
Minesweeper	0.049	0.004	0.068
Tetris	0.042	0.005	0.056
Cut the Rope	0.052	0.010	0.076
Portal 2	0.067	0.012	0.078
The Talos Principle	0.073	0.013	0.070
The Witness	0.080	0.012	0.071

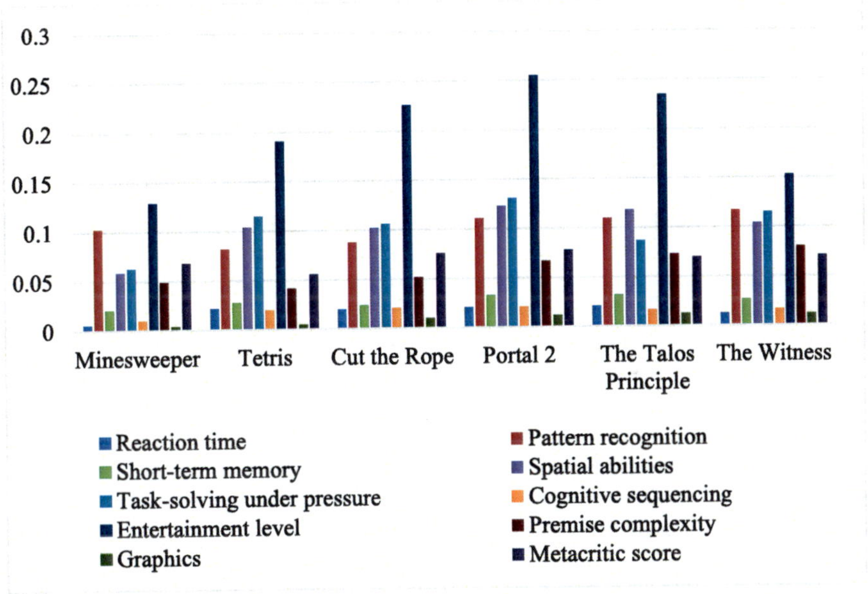

Fig. 2. Visual representation of the parameters by priority for each of the analyzed videogames

According to the coefficients presented in Tables 6, 7 and 8, a summed calculation of priorities is made for each of the videogames, which will clearly indicate which of the analyzed videogames has the most influence on logical reasoning of the player.

As shown on the Fig. 3, the results of the conducted analysis indicate that out of all analyzed videogames, Portal 2 has the most influence on logical reasoning, with a coefficient of 0.852. It is followed by The Talos Principle and Cut the Rope with coefficients of 0.776 and 0.727, respectively. The last three games are The Witness with a coefficient of 0.708, and Tetris and Minesweeper with coefficients 0.664 and 0.510, respectively. This analysis demonstrated that different aspects of games that are determined in the process of the game design can affect the players logical reasoning. However, the entertainment level and task-solving under pressure were the most important factors.

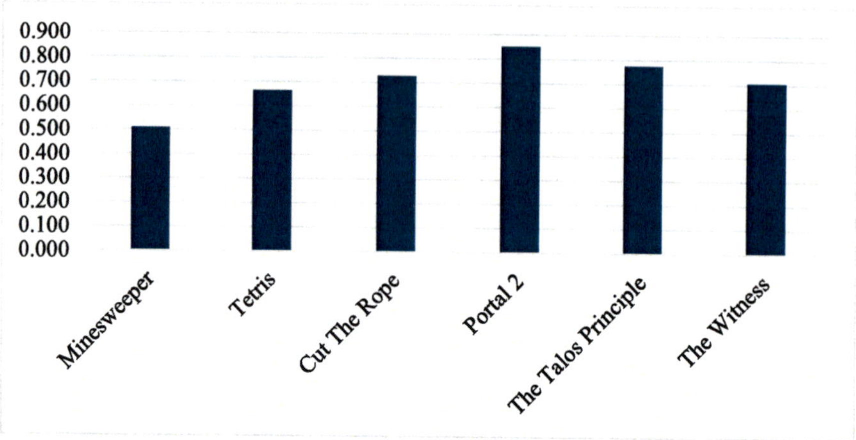

Fig. 3. Summed ranking of the analyzed videogames

5 Conclusion

Puzzle videogames represent a genre of videogames that have an appointed goal which is achieved through thinking, planning, analyzing the surrounding, strategizing, and thought-out decision making. By playing puzzle videogames, players often get a sense of accomplishment, which makes this genre appealing, and on the other hand, improves logical reasoning.

This paper contains an analysis of influences of puzzle videogames on player's logical reasoning. The analysis was conducted by using the AHP methodology. The results of the analysis have shown that modern puzzle games significantly influence logical reasoning of the players, and the calculation results for six different videogames are shown in this paper.

The conducted research may be useful in the process of designing puzzle games, i.e. the designers of videogames and game development companies may find this research helpful, in the way that the criteria presented serve as a guideline of which elements of the game should be optimally fulfilled.

This paper contributes to the already established knowledge base, by analyzing and comparing the influence of different puzzle games, which improve the logical reasoning by their own positive influences. This is very important in the early mental development of young users.

References

1. Casey, D.: Why Atari's PONG Matters 45 Years Later (2017). https://nerdist.com/article/atari-pong-45th-anniversary-history-video-game/. Accessed 31 Aug 2020
2. Alsop, T.: Percentage of households with home computers in the United Kingdom (UK) from 1985 to 2018 (2020). https://www.statista.com/statistics/289191/household-penetration-of-homecomputers-in-the-uk/. Accessed 31 Aug 2020

3. Weiss, B.: In: Classic Home Video Games, 1972–1984: A Complete Reference Guide. s.l.:s.n., p. 316 (2011)
4. Cobbett, R.: The most successful game ever: a history of Minesweeper (2009). https://www.techradar.com/news/gaming/the-most-successful-game-ever-ahistory-of-minesweeper-596504. Accessed 31 Aug 2020
5. Brilliant, T.D., Nouchi, R., Kawashima, R.: Does video gaming have impacts on the brain: evidence from a systematic review. Brain Sci. **9**(10), 251 (2019)
6. learningworksforkids.com: Cut the Rope - About Page (n.d.). https://learningworksforkids.com/playbooks/cut-the-rope/. Accessed 1 Sept 2020
7. Oei, A.C., Patterson, M.D.: Playing a puzzle video game with changing requirements improves executive functions. Comput. Hum. Behav. **37**, 216 (2014)
8. Shute, V.J., Ventura, M., Ke, F.: The power of play: the effects of portal 2 and lumosity on cognitive and noncognitive skills. Comput. Educ. **80**, 58–67 (2014)
9. Shaver, M.: What Was the World Like the Year Tetris was Born? (2017). https://tetris.com/article/128/what-was-the-world-like-the-year-tetris-wasborn. Accessed 31 Aug 2020
10. Valve Corporation: Portal 2 - About Page (2011). https://www.thinkwithportals.com/about.php. Accessed 1 Sept 2020
11. Croteam: The Talos Principle - Official Page (2014). http://www.croteam.com/talosprinciple/. Accessed 1 Sept 2020
12. ign.com: The Witness - IGN Page (2016). https://adria.ign.com/the-witness. Accessed 1 Sept 2020
13. Mayfield Clinic: Anatomy of the Brain (2018). https://mayfieldclinic.com/pe-anatbrain.htm. Accessed 1 Sept 2020
14. hr.divineinnerhealth.com: Što je prefrontalni korteks (n.d.). https://hr.divineinnerhealth.com/ventromedialnaja-prefrontalnaja-kora.php. Accessed 1 Sept 2020
15. Patel, A., Biso, G.M.N.R., Fowler, J.B.: Neuroanatomy, temporal lobe. In: StatPearls. StatPearls Publishing, Treasure Island (FL) (2020)
16. Brunelli, M.: Introduction to the Analytic Hierarchy Process. SpringerBriefs in Operations Research. Springer, Cham (2015). https://doi.org/10.1007/978-3-319-12502-2
17. Saaty, T., Vargas, L.G.: Models, Methods, Concepts & Applications of the Analytic Hierarchy Process, pp. 6–9. Springer, New York (2001)
18. Hunjak, T., Jakovčević, D.: Višekriterijski modeli za rangiranje i uspoređivanje banaka. In: The Proceedings of Zagreb Faculty of Economics and Business, p. 48 (2003)
19. Saaty, T.L.: Decision-making with the AHP: why is the principal eigenvector necessary. Eur. J. Oper. Res. **145**(1), 87–88 (2003)
20. google.com: Google Forms - About (n.d.). https://www.google.com/forms/about/. Accessed 11 Mar 2021
21. Goepel, K.: AHP Online System - AHP-OS (2018). https://bpmsg.com/ahp/. Accessed 6 Oct 2020
22. metacritic.com: Video Game Reviews - metacritic.com (n.d.). https://www.metacritic.com/game. Accessed 4 Oct 2020

Development of 3D Serious Game: Colorful Classroom to Effectively Teach Children Colors Letters and Numbers

Rijad Sarić[1(✉)], Mevlid Halilović[2], Damir Bajramović[3], Gabriela Raducan[1], and Edhem Čustović[1]

[1] School of Engineering and Mathematical Sciences, La Trobe Innovation and Entrepreneurship Foundry (LIEF), La Trobe University, Melbourne, Australia
{r.saric,g.raducan,e.custovic}@latrobe.edu.au

[2] Faculty of Engineering and Natural Sciences, International University of Sarajevo, Sarajevo, Bosnia and Herzegovina
mevlidhalilovic@student.ius.edu.ba

[3] Faculty of Electrical Engineering, University of Banja Luka, Banja Luka, Bosnia and Herzegovina
damir.bajramovic@etf.unrs.ba

Abstract. The application of serious games (SGs) has shown significant improvements in the education of both children without disorders and children with disorders such as autism. The major advantage of SGs over the traditional teaching approach is the possibility to boost motivation and engagement of children during the learning process. Currently, formal educational systems of many developed countries integrate SGs into primary and secondary schools. This re-search focuses on developing a 3D game called Colorful Classroom to enable children learn letters of the English alphabet, colors and numbers. The entire concept of the game is established using the Unity Engine, and all 3D models, including classroom and other objects, are modelled from scratch in Maya Autodesk. Evaluation of the game was conducted on two testing samples com-posed of typically developing, mentally delayed and autistic children. The findings show progressive learning of both tested groups and positive feed-back from teachers regarding the game design and logic. Finally, the game described in this research paper could be enhanced by designing more classes in different lessons.

Keywords: Serious games · Education · 3D modeling · Unity game engine

1 Introduction

The gaming industry has become one of the fastest-growing industry sectors as a result of progressive innovation. The extensive development of computer graphics has significantly improved the quality of video games and had allowed easy access including full implementation on mobile or online platforms (Santasärkkä 2017). As of 2019, the gaming market represents the core of entertainment, with more than 2.5Bn game

N. Ademović et al. (Eds.): IAT 2021, LNNS 316, pp. 248–263, 2022.
https://doi.org/10.1007/978-3-030-90055-7_19

enthusiasts over the world. In the same year, gaming consumers spent $151.21Bn on various video games, generating an increase of 9.6% in 2019. Console and mobile gaming (smartphone and tablet) have remained the two most used platforms for gaming in the last six years. However, total growth of 3.5% to $39.5Bn is assumed regarding the computer gaming market by 2022 (Wijman 2019; McDonald 2017). Generations encompassing the mind 1990's onwards, better known as Gen Z, iGen or Net Gen, feel comfortable in using the Internet and new technologies from a very young age. According to a report by NDP Group in 2009, responsible for global market research, 91% of the population under the age of eighteen intensively play video games. However, the highest growth of game users is noticeable in children aged 2–5, who play games mostly for entertainment purposes. In addition to that, it indicates that male and female teens between the age of 15–17, who play video games faced great attention problems while listening to school lessons (Wijman 2019; Resinger 2011). The sharp increase of games containing a balance between serious and playful stories is also noticeable in the past five years. These games require a compulsory assessment before production since their real impact has to be adequately understood (McDonald 2017). The rapid development of 3D technologies has also resulted in the integration of such technology into the gaming industry. 3D effects have almost entirely replaced limited 2D effects/technology, bringing qualitative change to video gaming experiences. Vivid scenes found in high-end Hollywood productions can now readily be implemented in video games. All video games can be classified with respect to the visual effect's implementation. In 2D games, graphics elements have only height and width and the camera has a fixed perspective, where buildings and other objects are constructed of single map elements. In 3D games, all graphics elements are created of several three-dimensional graphics polygons. The third category is the pseudo-3D game, also known as 2.5D game, where the needs of player are fulfilled using the features from 2D and adding the 3D with certain limitations (Resinger 2011; Zhang et al. 2015; Cao et al. 2015).

Early learning, as well as child education, can be significantly improved by including adequate modern technologies e.g., educational games. In developed countries, it is no longer considered a novelty to use computers to educate, empower and engage during the school lessons. In the last year, the outbreak of the COVID-19 pandemic resulted in almost every nation, enforcing self-quarantine measures on their citizens. Teaching and learning were forced to undergo rapid digitalization and online became the only way to deliver education. National television stations began broadcasting school lessons in some countries to substitute and complement other online teaching efforts. Evidently, distance learning and other forms of technology-based learning became crucial in both formal and informal education.

According to Backlund and Hendrix (2013), most recently, educational games became considerably more acceptable as an assistive tool in the education of children with disabilities. Educational games can be considered as "serious games" usually running on PCs or mobile devices to keep children motivated while learning. In general, the term serious game (SG) can be defined as specifically designed games whose main purpose goes beyond pure entertainment. Compared to traditional learning methods including textbooks, SGs achieve impressive results in terms of child motivation, better learning outcomes, engaging & entertaining experiences, and adjustment to different learning abilities which are especially useful for children with disabilities (Connolly et al. 2012; Durlak et al. 2011). Eventually, the critical role in many SGs has an assessment of learning performance. It is important to understand competencies as well as the learning progress of the children as learners. The performance rate is determined by using aspects such as measuring, gathering, analyzing, and interpreting achieved scores, task completion rates, completion times, success rates, etc. (Wiemeyer et al. 2016). The development of an effective SG requires considerable effort. The key requirements are long technical lifetime, high impact as well as a high number of users and in many cases large budget for serial production (Guttenbrunner et al. 2010). Hence, the implementation of one SG could have numerous challenges. During the implementation phase of the SG, it is necessary to obtain useful feedback to achieve a valuable user experience. The main challenge is the ability to prove the effectiveness as well as the practicality of the SG as an educational tool. Evaluation of the game effectiveness is determined by taking into account the learning performance of children. Measuring required parameters can assist teachers to understand the level of game effectiveness as a pedagogical tool. Additionally, it is important to address the issues of practical use of games while teaching children. There are a few issues that should be considered such as user & general acceptance that includes students, teacher and parents, technological restrictions, and appropriate adjustment to school curricula (Kirriemuir and McFarlane 2004; Barr et al. 2011).

By nature, children are playful and curious. They learn about our world via interaction. As they grow, they begin to settle, especially nowadays when having a mobile phone or PC is no longer a luxury. Gaming represents a way of entertainment that children eventually discover, sometimes much earlier than their parents would prefer. However, it is not something that should be feared, but rather an opportunity to combine entertainment with education. This form of education becomes particularly useful in scenarios where children with disabilities are failing to keep up with their peers. Finding a special school for such children is proving difficult, particularly in less developed regions that are lagging in the provision of programs and infrastructure for these children. The aim of this paper represents the development of an SG of educative nature. The game enables children with intellectual disabilities, developmental delays, and healthy preschoolers to quickly learn to recognize uppercase and lowercase letters of the English alphabet including fundamental colors in a 3D virtual classroom environment. The game design consists of progressive learning accomplished by increasing the game level difficulty. The major advantage can be tested by tracking the progress of children over a particular time-period.

2 Related Work

The concept of SGs proved to be successful in teaching children since the level of enjoyment in the classroom increases when learning the class material subconsciously. In the 21st century, computational thinking became a necessary skill for every individual. It represents successful identification of a problem including steps that lead towards problem-solving using a PC. In recent years, children in their early age have been introduced to programming in schools to enhance their computational reasoning (Barr et al. 2011;

Little et al. 2008). There is a myriad of research reviews investigating the integration and the impact of SGs into the classroom, to adequately enhance the computational skills of children (Cheng et al. 2015; Kempf and Martin 2013). For instance, Vahldick et al. (2015) proposed SGs for learning fundamental programming skills. The games, C++- or Java-based, are divided according to the type, platform, competency, and storyline. An important question that arises, is what limitations are of educational games as far as pedagogy is concerned. In particular, the findings of Kerres et al. (2019) acknowledged that almost every game can generate some valuable experience, suitable for making a connection with teaching and using it for reflection. It is also pointed out that SGs can most likely initiate better acquisition of declarative information together with standard motor learning. Cheng et al. (2015) conducted a systematic review of SGs focusing on science education from 2002 to 2013. According to this review, the adventure or role-playing is the most popular game type (29 studies) followed by simulation games, (11 studies) and puzzle games (6 studies). Adding virtual reality (VR) or augmented reality (AR) technology makes a significant impact on SGs by offering the more realistic experience of single gameplay (Lugrin et al. 2016; 2018).

Furthermore, SGs have become increasingly applicable in a wide range of medical therapies such as pain management and rehabilitation (De Freitas 2006). As stated by Bellani et al. (2011) such games combined with VR technologies offer great benefits for children with a serious neurodevelopmental disorder known as Autism Spectrum Disorder (ASD). Children suffering ASD have severe issues to socialize, communicate and speak in their surrounding community. SGs designed for children with special needs are usually combined with "play therapy" e.g., building LEGO blocks which allow children to interact with the virtual world in their own way and therefore improve their social and communication skills (Morgenthal 2015). For example, ComFiM (Ribeiro and Raposo 2014) and Open Autism apps (Hourcade et al. 2013) are developed to achieve positive interaction among children with ASD using a touchscreen tablet. Similarly, GameBook (Cunha et al. 2016) refers to a mobile app that tends to boost social and cognitive skills of children through imagination and motivation. In addition to that, Bossavit and Parson discuss the work where high functioning ASD teens participated in developing their own SG in order to better understand Geography lessons (Bossavit and Parsons 2016). Fletcher-Watson et al. (2016) described the workshop for enhancing communication skills where pre-school ASD children were designing the SG for iPad.

Developing a SG for children with ASD can be a challenging task because social interaction and communication problems need to be addressed. To improve social and communication skills during therapy exercises, different strategies are used. A large number of children with ASD respond much better to visual instructions, rather than verbal ones. Also, using visual feedback has been proven to be a successful strategy to improve social interaction during exercises (Morgenthal 2015; Hodgdon 1995; Lesley et al. 2008). Thus, two types of gaming systems were developed for ASD children namely,

- Graphical User Interface (GUI) game approach
- Traditional techniques that deploy tangible objects such as LEGO game sets.

Playing games, children learn as they play, solving puzzles while staying within the constraints of the virtual environment. In this way, conventional teaching methods like writing in the classroom can be transformed from the workspace to the virtual game space (Colby and Colby 2008). Hence, during the development of SGs, it needs to be ensured that intended learning outcomes and features (e.g., the level of difficulty, duration, aesthetic, and interaction modalities) do not conflict with each other (Clark 2003). De Loper et al. (2017) demonstrated an SG with a high story and educational impact, achieving an excellent balance between entertainment and learning components, during the development of the conceptual model. Overall, the common issue observed is the lack of teachers' feedback during the design and development of 3D serious games (Backlund and Hendrix 2013).

3 Materials and Methods

In general, game programming covers a wide area of expertise including gameplay implementation with simple gameplay triggers using visual scripting languages, GPU programming, and mathematics/optimization techniques. In this research study, the SG named "Colorful Classroom" was developed using the Unity Engine, including Standard Asset packages and other free packages available in the Unity Store. Almost all 3D assets were made using Autodesk's Maya 2016. Textures of the designed 3D models were generated using Cinema4D's noise generators, in conjunction with its material editors. Pronunciations of letters and numbers were recorded using the open-source digital audio workstation, called Audacity, with a standard microphone. All coding was undertaken in MonoDevelop available with the Unity game engine.

3.1 Game Design

The entire gameplay is composed of a 3D classroom environment with a single player in the first-person view. The virtual classroom was modelled in such a way to illustrate the same atmosphere as in the actual primary school classroom. Inspiration for classroom design was found in "Carpet teaching areas in primary school" (Charles 2017), published by Primary Education the largest education union in the UK. In Fig. 1, the picture of the primary school classroom taken as a reference is shown. Any element not visible in the classroom photo as indicedted in Fig. 1 is left to 3D modeler imagination.

Aside from the wall clock designed with synchronized local time in GMT+2 zone, which is also a free asset made by VIS Games available on Unity Store, the entirety of the classroom and all other objects such as chairs, tables, pens, abacus, boards, bins, balls, etc. were made from scratch, using 3D modelling software Maya, developed by Autodesk. Textures used for certain objects were created by generating a black and white noise map. They were layered with reduced opacity, on top of a solid color layer, resulting in shapes that can be seen on the floor throughout the classroom, and with a different color on a floor mat in front of the board. The optimization of the game is important since 3D models with a large number of polygons can drastically increase the computational complexity of a game. Figures 2a), b) and c) show the 3D model of the classroom together with all available school equipment from three different perspectives.

Fig. 1. The primary school classroom used as a reference for designing the 3D environment of the serious game (Primary Practice 2017)

3.2 Game Logic

The centerpiece of this high-level SG is the user-interface (UI) canvas that is activated when a player steps in front of the interactive board and presses the Enter key. Once the event is triggered, the first-person view is disabled, and the camera scene is switched to a UI of an onboard game. From there, the player is prompted to select the desired level before commencing to play the game. The goal of each level is to determine a letter-color, letter-letter or number-color pair that is randomly generated, with each pair corresponding to a certain audio notification and visual representation. The first three levels are sorted by difficulty from easiest, three by three (three uppercase letters, three colors) matching, to hardest nine by nine (nine uppercase letters, nine colors) matching. The fourth and fifth levels introduce matching between uppercase and lowercase letters as well as matching between the first nine natural numbers and the basic colors. When a player starts one of the first three levels, on the left side of the board there are 3 to 9 random circles with letters, whereas on the right-side the random color circles are generated. All colors, letters and numbers are generated from a predetermined set stored in the internal database. The correct answer of each level is shown on the top of the board where a circle with the correct color and letter, uppercase and lowercase letter or lastly number and color indicate the answer pair. By clicking a specific color, letter or number, the player will be given a pronunciation of that letter, color, or number in the form of a sound.

Forming a correct letter-color, letter-letter, or number-color pair results in a "correct" sound and a message stating so. In case of an incorrect answer, a "wrong" sound is generated with the corresponding message. Upon reaching a correct answer, the player received a point towards the next difficulty level (4/18). However, if the answer is incorrect, then points get deducted, and the game allows the player one more attempt for each mistake made, before going any further. To progress to the next level of difficulty, the

a) A 3D models of the virutual classroom in the first perspective corner

b) A 3D models of the virutual classroom in the second perspective corner

Fig. 2. continued

c) A 3D models of the virutual classroom in the third perspective corner

Fig. 2. The 3D model of the virtual classroom and school equipment developed to describe the real classroom environment.

player needs to accumulate enough points to pass a certain threshold. Each progressive level gets three new letters and colors to choose from, increasing the variety of options, up to nine letters, and nine colors. In Figs. 3a), b) and c) the general scheme of the UI design as well as developed UIs from the game are represented.

Levels four and five are experimental, purposely designed to examine how well children, after solving the same kind of problem, adapt to a new one in a similar setting. It is also necessary to describe the dynamic behavior of high-level SGs by identifying all actions of the main elements in the game as well as the point of interest. Establishing a starting reference point, in order to describe activities of how certain game aspects are connected, represents a designed sequence made using Unified Modelling Language (UML) (Marlon and Arthur 2001). When the gameplay starts, the player identifies elements related to each node in the game sequence. Each event of the game is divided into action, tools, or goals i.e. "Next level threshold". Figure 4 shows the activity diagram that describes one gameplay of "Colorful Classroom" SG where a player can choose between five levels and play a level until reaching a certain number of points to proceed to the next level. The player can return to the main menu at any point of the ongoing gameplay.

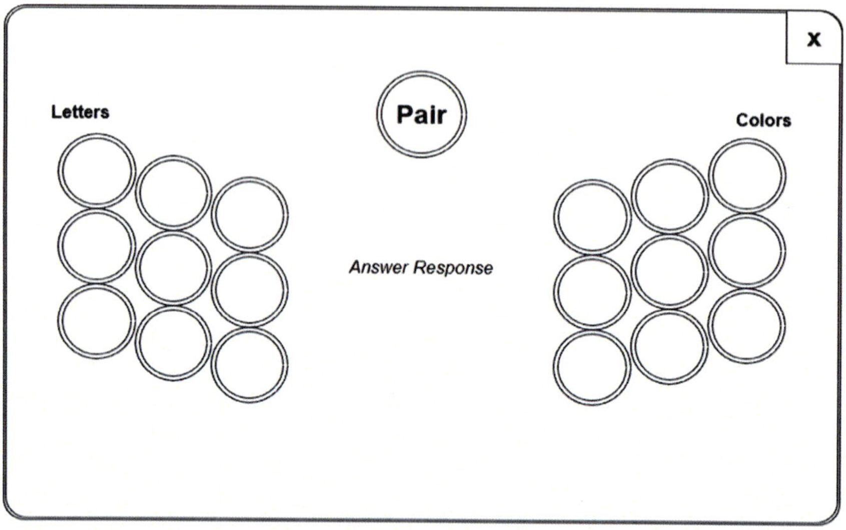

a) The basic UI scheme applied for game development

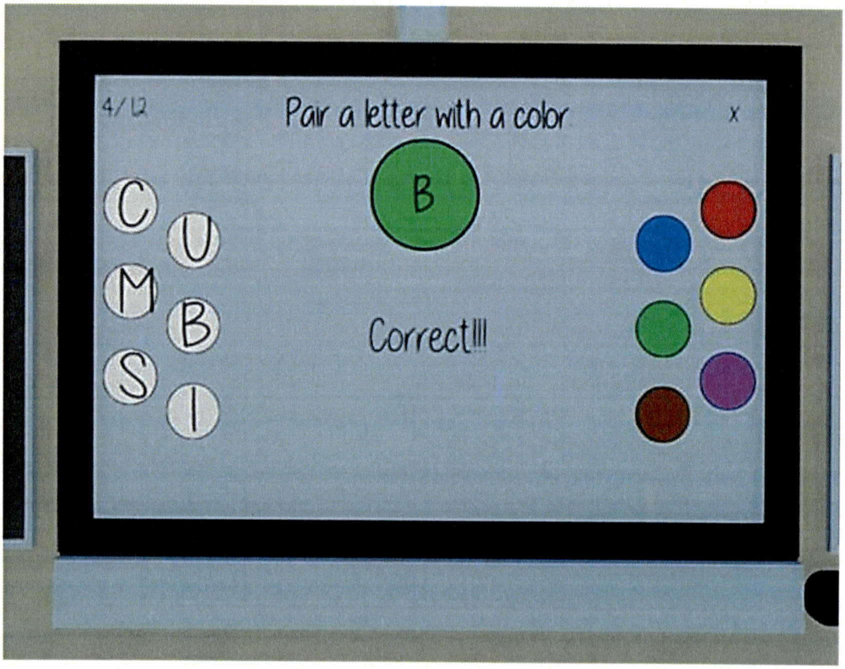

b) The example of interactive board second level developed based on the UI designed scheme

Fig. 3. continued

c) The example of interactive board fifth level developed based on the UI designed scheme

Fig. 3. UI developed for playing SG of matching letters, colors, and numbers on the interactive board in the classrom

3.3 Associative Learning Approach

Associative learning is the process of learning something in response to any two arbitrary stimuli. This method of finding meaningful relationships between words, pictures and their referents has been used for a long time in our society (Preissler 2008). When a child follows instructions and listens quietly while someone is talking, such behavior should lead to some certain rewards such as sweets, toys, words of affirmation, etc. Contrary to that, failure to abide by the provided rules results in an unpleasant situation or punishment. In the above-mentioned situations, a child learns not only the value of obeying the rules but also the price for breaking them. In the context of understanding pictures, letters, numbers, and colors, typically developing children understand the representational role of the picture or letter by 24 months of age and they are capable of using that information to identify objects in the real word by 30 months (Deloache and Burns 1994).

In the SG implemented in this research, the associative mapping approach was applied. Multiple stimuli were established, some of which were visual and audio cues – the shape of the letter/numbers or color including audio pronunciation. Moreover, another stimulus was implemented through correct/incorrect events, where, in case of a correct answer, the player was presented with a "reward" in the form of a pleasant chime-like sound. In contrast, the case of incorrect answer results in an unpleasant, irritating sound. This way, the first set of stimuli created associations between color or shape of the letter and its pronunciation, while the second one attempted to break unnecessary

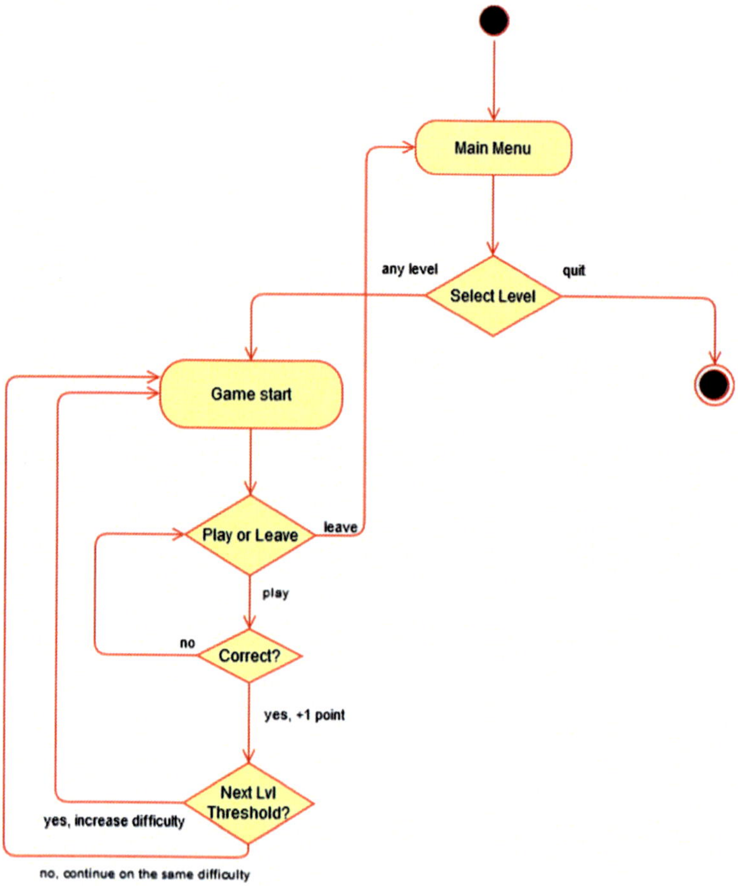

Fig. 4. UML activity diagram generated to demonstrate the dynamic behavior of the created game

associations between letters and colors e.g., letter "A" getting associated with the color red. This approach did not change the expectations of levels four or five, even though, in case of level four, where dealing with uppercase and lowercase letters, the necessity of second stimuli was much clearer than in the previous levels.

4 Results and Discussion

Testing of the developed SG was conducted on two different data samples. The first test consisted of 50 children, aged from 4 to 9, which were previously confirmed not to have any prior knowledge about the English alphabet. Among these, 20 were developmentally delayed and the other 30 were typically developing children. The testing period was two weeks, at most 30 min per day. The second test was made on children with autism, along with typically developing children with similar levels of development, to examine how well they progress through the developed SG. In this case, however, the sample was smaller (30 children), as the number of children with autism was not so large. The game was played by 19 typically developing children and 11 children with autism for three weeks (21 days), 30 min per day.

According to the teacher's report comments, the overall response of the children in the first test was very satisfying in terms of their performance. After testing, the teachers provided feedback about children experience during gameplay. In the report, teachers explained that the participants liked the concept as well as the design of the game, and the children found the game engaging and fun. The SG was supplied with an official manual document composed of all the guidelines for the teachers and what is expected from the children in terms of game interaction. The detailed instructions regarding the commands and how to proceed to the next level are available in the game main menu, after starting the game by running the.exe file. The analysis of the obtained average scores indicates that healthy children, aged 4 to 6, did not have any issues during the progression through the game playing day after day in the classroom, under the supervision of their teachers. Oppositely, referring to Perna (2013), children diagnosed with developmental delays, aged poorly developed their cognitive skills (IQ under 70). Analysing their average score, group of children were struggling during the gameplay and catching up slowly with children whose development was classed as normal. However, it was noticed that these children were consistently getting better and better. Figure 5 shows the average scores achieved over the three weeks testing period by healthy and delayed children in the first test.

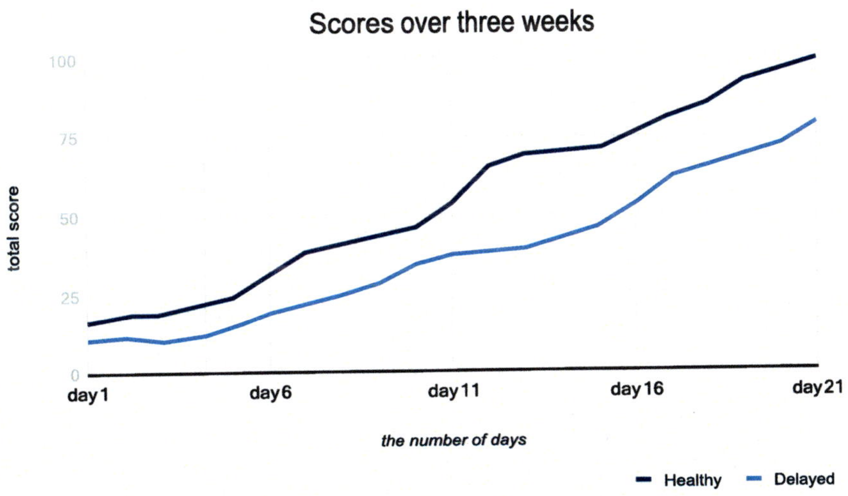

Fig. 5. Average scores achieved during the testing of typically developing children and children with development delays performance while playing implemented SG

Besides, the second test provided a similar pattern of the children responses concerning the game as in the first test. As for learning outcomes, a gap in learning speed between typically developing children and children with autism was considerably smaller, than the first test. The primary reason for this is that autism, as a neurodevelopment disorder, does not exactly impair their ability to learn, rather limits the effectiveness of traditional learning methods. According to the feedback of teachers, and achieved average scores within 21 days, the children with autism did not learn the symbolic nature of letters, numbers, and colors conventionally but rather through associative pairs. Figure 6 illustrates the average scores of typically developing children and children with autism performance over a three-week testing period.

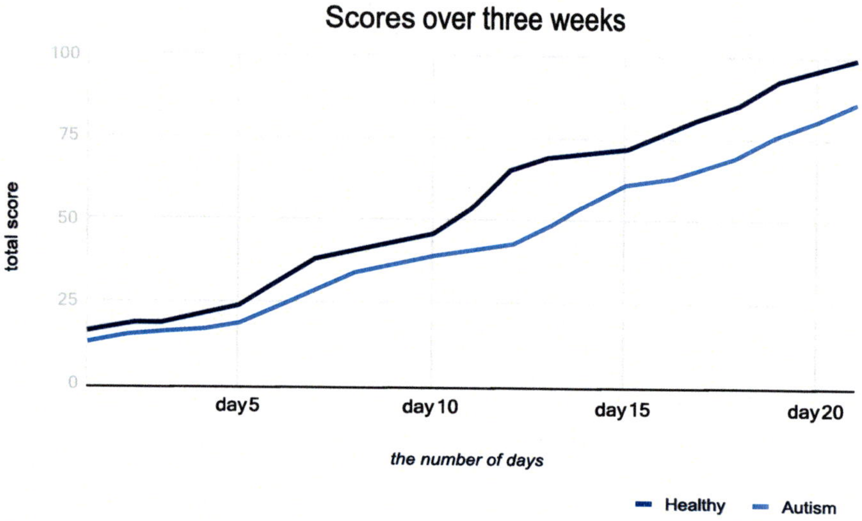

Fig. 6. Average scores achieved during the testing of typically developing children and children with autism delays performance while playing implemented SG

The primary goal of the created tests explained in this research is to see how well the SG will be received by children with different learning capacities and how well it serves as a learning tool. The results produced important insights into how this game could be further improved and/or to widen the variety of information that could be conveyed through such a medium, providing even better learning outcomes. One such improvement would be introducing a larger variance of shapes and sounds, simply by randomizing commonly used fonts for each letter, slightly different shades of colors, and having differently intonated sounds.

Similar research studies have shown SG implementation for the educational purposes where the primary groups were ASD children. For instance, Serret et al. (2014) developed SG called *JeStiMulE* to track ASD children's performance in recognizing different emotions such as sadness, happiness, and anger. After game testing, the ASD children were significantly better in recognizing basic human emotions. In a similar fashion, Bono et al. (2016) designed gaming platform called *GOLIAH* consisting of eleven SGs incorporating audio and visual stimuli principle which is also implemented in "Colorful Classroom" SG. Also, Husni and Budianingsih (2013) described the mobile-based implementation of SG to enable children to learn language vocabulary faster. Such implementation on Android/iOS mobile devices would be feasible for SG described in this research since it makes the game easily accessible for children and their parents. Additionally, implementing this associative mechanism of matching between letters, colors and numbers underline how such virtual environment of SGs could help children to spend their time in school though fun and interactive way while learning at the same time.

5 Conclusion

This research describes an implementation of a SG called "Colorful Classroom" developed for educational purposes as a hand-on solution to help children effectively adopt new learning material. In this SG, levels of the game were designed to boost cognitive functions of typically developing children and mentally delayed children, including children with diagnosed autism. This associative learning technique is implemented through the letter, color, and number matching, including visual and audio hints. As a stand-alone application, this game can be deployed not only in the formal education environment but also as a part of informal education, which would be effective for mentally delayed children and their parents. The findings of this research revealed that children with disabilities can progressively improve while playing the developed SG. This is also valid for autistic children whose performance was even better in the three-week testing of the game. Owing to the positive feedback from the teachers and children who participated in the experimental testing of the designed game, this research opens the possibility for further development of the game and creating a specific environment in the school with the class for different subjects such as mathematics, English language, fundamentals of programming and others.

References

Santasärkkä, S.: The digital games industry and its direct and indirect impact on the economy. Case study: supercell and Finland. Thesis, Metropolia University of Applied Sciences, Helsinki (2017)

Wijman, T.: The Global Games Market Will Generate $152.1 Billion in 2019 as the U.S. Overtakes China as the Biggest Market (2019). https://newzoo.com/insights/articles/the-global-games-market-will-generate-152-1-billion-in-2019-as-the-u-s-overtakes-china-as-the-biggest-market/. Accessed 04 Apr 2020

McDonald, E.: The global games market will reach $108.9 billion in 2017 with mobile taking 42% (2017). https://newzoo.com/insights/articles/the-globalgames-market-will-reach-108-9-billion-in-2017-with-mobile-taking-42/. Accessed 05 Apr 2020

Resinger, D.: 91 percent of kids are gamers, research says (2011). https://www.cnet.com/news/91-percent-of-kids-are-gamers-research-says/. Accessed 05 Apr 2020

Zhang, S., Zhou, H.J., Zhang, L.T.: Design and realization of key technology for body – building design of body piece legacy game based on unity3D and Kinec. J. Sanming Univ., 32–36 (2015)

Cao, Y., Guo, L.P., Du, H.Y., et al.: Cartoon style 3D game scene design production technology. J. Softw., 22–25 (2015)

Backlund, P., Hendrix, M.: Educational games - are they worth the effort? A literature survey of the effectiveness of. In: 5th International Conference on Games and Virtual Worlds for Serious Applications (VS-GAMES), pp. 1–8 (2013)

Connolly, T.M., Boyle, E.A., MacArthur, E., Hainey, T., Boyle, J.M.: Systematic literature review of empirical evidence on computer games and serious games. Comput. Educ. **59**(2), 661–686 (2012)

Durlak, J.A., Weissberg, R.P., Dymnicki, A.B., Taylor, R.D.: The impact of enhancing students' social and emotional learning: a meta-analysis of school-based universal interventions. Child Dev. **82**(1), 405–432 (2011)

Wiemeyer, J., Kickmeier-Rust, M.D., Steiner, C.M.: Performance assessment in serious games. In: Dörner, R., Göbel, S., Effelsberg, W., Wiemeyer, J. (eds.) Serious Games: Foundations, Concepts and Practice, pp. 273–302. Springer, Berlin (2016)

Guttenbrunner, M., Becker, C., Rauber, A.: Keeping the game alive: evaluating strategies for the preservation of console video games. Int. J. Digit. Curation **5**, 64–90 (2010)

Kirriemuir, J., McFarlane, A.: Literature review in games and learning, hal-00190453 (2004)

Barr, D., Harrison, J., Conery, L.: Computational thinking: a digital age skill for everyone. Learn. Lead. Technol. **38**(6), 20–23 (2011)

Little, P., Wimer, C., Weiss, H.B., et al.: After school programs in the 21st century: their potential and what it takes to achieve it. Issues Oppor. Out-of-Sch. Time Eval. **10**(1–12) (2008)

Cheng, M.T., Chen, J.H., Chu, S.J., Chen, S.Y.: The use of serious games in science education: a review of selected empirical research from 2002 to 2013. J. Comput. Educ. **2**, 353–375 (2015)

Kempf, K., Martin, S.: Autonomous exercise game use improves metabolic control and quality of life in type 2 diabetes patients – a randomized controlled trial. BMC Endocr. Disord. **13**(1), 57 (2013)

Vahldick, A., Mendes, A.J., Marcelino, M.J.: Analysing the enjoyment of a serious game for programming learning with two unrelated higher education audiences. In: Proceedings of the European Conference on Games-based Learning, pp. 523–53 (2015)

Kerres, M., Bormann, M., Vervenne M.: Didaktische Konzeption von Serious Games: Zur Verknüpfung von Spiel- und Lernangeboten. In: Medienpädagogik (2019). http://mediendidaktik.unidue.de/sites/default/files/kerres0908_0.pdf. Accessed 21 Mar 2020

Lugrin, J.L., Latoschik, M.E., Habel, M., Roth, D., Seufert, C., Grafe, S.: Breaking bad behaviors: a new tool for learning classroom management using virtual reality. Front. ICT **3**, 26 (2016)

Lugrin, J.L., Oberdorfer, S., Latoschik, M.E., Wittmann, A., Seufert, C., Grafe, S.: VR-assisted vs video-assisted teacher training. In: Proceedings of the 25th IEEE Virtual Reality (VR) Conference (2018)

De Freitas, S.: Learning in immersive worlds. Technical report, Joint Information Systems Committee, Bristol (2006)

Bellani, M., Fornasari, L., Chittaro, L., Brambilla, P.: Virtual reality in autism: state of the art. Epidemiol. Psych. Sci. **20**, 235–238 (2011)

Morgenthal, A.H.: Child-centered play therapy for children with autism: a case study, Ph.D. dissertation, Clinical Psychology, Antioch University New England, Keene, New Hampshire (2015)

Ribeiro, P.C., Raposo, A.B.: ComFiM: a game for multitouch devices to encourage communication between people with autism. In: 2014 IEEE 3rd International Conference on IEEE Serious Games and Applications for Health (SeGAH) (2014)

Hourcade, J.P., et al.: Evaluation of tablet apps to encourage social interaction in children with autism spectrum disorders. In: Proceedings of the SIGCHI Conference on Human Factors in Computing Systems. ACM (2013)

Cunha, P., et al.: Augmented reality for cognitive and social skills improvement in children with ASD. In: 13th International Conference on Remote Engineering and Virtual Instrumentation (REV), Madrid, pp. 334–335 (2016)

Bossavit, B., Parsons, S.: This is how I want to learn: high functioning autistic teens co-designing a serious game. In: Proceedings of the 2016 CHI Conference on Human Factors in Computing Systems, pp. 1294–1299. ACM (2016)

Fletcher-Watson, S., et al.: Designing for young children with autism spectrum disorder: a case study of an iPad app. Int. J. Child-Comput. Interac. **7**, 1–14 (2016)

Hodgdon, L.A.: Visual Strategies for Improving Communication Practical Supports for School and Home. QuirkRoberts Publishing, Troy (1995)

Lesley, L., Xie, et al.: Are tangibles more fun? Comparing children's enjoyment and engagement using physical, graphical and tangible user interfaces. In: Proceedings of the 2nd International Conference on Tangible and Embedded Interaction. ACM, pp. 191–198 (2008)

Colby, R.S., Colby, R.: A pedagogy of play: integrating computer games into the writing classroom. Comput. Comp. **25**(3), 300–312 (2008)

Clark, D.: Learning by playing: can computer games and simulations support teaching and learning for post-16 learners in formal, workplace and informal learning contexts? Computer games in education and training. Presentation at LSDA seminar London (2003)

Lope, R.P.D., Medina-Medina, N., Soldado, R.M., Garcia, A.M., Gutierrez-Vela, F.L.: Designing educational games: key elements and methodological approach. In: 9th International Conference on Virtual Worlds and Games for Serious Applications (VS-Games), vol. 00, pp. 63–70 (2017)

Primary Practice: Carpet teaching areas in Primary Schools (2017). http://primarypractice.co.uk/2017/05/08/carpet-teaching-areas-primary-schools/. Accessed 23 Mar 2020

Dumas, M., ter Hofstede, A.H.M.: UML activity diagrams as a workflow specification language. In: Gogolla, M., Kobryn, C. (eds.) UML 2001. LNCS, vol. 2185, pp. 76–90. Springer, Heidelberg (2001). https://doi.org/10.1007/3-540-45441-1_7

Preissler, M.A.: Associative learning of pictures and words by low-functioning children with autism. Autism **12**(3), 231–248 (2008)

Deloache, J.S., Burns, N.M.: Early understanding of the representational function of pictures. Cognition **52**, 83–110 (1994)

Perna, R.: Early Developmental Delays: A Cross Validation Study. Journal of Psychological Abnormalities in Children (2013)

Serret, S., Hun, S., Iakimova, G., et al.: Facing the challenge of teaching emotions to individuals with low- and high-functioning autism using a new Serious game: a pilot study. Mol. Autism **5**, 37 (2014)

Bono, V., Narzisi, A., Jouen, A., et al.: GOLIAH: a gaming platform for home-based intervention in autism - principles and design. Front. Psychiatry **7**, 70 (2016)

Husni, E., Budianingsih: Mobile applications BIUTIS: let's study vocabulary learning as a media for children with autism. Procedia Technol. **11**, 1147–1155 (2013)

Reliable Book Recommender System: An Evaluation and Comparison of Collaborative Filtering Algorithms

Aldin Kovačević(✉) and Zerina Mašetić

Faculty of Engineering and Natural Sciences, Department of Information Technologies, International Burch University, Sarajevo, Bosnia and Herzegovina
aldin.kovacevic@stu.ibu.edu.ba

Abstract. Nowadays, a vast number of online businesses and services use recommender systems to provide relevant product suggestions to their customers. Recommender systems can be defined as systems employing various data mining techniques to predict users' interest in certain items or topics, among a tremendous amount of available data. While there exist varying methodologies to construct a recommender system, one of the most potent and commonly used techniques is collaborative filtering (CF), which evaluates items based on the opinions of other users. This paper examines several commonly used collaborative filtering models with the goal of creating a reliable book recommendation system. The research focuses on training, evaluating and comparing several neighborhood-based and matrix factorization-based recommender algorithms, using a combination of two book datasets: the BookCrossing dataset compiled in 2004, and the Goodbooks "most popular books" dataset from 2017. Grid search with cross-validation is used to find the best performing combination of several hyper-parameters and different CF models. Ultimately, a matrix factorization-based SVD++ approach is selected as the most performant model, with MAE of 0.57 and RMSE of 0.752, making it a considerably good model based on the provided data. Moreover, a manual comparison between a user's book history and their recommended book titles shows that users are generally recommended books of the same or similar genres. Lastly, the paper discusses certain shortcomings of the proposed model and offers several suggestions for future improvements.

Keywords: Recommender systems · Collaborative filtering · Book recommendations · Neighborhood-based approaches · Matrix factorization-based approaches

1 Introduction

With the ever-growing amount of information and data available on the internet, users have an increasing need of tools to find and access appropriate information. Recommender systems are an example of one of the most successful tools for that purpose. In technical terms, a recommendation system (or recommendation engine) can be defined

N. Ademović et al. (Eds.): IAT 2021, LNNS 316, pp. 264–280, 2022.
https://doi.org/10.1007/978-3-030-90055-7_20

as a system which operates on a business environment, taking a user's online footprint as input and generating a probable output footprint consisting of item predictions for that user, which are as close to reality as possible [1]. Recommender systems have quickly become an important aspect of the information and e-commerce landscape, allowing users to quickly filter through large swaths of information and product spaces in order to find the relevant data [2].

Throughout history, humans have always relied on "recommendation systems", in one way or another. Before the information age took a considerable hold in society, people used recommendations or advice from their peers or experts in a particular field to guide their decision-making process and discover new material. However, this approach was limited in the sense that there might exist an "independent item" that a person would like, but if no one close to that person is familiar with the item, it would be unavailable to said person. This led to the creation of computer-based recommendation systems that can "expand" the circle of users from whom a person can obtain recommendations, and mine users' history and preferences to uncover new patterns, providing a more finely-tuned experience [2]. Over the past 20 years, there has been a multitude of various approaches to building recommender systems [3], with the most commonly used (and most performant) approaches being content-based algorithms and collaborative filtering (CF) algorithms [4]. Content-based algorithms attempt to recommend items similar to the items the user has liked in the past, correlating with the user's preferences [5]. On the other hand, collaborative filtering algorithms recommend items to users by evaluating the preferences, opinions and ratings that other users have expressed regarding those items [6]. While both of these approaches have their benefits and limitations, this paper is going to focus on collaborative filtering. Since content-based filtering relies on the properties of actual content to be recommended, its results might not transfer from one domain to another. On the other hand, collaborative filtering relies only on the user rating matrices for particular items, which are usually similar in form across multiple domains [8]. Moreover, collaborative filtering has been met with major success in both research and practice, being widely used in information filtering and e-commerce applications [7].

The topic of recommender systems has been the subject of numerous research papers for quite some time, primarily focusing on movie, music and product recommendations [8]. However, few papers have explored the field of book recommendations [1, 9]. This paper presents an exploratory study which aims to produce a satisfactory recommender system model that can be employed for relevant book recommendations on a bookshop or library website. After performing an exploratory analysis of two popular book datasets, BookCrossing and Goodbooks-10k, this research paper bases its approach around the collaborative filtering (CF) methodology. Namely, since CF recommender systems can depend on a variety of factors and initial parameters, grid search with cross-validation is applied in order to find the best performing set of model hyper-parameters. In short, the main contributions of this paper can be summarized as follows:

1. creating an acceptable book recommendation system by testing a variety of collaborative filtering approaches with a combination of parameters, and deducing the best performing model
2. fine-tuning of the recommender model and discussing possible future improvements.

2 Literature Review

The prototypes of recommender systems occurred fairly early in the history of computing, with Grundy [10], a computer-based librarian, being an early step towards automatic recommendation systems. Grundy functioned by grouping users into common "stereotypes" based on short interview questions and using hard-coded information about various stereotypes to generate favorable book recommendations. While fairly primitive in its execution (and according to today's standards), it represented one of the first major entries in the space of recommender systems.

First major strides regarding collaborative filtering started occurring in the early 1990s, when CF-based approaches began to arise as a solution to online information overload [2]. One of the first examples were Tapestry [11], a manual collaborative filtering system for relevant data retrieval, and GroupLens [12], which identified Usenet articles that could be interesting to a user, based on other users' ratings. Soon afterwards, various collaborative recommender systems, such as BellCore [13] for movies, Ringo [14] for music and Jester [15] for jokes followed. Moreover, recommender systems were also recognized in the marketing sphere for the purposes of increasing sales and improving customer experience [16], culminating in Amazon's adoption of recommender technology which could suggest relevant items to customers based on their previous purchase history, and the history of similar customers [3].

Due to the popularity and well-documented nature of collaborative filtering methods, there exists a plethora of relevant research papers on recommendation systems. In their 2005 paper, Adomavicius and Tuzhilin [3] made an extensive categorization of currently known recommender algorithms into content-based, collaborative and hybrid approaches, summarizing their possible extensions. On the other hand, Su and Khoshgoftaar [17] gave more focus to CF-based methods, including memory-based, model-based and hybrid methods. Their survey outlines some of the most sophisticated recommendation algorithms available as of 2009. Another excellent survey about collaborative filtering systems was produced by Ekstrand et al. [2], explaining the process which goes into creation and evaluation of CF algorithms in great detail. Apart from these theoretical surveys, a couple of experimental studies have also been conducted. A study by Breese et al. [18] compares the performance of two widely used memory-based methods – Pearson correlation and vector similarity – against two classical model-based approaches – clustering and Bayesian network. The comparison was done on three different datasets, indicating that Bayesian network and correlation-based methods feature a better performance. An experimental study into CF algorithms for e-commerce was conducted by Huang et al. [19], comparing user-based, item-based, dimensionality reduction-based and a few other model-based methods. This study uses precision, recall and F1 score as its evaluation metrics, ignoring the standard MAE and RMSE error scores commonly used in other papers. Recommender systems research was additionally reinvigorated in 2006, with the launch of Netflix's 1-million-dollar prize competition [22] to come up with an improved movie recommendation algorithm. The winning approach implemented a hybrid method which used temporal dynamics to account for the timestamps of movie ratings [23], and the Netflix dataset is still used today as a standard dataset for evaluating the performance of CF algorithms [8].

With the advent of the Netflix competition and the rise of e-commerce, most of the research into recommendation systems has been focused on movies, businesses and commercial products [2]. In terms of book recommender systems, a few relevant studies have been conducted. Kurmashov et al. [9] implemented a collaborative filtering model using the Pearson correlation as its similarity method. Based on user feedback, the quality of their recommender system was rated at 77%. Parvatikar and Joshi [30] combined collaborative filtering with association mining techniques to fill in the missing ratings, creating a system with an average MAE score of 0.7. Another collaborative filtering system with association mining was independently proposed and implemented by Patil et al. [31]. Rana and Deeba [32] attempted a slightly different collaborative filtering approach, by utilizing Jaccard similarity - the ratio of common book readers vs. the readers who rated the books individually. Their approach yielded an RMSE of 1.504 on a dataset with a 10-based rating scale. Uko et al. [33] proposed a collaborative filtering model with adjusted cosine similarity, averaging an RMSE of 0.9. On the other hand, Sase et al. [1] suggested a hybrid recommendation approach, combining collaborative filtering, relevant book properties and user demographics. Cho et al. [34] tested a variety of recommendation system approaches, including a k-NN collaborative filtering model with adjusted cosine similarity which yielded an RMSE of 1.29 on a 10-rating scale.

In recent years, recommender techniques other than collaborative filtering have begun to emerge, namely content-based approaches rooted in information retrieval, Bayesian inference and case-based reasoning methods [20]. Content-based methods work with the actual content or properties of items in the recommender system. These algorithms can use item content to supplement user rating patterns, or completely replace them. Apart from that, as various recommender strategies matured, hybrid recommender systems [21] emerged, combining multiple algorithms into composite systems that build upon the strengths of its individual parts. Despite all of that, collaborative filtering has remained an effective recommendation approach, both on its own and as a hybridized approach with additional content-based properties.

3 Materials and Methods

Among a variety of recommender system approaches, this paper focuses on building a collaborative filtering (CF) model. Collaborative filtering is a popular family of recommendation algorithms that gives out recommendations based on the ratings or behavior of other users in the system [2]. With CF, the ratings of other users in the system can be gathered and selected in order to predict another user's likely preferences. In other words, if a group of users agree about their preference for certain items, they are also likely to agree about other items which they might not have seen yet. To elaborate further, if a group of users prefers the same items as user A, then user A is likely to prefer the items that the group likes, which the user has not seen yet. This approach will be the key principle examined throughout the research paper. Figure 1 showcases an overview of the primary steps taken during the research process.

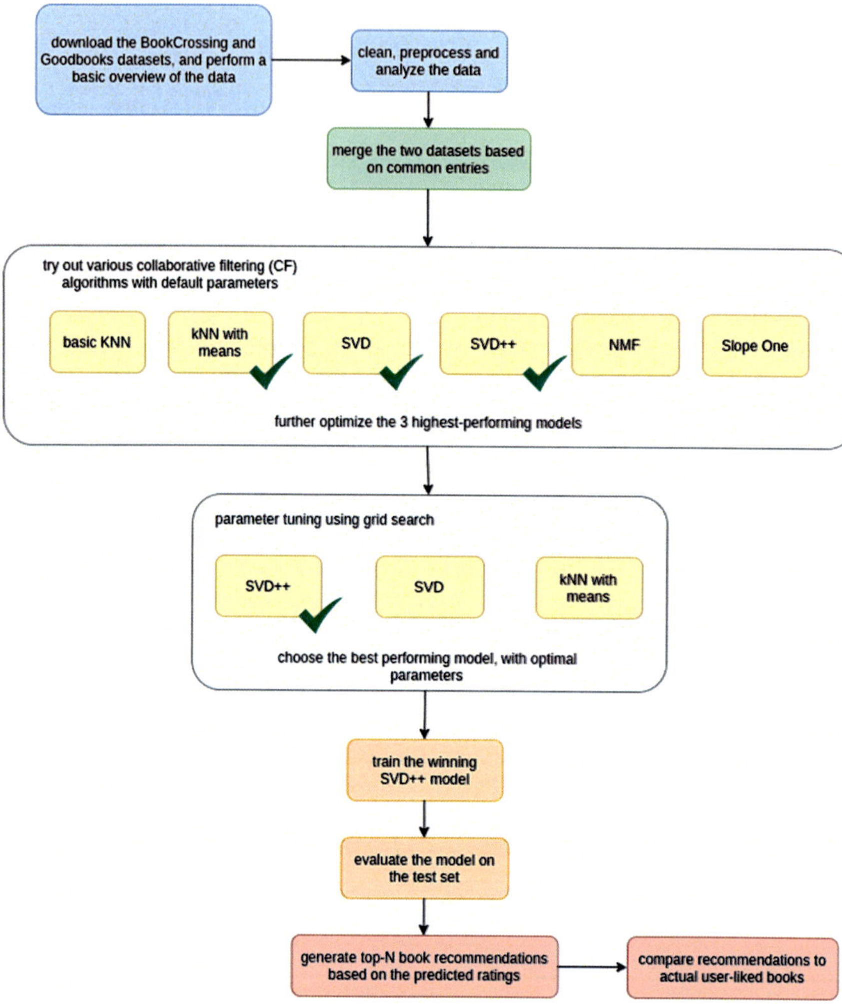

Fig. 1. Model diagram with key steps

3.1 Dataset Overview

Two different book datasets are used as the basis of this research: the BookCrossing [24] and Goodbooks-10k "top 10,000 books" [25] datasets. The BookCrossing dataset was compiled from the BookCrossing forum over a 4-week period in 2004, and comprises 271,379 books, 275,858 users with anonymized demographics and 1,149,780 implicit and explicit book ratings. On the other hand, the Goodbooks-10k dataset consists of 10,000 most popular books up to 2017 crawled from a book review website "Goodbooks", with around 6 million ratings and 53,424 users. Additionally, this dataset contains a list of user-defined book tags, which can be used to infer literary genres.

Both of the datasets contain several key files. The "books.csv" file of the BookCrossing defines the ISBN number, book title and author, publisher and year of publication, and links to the book cover image in three formats (small, medium and large). The user ratings are defined separately in a "ratings.csv" file, with user ID, book ISBN (foreign key) and the actual rating from 0 to 10. Additionally, the set contains a "users.csv" file with anonymized user demographics (user ID, location and age). The Goodbooks-10k "books.csv" equivalent defines the same columns as BookCrossing, with a few additions. The file contains a "book ID" (which is later used to join the file with ratings), and several other IDs, such as Goodreads ID and edition IDs. Moreover, the file has information about the book language, number of reviews and average rating. Goodbooks-10k's "ratings.csv" is similar to the BookCrossing's file, with the exception of using the book ID as the foreign key, not the ISBN number. Moreover, ratings are given on a scale from 1 to 5. Lastly, this dataset also contains a "tags.csv" file which comprises a collection of user-given tags for various books. With some preprocessing, this collection can be used to infer book genres.

3.2 Data Cleaning and Preprocessing

After performing an initial exploratory and statistical analysis of the two datasets, they are processed and prepared for use with the recommender model. Standard data cleaning and preprocessing steps are performed, including the removal of duplicate items, replacement of N/A and outlier values with median/mode values and standardization of column names (since the two datasets use different names for their attributes).

Special preprocessing steps are undertaken with the rating datasets. First of all, the BookCrossing ratings also contain implicit ratings, or books "rated" with a 0. Since a 0-rating is not indicative of a user's preference for that item, implicit ratings are removed from the dataset. Secondly, since BookCrossing ratings are presented on a scale from 1 to 10, they are transformed down to a rating scale of 0.5 to 5, to be in line with Goodbooks-10k ratings. The resulting rating distributions are showcased on Fig. 2. The distribution of book ratings for both datasets suggests that users are generally more likely to give a positive rating (3.0 and up) to a book than a non-favorable rating.

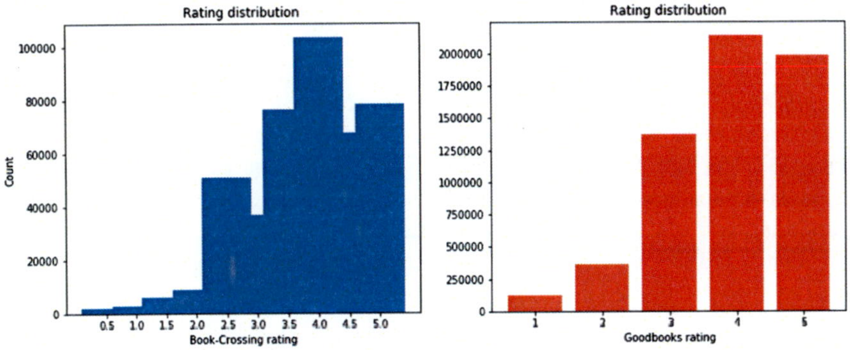

Fig. 2. Rating distribution across both datasets

Lastly, to have a balanced representation of both datasets in the final model, a combined set of ratings is generated and limited to around 30,000 ratings. To ensure statistical significance, the ratings are extracted from users who have rated at least 50 books and books which have received at least 25 ratings. One of the reasons for the dataset reduction is the computational constraints involved with subsequent model training and evaluation. Through repeated experimentation and testing, a dataset of ~30,000 ratings was found to be computationally feasible, producing satisfactory results.

3.3 Collaborative Filtering Techniques

Based on their approach, collaborative filtering (CF) recommender systems are generally categorized into two distinct families: memory-based and model-based methods [8]. Memory-based algorithms load in the entire rating matrix into memory when making recommendations. These algorithms try to find relevant recommendations based on the similarity between a given user/item and the remainder of the rating matrix. Memory-based algorithms can adapt to changes in the data, but their processing time can be large, depending on the size of the input dataset. On the other hand, model-based algorithms fit a parameterized model to the rating matrix, later providing recommendations based on the produced model. This family of algorithms has constant time regardless of input size, but the model needs to be recomputed whenever data changes.

Neighborhood-based methods are among the most used memory-based CF techniques. These k-NN inspired algorithms predict ratings by looking at the users with ratings similar to the input user, or at items that are similar to the input item. The main assumption behind neighborhood-based approach is that if two users have similar preferences regarding a certain set of items, they are likely to have similar preferences about the remaining items. Alternatively, if two items are rated similarly by a portion of the users, the two items are likely to be rated in a similar manner by the remaining users. Based on these two different approaches (user similarity and item similarity), neighborhood-based algorithms can be further classified into user-based CF methods [5] and item-based CF methods [18]. As the name suggests, the user-based approach identifies users that are similar to the input user, and estimates the user's preference for an item to be the average rating given out by these similar users. Alternatively, the item-based approach identifies items that are similar to the input item and predicts the item's rating to be the average rating of these similar items. The user/item similarity can be calculated in a variety of ways, with the most commonly used similarity measures being Pearson correlation, (adjusted) cosine similarity and mean-squared difference (MSD) [2].

On the other hand, model-based methods fit a parametric model to the training data. This model can then be used to predict unseen ratings and provide reasonable item recommendations. Model-based approaches are more varied than neighborhood-based algorithms, with some popular options being cluster-based CFs, Bayesian classifiers, regression-based methods and matrix factorization-based approaches [8]. In recent years, matrix factorization-based methods have become particularly popular, primarily due to the success of the SVD (singular value decomposition) algorithm during the "Netflix Prize" competition [23]. The main approach of matrix factorization-based algorithms is to factorize a rating matrix into a product of two low-rank matrices (user-profile and item-profile) which are then used to estimate the missing ratings [26]. In an SVD approach, each user is a row in a matrix structure, each column is an item and elements of the matrix are ratings given to the items by users. Factorizing this matrix using SVD results in the extraction of latent factors, mapping each user and item into a lower-dimensional latent space. This makes it easier to clearly represent the relationships between users and items, and calculate the expected user ratings [35].

In order to create a suitable book recommender system, this paper examines the most popular algorithms from both the memory-based and model-based algorithm families, giving a special focus to neighborhood-based and matrix factorization-based methods, as the most common representatives of their respective algorithm families. Moreover, the research also tests whether a user-based or an item-based approach yields better results for a neighborhood-based model.

3.4 Recommender Algorithms

The training and evaluation of recommender models was done using Surprise [27], a scikit-based Python library for building and analyzing recommender systems. Several representative models of major collaborative filtering approaches were tested, namely neighborhood-based and matrix factorization-based models. The neighborhood-based models work on the principle of finding similarities between items or users. Two neighborhood-based (memory-based) algorithms were selected, basic k-NN and k-NN with means, with the difference being that the latter takes the mean ratings of each user into account. Matrix factorization-based models extract features and correlation from user and item matrices. Out of matrix factorization-based (model-based) algorithms, the paper examined SVD [26], SVD++ [23], which is a variant of SVD that captures implicit ratings (the fact that a user rated an item, regardless of the rating value) in addition to explicit ratings and NMF [28], an algorithm similar to SVD, which prevents negative elements in factorized matrices. Lastly, a linear regression-based Slope One [29] algorithm was also tested.

All candidate models were fitted to the training set generated from relevant ratings using Surprise (80-20 split), and evaluated through a 5-fold cross-validation process. The best performing models were later fine-tuned and optimized using the grid search approach to select the optimal combination of parameters. Table 1 showcases the parameters used to build the initial models.

Table 1. Initial recommender model parameters

Recommender model	Chosen parameters
Basic k-NN	{k = 40, similarity = 'cosine', user_based = True}
k-NN with means	{k = 40, similarity = 'cosine', user_based = True}
SVD	{n_epochs = 20, lr_all = 0.005, reg_all = 0.02}
SVD++	{n_epochs = 20, lr_all = 0.007, reg_all = 0.02}
NMF	{n_epochs = 20, biased = False, reg = 0.06}
Slope one	No special parameters

4 Model Optimization and Evaluation

Various evaluation measures for prediction accuracy have been used throughout the available research into collaborative filtering systems. The most common ones are the mean absolute error (MAE) and the root mean squared error (RMSE). MAE is measured on the same scale as the original rating, and represents a measure of how much a predicted rating differs from the original rating:

$$MAE = \frac{1}{n}\sum\nolimits_{u,i} \left|p_{u,i} - r_{u,i}\right| \tag{1}$$

where $p_{u,i}$ is the predicted and $r_{u,i}$ the actual rating a user *u* has given to an item *i*. On the other hand, RMSE places a greater emphasis on larger errors, penalizing them more. Since errors are squared before being averaged, RMSE gives a higher weight to large errors [2].

$$RMSE = \sqrt{\frac{1}{n}\sum\nolimits_{u,i} (p_{u,i} - r_{u,i})^2} \tag{2}$$

4.1 Initial Model Evaluation

Since this paper aims to minimize potential large errors in predicted ratings, RMSE was used as the primary performance metric to evaluate the accuracy of the recommender model, as well as determine the best combination of parameters during the grid search phase. However, the MAE value was calculated for every algorithm as well, since it represents the average offset between predicted and actual ratings. The lower the value of RMSE/MAE is, the better accuracy is expected from the recommender model. The performance of the initial models was evaluated using a 5-fold cross-validation on the training set. Figure 3 shows a unified plot of MAE (red) and RMSE (blue) values for all six trained algorithms.

Based on the RMSE plot, SVD++ and SVD are the best performing algorithms, with RMSE values of 0.765 and 0.769, respectively. Table 2 showcases the full list of MAE and RMSE scores for all tested algorithms.

Looking at both performance metrics (RMSE and MAE), SVD++ and SVD algorithms are in the lead, with SVD++ having a slight advantage (0.769 vs. 0.765 for RMSE, and 0.582 vs. 0.59 for MAE). Out of the remaining models, the closest to them is k-NN with means, having an RMSE of 0.837 and MAE of 0.623. The remaining models (NMF, basic k-KNN and Slope One), while not displaying an overly bad performance, still lag behind the top algorithms. In general, it seems that matrix factorization-based recommendation models outperform neighborhood-based and regression-based algorithms, at least for the particular ratings dataset used in this paper.

4.2 Grid Search and Parameter Tuning

SVD++, SVD and k-NN with means achieved the best performance metrics during the initial testing, and therefore merit further research. Using grid search with 5-fold cross-validation, these three algorithms will be trained and evaluated with different parameter

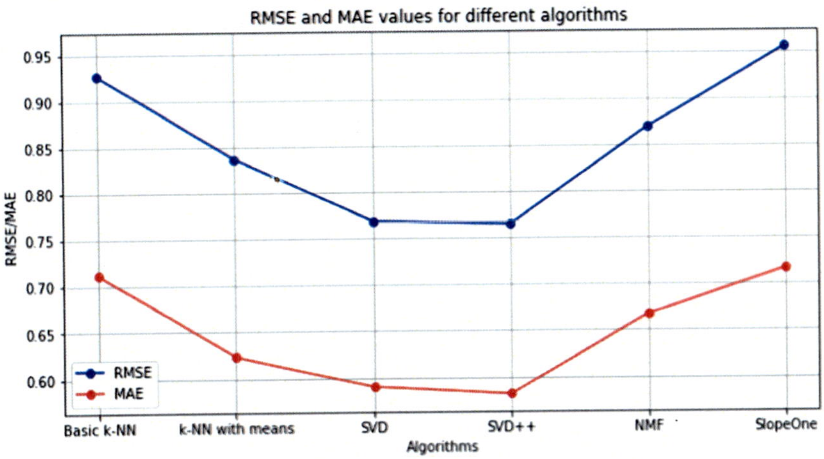

Fig. 3. MAE and RMSE scores for various algorithms

Table 2. RMSE and MAE scores for different models

Recommender model	RMSE	MAE
Basic k-NN	0.927	0.712
k-NN with means	0.837	0.623
SVD	0.769	0.59
SVD++	0.765	0.582
NMF	0.869	0.667
Slope one	0.955	0.717

grids, to see if hyper-parameter tuning can lead to better (lower) RMSE and MAE scores than those obtained in the initial iterations. The algorithms and their parameter grids are outlined in Table 3.

Table 3. Grid search parameter grids for different models

Recommender model	Parameter grid
SVD++	{n_epochs = [20, 25, 40], lr_all = [0.005, 0.007, 0.009, 0.01], reg_all = [0.2, 0.4, 0.6]}
SVD	{n_epochs = [20, 25, 40], lr_all = [0.005, 0.007, 0.009, 0.01], reg_all = [0.2, 0.4, 0.6]}
k-NN with means	{k = [15, 20, 25, 30, 40, 50, 80, 100], similarity = ['pearson', 'cosine'], user_based = [True, False]}

For SVD++ and SVD, the relevant parameters are the number of epochs (n_epochs), the learning rate (lr_all) and the regularization term (reg_all). In k-NN with means, there exist two different kinds of adjustable parameters. First of all, a parameter inherent to k-NN models is the number of neighbors to be examined (k). However, all memory-based models also depend on the similarity function and the approach to collaboration: user-based or item-based. Neighborhood-based models in Surprise use the cosine similarity and user-based filtering by default, so the grid search will examine the possibility of utilizing a different similarity function and/or item-based filtering to increase performance. Like in the initial phase, the algorithms were fitted and evaluated on the training set. RMSE was used as the metric for choosing the best subset of hyper-parameters. The optimal parameters for each model, along with RMSE and MAE values, are available in Table 4.

Table 4. RMSE, MAE and optimal parameters after grid search

Recommender model	RMSE	MAE	Optimal parameters
SVD++	0.759	0.581	{n_epochs = 25, lr_all = 0.007, reg_all = 0.2}
SVD	0.761	0.582	{n_epochs = 20, lr_all = 0.009, reg_all = 0.2}
K-NN with means	0.828	0.62	{k = 100, similarity = 'cosine', user_based = False}

Looking at the newly computed performance metrics, SVD++ once again shows the best results, with RMSE of 0.759 and MAE of 0.581. It is closely followed by regular SVD, with 0.759 and 0.581 for RMSE and MAE, respectively. k-NN with means, while performing better than the original model, does not stack up to SVD-based algorithms. However, it is worth noting that grid search evaluation found item-based filtering to be the more performant approach (user_based = False), suggesting that item-based filtering is the preferred option for models based on similarity (such as k-NN), given the contents of the examined book ratings dataset. In conclusion, after conducting a grid search hyper-parameter optimization, SVD++ is the best performing algorithm, so it will be used throughout the remainder of this paper.

5 Experimental Results

After initial tests and evaluation, the selected recommender model was the matrix factorization-based SVD++, with 25 epochs, learning rate of 0.007 and regularization term of 0.2. Experimental results obtained by using this model are outlined in the following sections.

5.1 Test Set Evaluation

Cross-validation and tuning of hyper-parameters were done on the previously extracted training subset of data, whereas 20% of the original data was withheld for the test set. The test set is used to get an unbiased estimation of the model's performance by applying it to unknown samples. After re-training the SVD++ model on the training set using the previously obtained optimal parameters, rating predictions were generated for the samples in the test set. Upon evaluation, RMSE for the test set was 0.752, and its MAE was 0.57. These metrics closely resemble the cross-validation metrics calculated via the training set, which indicates that the model is unbiased and can generalize well to unseen data. Figure 4 presents a visualization of actual (blue) versus predicted (red) ratings for 50 randomly sampled books, given by different users.

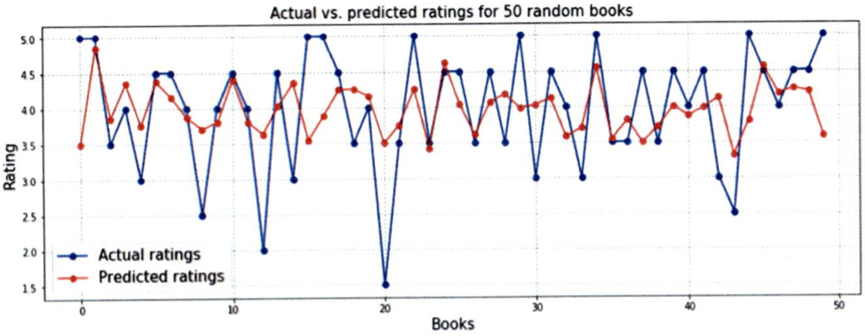

Fig. 4. Actual vs. predicted ratings for randomly sampled books

Based on the rating plot, it seems that the predicted ratings generally follow the trend of actual ratings: higher predicted ratings mostly correspond to higher actual ratings, and vice versa (except for a few outlier cases), sometimes even closely matching them.

5.2 Generating Top-N Recommendations

Since the chosen recommender model performs well on unseen data, it is suitable for use as the crux of the main recommender system which generates top-N recommended books for a given user. First, a so-called "anti-test set" is constructed. Since not every user has rated every book, a collection of all (user, book) pairs which are not in the original ratings dataset is constructed. This ensures that the user will not be recommended books which they have already read. SVD++ is then fitted to the anti-test set and rating predictions are generated for all (user, book) tuples. When a user wants to receive book recommendations, top N books (N being the number of books the user requests) with the highest predicted ratings by that user are selected to be displayed. Table 5 showcases an overview of the top-10 manually rated books (by the user) and the top-10 books recommended by the system for a randomly sampled user from the dataset, with an ID of 270713.

Judging from the top-rated books, user 270713 prefers books that are more fictitious in nature, such as sci-fi and fantasy genres (making up half of the selection). Additionally, they also enjoy an occasional drama or an autobiographical work. When it comes to the recommended books, it seems like the system picked up on the user's preferences, with half of the recommendations being sci-fi or fantasy books. Moreover, the remaining books are mostly related to drama and autobiographical genres, which are also present

Table 5. Top-rated and top-recommended books for a randomly selected user

Highest rated books by the user		Top books recommended by the system	
R	Title (genre)	R	Title (genre)
5.0	The Autobiography of Malcolm X (autobiography)	4.59	Weirdos From Another Planet! (sci-fi, comic)
5.0	Anne Frank: The Diary of a Young Girl (autobiography)	4.57	A Tree Grows in Brooklyn (semi-autobiography)
5.0	The World According to Garp (drama)	4.55	The Two Towers (fantasy)
5.0	Xenocide (sci-fi)	4.53	A Wrinkle in Time (sci-fi)
5.0	Ender's Game (sci-fi)	4.51	The Return of the King (fantasy)
5.0	Watership Down (adventure)	4.5	84 Charing Cross Road (drama)
5.0	The Red Tent (drama)	4.49	Where the Sidewalk Ends (poetry)
5.0	Speaker for the Dead (sci-fi)	4.47	Lolita (drama)
5.0	The Mists of Avalon (fantasy)	4.46	The Little Prince (novella)
5.0	Ender's Shadow (sci-fi)	4.45	Good Omens (fantasy)

in the books previously read by the user. All in all, it appears that the recommender system tends to produce books similar to the user's previous reading history.

6 Discussion

Based on the performance metrics evaluated using the final SVD++ recommender model, this paper produced satisfactory results. With RMSE of 0.752 and MAE of 0.57, the model seems to exhibit a better [30, 33] or comparable [32, 34] performance in comparison with related research. On the other hand, certain papers utilized different scoring metrics [9] or did not provide metrics at all [1, 31] and focused only on the proposal of the algorithm, making them difficult or impossible to adequately compare with the results obtained in this research. Moreover, even among the papers with usable performance metrics, different book datasets and rating scales are used, so one should be careful when analyzing these studies, making sure to examine them in context. For an easier overview, Table 6 showcases a comparison between previous research with relevant metrics and the results obtained during this study ("–" stands for "not reported or not applicable").

Of the two metrics, it is generally more difficult to interpret the meaning of RMSE in a practical sense (due to it stemming from the squared error). However, MAE is straightforward: it represents the average difference between predicted and original ratings. With a MAE of 0.57, the predictions generated by this model, on average, differ from the actual book ratings by ± 0.57. Since the ratings in the dataset are scaled from 0.5 to 5, with steps of 0.5, this can be considered a great metric, as it means that the average

Table 6. Metrics reported by previous research vs. this paper

Research	Method	RMSE	MAE	Other metrics
Kurmashov et al. [9]	CF with Pearson correlation	–	–	77% quality (user feedback)
Parvatikar and Joshi [30]	CF with association mining	–	0.7	–
Uko et al. [33]	CF with cosine similarity	0.9	–	–
Cho et al. [34]	CF using k-NN and cosine similarity (0–10 rating scale)	1.29	–	88.66% precision & 91.16% recall (using a content-based approach)
Rana and Deeba [32]	CF with Jaccard similarity (0–10 rating scale)	1.504	–	–
This paper	**CF using SVD++**	**0.752**	**0.57**	–

predictions do not deviate too much from the original ratings. While it is not feasible to manually inspect the recommended books for every user in the dataset, a showcase of predictions for a few random users, e.g. the user in Table 5, yields an interesting observation: even though ratings are the only feature used in the recommender model, the model still manages to suggest books of similar genres to the ones the user has enjoyed in the past. This fact, along with low error scores, indicates that the constructed recommender system has considerable quality and relevance.

Looking at Fig. 4, which plots the actual vs. predicted ratings for a random subset of books, a peculiar detail can be noticed: while the predicted ratings generally follow the trend of actual ratings (with a few outlier cases), the actual low ratings differ greatly from predicted low ratings. For several original ratings at the lower end of the scale, such as 2.5, 2 or 1.5, while the predicted ratings are lower than usual, they are nowhere near the actual rating. In fact, the lowest predicted rating on the plot is somewhere around 3.3, which is (at this scale) very far away from the actual lowest ratings. On the other hand, actual higher ratings (3.5+) generally have correspondingly high predicted ratings. This represents a drawback of the current model, and this disparity can be attributed to the uneven distribution of ratings in the input dataset. In Fig. 2, it can be seen that more than two-thirds of given ratings are neutral or positive (3.0+), meaning that users are generally more likely to submit positive ratings than unfavorable ones. Due to an insufficient number of low ratings in the training dataset, the model does not generalize well to them, leading to significant discrepancies between actual and predicted low ratings, as observed in Fig. 4. Using a dataset with more evenly distributed positive and negative ratings could be a way to mitigate this issue.

Despite the good performance and metrics exhibited by this model, there is still a lot of room for future work and improvements. First of all, the approach examined in this paper is a collaborative filtering algorithm relying solely on user ratings to recommend relevant books. Generally, CF algorithms use only the rating matrix and similarities between user/item ratings to generate predictions. In the previous paragraph, it was also discussed how this can lead to skewed results, if the ratings are not evenly distributed. However, the dataset used in the paper contains some additional interesting properties, such as the book author, genre, user demographics (age and location) and book publication year; all of which are factors that could contribute to the user's preference for a book. Another potentially useful (but perhaps not immediately obvious) factor could be a book's "page number/price" ratio. Even though a book might fit a user's preference based on genre and other factors, some users might also want to ensure that they would be getting "their money's worth" from the book; in other words, that the amount they are paying is proportional to the amount of content provided. Using this ratio would require enriching the current dataset with book prices (since book prices can vary between different outlets and bookstores, a base reference should be established) or finding a different input dataset containing price data. Meaningfully incorporating all of these properties into the recommendation algorithm could lead to the creation of a more reliable hybrid collaborative filtering/content-based recommender system which might yield better performance and offer improved recommendations.

Another aspect which could be improved is the size of the input dataset. Due to computational constraints, the ratings dataset was limited to ~30,000 relevant entries for the duration of this study. A richer and more extensive dataset (along with a better machine) could lead to increased performance metrics and more variety in recommendations. Lastly, one of the issues of collaborative filtering in general is trust: since the recommendation system relies on the ratings given by users, is the feedback given by a user genuine or not? However, as this is a broad and general issue, it was not examined during this research. Inclusion of additional content-based parameters (and the creation of a hybrid algorithm) into the recommender model could offset issues related to genuineness of user ratings.

7 Conclusion

Overall, this study managed to yield several interesting results. Firstly, it resulted in a collaborative filtering book recommender system, based on the matrix factorization approach using SVD++, with RMSE of 0.752 and MAE of 0.57. Since the ratings in the input dataset range from 0.5 to 5, with a rating step of 0.5, this basically means that, on average, predicted ratings differ from the original ratings by ~0.57, which is an acceptable deviation. Moreover, upon inspecting the list of predicted books for a user and comparing them to the user's previous book resume, it appears that the algorithm does indeed recommend books of similar genres and ratings to the ones previously enjoyed by the user. While the model presents good overall results, there is still significant room for improvement. Inclusion of a larger dataset with more balanced positive/negative ratings could result in better generalization and more accurate predictions for lower ratings. Additionally, collaborative filtering could potentially be combined with other hybrid recommendation approaches; for example, incorporating user demographics, book genres and prices into the recommendation algorithm for more relevant results. All in all, this paper presents an interesting insight into collaborative filtering algorithms and makes a case that book recommender systems based on SVD++ definitely merit additional research.

References

1. Sase, A., et al.: A proposed book recommender system. IJARCCE, 481–483 (2015). https://doi.org/10.17148/ijarcce.2015.42108
2. Ekstrand, M.D.: Collaborative filtering recommender systems (2011). https://doi.org/10.1561/9781601984432
3. Adomavicius, G., Tuzhilin, A.: Toward the next generation of recommender systems: a survey of the state-of-the-art and possible extensions. IEEE Trans. Knowl. Data Eng. **17**(6), 734–749 (2005). https://doi.org/10.1109/tkde.2005.99
4. Almazro, A., et al.: A Survey Paper on Recommender Systems. Concordia Institute for Information Systems Engineering, Montreal (2010)
5. Pazzani, M.J., Billsus, D.: Content-based recommendation systems. In: Brusilovsky, P., Kobsa, A., Nejdl, W. (eds.) The Adaptive Web. LNCS, vol. 4321, pp. 325–341. Springer, Heidelberg (2007). https://doi.org/10.1007/978-3-540-72079-9_10
6. Sarwar, B., et al.: Item-based collaborative filtering recommendation algorithms. In: Proceedings of the Tenth International Conference on World Wide Web – WWW 2001 (2001). https://doi.org/10.1145/371920.372071
7. Han, J., Kamber, M., Pei, J.: In Data Mining: Concepts and Techniques. Elsevier/Morgan Kaufmann, Amsterdam (2012)
8. Lee, J., Sun, M., Lebanon, G.: Comparative study of collaborative filtering algorithms. In: Proceedings of the International Conference on Knowledge Discovery and Information Retrieval (2012). https://doi.org/10.5220/0004104001320137
9. Kurmashov, N., Latuta, K., Nussipbekov, A.: Online book recommendation system. In: 2015 Twelve International Conference on Electronics Computer and Computation (ICECCO) (2015). https://doi.org/10.1109/icecco.2015.7416895
10. Rich, E.: User modeling via stereotypes*. Cogn. Sci. **3**(4), 329–354 (1979). https://doi.org/10.1207/s15516709cog0304_3
11. Goldberg, D., et al.: Using collaborative filtering to weave an information tapestry. Commun. ACM **35**(12), 61–70 (1992). https://doi.org/10.1145/138859.138867
12. Resnick, P., et al.: GroupLens. In: Proceedings of the 1994 ACM Conference on Computer Supported Cooperative Work, CSCW 1994 (1994). https://doi.org/10.1145/192844.192905
13. Hill, W., et al.: Recommending and evaluating choices in a virtual community of use. In: Proceedings of the SIGCHI Conference on Human Factors in Computing Systems, CHI 1995 (1995). https://doi.org/10.1145/223904.223929
14. Shardanand, U., Maes, P.: Social information filtering. In: Proceedings of the SIGCHI Conference on Human Factors in Computing Systems, CHI 1995 (1995). https://doi.org/10.1145/223904.223931
15. Goldberg, K.: In Eigentaste: A Constant Time Collaborative Filtering Algorithm. Electronics Research Laboratory, College of Engineering, University of California, Berkeley (2000)
16. Ansari, A., Essegaier, S., Kohli, R.: Internet recommendation systems. J. Mark. Res. **37**(3), 363–375 (2000). https://doi.org/10.1509/jmkr.37.3.363.18779
17. Su, X., Khoshgoftaar, T.M.: A survey of collaborative filtering techniques. Adv. Artif. Intell. **2009**, 1–19 (2009). https://doi.org/10.1155/2009/421425
18. Breese, J., Heckerman, D., Kadie, C.: Empirical analysis of predictive algorithms for collaborative filtering. In: Proceedings of Uncertainty in Artificial Intelligence (1998)
19. Huang, Z., Zeng, D., Chen, H.: A comparison of collaborative-filtering recommendation algorithms for E-commerce. IEEE Intell. Syst. **22**(5), 68–78 (2007). https://doi.org/10.1109/mis.2007.4338497
20. Schafer, J.B., Konstan, J.A., Riedl, J.: E-commerce recommendation applications. Appl. Data Min. Electron. Commerce **5**, 115–153 (2001). https://doi.org/10.1007/978-1-4615-1627-9_6

21. Burke, R.: Hybrid web recommender systems. In: Brusilovsky, P., Kobsa, A., Nejdl, W. (eds.) The Adaptive Web. LNCS, vol. 4321, pp. 377–408. Springer, Heidelberg (2007). https://doi.org/10.1007/978-3-540-72079-9_12
22. Bell, R.M., Koren, Y., Volinsky, C.: All together now: a perspective on the NETFLIX PRIZE. Chance **23**(1), 24 (2010). https://doi.org/10.1007/s00144-010-0005-2
23. Koren, Y.: Factorization meets the neighborhood. In: Proceeding of the 14th ACM SIGKDD International Conference on Knowledge Discovery and Data Mining, KDD 2008 (2008). https://doi.org/10.1145/1401890.1401944
24. Ziegler, C.-N., et al.: Improving recommendation lists through topic diversification. In: Proceedings of the 14th International Conference on World Wide Web, WWW 2005 (2005). https://doi.org/10.1145/1060745.1060754
25. Zajac, Z.: Goodbooks-10k: a new dataset for book recommendations, FastML (2017)
26. Pazzani, M., Billsus, D.: Learning collaborative information filters. In: Proceedings of the Fifteenth International Conference on Machine Learning, pp. 46–54 (1998)
27. Hug, N.: Surprise, a Python library for recommender systems, Surprise (no date). http://surpriselib.com/. Accessed 20 Jan 2021
28. Luo, X., et al.: An efficient non-negative matrix-factorization-based approach to collaborative filtering for recommender systems. IEEE Trans. Industr. Inf. **10**(2), 1273–1284 (2014). https://doi.org/10.1109/tii.2014.2308433
29. Lemire, D., Maclachlan, A.: Slope one predictors for online rating-based collaborative filtering. In: Proceedings of the 2005 SIAM International Conference on Data Mining (2005). https://doi.org/10.1137/1.9781611972757.43
30. Parvatikar, S., Joshi, B.: Online book recommendation system by using collaborative filtering and association mining. In: 2015 IEEE International Conference on Computational Intelligence and Computing Research (ICCIC) (2015). https://doi.org/10.1109/iccic.2015.7435717
31. Patil, A.E., et al.: Online book recommendation system using association rule mining and collaborative filtering. IJCSMC **8**(4), 83–87 (2019)
32. Rana, A., Deeba, K.: Online book recommendation system using collaborative filtering (with Jaccard similarity). J. Phys. Conf. Ser. **1362**, 012130 (2019). https://doi.org/10.1088/1742-6596/1362/1/012130
33. Uko, E., Eke, B.O., Asagba, P.O.: An improved online book recommender system using collaborative filtering algorithm. Int. J. Comput. Appl. **179**(46), 41–48 (2018). https://doi.org/10.5120/ijca2018917193
34. Cho, J., et al.: Book recommendation system. Bachelor's thesis. Carleton College (2017)
35. Kumar, V.: Singular value decomposition (SVD) in recommender system. Anal. India Mag. (2021). https://analyticsindiamag.com/singular-value-decomposition-svd-application-recommender-system/. Accessed 10 May 2021

Bidirectional LSTM Networks for Abstractive Text Summarization

Aldin Kovačević(✉) and Dino Kečo

Faculty of Engineering and Natural Sciences, Department of Information Technologies, International Burch University, Sarajevo, Bosnia and Herzegovina
aldin.kovacevic@stu.ibu.edu.ba

Abstract. With the constant increase in the amount of available information, especially textual, there exists a need for an efficient and automated text summarization mechanism. Two main text summarization algorithms have been the focus of numerous research studies: extractive and abstractive summarization. This paper focuses on implementing a variant of abstractive summarization using Long Short-Term Memory (LSTM) networks by modeling the summarization task as a Sequence-to-Sequence (Seq2Seq) problem. The dataset used in the research was Amazon's database of food and product reviews. The study employs the encoder-decoder architecture, with stacked LSTMs in the encoder phase and an attention layer mechanism. Moreover, the paper experiments with using bidirectional LSTMs instead of the standard unidirectional approach. This, combined with stacking, leads to increased sequence representation and better summarization performance. The trained model was evaluated using BLEU and ROUGE performance metrics, showing acceptable results. Moreover, the produced summaries generally seemed to correspond to original summaries, conveying the intended meaning. In the end, the paper outlines several possible improvements to be attempted in future models.

Keywords: Abstractive text summarization · Seq2Seq models · Encoder-decoder architecture · Stacked LSTMs · Bidirectional LSTM

1 Introduction

As an effect of the ever-evolving information gathering and processing industry, there is a plethora of data being generated every second, most of it textual. Whereas humans are generally proficient at analyzing, understanding and summarizing the meaning of textual documents, machines are lacking in this regard. In today's landscape of over-abundant data, and lack of manpower and time to properly interpret the incoming information, automatic text summarization algorithms are becoming increasingly crucial [1].

1.1 Text Summarization

Text summarization refers to the process of distilling the most important information from a source (or sources) in order to produce an abridged version for a particular

N. Ademović et al. (Eds.): IAT 2021, LNNS 316, pp. 281–293, 2022.
https://doi.org/10.1007/978-3-030-90055-7_21

user or task [2]. Text summarization aims to produce a fluent and concise summary of textual information, while simultaneously preserving key information content and overall meaning.

Text summarization is a common problem in natural language processing (NLP), and one that is becoming increasingly prominent today. According to the International Data Corporation (IDC) [3], the amount of digital data in annual circulation will rise to a voluminous 175 zettabytes by 2025. Hence, with such a volume of data in daily use, there exists a pressing need to develop efficient algorithms that can automatically and accurately shorten longer texts and derive fluent summaries [4]. Some of the main applications of automatic text summarization include:

- reducing reading and comprehension time
- facilitating the process of selecting relevant documents and researching for information
- improving the effectiveness of indexing
- creating personalized and less biased summaries
- increasing the amount of information that can fit in a medium

There are several ways of structuring a text summarization problem, but they are generally placed in the realm of supervised machine learning (future outcomes are predicted based on existing input data). Before summarization, the input data should be cleaned, preprocessed and possibly transformed in order to make it processable for a resulting summarization model. Moreover, the summarization model requires a way to recognize important "merited" words and keyphrases that will be used to train the machine learning algorithm. Some of the features that can be used are the keyphrase length and frequency, the most recurring words, parts of speech and sentence relevance, and similar [4].

1.2 Types of Text Summarization

When it comes to the classification of text summarization algorithms, they can be observed in the context of input type, purpose and output type [1]. Regarding the input type, summarization methods can be either single-document (which operate on texts of shorter length) or multi-document (which operate on long or multi-document inputs).

According to purpose, text summarization can be generic, domain-specific and query-based [1]. Generic summarization behaves the same for all kinds of input text. In contrast, domain-specific summarization uses specialized domain knowledge to try producing a more accurate summary and is generally used with scientific research papers. Query-based summarization produces summaries which answer specific questions about the input text.

Lastly, based on output type, summarization algorithms can be extractive and abstractive. Extractive summarization selects important, unedited sentences from the input text to build a summary. On the other hand, abstractive summarization tries to generate a natural-sounding summary consisting of new phrases and sentences, which do not necessarily appear in the original text. This approach can offer a more precise and coherent summary, but it is significantly more difficult to implement than its extractive counterpart [1].

1.3 Seq2Seq Modeling

Sequence-to-sequence (Seq2Seq) models are a type of deep learning models that take a sequence of items (e.g. letters, words, numerical data) as input and produce another sequence of items as output. First proposed in 2014 [5, 6], Seq2Seq models have since achieved tremendous success in various deep learning fields, namely image captioning, machine translation and text summarization [7].

This research paper focuses on the implementation of a Seq2Seq model for text summarization using Long Short-Term Memory (LSTM) networks. LSTMs are a special class of recurrent neural networks capable of learning long-term dependencies [8], which is especially useful in problems where sequences are interdependent, like in text summarization. The main contributions of this paper are as follows:

1. developing an effective abstractive text summarization model using an encoder-decoder architecture with bidirectional LSTM networks
2. evaluating the performance of the model using BLEU and ROUGE scores

2 Literature Review

Seq2Seq modeling has seen successful application in a variety of natural language processing tasks, namely machine translations, headline generation, text summarization and speech recognition [9]. One of the most significant breakthroughs in the field of abstractive text summarization occurred in 2015, when Rush et al. [10] applied a neural attention-based model to a news summarization (DUC-2004) problem. The model featured an attention-based encoder and a neural network language model (NNLM) in the decoder, achieving a considerable performance gain over the traditional methods at the time.

Stemming from the success of the neural network approach, Chopra et al. [11] improved upon the Rush et al. model by using a recurrent neural network (RNN) decoder instead of NNLM. They experimented with Elman and LSTM decoder architecture, and included a convolutional attention-based encoder. This model then proceeded to outperform recent state-of-the-art algorithms on a commonly used Gigaword testing corpus.

However, abstractive text summarization architecture still lacked in several key aspects, mostly related to modeling out-of-vocabulary words and detecting document hierarchies [9]. Nallapati et al. [12] sought to address these critical issues by introducing several new elements to the RNN encoder-decoder architecture, most prominently a feature-rich encoder to capture keywords, a switching generator-pointer to model OOV (out-of-vocabulary) words and the hierarchical attention necessary to capture document structures. Moreover, they also set a precedent by establishing their model benchmarks on the CNN/Daily Mail dataset, which consists of pairs of news article documents and multi-sentence summaries. Previously, most abstractive summarization models focused on examining short documents with single-sentence summaries.

Despite the success of Nallapati et al., models working on long documents with multi-sentence summaries still suffered from an inability to reliably reproduce the most important information contained in an input document. Additionally, these models often repeated words and sentences, which made the generated summaries sound unnatural [9]. Working to address these issues, See et al. [13] proposed a hybrid pointer-generator network that combines extraction with abstraction. This architecture is capable of copying words from source text via a pointer, while also generating novel words from a vocabulary using a generator. This approach takes care of both important information reproduction and OOV words. Numerous follow-up studies would prove the effectiveness of the pointing/copying mechanism by achieving state-of-the-art performance results [9].

Even though most of the prevalent Seq2Seq models rely on RNNs, significant success in the field of abstractive summarization has also been achieved using encoder-decoder models based on convolutional neural networks (CNN). CNNs can be parallelized during training and evaluation, and their computational complexity is linear with respect to sequence lengths. Thanks to these features, CNNs have a potential to solve a major shortcoming of RNNs: computational constraints and dependency on previous hidden states [9]. Kalchbrenner et al. [14] proposed a state-of-the-art ByteNet model which uses a one-dimensional convolutional neural network in both encoder and decoder components, resulting in exceptional performance. Gehring et al. [15] tried to apply a CNN-based approach to large-scale benchmark datasets by creating an encoder model comprised of convolutional layers and replacing the LSTM decoder with a CNN decoder [16], demonstrating strong performance.

3 Materials and Methods

The following section outlines an overview of the utilized dataset, the steps performed in data preprocessing, and the explanation of the proposed summarization model. Figure 1 showcases the key steps of the research process outlined in this paper.

3.1 Dataset Overview and Statistical Analysis

This paper attempts to build an abstractive summarization model using the "Amazon Fine Food Reviews" dataset. The dataset comprises a collection of 568,454 food and product reviews over a period of 10+ years, up to October 2012. Reviews include product and user information, a numerical rating, review text, review timestamp and a brief human-produced summary of the review. For this paper, only the review text and summary are relevant. On average, each review is composed of 96.25 words, spanning 4.98 sentences. Moreover, there are around 4.82 words per summary, averaged across 1.07 sentences.

3.2 Data Preprocessing

Data preprocessing is an important starting step in text summarization, since the input data has to be as "clean" as possible to facilitate smooth operation and a good resulting model. Data preprocessing in text summarization is a challenging problem that needs to be tailored to specific application needs. However, it usually boils down to simplifying the learning problem and reducing the size of the vocabulary.

The approach in this document uses the Python pandas library to first load each review and its corresponding summary into the memory. Afterwards, it removes empty,

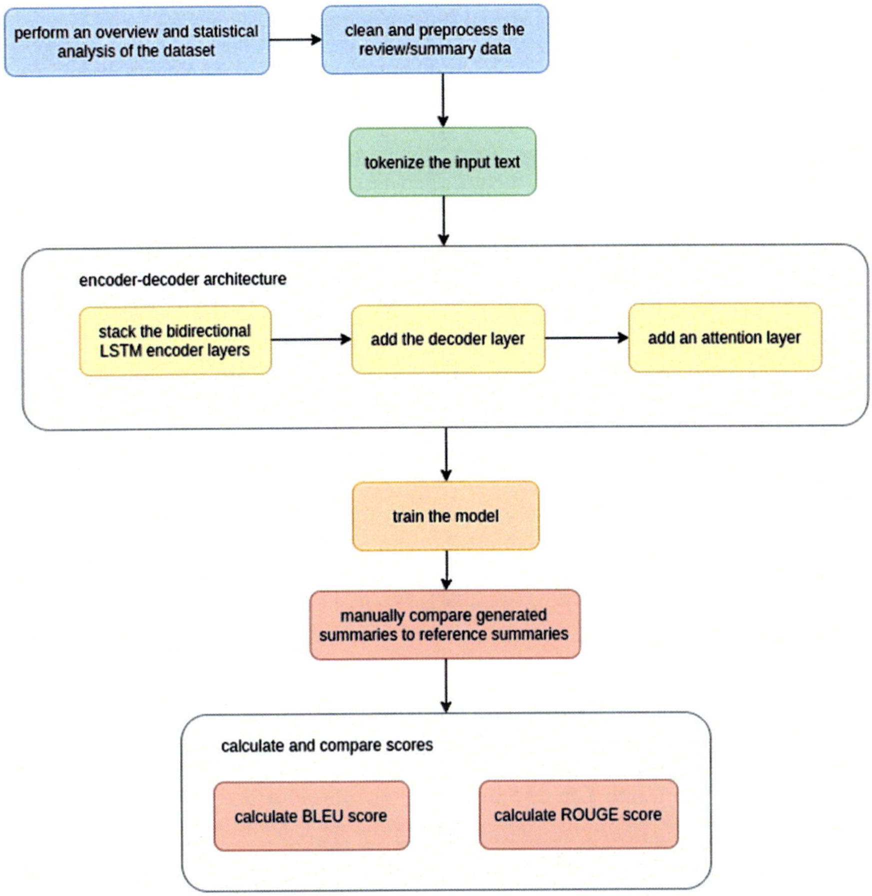

Fig. 1. Model diagram with key steps

duplicate and "N/A" lines, or lines with missing reviews/summaries. To further simplify the vocabulary, all words are turned into their lowercase counterparts. Next up, contents in the parentheses are removed, as they are most likely not relevant to the summarization. Since the reviews contain frequent contractions, a contraction mapping was made to "translate" all shortened forms into full words (e.g. "I've" to "I have"). In addition to that, all punctuation, Saxon genitives (possessive 's), and non-alphabetic characters have also been removed to simplify the model.

Afterwards, a removal of stopwords has been performed on the review texts. The collection of stopwords used in the document was provided by Python's NLTK (Natural Language Toolkit) open-source library. Stopword removal was not performed on summary data, in order to keep summaries more linguistically rich and flexible. Tables 1 and 2 show an example of an unprocessed (original) and cleaned reviews and summaries.

Another preprocessing method that was utilized is text normalization, or reducing the words to their root forms. This serves to simplify the input data and reduce training

Table 1. Original and cleaned review text

Original review	Cleaned review
I have bought several of the Vitality canned dog food products and have found them all to be of good quality. The product looks more like a stew than processed meat and it smells better. My Labrador is finicky and she appreciates this product better than most	Bought several vitality can dog food product found good quality product look like stew process meat smell well labrador finicky appreciates product well

Table 2. Original and cleaned summary text

Original review	Cleaned review
Good quality dog food	Good quality dog food

time. The form of text normalization performed on the review dataset was lemmatization, which uses vocabulary and morphological analysis to obtain word roots [17], based on word type. However, lemmatization was only performed on the input reviews, and not on the summaries, in order to keep summaries more natural-sounding and authentic. Additional data cleaning options include removal of low-frequency words (such as names) and removal of short words (length less than 2).

In the end, the cleaned dataset consists of 393,565 summary/review pairs. The dataset is then split into the training and testing sets, with 80% of the dataset being used as training data, and the remaining 20% as the testing data, resulting in 314,852 training samples and 78,213 test samples.

3.3 Word Tokenization

In abstractive text summarization, textual input needs to be converted into a numerical format. This stage of processing utilizes Keras, a deep learning Python library which runs as an interface for TensorFlow. Keras contains utilities for tokenization, which represents encoding a word sequence into a sequence, or vector, of integers. The Keras tokenizer uses a list of training texts to create an internal vocabulary index based on word frequency. Afterwards, it transforms each text in the text list to a sequence of integers, using the word frequency index generated in the first step.

Even though all input texts are of different lengths, using a constant sequence size for all samples will simplify and unify data for the model. Therefore, when utilizing integer sequences, the sequence size needs to be truncated to a certain length and smaller sequences need to be padded with zeros to the maximum length. Plotting the distribution of text lengths as histograms reveals that the average word count for cleaned reviews is 38.42, whereas the summary word count is 4.15. Moreover, most review lengths fall under the 0–100-word range, whereas summary lengths fall between 1 to 10 words (Fig. 2).

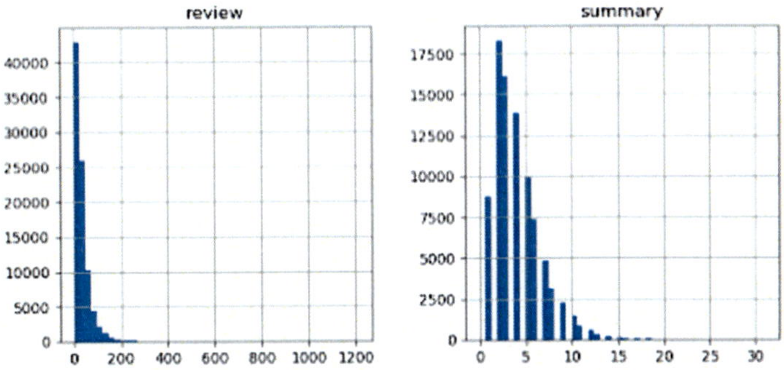

Fig. 2. Distribution of review and summary lengths

Based on this, the maximum length of a review will be taken as 80 words, and the maximum summary length will be 10 words

4 Model Building

As mentioned previously, a Seq2Seq model can be built for any problem that involves sequential information. In text summarization, the goal is to turn a long sequence of words (input) into a short summary sequence (output); in other words, this represents a typical many-to-many Seq2Seq problem.

Encoder-decoder architecture represents one of the most common methods for solving Seq2Seq problems where input and output sequences vary in length. It consists of two main components, an encoder and a decoder, which are typically modeled as either a Gated Recurrent Neural Network (GRU) or an LSTM, as is the case in this paper. The encoder captures the context of the input information using a hidden state vector. Using this state vector, the decoder produces the output sequence [7]. A visual representation of the encoder-decoder architecture is shown in Fig. 3.

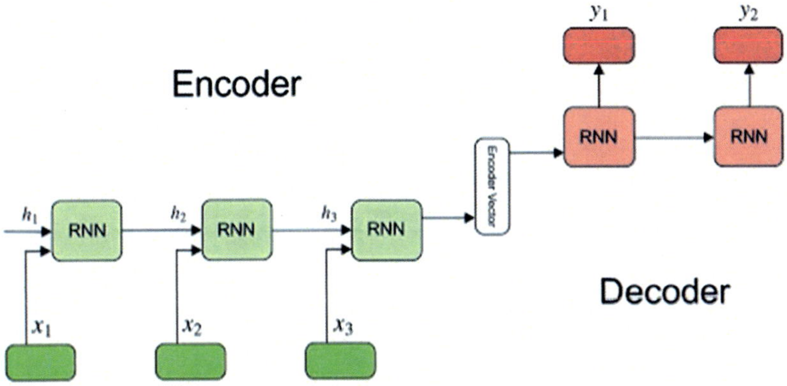

Fig. 3. Encoder-decoder architecture [25]

The encoder-decoder workflow is divided into two phases:

- training phase
- inference phase

4.1 Training Phase

The training phase begins by first setting up an encoder and a decoder, followed by the model training. At each timestep, one word from the input sequence is fed to the LSTM encoder. The input at each timestep is then processed, capturing the context of the input sequence. The context is defined in the final encoder output by the hidden state h_i and cell state c_i. These values are then used to initialize the decoder, which is also an LSTM component. Afterwards, the target sequence is read by the decoder, which then attempts to predict the same sequence with a one-timestep offset – essentially, predicting the next word based on the previous word in the sequence. In order to use a decoder on a target sequence, special *<start>* and *<end>* tokens are added to a sequence before processing. These tokens represent special words which do not appear anywhere else in the dataset (e.g. "starttok" and "endtok"). Since the target sequence is unknown while decoding, *<start>* and *<end>* tokens would be consistent and always signify the sentence boundaries.

4.2 Inference Phase

Inference, or testing, phase refers to trying out a model on new source sequences where the target sequence is unknown. In order to infer a test target sequence, the input sequence is encoded, and the starting state of the decoder is initialized using the final state of the encoder. Following that, the *<start>* token is appended to the beginning of the sequence. The decoder will run until it generates the *<end>* token or hits the maximum sequence limit, producing a probability output for the next word and appending it to the sequence.

4.3 Attention Mechanism

One flaw of the encoder-decoder architecture is that the entire input sequence needs to be transformed into a fixed-length vector during the encoding phase. This might make it challenging for the neural network to "remember" important information when dealing with long sentences, since the necessary information might not fit into the fixed-length vector. Because of that, the performance of a traditional encoder-decoder architecture deteriorates proportionately to the increase of sentence length [18].

This is a problem that the attention mechanism, first presented by Bahdanau et al. [18], tries to rectify. Instead of looking at the entire sequence, the attention mechanism focuses on a few specific parts of the sequence in order to predict a new word, an optimization achieved by increasing the importance of certain sequence parts. By utilizing an attention mechanism, the decoder can process different parts of the input sequence at each step of the output sequence generation. Rather than encoding the input sequence into a single fixed-length vector, the model learns how to generate a context vector at each step, and which sequence parts are more important for further processing [19].

4.4 Model Overview

The initial model described in this paper uses an LSTM-based encoder-decoder architecture and deep learning mechanisms provided by Keras. Moreover, in order to mitigate the issue of long sequences, it incorporates a global attention mechanism outlined by Bahdanau et al. [18] that aligns the model to parts of the input sequence that are relevant to the current prediction.

The encoder LSTM architecture is a stacked LSTM consisting of five layers, which makes the model deeper and in turn, should lead to a better sequence representation [20]. Moreover, this paper attempts to innovate by utilizing bidirectional LSTMs in the encoder phase. Since the meaning of a word can depend on both its following and antecedent lexemes, training a bidirectional network can lead to more accurate results that a standard unidirectional approach. Encoder layers are represented as bidirectional LSTMs, with the 300 latent dimensions, and dropout and recurrent dropout rate of 0.4. The decoder phase uses a single LSTM layer, with 600 latent dimensions (since it concatenates the outputs of the bidirectional encoder), dropout rate of 0.4 and recurrent dropout rate of 0.2. Following the encoder and decoder layer, the attention layer is attached to the model. Lastly, a time-distributed dense layer with softmax activation is added. Maximum number of iterations is set to 50, with early stopping defined if the validation loss, defined by sparse categorical cross-entropy, increases in one iteration. The model is trained on 314,852 samples, and validated on the remaining 78,213, with the batch size of 512.

5 Results and Discussion

On average, one epoch takes around 530 s to train using NVIDIA Tesla K80 GPU, Intel Xeon CPU and 25 GB of RAM (resources provided by Google Colaboratory). The training usually finishes early after roughly 20 epochs, having experienced a slight increase in validation loss, as shown in Fig. 4.

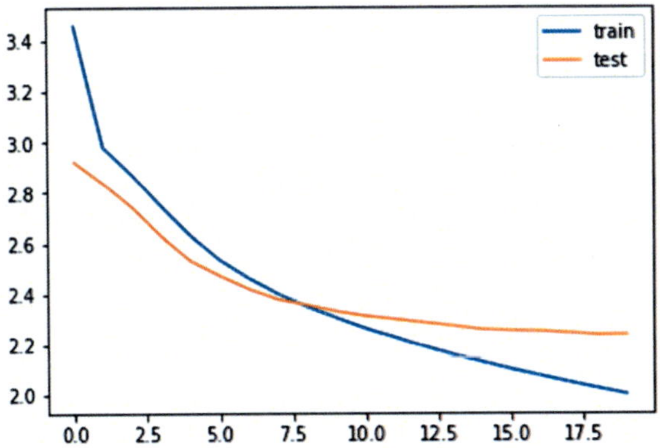

Fig. 4. Train vs. test error

5.1 Experimental Results

Target sequences were decoded using a greedy approach, with the *argmax* function. On first inspection, the generated summaries seem quite acceptable. In fact, some of them even convey a more relevant meaning than some of the human-provided summaries. A few examples of generated summaries are provided in Table 3. RS will denote "reference (human) summary" for short, whereas MS will stand for "model-generated summary".

Table 3. Comparison of original and generated summaries

Reference (original) summary (RS)	Generated summary (MS)
Love this syrup	Great syrup
Delicious	Great candy
Nothing special	Not good
Well on my hair	Great conditioner
Amazing popcorn	Great popcorn
Wonderful cracker	Great taste
Great taste	Delicious
Really great decaf	Good coffee
Best mayo ever	Best mayo ever
Paid extra	Rip off

By inspecting and comparing the model-generated summaries to reference summaries it can be seen that they are generally acceptable and convey an appropriate sentiment. In some case, they even provide more context than the human-provided summary. For example, "great candy" (MS) is a more accurate summary for a candy product that simply "delicious" (RS); or, "great conditioner" (MS) as opposed to "well on my hair" (RS). Moreover, the model seems to capture and connect inferred meanings, as evidenced by summarizing "rip off" (MS) where the original summary was "paid extra" (RS).

5.2 Scoring the Model

A commonly used scoring metric in text-based Seq2Seq problems is Bilingual Evaluation Understudy Score, or BLEU. It is a metric used to evaluate a generated sentence in comparison to a reference sentence, where a score of 1.0 signifies a perfect match and a score 0.0 a "perfect mismatch" [21]. It can be said that BLEU is a measure of text summarization precision: how many words in the model-generated summary appear in the human reference summary. When executed on the comparison of target and model-generated summaries from the test dataset, BLEU produced an average score of 0.814. A BLEU score above 0.50 is considered good [22], so this model seems to provide a satisfactory performance, which is backed up by manual inspection of the generated summaries.

Another commonly used metric for evaluating text summarization is ROUGE, or Recall-Oriented Understudy for Gisting Evaluation. Similarly, to BLEU, ROUGE compares an automatically produced summary against an original reference summary [23]. However, ROUGE is mostly used as a measure of text summarization recall: how many words in the human-produced summary appear in the model-generated summary. ROUGE features several different metrics, namely ROUGE-N which measures the N-gram overlap between summaries, ROUGE-L, which measures statistics based on the longest common sub-sequence (LCS) and ROUGE-W, which is a weighted version of ROUGE-L. Table 4 showcases the ROUGE scores for the model.

Table 4. ROUGE scores for the model

ROUGE metric	Score (* 100%)
ROUGE-1	14.25
ROUGE-2	2.72
ROUGE-L	15.34
ROUGE-W	14.06

The calculated ROUGE scores are all significantly lower than the BLEU metric. This implies the following: the proposed model generates a lot of words that are found in the human reference summaries (high BLEU). On the other hand, there are also a lot of words in the reference summary that do not appear in the generated summaries (low ROUGE). This can be correlated to the fact that human summaries include a lot of varied vocabulary and terms, whereas the generated summaries are mostly expressed through the different forms and variations of adjectives "good" and "bad". In short, the model did learn to convey proper meaning in its summarization, albeit with a limited vocabulary.

However, it has to be noted that ROUGE scores are generally low for short summaries [24], which happens to be the case in this dataset. A higher summary length might lead to increased ROUGE scores, as there is a higher possibility for words matches. Moreover, even though the generated summaries do not exactly reflect the human summaries in their structure, they convey the correct meaning most of the time, as evidenced by manual review.

6 Conclusion

Overall, based on manual inspection, the trained model seems to produce mostly relevant summarization, despite the BLEU and ROUGE score mismatch. The bidirectional LSTM approach yielded an increase in summarization precision, albeit at the cost of increasing the training time of the model. Some possible future improvements might include increasing the training dataset size and using varied data in order to increase the generalization capability. Moreover, the beam search strategy, which suggests several possible output sequences for sequence decoding, might produce more favorable predictions over the greedy *argmax* approach. Additionally, the model might be improved through the implementation of pointer-generator networks to handle OOV (out-of-vocabulary) issues and produce more novel and natural-sounding summaries. In any case, the bidirectional LSTM summarization approach shows promising results and warrants further research.

References

1. Chauhan, K.: Unsupervised text summarization using sentence embeddings, medium. Jatana (2019). https://medium.com/jatana/unsupervised-text-summarization-using-sentence-embeddings-adb15ce83db1. Accessed 4 Jan 2021
2. Mani, I., Maybury, M.T.: Advances in Automatic Text Summarization. The MIT Press, Cambridge (2001)
3. Patrizio, A.: IDC: Expect 175 zettabytes of data worldwide by 2025. Network World (2018). https://www.networkworld.com/article/3325397/idc-expect-175-zettabytes-of-data-worldwide-by-2025.html. Accessed 5 Jan 2021
4. Garbade, D.M.J.: A quick introduction to text summarization in machine learning, medium. Towards Data Science (2018). https://towardsdatascience.com/a-quick-introduction-to-text-summarization-in-machine-learning-3d27ccf18a9f. Accessed 11 Jan 2021
5. Cho, K., et al.: Learning phrase representations using RNN encoder–decoder for statistical machine translation. In: Proceedings of the 2014 Conference on Empirical Methods in Natural Language Processing (EMNLP) (2014). https://doi.org/10.3115/v1/d14-1179
6. Sutskever, I., Vinyals, O., Le, Q.V.: Sequence to Sequence Learning with Neural Networks (2014)
7. Dugar, P.: Attention - Seq2Seq Models, Medium. Towards Data Science (2019). https://towardsdatascience.com/day-1-2-attention-seq2seq-models-65df3f49e263. Accessed 31 Jan 2021
8. Olah, C.: Understanding LSTM networks, understanding LSTM networks -- colah's blog (2015). https://colah.github.io/posts/2015-08-Understanding-LSTMs/. Accessed 6 Jan 2021
9. Shi, T., et al.: Neural abstractive text summarization with sequence-to-sequence models. ACM/IMS Trans. Data Sci. **2**(1), 1–37 (2021). https://doi.org/10.1145/3419106
10. Rush, A.M., Chopra, S., Weston, J.: A neural attention model for abstractive sentence summarization. In: Proceedings of the 2015 Conference on Empirical Methods in Natural Language Processing (2015). https://doi.org/10.18653/v1/d15-1044
11. Chopra, S., Auli, M., Rush, A.M.: Abstractive sentence summarization with attentive recurrent neural networks. In: Proceedings of the 2016 Conference of the North American Chapter of the Association for Computational Linguistics: Human Language Technologies (2016). https://doi.org/10.18653/v1/n16-1012
12. Nallapati, R., et al.: Abstractive text summarization using sequence-to-sequence RNNs and beyond. In: Proceedings of the 20th SIGNLL Conference on Computational Natural Language Learning (2016). https://doi.org/10.18653/v1/k16-1028
13. See, A., Liu, P.J., Manning, C.D.: Get to the point: summarization with pointer-generator networks. In: Proceedings of the 55th Annual Meeting of the Association for Computational Linguistics (Volume 1: Long Papers) (2017). https://doi.org/10.18653/v1/p17-1099
14. Kalchbrenner, N., et al. Neural Machine Translation in Linear Time. ArXiv (2016)
15. Gehring, J., et al.: A convolutional encoder model for neural machine translation. In: Proceedings of the 55th Annual Meeting of the Association for Computational Linguistics (Volume 1: Long Papers) (2017). https://doi.org/10.18653/v1/p17-1012
16. Dauphin, Y.N., et al.: Language modeling with gated convolutional networks. In: International Conference on Machine Learning, pp. 933–941 (2017)

17. Singh, S.: How To Remove Stopwords in Python: Stemming and Lemmatization, Analytics Vidhya (2020). https://www.analyticsvidhya.com/blog/2019/08/how-to-remove-stopwords-text-normalization-nltk-spacy-gensim-python. Accessed 3 Jan 2021
18. Bahdanau, D., Cho, K., Bengio, Y.: Neural machine translation by jointly learning to align and translate. In: ICLR (2015)
19. Radhakrishnan, P.: Attention Mechanism in Neural Networks, Hacker Noon (2021). https://hackernoon.com/attention-mechanism-in-neural-network-30aaf5e39512. Accessed 24 Jan 2021
20. Brownlee, J.: Stacked Long Short-Term Memory Networks, Machine Learning Mastery (2019). https://machinelearningmastery.com/stacked-long-short-term-memory-networks/. Accessed 31 Jan 2021
21. Brownlee, J.: A Gentle Introduction to Calculating the BLEU Score for Text in Python, Machine Learning Mastery (2019). https://machinelearningmastery.com/calculate-bleu-score-for-text-python/. Accessed 2 Feb 2021
22. What is BLEU Score? (no date). KantanMT. https://kantanmt.com/whatisbleuscore.php. Accessed 1 Feb 2021
23. freeCodeCamp.org: An intro to ROUGE, and how to use it to evaluate summaries, freeCodeCamp.org. freeCodeCamp.org (2017). https://www.freecodecamp.org/news/what-is-rouge-and-how-it-works-for-evaluation-of-summaries-e059fb8ac840. Accessed 2 Feb 2021
24. Schluter, N.: The limits of automatic summarisation according to ROUGE. In: Proceedings of the 15th Conference of the European Chapter of the Association for Computational Linguistics: Volume 2, Short Papers (2017). https://doi.org/10.18653/v1/e17-2007
25. Kostadinov, S.: Understanding Encoder-Decoder Sequence to Sequence Model, Medium. Towards Data Science (2019). https://towardsdatascience.com/understanding-encoder-decoder-sequence-to-sequence-model-679e04af4346. Accessed 10 May 2021

Mechanical Engineering

Thermal Characterization of Straw-Based Panels Made Out of Straw and Natural Binders

Benjamin Rudalija and Benjamin Duraković(✉)

International University of Sarajevo, Sarajevo, Bosnia and Herzegovina
bdurakovic@ius.edu.ba

Abstract. This study investigates the possibility of using crop straw combined with natural binders as building construction material. The straw as a renewable material may have potential in reducing building energy demand. Four types of binders were used to produce composite panel samples. The Guarded Hot Plate method is used to characterize thermal properties and identify the best straw-binder mixture of the composite. The result showed that gypsum-based composite showed the best thermal properties suggesting that straw-based composite panels can be used in light weighed building construction.

Keywords: Straw composite panels · Thermal conductivity · Sustainability

1 Introduction

To reduce building energy demand and to reduce carbon footprint [1], differences approached were applied such as application of different environment-friendly materials and various energy storage techniques [2–5]. Latent heat storage have a huge potential in reducing building energy demand [6, 7]. In addition, some of the sustainable features of traditionally built houses can be used in today's buildings [8–10]. Renewable insulation materials have many advantages over synthetic insulation materials. Benefit of waste material can be used in energy saving [11, 12]. Crop straw is an agricultural by-product which, due to its advantages, is increasingly used as building material in 21st century. Some of the biggest advantages of this material are very low price, good thermal and acoustic insulations [13, 14].

In recent years there are several research and development project on environmentally-friendly building materials [15–18]. Experimental studies were conducted on concrete and clay masonry wall with units filled with coconut fiber [19]. Thermal and mechanical characterization of various composites is based on renewable material such as hemp shiv, sugar cane bagasse and cement composite, etc. [20, 21]. Due to favorable hygrothermal and acoustic properties, straw as by-product from agriculture recently drew researchers' attention [22, 23]. If straw is exposed to moisture, rapid degradation of straw is observed in a laboratory setting, which can be overcome by proper design of straw envelope [24]. Various preparation techniques and methods were developed to utilize straw bales as load bearing material [25]. Usually, straw is used in the form of compressed bales shaped as blocks or combined with a binding material

N. Ademović et al. (Eds.): IAT 2021, LNNS 316, pp. 297–304, 2022.
https://doi.org/10.1007/978-3-030-90055-7_22

to form composites. Thermal conductivities of straw bales reported in the literature are shown in Table 1.

Table 1. Reported thermal conductivity of straw bales

Straw orientation	Density 123 kg/m^3	Thermal conductivity W/mK	Reference
Perpendicular	75–90	0.052–0.056	[26]
Perpendicular-parallel	80	0.051–0.072	[27]
Parallel	75	0.066	[28]
Perpendicular-parallel	126–129	0.056–0.078	[29]
na	63–123	0.06–0.064	[29]

Depending on straw orientation in bale, thermal conductivity reported in the literature ranges from 0.052 to 0.078 W/mK, while the density goes from 63 to 129 kg/m^3. Thermal conductivities of straw-based composites reported in the literature are shown in Table 2.

Table 2. Reported thermal conductivity of straw based composite panels

Composite panel	Density 123 kg/m^3	Thermal conductivity W/mK	Reference
Wheat straw, hemp shiv	188	0.76	[30]
Straw and olive fibers	150	0.06	[31]
Miscanthus × *giganteus*	330–819	0.08–0.1	[32]
Hemp concrete	200–800	0.06–0.18	[33]
Straw and clay	200–800	0.062–0.18	[34]

Composites based on straw are a hot research topic mentioned in the literature by different researches. Mechanical and thermal properties of corn straw-based polyurethane composite panels were investigated by Jiang et al. [35]. Also, thermal and hygric properties of hemp shiv glued with wheat straw were investigated by Collet et al. They found that thermal conductivity of these panels is in the range of 0.071 to 0.076 W/mK [30]. Mechanical and physical properties of composite panels made from rice straws using two binding materials polymeric diphenylmethane diisocyanate and urea-formaldehyde binders were investigated and the results are reported in [36]. They found that static bending and internal bonding strength of the panels initially increase, then decrease with decreased particle size. Alkali treated rice straw with geopolymer has better binding effect than untreated rice straw [37]. Straw fibers combined with sodium silicate solution binder were used to create composite panels [31]. The investigated thermal properties of the composite and found the panel has good insulation properties. Commercially available synthetic insulation materials such as extruded and expanded polystyrene have thermal conductivity in the range of 0.029–0.040 W/mK. Biomass-based insulation material has thermal conductivity values in the vicinity of 0.060 W/mK [32].

The purpose of this research is to contribute to the investigation of straw-based composites, in particular, to investigate the impact of common binding materials to the thermal conductivity of the straw-based composite panels. These composite panels have the potential to be used in light-weighted building construction due to their availability, low cost, low thermal conductivity and overall, they are renewable materials. To achieve the goal, an experimental research is conducted as it is described below.

2 Materials and Methods

The Hot Guarded Plate method is used to determine thermal conductivity of straw-based panels. The scheme of the experimental setup is presented in Fig. 1, which show the cross section of the guarded hot plate setup.

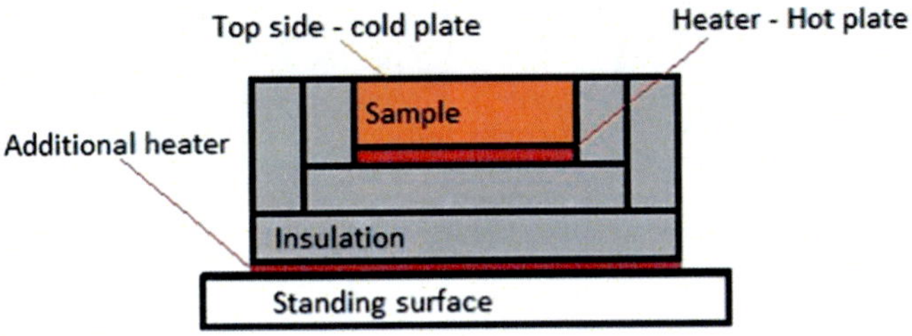

Fig. 1. Scheme of guarded hot plate

In Fig. 1, gray part shows thermal insulation while orange part represents test sample L × W × H = 200 mm × 200 mm × 80 mm. Hot plate is an electrical heater with the power of 9.62 W. The average temperature is measured at both sides of the test sample at know constant heat flow through the sample. Actual test setup with data logger is shown in Fig. 2.

Fig. 2. Experimental setup

Vernier LabQuest 2 data logger is used to measure the temperature. The accuracy of the surface temperature sensors is ± 0.2 °C at 0 °C, ± 0.5 °C at 100 °C, which is sufficiently accurate for this type of study. The data are recorded for each sample until the steady state is achieved, and the thermal conductivity is determined based on the recorded data.

2.1 Samples

Natural binding materials are used for sample fabrication such as: gypsum, sand and lime, lime, gypsum and lime, sugar (sweet water), which is shown in Fig. 3. Every sample used approximately same amount of straw and the same volumetric ration of binder. The weight of the specimens combined with binders varied from 1620 g–2500 g.

a) b) c) d)

Fig. 3. Straw-based test samples: a) gypsum binder; b) lime-sand binder; c) sugar water binder; d) cement-lime-gypsum binder

Prepared samples were dried and tested within temperature interval 60–80 °C. Mixing ratio for each sample is given in Table 3.

Table 3. Mixing ratio of each panel sample

Material	Mixing ratio
Gypsum	1:0.8 (water)
Lime, sand	1:3
Cement, gypsum, lime	1:1:3
Sugar-water	1:0.75

2.2 Thermal Conductivity

Based on recorded data from the experiments for the study state, thermal conductivities were determined using based on the heat flux equation:

$$q = \lambda \frac{t_1 - t_2}{L}$$

From where thermal conductivity is:

$$\lambda = q \frac{L}{t_1 - t_2}$$

where, q is heat flux [W/m^2], λ is thermal conductivity in W/mK, t_1 is temperature on the hot plate in °C, t_2 is temperature on the cold plate in °C, L is thickness of the sample in m. The heat flux is determined as the ration between the heater power and the hot plate area, which is 240.5 W/m^2.

3 Results and Discussion

Using recorded data for the steady state heat transfer and the Equation for thermal conductivity (λ) calculation, results for λ are obtained for each sample and shown in Table 4.

Table 4. Calculated thermal conductivity of each sample

Material	Sample number	Heat flux	Hot plate	Cold plate	Straw sample thickness	Thermal conductivity
		q[W/m^2]	t_1(°C)	t_2[°C]	L[m]	λ [W/mK]
Gypsum	1	240.5	84.9	22.3	0.08	0.31
	2	240.5	85.2	22.2	0.08	0.31
	3	240.5	86	20.4	0.08	0.29
Gyps, cem., lime	4	240.5	69.3	22.2	0.08	0.41
	5	240.5	62.8	22.8	0.08	0.48
Lime, sand	6	240.5	70.5	16.98	0.08	0.36
	7	240.5	67.3	22.67	0.08	0.43
Sugar	8	240.5	66.6	16.23	0.08	0.38
	9	240.5	71.2	15.8	0.08	0.35

In Table 4, the last column displays thermal conductivities of each sample. It is observed that there is no significant difference in the thermal conductivity variance among the different binders. Depending of the binding material type, sample density can significantly vary. The relationship between the sample thermal conductivity and the sample density are plotted in Fig. 4.

As expected, thermal conductivity increased with the increase of sample density. According to the French thermal regulations (RT2012), straw and mixtures below density of 450 kg/m^3 can be approvingly used in new types of buildings [34]. Although no sample falls into that category, samples can be used in some light weight construction, and definitely, there is a plenty of space for improvement. According to the results, it is observed that gypsum sample favorably has the lowest thermal conductivity, while other materials and their thermal conductivities are grouped on the right side of the graph with higher densities.

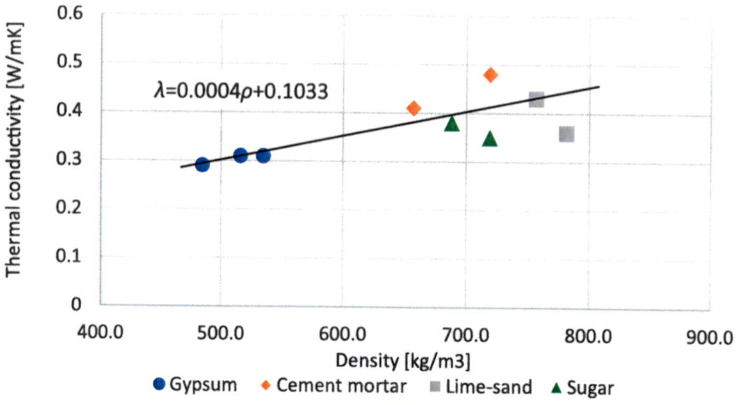

Fig. 4. Thermal conductivity versus density

4 Conclusion

This research showed that the straw-based composite panels have potential to be used in light-weighted building construction. The straw binders studied in this research are available, affordable and not harmful to the environment. The composite panel based on gypsum as binder provides the lowest thermal conductivity but structure of the panel has to be further investigated. Panels based on sugar provide the best structure while their thermal conductivity is increased and values are in the vicinity of sand and lime-based panels. As expected, the composite panel based on sand and cement has the highest thermal conductivity, the highest density, and as such it is less favorable as construction material. In addition, lime-sand is not a good binder, as the highest density is associated with these panels.

References

1. Yıldız, G., Çalış, B., Gürel A.E., Ceylan, I.: Investigation of life cycle CO2 emissions of the polycrystalline and cadmium telluride PV panels. Environ. Nanotechnol. Monit. Manag. **14**, 100343 (2020). https://doi.org/10.1016/j.enmm.2020.100343
2. Findik, F., Ermiş, K.: Thermal energy storage. Sustain. Eng. Innov. **2**(2), 66–88 (2020). https://doi.org/10.37868/sei.v2i2.115
3. Küçük, M., Findik, F.: Selected ecological settlements. Herit. Sustain. Dev. **2**(1), 1–16 (2020). https://doi.org/10.37868/hsd.v2i1.35
4. Halilovic, M.: Vernacular architecture sustainability principles: a case study of Bosnian stone houses in Idbar village. Period. Eng. Nat. Sci. **8**(4), 2564–2574 (2020)
5. Ceylan, I., Yildiz, G., Gürel, A., Ergün, A., Tosun, A.: The effect of malfunctions in air handling units on energy and exergy efficiency. Heat Trans. Res. **51**(11), 1007–1028 (2020). https://doi.org/10.1615/HEATTRANSRES.2020032895
6. Durakovic, B., Torlak, M.: Experimental and numerical study of a PCM window model as a thermal energy storage unit. Int. J. Low-Carbon Technol. **12**(3), 272–280 (2017)
7. Duraković, B.: PCMs in building structure. In: PCM-Based Building Envelope Systems. Green Energy and Technology, pp. 63–87. Springer, Cham (2020). https://doi.org/10.1007/978-3-030-38335-0_4

8. Gurel, A.E., Yıldız, G., Ceylan, I..: Use of air solar collectors in space heating applications. In: 14th National Plumbing Engineering Congress, pp. 1866–1875 (2019)
9. Örnek, E.: Sustainable characteristics of the vernacular house and its impact to the building physics. Herit. Sustain. Dev. **1**(1), 14–20 (2019)
10. Duraković, B.: Passive Solar Heating/Cooling Strategies. In: PCM-Based Building Envelope Systems. Green Energy and Technology, pp. 39–62. Springer, Cham (2020). https://doi.org/10.1007/978-3-030-38335-0_3
11. Uyanik, I., Özkan, O., Mihçiokur, H.: Waste management in a university campus. Sustain. Eng. and Innov. **3**(1), 49–53 (2021). https://doi.org/10.37868/sei.v3i1.id137
12. Aydin, N.: Electricity generation potential of municipal solid wastes produced in the province of Edirne. Sustain. Eng. Innov. **3**(1), 61–67 (2021). https://doi.org/10.37868/sei.v3i1.id138
13. Durakovic, B., Yıldız, G., Yahia, M.: Comparative performance evaluation of conventional and renewable thermal insulation materials used in building envelops. Tehnicki vjesnik - Technical Gazette **27**(1), 283–289 (2020)
14. Yildiz, G., Duraković, B., Almisreb, A.A.: Thermal insulation materials towards high thermal performance: a review. Int. J. Adv. Sci. Eng. Inf. Technol (2021, in Press)
15. Dhakal, U., Mark, G., Russell, R., Umberto, B.: Hygrothermal performance of hempcrete for Ontario (Canada) buildings. J. Clean. Prod. **142**, 3655–3664 (2017). https://doi.org/10.1016/j.jclepro.2016.10.102
16. Jami, T., Karade, S.R., Singh, L.P.: A review of the properties of hemp concrete for green building applications. J. Clean. Prod.**239**, 117852. https://doi.org/10.1016/j.jclepro.2019.117852
17. Basyigit, C., Alkayis, M.H., Kartli, M.I.: Environmental effects of utilization of sustainable building materials. Herit. Sustain. Dev. **3**(1), 64–70 (2021). https://doi.org/10.37868/hsd.v3i1.57
18. Baykal, F.Y.: The polluter should... pay? Herit. Sustain. Dev. **3**(1), 71–77 (2021). https://doi.org/10.37868/hsd.v3i1.58
19. Iwaro, J., Mwasha, A.: Effects of using coconut fiber-insulated masonry walls to achieve energy efficiency and thermal comfort in residential dwellings. J. Architect. Eng. **25**(1), 04019001 (2019). https://doi.org/10.1061/(asce)ae.1943-5568.0000341
20. Onésippe, C., Passe-Coutrin, N., Toro, F., Delvasto, S., Bilba, K., Arsène, M.: Sugar cane bagasse fibres reinforced cement composites: thermal considerations. Appl. Sci. Manuf. **41**(4), 549–556 (2010). https://doi.org/10.1016/j.compositesa.2010.01.002
21. Hussain, A., Calabria-Holley, J., Lawrence, R., Jiang, Y.: Hygrothermal and mechanical characterisation of novel hemp shiv based thermal insulation composites. Constru. Build. Mater. **212**, 561–568 (2019). https://doi.org/10.1016/j.conbuildmat.2019.04.029
22. Walker, P., Thomson, A., Maskell, D.: Straw bale construction. In: Nonconventional and Vernacular Construction Materials, pp. 127–155. Woodhead Publishing, Cambridge (2016). https://doi.org/10.1016/b978-0-08-100038-0.00006-8
23. Evola, G., et al.: Performance comparison between building insulating materials made of straw bales and EPS for timber walls. In: IOP Conference Series: Materials Science and Engineering. p. 072020. Institute of Physics Publishing (2019). https://doi.org/10.1088/1757-899X/609/7/072020
24. Thomson, A., Walker, P.: Durability characteristics of straw bales in building envelopes. Constru. Build. Mater. **68**, 135–141 (2014). https://doi.org/10.1016/j.conbuildmat.2014.06.041
25. Franco, W., Quaglia, G., Ferraresi, C.: Experimentally based design of a manually operated baler for straw bale construction. In: Boschetti, G., Gasparetto, A. (eds.) Advances in Italian Mechanism Science. MMS, vol. 47, pp. 307–314. Springer, Cham (2017). https://doi.org/10.1007/978-3-319-48375-7_33

26. Andersen, B.M., Munch-Andersen, J.: Halmballer og muslinger som isoleringsmaterialer - Sampling and testing report (2001)
27. Promis, G., et al.: Hygrothermal performance of a straw bale building: in situ and laboratory investigations optimization of the formulation of eco-aggregates concretes obtained by accelerated carbonation process of incinerator waste View project SAPICO2 View project Hygrothermal performance of a straw bale building: in situ and laboratory investigations (no date). https://doi.org/10.13140/RG.2.1.5115.9763
28. Conti, L., Barbari, M., Monti, M.: Steady-State Thermal Properties of Rectangular Straw-Bales (RSB). Building **6**, 6040044 (2016). https://doi.org/10.3390/buildings6040044
29. Platt, S., et al.: Manufacture and characterisation of prototype straw bale insulation products. Constr. Build. Mater. **262**, 120035 (2020). https://doi.org/10.1016/j.conbuildmat.2020.120035
30. Collet, F., Prétot, S., Lanos, C.: Hemp-straw composites: thermal and hygric performances. Energy Procedia **139**, 294–300. https://doi.org/10.1016/j.egypro.2017.11.211
31. Liuzzi, S., et al.: Characterization of biomass-based materials for building applications: the case of straw and olive tree waste. Ind. Crops Prod. **147**, 112229 (2020). https://doi.org/10.1016/j.indcrop.2020.112229
32. Savic, A., et al.: Thermomechanical behavior of bio-fiber composite thermal insulation panels. Energy Build. 229, 110511. https://doi.org/10.1016/j.enbuild.2020.110511
33. Arnaud, L., et al.: Hemp and the Construction Industry, pp. 239–259. CABI, Wallingford (2013)
34. Oudhof, N., et al.: Measurement of the hygrothermal properties of straw-clay mixtures. In: ICBBM, vol. 33, pp. 474–479 (2015)
35. Jiang, D., et al.: Environmentally friendly alternative to polyester polyol by corn straw on preparation of rigid polyurethane composite. Compos. Commun. **17**, 109–114 (2020). https://doi.org/10.1016/j.coco.2019.11.007
36. Li, X., Cai, Z., Winandy, J.E., Basta, A.H.: Selected properties of particleboard panels manufactured from rice straws of different geometries. Bioresource Technol. **101**(12), 4662–4666 (2010). https://doi.org/10.1016/j.biortech.2010.01.053
37. Huang, Y., et al.: Study on untreated and alkali treated rice straw reinforced geopolymer composites. Mater. Chem. Phys. 262, 124304 (2021). https://doi.org/10.1016/j.matchemphys.2021.124304

Influence of Fan Duty Point on the Performance of Mechanical Draught Wet Cooling Towers

Berina Delalić-Gurda[1](✉), Džana Kadrić[1], Almin Halač[1], Nijaz Delalić[1], and Elvedina Sikira[2]

[1] Faculty of Mechanical Engineering, University of Sarajevo, Sarajevo, Bosnia and Herzegovina
b.delalic@mef.unsa.ba

[2] Elektroprivreda Bosne I Hercegovine, Sarajevo, Bosnia and Herzegovina

Abstract. Mechanical draught wet cooling towers are widely used in HVAQ and industry for transferring heat from main process to the environment. With direct contact between hot (water) and cold fluid (air), which enables both sensible and latent heat transmission, it can meet high cooling demand even during hot summer period. However, with constant use of fans and pumps, it is very important to implement adequate control scheme in order to minimize energy consumption of system itself. This paper presents findings of a project focused on performance evaluation of a running thermal power plant cooling system, which was based on measurements of key parameters in cooling tower operation. The results showed that fan operating mode can strongly influence performance characteristics of large mechanical draught cooling towers, and that position of fan duty point is one of the important factors to enable implementation of energy efficient control strategy.

Keywords: Wet cooling towers · Mechanical draught · Fan duty point · Energy efficient control

1 Introduction

In many HVAQ systems, industrial facilities and large thermal power plants there is a constant or periodical need to discharge low potential heat to the environment. Mechanical draught wet cooling tower is system for indirect transfer of waste heat from the main process to the environment, widely used due to several advantages: it enables evaporative cooling so the water can be cooled near to the wet-bulb air temperature; it uses less space in comparison to natural draught cooling towers; it is easier to control its operation by regulating water and air flow through fans and pump operation. Figure 1 presents main elements of wet cooling tower through an example of use in a thermal power plant [1].

For large industrial or thermal power plant cooling systems, annual energy consumption of fans can reach up to 7 GWh, if considering example of a plant working with 6 cooling tower cells in average for 11 months in year, with fan power of 145 kW. In such cases even 1% reduction of energy consumption would result in 70 MWh of yearly energy saving and related CO_2 emission reduction depending on running fuel. Additionally, reduced use of fans can have positive effects in the aspect of water management,

N. Ademović et al. (Eds.): IAT 2021, LNNS 316, pp. 305–312, 2022.
https://doi.org/10.1007/978-3-030-90055-7_23

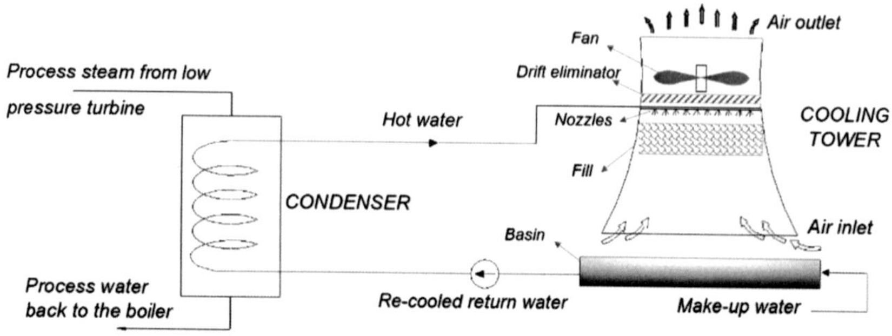

Fig. 1. Schematic representation of wet cooling tower within a thermal power plant

since in average 1–2% of total water flow rate through the tower is lost due to evaporation and drift. Hence, carefully optimised operation of large cooling system is highly important for improving its energy efficiency and reducing environmental impact.

However, design and implementation of energy efficient control scheme can be a challenging task when the plant is already in operation, especially in case if some elements of cooling system are not working as foreseen. Cooling tower's fans are one of the key factors in the cooling system control process because they are significant power consumer and represent a variable to be minimized. Fan's power consumption in general is related to operating conditions through fan performance curve, but it also depends on many design factors, such as shape and size of diffuser or plenum chamber. One of the most influencing parameters in this aspect is fan blade tip clearance which can create efficiency loss even with millimetre wide extra space [2]. Before investigating causes of change in fan efficiency, it is important to timely detect deviations in fan behaviour during the system operation because it can have strong influence on cooling tower's thermal performance and even hamper implementation of control scheme.

2 Objective and Methodology

In order to ensure proper implementation of control scheme which would lead to energy efficiency improvement, it is important to have deep knowledge about interactions between different elements of cooling system. This paper focuses on investigating correlation between cooling water flow rate through cooling tower unit and the operating mode of cooling tower's fan. The water flow rate is important parameter because it can be considered as a manipulated variable in control scheme, as it is generally adjustable based on plant cooling demand. Changes in water flow rate are expected to influence operation mode of tower's fan, but it is important to learn the details of this interaction in order to actually achieve energy savings and improve operation of cooling system at off-design conditions.

Investigation of cooling tower performance in regards to fan operating mode is based on observations and results from measurements performed on an operational unit of thermal power plant. The measurements covered monitoring and acquisition of fan motor power consumption for a cooling tower cell while the water flow rate through tower was gradually reduced from nominal flow rate (2423 m^3/h) to zero. Change in water flow rate was performed by manually closing valve on main distribution pipe for examined tower cell by 10% in every measurement step, from fully opened to fully closed valve. This method resulted in eleven measuring points for both fan motor power consumption and for water flow rate. Water flow rate through the tower was measured at the distribution pipe by digital ultrasonic flowmeter, utilising Z method for positioning the pair of probes. Fan motor power was measured by 3-phase energy and power disturbance analyser with flexible AC current sensors utilising Rogowski principle. Fan power was calculated based on measured fan motor power, reduced by motor and belt efficiency factors, both valued at 0.95. Observed fan can only work in on/off mode, so no changes in fan operating mode were made by plant operator during the measurements.

In parallel with measurement of fan power consumption in regards to changing water flow rate, the air static pressure drop across the examined tower was measured. This was performed by measuring pressure difference between surrounding air and the air in zone between drift eliminators and fan (see Fig. 1). Measurements were performed by digital differential manometer for the same eleven measuring points as for fan power consumption.

Finally, it was important to discover actual air flow rate provided by examined fan through the tower. Determining total air flow rate through observed cooling tower was performed by measuring its velocity at the entrance. Examined cooling tower unit has four rectangular shaped ports for air inlet at the lower half of tower construction. Measurements of air velocity were carried out across all four air inlet ports by using equal-area method for traverse reading, which resulted in total of 300 measuring points for only one tower operating mode. Hence, measurements of air flow rate through the tower, with operating fan, were performed only in two steady states i.e., for nominal and for zero water flow rate through the tower.

During the measurements, power plant unit was operating with near nominal load and the cooling system of eleven towers was engaged in full capacity. Since described measurements focused on hydraulic aspect of cooling tower, plant load had little influence on measurement results.

3 Results of Measurements and Determination of Actual Fan Duty Point

Figure 2 presents results of described measurements for fan power consumption in regards to changing water flow rate through the tower, for eleven measuring points. In terms of designing energy efficient control scheme, described measurements showed unexpected trends. It was assumed that in periods of lower cooling demand of power plant, it will be possible to reduce water flow rate through the towers and consequently the fan power consumption would decrease to create energy savings. However, measurements showed that while reducing water flow rate, the fan power consumption became higher, as visible in Fig. 2. This indicates that it would not be possible to achieve any energy savings by changing water flow rate through the tower, with current state of examined equipment.

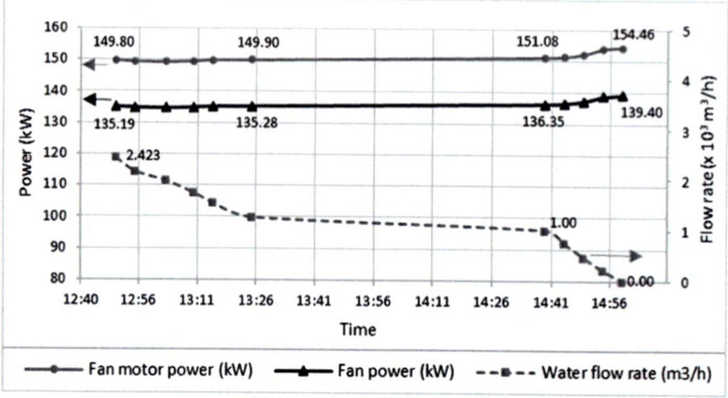

Fig. 2. Measurement results for fan power in regards to water flow rate through cooling tower

Figure 3 shows measurement results for relative air static pressure in the zone between drift eliminator and fan, which represents air static pressure drop across the tower. Trends of change in this set of measurements are as expected, having in mind that reduced water flow rate creates less resistance for the air to flow through the tower so the pressure drop of the air will also be reduced.

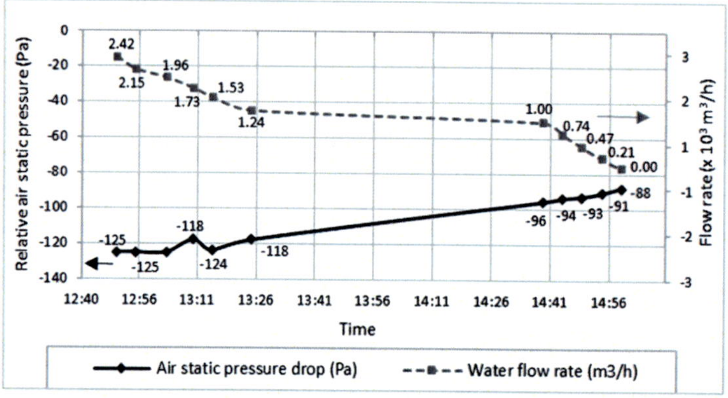

Fig. 3. Measurement results for air static pressure drop in regards to water flow rate

However, it is important to note here that the designed value of air static pressure drop is −175 Pa and it is foreseen to occur at examined cooling tower unit at full nominal water flow rate. Since actual air pressure drop at same conditions is measured to be −125 Pa, it implies that examined fan could overcome additional 50 Pa of resistance in comparison to current state. Additionally, this indicates a shift in fan's duty point in regards to design conditions. Fan's duty point is determined based on manufacturer's performance curves which provide corelation between air flow rate, fan power and air static pressure drop. Since the measurements for air flow rate were performed only for two different values of water flow rate, measurements for fan power and air static pressure drop are reduced to the same two measuring points, as presented in Table 1.

Table 1. Measured values of parameters related to fan performance at nominal and zero water flow rate

Measured parameter	At nominal water flow rate	At zero water flow rate
Air flow rate (m^3/s)	460	513
Static pressure drop (Pa)	−125	−88
Fan power (kW)	135.19	139.40

Based on the results of measurements from Table 1, duty point of observed fan was determined for the two tower operating points, as presented in Fig. 4. On left-side diagram, it can be observed that with reduction of water flow rate, there is less disturbance for air flow through the tower so static pressure drop is decreasing and air flow rate is on the rise, which can be considered expected. However, the right-side diagram shows that the same rise of air flow rate, caused by lower resistance from water and decreased air static pressure drop, is actually increasing the fan power consumption. Further analysis of this observation and its implications on cooling system performance is presented in following chapter.

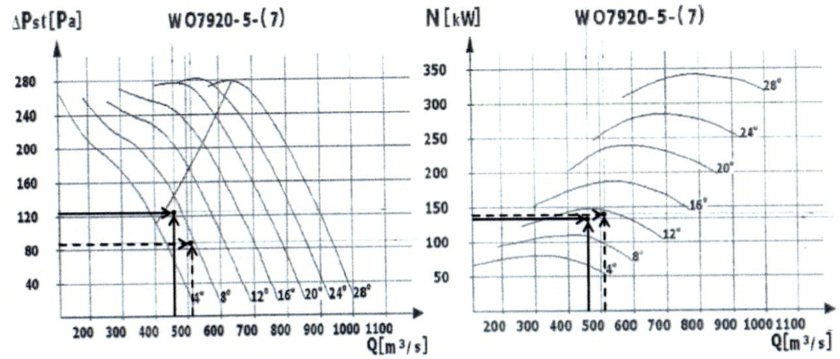

Fig. 4. Duty points of examined fan in real operating conditions for nominal water flow (solid line) and for zero water flow through the tower (dashed line)

4 Influence of Shifted Fan Duty Point on Tower Performance

Influence of fan's operating mode on the performance of cooling system will be investigated in two aspects: how it affects thermal performance of cooling tower, and why it is important for efficient energy management of system.

Thermal Performance of Cooling Tower

For discussion on thermal performance of cooling towers, comparison of certain parameters between design conditions and current operating conditions will be made. Table 2 presents key parameters related to fan operation and it can be seen that operating fan currently does not provide expected air flow rate, even though the static pressure drop is lower than in design conditions at nominal water flow rate.

Table 2. Parameters of fan performance and cooling performance at nominal water flow rate

Parameter	Design conditions	Current operating conditions
Air flow rate (m^3/s)	507	460
Static pressure drop (Pa)	−175	−125
Fan power (kW)	145.70	135.19
Merkel number $Me(-)$	1.62	1.53

For analysis of tower's thermal performance, Merkel number is calculated using empirical correlation [3]:

$$Me = m\left(\frac{\dot{m}_a}{\dot{m}_w}\right)^n$$

where $\dot{m}_a/\dot{m}_w$ is ratio between mass flow rates of air and water, and m and n are empirical coefficients which for newly installed cooling towers have values of $m = 1{,}82$ and $n = 0{,}61$, and parameter m decreases with time as a result of fouling. In order to isolate the influence of reduced air flow rate, fouling effects are neglected and empirical coefficients for newly installed cooling towers are used in calculation of Merkel number for both design and current operating conditions. Results are presented in Table 2 and it is evident that Merkel number is approximately 5% lower due to the reduced air flow rate provided by the fan.

On the other hand, widely used formulation of Merkel number is following [4]:

$$Me = \frac{h_d \cdot A}{m_w}$$

where, h_d ($kgm^{-2}s^{-1}$) is mass transfer coefficient and A (m^2) is the heat and mass transfer surface area. Comparing the conditions with same water mass flow rates, it yields that decrease in Merkel number is directly showing decrease in mass transfer coefficient which is one of the key parameters indicating energy transfer rate between water and air. Hence, reduced air flow rate directly influences intensity of heat transfer within the tower and hinders cooling performance of the system.

Correlation Between Air Flow Rate and Fan Power Consumption

Analysis of fan performance curve sheds some light on the unexpected rise in fan power consumption for reduced water flow rate. Figure 5 presents performance curve of examined fan, related to fan power. On the left side of Fig. 5, the diagram shows duty point of fan in design conditions for nominal water flow rate and design blade angle of 11.9°. The right-side diagram presents duty point of fan in real operating conditions, determined based on the results of performed measurements. Solid lines are related to nominal water flow rate through the tower, and the dashed line indicates conditions with zero water flow rate.

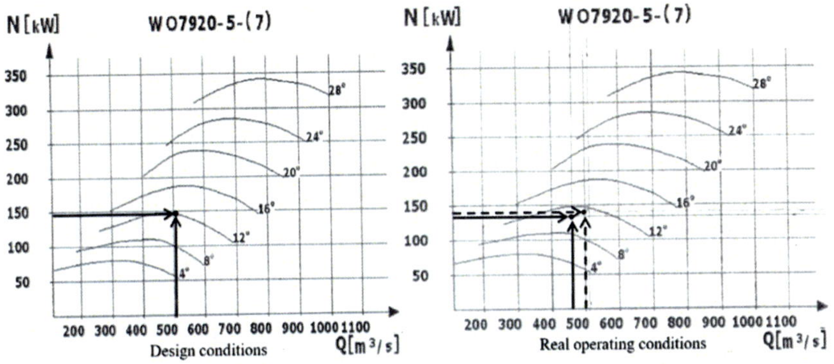

Fig. 5. Performance curve of examined fan in design and in real operating conditions

Based on the presented diagrams, the unusual behaviour of tower's fan can be explained as follows:

> In design conditions (left-side diagram of Fig. 5), duty point of fan for nominal water flow rate is positioned at the peak of curve so when water flow rate is reduced, it provides higher air flow rate resulting in decrease of fan power consumption, because the duty point moves to the right from curve's peak;

> In real operating conditions (right-side diagram of Fig. 5), duty point of fan for nominal water flow rate is slightly shifted to the left from curve's peak, so when the water flow rate decreases and air flow rate increases, duty point is still moving over the rising part of curve so the fan power will also increase.

Having this in mind, it is evident that current operating conditions of fan are inconvenient for implementation of control scheme which should improve energy efficiency of the system. Hence, it would be recommended to further investigate and try to resolve problems that are causing shift of fan duty point, prior to designing and implementation of an energy efficient control scheme.

5 Conclusions

Mechanical draught wet cooling tower is a complex system that needs to be carefully observed, analysed and tuned up in all its aspects in order to enable energy efficient operation. This paper shows by real example how important it is to bring more attention to evaluating performance of cooling tower fans, as those are large energy consumers and can strongly affect overall cooling tower performance. Presented measurements revealed that examine cooling tower fan is working with duty point different from conditions foreseen by design. Analysis showed two undesirable consequences of observed fan operating mode: 1) the fan is not providing expected air flow rate, which is important factor for heat and mass transfer process and it directly affects thermal performance of cooling tower; and 2) dependency between water flow rate and fan power consumption is reversed in comparison with design conditions data, which can completely undermine implementation of control scheme. Further investigation should be carried out in regards to condition of fan system elements such as plenum chamber, casing and diffuser, with the aim to target causes of shifted fan duty point.

It is important here to mention that performance of particular fan section was never fully evaluated after installation, since EU Standard for cooling tower acceptance testing [5] prescribes to measure only power consumption as an indicator of fan operation, and no other parameter related to fan performance was required to be monitored and examined. Fan units are in general evaluated only in testing facilities and there are rarely any comprehensive on-site checks of overall fan system performance after installation. Hence, it should be considered to introduce additional testing for performance evaluation of cooling towers with focus on fan system, in order to ensure most efficient operation.

References

1. Erens, P.J.: Cooling towers. In: VDI-Gesellschaft Verfahrenstechnik und Chemieingenieurwesen (ed.) Heat Atlas, pp. 1485–1501. Springer, Heidelberg (2010). https://doi.org/10.1007/978-3-540-77877-6_107
2. Kröger, D.G.: Air-Cooled Heat Exchangers and Cooling Towers –, vol. II. PennWell Corporation, Tulsa (2004)
3. Zemanek, I.: Heat and Mass Transfer in Cooling Tower Packing. National Research Institute for Machine Design, Prague (1989)
4. Kloppers, J.C., Kröger, D.G.: A critical investigation into the heat and mass transfer analysis of counterflow wet-cooling towers. Int. J. Heat Mass Transf. **48**, 765–777 (2005)
5. European Committee for Standardization: Heat Exchangers - Method of Measurement and Evaluation of Thermal Performances of Wet Cooling Towers, BS EN 14705:2005. European Committee for Standardization Publications, Brussels (2005)

Application of MEMS Accelerometers in Measuring Vertical Oscillations in Motor Vehicles

Jasmin Šehović(✉)

Mervik d.o.o. Sarajevo, Sarajevo, Bosnia and Herzegovina
jasmin.sehovic@mervik.ba

Abstract. Modeling complex oscillatory systems presents a constant challenge for researchers. Measurement data must be used in order to verify these models. In this paper, an accelerometer based on MEMS (micro electro - mechanical system) technology for measurement of vertical oscillations that occur on a vehicle is presented. A description of the device, methods of its use when measuring vertical oscillations and examples of measured oscillations on a passenger motor vehicle are presented. The practical application of the obtained results measured using this accelerometer is also shown.

Keywords: Acceleration · Accelerometer · Oscillations · Vehicle · MEMS

1 Introduction

Any mathematical modeling of oscillatory systems requires certain approximations. Therefore, mathematical models dealing with these analyses need to be verified with real oscillation values of a particular system. For this purpose, the oscillation parameters of the system must be measured with special devices [1]. The parameters to be measured are the acceleration, velocity and travel of a specific oscillatory system. Measurement methods can be contact and non-contact [2]. Contact methods include measurements with devices such as accelerometers, velocimeters and travel sensors, while non-contact methods include devices that work on the principle of laser measurement. One such example of oscillation measurement on a motor vehicle is shown in [3]. In this case, accelerometers were used to measure vertical oscillations, with a complex measurement setup, which can be seen in [3]. Given the complexity of the vehicle's oscillation system, there are also devices that can only measure the longitudinal oscillations of vehicles that occur due to the motion regime of the vehicle. Such an example is given in [4] where the longitudinal acceleration and deceleration of a vehicle are measured in order to model the vehicle's braking process. An example of measurement of road profile using laser beams is shown in [5]. All of the above examples of oscillation measurement methods require a complex measurement setup, expensive equipment for signal acquisition and processing, and a lot of space to accommodate a complete measurement setup. Figure 1 shows the general scheme of one measurement setup during oscillations measurement.

N. Ademović et al. (Eds.): IAT 2021, LNNS 316, pp. 313–320, 2022.
https://doi.org/10.1007/978-3-030-90055-7_24

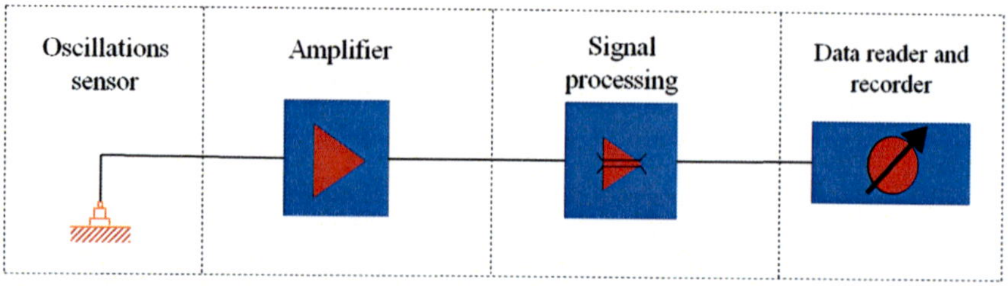

Fig. 1. Measurement setup for recording of the oscillations [1]

In order to simplify the measurement process and reduce the space and the measuring setup shown in Fig. 1, MEMS devices (micro electro-mechanical system) have recently been increasingly used [6]. These systems are very compact devices that contain an element that acts as an oscillation sensor but also electronics through which the recorded signal is processed, recorded and sent to the software on the computer for further work with the recorded data. One such device is described below and an example of oscillations measurement on a passenger motor vehicle with its practical application is also given.

2 Description of the Experimental Method of Oscillations Measurement

The biggest issue in setting up accelerometers for experimental research of vertical oscillations in vehicles is the limited space. In this paper, a simple device for measuring vehicle's vertical accelerations is proposed. It is a triaxial accelerometer SlamStick X, manufactured by Mide Technologies, USA. The accelerometer is shown in Fig. 2 with its dimensions presented both in inches and milimeters (milimeters shown in brackets).

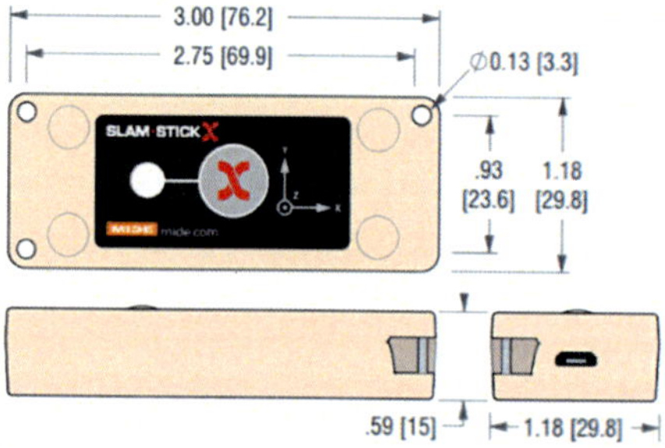

Fig. 2. Accelerometer with its dimensions [7]

All the above characteristics make this accelerometer ideal for use in the study of vertical oscillations on motor vehicles. This accelerometer has practical application both in laboratory test conditions, where the source of excitation of the oscillation system is controlled, and in road conditions where vehicles are driven over various obstacles.

The following is an example of measurement of vertical accelerations on a single passenger motor vehicle performed using accelerometers, both in laboratory conditions and in road conditions.

Measurements in laboratory conditions were performed at the vehicle's technical inspection station using the excitation device in the form of a vehicle's shock absorber tester. Measurements were performed at three characteristic positions: the excitation device (shock absorber tester, $\ddot{z}_p$), the vehicle wheel (lower arm, $\ddot{z}_t$) and the vehicle body ($\ddot{z}_v$). Figure 3 shows the installation of these devices at the given positions.

a) Shock absorber tester b) Lower arm c) Vehicle body

Fig. 3. Installation of the accelerometers at the characteristic positions

Figure 4 shows recorded results of vertical accelerations measurements at characteristic positions.

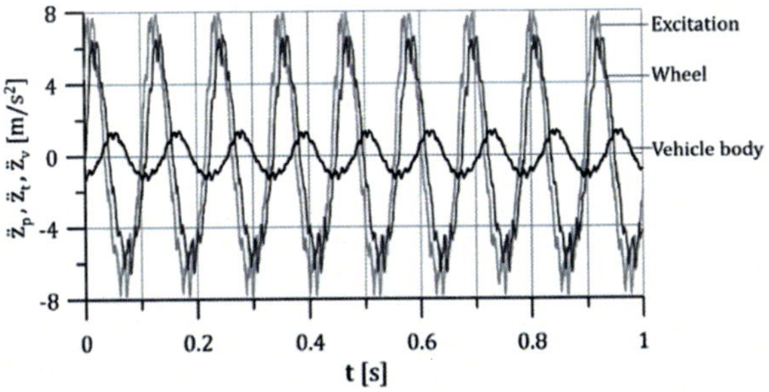

Fig. 4. Recorded vertical accelerations at characteristic positions

The following is a practical use of the accelerometers in researching vertical accelerations of vehicle wheel and body during motion in road conditions. Vehicle motion on city roads, highways and special test tracks is all considered movement in road conditions. Specifically, in this paper, the source of excitation was an obstacle on the road 5 cm high, which is shown in Fig. 5, and vehicle passing over the obstacle, which is shown in Fig. 6.

Fig. 5. Road obstacle as a source of excitation

Fig. 6. Vehicle passing over road obstacle

Two accelerometers were used on the vehicle to record the acceleration of the vehicle wheel and body when passing over the obstacle as shown in Fig. 6. The accelerometers were installed on the lower arm of the wheel and at the location of the vehicle's shock absorber top mount. Accelerometer placement locations are shown in Fig. 7.

a) Lower arm accelerometer

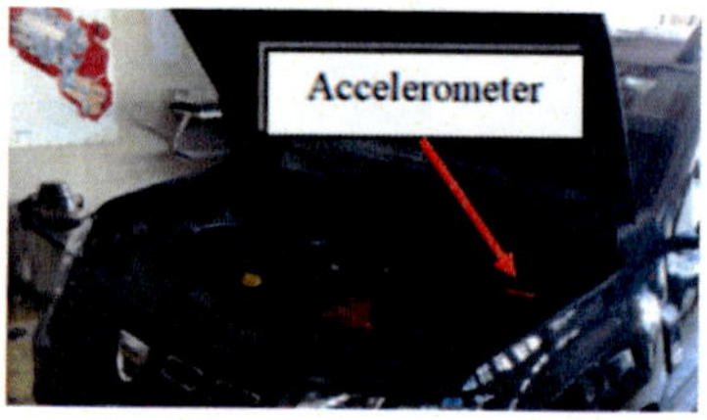

b) Vehicle body accelerometer

Fig. 7. Positioning of the accelerometers

After placing the accelerometers to the specified positions, the vehicle was driven on the test track passing over the obstacle shown in Fig. 5, and measurements of the vertical accelerations of the vehicle wheel ($\ddot{z}_t$) and the body ($\ddot{z}_v$) were performed. Figure 8 shows an example of the vertical acceleration recorded results at the vehicle wheel ($\ddot{z}_t$) and body ($\ddot{z}_v$).

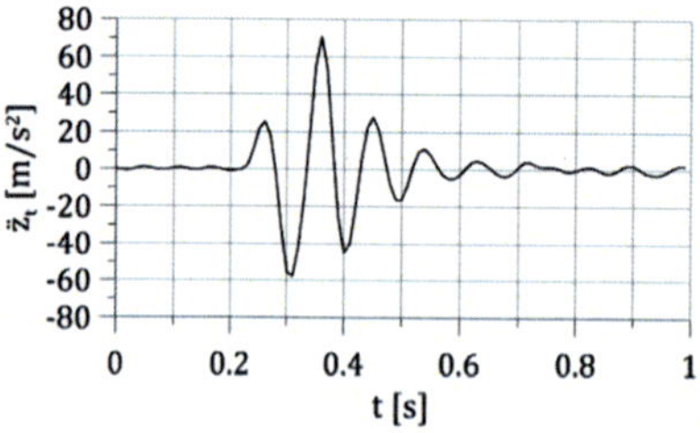

a) Wheel

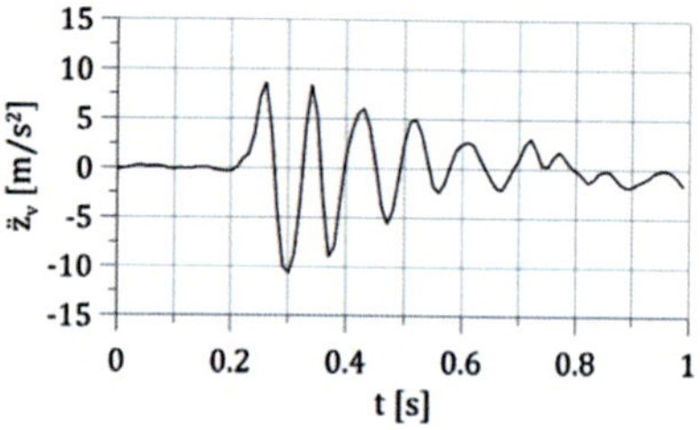

b) Vehicle body

Fig. 8. Recorded vehicle vertical accelerations

Further analysis of the recorded values (both in laboratory and road conditions) of the vehicle vertical accelerations and their practical use is shown next.

3 Analysis of the Recorded Results

The recorded vertical accelerations in laboratory conditions, as shown in Fig. 4, can be further used to define the kinematic values of velocity and travel in vertical direction of the characteristic positions from Fig. 3. To analyze the recorded results, one part of the system excitation for 0.118 s was chosen. Since the excitation is of harmonic character, the results of acceleration were observed as harmonic functions of the given excitation frequency. The obtained harmonic functions were created by the harmonic analysis of the experimental results recorded at the characteristic positions. They are shown by solid lines in the following Figures. In given Figures, the experimental results recorded with accelerometers are also shown using dotted lines. Real oscillations recorded on the vehicle (dotted lines) are the result of oscillations of a very complex vehicle system with variable stiffness in different cross sections of the system. Figure 9 shows a diagram of the acceleration of the excitation device for one period of oscillation of the tested vehicle. Figure 10 shows a diagram of the wheel acceleration, and Fig. 11 shows a diagram of the vehicle body acceleration for one period of oscillation.

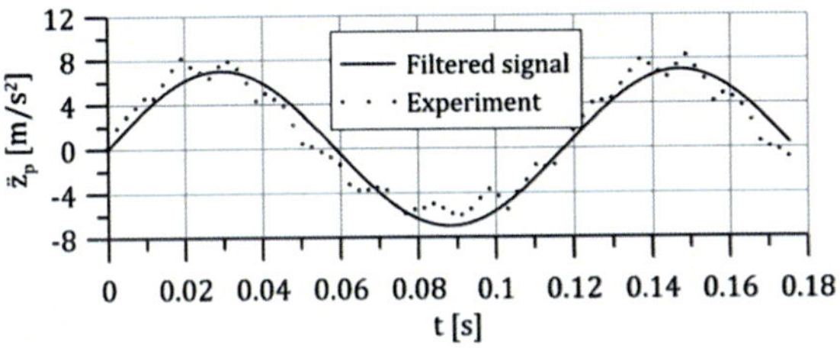

Fig. 9. Acceleration diagram of the excitation device for one ocillation period

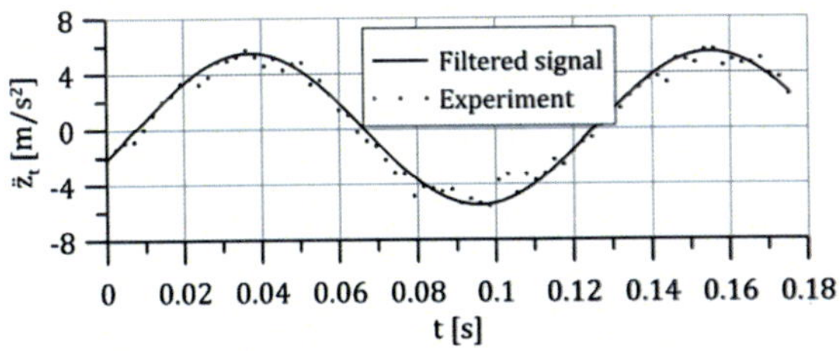

Fig. 10. Acceleration diagram of the wheel for one ocillation period

Figure 12 shows the acceleration results at characteristic positions after a harmonic analysis of experimental results.

After integrating the acceleration functions obtained by the harmonic analysis, shown in Fig. 12, the obtained results of velocity and travel of the excitation device, wheel and vehicle body are shown in the following Figures. Figure 13 shows the velocity in vertical direction at the characteristic positions, and Fig. 14 shows the travel of the characteristic positions in the vertical direction.

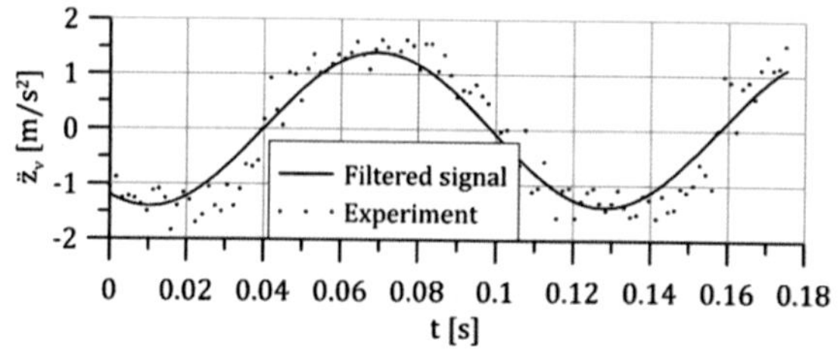

Fig. 11. Acceleration diagram of the vehicle body for one ocillation period

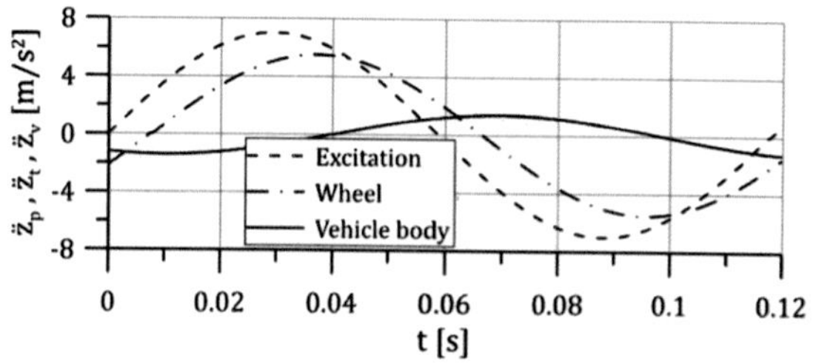

Fig. 12. Acceleration diagram at characteristic positions

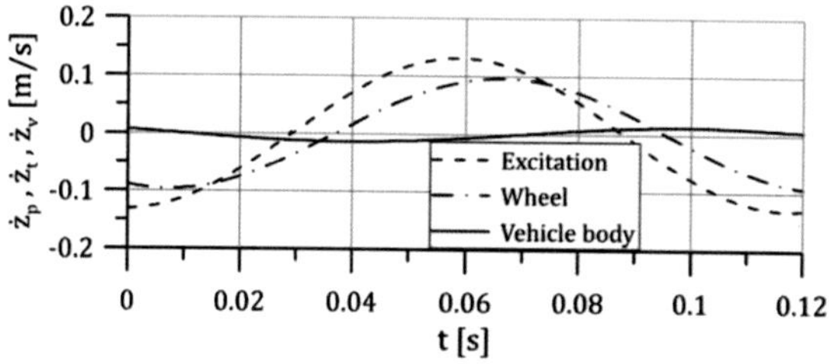

Fig. 13. Velocity diagram at characteristic positions

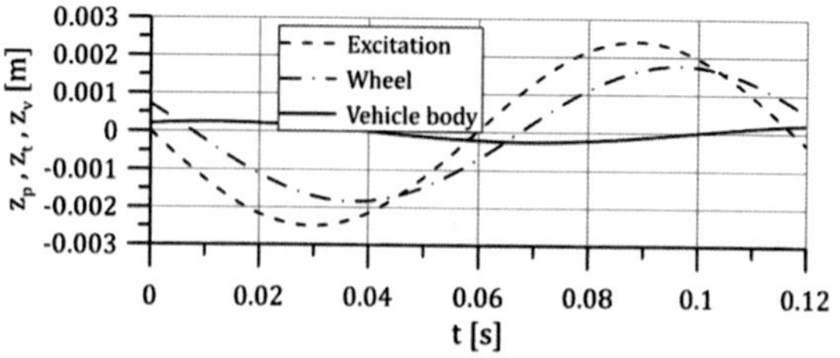

Fig. 14. Travel diagram at characteristic positions

The results shown in Figs. 12, 13 and 14 along with the use of the well-known 1D vehicle oscillating model [8] were used to define the damping characteristics of the vehicle suspension system, i.e. damping force in the damper (F_p). Figure 15 shows the damping force as the final result of the experiment being performed using the accelerometers. A detailed determination of the damping force and other damper's characteristics can be found in [9].

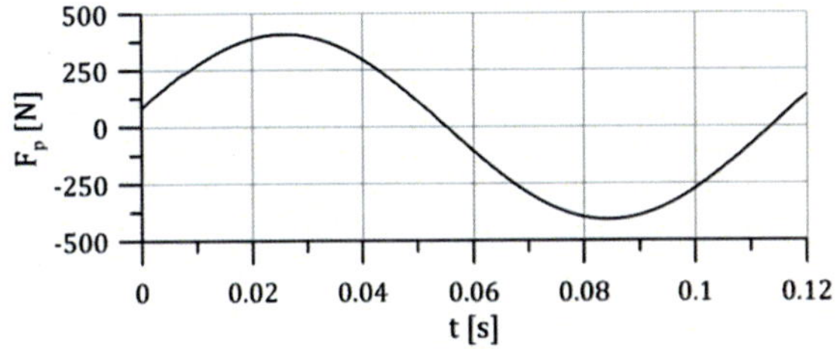

Fig. 15. Damping force diagram

The recorded vertical wheel ($\ddot{z}_t$) and vehicle body ($\ddot{z}_v$) accelerations of the vehicle experiment in road conditions, shown in Fig. 8, can be used in defining the suspension system damper piston kinematics. The acceleration of the damper piston ($\ddot{z}_k$) is defined as the difference between the vertical accelerations of the vehicle body ($\ddot{z}_v$) and the wheel ($\ddot{z}_t$). The obtained results can be further used to verify any numerical model that deals with the research of vehicle oscillations. Figure 16 shows the practical use of these results to compare with the calculation results of one hydrodynamic damper model.

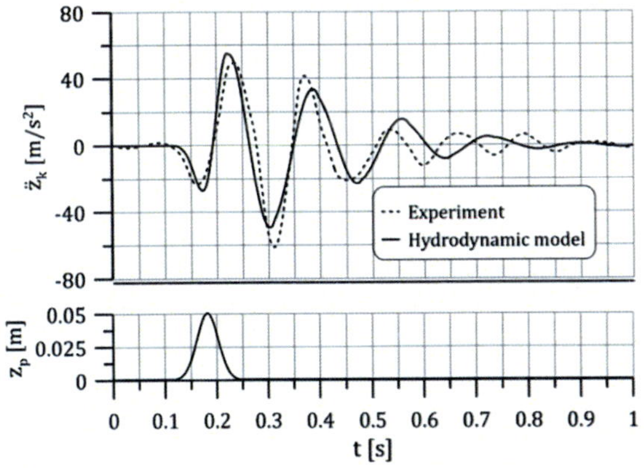

Fig. 16. Comparison of the experimental and hydrodinamic damper model results

Figure 16 shows, in dotted line, the damper piston acceleration. It is obtained by processing experimental data of the vertical acceleration recording (Fig. 8) using accelerometers. Result of the numerical calculation of a hydrodynamic damper model (solid line)

with the excitation brought to the system (z_p) is also shown. This way, the hydrodinamic numerical damper model is verified using the experimental data recorded by the proposed accelerometer, without much investment in expensive experiments.

4 Conclusion

In this paper, a MEMS accelerometer for measuring vertical accelerations on a passenger motor vehicle, manufactured by Mide Technologies, is presented. Several measurements were performed using these devices in laboratory and road conditions. The measurement results are shown, as well as the analysis of the measurement results of vertical oscillations. It has been shown that the use of this device can contribute to quality measurements of vertical accelerations on vehicles. The contribution of this work and application of these devices is to prove that it is not necessary to use complex measuring equipment and measurement setups, but that this device can be used in experimental measurements due to its compactness and ease of use, both on vehicles and other complex mechanical systems.

References

1. Doleček, V., Voloder A., Isić, S.: *Vibracije.,* Sarajevo: Mašinski fakultet Sarajevo (2009)
2. Schmitz, T.L., Smith, K.S.: Mechanical Vibrations – Modeling and Measurement, 2nd edn, Springer, Cham (2021)
3. Surblys, V., Žuraulis, V., Sokolvskij, E.: The influence of semi-active suspension adjustment on vehicle body pitch oscillations. Transp. Telecommun. **20**(2), 107–113 (2019)
4. Pečeliunas, R., et al.: Experimental research into motor vehicle oscillations in the case of changeable deceleration. Transport **20**(5), 171–175 (2005)
5. http://www.pavetesting.com. Accessed 1 Feb 2021
6. Looney, M.: MEMS vibration monitoring: from acceleration to velocity. Analog. Dialog. **51**(06), 1–5 (2017)
7. https://www.mouser.com/datasheet/2/606/slam-stick-x-datasheet-466833.pdf. Accessed 5 Feb 2021
8. Jazar, N.R.: Vehicle Dynamics – Theory and Application, 3rd edn, Springer International Publishing AG, Cham (2017)
9. Šehović, J., Filipović, I., Pikula, B.: Experimental determination of non-linear characteristics of the passenger vehicle suspension system. Trans. Famena **XLIV**(2), 13–22 (2020)

Oscillatory Behaviour of Bubbles

Mirela Alispahić(✉) and Šefko Šikalo

Mechanical Engineering Faculty, University of Sarajevo, Vilsonovo šetalište 9, 71000 Sarajevo, Bosnia-Herzegovina
hajro.m@mef.unsa.ba

Abstract. Results of an experimental study of pool boiling of distilled water on copper, brass, and SS surfaces are presented in this paper. Visualization of the boiling process has performed in the condition of high subcooling ($\Delta T_{\text{sub}} =$ 45 K) with multiple bubbles randomly appearing onto the surface. The parameters related to bubble growth characteristics (bubble formation, base diameter, equivalent diameter, bubble growth time) are qualitatively and quantitatively presented for low surface superheat. The total bubble growth time is divided into initial bubble growth time and oscillatory growth time. For a low degree of wall superheat, oscillatory growth time is quite long (two orders of magnitude larger than the initial bubble growth time) and directly related to bubble size. It appears that the bubbles with a lower ratio of equivalent diameter to base diameter have a prolonged transition from hemispherical to an irregular spherical shape and higher volume before departure.

Keywords: Subcooled pool boiling · Base bubble diameter · Equivalent bubble diameter · Initial growth time · Oscillatory growth time

1 Introduction

Nucleate boiling has been a subject of study for several decades because of its very efficient heat transfer mechanisms encountered in many industrial applications (e.g. steam generators, heat exchangers, cooling equipment, etc.). Previous researches, based on the mechanisms of the process, has shown that high efficiency of heat transfer by boiling is achieved by the movement of liquid near the formed bubble (formation of bubbles create intense convection of liquid along the heated surface), and also by transient conduction and evaporation in the microlayer. A good overview of previous research based on heat transfer mechanisms, as well as a summary of previous experimental and numerical research of nucleate boiling, was given by [1]. Regardless of the significant scientific interest and practical engineering interest in the nucleate boiling regime, there is still no general model which could adequately describe the boiling process. The relative contributions of individual heat transfer mechanisms to the total heat transfer are very sensitive to the conditions in which boiling occurs (surface topography, physical properties of liquid and vapour, surface superheat, fluid subcooling, surface wettability, etc.), so they are difficult to predict. Recent works are more focused on the experimental

N. Ademović et al. (Eds.): IAT 2021, LNNS 316, pp. 321–328, 2022.
https://doi.org/10.1007/978-3-030-90055-7_25

investigation of those conditions on the bubble growth and departure characteristics, [2–7].

The present work aims to improve the understanding of the effect of high subcooling on bubble growth characteristics in the case of high bubble surface densities.

2 Experimental Setup and Measurement Techniques

The boiling vessel is made of stainless steel and glass; it is steel on two sides, and the front and back are made of glass to visualize the boiling process. The dimensions of the vessel are 150 × 150 × 250 mm with an open top so that boiling is at atmospheric pressure. The test block is placed inside the vessel and the boiling surface is horizontal. The sides of the block are thermally insulated with silicone sealant to reduce heat losses and to avoid the appearance of unwanted nucleation sites. The heat transfer area is 20 × 20 in mm bare top surface of the block. A DC power supply is connected to the 5.2 mm diameter cartridge heater mounted at the bottom of the test block. The vessel and the test block were insulated with silicone sealant on the underside, to prevent leakage.

A schematic of the set-up used in the experiment is shown in Fig. 1.

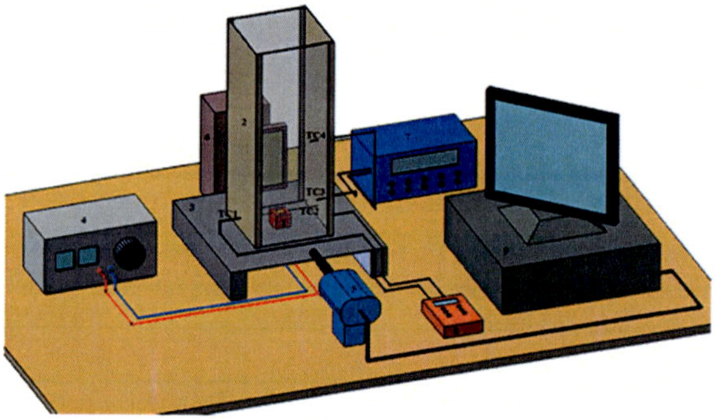

Fig. 1. Experimental setup: 1-test block; 2-boiling vessel; 3-metal support; 4-DC-power supply; 5-HS camera; 6-LED backlight; 7,8- data loggers; 9-PC; TC-thermocouple.

A high-resolution camera (pco.1200 s) is used to visualize the boiling process. The recording speed of the camera is set at 520 fps and a spatial resolution of 1280 × 1024 pixels. The image calibration is carried out using the known dimensions of the etalon plates with the specified magnification of the imaging Navitar lens. The camera is set up laterally to better monitor the boiling process. A backlighting technique with a flat LED light is used.

Four K-type thermocouples are connected to the temperature data loggers. The thermocouples TC1 and TC2 are inserted into the test block close to the boiling surface (~3 mm under the boiling surface) and thermal paste was used to reduce the thermal contact resistance between the thermocouples and the test block. To monitor the water temperature, another thermocouple TC3 is placed next to the boiling surface. The bulk liquid temperature is measured with TC4 located at the lower part of the boiling vessel.

Table 1. Description of used devices and sensors.

Sensors/Device	Description
Thermocouples Type K	Accuracy ±1 °C to 250 °C
4-channel temperature data logger	Type K sensors up 1370 °C (resolution 0.1 °C to +200 °C; accuracy ±0.15%; >200 °C, ±1 °C), Type J sensors up to 1100 °C;
Fluxmeter	Built-in TC type K up to +200 °C; sensor thickness 0.18 mm; sensitivity 0.981 $\mu V/(W/m^2)$
Digital multimeter	Rigol (model 3058E), with 5 significant digits; it is possible to select the type of thermocouple
DC power supply	0–30 V, 0–15 A (resolution 0.1 V, 0.1 A)

A short description of used devices and sensors is given in Table 1.

The surface temperatures are tested for each block placed inside the vessel, first in air ("dry") and then in water. The surface temperature and the heat flux are monitored using the flux meter attached to the heated surface in ambient air. The temperature values ware compared with mean temperatures obtained from TC1 and TC2. Also, heat flux values obtained on the flux meter are compared with heat flux calculated from applied power input to the heater. The power supplied to the heater from the voltage device is corrected for the values of losses estimated in the system (heat losses due to heating of the vessel, and losses through contacts with the heater and the block).

The test surface temperatures are calculated by averaging the temperature values measured with the thermocouples TC1 and TC2.

Boiling surfaces are mechanically polished, and roughness is measured on all the surfaces (in terms of average roughness). Also, the wetting properties of the boiling surfaces are determined under ambient conditions, as listed in Table 2.

Table 2. Wetting properties of boiling surfaces on $t_{amb} = 22$ °C.

Material	θ_o,°	θ_a–θ_r,°	Ra, μm
Copper	87	>60	1.29
Brass	77	>60	1.64
Stainless steel	65	>60	1.08

The temporal evolution of bubble growth is evaluated for each test surface, from the entire bubble growth period until detachment. Mean values are calculated taking into account the diameters taken from different nucleation sites (from the same nucleation at least two consecutive bubbles, in the case of slow-growing bubbles).

The equivalent bubble diameter is calculated from:

$$D_{eq} = \sqrt[3]{D_x^2 \cdot D_y} \tag{1}$$

where D_x and D_y are measured bubble dimensions in the horizontal and vertical directions, respectively, similar as [6].

Uncertainties associated with bubble equivalent diameter and base diameter are shown in Table 3.

Table 3. Uncertainties associated with bubble equivalent diameter and base diameter.

Parameter	Estimated as:	The uncertainty associated with the identification of the boundary of the bubble
Equivalent diameter, mm	$\frac{\Delta D_{eq}}{D_{eq}} = \pm\sqrt{\left(\frac{2\Delta D_x}{3D_x}\right)^2 + \left(\frac{\Delta D_y}{3D_y}\right)^2}$	ΔD_x is estimated to be 4 pixels; ΔD_y is estimated to be 8 pixels, for the worst case
Base diameter, mm		8 pix for brass and coper surfaces; 4 pix for stainless steel

The number of active nucleation sites is determined by counting it on the surface from images extracted from the high-speed video. The uncertainty associated with the number of the active sites is estimated by comparing counted active sites in the single image and between the first and tenth image, for example, to obtain a more accurate value for this parameter. Thus, the estimated uncertainty is up to 10% on the steel surface and up to 14% on copper and brass surfaces. Because of large values of bubble base on copper and brass, those surfaces look saturated with bubbles.

3 Results and Discussion

The behaviour of the bubbles on all observed surfaces is similar: it could be described by the rapid initial growth of the bubbles and the subsequent long period of oscillatory growth until the final detachment from the heated surface. The appearance of a new bubble is observed immediately after the departure of the previous bubble, in the observed time, which means that there is no waiting time between successive bubbles. That means that the departing bubble pulls the next one, thus forming a small vertical vapour column. This phenomenon is easily observed at the lowest wall superheat, as illustrated in Fig. 2.

The appearance of bubbles on the surface with and without waiting time is also observed by [2] and [8, 9] stated that the waiting time is very short, up to 1 ms.

The visible growth of the new bubble begins immediately in the vicinity of the departed bubbles. In general, the initial bubble growth time lasts until the base diameter reaches its maximum value. During this period, the base diameter grew to a mean value of 1.65 mm for brass and steel surfaces and 1.85 mm for the copper surface in 4 to 10 ms. After this initial period of surface drying, the base diameter begins to decrease; there is a narrowing of the dry zone, no departure of the bubbles was observed. This phenomenon is also observed by [2] in the case of bubbles departing without waiting time.

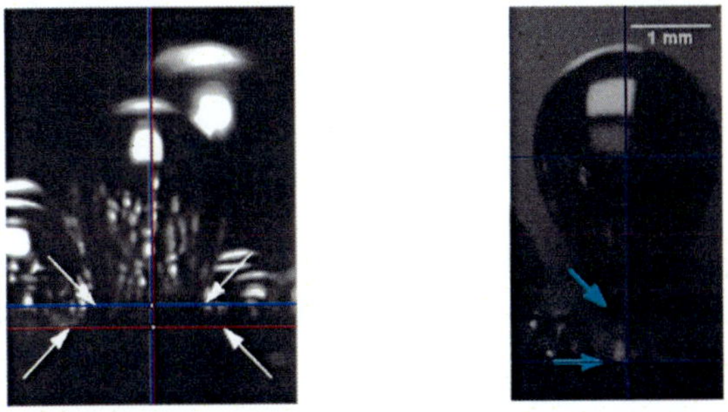

Fig. 2. Bubble departure at different nucleation sites on copper: a) the position where the vapour film breaks up at 8 K of surface superheat and b) 4 ms before the departure at 5.4 K.

Figure 3 shows the variations in the dimensions of the bubble growing on the copper at 5.4 K of surface superheat, from the moment they become visible on the surface until the final detachment from it. All values represent physical values obtained from a series of images for different active nucleation sites.

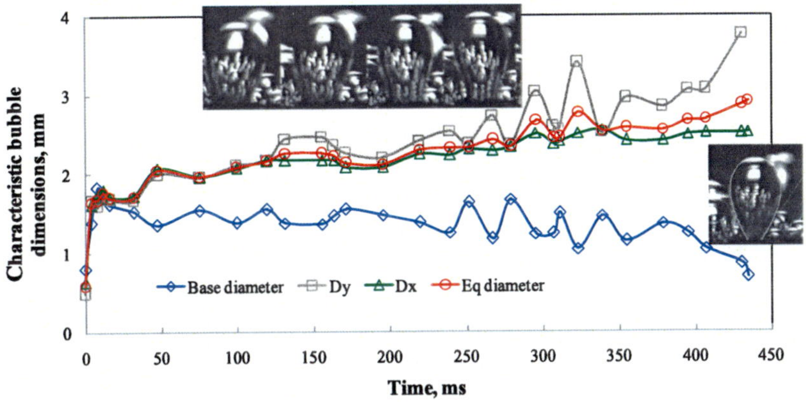

Fig. 3. Oscillatory behaviour of bubbles on copper surface at superheat of 5.4 K (behaviour of bubbles on copper and brass surfaces are similar).

At the beginning of the oscillatory period in brass and copper, only a slight increase in the equivalent diameter has observed in the first 100 ms, as shown in Fig. 4.

The variations in bubbles dimensions are significant in the later period of oscillatory growth and are described as changes in the vertical direction followed by changes in the base, as illustrated in Fig. 3 and 5. Therefore, the bubbles on the copper and brass surfaces are more balloon-shaped rather than a regular sphere. However, the ratio of equivalent diameter to the base diameter is about 1.5 in the first 250 ms, than increases to 2.2 in the later period (increases to 3 before the detachment from the surface). Thus, the equivalent diameter does not describe behaviour of the bubbles adequately. Interestingly, the base diameter during the oscillatory period does not exceed its maximum value, reached during the initial growth.

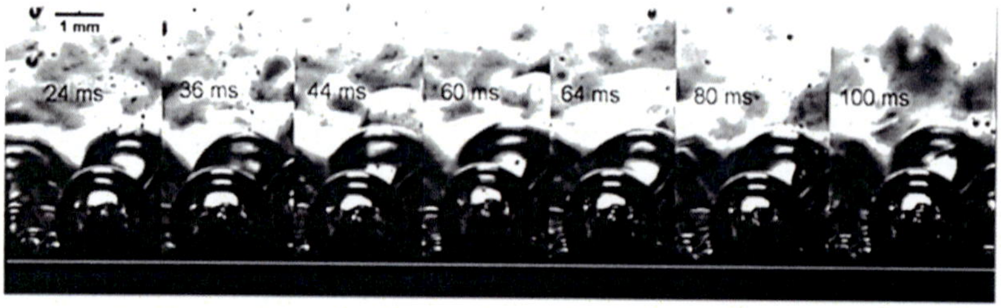

Fig. 4. Typical bubble behaviour for the first 100 ms (from the moment the bubble is visible on the heated surface).

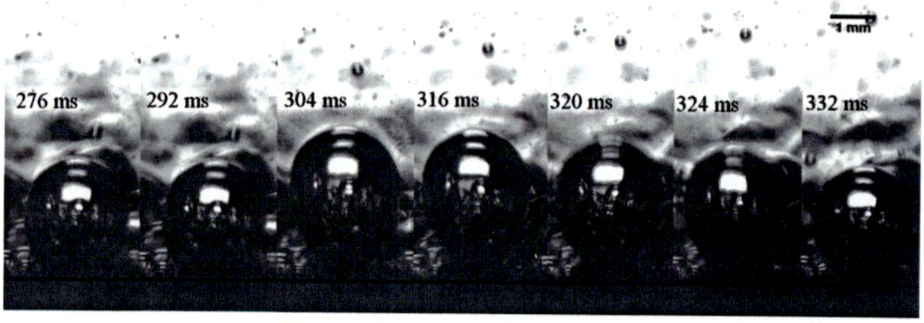

Fig. 5. Oscillatory behaviour of bubbles on brass surface at superheat of 6.1 K.

Equivalent diameter better reflects the changes in volume on steel in comparison to the other two surfaces with milder variations of the volume in the vertical direction. The base diameter on the steel surface decreases continuously in the first 250 ms; later oscillates around the same values which causes a slower growth of the equivalent diameter with slightly lower values compared to the bubble diameter on copper and brass surfaces (it grows about 2 mm in 300 ms, then it oscillates around the same value), as illustrated in Fig. 6.

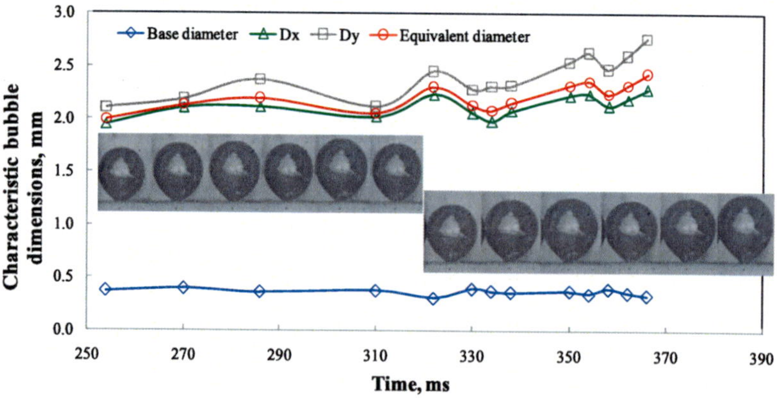

Fig. 6. Changes in characteristic bubble dimensions at SS during oscillatory growth at 7 K.

The bubble base behaviour is very similar to that presented by [10], but there are differences in the equivalent bubble diameter. Gradual but constant increase of the bubble equivalent diameter is observed during the oscillatory regime. Bubble departure diameters are up to 3 mm for copper and brass surfaces, whereas for the steel they are slightly smaller, up to 2.5 mm. However, [10] recorded a decrease of the equivalent diameter in time: after 300 ms, the equivalent diameter is about half the value of that obtained for brass and copper.

Starting from the assumption that the transient conduction occurs only after the bubble departure [11], it is possible to neglect it, since the waiting time between two bubbles from the same nucleation site was not observed. There is no disturbance of the superheated layer since the departure of the bubbles from the surface during the oscillatory period was only occasional. It follows that the heat transfer is limited to the initial evaporation at the base of the bubble and later to the evaporation at the top of the bubble and re-evaporation of the wetted microlayer in the oscillatory mode (heat transfer related to bubble activity on the heated surface excluding natural convection). However, the volume of the bubbles increases intensively when the base begins to narrow significantly. That means that the liquid reaches the heated surface, although the bubbles depart seldom.

If only the mechanism of evaporation in the microlayer had been active during the entire growth period, then the bubbles would not grow significantly from the achieved maximum base diameter during the initial period. Thus, each subsequent bubble at the same nucleation site would be smaller because less liquid remains trapped in the microlayer (no waiting time). However, variations in the equivalent diameter between successive bubbles are $\pm 10\%$ and do not have a decreasing trend. There are some variations between bubbles from different nucleation sites, but the overall behaviour is consistent across the evaluated bubbles.

Additionally, the increase in bubble volume over time may have contributed by the high initial density of active nucleation sites (the most initially active nucleation sites are observed on brass and the least on steel surface), as well as surface wetting properties. Thus the simultaneous oscillatory behaviour of more bubbles on heated surfaces could lead to thickening of the boundary layer and heating of the liquid near bubbles due to the improved heat transfer by evaporation at the bubble top. That could explain why [10] obtained different results in their work.

4 Conclusion

The provided investigation of subcooled boiling is shoved the bubble dimensions significantly depend on the oscillatory growth time and activation of different heat transfer mechanism during this period of growth (evaporation at the bubble top, transient conduction, and convection due to the bubble movement). Evaporation at the bubble base is limited by the bubble departure without the waiting time. Also, it appears that sphericity of the bubbles affects the volume before the departure: the bubbles with a lower ratio of the equivalent diameter to the base diameter (less than 2.5) during the growth on the heated surface have a prolonged transition from hemispherical to irregular spherical shape and higher volume before the departure. However, further investigation is needed in the oscillatory regime to offer sufficient data for a more reliable assessment of the influence on bubble volume and a possible correlation with heat transfer mechanisms. In those terms, it is required to investigate the oscillatory behaviour for both spherical and non-spherical bubbles with respect to increased surface superheat.

References

1. Kim, J.: Review of nucleate pool boiling bubble heat transfer mechanisms. Int. J. Multiphase Flow **35**, 1067–1076 (2009)
2. Demiray, F., Kim, J.: Microscale heat transfer measurements during pool boiling of FC-72: effect of subcooling. Int. J. Heat Mass Transfer **47**, 3257–3268 (2004)
3. McHale, J.P., Garimella, S.V.: Nucleate boiling from smooth and rough surfaces – Part 1: fabrication and characterization of an optically transparent heater–sensor substrate with controlled surface roughness. Exp. Thermal Fluid Sci. **44**, 456–467 (2013)
4. Teodori, E., Moita, A.S., Moreira, A.L.N.: Influence of surface topography in the boiling mechanisms. Int. J. Heat Fluisd Flow **52**, 50–63 (2015)
5. Goel, P., Nayak, A.K., Kulkarni, P.P., Joshi, J.B.: Experimental study on bubble departure characteristics in subcooled nucleate pool boiling. Int. J. Multiphase Flow **89**, 163–176 (2017)
6. Wang, X., Wu, Z., Wei, J., Sundén, B.: Correlations for prediction of the bubble departure radius on smooth flat surface during nucleate pool boiling. Int. J. Heat Mass Trans. **132**, 699–714 (2019)
7. Chang, Y.H., Ferng, Y.M.: Experimental investigation on bubble dynamics and boiling heat transfer for saturated pool boiling and comparison data with previous works. Appl. Therm. Eng. **154**, p284-293 (2019)
8. Moghaddam, S., Kiger, K.: Physical mechanisms of heat transfer during single bubble nucleate boiling of FC-72 under saturation conditions – I. Experimental investigation. Int. J. Heat Mass Trans. **52**, 1284–1294 (2009)
9. Siedel, S., Cioulachtjian, S., Bonjour, J.: Experimental analysis of bubble growth, departure and interactions during pool boiling on artificial nucleation sites. Exp. Thermal Fluid Sci. **32**, 1504–2151 (2008)
10. Narayan, S., Singh, T., Singh, S., Srivastava, A.: Experiments on the effects of varying subcooled conditions on the dynamics of single vapor bubble and heat transfer rates in nucleate pool boiling regime. Int. J. Heat Mass Trans. **134**, 85–100 (2019)
11. Mikic, B.B., Rohsenow, W.M.: A new correlation of pool boiling data including the effect of heating surface characteristics. Int. J. Heat Mass Trans. **91**, 245–250 (1969)

Using Hydraulic Model Tests for Water Intake Structure Redesign of Hydro Power Plant

Hajrudin Džafo[1](✉), Sadžit Metović[2], and Edin Kasamović[2]

[1] Investment Department, JP Elektroprivreda BiH, Vilsonovo Setaliste 15, 71000 Sarajevo, Bosnia and Herzegovina
h.dzafo@epbih.ba

[2] Faculty of Mechanical Engineering, University of Sarajevo, Vilsonovo Setaliste 9, 71000 Sarajevo, Bosnia and Herzegovina

Abstract. The paper presents the interlaced concept as well as the main finding of the investigation, which lead to a successful adaptation of the intake structure existing hydropower plant Una Kostela, Bosnia and Herzegovina, built in 1954. After conducting hydraulic model tests on a physical model and analysis performed on mathematical models, many dilemmas were clarified (or removed). This caused the need of a total redesign of the existing water intake. The results obtained by these hydraulic laboratory tests represents a basis for the design and construction of intake structures envisaged for the innovation of this hydropower plant. A mathematical model was applied, an experimental hydraulic tests in the laboratory done, with calibration and verification of this model. Namely, with these hydraulic tests on a physical model in the laboratory, it was necessary to test and confirm the designed technical solutions from the preliminary design. The construction of new hydropower plants is under great scrutiny and opposition from green NGOs for environmental protection. Reconstruction and expansion of existing hydropower plants is a solution for increasing electricity generation from renewable energy sources. One possibility is the reconstruction and upgrade of existing hydropower plant Una Kostela, Bosnia and Herzegovina, built in 1954.

Keywords: Hydraulic model tests · Mathematical models · Reconstruction · Preliminary design

1 Introduction

Bosnia and Herzegovina (BiH) is the rare country in the region with a positive balance of electricity exports and the eighth country in Europe in terms of hydro power - just over one third (37%) of the hydropower potential is being used. Also, it is estimated that BiH has good potentials for energy production from renewable sources in the Balkans, and they are 30% higher than the EU average. As regards renewable energy, the country's 2020 target of 40% is not likely to be met. Market-based support schemes are still missing, [1]. Bosnia-Herzegovina, a country of around 3.5 million people, is currently a net exporter of electricity. More than half of its electricity generation capacity is made

N. Ademović et al. (Eds.): IAT 2021, LNNS 316, pp. 329–336, 2022.
https://doi.org/10.1007/978-3-030-90055-7_26

up of hydropower, while the remainder is made up of five lignite power plants. As of 2020 the country had around 2.076 MW net installed hydropower capacity larger than 10 MW, 2065 MW of lignite, 159 MW of small hydropower, 100 MW wind power, and about 20 MW solar. Increasing electricity generation from renewable energy sources is a projected scenario in the BIH energy sector. The construction of new hydropower plants is under great scrutiny and opposition from green NGOs for environmental protection. In BiH in particular, there have been significant variations in the climate, as well as the occurrence of extreme climatic and hydrological events in recent years, [2]. When it comes to planning new hydropower projects in line with ecological needs, environmental flow should be one of the primary factors for investors and regulators. There are over 200 methods used globally for determining environmental flow, with methodologies varying from country to country and from project to project [3]. Historically and presently, the planning and construction of hydropower plants (HPPs) in Bosnia and Herzegovina have not been preceded nor sufficient understanding of the impact these objects on the environment, [4–6].

Reconstruction and expansion of existing hydropower plants is a solution in this direction with this goal. One possibility is the installation of new capacity and the completion of the reconstruction and expansion of the existing hydropower plant Una Kostela, Bosnia and Herzegovina, built in 1954. During the years of exploitation, certain reconstructions were performed for the purpose of improving the energy characteristics by the reconstruction of the water intake structure, Fig. 1.

Fig. 1. Water intake structure - existing condition before the upgrade [7]

2 Reconstruction Project of the HPP Una Kostela

The start of the exploitation is planned for 2023 when it is expected for the installed power to be increased by about 5,0 MW with an additional production of about 23 GWh. The project of reconstruction and expansion of HPP Una Kostela represents a project on the already existing HPP, the reconstruction in riverbed of Una River and the upgrade through the construction of the so-called Aneks, with the aim of increasing the electricity generation, the optimal use of Una River and the extension of the working life of the plant. Considering that these works are performed on the already existing plant or in its immediate environment, they will not change the existing state of the environment and the flow of water. HPP Una Kostela was put into operation back in 1954, and its optimum long-term operation was provided through programs of maintenance, revitalization, and reconstruction on several occasions. Reconstruction of the plant in this HPP started back in year 2000 within the first phase, and it included the replacement of the mechanical-electrical equipment and certain parts of the plant were reconstructed as well.

The existing HPP Una Kostela is a flow power plant of the derivation type, located on the left side of the Una river, downstream from Bihac city. The hydroelectric power plant was put into operation in 1954. It was originally equipped with four units each with installed capacity of 1,5 MW. In order to preserve the tufa on the stretch from the overflow threshold to the drainage outlet, regulation was performed troughs with concrete cascades at a length of about 300 m. There is an obligation in this part of the Una discharges/overflows of 2,0 m^3/s, in order to preserve the natural environment. Table 1. contents statistical characteristic flows on the water intake profile, [7].

Table 1. Basic vulnerability index values for various types of masonry structures, [17]

Mean annual flow	97,6 m^3/s
High water return period 5 years, $Q_{20\%}$	611,7 m^3/s
High water return period 20 years, $Q_{5\%}$	787,3 m^3/s
High water return period 100 years, $Q_{1\%}$	981,9 m^3/s
High water return period 1000 years, $Q_{0,1\%}$	1257,5 m^3/s
Minimum mean annual flow	20,1 m^3/s

Reconstruction with replacement of mechanical and electrical equipment (2001) gave new characteristics of the power plant through the increase of the installed flow from 64,0 m^3/s to 88,0 m^3/s and installed capacity to $4 \times 2{,}53 = 10{,}14$ MW. The reconstruction project of the water intake structure on the hydropower plant Una Kostela involves: redesign of intake structure, supply tunnel capacity testing, overflow and flood spillway on the water chamber.

An effective solution it was necessary that is acceptable according to technical regulations and at the same time, economically acceptable. the data collected by this research was useful for decision makers. In real operational conditions, due to the lack of realization of the threshold exceedance, it was not possible to achieve an increase in the value of the installed flow and power, so this issue remained for further processing and analysis. In parallel, based on the still low degree of installation, an increase in capacity was considered by building an additional (fifth) unit - Una Anex. As it is not possible to place the fifth unit in the space of the existing machine building, it is planned to place it in a special machine building, with a command connected to the existing one, [8].

Table 2. Basic parameters after the reconstruction in 2001, [8]

Hydro power plant characteristics:	
Installed power plant flow	88,0 m^3/s
Minimum operating flow of one unit	8,6 m^3/s
Maximum upper water level in the chamber	211,60 m.a.s.l
Minimum allowed elevation in the chamber	209,15 m.a.s.l
Lower water level for flow of 88,0 m^3/s	197,55 m.a.s.l
Maximum gross head	14,97 m
Maximum net head	13,70 m
Minimum net head	9,60 m
Number of turbine-generator units	4
Turbine type	Kaplan vertical
Installed capacity of power plant	10,14 MW
Possible annual production	59,50 GWh

3 Hydraulic Model Tests of Water Intake Structure

Program of Hydraulic Model Tests: A physical model was built in the hall of the Hydraulic Laboratory of the Institute of Water Management "Jaroslav Černi", in Belgrade, [8]. Hydraulic model tests realized in more phases. Phase 2-testing of the water intake structure (existing and new power plants) was explained in this paper. Hydraulic model tests of the water intake structure performed for the following input mean-average flows: 52; 63; 74; 90; 96; 97; 118; 140 and 168 m^3/s, as well as flows ordered values as $Q_{20\%}$, $Q_{5\%}$, $Q_{1\%}$, $Q_{0,1\%}$. For each of the scenarios, the following task was done: (1) determination of flow conditions and flow conditions; (2) checking whether vortex and air flows into the tunnel occurs during operation, as well as proposing possible measures to prevent these phenomena; (3) checking the capacity, dimensions and hydraulic characteristics of the water intake structure. Phase 3 was testing with rubber dam. Hydraulic model tests and analysis of overflow over the threshold with a rubber dam done for two positions: raised (full dam) and lowered (empty dam).

The main goals of hydraulic model tests was: (a) verification of the designed solution of the water intake: contact conditions, current image, shape, dimension and position, detection of possible adverse events and proposed measures for prevention of these phenomena; (b) checking the capacity, dimensions and hydraulic characteristics of the entrance structure in stationary and transient modes of operation (abrupt closing and opening of turbines for existing and new aggregate; (c) checking the designed solution of the sill and rubber dam: contact conditions, flow conditions on overflow, capacity, shape and dimensions of dividing columns on overflow, determination places where under-pressures may occur, proposed measures to eliminate these phenomena, determining the flow curve over a rubber dam.

Physical Model Project: The project of the physical model, for the defined subject and the goal of the examination, determined the precise limits and dimensions of the model (according to the area that is influenced by the flow conditions on the considered objects - entrance structure overflow threshold, riverbed upstream and downstream). The proposed size of the physical model is 1:30, which means that the length of the model will be about 33 m. In accordance with the hydrological data, it will be possible to test the entire flow range up to $Q_{0.1\%} = 1257{,}5\ m^3/s$ on the model.

Hydraulic Similarity Conditions and Choice of Hydraulic Model Size: In the area covered by the physical model [8], free-surface flow prevails for the entire range of considered flows. In open flows, there are forces of friction, gravity and inertial fictitious forces, so the similarity of phenomena in relation to nature is subject to Froude's law, while satisfying the influence of frictional forces. The ratio of inertial and gravitational forces on the model and in nature is the same, which means that the Froude number, Fr, must be the same in nature and on the model, i.e. according used in [8]:

$$F_{r_p} = F_{r_m} \tag{1}$$

$$F_r = \frac{V}{\sqrt{gL}} \tag{2}$$

where:
V - velocity, m/s
L - length, m
g - acceleration due to gravity, m/s^2.

Physical Hydraulic Model: The physical hydraulic model was made in the scale for lengths R = 1: 30 and covers an area of approx. 33 × 10 m. Froud's similarity was used on the model. For measuring levels, pressures, speed and water flow, both classical and state-of-the-art instrumentation techniques were used. Numerical analyzes are based on line models, described by the corresponding equations for non-stationary (unsteady) water flow, with initial and boundary conditions.

Testing of the Water Intake Structure (Existing and New-Redesigned): The aim of these tests was to check the designed capacity of the water intakes, [8]. Water level elevations were measured for each of the scenarios. For the river flows of 74, 96, 118 and 140 m^3/s, in the case of a raised rubber dam, the model showed that upgraded hydropower plant with fifth turbine-generator unit can operate with the installed capacity with simultaneous operation of existing power plan respectively with one, two, three and

Fig. 2. Physical hydraulic model of redesigned water intake structure, [8]

four turbine-generators. For a river flow of 90 m^3/s, in the case of a raised rubber dam, existing hydro power plant can operate with the installed capacity. For the river flow of 168 m^3/s, with the rubber dam lowered, the water level in the entire flow tract of the existing power plant is noticeably low (in the chamber the elevation is 208,92 m.a.s.l., so the capacity of the supply tunnel is less than the installed flow and is approximately 79 m^3/s, according to [8].

Fig. 3. Physical hydraulic model of redesigned intake structure - the upgrade solution, [8]

Figure 5(a) presents the current state of the water intake, and the future state that was tested on a hydraulic model in the laboratory, Fig. 5(b), according to [8].

Fig. 4. Physical hydraulic model of threshold in the riverbed. A model of a full rubber dam was installed on one of three segments, [8]

(a) (b)

Fig. 5. (a) – real situation of water intake structure; (b) Physical hydraulic model with upgrade redesign of the water intake structure, [8]

4 Conclusion

The physical hydraulic model confirmed the technical solutions of the conceptual design and answered some questions of the design of the water intake structure. After evaluating the physical model tests, the main conclusions are as follows:

a) A favorable flow at the water intake structure has been confirmed;
b) The water intake structure in conditions with a lowered rubber dam cannot achieve full capacity (simultaneous operation of fifth units of power plants at installed flow) for natural flows less than 300 m^3/s;

c) It is not rational to plan for an existing HPP to operate with a higher installed flow than originally provided. Therefore, we believe that subsequently installed turbines ($4 \times 22 = 88\ m^3/s$) should operate only up to the originally intended installed flow;
d) No air was drawn into the tunnel at the water intake structure.

References

1. Bosnia and Herzegovina Annual Implementation Report, State of Energy Sector (2020)
2. Final Report on Hydraulic Model Test Performed on the Physical Model, JP EP BiH (2019)
3. Milosavljevic, D., et al.: Climate changes and renewable sources of energy in Bosnia and Herzegovina. Contemp. Mater. **9**(1), 58–69 (2018)
4. Tesnjak, S., et al.: 3R Model from Small HPP Design (2012)
5. Vucjak, B., et al.: Applicability of multi-criteria decision aid to sustainable hydropower. Appl. Energy **101**, 261–267 (2013)
6. Vucjak, B., Smolar-Zvanut, N., Antonelli, F.: Development of environmental flow assessment procedure for Bosnia and Herzegovina (n.d.)
7. Preliminary design of HPP Una Kostela -reconstruction- Energoinvest, Sarajevo (2016)
8. Hydraulic model tests for intake structure and threshold in the river level for HPP "Una Kostela" – Reconstrution and expansion", done by Hydraulic Laboratory of the Institute of Water Management "Jaroslav Černi", in Belgrade (2019)

Experimental Analysis and Mathematical Modeling of Deformation for a Frame of a Machine for Sheet Metal Design Manufacturing Process

Ermin Bajramović(✉), Milan Jurković, Dženana Gačo, Emir Bajramović, and Esad Bajramović

Technical Faculty Bihac, University of Bihac, Bihać, Bosnia and Herzegovina
ermin.bajramovic@unbi.ba

Abstract. In this paper, based on the experimental investigations of the cold forming line of the sheet metal using rollers, the deformations of the supporting frame of the machining module were analyzed. After that, a comparison of experimental and modeled values was performed. Experimental studies have included the measurement of deformation of the frame in the cold rolling process on a single machining module. The following resources were used to perform the experiment: a cold forming profile line, three types of sheet metal of different thicknesses, widths and mechanical characteristics, measuring straps for measuring frame deformation, signal amplifiers and two computers with appropriate programs.The goal of the paper is to develop a mathematical model to obtain a model which can be used for analysis of deformation of the supporting frames of the processing modules. The resulting mathematical model could be important for engineers to develop new production lines for cold forming sheet metal using rollers.

Keywords: Production line · Machining module · Supporting frame · Sheet metal · Deformations · Mathematical modeling

1 Introduction

1.1 Cold Forming Process with Rollers

Profile fabrication by cold forming using rollers is a highly productive method used in mass production. The basic characteristic of cold forming of profiles by rollers is the gradual formation of a given shape in several stages which is achieved by passing the workpiece through a number of shaping modules, as shown in Fig. 1.

1.2 Literature Review

The process of the profiles' cold forming by rollers is influenced by several parameters in the process of their obtaining such as tool and profile geometry, characteristics of

N. Ademović et al. (Eds.): IAT 2021, LNNS 316, pp. 337–347, 2022.
https://doi.org/10.1007/978-3-030-90055-7_27

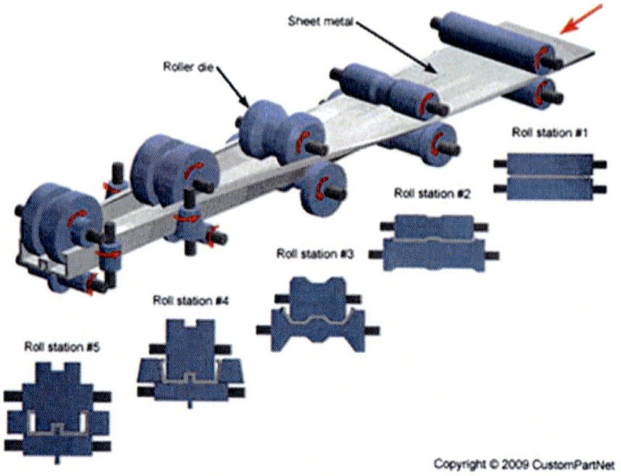

Fig. 1. Pri Scheme of the profile cold forming process

the material from which the profile is obtained, forces required to overcome resistance, friction occurring between tools and materials, installed driving force systems etc. Therefore, new process knowledge and determination of optimal parameters is a continuous engineering task in order to increase product quality while reducing production costs. Some of the authors cite research in their papers and provide guidance for improvement in the process of cold rounding of profiles using rollers. Thus, Gehring [1] in his doctoral dissertation discussed the suitability of metal strips and sheets for profiling by roller and the parameters in the process that are influenced by the material characteristics. Hellborg [2] in his doctoral thesis investigated the geometry of rollers and performs a simulation of rollers in the process of the profiles' cold rolling. In his master's thesis Jušić [3] dealt with theoretical research and experimental diagnostics of the load of a production line for forming open profiles made of sheet metal using rollers.

Authors Traub et al. [4] investigated the optimisation of energy efficiency in cold forming by rolling with a new sensor concept, Han and Wang [5] in their work investigated and analysed the deformations of the QP980 steel rolling mechanism based on finite element simulation. Abdollahi et al. [6] talked about friction irregularities during cold rolling of material. Safari and Joudaki [7] investigated the roundness and slope of arches from previously drilled sheets in circular trapezoidal section shaping. Traub et al. [8] experimentally and numerically investigated bending zones in rolling design. Paralikas et al. [9] examined the influence of the roll forming process main parameters on the quality of the V-section profile of AHSS. In their work, Jurković et al. [10, 11] through experimental research, demonstrated the diagnostics of a roll forming line using a roller and an experimental and modeling approach to improving the utilisation rate of a cold rolling production line.

A real-time experimental study provided a series of input parameters for the analysis of a cold sheet metal forming line. The parameters will certainly serve to further research and improve the production line and the process of cold forming using rollers.

2 Experimental Research and Analysis of the Cold Design Production Line

2.1 Display of Cold Forming Production Line

The production line used for research and analysis was 10 m long and had 20 processing modules for profile formation. The profile was obtained gradually from the first to the twentieth module by the flower pattern principle. The experimental research was carried out in the actual production process (Fig. 2). Each machining module has an upper and a lower roller (Fig. 4). The power of the electric motor is 11 kW, the output speed is 41 min-1, the transmission drive from the electric motor to the axles was realised by a chain. The production line consisted of a 10 000 kg payload, a straightening device at the inlet, a profile cutting off device at the outlet of the line and a control unit with appropriate management software.

Fig. 2. Production line used in experimental research

Schematic presentation of the production line with labels and measuring points in the experiment (Fig. 3).

When measuring the line with the measuring equipment the values of force, torque and elastic deformation were measured at several measuring points in three different sets (Fig. 3).

Elastic deformation of the frame (I set – II machining module, left side), where eight values for elastic deformations were obtained for each sheet passage at eight frame locations, where strain gauges were installed to measure deformation (Fig. 5 and Fig. 6).

The experiment resulted in a series of findings regarding structural and other parameters that should serve as well as the basis for designing new generation production lines. In this paper, emphasis was placed on the deformations of the workpiece frame. The experimental values of frame deformations for specific material thicknesses provided guidance for modeling and comparing results. The goal was to develop a mathematical model that would describe the deformations of the supporting frames of the machining modules.

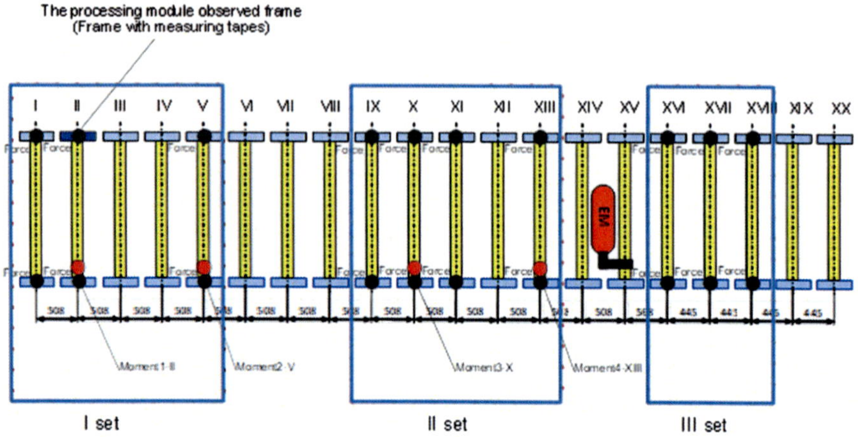

Fig. 3. Schematic presentation of production line with labels and measuring points

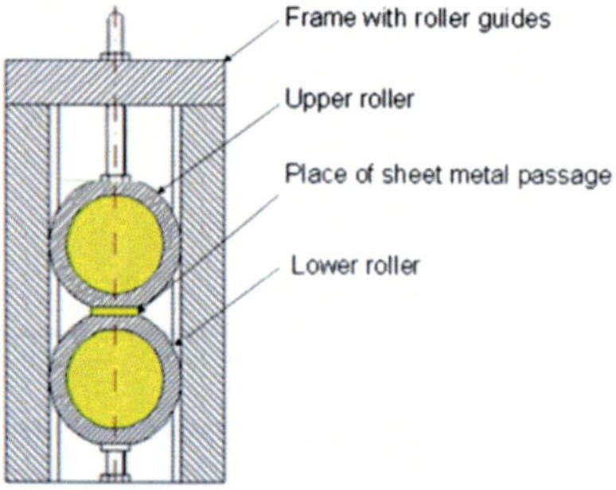

Fig. 4. Processing module frame (upper and lower roller)

2.2 Influence Parameters on Deformations of the Workpiece Frame

The size of the deformation of the workpiece frame is influenced by several parameters, the force exerted by the workpiece module due to the rolling of the sheet metal which is a function of the characteristics of the sheet being profiled $F = f(s,b,R_m)$, the friction force, then the cross section of the frame. The friction force and the cross-section of the frame in the observed process are constants, and only the load of the workpiece module on the frame or the input variables that affect the load due to rolling, namely thickness (s), width, will be considered (b) and material strength (R_m).

Figure 5 shows the frame of the second machining module observed in the experiment with the marked locations of the strain gauges. Figure 6 shows the strain gauges on the same frame.

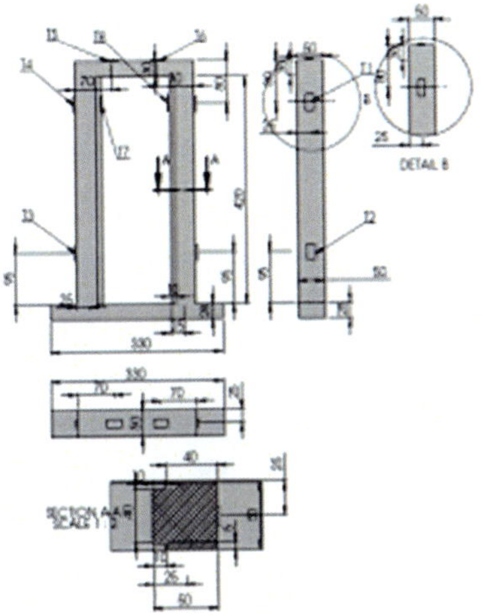

Fig. 5. Frame - the locations the strain gaguge

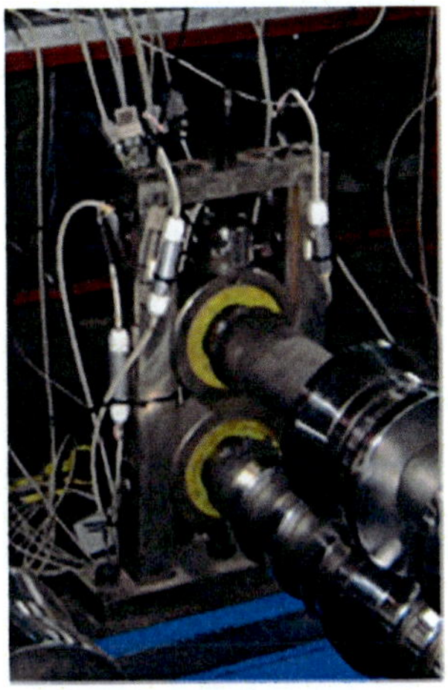

Fig. 6. Frame – display of the strain gauges

2.3 Materials Used in the Experiment

For the aforementioned experiment, metal sheets in strips, three sheet widths (950 mm, 1100 mm and 1250 mm), three sheet thicknesses (0.50 mm, 0.60 mm and 0.70 mm) were used and the sheet length 2500 mm with the following features:

- steel sheet DX51D (DIN 17162–1, EN 10327), mechanical properties: $R_m = 383$ MPa, $R_e = 278$ MPa,
- steel sheet DX53D (DIN 17162–1, EN 10327), mechanical properties: $R_m = 270$ MPa, $R_e = 140$ MPa,
- aluminum sheet Al 99,5 (EN 1050), mechanical properties: $R_m = 130$ MPa.

2.4 Measuring Devices and Data Transmission

For measurements, because multiple output quantities were measured in real time, it was necessary to provide measuring equipment that was able to simultaneously perform all of the above measurements. The equipment contained the following: eight specially designed force transducers according to the design requirements on the line itself, eight strain gauges which were glued to the frame of the line and by means of cables connected to an amplifier using ap-propriate connectors. In addition, two laptops were provided with catmanEasy 3.3.5 installed, recording outputs of all measured sizes as well as three amplifiers DMCPlus, Spider8 and QuantumX [12].

2.5 Experiment and Results Analysis

The experiment was performed with 3 input variables, sheet thickness (s), sheet width (b) and material tensile limit (R_m). Experiment plan:

$$N = 2^3 + n_o = 8 + 4 = 12 \quad (1)$$

Physical and coded values of experiment plan parameters are shown in Table 1.

Table 1. Physical and coded values of experiment plan parameters

Influential parameters		Coded and physical values of the process input parameters		
Coded parameters	X_i	−1	0	1
Physical parameters	$x_1 = R_m$ [MPa]	130	270	383
	$x_2 = s$ [mm]	0.5	0.6	0.7
	$x_3 = b$ [mm]	950	1100	1250

The deformations of the frames were measured on module number two, on the left side, where eight strain gauges were glued to eight measuring points, the change of deformations depending on the material used was monitored and the results obtained are shown in Table 2.

Table 2. Experiment plan with deformation values of the processing module frame

No	R_m [MPa]	s [mm]	b [mm]	X_1	X_2	X_3	Frame deformations Δli [μm/m]							
							T1	T2	T3	T4	T5	T6	T7	T8
1	130	0.5	950	−1	−1	−1	1.019	1.609	−1.637	- 0.418	−0.569	0.342	0.500	1.078
2	383	0.5	950	1	−1	−1	1.867	2.333	−1.120	0.588	0.555	0.941	0.326	1.286
3	130	0.7	950	−1	1	−1	3.732	1.741	1.799	−1.857	0.161	−0.141	0.890	0.174
4	383	0.7	950	1	1	−1	2.085	2.646	−1.644	0.516	1.008	1.670	0.674	1.730
5	130	0.5	1250	−1	−1	1	3.822	1.386	0.844	−1.785	−1.075	−0.256	1.332	0.786
6	383	0.5	1250	1	−1	1	4.507	2.499	2.536	−2.414	−0.196	1.221	1.819	0.694
7	130	0.7	1250	−1	1	1	4.190	2.798	2.159	- 1.727	−0.817	0.535	1.878	0.964
8	383	0.7	1250	1	1	1	2.269	3.306	−1.724	−0.160	2.815	3.098	1.514	3.167
9	270	0.6	1100	0	0	0	3.664	2.159	1.650	−1.855	−0.748	−0.125	1.417	0.458
10	270	0.6	1100	0	0	0	2.470	3.592	1.944	1.431	−2.120	−0.664	−0.080	1.295
11	270	0.6	1100	0	0	0	2.066	1.438	−2.377	−0.584	0.099	1.212	0.426	0.735
12	270	0.6	1100	0	0	0	2.024	1.547	−2.456	−0.649	−0.166	1.008	0.551	1.027

Table 2 shows the experimental values obtained for frame deformations for twelve experiments at eight measurement sites [12]. According to the reported values, it can be concluded that the deformations of the frame were very small.

3 Mathematical Model Modeling

On the basis of the experimental data, it was necessary to develop p mathematical model for modeling the deformations of the frames, to calculate the modeled values of the deformations of the frames and to compare them with the experimental values.

3.1 Mathematical Model

Input parameters: Sheet width, b [mm]; Sheet thickness, s [mm]; Sheet material, R_m [MPa].

Output parameters: Frame deformations, Δl_{ok} [µm/m].

The process state function is:

$$\Delta l_{ok} = f(s,b,R_m) \tag{2}$$

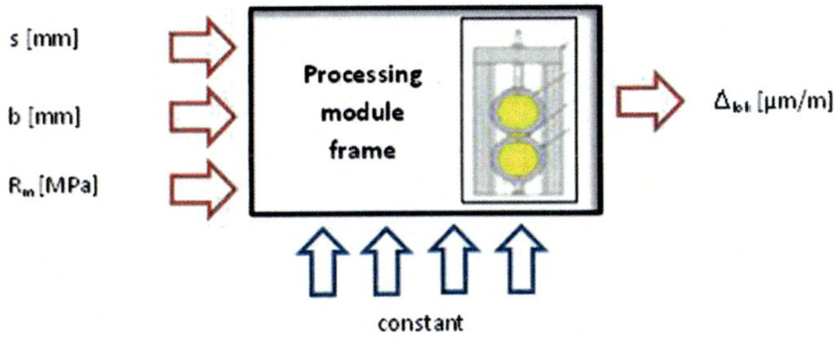

Fig. 7. Input-output values of the rolling process for modeling the deformations of the workpiece frame.

3.2 Selecting a Starting Model

For an unknown process mechanism, a mathematical model based on experimental and regression analysis with real regression coefficients b_i, b_{ii}, b_{im}, b_{imk}, takes the form:

$$Y = \sum_{i=0}^{k} b_i X_i + \sum_{1\le\langle m}^{k} b_{im} X_i X_m + \sum_{i=1}^{k} b_{ii} X_1^2 + \sum_{1\le i\langle m\le k}^{k} b_{imk} X_i X_m X_k \tag{3}$$

The initial mathematical model chosen is a three-factor polynomial:

$$\begin{gathered} Y = b_0 + b_1X_1 + b_2X_2 + b_3X_3 + b_{12}X_1X_2 \\ + b_{13}X_1X_3 + b_{23}X_2X_3 + b_{123}X_1X_2X_3, \end{gathered} \tag{4}$$

Mathematical regression coefficients:

$$b_0 = \frac{1}{N}\sum_{j=1}^{N} X_{0j}Y_j = \frac{1}{N}\sum_{j=1}^{N} Y_j \quad \text{because\_it\_always\_is\_}X_{0j=1}$$

$$b_i = \frac{1}{N-n_0}\sum_{j=1}^{N} X_{ij}Y_j, \quad \text{for\_i} = 1, 2, 3, \ldots \tag{5}$$

$$b_{im} = \frac{1}{N-n_0}\sum_{j=1}^{N} X_{ij}X_{mj}Y_j \quad \text{for } 1 \le i\langle m \le k$$

Table 3. Experimental and modeled values

Exp	Physical values			Coded values			Experimental values	Modeled values
Nj	R_m[MPa]	s [mm]	b [mm]	X_1	X_2	X_3	$Y_i^E = \Delta l_i^E$ [μm/m]	$Y_i^R = \Delta l_i^R$ [μm/m]
1	130	0.5	950	−1	−1	−1	0.2405	−0.0675
2	383	0.5	950	1	−1	−1	0.8470	0.4638
3	130	0.7	950	−1	1	−1	0.8124	0.4021
4	383	0.7	950	1	1	−1	1.0856	0.9334
5	130	0.5	1250	−1	−1	1	0.6318	0.3825
6	383	0.5	1250	1	−1	1	1.3333	0.9138
7	130	0.7	1250	−1	1	1	1.2475	0.8521
8	383	0.7	1250	1	1	1	1.7856	1.3834
9	270	0.6	1100	0	0	0	0.8275	0.6863
10	270	0.6	1100	0	0	0	0.9835	0.6863
11	270	0.6	1100	0	0	0	0.3769	0.6863
12	270	0.6	1100	0	0	0	0.3608	0.6863

Testing the significance of the mathematical model coefficients yields the results of checking the significance of the coefficients and verified by the Fisher test. After checking the significance, the coefficients b_0, b_1, b_2, b_3 were significant while the coefficients b_{12}, b_{13}, b_{23}, b_{123} were insignificant, so that the mathematical model took the form:

$$Y = F_i = 0.8777 + 0.2649X_1 + 0.2348X_2 + 0.2516X_3 \quad (6)$$

The mathematical model of adequacy checking according to the Fisher criterion for the condition $F_a \le F_t\ (f_1, f_2) = F_t\ (f_a, f_0)$. The condition is met with $F_a = 6.66 \le F_t = 9.01$ and the mathematical model is adequate. After determining the adequacy of the model and decoding it, the final expression in physical values is:

$$y = \Delta_{sr} = 0.0021 \cdot R_m + 2.348 \cdot 0.0015 \cdot s + b - 2.9393 \quad (7)$$

With 95% accuracy, the multiple regression coefficient $R = 0.97$ indicated that the model described the process well.

3.3 Comparison of Experimental and Modeled Values

Table 3 shows the mean experimental and modeled left frame deformation values of the second production line of the cold forming sheet metal using rollers.

Figure 8 shows a diagram of the mean values of the experimental and modeled deformations where it can be seen that the model describes well the state of the frame deformation process.

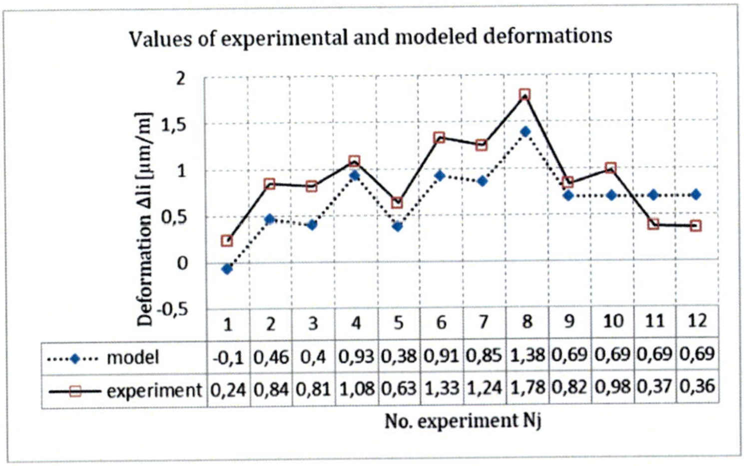

Fig. 8. Diagram of the values of experimental and modeled deformations

3.4 Analysis of the Frame Deformations Obtained Values

The experimental deformation values of the frame on the second machining module (left side) shown in Table 2 show that their absolute values, maximum and minimum values (Δl) do not follow any logical sequence, either by the magnitude of the deformation or by the sign positive (+) or negative (−) value.

The magnitude and sign of the stress depends on the profile tension in the space between the two machining modules (and if the rolling speed of the profile was constant at each succeeding module, the bending angle was increased according to the flower pattern or an adjacent roller engages the material) and the friction force between the rollers of the machining module and the material. The contact between the sheet and the rollers of the workpiece module was not constant, the reason was the withdrawal of the sheet from the rollers (flower pattern) and if there was no change in the thickness of the material during the rolling process. The deformations of the workpiece frame were very small and satisfied the rigidity and accuracy of the profiled sheet in the tested thicknesses.

The smallest deformations of the frame were caused by rolling of the material with the smallest input parameters R_m, s, b, experiment 1, ($\Delta l_{sr.exp.} = 0.2405$ [μm/m]), while the largest deformations occurred in experiment 8, ($\Delta l_{sr.exp.} = 1.7856$ [μm/m]) where materials with the highest input parameters R_m, s, b were used. The same was shown by the modeled data in experiment 1, the smallest values of the input parameters R_m, s, b, ($\Delta l_{sr.mod.} = -0.0675$ [μm/m]), that is, in experiment 8 ($\Delta l_{sr.mod.} = 1.3834$ [μm/m]) where the materials with the highest values of the input parameters R_m, s, b were used. The data presented confirmed that the deformations of the workpiece frame were influenced by the thickness, width and mechanical properties of the material being profiled.

3.5 Application in Practice

The experimental researches and deformation modeling of the frame processing module of the production line for cold forming sheet metal have provided new insights that can be of great importance for practical application. The production line has very low loads on the workpiece frame, as shown by the deformations of the frame.

The obtained experimental and modeled data on the frame deformations of the machining modules have a practical value that could be used by the engineers when making new production lines for cold sheet metal forming using rollers.

Future directions of research, when it comes to frame structures of machining modules of such systems, should go in the direction of choosing as good and optimal sheet metal profiling lines as possible with supporting frames that have grip for rapid modification of machining modules with rollers.

This would make it possible to produce more geometric shapes than sheet metal on a single machining system.

4 Conclusion

Based on the experimental research, deformation modeling of the workpiece frame was performed in order to modernise and improve the process and production line for the production of cold sheet shaping profiles in order to achieve optimal structural values. All the measurements carried out showed that the deformations in the mentioned operating and measurement conditions gave very small values.

Although deformation measurements were performed on only one frame, the assumption was that the load of the frame and its deformation also had an increase in their values from the first to the twentieth module, and the deformation of each frame depended on the tensile strength (R_m), thickness (s), and width (b) of the material used in the rolling process. With the multiple regression coefficient $R = 0.97$, model (7) obtained by modeling accurately and reliably showed that it was within the space covered by the experiment as a function of the mentioned parameters.

References

1. Gehring, A.: Beurteilung der Eignung von metallischen Band und Blech zum Walzprofilieren. PhD Thesis, Karlsruhe: Universitätsverlag Karlsruhe (2008)
2. Hellborg, S.: Finite element simulation of roll forming. PhD Thesis, Linköping: Linköping University (2007)
3. Jusic, A.: Teorijsko istraživanje i eksperimentalna dijagnostika opterećenja proizvodne linije za oblikovanje otvorenih profila iz lima pomoću valjaka. Master Thesis, Bihac: University of Bihac (2013)
4. Traub, T., Güngör, B., Groch, P.: Measures towards roll forming at the physical limit of energy consumption. Int. J. Adv. Manaf. Techol **104**(5–8), 2233–2245 (2019)
5. Yun, H., Wang, N.: Deformation mechanism analysis of roll forming for QP980 steel based on finite element simulation. J. Iron Steel Res. Int. **26**, 1178–1187 (2019)
6. Abdollahi, H., Beik, A., Dehghani, K.: Irregularity in friction hills during the cold rolling of materials. Int. J. Mater. From **1**, 343–346 (2008)

7. Safari, M., Joudaki, J.: Ovality and bow defect of pre-punched sheets in roll forming of trapezoidal sections. Int. J. Eng. **A31**(7), 1123–1128 (2018)
8. Traub, T., Chen, X., Groche, P.: Experimental and numerical investigation of the bending zone in roll forming. Int. J. Mehan. Sci. **131–132**, 956–970 (2017)
9. Paralikas, J., Salonitis, K., Chryssolouris, G.: Investigation of the effects of main roll-forming process parameters on quality for a V-section profile from AHSS. Int. J. Adv. Manaf. Technol. **44**(3–4), 223–237.
10. Jurkovic, M., Jusic, A., Hasanagic, N.: Experimental diagnostics line profiling sheet metal using rollers. In: RIM 2013: 9th International Scientific Conference on Production Engineering, Budva, pp. 26–29 Septembar 2013
11. Jurkovic, M., Jurkovic, Z., Buljan, S., Obad, M.: An experimental and modelling approach for improving utilization rate of the cold roll forming production line. Adv. Prod. Eng. Mang. **13**(1), 57–68 (2018)
12. Testing Laboratory TRCpro doo: Roller Frame Test Repot. Report no. ZA_TRC_4.02.06. CAZIN_003 doc. Petrovaradin (2012)

Kinetics of Internal Stresses of Water Varnishes

Izet Horman[1](✉), Esad Azemović[2], and Amina Pandžo[3]

[1] Faculty of Mechanical Engineering, University of Sarajevo, Vilsokovo šetalište 9, 71 000 Sarajevo, Bosnia and Herzegovina
horman@mef.uns.ba
[2] Saplast Sarajevo, Nikole Sopa 245, 71000 Sarajevo, Bosnia and Herzegovina
[3] Secom Visoko, Arnautovici bb, 71300 Visoko, Bosnia and Herzegovina

Abstract. The paper presents the results of an experimental investigation of the kinetics of internal stresses in aqueous and UV aqueous varnish films formed on a solid substrate. Internal stresses occur as a result of volume change during the curing and adhesion period of the film and substrate. The cantilever method was used for the analysis of internal stresses. This method is based on the principle of a clamped thin elastic narrow plate at one end. The influence of film thickness as well as the kinetics of internal stresses were controlled. In the kinetics of internal stresses of the investigated water varnish, a period in which internal stresses do not occur at the beginning of drying, a period of occurrence of internal stresses and their intensive growth, a period of decreasing intensity of internal stresses to maximum and a period of retention at the achieved level can be observed.

Keywords: Kinetics of internal stresses · Water-based varnish · Cantilever method · Film

1 Introduction

During the drying process of aqueous materials, water evaporates and molecules coalesce, caused by an accelerated drying process using ultra violet rays. In aqueous varnishes, the process of evaporation and binding of molecules takes place at room temperature. These vapors and the chemical reaction of the molecules cause the film to shrink, which is simultaneously bound to the substrate by adhesive forces. The result of these changes is an internal strain in the film.

All polymeric materials, which are used in the surface treatment of wood, are subject to changes in volume (shrinkage) during the hardening period. In the case of a film of material applied to a rigid substrate, stretching is only possible perpendicular to the substrate, while changing the dimensions parallel to the substrate is prevented by the adhesion of the film and the substrate. This leads to the appearance of internal stresses in the film which is stretched. Shear stress occurs at the film-substrate boundary. Internal stresses are particularly pronounced on dimensionally unstable substrates, such as wood and wood products [1].

During the curing of the film of polymeric materials, the internal stresses follow a known order of changes characteristic of all macromolecular systems, ie three periods in their development differ, as shown in Fig. 1.

N. Ademović et al. (Eds.): IAT 2021, LNNS 316, pp. 348–358, 2022.
https://doi.org/10.1007/978-3-030-90055-7_28

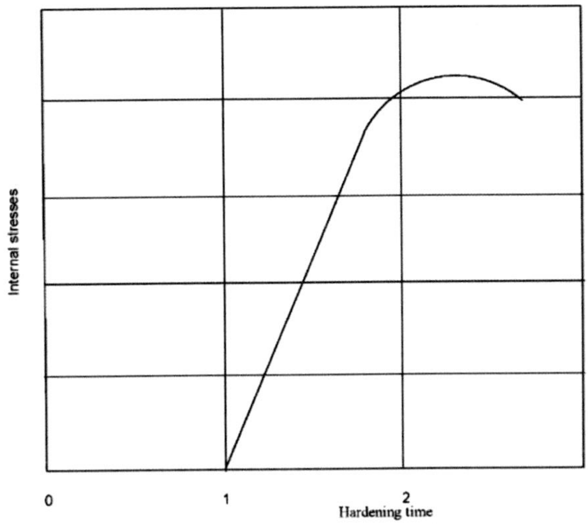

Fig. 1. Kinetics of development and relaxation of internal stresses in polymeric materials

I period (0–1), the film is in a liquid state, there is an intensive loss of volatile components, intermolecular forces still allow unimpeded mobility of molecules in all directions and internal stresses do not develop. In some cases, in the case of varnishes applied to a wooden base, negative internal stresses occur during this period as a result of swelling of the surface layer of the base.

II period (1–2), in which internal stresses occur in the film and reach the maximum value. If these values are greater than the mechanical properties of the film or the adhesion of the substrate, cracks or separation from the substrate occur.

II period (after 2), in which there is a decrease in internal stresses. This is a period of relaxation processes in the film that take place faster or slower depending on the type of film, the substrate, external conditions, etc. Internal stresses are reduced, but always maintain a certain level. These are residual stresses that are significant for the assessment of the durability of the surface treatment film, because they negatively affect the exploitation properties.

In stress analysis, there are various influencing factors, which increase or decrease the internal stresses in the film. One of the more important, influential factors is the dry film thickness.

The transition of liquid materials for surface treatment applied to wood, the surface from the liquid state to hard covers, is achieved:

- cooling,
- evaporation of volatile ingredients,
- chemical reaction and
- simultaneous evaporation and chemical reaction.

Film formation by evaporation of volatile ingredients is present in many modern wood surface treatment materials. The transition from liquid to solid state in the case of evaporation of volatile components is accompanied by a continuous reduction in the thickness of the film. The whole process can be divided into three characteristic stages which are clearly seen in Fig. 2.

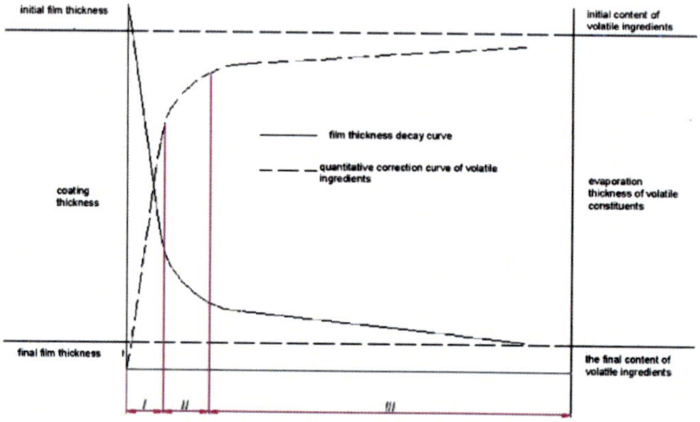

Fig. 2. Dependence of the thickness of the coating (film) and the amount of evaporated volatile components in the hardening of the coating by evaporation of volatile components [2].

The first stage occurs immediately after the application of the liquid material, and the evaporation takes place intensively as from the free surface of the liquid. This stage, in addition to intensive evaporation, is characterized by an intensive reduction in the thickness of the coating, which is accompanied by a rapid increase in the viscosity of the applied material.

The second stage is the most important, the most important for the internal stresses in the film and starts from the moment of scrum formation and coincides with the beginning of the drying stage called "dust dry", and the evaporation curve and reducing film thickness have a slight transition. The reduction of the coating thickness at this stage takes place with the deformation of the film surface. The shrinkage of the film is opposed by its force of adhesion with the wooden surface, so there are weighting stresses that are not particularly large and it is reduced by relaxation phenomena in the still uncured film.

The third stage is characterized by a decrease in the intensity of evaporation and a slowdown in the decrease in film thickness. At this stage, the final thickness of the film is formed. The cover turns into a solid phase, although the remains of volatile ingredients remain in it for a long time.

2 Methodology for Testing the Kinetics of Internal Stresses

All wood surface treatment materials have appropriate properties that must be controlled, tested, finished and prepared for application to the treated surface. Testing, measuring and controlling the properties of polymeric and all other surface treatment materials is divided into three groups:

- measuring the roughness of wooden surfaces before surface treatment,
- testing of liquid materials for surface treatment,
- testing of decorative protective covers.

2.1 Testing of Decorative Protective Covers – Films

Testing of decorative protective covers includes testing of physical, mechanical, operational properties of the cover - film. These tests are primarily significant from the point of view of final product quality control, of which the quality of surface treatment is an integral part, and in some cases the most important part of the overall quality of the final product. The group of these tests includes:

- thickness of dry films,
- hardness,
- resistance to mechanical damage,
- adhesion,
- elasticity,
- shine,
- resistance to the action of liquid agents,
- resistance to the effects of low and high temperatures

2.2 Measurement of Dry Film Thickness

Film thickness measurement consists of removing (separating) a narrow strip of film from the substrate and directly measuring the film thickness at that location using a comparator as shown in Fig. 3.

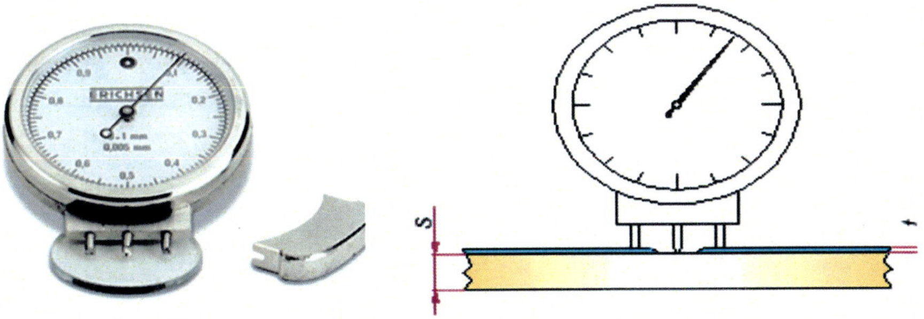

Fig. 3. Mechanical measurement of dry film thickness

Before the comparison, the comparator is calibrated on the glass plate, ie the hands of the comparator are brought to zero in the position when the tip of the comparator needle is at the same level with the feet.

The film thickness of the water-based varnish was measured on a glass plate. Before applying the varnish on the tile, an adhesive narrow strip is placed on the tile. It is important to note that the strip does not have a large width due to the limited spacing of the feet on the comparator. Removal of the tape of the adhesive tape together with the film formed on it is performed when the coating becomes gelatinous, and the thickness is measured when the film dries.

2.3 Console Deflection Measurement Method

The method consists in the following: On one side of a thin, narrow and long plate of elastic material, the dimensions and elastic properties of which are known, an appropriate amount of film-forming material is applied and after curing the film is formed.

During the curing process, the film tends to shorten, but this is prevented by the adhesion forces to the substrate. The result of this process within the film is an internal tension in it. As in the case of the cantilever method, the substrate is flexible, it follows every dimensional change of the film by bending [3].

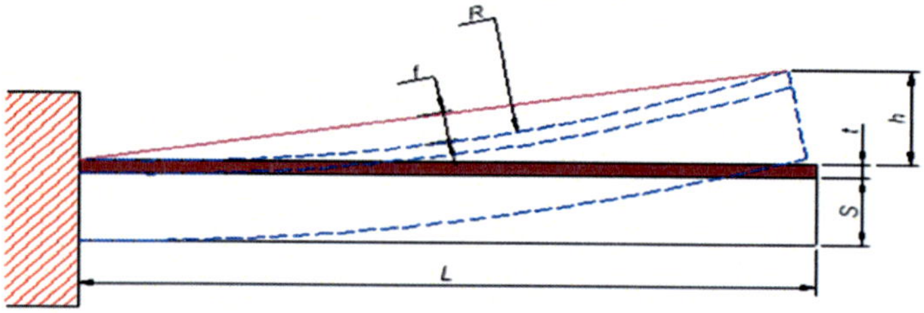

Scheme of clamping of the elastic plate with the applied film before and after hardening. S - plate thickness; t - film thickness; L - plate length; R - radius of curvature; f - deflection; h - deflection of the console.

2.4 Console Deflection Measurement Method

The method of measuring the deflection of the bracket [h] is the same for aqueous and UV aqueous varnish. The main difference is in the number of deflection measurements for water and UV water varnish. According to this difference, they have two types of console deflection measurements [3, 4]:

- measuring the deflection of the bracket [h] when drying naturally,
- cantilever deflection measurement [h] for forced drying.

When measuring the cantilever deflection for V-1 watercolor has five measurement time intervals, from 0; 0.5; 1; 24; 48 [h]. There are several time intervals (nine in total) for measuring the deflection of V-2 water-based varnish, so the kinetics of internal stresses can be analyzed in more detail. Before applying the water-based varnish on the console, it is necessary to measure the deflection of the console [h] as in Fig. 4 and 5. The measurement of the deflection of the console [h] is by means of a movable scale. Positioning the moving scale at the required location. The first measurement is called zero measurement, ie measurement before dispersion of the varnish. The deviation of the console [h] is measured by applying the varnish to the surface, by the same procedure, with an instrument in the first, second, third, fourth interval. 50 tubes were used to test the water-based varnish.

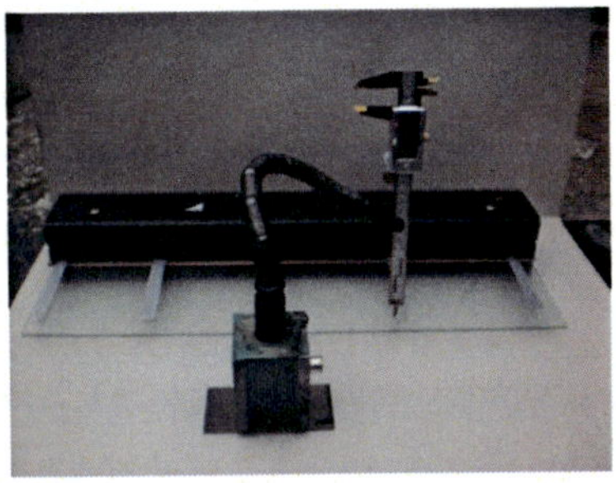

Fig. 4. Cantilever deflection measurement [h]

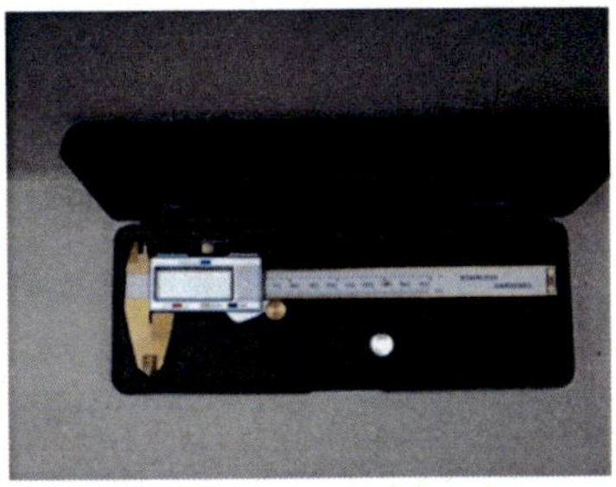

Fig. 5. Movable scale accuracy 0.01 [mm]

2.5 Measuring the Deflection of the Console During Forced Drying

The procedure and measurement of cantilever deflection is exactly the same for forced drying. In this procedure, too, there is a zero measurement, before the console enters the electronic station. By applying the varnish, the test stations enter a drying channel in which there is no access to measure the deflection of the cantilever [h]. The total drying process from application to exit takes 13 [min]. When the console exits the channel, the deflection of the console is measured by the same procedure. The differences between the initial and final measurements of the console give the actual value of the deviation of the console [h]. 50 tubes were christened to test the UV aqueous varnish (Fig. 6).

Fig. 6. Consoles do not expose from UV channels

3 Results

Basic properties of water and UV water varnish. Two types of varnishes were used for the analysis of internal stresses in cover materials: water and UV water varnishes. In order to properly analyze the internal stresses in water-soluble materials, the basic technical properties of these materials must be examined, which are:

- Viscosity,
- Dry Matter Content,
- quantities of liquid materials applied.

3.1 Viscosity

In the practice of surface treatment of wood, especially in plant laboratories, conditional viscosity is used much more often. However, in higher-level laboratories and research papers, absolute viscosity is measured. Conditional and absolute viscosity were measured in this paper. Conditional viscosity was measured in power plants for surface treatment of wood, and absolute viscosity in the laboratory of the Faculty of Mechanical Engineering in Sarajevo.

3.2 Conditional Viscosity

The conditional viscosity test was performed according to FORDU JUS B.H8.021 [5]. The basic characteristics of the instrument are: the volume of the funnel 100 ± 1 ml with a protrusion opening of 4 [mm]. Viscosity measurement is at room temperature 21 °C and relative humidity 51% (Tables 1 and 2).

Table 1. Experiment results according to Ford for conditional viscosity

Type of varnish	FORDU-u JUS B.H8.021			Volume [ml]	Time [sec]
	[4 mm]	[6 mm]	[8 mm]		
Water-based varnish	4 mm	–	–	100	**240"**

Table 2. Experiment results according to Ford for conditional viscosity Experiment results according to Ford for conditional viscosity

Type of varnish	FORDU-u JUS B.H8.021			Volume [ml]	Time [sec]
	[4 mm]	[6 mm]	[8 mm]		
UV water-based varnish	4 mm	–	–	100	**117"**

The results presented in the table are satisfactory and are close to the catalog results of the manufacturers of this material.

3.3 Absolute Viscosity

The absolute viscosity was determined by Brookfield prescribed by JUS H.K8.022 (Table 3).

Table 3. Experiment result according to Brookfield for absolute viscosity

Type of varnish	Viscosity [mPa·s]
Water-based varnish	750
UV water-based varnish	1230

3.4 Dry Matter Content

The dry matter content of aqueous and UV aqueous varnish is determined by forced drying (in a laboratory dryer) [5]. The test was performed using a 100 × 100 × 0.18 [mm] aluminum plate and a digital scale (of the order of 0.1 [g]).

The following tables show the results of dry matter content for aqueous and UV aqueous varnish (Table 4).

Table 4. Experiment results with dry material content

Type of varnish	m_1 [mm]	m_2 [mm]	Si [%]	Sm [%]
Water-based varnish	0,67	1,76	61,93	**38,07**
Type of varnish	m_1 [mm]	m_2 [mm]	Si [%]	Sm [%]
Water-based varnish	0,74	1,86	60,21	**39,78**

Discussion of the Obtained Results

The percentage of dry matter content of UV water-based varnish is higher than the percentage of dry matter content of water-based varnish. The percentage of volatile substances in water-based varnish is higher than the percentage of volatile substances in UV water-based varnish.

3.5 The Amount of Application of Liquid Material

In the surface treatment of wood, there are different methods of applying varnish to the treated surface. During the testing of varnishes on internal stresses, two methods of varnish application were used:

- manual application of varnish (airless dispersion),
- automatic application ("robotic arm").

3.6 Manual Varnish Application (Airless Dispersion)

Applying water-based varnish on the treated surface is done manually by the method of airless dispersion. During application, the light ambient temperature is 21 °C and the relative humidity is 51%. The water varnish was applied on an aluminum base, measuring 100 × 100 × 0.18 mm (Table 5).

Table 5. Experimental result of the amount of water varnish applied to the surface

Type of varnish	m_2 [g]	m_1 [g]	A [m^2]	Q $\left[\frac{g}{m^2}\right]$
Water-based varnish	123,5	122,3	0,01	**120,0**
	123	122,3	0,01	**70,0**

3.7 Internal Stresses of Water-Based Varnish

Internal stresses in water-based varnish were analyzed for two groups:

- water-based varnish with a dry film thickness of 60 [μm]
- water-based varnish with a dry film thickness of 30 [μm]

Experimental results for water-based varnish, dry film thickness 60 [μm].

The varnish is applied to the console by hand, by the method of airless dispersion, with the amount of application Q = 120 [g/m^2], relative humidity 51% and temperature 21 °C. The console deflection measurement time is at 0 intervals; 0.5; 1; 24; 48 [h]. The process of drying water-based varnish is natural without the action of other accelerators to reduce the drying time interval. In the analysis of water-based varnish, there were a total of five test stations. There are five consoles at each test station, which means that the total number of tested consoles is 25 pieces.

Table 6 shows the cumulative results of internal stresses at time intervals with a film thickness of 60 [μm].

Table 6. E Cumulative internal stresses in water-based varnish, dry film thickness 60 [μm]

No	Console materia	S-Console thickness [mm]	t-Film thickness [μm]	L-Length free canopies [mm]	Cantilever deflection time [h]	h-Cantilever deflection [mm]	y1-Poisson's ratio for plate	E1-Modul els. plates [MPa]	δ-Internal stresses [MPa]
Application method: MANUAL SPRAY					Drying method: NATURAL DRYING				
1	Aluminum	0,18	60	100	0	0,000	0,345	70000	0,000
2	Aluminum	0,18	60	100	0,5	0,571	0,345	70000	0,486
3	Aluminum	0,18	60	100	1	0,725	0,345	70000	0,998
4	Aluminum	0,18	60	100	24	0,954	0,345	70000	1,314
5	Aluminum	0,18	60	100	48	0,954	0,345	70000	1,314
Relative humidity 51%; Room temperature 21 °C; Varnish application amount is: 120 [g/m^2]									

In the diagram shown in Fig. 7, are the cumulative results of the internal stresses in a film of thickness t = 60 [μm]. There were 25 consoles for this internal stress analysis. The maximum mean stress for this coating thickness is σs max = 1.314 [MPa], and the vertical mean displacement of the cantilever is hs = 1.31 [mm].

Experimental results for aqueous varnish, dry film thickness 30 [μm].

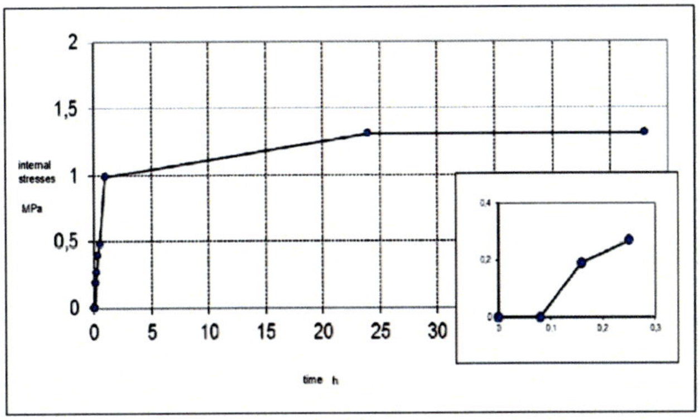

Fig. 7. Aggregate internal stresses in water-based varnish for film thickness t = 60 [μm]

The varnish is applied to the console by hand, by the method of airless dispersion, with the amount of application Q = 70 [g/m^2], relative humidity 51% and temperature 21 °C. The console deflection measurement time is at 0 intervals; 0.08; 0.16; 0.25; 0.33; 0.5; 1; 24; 48 [h]. In this analysis, the dry film thickness is t = 30 [μm] and this is the main difference compared to the previously presented results (Table 7).

Table 7. Internal stresses in water-based varnish, dry film thickness 30 [μm]

No	Console material	S-Console thickness [mm]	t-Film thickness [μm]	L-Length free canopies [mm]	Cantilever deflection time [h]	h-Cantilever deflection [mm]	y1-Poisson's ratio for plate	E1-Modul els. plates [MPa]	δ-Internal stresses [MPa]
Application method: MANUAL SPRAY					Drying method: NATURAL DRYING				
1	Aluminum	0,18	**30**	100	0	0,000	0,345	70000	**0,000**
2	Aluminum	0,18	**30**	100	0,08	0,000	0,345	70000	**0,000**
3	Aluminum	0,18	**30**	100	0,16	0,680	0,345	70000	**1,460**
4	Aluminum	0,18	**30**	100	0,25	0,820	0,345	70000	**1,770**
5	Aluminum	0,18	**30**	100	0,33	0,940	0,345	70000	**2,030**
6	Aluminum	0,18	**30**	100	0,5	1,080	0,345	70000	**2,330**
7	Aluminum	0,18	**30**	100	1	1,240	0,345	70000	**2,670**
8	Aluminum	0,18	**30**	100	24	1,560	0,345	70000	**3,360**
9	Aluminum	0,18	**30**	100	48	1,560	0,345	70000	**3,360**
Relative humidity 51%; Room temperature 21 °C; Varnish application amount is: 120 [g/m^2]									

In this test, the results of the kinetics of internal stresses of water-based varnish were specified. From Fig. 8, four periods can be observed. Period I in which no internal stresses occur at the beginning of drying of 8 min, in which the intense evaporation of volatile components and the unimpeded mobility of molecules.

Period II in which internal stresses occur and their intensive growth up to 1 [h] after the start of drying, period III or decrease in the intensive increase of internal stresses

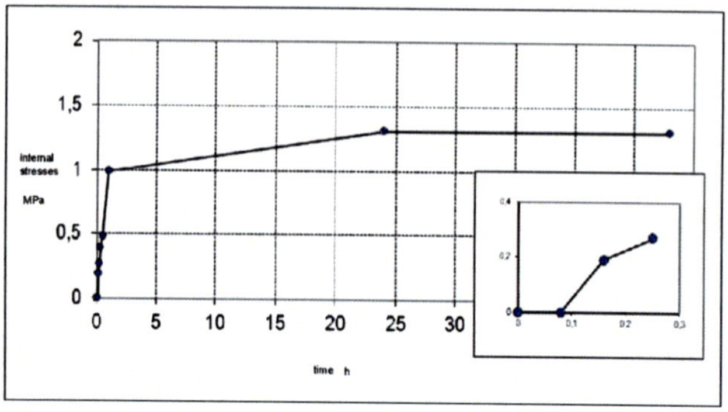

Fig. 8. Internal stresses in water-based varnish, film thickness 30 [μm]

until reaching a maximum of 3.36 [MPa] 24 [h] after the start of drying and the final period in to whom the stresses are kept at the achieved level. The maximum vertical displacement of water-based varnish with dry film thickness is t = 1.56 [mm].

References

1. Horman, I., Vukas, N., Azemović, E.: CNC tehnologije u finalnoj obradi drveta. Mašinski fakultet, Sarajevo (2014)
2. Jaić, M., Živanović, R.: Površinska obrada drveta, Beograd (2000)
3. Alić, O.: Površinska obrada drveta. Mašinski fakultet, Sarajevo (1997)
4. https://www.ivje.hr/index.php/tehnologija/nanosenje-mokrih-premaza
5. http://automatizacija1.etf.rs/poglavlja/Robotika%2010.htm

Civil Engineering

The Impact of Pandemic on the Survival of Construction Industry: A Case of COVID-19 in Bosnia and Herzegovina

Nerma Smajlović Orman(✉) and Ahmed El Sayed

Faculty of Natural Sciences and Engineering, International Burch University, Sarajevo, Bosnia and Herzegovina

nerma.smajlovic.orman@ibu.edu.ba

Abstract. COVID-19 has changed the world overnight and the question arises as to how much the course of the construction industry has changed since this industry largely depends on the dynamics of work, teamwork, but also the daily meetings of different teams. The objectives of this study are to learn about the impact of the pandemic on the construction industry in Bosnia and Herzegovina, but also to compare Bosnian-Herzegovinian trends with growing or declining trends in the construction industry and the workforce in Europe and other regions. Qualitative research methods have been employed to study this topic, meaning that after identifying the scope, problems, objectives of the study, and research goals, in-depth interviews have been conducted with leading experts from the construction industry in Bosnia and Herzegovina. Furthermore, experiences and statistics from other target countries are reviewed in this paper. The aforementioned interviews provided information on experts' experiences in the time of the pandemic as well as their analysis of the impact of COVID-19 on the construction industry in Bosnia of Hercegovina. The findings of this research will provide an objective picture of trends in the construction industry in Bosnia and Herzegovina during the caused by COVID-19 pandemic. Moreover, the paper includes expert recommendations on how to deal with such issues as pandemics in the future and how to become more resilient.

Keywords: Construction industry · COVID-19 · Pandemic crisis · Bosnia and Herzegovina

1 Introduction

On 31st December 2019, the WHO China Country Office announced cases of pneumonia of unknown etiology detected in Wuhan City, in the Hubei Province of China [1]. After just over a month, the WHO named the virus COVID-19, and a worldwide pandemic was declared on March 11th, 2020 by the Director of WHO [2].

New terms have emerged as new items in the lexical inventories of the entire world where the construction industry is not excluded [3]. These terms are COVID-19, social

N. Ademović et al. (Eds.): IAT 2021, LNNS 316, pp. 361–373, 2022.
https://doi.org/10.1007/978-3-030-90055-7_29

distancing, self-isolation, quarantine, furlough, palliatives, and many more [3]. The construction industry is far different from other industries, primarily because it typically requires the presence on-site of all the project members such as workers, engineers, supervisors but also others who are part of the project [4]. It is crucial to understand how this industry addresses unforeseen situations such as pandemics [4].

The first confirmed case of COVID-19 infection outside of China was detected in Singapore on January 23rd, 2020, and this information was announced by the Ministry of Health of Singapore (MOH-Sg), when 25 people were infected by COVID-19, including staff in the tourism sector, retail and hospitality industry, transport and security workers, and construction workers [5]. Many construction stakeholders have become familiar with the fact that construction project success considers factors such as time, cost, and quality but also through the effect of health and safety [6].

Due to the above, certain questions arose regarding the risks in the field of the construction industry. This paper examines two main issues, namely how to preserve the safety of workers and how the pandemic affected construction processes and industry in general. The objectives of this study are therefore to provide more information about the impact of the pandemic on the construction industry in Bosnia and Herzegovina and to compare Bosnian-Herzegovinian construction industry trends with growing or declining trends in the construction industry and the workforce in Europe and other regions. The first contribution of this study is the actual knowledge about the impact of the COVID-19 pandemic on the construction industry in Bosnia and Herzegovina, supplemented by information about potential delays in large infrastructure projects and the opinions of domestic experts on the current state of the construction industry in Bosnia and Herzegovina. The second contribution is the actual analysis of workforce statistics in the construction industry in Bosnia and Herzegovina and other countries in Europe and the region, supplemented by recommendations for recovering from the identified impact of the pandemic on the construction industry in Bosnia and Herzegovina.

2 Literature Review

The impact of the pandemic crisis on the survival of the construction industry was not in the focus of interest of authors in Bosnia and Herzegovina and therefore this research study relies on the review of the literature on this topic published in other countries or literature that speaks generally about the impact of COVID-19 on the construction industry. The construction industry is a sector of the economy which meaning is the same in both, developed but also developing countries [7]. This sector transforms different resources into constructed facilities through the process which includes planning, design, and construction [7].

One of the studies on the impacts of the global recession from the first decade of the new millennium on Southeast European Countries is mentioned that high trade and budget deficits were one of the main facts about the economies before the global recession of southeast European countries [8]. There are more facts such as modest inflow of foreign direct investments, and a macroeconomic imbalance which was reflected in higher domestic consumption concerning the value of production, as well as increasing foreign debt [8]. Economic growth in almost all southeast European countries related to the time before the global recession was based on increasing public consumption, exports, and personal consumption [8]. The very first sign of economic recession in these countries was a weighty reduction in exports as a direct consequence of a strong decline in foreign demand, and soon after, sectors as construction and industry experienced the same in form of the stagnation and reduction of activities [8].

The world is now facing a new uncertain situation caused by the coronavirus COVID-19, which has resulted in global blockades as a declining factor in economic activities [9]. The situation deteriorated significantly during the pandemic in the construction industry, which was visible through the reduced supply of construction materials [4]. Following the spread of the virus, many countries started implementing severe measures to reduce the movement of people, thus significantly obstructing the construction processes because those require on-site work and the presence of every project member who is tasked with checking and monitoring all the work activities [4]. After investigation and assessing the consequences of pandemic COVID-19 and it has been proved that the suspension of projects, labor impact and job loss, and transgression of time and cost were statistically the most impacting factors in the time of pandemics [4].

During the lockdown it has been difficult to manage projects since working from home was one of the measures implemented for the staff members what led to delays in project activities, especially because many staff members were not able to be present physically on the site and work on their necessary activities [10]. Moreover, this was challenging for the managers to manage their teams and in this manner, project teams have used technological tools like video meetings which are proven as the most effective way of communication in pandemics [10]. All construction design offices and ongoing construction projects were tendered to abruptly suspend construction work on-site due to a lockdown caused by the pandemic, and design engineers were relocated to do their work from home [11].

One of the studies which speak about the construction industry in Nigeria shows that "transportation problem (for both materials and workers)", "project abandonment", "delay in construction activities", "high cost of construction materials", "reduction in working hours per day", "lack of funding" and "shortage of workforce" were the main disruptions in the construction industry caused by COVID-19 pandemic [12]. Furthermore, the pandemic revealed safety management challenges and some of them were also present before the pandemics and they are related to the lack of compliance by workers on site such as ignorance of workers about risk factors proven by their incorrect usage of safety equipment on construction site, theft of safety materials but also a very poor supply of this previously mentioned safety equipment by contractors [13]. A new health and safety management challenges are the sanitation of construction materials and difficulties with sharing tools and equipment on-site and these challenges have appeared with the social distancing rules, offsite behavior and public transportation of workers, during material delivery and on-site behavior of workers who were superstition concerning the COVID-19 safety hazard [13].

In the first few months of the pandemic were orders to stay home enforced across several US states, to stop the spread of the virus [14]. The construction was considered crucial in most states, while several states have found that construction work is at least partially irrelevant [14]. Some of the participants in the study mentioned that there was a smaller impact on their projects and operations were continued like normal even new safety measures were adopted, while others mention that construction work at their workplaces had completely stopped. The study found that the construction industry experienced several adverse effects such as delayed material delivery, material shortages, delayed permits, lower productivity rates, cash flow challenges, project suspension, price escalation, and potential conflicts and disputes [14].

Although there are several research studies based on the impact of COVID-19 on the construction industry, data are scarce for much of the world, particularly for the region of Southeast Europe. According to the literature listed above, three main conclusions are drawn. Firstly, in almost all parts of the world, with special reference to the US, UK, Jordan, South Africa, and Nigeria, a slowdown in the construction industry was observed at the beginning of the pandemic, but it did not have a significant impact on the industry's survival. Secondly, delays or suspension occurred due to the material delays when is a notable material shortage at the same time with price escalation, but also potential conflicts caused by project suspensions or decreased financial flow. Lastly, construction sites continued to work, but the designers worked from home.

3 Methodology

The research is based on qualitative research methodology. The qualitative methodology is based on running in-depth interviews with leading experts from the construction industry in Bosnia and Herzegovina, as well as on reviewing experiences and case studies from other targeted countries (neighboring countries, EU countries). In-depth interviews are one of the most frequently used tools for qualitative research studies, primarily because of their open-ended character and discovery-oriented nature, which allow the researcher to find out and explore the deep feelings and attitudes of interviewees about the subject in question [15]. Questions in in-depth interviews were formulated in the way to be open-ended and easy to understand [16].

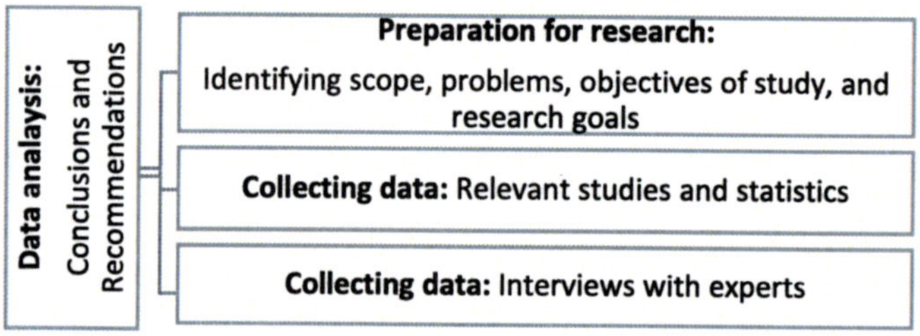

Fig. 1. Methodology flowchart made by authors

After gathering information on the participants' backgrounds such as professional role and experience, the responses to the following questions were targeted for solicitation. The first research question pertains to the analysis of the current situation in the construction industry since the beginning of the pandemic, including the global situation, which could depend on the workplace or project scope. The next research question seeks to investigate the impact of COVID-19 on construction companies – smaller firms/subcontractors or start-ups as well as companies with high debt levels and low cash reserves – as an important part of the economy and the stronghold of the real sector. The third research question investigates the impact of lockdown and other measurements imposed by the COVID-19 pandemic on already signed construction contracts – whether they have been revised or their implementation continued according to the initial plans – at the same time attempting to identify the potential delays and their causes.

The main statement is that the construction industry in Bosnia and Herzegovina was significantly affected by the COVID-19 pandemic.

With regard to the research questions, the authors conducted in-depth interviews with six experts from Bosnia and Herzegovina. The targeted population was generally experts with 15 to 40 years of experience in the construction industry in Bosnia and Herzegovina and with some international experience. Interviews were conducted with civil and traffic engineers with or without academic references, with engineers who work in managerial positions, and engineers with experience in public or private companies during the economic recession.

The diversity of qualitative research is reflected in the variety of approaches to criteria for evaluation [17]. In addition, different criteria such as research ethics, responsibility, and consequences of research must be considered [18]. Given that the authors selected participants who are recognized experts in their fields and work in positions that require supervision of all segments of the construction sector, the evaluation criteria are set so that credibility is measured by the frequency of the same responses.

4 Results and Discussion

The interviews were conducted in the summer of 2020, four months after the pandemic was pronounced. Then, it was too early to take any data as an official depiction of the real impact of COVID-19 on the construction industry in Bosnia and Herzegovina. Two interviews were conducted in February 2021, and at this point, it was also possible to consider the official statistical data when describing the subject of the study.

Interview Results

The first interview was conducted at the beginning of summer 2020 with the expert who is also an associate professor from Bosnia and Herzegovina. His biography is quite respectful, with 26 years of work experience in the academy and construction industry itself. It is important to mention that he participated in the development and implementation of projects whose worth amounts to more than 100 million dollars, including projects of infrastructure, roads, bridges, water supply systems, drinking water treatment plants, wastewater systems, hospitals, schools, power facilities, etc. His greatest professional achievement is the introduction of EURO codes in the standardization of Bosnia and Herzegovina in cooperation with other leading experts. According to him, when the economy falls into crisis, the construction is first to suffer. Furthermore, he said that when the economy emerges from the crisis, it is again felt first in construction. He thinks, in the specific case of COVID-19, in which taxes are reduced, the amount of revenue is reduced, the quantity of tenders is reduced, and then the amount of new product produced by the construction operative is reduced. He also said that when there is a revival at one point, the number of tenders, investors, and money increases automatically, which can be seen in the field with increased construction investments. According to this expert, this has been evident in the last global economic crisis before more than a decade, at least from his experience, and this crisis will confirm that. At this particular moment, according to him, there is no significant slowdown in construction works in Bosnia and Herzegovina caused by COVID-19, and there is no suspension of contracted works. However, he thinks that in the coming period there may be some slowdowns in the dynamics of work, due to the procurement of materials, availability of warehouses, shops, import of materials into Bosnia and Herzegovina, and the like. He concluded that this will certainly not be significant when it comes to larger infrastructure projects currently underway in Bosnia and Herzegovina. It is important to mention that this expert also referred to the increased number of skilled labor that worked in the construction industry in the European Union, and returned to Bosnia and Herzegovina with the pandemic.

The second interview was conducted with a civil engineer with 23 years of experience in the construction industry and five years of international experience in the Middle East region in the turbulent period 2008–2010, at the time of a global economic crisis. He shared his perspective as the head of a consultant company in the field of the construction industry. According to him, the pandemic did not significantly affect the scope of activities, because started projects continued during the declared pandemic and emergency. He not only added that the construction industry reacted very well to the new conditions imposed by the COVID-19 pandemic, but he also agreed with the aforementioned opinion regarding the continued work on big infrastructural project and delays with the imports of goods, materials, and raw materials, equipment, and highly sophisticated goods, all due to the blockade of borders and the occasional complete lock-down of certain countries and regions. He mentioned that there has been a significant increase in the number of infrastructure projects in recent years, early 2020 included; that most smaller firms have survived for that reason only, and that the construction sector as a whole has maintained a high level of activity. He thinks that state-owned construction companies have not been at their pre-war level of activity for a long time and that the focus of construction activities, as well as the majority of the workforce, has shifted to private companies. Therefore, he emphasizes, only infrastructural works in such times can offer certain security in providing work activities and income at the state level, and that regardless of the current problems, which have been going on since March 2020, the reaction of contracting authorities and construction companies has been quite good. He concluded that the state must help with additional actions in overcoming and mitigating the consequences of the pandemic.

The authors also spoke with one more associate professor, who has over 40 years of work experience, starting as a chief engineer at facilities in the then giant company of the former Yugoslavia – a company that is currently running out of maintenance. His career extends from the construction site, through the academy, and he has been permanently employed since 2003 in the Labor and Employment Agency of Bosnia and Herzegovina, at one of the leading positions. According to him, the current situation in the high-rise construction industry is reflected in the reduced use of production capacity caused by reduced demand for construction materials. He mentions also that the situation in low-rise construction is much better because the construction of highway sections on Corridor Vc is running according to the planned dynamics and there is increased delivery of products from the construction industry. He believes that only the first month of the appearance of the coronavirus had a greater impact on the delays in the work process until the reorganization of work and the procurement of protective equipment have been completed. He added that the companies mostly managed to adapt to the new working conditions. He said that he has no information that there now exists a construction company with state capital and that "Hidrogradnja" is in the process of bankruptcy and liquidation. This participant is completely sure that the existing contracts for design and construction will be revised to the mutual satisfaction of investors and contractors, but that it is to be expected that the volume of business in the complete construction industry will decrease because purchasing power decreases drastically from month to month.

The interview has also been conducted with a group of experts from one of the leading domestic companies, with more than 25 years of experience in structural design and consulting on structural design projects in Bosnia and Herzegovina and abroad. They said that it is difficult to answer the general question about the construction industry because the response to the COVID-19 pandemic differs depending on the status of the company and the type of work they are doing, as well as several other parameters. However, they believe that small companies that are mainly engaged in craftwork or rent labor to other companies did not pay attention to the situation with the appearance of the COVID-19 virus because those companies are solely dependent on investors; these companies did all the work wherever the investor required it. They added that the designers mostly did the already contracted work from home while the owners and managers of small companies in the struggle for survival simply worked on all construction sites where it was allowed, following recommendations to employees to use protective equipment. They also believe that employees only formally used the equipment without any major objections, aware that they will only get the payment if they perform the assigned tasks. In this review of state-owned construction companies, they point out that such companies almost do not exist if we exclude the construction operations of utility companies such as CPUC Rad, various cleanings in smaller places, construction operations in forestry companies, and the like, while in these state companies construction capacities are used as secondary services and were generally active only to the extent necessary for the primary activity. According to them, the already signed agreements with companies that have credit funds as a source of financing will probably continue as soon as the conditions are met, i.e. the state will not stop those investments for now, while in cases where the investments come from private capital there will probably be a serious slowdown, freezing of the situation and, depending on the degree of realization of the investment – if the investment is not in an advanced stage – a complete suspension of works and withdrawal can be expected. The situation is unpredictable, they said. Their opinion is that the higher degree of participation of private concerning state investments, the more investments will continue and the more construction operations will be engaged. They added that this will produce less demand and higher supply, which results in lower prices and lower profits of the already endangered construction industry, and in such relations, the profit for owners suffers first, and at the same time depreciation coverage and reduced investment in equipment and technology. They concluded that it is a matter of reducing the conjuncture of construction as an economic branch.

After six months two more interviews were conducted, precisely in February 2021. The first one was with a civil engineer, with experience in the construction sector of over 25 years. He is currently the team leader of the project whose objective is the development of sustainable water and sanitation services (wastewater collection and treatment) in the Federation of BiH, through the construction of necessary facilities, development of adequate infrastructure of the sector, and strengthening of municipal operators – public utility companies (PU). At the same time, he is the owner of a consulting company which deals with the design and consulting work in the field of construction in Bosnia and Herzegovina and abroad. From his experience, the only stagnation was evident in the period from mid-March to the end of May 2020, when a state of disaster due to a pandemic was declared, in such a way that the works were suspended until the procedures for the continuation of works were defined. According to him, the pandemic mostly affected the beginning of the realization of activities on new components, which was the case with the project on which he was engaged. The beginning of the mentioned project was extended for the spring of 2021, and the reason is that these are foreign contractors. He stated that the implementation of these contracts is based on the Yellow FIDIC contract terms so that in case of delays the investor faces additional claims by the Contractor, which potentially poses a risk of increasing the cost of implementation of these contracts between 10 and 30%. When it comes to local companies, he believes that they have adapted to the working conditions in the pandemic and he mentioned that the only requests of smaller construction companies were related to the extension of the deadlines for the completion of works. He believes that the unsystematic solution of the issue of vaccination of the population can lead to problems in the realization of activities that depend on foreign companies in the future, and that there are possible problems with the import of necessary equipment related to the realization of given contracts.

The last interview was done with a traffic engineer who has 16 years of experience in design and supervision of road construction projects, with special reference to the current position as a chief supervising route engineer on one section on the highway in Bosnia and Herzegovina – currently considered the most significant infrastructure project in this country. His review on the situation in the construction industry caused by COVID-19 started with the statement that the biggest problem before the pandemic was the lack of skilled labor because people were going abroad in search of better and safer jobs. According to him, this trend slowed down a bit during the pandemic. He said that the biggest problems that arose during the pandemic were ensuring the continuous payment on performed works, procurement and delivery of materials, and maintaining the existing number of employees. From his experience, it was necessary to perform work in small groups to comply with epidemiological measures, which caused a decrease in the intensity of work as well as an attempt to harmonize with investors the problem of increasing market prices of materials and their transport to the construction site. He also mentioned that the smaller companies faced a much bigger challenge – the challenge being that due to epidemiological measures they were unable to meet their contractual obligations and he noticed that some companies withdrew their employees from the construction sites, thus reducing costs and trying to keep them healthy. A significant part of the interview with him was about signed but unrealized contracts, with which the interlocutor is familiar. According to him, as far as designing is concerned, the matter is somewhat more favorable because a lot of projects are being realized. As an example, the respondent stated that in 2020 was prepared technical documentation for the next 30 km of highway, whose construction is scheduled to begin in 2021. He mentioned that contracts for works depend more on the applicability of technical solutions and the willingness of the Contractor to implement them than on the method of payment. He added that these projects are financed through the international and European Banks and their implementation is unquestionable. Respondent concluded that a big problem in the implementation of projects today is incomplete technical documentation, insufficient training of contractors, and lack of coordination between institutions responsible for the implementation of certain projects.

Statistical Data

According to the Agency for Statistics of Bosnia and Herzegovina (BHAS), in January 2020 the number of persons employed in B&H was 833.909. This is increased by 0.4% compared to December 2019 [19]. Following the statistics on BHAS, the number of persons employed in BiH amounted to 813.942 in December 2020 [20] (Table 1).

Table 1. The number of registered employed persons in the construction industry in B&H. Source: Agency for Statistics of Bosnia and Herzegovina [21]

Timeline	The number of registered employed persons in B&H	The number of registered employed persons in the construction industry in B&H
2019	820.974	39.582
01/2020	833.909	39.582
02/2020	832.200	39.887
03/2020	830.421	40.369
04/2020	808.894	40.152
05/2020	808.868	40.589
06/2020	804.028	40.515
07/2020	803.326	40.826
08/2020	799.207	40.909
09/2020	806.350	40.976
10/2020	807.985	41.011
11/2020	812.691	41.403
12/2020	813.942	40.446

Following the statistical data given in the table below, taken from the Agency for Statistics in Bosnia and Herzegovina, it is evident that the number of employees varied concerning the course of the pandemic in Bosnia and Herzegovina and the world, but the number of employees in the construction industry increased from the beginning to the end of 2020.

Using the official statistics of Germany, as one of the European leaders in the field of construction industry expansion, and neighboring countries, numerical data about the employment rate during 2020 were extracted and presented in Table 2. The comparison was made of the number of employees in the construction sector. It should be noted that these are countries with smaller to large differences in population, as compared to Bosnia and Herzegovina, but results are shown in the employment rate in the construction industry for every country.

Table 2. The number of registered employed persons in the construction industry in B&H compared with European and neighboring countries: Croatia, Serbia, and Germany [22–25]

	Employment in construction industry 2019	Employment in construction industry 2020	Employment rate
Bosnia and Herzegovina	*(12/2019)* 39 582	*(12/2020)* 40 446	+2,2%
Croatia	*(12/2020)* 93 679	*(12/2020)* 101 174	+8%
Serbia	*(qr.4/2019)* 110 461	*(qr.4/2020)* 118 461	+7,2%
Germany	2 552 000	2 570 000	+0,71%

Discussion

The impact of the pandemic caused by coronavirus COVID-19 on the construction industry in the world and Bosnia and Herzegovina since the beginning of the pandemic period has been very similar.

1) 6/6 respondents answered that COVID-19 did not significantly affect the construction sector in the previous period of one year.
2) 6/6 respondents answered that the companies (smaller and larger) provided a good and fast response to COVID-19 and epidemiological measures which came with the pandemic.

The first two conclusions of respondents' answers are supported by the first conclusion from the literature review.

3) 5/6 respondents commented and agreed that the occurred delays were due to the material delays or shortage according to the border blockades or occasional lock-downs and the like.

These reasons are also mentioned in the second conclusion from the literature review. Within the literature review, it was added that price escalation and potential conflicts caused by project suspensions or decreased financial flow are other reasons.

4) 6/6 respondents answered that the slowdown occurred at the beginning of the pandemic period, after which the contracts were revised and work in this industry continued.

Their responses supported the third conclusion from the literature review section of this paper which refers that construction sites continued to work, but the designers worked from home.

5) 2/6 respondents commented and agreed that the labor force was prevented from going abroad during the pandemic, so the growth of skilled labor is evident.

This fact was confirmed by statistics, where it can be seen that there was an increase in the employment rate of the labor force in the construction industry even in the months when there was an evident decline in employment rate generally in Bosnia and Herzegovina.

The main statement from the beginning of this research that the construction industry in Bosnia and Herzegovina was significantly affected by the COVID-19 pandemic has not been supported.

5 Conclusion

Considering that the topic is currently of global interest and that the governments of all world countries have had different responses and results during pandemics in the construction industry, this research paper aimed to answer how the coronavirus pandemic COVID-19 affected the construction sector in Bosnia and Herzegovina.

It is witnessed that various types of activities were banned during the pandemic in Bosnia and Herzegovina, but also in the whole world, or they were restricted due to certain reasons related to the restrictions of the crisis headquarters. There were prohibitions or restrictions on the work of all activities wherever, among other things, there was a possibility of contact of infected persons with healthy people, whether among employees themselves or employees and clients. Construction is a profession that must function on the site. However, construction works continued actively and were not affected by the special restrictions of the crisis staff, although construction workers work together, inside and out, using the same means of labor, and thus there was a high probability of clusters coming from construction sites. From the research findings, it can be concluded that COVID-19 has not significantly affected the construction industry in Bosnia and Herzegovina, and therefore the construction industry workers in offices but also on-site worked continuously almost all the time. It did not go unnoticed that the COVID-19 pandemic has resulted in the slowdown of works at the beginning of the pandemic since there were delays with material supply mostly, but economic losses caused by the revision of the signed contracts have not been emphasized by interview participants. They emphasized the increase in the workforce that was prevented from leaving Bosnia and Herzegovina during the pandemic.

The research was limited with the lack of information, and the negligible focus on the impacts of the COVID-19 pandemic on the construction industry in Bosnia and Herzegovina, which the authors faced at the beginning of this research.

It is recommended by authors for future researchers on the same or similar topics is to conduct a survey on the labor force leaving Bosnia and Herzegovina, and in particular to look at the level of their education or qualifications and age. The authors also recommending improving the results of this with the samples of smaller communities or more similar municipalities, because the economy in the country differs significantly depending on the region and the size of municipalities/cities. One of the limitations was a lack of public or official statistics on the number of infected construction workers and employees in the construction sector during the pandemic, which should be improved. The last recommendation is to research how slow Bosnia and Herzegovina is when it comes to following trends from the European Union, standardization in the construction industry, as well as the protection of workers' rights and their safety at work. The question is how this trend in Bosnia and Herzegovina managed to respond to the effects of the pandemic.

References

1. World Health Organization: Pneumonia of unknown cause – China (2020). https://www.who.int/csr/don/05-january-2020-pneumonia-of-unkown-cause-china/en/. Accessed 5 Feb 2021
2. Tedros, A.: General's opening remarks at the media briefing on COVID-19, 11 March 2020. https://www.who.int/director-general/speeches/detail/who-director-general-s-opening-remarks-at-the-media-briefing-on-covid-19---11-march-2020. Accessed 4 Feb 2021
3. Ogunnusi, M., Hamma-Adama, M., Salman, H., Kouider, T.: COVID-19 pandemic: the effects and prospects in the construction industry. Int. J. Real Estate Stud. **14**(Special Issue 2), 120–128 (2020)
4. Gamil, Y., Alhagar, A.: The impact of pandemic crisis on the survival of construction industry: a case of COVID-19. Mediterr. J. Soc. Sci. **11**(4), 122 (2020)
5. Koh, D.: Occupational risks for COVID-19 infection. Occup. Med. **70**(1), 3–5 (2020)
6. Smallwood, J., Haupt, T.: Impact of the South African construction regulations on construction health and safety: architects' perceptions. J. Eng. Des. Technol. **5**(1), 23–34 (2007)
7. Isa, R.B., Jimoh, R.A., Achuenu, E.: An overview of the contribution of construction sector to sustainable development in Nigeria. Net J. Bus. Manag. **1**(1), 1–6 (2013)
8. Buturac, G., Teodorović, I.: The impacts of the global recession on Southeast European Countries. East. Eur. Econ. **50**(1), 78–97 (2012)
9. Helm, D.: The environmental impacts of the coronavirus. Environ. Resour. Econ. **76**(1), 21–38 (2020). https://doi.org/10.1007/s10640-020-00426-z
10. Jallow, H., Renukappa, S., Suresh, S.: The impact of COVID-19 outbreak on United Kingdom infrastructure sector, 25 June 2020. https://doi.org/10.1108/SASBE-05-2020-0068. Accessed 20 Dec 2020
11. Bsisu, K.A.: The impact of COVID-19 pandemic on Jordanian civil engineers and construction industry. Int. J. Eng. Res. Technol. **13**(5), 828–830 (2020)
12. Osuizugbo, I.C.: Disruptions and responses within Nigeria construction industry amid COVID-19 threat. Covenant J. Res. Built Environ. (CJRBE) **8**(2), 37–48 (2020)
13. Amoah, C., Simpeh, F.: Implementation challenges of COVID-19 safety measures at construction sites in South Africa. J. Facil. Manag. **19**(1), 111–128 (2020)
14. Alsharef, A., Banerjee, S., Uddin, S.M.J., Albert, A., Jaselskis, E.: Early impacts of the COVID-19 pandemic on the United States construction industry. Int. J. Environ. Res. Public Health **18**(4), 1559 (2021)

15. Guion, L.A., Diehl, D.C., McDonald, D. (ed.): Conducting an In-Depth Interview, pp. 1–3 (2001)
16. Legard, R., Keegan, J., Ward, K.: In-depth interviews. In: Qualitative Research Practice: A Guide for Social Science Students and Researchers, vol. 6, no. 1, pp. 138–169. SAGE (2003)
17. Bitsch, V.: Qualitative research: a grounded theory example and evaluation criteria. J. Agribus. **23**(1), 75–91 (2005)
18. Lincoln, Y.S.: Emerging criteria for quality in qualitative and interpretive research. Qual. Inq. **1**(3), 275–289 (1995)
19. Agency for Statistics of Bosnia and Herzegovina: Demography and Social Statistics: Persons in pain employment by activity, January 2020, 20 March 2020. http://www.bhas.gov.ba/data/Publikacije/Saopstenja/2020/LAB_02_2020_01_0_BS.pdf. Accessed 4 Feb 2021
20. Agency for Statistics of Bosnia and Herzegovina: Demography and Social Statistics: Persons in paid employment by activity, December 2020, 22 February 2021. http://www.bhas.gov.ba/data/Publikacije/Saopstenja/2021/LAB_02_2020_11_0_BS.pdf. Accessed 22 Feb 2020
21. Agency for Statistics in Bosnia and Herzegovina (2020). https://bhas.gov.ba/Calendar/Category/17. Accessed 21 Feb 2021
22. Croatian Bureau of Statistics, 20 January 2020. https://www.dzs.hr/Hrv_Eng/publication/2019/09-02-01_12_2019.htm. Accessed 22 Feb 2021
23. Croatian Bureau of Statistics, 19 January 2021. https://www.dzs.hr/Hrv_Eng/publication/2020/09-02-01_12_2020.htm. Accessed 22 Feb 2021
24. Deutchland Statistisches Bundesamt, 18 February 2021. https://www.destatis.de/EN/Themes/Labour/Labour-Market/Employment/Tables/persons-employment-sectors-economic. Accessed 22 Feb 2021
25. Republički zavod za statistiku, Republika Srbija (2021). https://www.stat.gov.rs/sr-latn/oblasti/trziste-rada/registrovana-zaposlenost/. Accessed 11 May 2021

Environmental Impact of Different Types of Intersections in Urban Areas

Ammar Šarić(✉), Sanjin Albinović, Anisa Krnjić, Mirza Pozder, Suada Sulejmanović, and Žanesa Ljevo

Faculty of Civil Engineering, Department of Roads and Transportation, University of Sarajevo, Sarajevo, Bosnia and Herzegovina

Abstract. Cities in Bosnia and Herzegovina face a high concentration of small particles in the air that are harmful to human health. One of the key factors influencing environmental pollution is pollution caused by motor vehicle emissions. As the traffic load increases, the emission of harmful gases also increases, directly affecting the quality of life. This paper studies how different types of intersections in urban areas can affect the level of air pollution. Two existing signalized intersections in Sarajevo and variants of the planned roundabout were analyzed. All solutions are modeled in the PTV Vissim software package. A comparative analysis was performed for a signalized intersection and a roundabout at the same location in terms of environmental impact and basic emission factors. Different results were obtained at both locations. At the first location, the roundabout proved to be a partially better solution, while at the second location, the classic signalized intersection is a better solution regarding environmental pollution. Obtained results suggest that the level of service of the intersection significantly affects the emission of pollutants.

Keywords: Traffic emission · Air quality · Signalized intersection · Roundabout · Traffic simulation

1 Introduction

Emissions of toxic components from motor vehicle exhaust are one of the leading air pollutants in urban areas. Emissions of carbon monoxide (CO), hydrocarbons (HC), nitrogen oxides (NOx), and carbon dioxide (CO_2) have the most significant impact on air pollution in urban centers. In these areas, pollutant emissions are higher due to higher traffic intensity. The amount of emissions is directly affected by the number of motor vehicles, a technology used by motor vehicles, geometric elements of the road, driver behavior, type of intersection, etc. As the number of motor vehicles continues to grow, so does the exhausted emissions, which is a constant threat to air quality and thus human health.

N. Ademović et al. (Eds.): IAT 2021, LNNS 316, pp. 374–386, 2022.
https://doi.org/10.1007/978-3-030-90055-7_30

At classic signalized and non-signalized intersections, vehicles generally slow down or stop, leading to traffic interruptions. When stopping and then restarting the vehicle, there is higher fuel consumption and higher exhaust emissions. The general assumption is that roundabouts reduce the number of stops due to better traffic flow behavior, which means lower fuel consumption and lower exhaust emissions.

This paper aims to determine the amount of gas emissions at the existing signalized intersection and compare them with the emissions at the planned one or two-lane roundabout at the same location. Exhaust gas analysis can be done in several ways, at the macro (network) or micro (specific locations) level. In this paper, the approach using traffic microsimulation by PTV Vissim 11 is chosen.

In Europe, many air pollutants have decreased significantly over the last decades, resulting in improved air quality throughout the region. However, concentrations of air pollutants are still too high, and air quality problems persist. A significant proportion of Europe's population lives in cities where air quality standards are exceeded: ozone, nitrogen dioxide, and particulate matter (PM) pollution pose serious health risks [1]. Most of this pollution comes as a result of traffic. The following figure represents a diagram showing trends in air pollutant emissions from transport.

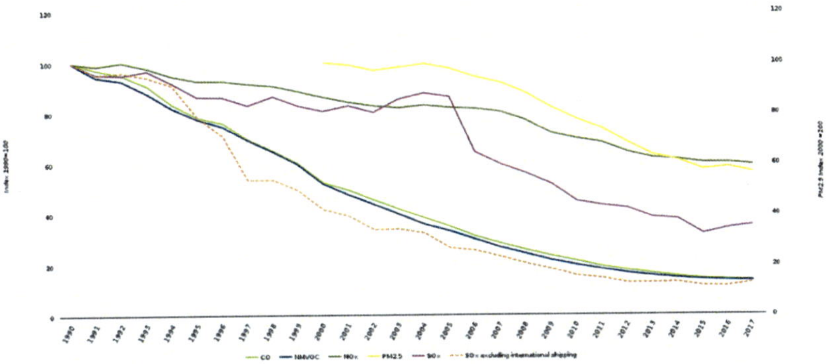

Fig. 1. Trends in air pollutant emissions from transport [1]

Based on Fig. 1, it can be seen that the emissions of all pollutants have been reduced despite the increase in activity within the transport sector. Between 1990 and 2017, emissions of nitrogen oxides (NOx) from the transport were reduced by 40%, sulfur oxides (SOx) were reduced by 66%, and carbon monoxide (CO) and non-ethane volatile organic compounds (NMVOC) were reduced by 87%. Between 2000 and 2017, emissions of PM2.5 decreased by 44% [1].

In Sarajevo Canton, air quality is monitored by the Federal Hydrometeorological Institute - FHMZ (Bjelave and Ivan Sedlo stations) and the Sarajevo Canton Public Health Institute (Otoka, Ilijaš, Vijećnica and Ilijaš stations). The following diagrams show the measurements of sulfur dioxide (Fig. 2), nitric oxide (Fig. 3), and suspended particles PM10 and PM2.5 (Fig. 4). The results are presented in percentile values because the number of allowed overruns is expressed concerning the absolute number of overruns from the obtained data set [2].

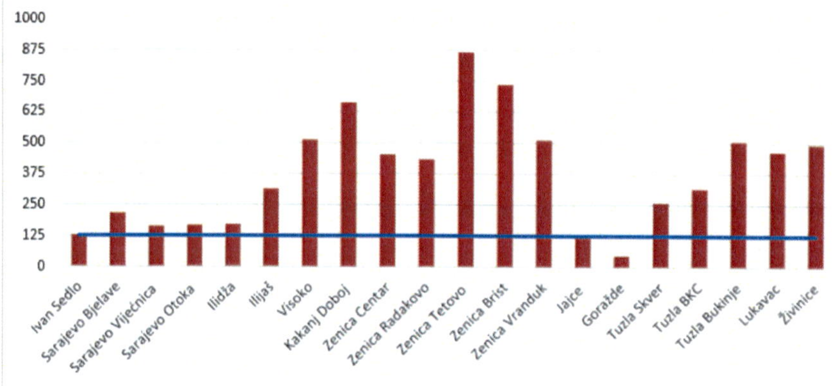

Fig. 2. Value of 99.73% of daily values of SO2 concentrations in 2019. Exceeding the value of 125 $\mu g/m^3$ corresponds to exceeding the number of allowed days in the year with a concentration above the limit value [2]

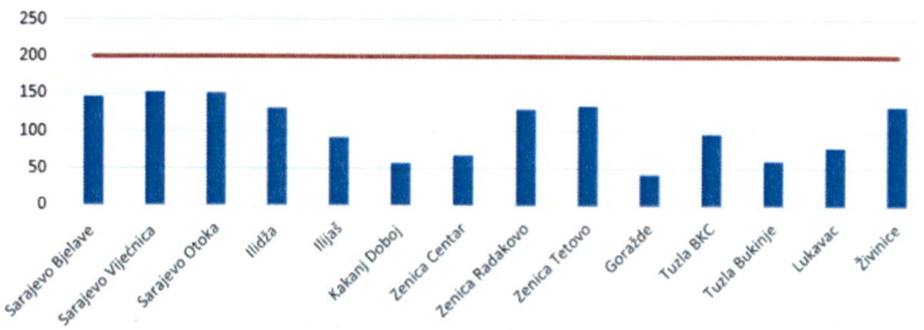

Fig. 3. Value of 99.79% of daily values of NO2 concentrations in 2019. Exceeding the value of 200 $\mu g/m^3$ corresponds to exceeding the number of allowed days in the year with a concentration above the limit value [2]

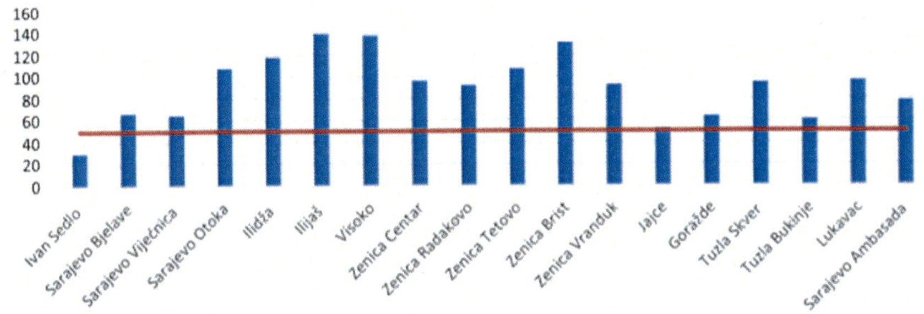

Fig. 4. Value of 90.4% of daily average values of the concentrations of suspended particles PM10 and PM2.5 in 2019. Exceeding the value of 50 μg/m^3 corresponds to exceeding the number of allowed days in the year with a concentration above the limit value [2]

Based on the previous diagrams, it can be concluded that exceptionally high concentrations of suspended particles dangerously impair the air quality in Sarajevo to the extent that it can seriously impair human health. This applies to both annual averages and the number of permitted exceedances of hourly or daily concentrations. High concentrations of sulfur dioxide were recorded in the western parts of the city and suburban settlements where the number of days was exceeded with the concentration of sulfur dioxide above the prescribed limit value, while the average annual concentration was within the prescribed values. High concentrations of nitrogen dioxide occur at measuring points in the most densely populated and busiest parts of the city (Otoka and Ilidža stations) [2].

Apostolos and Fotini [3] investigated the impact of transforming a small urban two-lane roundabout into a turbo roundabout on air pollutant emissions. Based on the traffic simulation conducted in Vissim and Enviver Pro, the emission of toxic exhaust gases (CO_2 (carbon dioxide), NOx (nitrogen oxides), and PM10 (particles)) was calculated. The results showed an average reduction of 32% on CO_2 emissions, 36% on NOx emissions, and 31% on PM10 emissions for the turbo roundabout compared to the classic roundabout.

A similar study was conducted by Jaworski, Lejda, and Mądziel [4] in Rzeszów (eastern Poland). The emission results of the selected exhaust gas components (NOx and PM10 particles) were obtained based on the VERSIT + Enviver emission model calculation. A traffic flow microsimulation was performed in the Vissim software using the field obtained traffic. A total of four scenarios were compared, representing two times of the day with different traffic volumes. Verification of emission results obtained with Enviver software was done in the COPERT 5 model.

Two vehicle classes were used, namely cars (which make up 80% of all vehicles) and trucks (20%). The car class consisted of 40% of vehicles with gasoline engines, 47% of vehicles with diesel engines, and 13% of LPG vehicles. The average age of car vehicles was set at eight years. For the class of trucks and buses, it was assumed that

82% of vehicles had diesel engines and 18% buses with CNG (compressed natural gas) drive. The average vehicle age for this vehicle class was set at nine years.

According to the study results, it can be concluded that the emission of exhaust gases changes significantly during the day. For selected exhaust components, i.e. NOx and PM10, the total emission is 3–4 times higher for peak hours compared to other parts of the day. During the morning peak hours, the NOx emission factor increases by 35–45%, and the PM10 emission factor by 32–43%. Considering the comparison between the existing two-lane roundabout and the modeled turbo rotor's alternative possibilities, an average speed increase of 38% can be noticed in the average morning congestion. This means a reduction in NOx emissions by 31% and PM10 by 26% in favor of turbo rotors. In the evening, when the traffic load is less, an increase in the average speed by 8% can be seen for the turbo rotor and a decrease in NOx emissions by 21% and PM10 by 13% compared to the two-lane roundabout.

Meneguzzer et al. [5] investigated the benefits of reducing emissions by replacing a signalized intersection with a one-lane roundabout in Vicenza, Italy. CO_2, CO, and NOx emissions were monitored using PEMS technology installed in a mobile passenger vehicle. The general conclusion is that CO_2 emissions decreased after the construction of the roundabout, while at the same time, NOx emissions increased. Regarding CO emissions, no statistically significant differences were obtained. It was concluded that different driving conditions have a significant impact on pollutant emissions in both variants. The obtained results indicate that the largest emissions are in the phase of vehicle acceleration.

2 Traffic Flow Microsimulation

This research aims to determine the impact of different types of intersections on the environment and basic emission factors using a traffic microsimulation tool. Traffic modeling was performed in the microsimulation software PTV Vissim 11. The research was done at two locations in Sarajevo, Bosnia and Herzegovina, on existing signalized intersections and planned roundabouts. At location 1, during the research, a two-lane roundabout was built instead of the original intersection.

The values of the following variables were analyzed in the research:

- number of vehicle stops to the number of non-stop vehicles,
- quantities of basic emission factors (CO, NOx, VOC),
- fuel consumption.

The layout of the original signalized intersection at Location 1 is given in Fig. 5, while Fig. 6 shows a new two-lane roundabout at the same location.

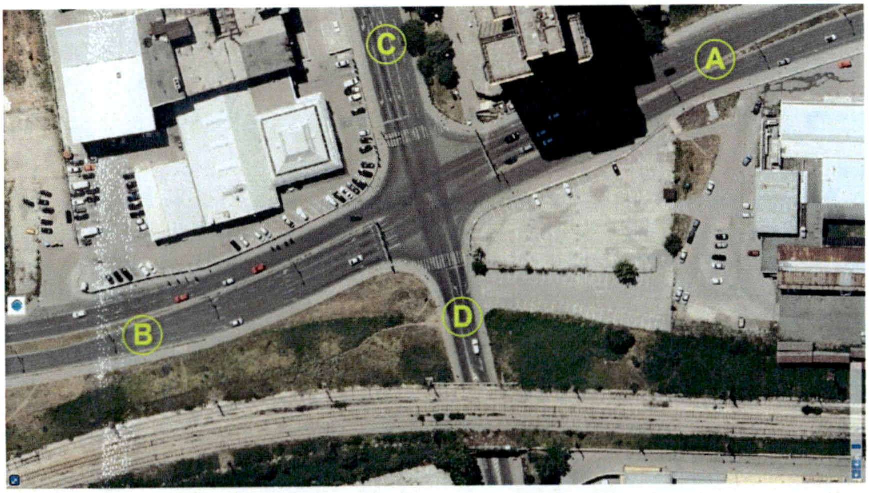

Fig. 5. Geometry of signalized intersection at Location 1

Fig. 6. Geometry of two-lane roundabout at Location 1

The traffic load at Location 1 measured during the period of maximum hourly traffic (afternoon peak hour) is shown in the following Table 1:

Table 1. Traffic load at Location 1

Entry	Left			Straight			Right		
	Car	HV	$\sum$	Car	HV	$\sum$	Car	HV	$\sum$
A	252	2	254	659	4	663	206	18	224
B	61	1	62	554	8	562	47	0	47
C	143	3	146	203	8	211	253	8	261
D	355	16	371	328	12	340	120	3	123

The following Fig. 7 shows the existing layout of the intersection at location 2. A one-lane roundabout is planned to be built at this location.

Fig. 7. Geometry of signalized intersection at Location 2

The traffic load at Location 2 measured during the period of maximum hourly traffic (afternoon peak hour) is shown in the following Table 2:

Table 2. Traffic load at Location 2

Entry	Left			Straight			Right		
	Car	HV	Σ	Car	HV	Σ	Car	HV	Σ
A	253	10	263	17	0	17	196	1	197
B	263	10	273	48	1	49	51	0	51
C	334	2	336	231	2	233	157	3	160
D	377	1	378	127	1	128	67	0	67

In the PTV Vissim 11 software, two separate models were formed at both locations. First, the existing signalized intersections with actual cycle lengths and phase arrangements were modeled, and then alternative solutions of roundabouts were modeled (Fig. 8 and 9). Since no roundabouts were built at the time of the research, their calibration was done based on priority rules and values of minimum gap time and minimum headway time at each approach. Each model was simulated for a peak traffic load of 1 h, with a simulation warm-up period of 600 s. Each simulation was repeated ten times with a different random seed number.

Fig. 8. Vissim model of the two-lane roundabout at Location 1

Fig. 9. Vissim model of the one-lane roundabout at Location 2

Along with the traffic simulation, a capacity analysis was performed for both types of intersections at both locations. In all variants, average vehicle delays and level of service were determined using the HCM 2010 methodology. Comparing the results for Location 1 (Tables 3 and 4) with the construction of the two-lane roundabout, the level of service of the whole intersection is still F, although the total delay of the whole intersection has decreased. It can be concluded that the two-lane roundabout did not solve all the problems of the traffic light intersection, but it still represents an increase in capacity.

On the other hand, at Location 2, the existing signalized intersection has a LOS F on only one approach, while the proposed one-lane roundabout has such a level of service on three approaches. From the aspect of capacity analysis, such a solution is unacceptable and does not improve the existing situation. The topic of this research is not primarily the traffic analysis at the considered intersections, so it will not comment in detail on the results obtained or propose measures to correct the results.

Table 3. Results of traffic analysis for signalized intersection at Location 1 [6]

Year	2018	Phase I (sec)	30	Protected left movement (sec)		10
Entry	A			B		
Lane	L	T	R	L	T	R
Average delay per lane (sec/veh)	246.5	27.3	56.2	19.2	24.4	20.4
Level of service per lane	F	C	E	B	C	C
Average approach delay (sec/veh)	76.5			23.4		
Level of service per approach	E			C		
Intersection level of service	F					
Year	2018	Phase II (sec)		35		
Entry	C			D		
Lane	L	T	R	L		T/R
Average delay per lane (sec/veh)	148	19.4	23.6	116		39
Level of service per lane	F	B	C	F		D
Average approach delay (sec/veh)	51.3			73.3		
Level of service per approach	D			E		
Intersection level of service	F					

Table 4. Results of traffic analysis for a two-lane roundabout at Location 1 [6]

Entry	A		B		C		D	
Lane	L	R	L	R	L	R	L	R
Degree of saturation (v/c)	0.61	0.63	1.04	1.06	0.71	1.01	0.67	0.79
Average approach delay (sec/veh)	16.7	16.9	78.4	73.7	31.9	78.4	21.0	28.4
Level of service per approach	C	C	F	F	D	F	C	D
Average approach delay (sec/veh)	16.8		75.8		60.0		25.1	
Level of service per approach	C		F		F		D	
Intersection level of service	F							

Table 5. Results of traffic analysis for signalized intersection at Location 2 [7]

Year	2017	Phase I (sec)	29	Phase II (sec)	38	Cyclus (sec)	77
Entry	A		B		C	D	
Degree of saturation (v/c)	0.30	0.55	1.16	0.92	1,16	0,76	0,68
Average delay per lane (sec/veh)	21.9	22.2	112.3	45.7	112,3	30,8	19,7
Level of service per lane	C	C	F	D	F	C	B
Average approach delay (sec/veh)	22.2		43.7		112.3	23.3	
Level of service per approach	C		D		F	C	
Intersection average delay (sec/veh)	50,3						
Intersection level of service	D						

Table 6. Results of traffic analysis for a one-lane roundabout at Location 2 [7]

Entry	A	B	C	D
Degree of saturation (v/c)	0.65	1.00	0.96	1.21
Average approach delay (sec/veh)	19.55	67.24	54.64	132.25
Level of service per approach	C	F	F	F
Intersection average delay (sec/veh)	77.78			
Intersection level of service	F			

3 Evaluation of Traffic Emissions

After all, simulations were performed at both observed locations, data on the main indicators of harmful gas emissions were collected. The data were obtained for each movement, but they are presented in the form of average values for each approach for a more straightforward presentation. In this way, the traffic simulation gives data on pollution exclusively based on general indicators such as traffic load, delay, and representative vehicle.

Figure 10 presents a diagram with the results for both types of intersections at Location 1. The average fuel consumption at the signalized intersection is 101.4 l, while at the roundabout is 85.6 l. The average number of stops is identical at both intersections. The obtained values indicate that the CO emissions at both locations are almost identical, while VOC concentration and NOx concentration is higher by 16% at the signalized intersection.

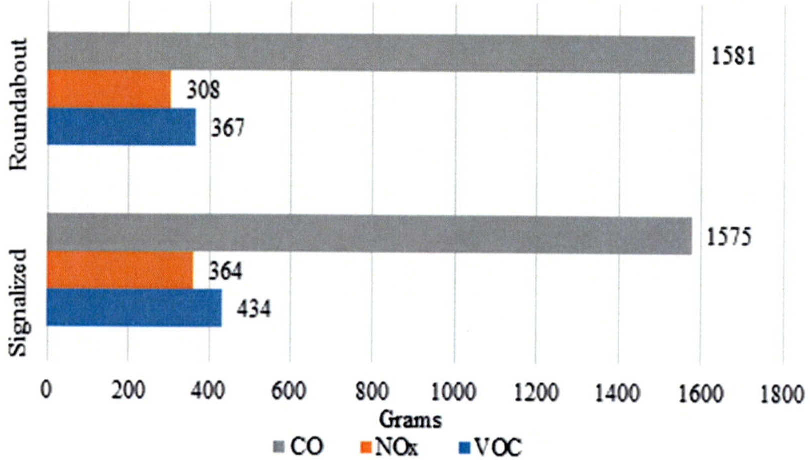

Fig. 10. Main emission indicators at Location 1

The results show that at Location 1 both types of intersections have almost equal pollutant emissions. A signalized intersection is a slightly less favorable solution in terms of VOC and NOx emissions. It is imperative to compare these results with the results of the level of service analysis (Tables 3 and 4). The delays and LOS results correspond to the results of pollutant emissions and show that these two solutions provide almost the same quality of traffic. Also, in that case, there is a small and not so significant advantage of a two-lane roundabout related to a signalized intersection.

Figure 11 shows the results of emissions at Location 2. The average fuel consumption at the traffic light intersection is 66.88 l, while at the roundabout is 192.25 l. The average number of stops at a signalized intersection is ten and at a one-lane roundabout is 29.

The obtained results show that the signalized intersection at Location 2 is a significantly better solution than the one-lane roundabout in all emission indicators. This result contradicts the general assumption that roundabouts are more favorable solutions in terms of environmental impact. However, if we take a closer look at the LOS results in Tables 5 and 6, it is clear why such results were obtained. Namely, the existing solution with a signalized intersection has a LOS F only on one approach, while in the case of a one-lane roundabout, the LOS F is on three of the four approaches. Also, the total delay is increased by 27 s on the one-lane roundabout. Thus, significantly worse operational characteristics of the one-lane roundabout have led to vehicles spending more time at intersections, stopping more times, consuming three times more fuel, and ultimately, releasing more pollutants into the atmosphere.

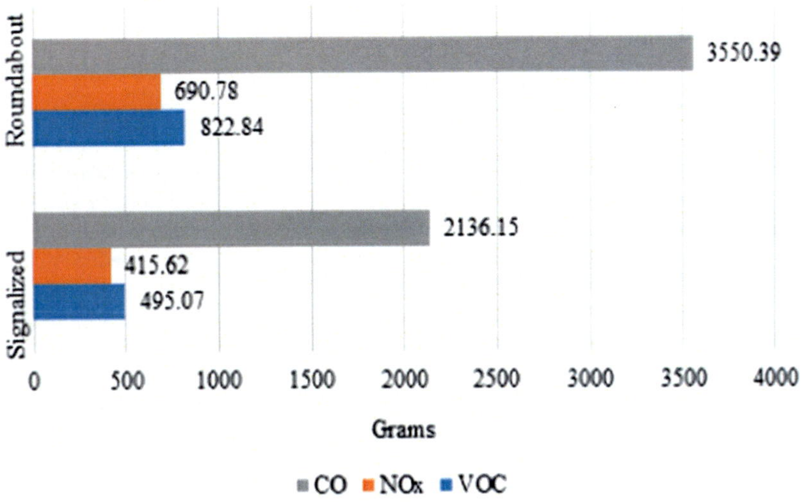

Fig. 11. Main emission indicators at Location 2

4 Discussion and Conclusion

This paper presents a comparative study at signalized intersections and roundabouts regarding their impact on the environment. The analysis was performed at two locations in Sarajevo; Location 1 with a large signalized intersection and a two-lane roundabout as an alternative solution, which has been built in the meantime, and Location 2 with a smaller signalized intersection and an alternative in the form of a one-lane roundabout.

The emission values of individual pollutants were determined based on traffic microsimulation in the PTV Vissim 11 software. All traffic simulations were done with real traffic load and existing geometric characteristics of both locations.

This research's initial assumption is that by changing the way traffic is regulated and the intersection's geometry, lower emission of pollutants into the atmosphere is achieved. However, the obtained results indicate that such an assumption cannot be definitive without a prior level of service analysis of different intersection types. Namely, at Location 1, both types of intersections have similar operational characteristics, with a slight advantage of the roundabout. The results of pollutant emissions showed the same trend.

On the other hand, at Location 2, the alternative solution has a significantly higher delay and a lower level of service than the existing signalized intersection, thus higher emissions of pollutants. Therefore, the amount of emissions is directly proportional to the indicators of traffic quality.

This type of research can be extended to more detailed analysis and traffic simulation that includes traffic flow characteristics such as vehicle type, vehicle age, fuel type, etc.

In less developed countries, such as Bosnia and Herzegovina, no significant increase in electric vehicles' use can be expected in the coming period, and traffic will continue to be one of the key factors of pollution in urban areas. Therefore, detailed analysis and evaluation of different criteria are needed when choosing the optimal type of intersection.

References

1. Europian Environment Agency (EAA): National emissions reported to the Convention on Long-range Transboundary Air Pollution (LRTAP Convention). https://www.eea.europa.eu/data-and-maps/daviz/trend-in-emissions-of-air-pollutants-6#tab-chart_3. Accessed 1 Feb 2021
2. Federal Hydrometeorogical Intstitute: Annual report on air quality in the Federation of Bosnia and Herzegovina for 2019 (Bos. 'Godišnji izvještaj o kvalitetu zraka u Federaciji Bosne i Hercegovine za 2019. Godinu'), Sarajevo, Bosnia and Herzegovina (2020)
3. Apostolos, A., Fotini, K.: Turbo-roundabouts as an alternative to roundabouts in terms of traffic safety, capacity and pollutant emissions. In: 7th Pan-Hellenic Road Safety Conference, Larissa, Greece (2018)
4. Jaworski, A., Lejda, K., Mądziel, M.: Emission of pollution from motor vehicles with respect to selected solutions of roundabout intersections. Combust. Eng. **168**(1), 140–144 (2017). https://doi.org/10.19206/CE-2017-122
5. Meneguzzer, C., Gastaldi, M., Rossi, R., Gecchele, G., Prati, M.V.: Comparison of exhaust emissions at intersections under traffic signal versus roundabout control using an instrumented vehicle. Transp. Res. Procedia **25**, 1597–1609 (2017). World Conference on Transport Research - WCTR 2016 Shanghai, 10–15 July 2016
6. Čolaković, A.: Idejno rješenje raskrsnice Hamdije Čemerlića-Put života (Energoinvest). Master thesis, Faculty of Civil Engineering, University of Sarajevo (2018)
7. Đedović, H.: Idejno rješenje raskrsnica Patriotske lige – Bolnička-StjepanaTomića. Master thesis, Faculty of Civil Engineering, University of Sarajevo (2017)

Energy Efficient Building in Bosnia and Herzegovina

Berina Sejdinović(✉)

Travnik, Bosnia and Herzegovina

Abstract. EU countries' commitments "20/20/20" are the commitments that place the most tasks on the construction industry. By 2035, Bosnia and Herzegovina's Framework Energy Strategy provides the context and direction of energy development in Bosnia and Herzegovina and seeks the right balance in the context of "energy trilemma". Typology of public buildings in Bosnia and Herzegovina is necessary prerequisite for defining the appropriate strategy for managing public buildings in all aspects. The EPBD/Energy Performance of Buildings Directive is, the main legislative instrument for promoting the energy efficiency of buildings and encouraging renovation within the EU. Energy efficiency model represents a universally applicable, transparent, non-discriminatory and socially sensitive framework for the promotion, implementation and support of energy efficiency projects in Sarajevo Canton, and its main goal is long-term and systemic support to citizens in improving energy characteristics, i.e. warming of housing, under more favorable conditions. Low-energy houses are the foundation of applications of sustainable construction throughout their lifetime, ranging from building materials whose production has minimal impact on the environment, through their energy efficiency over the lifetime, to adequate waste management. Currently, there are only a few products on the BiH market for households with a special intention of financing energy efficiency measures: loans with special intent - energy efficiency.

Keywords: Energy efficiency · EPBD directive · Energy efficient building · Low-energy houses · Passive houses · Zero energy houses · Independent or autonomous houses · Houses with surplus energy

1 Introduction

"The construction industry is responsible for the most tasks required under the "20/20/20" commitments of the EU Member States, which include:

- *20% cut in greenhouse gas emissions from 1990 levels (even 30% if the preconditions are met);*
- *20% of EU energy from renewables;*
- *20% improvement in energy efficiency.*

The European Union countries are not set only on saving 20% of the energy used in buildings by 2020 but also introduced a long-term goal to reduce emissions from the

N. Ademović et al. (Eds.): IAT 2021, LNNS 316, pp. 387–402, 2022.
https://doi.org/10.1007/978-3-030-90055-7_31

sector by between 80 and 95% by 2050. In order to achieve that, all Member States but also potential candidate and candidate countries must adopt National Energy Efficiency Action Plans and individual strategies for each of the affected sectors [1].*"*

Many EU countri es now have a significant backlog in 2021, the full implementation of the obligations has not been carried out on time and therefore are not dynamic national commitments fulfilled by 2020.

"As part of the European Green Deal, the Commission proposed in September 2020 to raise the 2030 greenhouse gas emission reduction target, including emissions and removals, to at least 55% compared to 1990.

Key targets for 2030:

- *At least 40% cuts in greenhouse gas emissions (from 1990 levels);*
- *At least 32% share for renewable energy;*
- *At least 32.5% improvement in energy efficiency.*

The EU aims are to be climate-neutral by 2050 – an economy with net-zero greenhouse gas emissions [2].*"*

BiH needs to seize the opportunity to develop through clean technologies in those sectors with which it can be competitive in the demanding EU market. Pure natural building materials and elements can be the strength of the future sustainable socioeconomic development of our state.

In addition to all of the above, it is especially necessary to point out that BiH has the opportunity to start using IPA funds in a much better way than in the previous period. IPA funds would be available to all citizens who align their projects with adopted and adopted strategies. Funds from these funds are allocated in the form of grants, which is a special benefit for development.

In Fig. 1, all stakeholders were identified as well as their level of interest and influence, which should be used in a considered way for development of strategy. The major target stakeholders are The Council of Ministers, The State Owned Enterprises, the construction industry, the wood industry, the energy industry, the transportation industry, employee unions, the wider public and consumer groups, Small and medium sized enterprises, craft associations, and technology clusters [1].

"By 2035, Bosnia and Herzegovina's Framework Energy Strategy provides the context and direction of energy development in Bosnia and Herzegovina and seeks the right balance in the context of "energy trilemma". Launching real investments, market and regulatory reforms in all segments of energy, supported by solid and structured implementation from all key participants, is very important for Bosnia and Herzegovina [3].*"*

"*Energy* trilemma" has dimensions:

- Energy security - Reflects a nation's capacity to meet current and future energy demand reliably, withstand and bounce back swiftly from system shocks with minimal disruption to supplies;
- Energy equitity - Assesses a country's ability to provide universal access to aordable, fairly priced and abundant energy for domestic and commercial use;

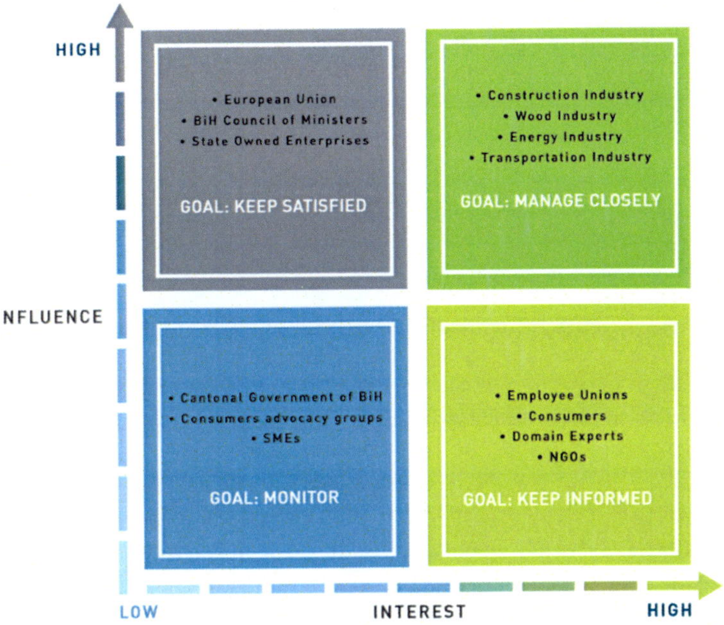

Fig. 1. Map of stakeholders [1]

- Envoronmental sustainability of energy systems - Represents the transition of a country's energy system towards mitigating and avoiding potential environmental harm and climate change impacts [4].

Secure and affordable energy – are the two extremely important components of the energy trilemma in the context of the economic situation and geopolitical position of Bosnia and Herzegovina. In terms of energy security, it is important to understand that, at present times, Bosnia and Herzegovina cannot achieve the energy security in all segments on its own, primarily because of non-existence of its own oil and gas production. With this regard, it is primarily important to actively manage the physical market integration with the neighbouring countries, understand and implement the options of supply routes diversification develop partnership with business entities which supply the domestic market. In the oil and petroleum derivatives segment, an emergency stocks system should be established. Likewise, the adequate domestic infrastructure management towards promotion of quality and security of supply should not be disregarded. When it comes to electricity, in the context of Bosnia and Herzegovina, there is relatively high level of security of supply that need to be maintained in the forthcoming period, balancing its high share of domestic generation, integration with regional markets, energy efficiency trends, decentralised generation, and clean energy competitiveness.

Directive 2002/91/EC of the European Parliament and of the Council of 16 December 2002 on the energy performance of buildings has objective to promote the improvement of the energy performance of buildings within the Community, taking into account outdoor climatic and local conditions, as well as indoor climate requirements and cost-effectiveness [5].

As a great step towards the typology of the non-residential fund in Bosnia and Herzegovina, this typology of public buildings has been developed, which in a way represents part of the response to Article 5. Directive 2012/27/EU, which highlights the priority need to renovate buildings by public authorities as a role model for other energy consumption sectors.

As described in the previous chapters, the results of the census showed that there are a total of 7600 public buildings in Bosnia and Herzegovina [6].

The EPBD is, together with the Energy Efficiency Directive, the main legislative instrument for promoting the energy efficiency of buildings and encouraging renovation within the EU.

EPBD (2010/31/EU) has been in place since 2010 and helps consumers make informed choices by enabling them to save both energy and money.

This has also resulted in a positive change in trends in the energy properties of buildings; After EPBD's introduction of energy efficiency requirements in national building laws, today's buildings spend only half as much as typical buildings from the 1980s [7].

The revised EPBD (2018/844/EU), which complements parts of the 2010 EPBD and introduces new elements, is an important part of implementing the Juncker Commission's priorities for building a "resilient Energy Union and a promising climate change policy".

The Commission launched a public debate in June 2015 to help support the revised directive, including for example targets to accelerate cost-effective renovation of existing buildings, with a vision of a decarbonised building stock by 2050, and mobilise investment [7].

For the purpose of calculation, buildings should be classified into one of the following categories:

(a) single-family houses of different types;
(b) apartment blocks;
(c) offices;
(d) education buildings;
(e) hospitals;
(f) hotels and restaurants;
(g) sports facilities;
(h) wholesale and retail trade services buildings;
(i) other types of energy-consuming buildings (Table 1).

Table 1. Typology results - type object tabs [8]

	I Pre-education	II Education	III Health	IV Sport	V Culture	VI Administrative	VII Whole-day stay
A Until 1945							
B 1946 - 1965							
C 1966 - 1973							
D 1974 - 1987							
E 1988 - 2009							
F After 2010							

2 Energy efficient construction

"Increasing energy prices as well as global climate change are the main reasons to reflect on our energy consumption habits. Given that it is the buildings that consume about 40% of the total energy consumption and are responsible for 36% of carbon dioxide emissions, energy efficient residential buildings and houses are becoming more interesting and their concepts are becoming more diverse [9]*."*

2.1 Types of Energy Efficient Houses

Energy efficient houses are called:

- low energy houses
- passive houses (ultra-low energy houses)
- zero-energy houses or net zero energy houses
- independent or autonomous houses (autonomous buildings, houses with no bills)
- energy-plus houses (houses with surplus energy) [10].

2.1.1 Low Energy Houses

"Low-energy houses are the foundation of the application of sustainable construction throughout their lifespan, starting with building materials whose production does not

burden the environment, through their energy efficiency and rational consumption of energy throughout their lifespan, to rational waste management. In addition, low-energy houses (whose group includes passive houses) provide high comfort housing with a comfortable climate throughout the year without standard heating and cooling systems, with very low energy costs and are less (or not at all) energy dependent [11]."

Figure 2 shows a low-energy home in California. Under recent laws and regulations in California, new and refurbished homes will have to use energy far more efficiently in coming years, to lower consumption and help reduce the state's carbon fooftprint. Low energy houses will include high-performance features.

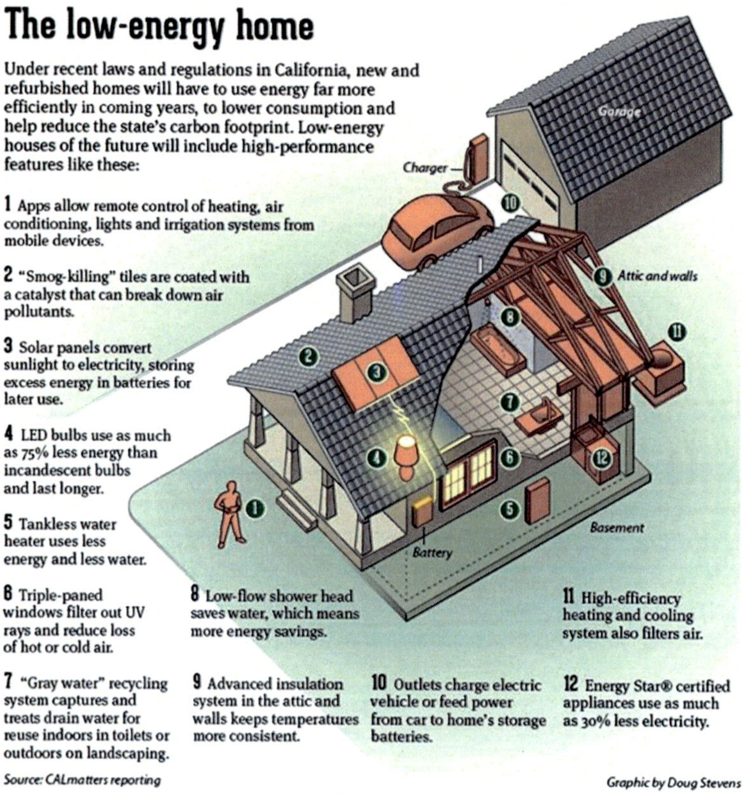

Fig. 2. Low energy house [12]

2.1.2 Passive Houses/Ultra-Low Energy Houses

"A passive house, according to the definition of the Institute for Passive Houses in Darmstadt, is considered to be a building that annually does not exceed the need for energy of 15 kilowatt hours (kWh) per square meter. An additional characteristic typical of this type of house: Much of the need for heating is covered with solar energy or waste heat of the tenants and technical equipment.

Thus, in this context, the name "passive" house was created. An additional characteristic of the standard for passive houses is the high efficiency of thermal insulation of such buildings [13]."

2.1.3 Zero-Energy Houses/Net Zero Energy Houses

"*As with passive houses, these houses are characterized by minimal energy consumption and the use of renewable sources, which is why the annual amount of energy taken from the grid and handed back to the grid equalizes and amounts to zero. These houses are characterized by energy efficiency and the use of renewable energy sources resulting in a high energy class in case of energy certification* [14] (Fig. 3)."

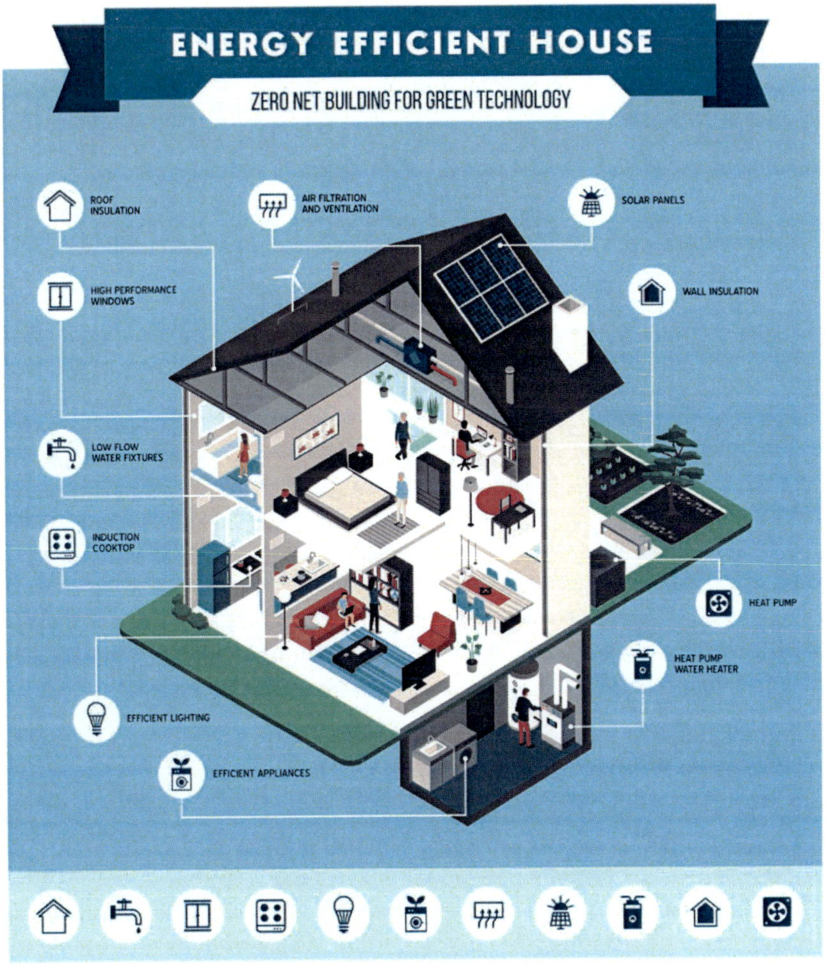

Fig. 3. Zero net building for green technology [15]

2.1.4 Autonomous Buildings/Houses with No Bills

"The autonomous house with no bills is designed to function normally independently of outside infrastructure support. Therefore, there is no connection to the electricity distribution network, plumbing, sewage, communication network, and in some cases there is no connection to public roads.

An autonomous house is much more than an energy-efficient house – energy in this case is just one of the resources that needs to be obtained from nature [16]."

2.1.5 Energy-Plus Houses/Houses with Surplus Energy

"A house with surplus energy is a house that produces more energy on average throughout the year using renewable energy sources than it takes from external systems. This can be achieved by using renewable sources, small electricity generators, low-energy construction techniques, excellent insulation and use of energy efficient systems and very careful choice of location for the house. As a rule, such houses use the most energy efficient devices and systems for heating, cooling and the like. Such houses, in addition to contracts for the sale of surplus energy to distribution companies, can represent permanent income to their owners instead of the constant cost of energy [17]."

2.2 Energy Efficient Construction – The Need for Energy and U-Values

"Energy standards of buildings are reflected in different values. The following list shows the maximum need for energy of different types of buildings. The lower the value, the less energy must be consumed to heat the building. 10 kWh corresponds to approximately the energy obtained from one liter of fuel oil. These amounts are considered benchmarks, except for values specified for passive houses [13]."

When it comes to the need for energy, insulation of different parts of the house plays a significant role. The following table shows the average heat transfer coefficient (U-value) of individual parts of different types of building. The lower the U-value, the better the energy efficiency situation. Because a lower U-value means less heat comes out of different parts of the building's sheath. The following figures can also be considered benchmarks [13]. U-value depends mostly on layers thickness and used materials [18].

They vary depending on the definition and depending on the type and age of the parts of the building displayed [13].

2.3 Profitability of Increased Construction Costs

"Initially, the cost of construction per m^2 is about 5–12% higher in low-energy standard and by about 15–22% in passive standard compared to classical construction.

But the house is not built for a short shelf life and use.

The calculation shows that in the long term the investment in the construction of a low-energy house pays off over 3 to 5 years, while the profitability of investing in the construction of a passive house is seen in 12 to 17 years, sometime even after 8 years [19]."

2.4 KfW Promotional Programs for Energy Efficiency in Buildings - Main Elements and Success Factors

KfW is promotional bank of the Federal Republic of Germany. It was established in 1948 as Kreditanstalt für Wiederaufbau. Main stakeholders are: 80% Federal Republic, 20% federal states with famous headquarters in Frankfurt am Main Branches: Berlin, Bonn and Cologne. KfW have about 80 offices and representations worldwide [20].

3 Energy Efficient Building in Bosnia and Herzegovina

"Bosnia and Herzegovina, by signing the Agreement on the Establishment of the Energy Community, as well as the Stabilisation and Association Agreement, undertook the obligation to harmonise its own legislation with that of the EU and the adoption of European Directives and standards in the field of energy. One of the twists of the above Agreement is the enactment of the legislative framework. The area of energy efficiency is regulated by the Federation of BiH Energy Efficiency Law, adopted in April 2017. (Official Gazette of FBiH 22/17.) [21]." The Law assumes all requirements of EU Directives.

3.1 Analysis of the Current Situation in Sarajevo Canton from the Point of View of Energy Efficiency in the Building Sector

An analysis of the current situation in Sarajevo Canton from the point of view of energy efficiency in the building sector showed that this sector is the largest energy consumer with a share of about 55% in final consumption. - Poor heat protection of buildings, irrational energy consumption and underdeveloped awareness of citizens about the need for efi-later energy consumption are the basic characteristics of heating in our Canton. - Heat consumption per unit area is very high, while in EU countries with colder winters this consumption is several times lower.

The total housing stock in Sarajevo Canton consists of 70% of individual housing, while the rest consists of collective housing facilities. In order to achieve a reduction of energy consumption in buildings up to about 50%, it is necessary to apply energy efficiency measures to about 150 thousand units, or about 14 million m^2 of living space.

Reducing the energy needed to heat individual and collective housing facilities as well as public buildings will result in lower heating costs as well as a reduction in air pollution.

Taking into account these facts, the Ministry of Spatial Planning, Construction and Environmental Protection of Sarajevo Canton, in cooperation with the Sarajevo Regional Development Agency (SERDA), has developed a Model for Improving Energy Efficiency in Building Efficiency in Sarajevo Canton in the function of increasing the number of users (omassion), (Model EE). The Sarajevo Canton Government adopted the Model EE in December 2017.

The EE model represents a universally applicable, transparent, non-discriminatory and socially sensitive framework for the promotion, implementation and support of energy efficiency projects in Sarajevo Canton, and its main goal is long-term and systemic support to citizens in improving energy characteristics, i.e. warming of housing, under more favorable conditions.

It is important to note that energy efficiency measures also imply a shift from sub-standard to more environmentally friendly fuels, especially in urban environments and sustainable use of natural gas for heating. The realization of energy efficiency projects according to the established Model EE will ultimately contribute to a more beautiful urban picture, with better housing comfort, higher value of real estate and increasing economic activities in Sarajevo Canton [22].

The basic principle on which the EE Model is based is the repayment of energy efficiency (EE) costs from the achieved savings, which means that for owners of buildings where energy efficiency measures are implemented there is no increase in costs due to investments in EE measures, so the project slogan is: - half the account [23].

Users of the Energy Efficiency Model can be:apartment owners in collective housing buildings, owners of individual residential buildings (houses) and public buildings.

Beneficiaries-applicants of the Model EE project are individual residential buildings and collective housing buildings from the area of the municipalities of the Sarajevo Canton, which includes the city urban core and the sloping parts of the municipalities of the Sarajevo Canton.

Benefits for project users are [23].

- Covered costs of energy audits and project documentation for the implementation of energy efficiency measures;
- Possibility to use a dedicated credit line with banks that are partners in the implementation of the EE Model;
- Subsidizing part of the interest rate for beneficiaries who have been granted credit funds within the project;
- Co-financing of a part of costs in the amount of up to 45% for heating of buildings and other energy efficiency measures envisaged for a given building;
- Costs of supervision over the execution of works in order to improve energy efficiency in buildings are covered [23].

The model envisages activities for the promotion and education of condominium owners and owners of individual buildings on the benefits of applying energy efficiency measures in residential buildings, primarily energy savings and air improvement, but also some others that are in line with the commitment of the EU and the world.

Warming of buildings in order to save energy is realized through, for each user, binding phases as follows:

- *PHASE I implies conducting an energy audit of the facility, preparation of economically justified measures and investment technical documentation in accordance with applicable regulations;*
- *PHASE II implies the performance of construction - craft works on the warming of buildings in order to save energy* [23].

3.2 The First Passive House in BiH Was Built on Kovači

The first "house of the future" in BiH, which with its superior thermal insulation and use of energy from renewable sources significantly reduces energy consumption, was built on Kovači.

Passive house, located in one of the traditional old town streets, will serve as a model example to anyone who would like to look at what it looks like, and possibly opt for this kind of construction.

The house was built by the Municipality of Stari Grad, which gave up the location, and issued the necessary permits, so the Municipality of Stari Grad is the first to boast the so-called "house of the future".

The construction of the passive house began in May this year, and a complete interior has been arranged, and the house is equipped with the most modern furniture. The total area of the passive house is about 103 m^2 [24] (Fig. 4).

Fig. 4. Exterior of the first passive house in BiH [24]

3.3 Manufacturers of Low-Energy Houses in BiH

"Low-energy houses are the foundation of the application of sustainable construction throughout their lifespan, starting with building materials whose production has minimal impact on the environment, through their energy efficiency throughout their lifespan, to adequate waste management. In addition, low-energy houses, which include passive houses, provide high housing comfort with a comfortable climate throughout the year without standard heating and cooling systems, at very low cost at the expense of energy [25]."

3.3.1 PROMO d.o.o. Donji Vakuf

Modern house is an object that provides a sense of harmony of space and its purpose. It is designed according to the U line in which most of them occupy the living room with dining room and kitchen, and the second part is intended for bedrooms or study rooms. It is characterized by modern lines on the façade, contrasting wooden-dark façade and interesting openings. The horizontal upper openings on the front façade provide additional lighting and a sense of spatiality. This house also has an unusual roof shape and gives the entrance façade a special look [26].

3.3.2 Krivaja TMK d.o.o. Zavidovići

Krivaja low-energy houses are carefully designed and made for absolute comfort, while guaranteeing a minimum amount of energy for heating, illumination, etc.

Minimal energy consumption can be guaranteed thanks to special construction system, which is based on a large level of thermo-insulation, steam-permeable sheath, controlled ventilation system and maximization of the utilization of heat gains from solar radiation through optimal design.

The annual need for energy for heating in low-energy houses ranges between 15 and 50 kWh/m^2. Low-energy houses are slightly less energy efficient compared to passive houses, but the initial construction costs are much lower, and the investment pays off sooner [27].

3.3.3 SAVOX d.o.o. Milići

Savox ltd Milišići produce prefabricated houses and prefabricated buildings of all purposes, roh-bau (rough construction), taking into account the needs, desires and possibilities of customers.

Prefabricated houses are significantly cheaper than masonry construction. The fact that these houses manufactured in the factory and then transported and assembled, reduces costs regarding the construction, increases the efficiency of workers and drastically reduce duration of construction [28].

3.4 The Most Important Institutions for Supporting and Financializing Sustainable Development and Energy Efficiency in BiH

"The most important institutions for supporting and financializing sustainable development and energy efficiency in BiH are the European Bank for Reconstruction and Diversity (EBRD), the American Agency for International Development (USAID), the United Nations, United Nations Development Programme (UNDP), Global Environmental Fund (GEF), European Commission, Organisation for Economic Co-operation and Development (OECD), International Energy Commission (IEA), World Bank (WB), FBiH Environmental Protection Fund.

The largest organization that finances projects in the field of energy efficiency in Bosnia and Herzegovina is the EBRD. The EBRD has developed a sustainable energy credit line for the Western Balkans (WeBSEFF II-Western Balkans Sustainable Energy Financing Facility II), which finances projects in Bosnia and Herzegovina. The WeBSEFF II sustainable energy credit line for the Western Balkans operates through participating local banks that provide credit benefits for companies and local self-government units that want to invest in energy efficiency projects and small renewable energy projects [29]."

3.5 Special Purpose Loans - Energy Efficiency

Currently, there are only a few household products on the Market of BiH with a special purpose of financializing energy efficiency measures: loans with special purpose - energy efficiency.

It is important to mention that each bank in BiH uses its own prescribed methodology and has established its own ways of forming and changing interest rates for loans for physical faces. Work is being done on setting parameters for the formation of interest rates and setting the conditions under which the interest rate can be changed. Banks in BiH negotiate loans and give them to individuals at fixed or variable interest rates [30].

3.5.1 Bosna Bank International (BBI)

With the project of warming individual buildings in the municipality of Centar Sarajevo, citizens will be able to receive interest-free funds for this purpose in BBI Bank. The municipality will subsidize the profit margin, and in order to encourage citizens to get involved in energy efficiency projects, the municipality will, in addition to subsidizing the profit margin, depending on the number of squares that are being heated [31].

3.5.2 Sberbank BH

By providing a new credit line for Sberbank BH, successful cooperation between Sberbank, the largest Russian bank and the EBRD continues with the aim of supporting small and medium enterprises. The EBRD is already successfully cooperating with Sberbank in Belarus, Kazakhstan and Turkey through support for SME and energy efficiency projects [32].

3.5.3 UniCredit Bank

Unique credit products on the market with a grant cash payment to current account as an incentive for a realized project!

Installing new windows, thermal façade, roof or implementing another project energy efficiency in the household person is entitled to an incentive bonus – one-time grant amount granted!

Data on energy efficient products that can be financed from this credit line can be found on the Technology Selector of Unicredit Bank [33].

3.5.4 Raiffeisen Bank

Raiffeisen BANK dd Bosnia and Herzegovina, with the support of KfW Development Bank, has introduced a new credit line for financing investments in energy efficiency and projects that generate energy savings. The aim is to promote the efficient use of energy in Bosnia and Herzegovina in a sustainable way.

Through this credit line, Raiffeisen Bank contributes and supports long-term environmental protection, instiates the purchase of products and devices with efficient use of energy [34].

4 Conclusion

Given that buildings consume about 40% of total energy consumption and are responsible for 36% of carbon dioxide emissions, we need to think about our energy consumption habits.

Today there are five types of energy efficient houses. These are a low-energy house, passive house, zero energy house, independent or autonomous house and houses with superplus energy.

Building a low-energy house costs about 5–12% more expensive, and the construction of a passive house is about 15–22% more expensive compared to classical construction, but the construction of a low-energy house pays off over 3 to 5 years, while the profitability of investing in the construction of a passive house is seen in 12 to 17 years, sometime even after 8 years.

The benefits of building low-energy and passive houses are: reduced bills for heating, cooling and electricity; more comfortable and quality housing; longer life expectancy of the house; negative impact on the environment, i.e. reducing carbon dioxide emissions.

Citizens in Bosnia and Herzegovina need to develop awareness of energy-efficient construction that can be achieved through the Model of Improvement of Energy Efficiency in Building Efficiency in Sarajevo Canton in the function of increasing the number of users (affinity), developed by the Ministry of Spatial Planning, construction and environmental protection of Sarajevo Canton, in cooperation with the Sarajevo Regional Development Agency (SERDA) or a new credit line for financing investments in energy efficiency and projects that generate energy savings,introduced by some banks in Bosnia and Herzegovina and this must be a priority.

References

1. Klarić, S., Šamić, D., Katica, J., Kurtović, A., Duerod, M., Roso Popovac, M.: Energy efficiency in buildings as basis for sustainable social and economic development in Bosnia and Herzegovina. Green Council, Sarajevo (2016). https://green-council.org/publik/Vodic_EE_in_bild_eng.pdf. Accessed 20 Jan 2021
2. European Comission: Climate action (2021). https://ec.europa.eu/clima/index_en. Accessed 13 May 2021
3. Ministry of Foreign Trade and Economic Relations of Bosnia and Herzegovina (MOFTER), Federal Ministry of Energy, Mining and Industry (FMERI), Ministry of Industry, Energy and Mining (MIER) and "Komunalno Brčko": Framework Energy Strategy of Bosnia and Herzegovina until 2035 (2018). http://www.mvteo.gov.ba/data/Home/Dokumenti/Energetika/Framework_Energy_Strategy_of_Bosnia_and_Herzegovina_until_2035_ENG_FINAL....pdf. Accessed 2 Feb 2021
4. World Energy Council: World Energy Trilemma Index (2019). https://www.worldenergy.org/assets/downloads/WETrilemma_2019_Full_Report_v4_pages.pdf. Accessed 13 May 2021
5. Official Journal of the European Communities: Directive 2002/91/EC of the European Parliament and of the Council (2003). https://eurlex.europa.eu/LexUriServ/LexUriServ.do?uri=OJ:L:2003:001:0065:0071:EN:PDF. Accessed 12 May 2021
6. Nišandžić, M.: Typology of public buildings in Bosnia and Herzegovina (2018). https://issuu.com/unitednations_bih/docs/tipologija_javnih_zgrada_u_bosni_i_. Accessed 10 May 2021

7. European Comission: Energy Performance of Buildings Directive(2018). https://ec.europa.eu/energy/topics/energy-efficiency/energy-efficient-buildings/energy-performance-buildings-directive_en. Accessed 2 Feb 2021
8. Hadžikadić, E.:Typology of public buildings in Bosnia and Herzegovina. Fourth Energy Summit, Neum, 25–27 April 2018 (2018)
9. Direkt inženjering j.d.o.o.: Energy efficient construction (2016). http://direkt-inzenjering.com/. Accessed 5 Feb 2021
10. Ytong porobeton d.o.o.: Types of energy efficient houses (2021). https://gradnjakuce.com/energetski-ucinkovite-kuce/vrste-energetski-ucinkovitih-kuca/. Accessed 5 Feb 2021
11. Fond za zaštitu okoliša i energetsku učinkovitost: Energy efficient construction (2021). https://www.fzoeu.hr/hr/energetska_ucinkovitost/enu_u_zgradarstvu/energetski_ucinkovita_gradnja/. Accessed 8 Feb 2021
12. Capradio: Buildings in California Must Use Energy More Efficiently (2016). https://www.capradio.org/articles/2016/05/03/buildings-in-california-must-use-energy-more-efficiently/. Accessed 8 Feb 2021
13. Planradar: Energy efficient construction (2021). https://www.planradar.com/hr/energetski-ucinkovita-gradnja. Accessed 8 Feb 2021
14. Portal energetskocertificiranje.com.hr: Zero energy house (2015). https://www.energetskocertificiranje.com.hr/kuce-nulte-energije/. Accessed 9 Feb 2021
15. Ecomena: All About Building a Net Zero Home (2019). https://www.ecomena.org/building-a-net-zero-home. Accessed 9 Feb 2021
16. Kancelarija za energetski menadžment Opština Vrbas: Autonomous house (2011). http://www.eevrbas.org/korisni-saveti/ventilacija/27-korisni-saveti/niskoenergetska-gradnja/69-autonomne-ku%E7%9D%80e. Accessed 10 Feb 2021
17. Portal energetskocertificiranje.com.hr: House with surplus energy (2015). https://www.energetskocertificiranje.com.hr/kuce-s-viskom-energije/. Accessed 10 Feb 2021
18. NBS for specifiers: What is U-value (2015). https://www.thenbs.com/knowledge/what-is-a-u-value-heat-loss-thermal-mass-and-online-calculators-explained. Accessed 11 May 2021
19. Ytong porobeton d.o.o.: Profitability of increased construction costs (2021). https://gradnjakuce.com/pasivna-niskoenergetska-gradnja/niskoenergetska-ili-trolitarska-kuca/. Accessed 12 Feb 2021
20. Dorendorf, B.: KfW promotional programs for energy efficiency in buildings. National Roundtable on Financing Energy Efficiency in Latvia, Riga, 26 April 2018 (2018)
21. Office for Energy Efficiency Canton Sarajevo: BiH Energy Efficiency Legislation (2017). https://uredee.ba/bs/energetska-efikasnost/zakonska-regulativa-o-energetskojefikasnosti-u-bih/4. Accessed 12 Feb 2021
22. Office for Energy Efficiency Canton Sarajevo: About the Energy Efficiency Model (2017). https://uredee.ba/bs/energetska-efikasnost/o-modelu-energetske-efikasnosti/3. Accessed 12 Feb 2021
23. Serda: Model of improving energy efficiency in buildings in the Sarajevo Canton area (2021). https://serda.ba/bs/projekti/model-poboljsanja-energetske-efikasnosti-u-zgradarstvu-na-podrucju-kantona-sarajevo/20. Accessed 13 May 2021
24. Bim Bau d.o.o. Sarajevo: The first passive house in BiH was built on Kovači (2021). https://bimbau.ba/kuca-buducnosti-na-kovacima-sagradena-prva-pasivna-kuca-u-bih-akta-ba//. Accessed 13 Feb 2021
25. REIC Sarajevo: Low-energy houses (2009). https://www.yumpu.com/xx/document/view/35129775/niskoenergetske-zgrade-centar-za-energetsku-efikasnost. Accessed 13 Feb 2021
26. PROMO d.o.o. Donji Vakuf: Modern house (2021). http://www.montazneidrvenekuce.info/vijesti/moderna-niskoenergetska-prizemnica-78-m2/429. Accessed 13 Feb 2021
27. Krivaja TMK d.o.o. Zavidovići: Krivaja low-energy houses (2021). http://www.krivajahomes.com/ba/montazne-kuce/niskoenergetske-kuce/. Accessed 13 Feb 2021

28. SAVOX d.o.o. Milići: Prefabricated houses (2021). https://www.savox.ba/en/prefabricated-houses. Accessed 14 Feb 2021
29. Reelih BiH: Special available loans and funds (2021). https://topaodom.ba/kako/financirajte-projekt/posebni-dostupni-krediti-fondovi. Accessed 14 Feb 2021
30. Reelih BiH: Special purpose loans - energy efficiency (2021). https://topaodom.ba/kako/financirajte-projekt/krediti-sa-posebnom-namjenom/. Accessed 14 Feb 2021
31. Bosna Bank International (BBI): Joint project between BBI Bank and Municipality centre to reduce pollution in Sarajevo (2019). https://www.bbi.ba/bs/novosti/zajednicki-projekt-bbi-banke-i-opcine-centar-na-smanjenju-zagadenosti-u-sarajevu. Accessed 15 Feb 2021
32. Sberbank BH: EBRD loan of EUR 10 million for SME sector (2013). https://www.sberbank.ba/bs/ebrd-kredit-u-iznosu-od-10-miliona-eur-za-sme-sektor. Accessed 15 Feb 2021
33. Unicredit Bank: Grant cash bonus for purchasing an energy efficient product (2021). https://www.unicredit.ba/ba/stanovnistvo/kreditiranje/gotovinski_krediti/EBRD.htm. Accessed 15 Feb 2021
34. Raiffeisen BANK dd Bosnia and Herzegovina: Credits for energy efficiency (2021). https://raiffeisenbank.ba/stanovnistvo/krediti-za-energetsku-efikasnost. Accessed 15 Feb 2021

Factors Affecting the Process of Decision-Making at Construction Site; Case Study of Bosnia and Herzegovina

Almedina Mustafić(✉), Ahmed El Sayed, and Hajrudin Džafo

Faculty of Engineering and Natural Sciences, International Burch University, Ilidža, Bosnia and Herzegovina

Abstract. The need for efficient solutions and an efficient decision-making process is growing dizzyingly along with the fast pace of business in today's world. Effective project management requires decisions to be made in accordance with the importance and urgency of the situation and acted accordingly. The decision-making process at construction site is more complex due to certain conditions specific to construction engineering. This paper presents the results of research conducted in Bosnia and Herzegovina, which deals with the analysis of factors that have a direct impact on the decision-making process of the construction site, conducted for the purpose of developing a master's thesis. In-depth interviews were conducted with construction engineers in Bosnia and Herzegovina and the factors that have a direct impact on the decision-making process on the construction site in Bosnia and Herzegovina were identified. This paper presents the results of comparison of factors that have a direct impact on the decision-making process – known from literature-and those identified in the construction industry of Bosnia and Herzegovina. In addition, it represents the factors that have an impact on the decision-making process on the construction site in Bosnia and Herzegovina, listed by construction engineers, which do not appear in case studies from around the world.

Keywords: Construction project management · Decision-making process · Construction industry · Construction site

1 Introduction

What makes a construction site a unique working environment are its specific requirements [1] such as; construction site layout planning [2, 3], organization and management of project execution, choosing the subcontractor [4], positioning of machinery, plan of internal traffic movement and material circulation [5], material supply [6, 7]. Project management is a set of organized groups of skills, tools and activities necessary to achieve project goals [8]. The key success factor in construction project management is precisely decision-making process [9]. Many researches deal with the life of the project, the involvement of factors in the planning phases, decision making and

N. Ademović et al. (Eds.): IAT 2021, LNNS 316, pp. 403–409, 2022.
https://doi.org/10.1007/978-3-030-90055-7_32

functional plans. There are also studies on the owner's, designer's and contractor's participation in decision-making on the construction site [10]. Efficient decision-making is a schematic procedure in which the elements with the planned performance steps are clearly defined. All of this is a result of effective decision-making becoming a major link in the value chain [11]. Much depends on the strategy chosen. The encouragement and internalization of identity is explained by the example that the more experienced and successful decision maker is the more socially reliable in making final decisions [12]. Each decision is crucial for success or failure of the project, and each must be justified. They are made in accordance with the urgency and importance of the situation. Poor performance, breakdowns or unproductiveness on the construction site can lead managers to make critical decisions [13]. A great responsibility is placed on engineers supervising the construction site and leading the construction of the project [14]. Complex situations that go beyond the experience of the responsible person or the reliability of a joint decision - require the use of tools and techniques to aid in decision-making, to ensure the selection of the most effective solution [15]. Decisions optimal at the time they are being adopted could become unsatisfactory after a certain period. The decision-maker must consider what the acceptance of that decision will be, by the environment and community. Both, literature and practice recommend use of different methods to support the decision-making process on the construction site [16]. Multi criteria decision making techniques for engineering problems have developed very rapidly in recent years, and offer a wide range of facilitating decision-making methods [17]. Decision support systems are designed to process and manage structural data, to support not to replace the decision-making process [4].

An effective solution is a quick solution that is acceptable according to technical regulations and at the same time, economically acceptable. The data collected by this research can be useful for finding an effective solution for dynamic situations. By analyzing the state of the project management system in the construction industry of Bosnia and Herzegovina and comparing it with existing factors and methods that support the decision-making process, this paper is expected to answer whether the construction industry in Bosnia and Herzegovina is a suitable platform for applying advanced support methods in the decision-making process.

2 Methodological Approach

The important research questions for this study project related us follows:

- What is the current state of the construction industry in Bosnia and Herzegovina in terms of developed methods to support the construction project management process?
- How is the construction industry of Bosnia and Herzegovina a suitable/unsuitable platform for the application of developed methods to support the decision-making process on the construction site?

In this study assumes that:

a) Managerial and experiential skills of the supervisors and site engineers have a direct impact on the decision-making process at the construction site;

b) Architectural design documentation completeness can affect a decision-making process at construction site;
c) Advanced methodologies that support decision-making process are applicable in the field of construction industry in Bosnia and Herzegovina.

Qualitative analysis was applied in processing of information as a result of a direct interview with construction engineers in Bosnia and Herzegovina. In order to get the professional opinion and statements related to the impact of particular factors and their frequency of occurrence. After information being analyzed and compared, the real situation shows that is possible to apply decision-making process methodologies in construction industry of Bosnia and Herzegovina, in order to improve projects managing.

In-depth interview method applied. An interview is defined as a conversation between two people on a topic of common interest, according to a system based on questions and answers [18]. Qualitative research opens the possibility to achieve new knowledge, knowledge of new facts with an explanation of existing theories or hypotheses from the real life of the person we are examining [19]. In-depth interviews are among the most commonly used tools of qualitative research methodology. The reason for this is their open characteristics and discovery-oriented nature. They allow the researcher to discover the respondent's deep relationship to the topic of conversation [20].

In-depth interview answering to the question "*What are the factors that have a direct impact on project management at construction project?*" leaded the respondents to share their experience about what influenced their project management process and the importance of the engineer's abilities and experience. Answering to the question "*How many times in your experience, has the completeness of design documentation emerged as a factor influencing your project management of the construction project?* We could find out from the respondents which this factor affects the decision-making process on construction sites in Bosnia and Herzegovina. We learn about engineer skills through answering to the question "*What systems and methods do you use, when deciding or solving dynamic situations that affect the functionality of the work*?

The state of representation of developed methods for decision support in the construction industry of Bosnia and Herzegovina, was explored by obtaining answers to questions:

- *What systems and methods do you use, when deciding or solving dynamic situations that affect the functionality of the work?*
- *What software do you use to follow work dynamics at the construction site?*

Criteria for Evaluation the Decision-Making Process at the Construction Site in Bosnia and Herzegovina: From literature review identified factors affecting the process of decision-making at construction site. Through the analysis of case studies worldwide, research done about the factors that have an impact on the decision - making process on the construction site, presents the factors and lists them according to the strength of the impact and frequency of occurrence, [21]:

(1) Construction site layout planning,
(2) Supplier selection process;

(3) Materials management;
(4) Subcontractor selection;
(5) The tender process;
(6) Outcomes of stakeholder's management process;
(7) Site location;
(8) Appropriate choice of construction methods;
(9) The quality of design documentation

Target Population Sample: The population selected for research purposes are engineers employed in construction industry of Bosnia and Herzegovina. The main professions of engineers who throughout the process participate in the implementation of the construction project: project manager, supervision of the construction project and supervisors on the construction site. All respondents have minimum five years of work experience. Business owners, although eligible, are exempt from the target population.

3 Results and Discussion

Comparing the Results from the Interviews with the Literature Findings
From the Table 1 it can be concluded that construction engineers in Bosnia and Herzegovina do not consider the choice of location for the construction site and its layout planning as factors that have a direct impact on the decision-making process at the construction site.

Table 1. Factors that have a direct impact on the decision-making process at construction site; case study of Bosnia and Herzegovina

	Factors:	direct impact on the decision-making process at construction site; case study of Bosnia and Herzegovina
1	Construction site layout planning	□
2	Supplier selection process	■
3	Materials management	■ ■
4	Subcontractor selection	■ ■ ■
5	The tender process	■ ■
6	Outcomes of stakeholder's management process	■ ■ ■ ■ ■
7	Site location	□
8	Appropriate choice of construction methods	■ ■ ■ ■
9	The quality of design documentation	■ ■ ■ ■ ■ ■ ■ ■

The least mentioned factor that coincides with the factors from the literature review is the choice of construction material supplier. This group also includes site management materials and the tender process. The factors most often mentioned in almost every interview are the quality and completeness of the design documentation and the investor/stakeholder's management process. By comparing the factors, it was proven that most of the factors found in the construction industry of Bosnia and Herzegovina agree with the factors listed in the literature review section. The interviews proved that there are factors that have a direct impact on the decision-making process on the construction site in Bosnia and Herzegovina, which did not appear in the literature review section. These factors, as well as the others mentioned, cause delays in the execution of certain works on the construction project, which leads to an increase in costs or loss of earnings; and are listed below:

1. Bureaucratic system
2. Construction of facilities in phases where a new tender is announced for each phase and another contractor is selected
3. Project value - affects the quality of technical documentation
4. Lack of skilled workforce
5. Legal documentation - collecting urban and construction permits
6. Education in the construction profession.

Situations encountered by engineers in the construction industry in Bosnia and Herzegovina during the implementation of construction projects, cause frequent delays in the execution of works, increased costs or economic losses. The reasons for that, as they state in the interviews, are most often problems with finances, small investments or technical documentation. Problems in resolving dynamic situations on the construction site occur due to the great influence of investors, according to answers gathered from construction engineers in construction industry of Bosnia and Herzegovina. It is often stated in interviews that investors sometimes have a blackmailing approach to prices, which calls into question both; the quality of design documentation that is forced to be completed in a minimum time, and the quality of construction work that is also performed in incorrect conditions.

The emergence of foreign investors in Bosnia and Herzegovina is frequent. Construction engineers have stated that there has been a great demand lately for the materialization of buildings that is not in line with our climatic conditions, which calls into question the quality of the building and the entire life cycle. Unfortunately, as proved in the second hypothesis, the design of the project and the entire architectural documentation and its quality are often the cause of problems on the construction site. As reasons for this, engineers cite low investor prices and demands for the project to be done in a very short time. In these circumstances, lower quality of project documentation is expected, as well as incompleteness or lack of certain positions and details, caused by lack of time. The mentioned topic is an increasingly frequent topic of discussion and it is necessary to thoroughly investigate in order to determine and eliminate the causes of this problem.

The tender process, according to the interview answers, is not a big problem. While construction in phases is mentioned as an aggravating factor, where a tender is announced for each phase of construction of the facility and another contractor is selected. This leaves the possibility of various errors by the first contractor, which the next contractor has to deal with. Concerns were expressed about the competence of companies engaged in the construction of buildings in the Bosnian-Herzegovinian construction industry. The problems of the decreasing number of skilled craftsmen for construction jobs were commented, saying that those who are still there are "universal" masters who, when they come to the construction site, must know how to do every part of the job. The construction industry in Bosnia and Herzegovina should take into account the consideration of this problem and create legal provisions for this way of project implementation. As well as considering launching social projects that will present construction jobs in a good way and training young people for masters in construction.

4 Conclusion

The construction industry in Bosnia and Herzegovina is not as developed as in most European countries, for this reason, various methods to support the decision-making process on the construction site have not found their use there. Despite this fact, through this research, it is concluded that the construction industry/construction companies in Bosnia and Herzegovina, have engineers interested in improving the work quality. Companies also have access to more advanced software solutions to support work, and it is up to them to decide to implement this into the day-to-day work process. Also, in order to improve the way of doing business, companies that are active in the construction industry should invest in educating their staff on advanced methods to support the decision-making process on the construction site, as well as more advanced use of software solutions.

References

1. Scherer, R.J., Schapke, S.E.: A distributed multi-model-based management information system for simulation and decision-making on construction projects. Adv. Eng. Inform. **25**(4), 582–599 (2011)
2. Ning, X., Lam, K.-C., Lam, M.C.-K.: A decision-making system for construction site layout planning. Autom. Constr. **20**(4), 459–473 (2011)
3. Dhanure, Y., Pathak, S.: Optimization of site layout planning for multiple construction stages with safety consideration and requirements. Int. Res. J. Eng. Technol. (IRJET) **3**(5), 3173–3176 (2016)
4. Rosmayati, M., Hamdan, A.R., Otoman, Z.A., Mohamad Noor, N.M.: Decision support systems (DSS) in construction tendering processes. Int. J. Comput. Sci. Issues **7**(2), 35–45
5. Madjoubi, L., Yang, J.L.: An intelligent materials routing system on complex construction site. Logist. Inf. Manag. **14**(5/6), 337–344 (2001)
6. Kar, A.K.: Modeling of supplier selection in e-procurement as a multi-criteria decision making problem. Sprotus **9**(40). Modeling of Supplier Selection in e-Procurement as a Multi-Criteria D" by Arpan Kumar Kar. https://aisnet.org/. Accessed 27 June 2021
7. Cengiz, A.E., Aytekin, O., Ozdemir, I., Kusan, H., Cabuk, A.: A multi-criteria decision model for construction material supplier selection. Procedia Eng. **192**, 294–301 (2017)
8. Kerzner, H.: Project Management: A Systems Approach to Planning, Scheduling and Controlling, 8th edn. Wiley, Ohio (2002)

9. Marovic, I., Hanak, T.: Selection of adequate site location during early stages of construction project management: a multi-criteria decision analysis approach. In: 3rd International Conference on Innovative Materials, Structures and Technologies, Riga, Latvia, 27–29 September 2017 (2017)
10. Shapira, A., Laufer, A., Shenhar, A.J.: Anatomy of decision making in project planning teams. Int. J. Project Manag. **12**(3), 172–182 (1994)
11. Drucker, P.F.: The Effective Decision, pp. 1–91. Harvard University, Graduate School of Business Administration, A Harward Business Review Papaerback
12. March, J.G.: 'A Primer on Decision Making: How Decisions Happen', 1st edn. Simon & Schuster, New York (1994)
13. Ashley, D.B., Uehara, K., Robinson, B.E.: Critical decision making during construction. J. Constr. Eng. Manag. **109**(2), 146–162 (1983)
14. Poon, S.W., Price, A.D.F.: Decisions made on construction sites. In: 15th Annual ARCOM Conference, Liverpool, UK, 15–17 September 1999 (1999)
15. Espinoa, D.J., Castillo-Lopez, E., Hernandez, J.R., Canteras-Jordana, J.C.: A review of application of multi-criteria decision making methods in construction. Autom. Constr. **45**, 151–162 (2014)
16. Szafranko, E.: Decision problems in management of construction projects. In: 3rd International Conference on Innovative Materials, Structures and Technologies, Riga, Latvia, 27–29 September 2017 (2017)
17. Zavadskas, E.K., Antucheviciene, J., Adeli, H., Turskis, Z.: Hybrid multiple criteria decision-making methods: a review of applications in engineering. Int. J. Sci. Technol. **23**(1), 1–21 (2016)
18. Kvale, S., Brinkmann, S.: Interviews: An Introduction to Qualitative Research Interviewing, 2nd edn. Sage Publications, Thousand Oaks (2009)
19. Tracy, S.J.: Qualitative Research Methods: Collecting Evidence, Crafting Analysis, Communication Impact. Wiley, Oxford (2013)
20. Guion, L., Diehl, D., McDonald, D.: Conducting an In-Depth Interview. University of Florida, Florida (2001). https://journals.flvc.org/edis/article/download/127025/126671
21. Mustafić, A., Špago, S., Novalić, A., El Sayed, A.: Evaluation of factors affecting the process of decision making in construction site. In: Karabegović, I. (ed.) NT 2020. LNNS, vol. 128, pp. 888–896. Springer, Cham (2020). https://doi.org/10.1007/978-3-030-46817-0_100

Quality in Construction Project Management Process

Žanesa Ljevo(✉), Suada Sulejmanović, Mirza Pozder, Ammar Šarić, and Sanjin Albinović

Faculty of Civil Engineering, University of Sarajevo, Sarajevo, Bosnia and Herzegovina

Abstract. The construction sector, and thus the construction projects, are an essential factor in each country's economy. In Bosnia and Herzegovina (BIH), project management is still at a low level of performance. Over fifteen million new project management roles globally across seven project-intensive industries (construction is among them) were created between 2010–2020. This paper will present the research results in BIH, aiming to demonstrate the possibility of identifying the influence that the project management processes have on the quality of the end product at delivery. Through a survey and case studies, 79 (75) construction projects were examined. The survey covered clients, designers, contractors, and consultants. The obtained results showed differences in results through exploratory factor analysis models in the same project phase, and those models will help all participants in the project. This research will help participants in construction projects focus on the key factors that direct the project towards achieving its final quality.

Keywords: Construction · Project · Quality · Project management process

1 Introduction

The formal usage of project management (PM) leads to better and satisfying results (meet goals and business intent) in more than 80% of projects. However, the construction industry shows traditionally poor project management performance in their projects. Disciplined project management is the best way to ensure project deliverables or benefits [1].

It is estimated that over 15.7 million new PM roles were added globally across seven project-intensive industries by 2020, and construction is one of those industries [2]. By 2027, employers will need 87.7 million individuals working in PM-oriented roles [3].

Without the support of the top management, PM will not be adopted. Project management, for them, does not represent almost anything or very little. As a result, there are shortcomings in projects [4].

Research shows that small and medium-sized companies make up 99.8% of all companies in the E.U., generate 56% of GDP, and employ 70% of private ownership. However, no tendency to use known tools and techniques for PM has also been shown in small and micro-companies [4].

N. Ademović et al. (Eds.): IAT 2021, LNNS 316, pp. 410–420, 2022.
https://doi.org/10.1007/978-3-030-90055-7_33

Project management has evolved in the last few decades. Traditional PM skills were developed from construction industries' requirements to plan, control, and manage large and complex "tangible" projects [5].

Project management is a relatively recent professional endeavor rapidly growing to keep pace with the increasingly complex job market. The dedicated or standardized PM ensures that a civil engineering project (structural, transportation, and hydraulic engineering) has the time and resources to deliver all outputs and that outputs at every phase are quality tested. In practice, without project management, clients have a significant opportunity to be exposed to unclear objectives, a low-quality project, and late delivery of project, over budget, high risk. For quality assurance, project management is very important [6].

Construction companies that are poorly managed can spend up to 40% of their income because they had poor quality. Low quality is one reason for defects during work, outages, and wastages, and those reasons are significant areas of project management in construction projects [7].

2 Overview of Literature

By definition, projects are temporary organizations, limited by a particular scope, and implemented within a certain amount of time [3].

The quality of the project is defined as meeting the agreed project requirements, and it represents the architects' standard of a project and the client's stat-ed specification and instruction [8].

The quality of the project is defined by two dimensions: product quality and process quality [9].

The quality of construction is improving because of the supervisory system's implementation during the year. For the quality of construction, the availability of resources is essential, and more significant productivity is directly linked. This factor showed the product quality, but the process quality factor is not clearly defined [10].

In the literature, product quality is very well defined, but process quality is very poorly defined. The factors that affect quality management are found in the literature. However, this is not the case with the quality factors of the project management process. They are analyzed by university professors and separated as factors that directly affect the quality of the project management process. These quality factors (all project phases) are the focus of this research [11–16].

Three criteria or objectives for assessing the project's success are time, cost, and quality (iron triangle). Most people consider the iron triangle the project's principal target, but others suggest something more complex, defined in the literature on project management. The quality of the project, being the third dimension of the "iron triangle" is much more difficult to define, assess, plan, and control [9, 17].

In the literature, there are defined methods and techniques for time and cost planning and control (for example, Gantt chart, Critical Path Method (CPM), Earned Value Management method (EVM), Earned Schedule (ES) method). The Earned Schedule and Earned Value (project management methods) are used in construction projects for time and cost control. Through the cost breakdown on the different levels, structure cost may be organized [18].

The methods and techniques for planning and controlling the quality project management process for project life were not found in the literature.

A successful construction project requires quality control and quality policy, which is important for the project life cycle (from the conception phase to the execution phase). That will reduce and prevent costly repairs [19].

3 Research Methodology

In the first step, intense data collection from the literature (books, papers, online materials, etc.) about the project management process and the factors that affect his quality was made. In the next steps, when more information about a topic was known, a literature review and selection were made (Fig. 1) [16].

The key quality factors for the project management process (Q1 - planning and control; Q2 - involvement teamwork; Q3 - expertise, knowledge; Q4 - focus on the customer; Q5 - top management support; Q6 - communication; Q7 - continuous improvement; Q8 - coordination among project participants; Q9 - quality policy; Q10 - availability of resources; Q11 - supplier's quality management) were selected. Questions were created with all available information, and an online survey was sent to construction companies in Bosnia and Herzegovina by email. The accuracy of the data was expected, although the survey was anonymous but targeted [16, 20].

After that, face-to-face interweaves on a case study with targeted interlocutors were made. Due to the research's relevance and quality, project participants all over Bosnia and Herzegovina on different construction projects (structural engineering or water and traffic infrastructure, roads, bridges, buildings) were interviewed.

The variables - quality factors of the project management process of each project phase are numerically reduced by grouping them into new factors, using exploratory factor analysis (EFA). EFA examines the interdependencies within many variables and tries to explain them with a small number of common factors. Based on several criteria, the number of factors is determined. One of the criteria that were used is the root's characteristic size – Kaiser. The factors that have the root's characteristic size larger than ($\lambda \geq 1$) will remain. The variance explained by the variable on the corresponding factor is represented by the factor loading. The factor loading expresses the relationship of each variable to the corresponding factor. It is necessary to determine if the factor loadings are significant enough. To achieve significance for the 79 sample size (survey) factor loading should be 0.548 and 0.633 for the 75 sample size (case study). The highest factor loading indicates the variables that have the strongest correlation with a given factor [21, 22].

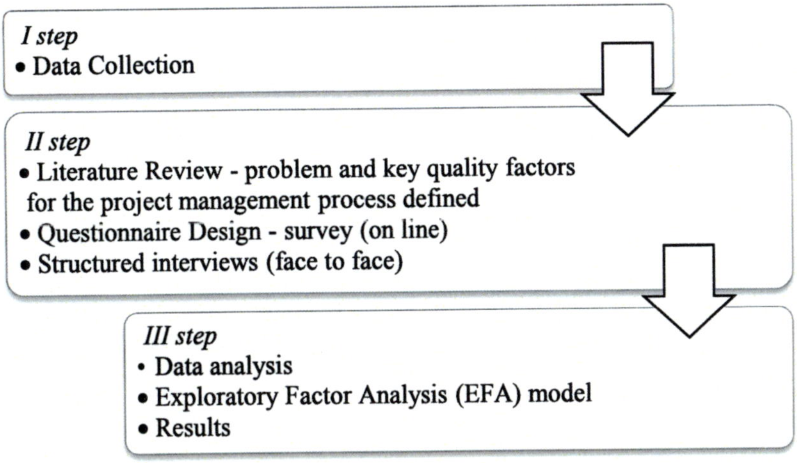

Fig. 1. Methodological approach

4 Results

A measure of sampling adequacy (MSA) quantifies the degree of correlation between variables and the justification of factor analysis. The index ranges from 0 to 1. If the MSA value is over 0.80, there is a very strong correlation, MSA between 0.70 and 0.80 - strong correlation, 0.60 and 0.70 - medium, and 0.50 and 0.60 - weak correlation. A measure of sampling adequacy presented a medium to very strong correlation between variables in every project phase (Table 1).

The value of the Kaiser–Meyer–Olkin (KMO) test measures sampling adequacy for each variable in the model as the following prerequisite, and it should be more than 0.50 in order to be included in the EFA. The minimum values of KMO are 0.721 (case study) and 0.823 (survey), which was the condition for the variable to be included in the factor analysis (Table 1).

Bartlett's test (Chi-Square) is used to test the null hypothesis that there is no significant correlation between the original variables. Based on the p (less than 0.05) value, the null hypothesis is rejected, which means a significant correlation between the original variables exists. The values less than 0.05 of the significance level indicate that factor analysis may be useful with data. In this research, $p = 0.00$ for both cases in every project phase, and the factor analysis is useful with data.

Table 1 shows all EFA (the values MSA, KMO, Chi-Square, communality, factor loadings, number of factors, Squared Loadings) results for surveys and case studies in each project phase.

Communality (the sum of the squares of the factor loadings) is a definition of common variance that ranges between 0 and 1. A particular variable's commonality tells how many variables are explained by common factors (the proportion of a standard variable's variance to all other variables together shows) and should be $\geq$0.50 [21, 22]. Communalities for all project phases are over 0.50; for example, in the execution phase, the minimum value is 0.511 (survey) and 0.643 (case study).

Table 1. EFA results for project phases – surveys and case study

		Conceptual	Defining and planning	Execution
MSA min.-max	Survey	0.632–0.908	0.710–0.887	0.722–0.880
	Case study	0.652–0.908	0.629–0.847	0.706–0.917
KMO	Survey	0.831	0.827	0.823
	Case study	0.787	0.721	0.777
Chi-Square	Survey	425.875	481.195	566.975
	Case study	392.815	574.188	493.622
p	Survey	0.00	0.00	0.00
	Case study	0.00	0.00	0.00
Min. – max. communality	Survey	0.521–0.775	0.520–0.747	0.511–0.821
	Case study	0.523–0.823	0.532–0.964	0.643–0.906
Min. – max. loading factors	Survey	0.564–0.842	0.559–0.853	0.616–0.891
	Case study	0.653–0.880	0.665–0.963	0.660–0.920
Number of factors	Survey	3	3	3
	Case study	3	3	3
Rotation sums of squared loadings - cumulative % (>60%)	Survey	64.057	66.605	70.098
	Case study	69.062	77.046	72.955

All factor loadings (Table 2) should be greater than 0.548 (survey, 79 samples) or 0.633 (case study, 75 samples). Minimum factor loadings for the new factors are 0.564 (conceptual), 0.559 (defining and planning), and 0.616 (execution) for surveys in every phase, and they are all above 0.475. Minimum factor loadings for the new factor (case study) are 0.684, 0.665, and 0.660, and they are all above 0.633.

Models with new factors and their factor loadings for survey and case study for the conceptual phase are presented below (Fig. 2, Table 1 and 2).

The conceptual phase factor models (survey, case study) are defined with three factors: planning, market conditions, and top management commitment. These models describe 64.067% (survey) and 69.062% (case study) of the variance of the basic set of variables, according to the criterion which is defined in Table 1. EFA results (Table 1) for the survey and case study are models with three factors in the conceptual phase.

The first factor is R1, called planning, with five variables: planning and control (0.806), expertise/knowledge (0.564), communication (0.696), coordination among project participants (0.619), and quality policy (0.703). As opposed to the survey, in the case study, the first factor is top management involvement with four variables: involvement teamwork (0.794), focus on the customer (0.844), top management support (0.684), and quality policy (0.706), see Table 2, and Fig. 2.

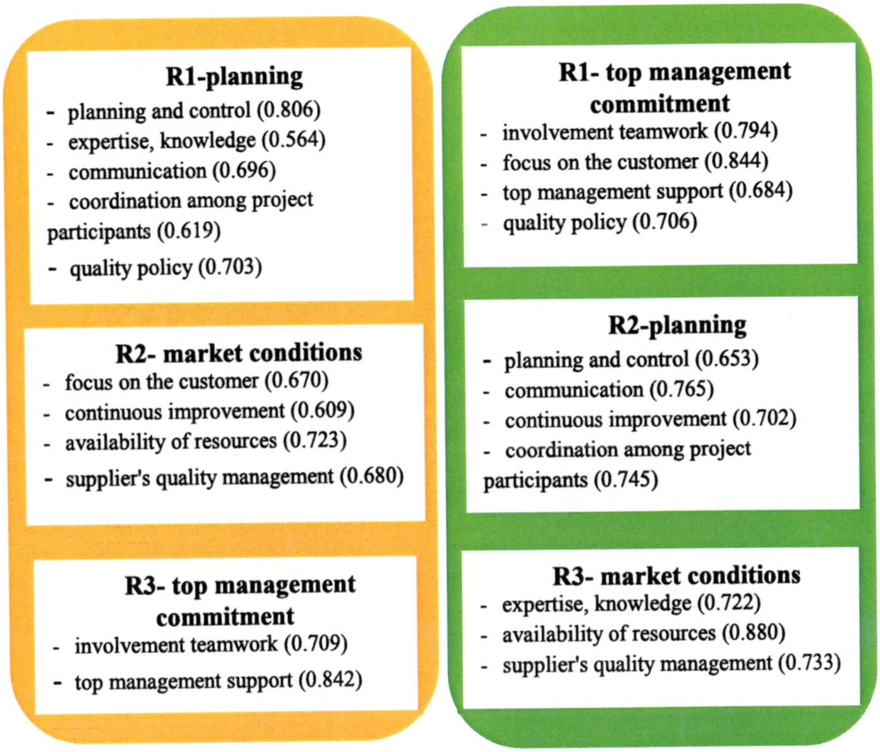

Fig. 2. Models for conceptual phase survey and case study

The second factor is R2, called market conditions, with four variables: focus on the customer (0.670), continuous improvement (0.609), availability of resources (0.723), and supplier's quality management (0.680). As opposed to the survey, in the case study, the second factor is planning involvement with four variables: planning and control (0.653), communication (0.765), continuous improvement (0.702), and coordination among project participants (0.745), see Table 2, and Fig. 2.

The third factor is R3 top management commitment with two variables: involvement teamwork (0.709) and top management support (0.842). As opposed to the survey, in the case study, the third factor is market conditions involvement with three variables: expertise, knowledge (0.772), availability of resources (0.880), supplier's quality management (0.733), see Table 2, and Fig. 2.

The execution phase factor model (survey) with three factors: planning, market conditions, and supplier involvement, described 70.098% of the variance of the basic set of variables. Furthermore, 72.955% (case study) of the variance of the basic set of variables described three factors: knowledge and resources, management, and quality policy see Table 1, and Fig. 3.

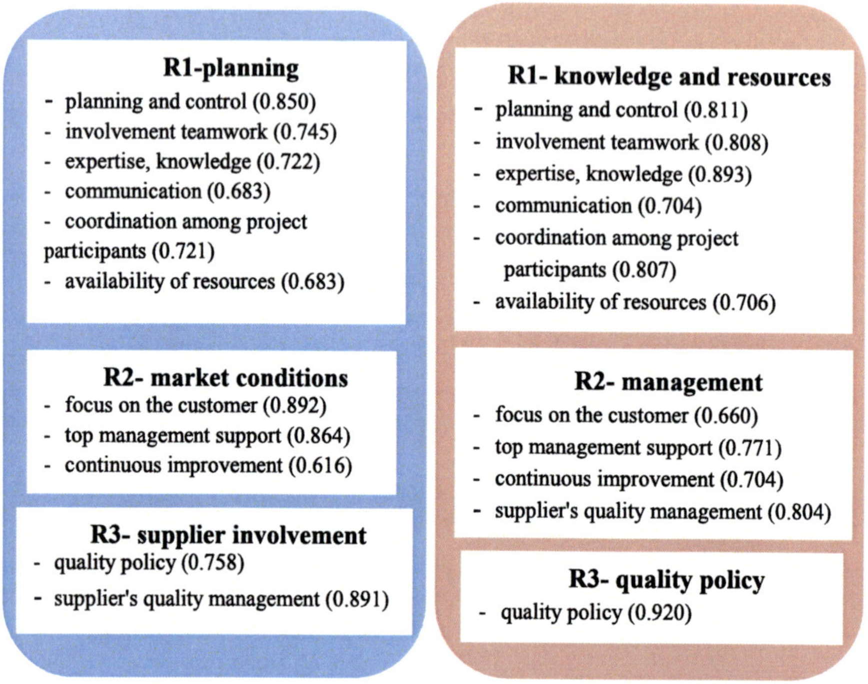

Fig. 3. Models for execution phase survey and case study

The first factor in the execution project phase (survey) is R1 called planning with six variables: planning and control (0.850), involvement teamwork (0.745), expertise, knowledge (0.722), communication (0.683), coordination among project participants (0.721), and availability of resources (0.683). As opposed to the survey, in the case study, the first factor is knowledge and resources, which has six variables: planning and control (0.811), involvement teamwork (0.808), expertise, knowledge (0.893), communication (0.704), coordination among project participants (0.807) and availability of resources (0.706), see Table 2, and Fig. 3.

The second factor in this project phase is R2, called market conditions, with three variables: focus on the customer (0.892), top management support (0.864), and continuous improvement (0.616). In the case study, the second factor is management involvement with four variables: focus on the customer (0.660), top management support (0.771), continuous improvement (0.704), and supplier's quality management (0.804), see Table 2, and Fig. 3.

The third factor is R3 supplier involvement with two variables: quality policy (0.758) and supplier's quality management (0.891). In the case study, the third factor is a quality policy with one variable quality policy (0.920), see Table 2 and Fig. 3.

Table 2. Factor structure matrix after varimax rotation for phase

	Main components after rotation			Main components after rotation		
	1	2	3	1	2	3
Factor	**Conceptual - Survey**			**Conceptual - Case study**		
Q1	**0.806**	0.098	0.229	0.286	**0.653**	0.119
Q2	0.509	0.007	**0.709**	**0.794**	0.296	−0.100
Q3	**0.564**	0.427	0.274	0.358	0.087	**0.722**
Q4	0.013	**0.670**	0.267	**0.844**	0.012	0.283
Q5	0.034	0.255	**0.842**	**0.684**	0.391	0.000
Q6	**0.696**	0.048	0.140	0.296	**0.765**	0.055
Q7	0.546	**0.609**	−0.028	0.010	**0.702**	0.260
Q8	**0.619**	0.596	0.104	0.162	**0.745**	0.483
Q9	**0.703**	0.366	−0.070	**0.706**	0.156	0.390
Q10	0.268	**0.723**	−0.085	−0.106	0.196	**0.880**
Q11	0.107	**0.680**	0.259	0.210	0.380	**0.733**
Factor	**Definition and planning - Survey**			**Definition and planning - Case study**		
Q1	**0.804**	−0.085	0.090	**0.829**	0.200	0.143
Q2	**0.771**	0.175	0.226	**0.790**	0.157	0.193
Q3	**0.743**	0.323	0.067	0.173	0.140	**0.950**
Q4	0.109	−0.088	**0.853**	0.442	**0.778**	−0.061
Q5	0.144	0.228	**0.668**	0.355	**0.847**	0.090
Q6	**0.581**	0.236	0.551	0.584	0.079	**0.665**
Q7	0.190	**0.628**	0.433	MSA < minimum excluded		
Q8	**0.720**	0.422	0.201	0.156	0.111	**0.963**
Q9	0.489	**0.605**	0.242	0.119	**0.665**	0.275
Q10	0.073	**0.846**	−0.104	**0.805**	0.086	0.173
Q11	0.232	**0.559**	0.498	−0.120	**0.821**	0.064
Factor	**Execution - Survey**			**Execution - Case study**		
Q1	**0.850**	0.063	0.031	**0.811**	0.091	−0.075
Q2	**0.745**	0.274	0.170	**0.808**	0.329	0.004
Q3	**0.722**	0.053	0.194	**0.893**	0.171	−0.020
Q4	0.091	**0.892**	0.131	0.466	**0.660**	0.091
Q5	0.242	**0.864**	−0.019	0.154	**0.771**	0.268
Q6	**0.683**	0.405	0.215	**0.704**	0.057	0.573

(continued)

Table 2. (*continued*)

	Main components after rotation			Main components after rotation		
	1	2	3	1	2	3
Q7	0.395	**0.616**	0.378	0.407	**0.704**	0.061
Q8	**0.721**	0.265	0.371	**0.807**	0.112	0.241
Q9	0.299	0.275	**0.758**	0.013	0.242	**0.920**
Q10	**0.683**	0.191	0.091	**0.706**	0.320	0.206
Q11	0.109	−0.010	**0.891**	−0.023	**0.804**	0.054

5 Discussion and Conclusion

The factor models for conceptual and execution project phases are the results of this research. The factor models (conceptual phase) are defined with three factors: planning, market conditions, and top management commitment. These models describe 69.062% (case study) and 64.067% (survey) of the variance of the basic set of variables (Table 1, Table 2, Fig. 2). The execution phase factor model (case study) with three factors: knowledge and resources - six variables, management - four variables, and quality policy - one variable described 72.955% of the variance of the basic set of variables. Furthermore, 70.098% (survey) of the variance of the basic set of variables described three factors: planning - six variables, market conditions - three variables, and supplier involvement - two variables (Table 1, Table 2, Fig. 3).

The quality models of the project management process of the phases throughout the project life cycle (construction projects) are elaborated in Bosnia and Herzegovina.

The survey covered investor, contractor/subcontractor, designer, project manager, and consultant like different management perspectives included in the project management process, but in the case studies, interviews were face-to-face with participants of that project phase.

Defined models are sets of the key factors of quality project management processes for a survey and a case study. These models can help participants in construction projects (Investor, Contractor/Subcontractor, Designer, Project Manager, Consultant) to focus on the quality (project management process) factors that were marked as important. They are then used as indicators and enable to direct the project towards achieving the product's final quality. It is possible to identify factors critical to quality project management processes (which will be a collection of previously adopted factor variables).

The differences in the models (survey and case studies) are visible. One reason for this is that the case study was based on interviews performed with carefully chosen participants, while the survey was conducted by email correspondence. The variables were evaluated with scores from one to six in the survey by the participants involved in the survey, but through face-to-face interviews with the chosen project participants (direct participants of each phase), the variables were precisely determined.

Models resulting from case studies can help project participants in certain phases to more easily identify and adequately respond to possible factors that will affect the project management process's quality.

Planning and control, like a variable of factor R1 in the execution phase, is measured by whether the plans created in the previous phase are used and controlled continuously and wherreted they are harmonized with the actual stated of the project implementation.

Customer satisfaction (focus on the customer-like variable factor market conditions) like variable of factor market conditions-survey (management-case study) needs to be evaluated with performance objectives [19].

Using some of the project management software packages by participants of a project, planning, and control is very easy to achieve, as part of the quality project management process, and for a better quality of project management process, and all participants need to work on soft skills for the better quality PM process.

Some of the companies in BIH are still using spreadsheets and manual processes. However, they must use enterprise-level project management solutions for scheduling, costing, management, and reporting to drive enterprise-level collaboration and visibility. Companies should combine standardized project management processes with the right technology and achieve quality project management processes (models quality PM process – research) to achieve better outcomes than their competitors.

References

1. Radujković, M., Sjekavica, M.: Project management success factors, creative construction. In: Creative Construction Conference 2017, CCC 2017, Primošten, Croatia, 19–22 June (2017)
2. Project Management Institute – PMI: The Strategic Impact of Projects (2015). https://www.pmi.org/-/media/pmi/documents/public/pdf/learning/thought-leadership/pulse/identify-benefits-strategic-impact.pdf?v=ba31966c-6d1a-4016-b43f-a0ccd34b9af5. Acceseed 20 Jan 2021
3. Project Management Institute – PMI: Project Management Job Growth and Talent Gap 2017–2027 (2017). https://www.pmi.org/learning/careers/job-growth. Acceseed 20 Jan 2021
4. Turner, R., Ledwith, A., Kelly, J.: Project management in small to medium-sized enterprises: matching processes to the nature of the firm. Int. J. Project Manage. **28**, 942–957 (2010)
5. Bourne, L., Walker, D.H.T.: Advancing project management in learning organizations. Learn. Org. **11**(3), 226–243 (2004)
6. Atson, B.: Why Is Project Management So Important to An Organization? (2017). https://thedigitalprojectmanager.com/why-is-project-management-important/. Acceseed 21 Jan 2021
7. Mashwama, N., Aigbavboa, C., Thwala, D.: An assessment of the critical success factor for the reduction of cost of poor quality in construction projects in Swaziland. Proc. Eng. **196**, 447–453 (2017)
8. Seeley, I.H., Murray, G.P.: Civil Engineering Quantities, Revised Jossey-Bass Publishers, San-Francisco (2001)
9. Turner, J.R.: Managing Quality. Gower Handbook of Project Management. Gower Publishing (2002)
10. Yung, P., Yip, B.: Construction quality in China during transition: a review of literature and empirical examination. Int. J. Project Manage. **28**(1), 79–91 (2010)
11. Ogwueleka, A.C.: A review of safety and quality issues in the construction industry. J. Constr. Eng. Project Manage. **3**(3), 42–48 (2013)

12. Husin, N.H., Adnan, H., Jusoff, K.: Management of safety for quality construction. J. Sustain. Dev. **1**(3), 41–47 (2008)
13. Joaquin, C.D., Hernandez, D., Aspinwall, E.: A framework for building quality into construction projects - Part I. Total Qual. Manage. **19**(10), 1013–1028 (2008)
14. Hoonakker, P.L.T.: Quality management in Construction Industry. In: Proceedings of ASQ World Conference on Quality and Improvement, Milwaukee, WI, USA, 1–3 May (2006)
15. Jraisat, L.: Quality in construction management: an exploratory study. Int. J. Qual. Reliab. Manage. **33**(7), 920–941 (2016)
16. Ljevo, Ž.: Uticaj upravljanja projektima na ostvarenje kvalitete kod građevinskih poslovnih sistema, Faculty of Civil Engineering at University of Sarajevo (2017)
17. Agarwal, N., Rathod, A.: Defining success for software projects: an exploratory revelation. Int. J. Project Manage. **24**(4), 358–370 (2006)
18. Solis, R., Carmen Sarahí, M.G., Jesús, Z.G.: Time and cost control in construction projects in southeast Mexico. Ingeniería Investigación y Tecnología **18**(4), 411–422 (2017)
19. Emiedafe, W.: Reasons Why Quality Control is Important for a Successful Construction Project (2017). https://sapientvendors.com.ng/construction-quality-control/. Acceseed 25 Jan 2021
20. Ljevo, Ž, Vukomanović, M., Džebo, S.: Assessing the influence of project management on quality during the early phases of construction projects. Org. Technol. Manage. Constr. **9**, 1584–1592 (2017)
21. Hair, F., Babin, C., Black, B., Anderson, R.: Multivariate Data Analysis, 7th edn. Cengage Learning, EMEA, Hampshire (2018)
22. Hair, F., Babin, C., Black, B., Anderson, R.: Multivariate Data Analysis - A Global Perspective. Pearsib, New Jersey (2010)

Seismic Vulnerability Analysis in Urban and Rural Regions of Visoko, BIH

Naida Ademović[1] (✉), Nermina Zagora[2], and Marijana Hadzima-Nyarko[3]

[1] Faculty of Civil Engineering, University of Sarajevo, Patriotske lige 30, 71 000 Sarajevo, Bosnia and Herzegovina
naida.ademovic@gf.unsa.ba

[2] Faculty of Architecture, University of Sarajevo, Patriotske lige 30, 71 000 Sarajevo, Bosnia and Herzegovina

[3] Faculty of Civil Engineering and Architecture Osijek, University J.J. Strossmayer Osijek, Osijek, Croatia

Abstract. Seismic vulnerability assessment of existing structures is of extreme importance as it provides information on the possible damage of structures if earthquakes of various intensities are to happen. In the case of scarce information, rapid vulnerability analysis with the application of the macroseismic model with vulnerability index can be implemented. This method is efficient for large-scale analysis as opposed to the detailed approaches which are used for the assessment of individual buildings. In the present study 94 buildings were elaborated of which 56.4% are located in the urban and 43.6% in the rural area. Three types of masonry structures were identified, unreinforced masonry (URM) with flexible floors, URM with rigid floors, and confined masonry. The average vulnerability index is in the range from 0.571 to 1.08 and the average value for the whole database set was 0.733. The most probable earthquake intensities in this region are VI and VII. For the higher intensity, the largest damage is expected for URM with flexible floors where 17.1% would experience significant to severe damage, and 33.2% moderate damage. Confined masonry would have the least damage, 56.8% of the structures would experience no damage at all, 34% slight damage, while only 8.1% moderate damage would be noted.

Keywords: Seismic analysis · Masonry structures · Vulnerability index · Behavior modifiers

1 Introduction

Visoko is located 22.42 km from Sarajevo (the capital of Bosnia and Herzegovina) and is generally considered a medium seismic zone with a peak ground acceleration (PGA) of 0.17 g.

Several seismic vulnerability assessment methods exist that can be used in the analysis of buildings [1–4]. Generally, there are three groups of these methods: empirical, analytical, and hybrid methods. Relying on expert judgment is common for all methods. In the process of the analysis, at certain points, assumptions have to be made, which are provided by selected experts. The selection of an adequate vulnerability method, which

N. Ademović et al. (Eds.): IAT 2021, LNNS 316, pp. 421–429, 2022.
https://doi.org/10.1007/978-3-030-90055-7_34

will be used plays a crucial role in the process of seismic risk assessment. Which method will be selected as the most convenient one is a function of several features, first of all on the obtained information, buildings' features, management of the collected data, and decision-makers. To perform the seismic vulnerability in this specific case vulnerability index method (VIM) was implemented. This method was a part of the Risk-UE project, which started in 2001 and was finalized in 2004 [5]. Until now seismic risk assessment of buildings in various European and African cities was done by applying this method. In their work, Lantada et al. [6] obtained seismic risk results for cultural historic structures in Barcelona. Pitikalis et al. [7] conducted a seismic assessment of the buildings located in Thessaloniki in Northern Greece, in two cities in Marocco seismic assessment with the application VIM method was done by Cherif et al. [8, 9] and in one of them, site effects were taken into account. The influence of the site effects on damage to the buildings was investigated for the Molise Region of Italy [10]. In Bosnia and Herzegovina, it was chosen to do the first assessment for buildings in Banja Luka due to its possible high-intensity levels of ground motion and the capital city of Bosnia and Herzegovina, Sarajevo [3].

2 Typological and Structural Survey of the Built-Up Area

The data obtained from the TABULA project [11] had joint characteristics as the ones defined in GEM required for seismic assessment which enabled the use of this data, which predominant purpose was for obtaining information on the energy efficiency of the buildings. From the whole database, information regarding the buildings in Visoko was extracted. In total 94 buildings were elaborated of which 56.4% are located in the urban and 43.6% in the rural area. Concerning the level of earthquake-resistant design, the buildings, according to EMS-1998 [12], are grouped into M5 U Masonry (old bricks) and M6 U Masonry-reinforced concrete floor. The influence of the slab type has a direct impact on the seismic behavior of structures exposed to seismic actions [3]. Wooden floors due to the lack of connection with the walls and wall interconnection exhibit weaker responses in relation to the URM with rigid floors which show a "box" behavior [3, 13, 14]. An example of the elaborated buildings is shown in Fig. 1.

a) b)

Fig. 1. a) Building in the urban area b) building in the rural area

Figure 2 shows the statistical elaboration of the analyzed buildings in the urban and rural regions. The construction age goes from 1946 to 2014 covering the existence of different construction rules which were enforced at a particular time. Three types of masonry structures were analyzed, and as seen from Fig. 2b) 60.6% are confined masonry structures, 23.4% are URM with flexible floors and the remaining 16% are URM with rigid floors. The majority of structures (87.2%) have an average surface area of up to 100 m^2 (Fig. 2c). Over half of the building stock is represented by two-storey buildings (66%), while 25.5% are three-storey buildings, and only a small percentage are one, four, and six-storey buildings (Fig. 2d).

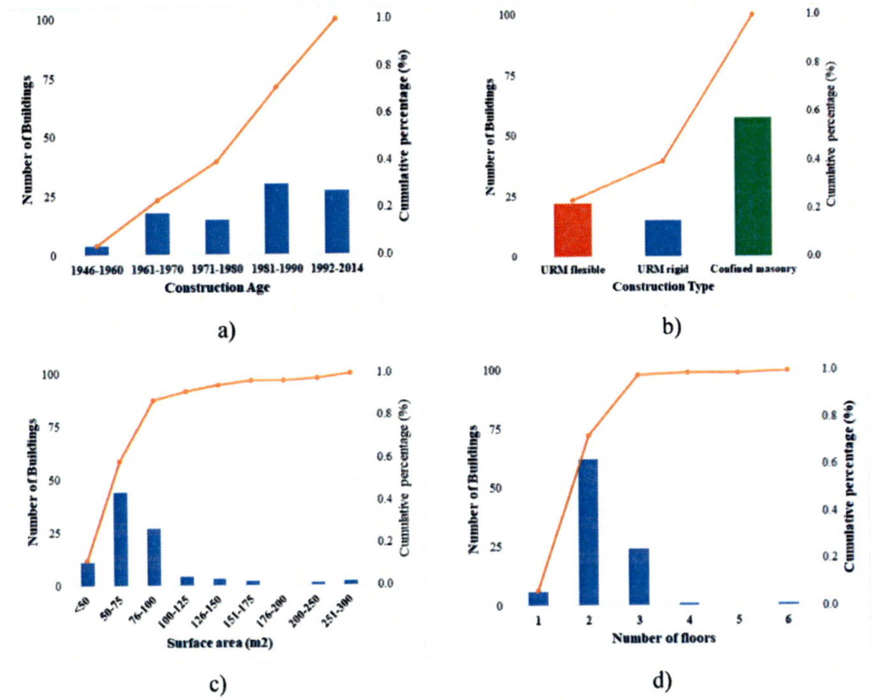

Fig. 2. Characteristics of buildings in Visoko

3 Vulnerability Assessment

Giovinazzi and Lagomarsino [15] developed the macroseismic method, with the application of the probability theory and fuzzy-set theory on the classes which define the vulnerability of structures defined by Grünthal [12] in the EMS-98. Hadzima-Nyarko et al. [16] listed the advantages and benefits of using the EMS-98. In this way, the vulnerability index method (VIM) provides a connection between the macroseismic intensity and the damage that was detected in the aftermath of earthquakes. It was noted that structures of similar typology will exhibit indistinguishable forms of damage.

Table 1. Basic vulnerability index values for various types of masonry structures [17]

Typology	Description	Vulnerability index V_I^*
M3.1	URM with flexible floors	0.740
M3.4	URM with rigid floors	0.616
M4	Confined masonry	0.451

Vulnerability index values for various types of masonry structures are taken into account as defined by Milutinovic and Trendafiloski [17] and presented in Table 1.

The vulnerability index composes of the basic vulnerability index connected to the typology, V_I^*, behavior modifiers V_m, and regional modifiers V_r [18]. The vulnerability index is calculated from Eq. (1)

$$V_I = V_I^* + V_m + V_r \tag{1}$$

The mean damage grade μ_D is calculated from Eq. (2)

$$\mu_D = 2.5\left[1 + tanh\left(\frac{I + 6.25 \times V_I - 13.1}{2.3}\right)\right] \tag{2}$$

where:

I – the macroseismic intensity,
V_I – the vulnerability index.

By applying the binomial probability distribution damage probability matrices are calculated.

Once building typology has been defined the next step is to define the average vulnerability indices which are correlated with certain vulnerability classes [17] and extended by [6].

The damage states and mean damage index values were employed as defined by [19]. They define five classes starting from no damage in the range from 0 to 0.5, and then each class has a full range of one complete value, from slight, moderate, extensive to complete damage having a maximum value of 4.5.

The most appropriate behavior modifiers were the ones proposed by Milutinovic and Trendafiloski [17] and Lantada et al. [6] (Table 2) with an expert judgment due to lack of information regarding damage after past earthquakes.

Table 2. Adopted behavior modifiers (after [17], and extended by [6])

Parameter	-0.04	-0.02	0	+0.02	+0.04	+0.06
Position of the building	Middle		Isolated		Corner	Header
Plan irregularity			Regular		Irregular	
Number of floors		Low (1-2)		Medium (3-5)		High (=6)
Roof			Light		Heavy	
Façade length			L=15 m		L>15 m	
State of preservation*	Very good <10 years		Good		Bad >40 years	

Regarding the state of preservation for the buildings constructed in the period from 1992 to 2014, it was decided to take an intermediate value of −0.02. Regional level typologies, which take into account the years of construction which is directly connected to the enforced structural codes and the building typology are considered by the regional vulnerability factors V_r. In this specific case, the upper limit proposed by Feriche et al. [20] was respected (+0.16) related to the level of the construction code, and e.g. for confined masonry, the value of +0.10 was taken for medium-level code in rural areas and +0.08 for urban areas. The distinction between the construction in urban and rural areas is reasonable and acceptable as better building control was provided in urban than in rural areas. The proposed regional factors only for this specific case are presented in Table 3. In the case of confined masonry besides low-rise buildings, there are medium-rise structures, so this was taken into account and noted with * in Table 3.

Table 3. Proposed regional indices for various types of masonry structures

Code Level	Typology	V_r Urban	V_r Rural
Low	M3.1	0.12	0.16
	M3.4		
	M4		
Medium	M3.1		
	M3.4	0.10	0.12
	M4	0.08	0.10
High	M3.1		
	M3.4		
	M4	0.04 0.05*	0.06 0.07*

4 Results

The inclusion of behavior modifiers has a direct impact on the vulnerability class of buildings meaning that it may to great extent influence the vulnerability class, and result in higher values during the seismic vulnerability assessment [16]. After taking into account the modifiers 38.3% of buildings passed to class A, 18.1% passed to B, and 20.2% passed to C. Interesting in this specific case is that 38.3%, as for class A was the transfer for both B and C together. Buildings of confined masonry type stayed in the same class (only 3 passed from class C to class B), URM with rigid floors skipped two classes (only one went from C to B), while all URM with flexible floors passed from class B to A, regardless of the location of the built-up area (Fig. 3) .

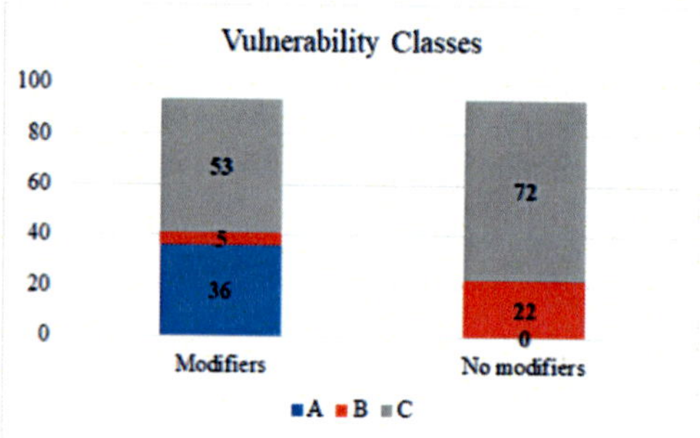

Fig. 3. Vulnerability classes as a function of (no)modifiers

The most probable intensities of earthquake ground motion for this region are VI and VII. In the case of VI intensity, 32.4% and 33.2% of M3.1 would experience slight and moderate damage respectably, 40.6% and 17.5% for M3.4 respectably, and 19.6% and 2.0% M4 respectably. In the case of earthquake ground shaking of VII intensity 33.6% (M3.1), 17.1% (M3.4), and only 1.0% (M4) would experience significant to severe damage (Fig. 4). The line of transfer between lower to higher damage levels is different depending on the building typology. In the case of URM structures with flexible floors, intensity level VII is marked (Fig. 4a), while for structures with rigid floors (Fig. 4b) one higher intensity is observed, and for confined masonry (Fig. 4c), intensity X is marked.

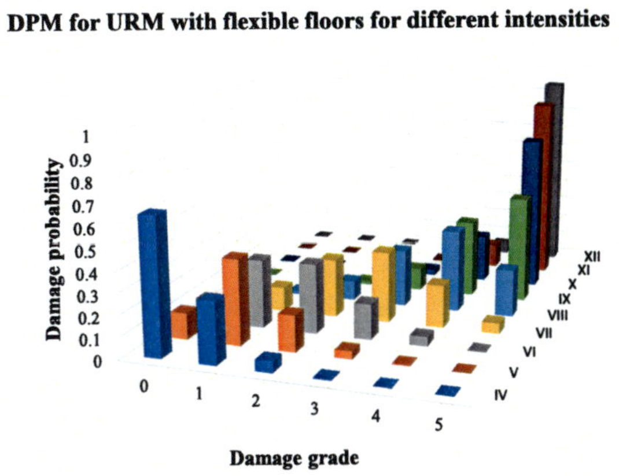

a)

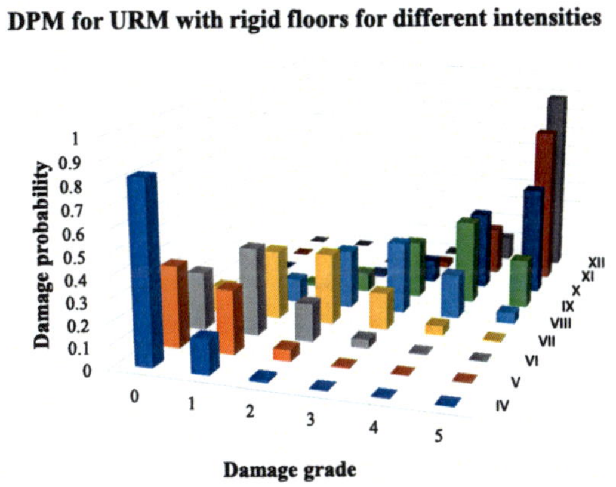

b)

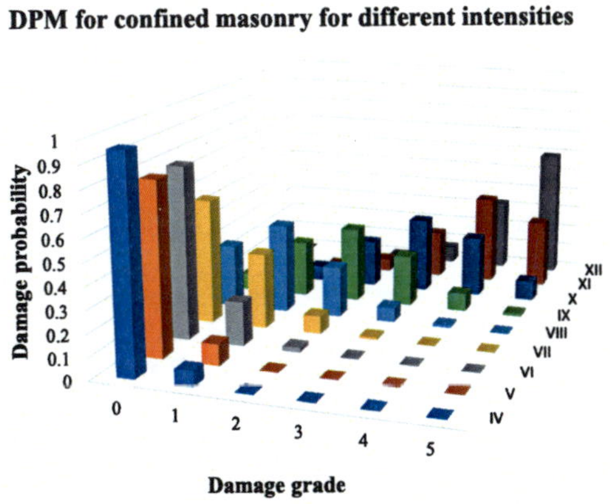

c)

Fig. 4. Damage probability in relation to damage grade and different intensities for a) URM with flexible floors; b) URM with rigid floors; c) confined masonry

5 Conclusion

The macroseismic method proposed by Milutinovic and Trendafiloski [17] and Giovinazzi [18], and extended by Lantada et al. [6] was used to access the vulnerability of urban and rural areas in Visoko. Three types of buildings were recognized and regional modifiers taking into consideration the country type of the located building (urban or rural) were proposed. In this specific case, the selected behavior and regional modifiers did not depend on the country type where the buildings were located. However, their impact was notable regarding the vulnerability classes. Once classification, as defined in EMS-98, was applied 72% of all buildings were in class C, 22% in class B, and 0% in class A. This percentage dramatically changed once the regional and behavior modifiers were implemented, transferring URM buildings with rigid floors from level C to A. This directly implies how these modifiers have a significant effect on the vulnerability class of buildings.

Acknowledgments. We gratefully acknowledge the Deutsche Gesellschaft für Internationale. Zusammenarbeit (GIZ) GmbH for giving their approval to use the data obtained through the project "Typology of residential buildings in Bosnia and Herzegovina".

References

1. Calvi, G.M., Pinho, R., Magenes, G., Bommer, J.J., Restrepo-Vélez, L.F., Crowley, H.: Development of seismic vulnerability assessment methodologies over the past 30 years. J, Earthq. Technol. **43**(3), 75–104 (2006)
2. Hadzima-Nyarko, M., Kalman Šipoš, T.: Insights from existing earthquake loss assessment research in Croatia. Earthq. Struc. **13**(4), 401–411 (2017)
3. Ademović, N., Hadzima-Nyarko, M., Zagora, N.: Seismic vulnerability assessment of masonry buildings in Banja Luka and Sarajevo (Bosnia and Herzegovina) using the macroseismic model. Bull. Earthq. Eng. **18**(8), 3897–3933 (2020). https://doi.org/10.1007/s10518-020-00846-8
4. Pavić, G., Hadzima-Nyarko, M., Bulajić, B., Jurković, Ž: Development of seismic vulnerability and exposure models-a case study of Croatia. Sustainability **12**(3), 1–24 (2020)
5. Mouroux, P., Bertrand, M., Bour, M., Brun, B.L., Depinois, S., Masure, P.: Risk-UE Team The European Risk-UE project: an advanced approach to earthquake risk scenarios. In: Proceedings of the 13th World Conference Earthquake Engineering, Vancouver, British Columbia, Canada, 1–6 August 2004, (paper no. 3329) (2004)
6. Lantada, N., et al.: Seismic hazard and risk scenarios for Barcelona, Spain, using the Risk-UE Vulnerability Index Method. Bull. Earthq. Eng. **8**(2), 201–229 (2010)
7. Pitilakis, K., Alexoudi, M., Argyroudis, S., Anastasiadis, A.: Seismic risk scenarios for an efficient seismic risk management: the case of Thessaloniki (Greece). In: Wasti, S.T., Ozcebe, G. (eds) Advances in Earthquake Engineering Urban Risk Reduction, vol. 66, pp. 229–244. Springer, Dordrecht (2006). https://doi.org/10.1007/1-4020-4571-9_15
8. Cherif, S.-E., Chourak, M., Abed, M., Pujades, L.: Seismic risk in the city of Al Hoceima (north of Morocco) using the vulnerability index method, applied in Risk-UE project. Nat. Hazards **85**(1), 329–347 (2016). https://doi.org/10.1007/s11069-016-2566-8
9. Cherif, S.-E., Chourak, M., Abed, M., Douiri, A.: Potential seismic damage assessment of residential buildings in Imzouren City (Northern Morocco). Buildings **8**(12), 179 (2019)

10. Chieffo, N., Formisano, A.: Induced seismic-site effects on the vulnerability assessment of a historical centre in the Molise Region of Italy: analysis method and real behaviour calibration based on 2002 earthquake. Geosciences **10**(1), 21 (2020)
11. IWU, I.W. TABULA (2017). http://episcope.eu/iee-project/tabula/
12. Grünthal, G.: European macroseismic scale 1998. Cahiers du Centre Européen de Géodynamique et de Séismologie **15**, 1–99 (1998)
13. Ademović, N., Hrasnica, M., Oliveira, D.V.: Pushover analysis and damage assessment of a typical masonry residential building in Bosnia and Herzegovina. Eng. Struct. **50**, 13–29 (2013)
14. Ademović, N., Oliveira, D.V.: Damage indicators for unreinforced masonry building walls subjected to seismic actions. Mater. Struct. **60**(4), 17–32 (2017)
15. Giovinazzi, S., Lagomarsino, S.: A macroseismic method for the vulnerability assessment of buildings. In: Proceedings of the 13th World Conference Earthquake Engineering, Vancouver, British Columbia, Canada, 1–6 August 2004 (paper no. 896) (2004)
16. Hadzima-Nyarko, M., Pavić, G., Lešić, M.: Seismic vulnerability of old confined masonry buildings in Osijek, Croatia. Earthq. Struc. **11**(4), 629–648 (2016)
17. Milutinovic, Z.V., Trendafiloski, G.S.: RISK-UE, an advanced approach to earthquake risk scenarios with applications to different European towns, Report to WP4: vulnerability of current buildings (2003)
18. Giovinazzi, S.: The vulnerability assessment and the damage scenario in seismic risk analysis, Ph.D. thesis, University of Florence and TU-Braunschweig (2005)
19. Barbat, A.H., Carreño, M.L., Pujades, L.G., Lantada, N.C., Omar, D., Marulanda, M.C.: Seismic vulnerability and risk evaluation methods for urban areas. A review with application to a pilot area. Struc. Infrastruc. Eng. **6**(1), 17–38 (2009)
20. Lantada, N., et al.: Seismic hazard and risk scenarios for Barcelona, Spain, using the risk-UE vulnerability index method. Bull. Earthq. Eng. **8**, 201–229 (2010)
21. Feriche, M., Vidal, F., Jimenez, C., Navarro, M.: A straightforward method applicable to Earthquake Damage Scenarios and Early Loss Assessment in urban areas of Southern Spain. In: 31st General Assembly of the European Seismological Commission ESC, Hersonissos, Crete, Greece, 7–12 September 2008 (2008)

Assessment of Regional Analyses Methods for Spatial Interpolation of Flood Quantiles in the Basins of Bosnia and Herzegovina and Serbia

Ajla Mulaomerović-Šeta[1](✉), Borislava Blagojević[2], Šemsa Imširović[3], and Bojana Nedić[4]

[1] Faculty of Civil Engineering, University of Sarajevo, Sarajevo, Bosnia and Herzegovina
ajla.mulaomerovic@gf.unsa.ba
[2] Faculty of Civil Engineering and Architecture, University of Niš, Niš, Serbia
[3] Higracon d.o.o. Sarajevo, Sarajevo, Bosnia and Herzegovina
[4] Hydro-Engineering Institute Sarajevo, Sarajevo, Bosnia and Herzegovina

Abstract. Regional analyses of streamflow attempt to link discharge to some catchment characteristics. This paper deals with regional analysis of flood discharge in Serbia and Bosnia and Herzegovina, based on at-site statistical analysis of 74 hydrologic stations with small to medium catchment area. Flood quantile estimation is performed according to Bulletin 17C procedure on the annual maxima records. Regionalization of catchments is performed in two ways: 1) subjectively, according to catchment location in large river basin and within one region encompassing all stations, and 2) clustering, by both k - means and Ward's algorithm. The catchment clustering methods use similarity matrix comprising morphologic catchment characteristics. In each of the defined regions, flood quantiles are linked to the catchment area by regression using jack-knife procedure. Regionalization methods performance is assessed by absolute error measures and method rank frequency. It is found that the results of flood quantile spatial interpolation are influenced by the flood record length and number of stations comprising the region i.e. region size. Larger regions give unsatisfactory results, as well as those containing a significant number of stations with potentially influential low floods. Regardless of a detected difference of regionally estimated quantiles to the reference ones at all stations, the clustering method according to the Ward's algorithm performs best.

Keywords: Flood quantiles · Bulletin 17C · Regression · Cluster analysis · Regional analysis

1 Introduction

Worldwide, floods pose a significant problem due to damage costs, often along with loss of lives. In order to harmonize the framework for evaluating the impacts and mitigating the flood damage, the European Parliament adopted the 'Directive on the assessment

N. Ademović et al. (Eds.): IAT 2021, LNNS 316, pp. 430–456, 2022.
https://doi.org/10.1007/978-3-030-90055-7_35

and management of flood risk' in 2007 [1]. On the footsteps of the EU, over the last few years, efforts have been made in the West Balkan region by non-EU member countries, to produce the flood hazard and flood risk maps [2–4]. One of the key input data for generating the fluvial flood hazard map is the estimate of the flood flow of a certain exceedance probability, a flood quantile. In Bosnia and Herzegovina (BiH), the selected exceedance probabilities i.e. return periods of floods are 500, 100 and 20 years, while in Serbia it is 1000, 100 and 50 years [5].

The flood quantiles are often required at the location without gauged data, or in the case gauging/hydrologic station matches the location of interest, the data record is short (less than 30 years). In such cases, either rainfall-runoff methods or regional analysis methods are used for flood quantile estimation.

The rainfall-runoff methods transform either design or recorded rainfall to runoff. In the design rainfall approach, the underlying assumption is equality of rainfall and runoff return period, while other assumptions are related to the design storm duration and excess rainfall estimation. Although widely used in practice, the setbacks of this approach are recognized [6]. The recorded rainfall approach usually implies simulation models, while the resulting runoff is further statistically analyzed and considered tentative [7].

The statistical regionalization procedures attempt to link flood flow parameters or quantiles to morphologic catchment characteristics and/or meteorological parameters. These methods are employed to estimate flood quantiles at ungauged locations, but also for comparative assessment of regional behavior of floods. Because flood flows as extreme events vary in wider range compared to mean flows, longer data records are required for reliable estimate of their statistical properties [8]. In Germany, T – year flood quantile estimation is required from the record of at least T/2 size [7], while in former Yugoslavia, the recommended record size was from T/5 to T/3 [9] as a compromise between existing record lengths and statistical considerations. In general, when adequate record length is not available, auxiliary data and procedures are suggested. One of such procedures is regional analysis, aiming at flood estimation reliability improvement.

A global effort in regional hydrology has been initiated by UNESCO (United Nation Education, Science and Culture Organization) in 1985 within its International Hydrology program (IHP) by launching a research program FRIEND (Flow Regimes from International Experimental and Network Data). The FRIEND program has included and coordinated regional research in 74 countries from all continents [10].

In the year 2010, COST Action ES0901 'European Procedures for Flood Frequency Estimation' has started, aiming at comparison and evaluation of different methods for flood frequency estimation across EU member states [11]. Seventeen participating countries have provided their reports related to both statistical and regional analyses for flood frequency, methods under development and situation with data. In the roof report [11] it is shown some of the countries do not use regional methods due to its specific conditions: Cyprus due to small territory, Greece because of significantly modified natural regime by reservoirs, Finland considers its existing hydrologic stations network dense enough, while Slovenia has not developed regional background for regional methods application at the time.

Regional approach to flood frequency assessment has been incorporated into regulations, standards, or guidelines of some countries a long time ago. The countries that support regional analysis include the USA [12], the UK [12, 13] Russia [15], Germany (individual provinces), Italy, Lithuania, Slovakia, Spain. Various methods for defining homogeneous regions have been proposed in the literature [16], but in practice, there is no generally accepted procedure for this [7]. These countries use different approaches to regionalization including region of influence (ROI) [13, 17], cluster [18–21], and within it a subjective method, according to different attributes used to calculate distance measure. Common attributes include morphological, climatological (average annual precipitation [22], time of occurrence of flood [23]); statistical properties (variation coefficient - *Cv*, skew - *Cs* [24]), or shape of distribution function (unimodal, multimodal [25]), and causes of flood occurrence [26]. Various tests are used to test homogeneity, most commonly by examining quantile or statistical parameters variability. Different methods are also used for information transfer including geostatistical methods [27], regression equations [13, 14, 28], index-flood method [18], as well as an approach that relies on the ROI method [13, 29, 30].

The multivariate exploratory techniques are used to define a region, because the similarity of the basins is determined in the multidimensional attribute space. Another way to group basins is called subjective, although it is performed according to the property of the main basin, geographical location, and climatologic area. Among multivariate techniques, cluster analysis [31] and principal component analysis (PCA) [32] are common.

In the West Balkans region, various regional analyses have been carried out in the 1970s and 1980s. The relation of the flood quantile to the average of maximum floods with the variation coefficient - *Cv* of maximum floods, was considered adequate for predicting flood quantiles in the catchments in Yugoslavia. The paper entirely dedicated to this topic was given by Popović and Andrejević [33] who, based on the data from 46 hydrological stations, defined envelopes of specific runoff (discharge per unit drainage area) for various return periods in the territory of Serbia. Janković and Malošević [34] defined the upper envelopes of specific runoff for the maximum observed floods and quantiles of 10-, 50-, 100-, 1000- and 10000-year return periods for Serbia without the province of Vojvodina (on the South of the river Danube and river Sava). The most recent regional analysis published for BiH dates from 2005 [35], where Hrelja has provided formulae for estimating floods of a given return period for the catchments in the Bosna river basin. The underlying frequency distribution is Gumbel. The regional formulae enable flood quantile estimation based on catchment area, simply by linking mean flood flow to the catchment area, and ratio of flood quantiles and mean flood flow to return period. A more powerful statistical methods are employed to regionalize flood quantiles of catchments in Bosnia and Herzegovina and Serbia in the recent research by Imširović [36], using cluster analysis based on distance matrix defined by physiographic characteristics and flood dataset *Cv*.

The aim of this research is to determine the potential for flood data transfer from basins in Serbia to basins in BiH by regionalization. The motivation for the research is lessening uncertainty in the flood studies in BiH due to the lack of systematic (gauged) flow data at most of the hydrologic stations in the period 1991–2000.

The basis for establishing regional relationships are data on peak flows (floods) from the standard normal World Meteorological Organization period 1961–1990 in the small and medium-sized catchments - with drainage area up to 4000 km^2. The applied methodology includes a description of the study area with available physiographic/morphologic catchment characteristics and hydrological data. The applied procedure for determining the flood quantiles includes the methods of regionalization with a focus on cluster methods. Upon the definition of the hydrologic regions, the flood quantile transfer is performed using the regression equations. These equations express flood quantiles as a function of a catchment area. The evaluation of the regionalization methods efficiency is examined using several common criteria, and regionalization methods are rated accordingly. In the concluding section, recommendations for the next research phase can be found, specifically pointing out the issues related to the size of the region, stations with data gaps and potentially influential low floods, and the best performing regionalization method.

2 Methodology

Figure 1 illustrates the paper structure (blue boxes) and applied methodology (white boxes) with detailed presentation of the core subchapter on regionalization (Fig. 1 - right). In the subchapter on study area, the locations of the basins, morphological and statistical characteristics of maximum annual flows are presented. The next step explains a procedure that is the current standard in the USA for flood frequency analysis, after which the method of regionalization is presented. Subjective and cluster methods used in pooling groups follow, with data transfer performed by regression methods. The methodology chapter ends with explanation of three methods applied for the evaluation of regionalization methods.

2.1 Study Area

The input data for this analysis are the maximum annual flows at gauging/hydrologic stations (HS) in BiH and Serbia, in the period 1961–1990. All considered basins belong to the Black Sea basin and temperate continental climate. There are 32 stations from BiH and 42 from Serbia.

The focus of research in BiH are stations from the Una river and the Bosna river basins. The locations of the stations and the catchment boundaries are shown in Fig. 2 and Fig. 4. Table 1 shows the number of stations per basin of the larger rivers illustrated in Fig. 2. Data gaps i.e. years without annual flow maximum at 22 of the 74 stations are shown in Fig. 4. Fifty-two considered stations have a complete data records in the period selected for research.

A 20-m (resolution 20 m × 20 m) Digital Elevation Model (DEM) of BiH was used to define (orographic) river catchments in the Bosna and Una river basins. The major catchment morphologic data in Serbia are adopted from previous research [37], while a smaller part was determined from the 30 m DEM of Serbia. The tool for determining catchment geomorphological parameters was free and open- source software Quantum Geographic Information System (QGIS) [38].

The physiographical/geomorphologic characteristics of the studied catchment sample are shown as box plots representing statistical data in Fig. 5, together with statistical properties of the transformed (logarithmic) peak annual flow datasets.

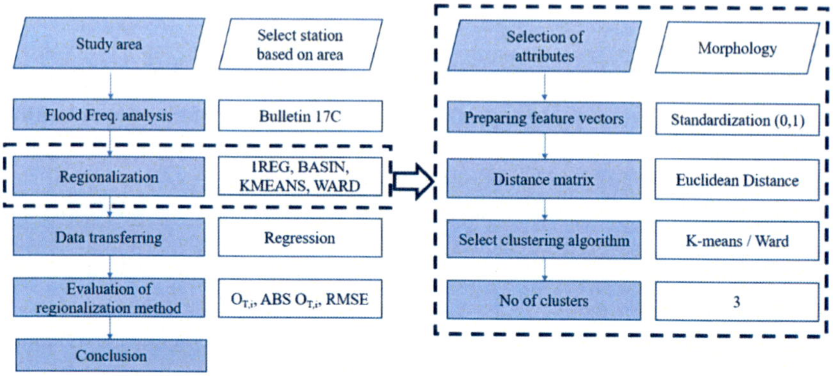

Fig. 1. Steps in research (left) and sub-steps in regionalization by cluster analysis (right)

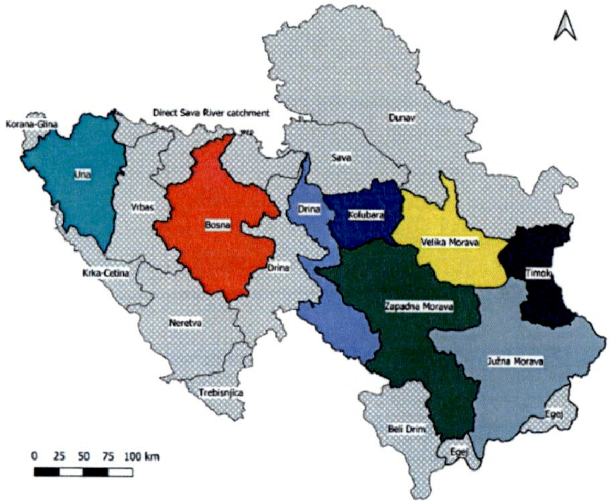

Fig. 2. Study area. Large river basins division.

Table 1. Number of sampled locations in each large river basin

River basin	Una	Bosna	Drina	Kolubara	Zapadna Morava	Velika Morava	Južna Morava	Timok
No. HS	15	17	3	6	9	5	15	4

2.2 Flood Frequency Analysis

The estimation of flood quantiles was performed by a procedure that is the current standard in the USA - Bulletin 17C [12], incorporated in the U.S. Army Corps of Engineers, Hydrologic Engineering Center's Statistical Software Package (HEC-SSP) version 2.1 [39]. In this version, flood frequency analysis (FFA) according to the pervious procedure

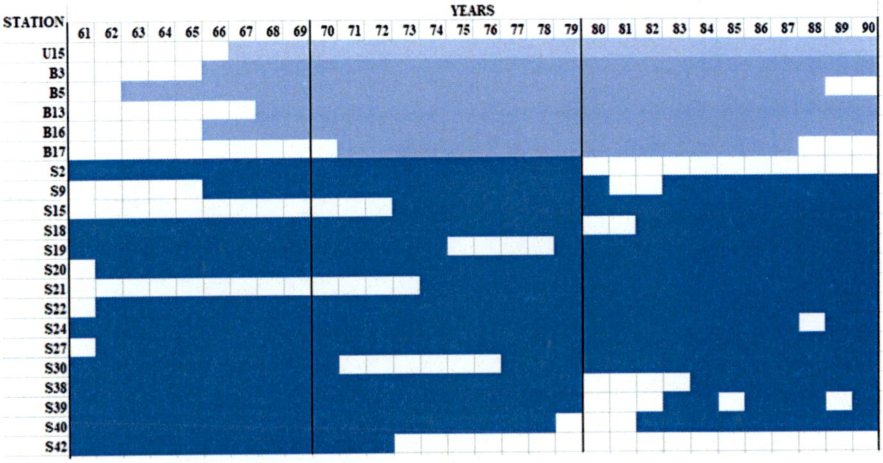

Fig. 3. HS with gauged flow data gaps (white color). The station labels correspond to the legend in Fig. 4.

of the Bulletin 17B [40] is also enabled. The main difference among the two procedures is the methodology for assessing the log-Pearson Type III (LPTIII) distribution moments - the Expected Moments Algorithm (EMA), introduced in the Bulletin 17 C. Along with the EMA came the Hirsch-Stedinger plotting position formula, capable of accommodating interval data. This novelty allows for the assessment of statistical parameters in the incomplete data records, where missing data is given as a range, e.g., through perception thresholds. Perception thresholds are described by lower and upper bound, where an upper bound in the systematic records is usually assumed to be infinite, as adopted for this research. The lower perception threshold bound here is set to the value of 1.2 *Qmax* at all stations shown in Fig. 3, where *Qmax* is the maximum gauged flow in the incomplete data record at each station. This is a slightly higher value of the lower perception threshold bound compared to those examined in [41], where the influence of the lower perception threshold bound value on the flood quantile estimates was analyzed for the catchments of medium and large drainage area in the territory of Serbia.

Other advancements of the flood frequency analysis in the Bulletin 17C include the Multiple Grubs-Beck test (MGBT) for low outliers, which can detect several outliers called Potentially Influential Low Flood (PILF). The aim of MGBT is to identify all observed floods that may significantly affect the form of annual maxima distribution function. Within the procedure of quantile assessment, PILFs are rejected (censored) from the dataset. In Fig. 6. the probability plots at two stations are shown, one without and the other with PILFs detected. Because this research is based on the 30 year data records, stations with a larger number of detected PILFs require attention due to a high uncertainty in quantile estimation based on reduced data record. Table 2 shows the number of detected PILFs at sample stations.

Seven return periods are selected in this research for flood quantile assessment: 10, 20, 50, 100, 200, 500 and 1000 years.

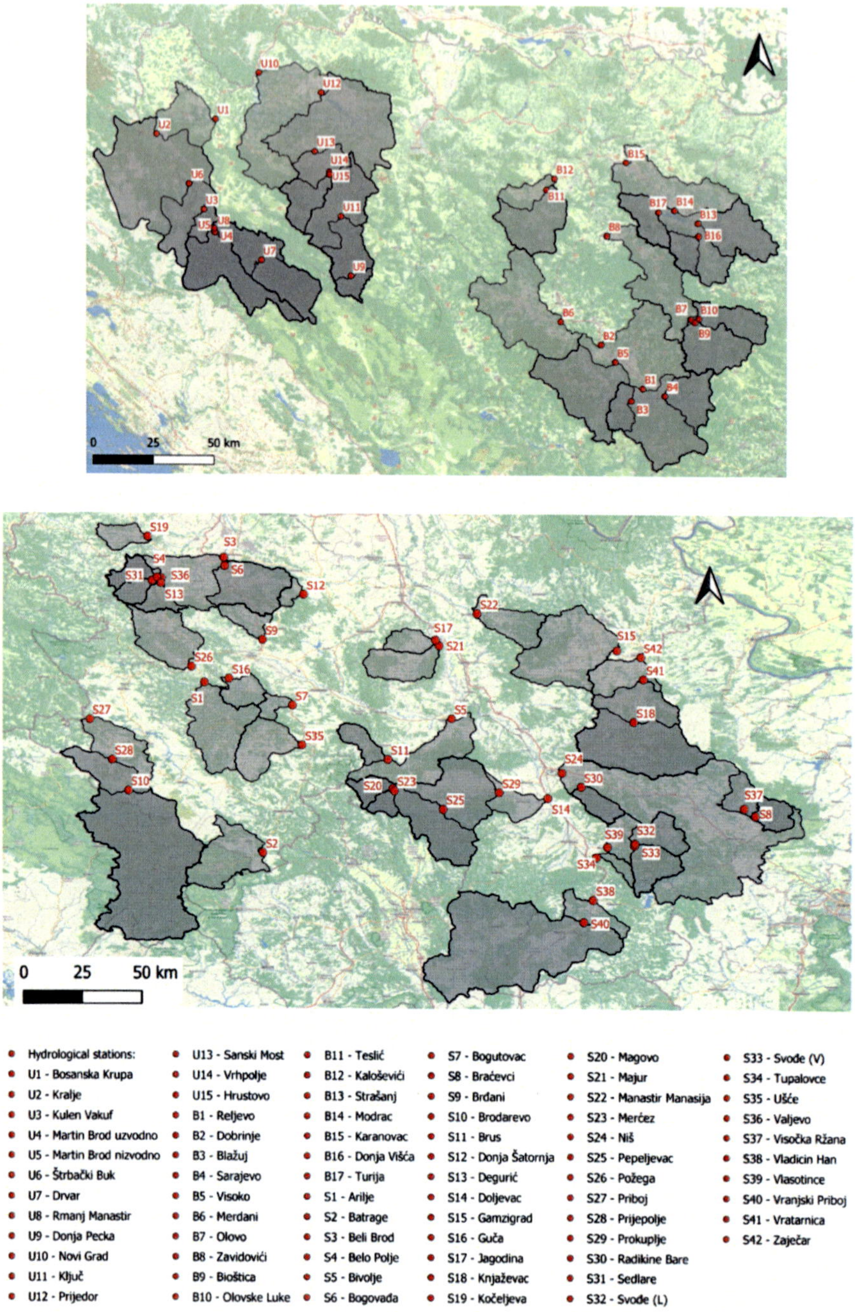

Fig. 4. Location of analyzed basins and hydrologic stations in BIH (up) and SRB (down)

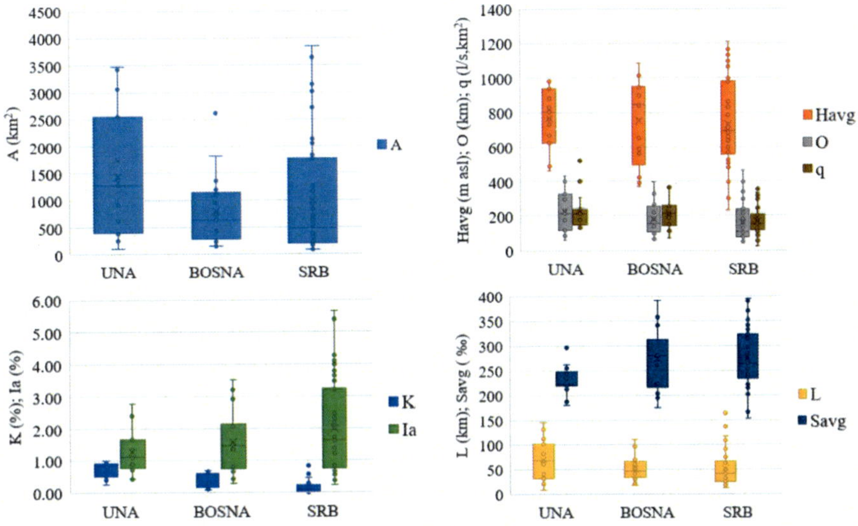

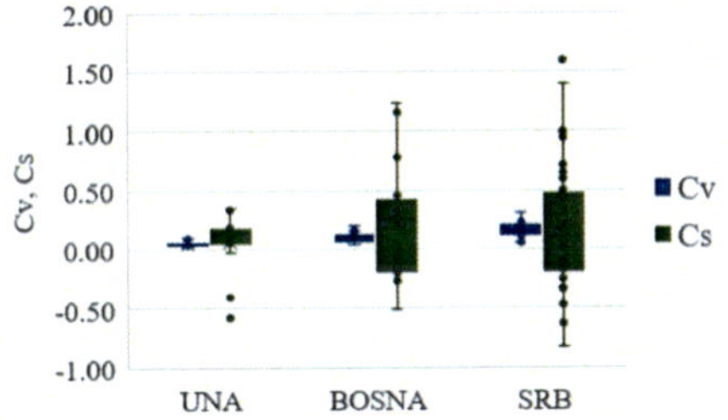

Key:
A - catchment area (km²), *Havg* - average catchment altitude (m asl), *O* - catchment perimeter (km), *Savg* - average catchment slope (%), *K* - percent of karst in catchment (%), *Ia* - absolute streamflow slope (%); *L* - the main streamflow length (km), *q* - specific flow (mean flood flow per drainage area) (l /s·km²), *Cv* - variation coefficient of the logarithmic dataset, *Cs* - skew of the logarithmic dataset.

Fig. 5. Box-plots of physiographical characteristics and statistical properties of the catchment sample shown separately for the large basins of BiH rivers (UNA and BOSNA) and together for all catchments in Serbia (SRB).

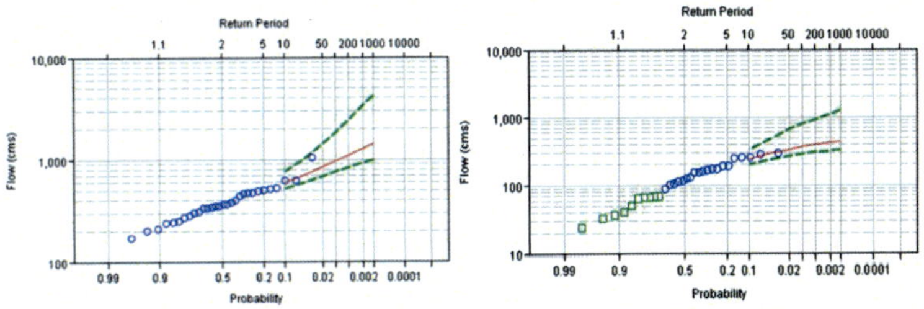

Fig. 6. Probability plots from the HEC – SSP at-site flood frequency analysis. HS Dobrinje (Bosna) – B2 without PILF (left), HS Bivolje (Serbia, Z. Morava) – S5 with PILF (right)

Table 2. PILF frequency in catchment data sample – HS

No. PILFs	No. station	HS
0	53	
1	7	U13, U14, B5, S15, S35, S36, S42
2	4	B13, S3, S6, S21
3	3	U12, S9, S16
5	4	B3, B16, B17, S19
6	2	U15, S41
10	1	S5
Sum	74	

2.3 Regionalization Methods

Hydrological regions are formed subjectively and by the cluster analysis. Subjectively formed regions are labeled in the following way:

- REG1 - considering all catchments as one cluster,
- BASIN - according to the main basin of larger rivers as presented by hydro-meteorological services (Fig. 2): UNA, BOSNA and SRB, where all catchments from Serbia comprised one region in the research.

By applying the cluster method based on physiographic similarity, the regional labels are:

- WARD – formed by the hierarchical method according to the Ward's algorithm,
- KMEANS – formed by the *K*-means partition clustering method.

2.3.1 Cluster Analysis

Cluster analysis is a generic name for various multivariate statistical procedures used to study, interpret, and classify data into similar groups or clusters, which may or may not overlap. The analysis conducted in this research comprises several steps shown in Fig. 1 - right.

The first step is to select attributes for defining the distance between catchments. In this research, attributes that are believed to be crucial for generating runoff are selected, comprising feature vectors in the next step. Most often in flood studies, attributes include catchment area, geological characteristics, average catchment slope, absolute streamflow slope, coordinates of catchment gravity centers, flood flow statistics (*Cv*, *Cs*) and specific flow (mean flood flow or a specific flood quantile per unit catchment area).

The attributes are selected here from different combinations of seven physiographic characteristics shown in Fig. 5. The final selection contains three catchment attributes: drainage area (*A*), average altitude (*Havg*), and average slope (*Savg*). The correlation significance between selected attributes is examined by the *t*-test.

Considering the units and order of magnitude of the selected attributes, the usual standardization was done to continue research with standardized variables with a mean of 0 and a standard deviation of 1.

The distance matrix may be composed from many measures of distance or similarity [16]. The one selected for this research is Euclidean distance matrix. The distance between each pair of stations is:

$$d_{ij} = \sqrt{\sum\nolimits_{k=1}^{n} \left(x_{ik} - x_{jk}\right)^2} \tag{1}$$

where:

d_{ij} – distance measure between feature vectors i and j (catchments i and j).

x_{ik} – standardized value of attribute k in the feature vector i.

xjk – standardized value of attribute k in the feature vector j.

n – number of attributes in feature vector.

The color coded Euclidian distance matrix is shown in Fig. 7.

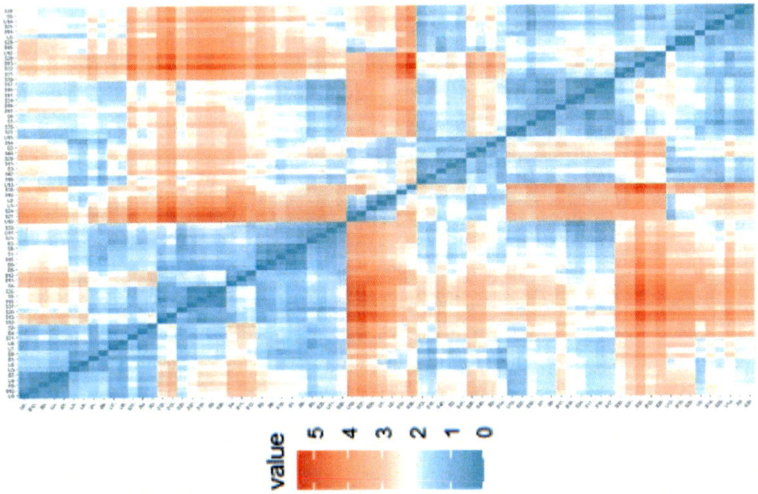

Fig. 7. The Euclidean distance matrix based on adopted morphological attributes. The colors denote: smallest distance value (0) - dark blue, the largest (~5) dark intestinal, and ~2 is white.

The next step in the procedure was clustering catchments. The two clustering algorithms were studied in this research: k-means and Ward's algorithm (Fig. 8 left).

K Means Algorithm

At the start of k-means clustering, the number of centers should be specified. Based on scree plot, the number of centers was selected. The basic idea is to define number of centers such that the total intra-cluster variation (total within-cluster variation or total within-cluster sum of square) is minimized. This involves steps [42]: Compute clustering algorithm (e.g., k-means clustering) for different values of k; For each k calculate the total within-cluster sum of square (wss); Plot the curve of wss according to the number of clusters k. The location of a bend (knee) or decrease in slope of wss is generally considered as an indicator of the appropriate number of clusters.

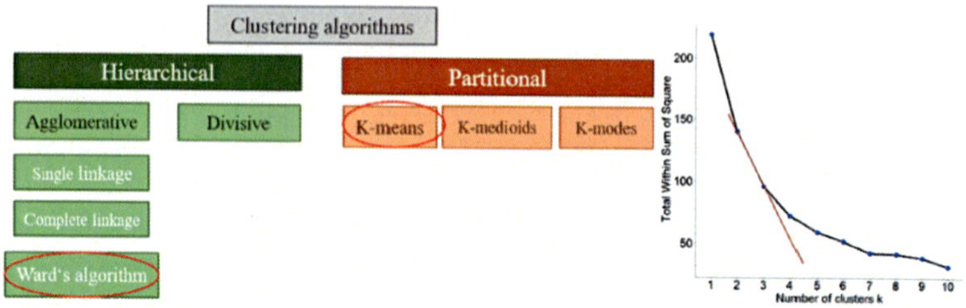

Fig. 8. Classification of the cluster methods (according to [16]) and Scree plot (right)

After selecting the number of clusters, the k-means algorithm randomly selects the initial center from the available data set. Then, each of the remaining objects is assigned to the nearest center, where the nearest implies the smallest squared Euclidean distance from object to the center of the cluster to which the object is assigned. After each object is assigned to the nearest center, the algorithm calculates the new center of each cluster and the procedure is iteratively repeated until the clusters formed in two consecutive iterations are equal. An illustration of the process is shown in Fig. 9. The objective function to be minimized when grouping is [16]:

$$F = \sum_{k=1}^{K} \sum_{j=1}^{n} \sum_{i=1}^{N_k} d^2\left(x_{ij}^k - x_{\cdot j}^k\right)^2 \tag{2}$$

where:
K – number of cluster (cluster's centers).
N_k – number of feature vectors (basins) in cluster k.
${x_{ij}}^k$ – standardized value of attribute j in the feature vector i assinged to cluster k,
d (.) – distance measure (squared Euclidean distance here).
n – number of attributes in feature vector.
$x_{\cdot j}^k$ – mean value of attribute j for cluster k, computed as:

$$x_{\cdot j}^k = \frac{\sum_{i=1}^{N_k} x_{i,j}^k}{N_k} \tag{3}$$

Ward's Algorithm
At the initial step of agglomerative hierrarhical clustering, all clusters are singletons (clusters containing a single point). The following steps analyze the merging of all possible pairs of clusters, where the criterion for choosing the pair of clusters to merge at each step is based on the optimal value of an objective function. Criterion function depends on the applied algorithm. Calculation steps and criterion function in agglomerative hierrarhical slustering algorithm are shown in Fig. 10.

Among hierarchical clustering algorithms, Ward's algorithm is most commonly used in in hydrology [18, 21, 44]. It is defined as an algorithm that merges the groups that will cause the smallest increase in the sum of the squares of the differences calculated according to (Eq. 2) (where $x_{\cdot,j}$ is mean value of attribute j after merging two clusters).

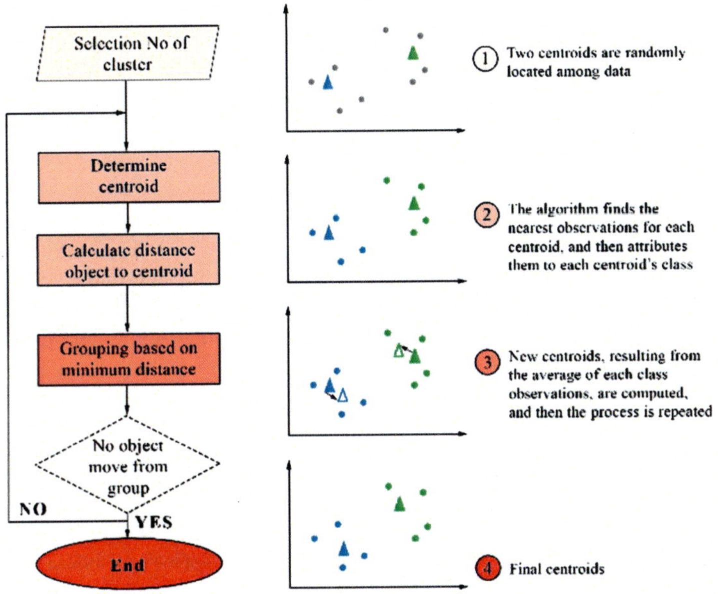

Fig. 9. *K*-means clustering algorithm (left), graphical interpretation of algorithm (right). Modified according to [43].

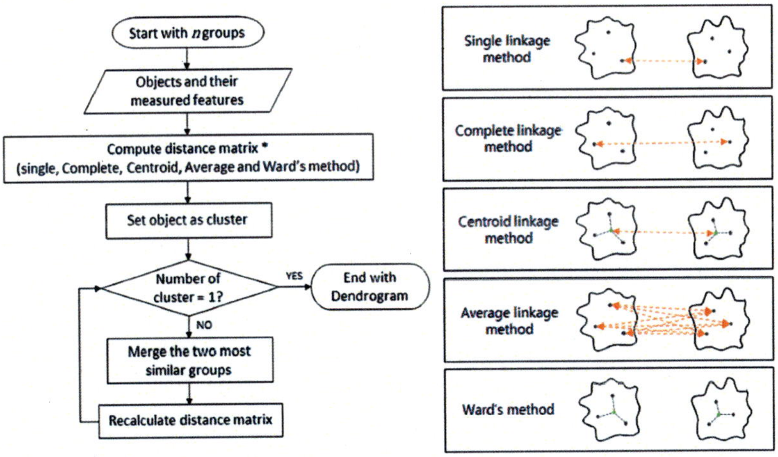

Fig. 10. Agglomerative hierarchical clustering algorithm (left), graphical interpretation of algorithm (right) [45]

2.4 Flood Quantile Transfer Model

In the regional analysis, physiographic attributes are usually associated with flood quantiles or statistical moments by regression. Rodda and Robinson [46] argue that moments of higher order and quantiles from the distribution tails cannot be adequately described by regression, because it is difficult to find independent variables to describe them [14, 47]. The improvements may include multiple regression taking into account the average annual precipitation or some other attributes [13, 14].

Still, there is a long tradition in using a single variable regression as a flood quantile transfer model:

$$Q_{T,REG} = a \cdot A^b \tag{4}$$

where:

$Q_{T,\ REG}$ – flood quantile of return period T in a cluster/region *REG*,

a, b - regression coefficients, A - basin area.

This transfer model is used in the research. Regression coefficients a and b were determined by ordinary last square (OLS) method for each station i, return period T and method of regionalization.

2.5 Assessment of Regionalization Methods Efficiency

Efficiency of regionalization methods is here based on comparison (measures of absolute error) of the flood quantiles $Q_{T,\ REG,\ i}$ calculated from the Eq. (4), to the reference ones - $Q_{T,\ i}$ estimated by the at-site flood frequency analysis from the data record.

The jack-knife procedure is applied for the $Q_{T,\ REG,\ i}$ regression relationships definition. The procedure steps are shown in Fig. 11.

The jack-knife procedure is also called 'take one out', which in this case means regression coefficients a and b in the Eq. (4) for station i are estimated from the set of stations comprising hydrologic region (N_k stations total), without one station - i. This procedure is repeated for each station, return period T and regionalization method. The flood quantiles $Q_{T,\ REG,\ i}$ at station i are then calculated from Eq. (4) for known catchment area of station (A_i).

Considering that 74 stations, 7 return periods and 4 regionalization methods (1REG, BASIN, KMEANS and WARD) were analyzed, 74 × 7 × 4 values of regression parameters a and b are estimated.

In one hydrologic region comprising a total of N_k stations with different magnitudes of flow, it is desirable to compare flood quantiles to reference values, both in the terms of magnitude and direction (underestimate, overestimate). For such an insight, at each station i and a set of return periods T, percent error (*PE*) formula, also known as percent bias, is used:

$$PE_{T,i} = \frac{Q_{T,REG,i} - Q_{T,i}}{Q_{T,i}} \cdot 100(\%), \tag{5}$$

while a sole magnitude comparison is perceived from its absolute value (*ABSPE*):

$$ABSPE_{T,i} = \left|PE_{T,i}\right|(\%). \tag{6}$$

An overall measure of regionalization method (RM) efficiency resembling root mean squared error formula is proposed:

$$RMSE_{T,RM} = \sqrt{\frac{\sum_{i=1}^{N}\left(ABSPE_{T,i} - \overline{ABSPE_T}\right)^2}{N}}\ (\%) \tag{7}$$

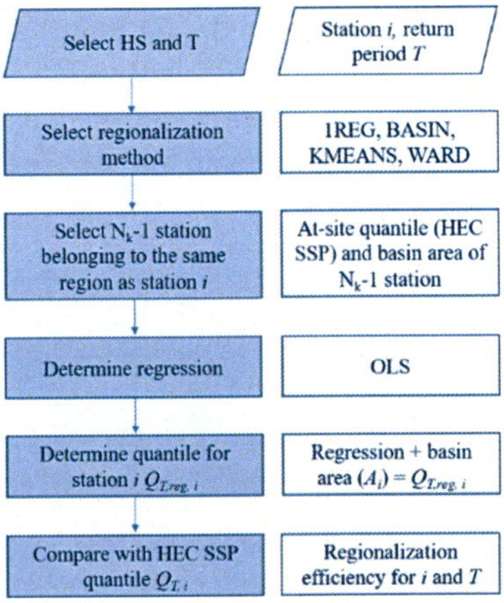

Fig. 11. Jack-knife procedure in estimating regionalization efficiency

where:

$RMSE_{T,RM}$ - root mean squared error of $ABSPE$ for return period T and regionalization method RM

N – total number od stations

$ABSPE_{T,i}$ - absolute percent error of station i for return period T in the considered RM

$\overline{ABSPE}_{T,RM}$ - mean absolute percent error for return period T in the considered RM.

3 Results and Discussion

3.1 Regionalization

In the initial regionalization phase, all physiographic attributes were selected for a distance matrix determination. Such an approach has led to grouping the catchments within the Una river basin in one cluster. This happened because the Una catchments are predominantly karstic and geographically furthest away from the ones in Serbia, resulting in quite large distances. In this way, subjective approach to regionalization according to the main river basin affiliation was indirectly justified.

In the next regionalization phase, the attributes were selected according to recommendations to exclude highly correlated attributes, and to use three attributes at most [7]. Table 3 shows both the degree of correlation between the attributes measured by correlation coefficient, and its significance according to t-test. Based on the correlation coefficient significance test results, some attributes are excluded in further analysis. The adopted attributes are: catchment area (A), catchment average altitude ($Havg$) and catchment average slope ($Savg$), shown in bold in Table 3.

The final number of clusters was determined upon the scree plot drawn using Package *stats* version 4.0.2. *screeplot* function, as shown in Fig. 8 on the right. The cases from 2 to 10 clusters were analyzed. Three clusters were finally chosen, because the slope rate of change in the scree-plot function is milder for more than 3 clusters, meaning that adding centers does not significantly reduce the total variance (*wss*).

The other considerations related to cluster centers included testing 25 random combinations of three centers, with 10 iterations each, by the Package *stats, kmeans* function, because the result of the *k*-means cluster may depend on the initially assumed centers. Likewise, Ward's algorithm was executed using Package *stats,* with *hclust* function, and the graphical representation is given as dendogram (Package *factoextra* version 1.0.7, *fviz_dend* function) (Fig. 12). From the dendogram it is clear which clusters would merge if a greater distance between them is applied (greater distance = smaller number of cluster centers) or in what way the existing ones would split if the distance was reduced (smaller distance = more cluster centers).

Table 3. Coefficient of correlation between attributes (upper table part -*r*) and its significance (lower table part *p*-values). Red color indicates significant correlation (*p*-values < 0.05). Selected attributes for distance matrix are bolded.

r	***A(km2)***	*O(km)*	*K(%)*	*L(km)*	*Ia(%)*	***Havg(m asl)***
O(km)	0.951					
K(%)	-0.015	0.043				
L(km)	0.882	0.901	-0.067			
Ia(%)	-0.472	-0.542	-0.038	-0.597		
Havg(m asl)	0.042	0.08	0.107	0.008	0.32	
Savg(%)	-0.071	-0.05	-0.365	-0.013	0.431	0.511
p values	*A(km2)*	*O(km)*	*K(%)*	*L(km)*	*Ia(%)*	*Havg(m asl)*
O(km)	0.0000					
K(%)	0.8983	0.7180				
L(km)	0.0000	0.0000	0.5720			
Ia(%)	0.0000	0.0000	0.7475	0.0000		
Havg(m asl)	0.7236	0.4955	0.3627	0.9469	0.0054	
Savg(%)	0.5466	0.6709	0.0014	0.9154	0.0001	0.0000

The final clustering of catchments is mapped in Fig. 13, according to both algorithms. A total of ten regions or clusters is defined in the regionalization process by two approaches. Catchment morphological attributes for each group of catchments is shown through statistical properties in Table 4.

According to research [18], an increase in the number of stations over 20 does not affect the accuracy of flood quantile estimates, but increases the chance of creating inhomogeneous region. As shown in the last column of Table 4, the following regions have more than 20 stations: 1REG, BASIN SERBIA, KMEANS 2, KMEANS 3, WARD 2 and WARD 3. From the standpoint of number of stations comprising the region, a better flood assessment results could be expected from BASIN BOSNA, BASIN UNA, KMEANS 1 and WARD 1 clusters.

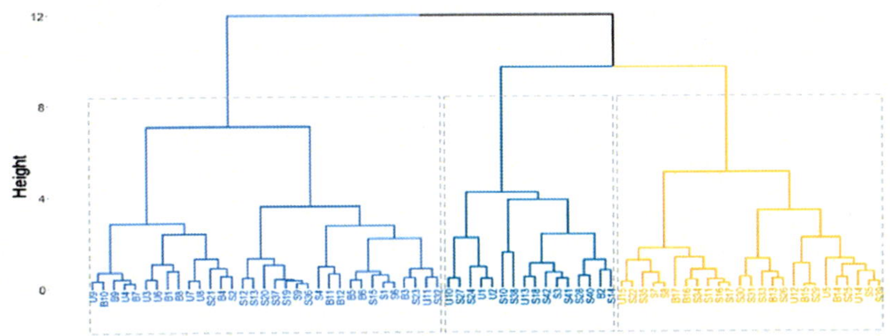

Fig. 12. Dendogram for Ward's algorithm. The dashed grey boxes show HS belonging to the center of each cluster.

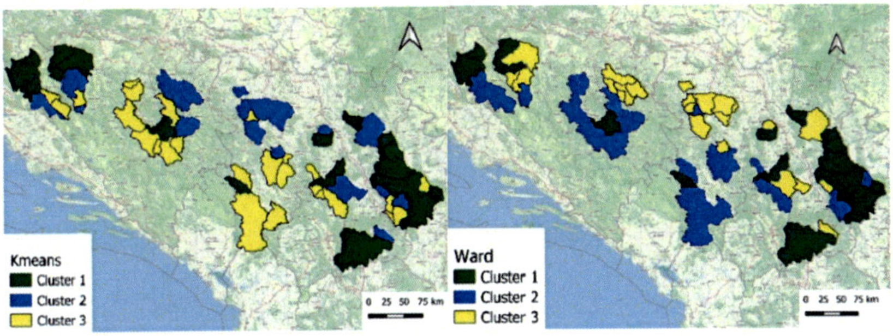

Fig. 13. Location of catchments assigned to K-means clusters (left) and Ward's clusters (right). Centers of clusters are indicated by color.

3.2 Regression

The potential of applied regression model (Eq. 4) is tested within each defined region on the reference values of flood quantiles (Q_T) assessed by FFA. Figure 14 shows the coefficient of determination (r^2), a measure of how well reference quantiles are replicated by the selected model, and a value of $r^2 = 0.5$, delimiting good (>0.5) from poor regression (<0.5).

According to the r^2 propagation through return periods (Fig. 14), better quantile estimates could be expected in the regions: BASIN UNA, BASIN SERBIA and WARD 3 for all return periods; 1REG, WARD 1, and KMEANS 1 for all but short T; KMEANS 2 for $T \leq 200$ years, compared to WARD 2, KMEANS 3 and BASIN BOSNA which could perform worst.

A cross-check of stations origin in the poor performing clusters for the selected regression model (Eq. 4), WARD 2 and KMEANS 3, reveals that in WARD 2 half of the catchments come from the worst performing region BASIN BOSNA, while in the KMEANS 3, they comprise a fair share of the catchments – one third.

Another important consideration relates to the uncertainty in the reference flood quantile estimates Q_T originating from the input data. This uncertainty is inherited in the regionally assessed $Q_{T,REG}$ i.e. spatially interpolated by the considered regression model. In the Bulletin 17C FFA, the flood record length is an important element influencing

Table 4. Basic statistical properties of catchment morphological attributes in all defined regions/clusters.

Region/Cluster		Attribute	MIN	MAX	AVG	MED	N_k
Subjective approach	1REG	*A*	84	3879	1104	702	**74**
		Havg	234	1210	743	721	
		Savg	15.3	39.3	26.4	25.9	
	BASIN UNA	*A*	117	3879	1708	1459	**15**
		Havg	463	980	769	805	
		Savg	18.0	29.5	23.0	22.2	
	BASIN BOSNA	*A*	155	2587	799	633	**17**
		Havg	369	1083	748	842	
		Savg	17.5	39.0	27.2	27.9	
	BASIN SERBIA	*A*	84	3870	1012	484	**42**
		Havg	234	1210	732	694	
		Savg	15.3	39.3	27.4	26.4	
Clustering approach	KMEANS 1	*A*	227	3879	2207	2150	**17**
		Havg	401	1179	741	709	(6 U; 1 B, 10 S)
		Savg	21.8	34.0	25.2	24.7	
	KMEANS 2	*A*	84	3052	780	400	**31**
		Havg	234	1010	593	576	(6 U; 8 B, 17 S)
		Savg	15.3	28.6	22.4	22.2	
	KMEANS 3	*A*	98	2762	652	540	**26**
		Havg	590	1210	918	902	(3 U; 8 B, 15 S)
		Savg	25.4	39.3	32.1	32.2	
	WARD 1	*A*	227	3879	2079	2079	**16**
		Havg	401	1179	728	706	(4 U; 1 B, 11 S)
		Savg	21.8	34.0	25.5	25.9	
	WARD 2	*A*	98	3684	768	540	**33**
		Havg	590	1210	929	945	(7 U; 22 B, 15 S)
		Savg	19.4	39.3	29.9	30.0	
	WARD 3	*A*	84	3191	924	466	**25**
		Havg	234	730	507	503	(4 U; 5 B, 16 S)

Key: MIN - minimum, MAX - maximum, AVG - average, MED - median, *Nk* - number of stations in cluster; Origin of stations in the cluster: U - Una, B - Bosna, S - Serbia

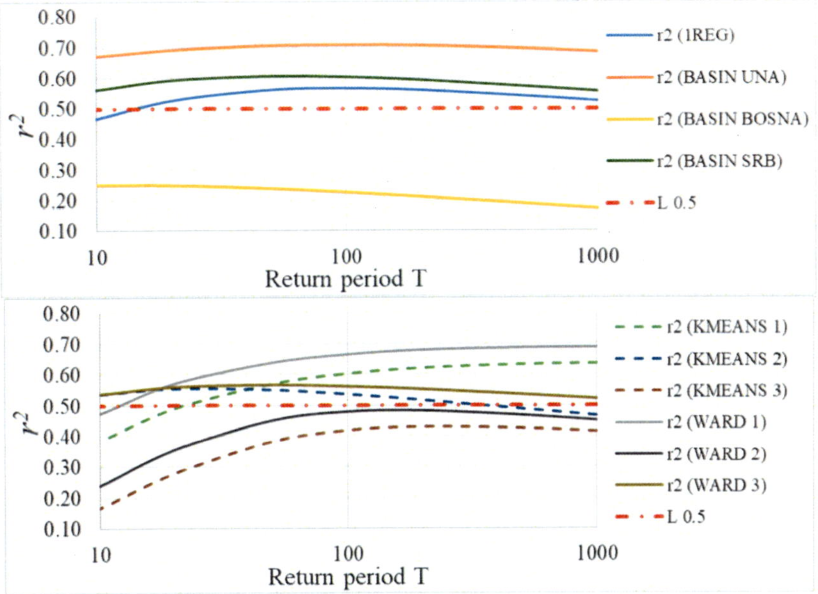

Fig. 14. Determination coefficient change across return periods T (years) for subjective (above) and clustering (below) regionalization approach.

uncertainty stemming from the flood input data, together with the number of detected PILF due to the outlier treatment approach. Table 5 presents situation within each defined region where the flood record length is shown per station, PILF as a total number detected in the stations comprising region, and number of stations with detected PILF.

Table 5. An average length of data series (N_{avg}) and number of PILF according to regionalization methods and cluster centers with number of stations with detected PILF.

REG	N_{avg}	*PILF* (no.stat.)	REG	N_{avg}	*PILF* (no.stat.)	REG	N_{avg}	*PILF* (no.stat.)
1REG	27.4	66 (21)	KMEANS 1	27.7	12 (4)	WARD 1	27.6	10 (4)
UNA	28.9	11 (4)	KMEANS 2	27.5	34 (9)	WARD 2	28.0	20 (8)
BOSNA	26.9	18 (5)	KMEANS 3	27.4	20 (8)	WARD 3	27.2	36 (9)
SERBIA	27	37 (12)						

Low number of PILFs and high value of N_{avg} would overall yield better (less uncertain) flood quantile estimation results in the region. The lowest ratio $PILF/N_{avg}$ is in the region WARD 1 (0.362), BASIN UNA (0.381) and KMEANS 1 (0.433). These are the regions with the highest r^2 at the same time. However, in other regions, this is not the case – the regions with high $PILF/N_{avg}$ ratio perform well according to coefficient of determination and vice versa. Therefore, the most reliable regionally assessed flood quantiles may come from the regions WARD 1, BASIN UNA and KMEANS 1, while quantiles estimated from the other regions should be closely examined.

3.3 Regionalization Efficiency

A comparison of the regionally assessed flood quantiles $Q_{T,REG}$ from the adopted regression model and reference values (Q_T) obtained through FFA is shown in Fig. 15 and Fig. 16 for the 100 years return period. Results for stations with and without gaps are shown separately. The percentage difference of quantiles – $PE_{T,i}$ (Eq. 5) for all stations and each regionalization method is split into two charts (Fig. 15 and Fig. 16) due to a wider range of differences in the subjective regionalization approach with one region comprising all stations (1REG).

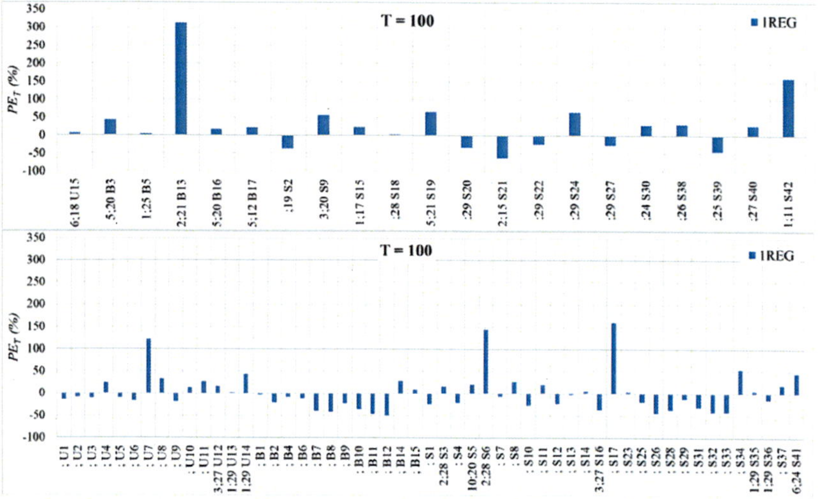

Fig. 15. Percentage difference of 100-year flood quantiles to reference ones in subjective regionalization method for each station. Adjacent to station labels on x-axes, number of detected PILFs and record length without PILF (years) is shown. Stations with gaps (above), and without gaps (below)

There are 5 stations in the cluster 1REG with significantly overestimated $Q_{100,REG,i}$ ($PE_{100,i}$ > 100%) (Fig. 15) and 4 stations with $PE_{100,i}$ > 50% as shown in Fig. 16. Four stations in Fig. 16 with $PE_{100,i}$ estimated around 50% come from KMEANS cluster centers 1 and 2 and do not match any of 9 stations from 1REG with significant overestimation of $Q_{100,REG,i}$. In general, both significant over– and underestimation in volume and occurrences are noted in 1REG compared to three other regionalization methods.

An indication of $PE_{T,i}$ spread out is given in Fig. 17 for four return periods and three regionalization methods.

It can be noticed from Fig. 17 that BASIN quantiles are systematically underestimated in the Una river catchments for all return periods (mean value of $PE_{T,i} < 0$), while in the Bosna river catchments, they are overestimated. In the Bosna river basin, WARD method has an advantage because the interquartile range is the narrowest and the mean value of $PE_{T,i}$ is approximately 0. For the catchments in Serbia, the BASIN method with the lowest interquartile range has an advantage.

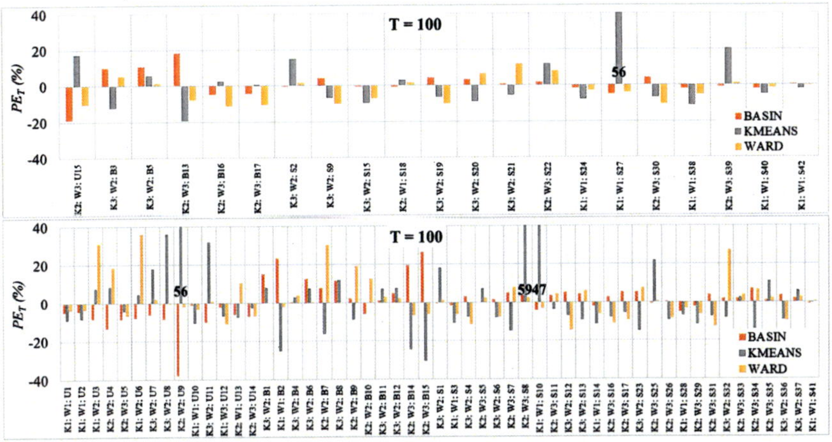

Fig. 16. Percentage difference of 100-year flood quantiles to reference ones in subjective regionalization method BASIN and clustering methods KMEANS and WARD for each station. Adjacent to station labels on x-axes, KMEANS (K) and WARD (W) clusters centers is shown. Stations with gaps (above), and without gaps (below).

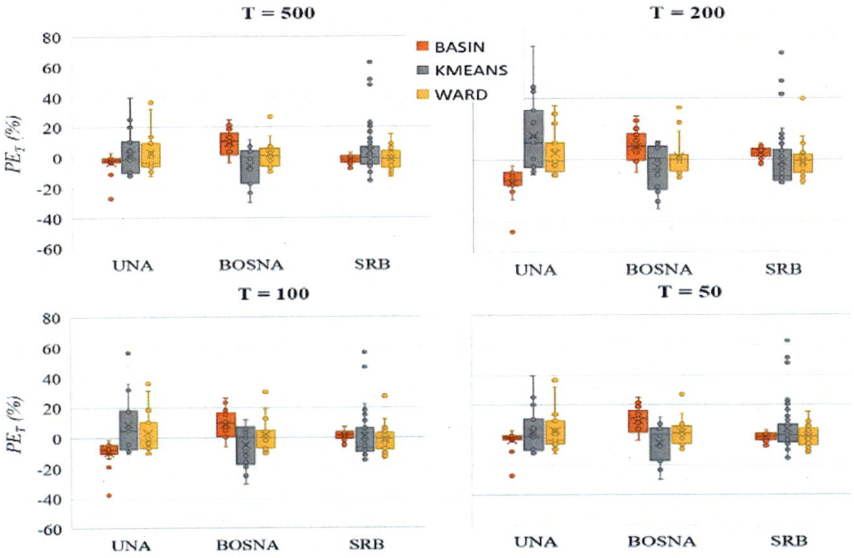

Fig. 17. Box-plot of PE_T for $T = 500$, 200, 100 and 50 years according to regionalization methods, disaggregated by basins

While $PE_{T,i}$ as a difference measure was selected for orientation regarding direction of the quantile difference, its absolute value $ABSPE_{T,i}$ (Eq. 6) allows for comparison of the difference volume.

A separate insight into both *ABSPE* and *RMSE* for 100-year quantile estimates for stations with and without gaps is shown in Table 6. Regionalization 1REG exibits the largest *ABSPE* values among all regionalization methods for both stations with and without gaps. The largest *ABSPE* (312%) is detected in station B13 with record length of 23 years, although there are stations with shorter records. The average value of $ABSPE_{T=100}$ for stations with gaps is 51.9% while for stations without gaps it is 29.8%. For other regionalization methods, $ABSPE_{T,i}$ values for statitons with gaps are slighty better compared to stations without gaps. In both cases, BASIN and WARD method perform better compared to other two methods.

Table 6. *ABSPE* and *RMSE* for return period T = 100 years according to regionalization methods

Statistics		Stations with gaps ($N = 21$)				Stations without gaps ($N = 53$)			
		1REG	BASIN	KMNS	WARD	1REG	BASIN	KMNS	WARD
ABSPE	MIN	1.7	0.1	0.5	0.0	1.1	0.0	0.1	0.0
	MAX	312.7	18.8	56.5	11.9	160.9	37.8	58.6	35.9
	AVG	51.9	4.7	11.2	6.1	29.8	6.3	13.3	7.6
RMSE		84.6	7.0	16.0	7.2	43.0	9.4	18.2	11.1

Based on the $ABSPE_{T,i}$ at each station for flood quantiles obtained by four regionalization methods, the method rank was determined. Rank 1 implies the method with the lowest $ABSPE_{T,i}$ (the best), while rank 4 is assigned to the highest $ABSPE_{T,i}$ value (the worst).

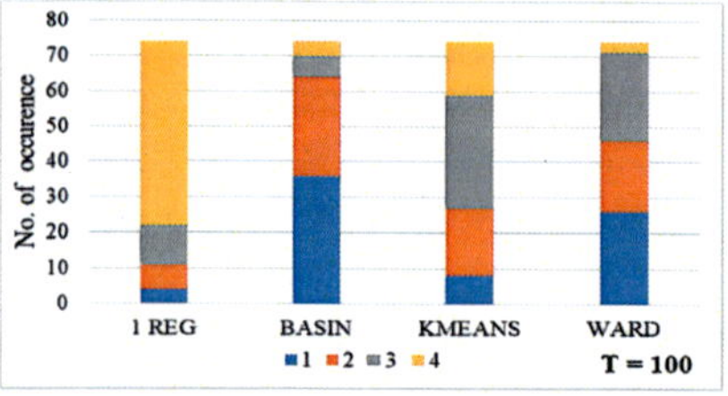

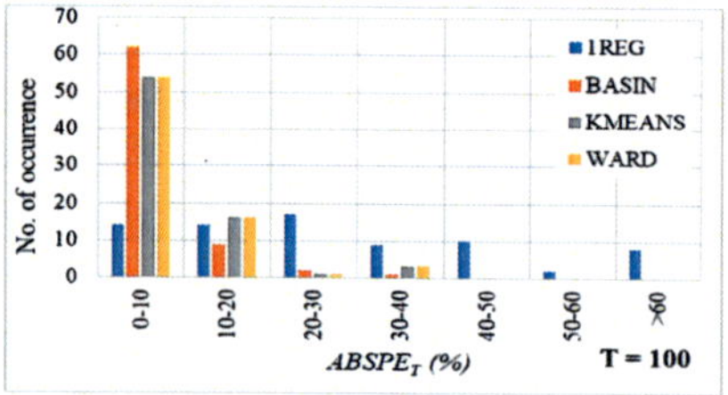

Fig. 18. Left - Frequency of ranks for $Q_{100,\,REG,i}$ per regionalization method (1. best, 4. worst). Right – Frequency of $ABSPE_{100,i}$ in 10% intervals for $Q_{100,\,REG,i}$ assesed by each regionalization method.

Within each method of regionalization and 100 year return period, the total number of ranks among 74 stations is shown in Fig. 18-left. Colors indicate individual rank while bar height indicates a frequency of rank occurrence – the number of stations holding a particular rank.

The left charts in Fig. 19, Fig. 20 and Fig. 21 represent the rank frequencies for $T = 100$ years disaggregated according to the large river basins. The advantage of the WARD over the other methods stands out for the Bosna river basin, and the BASIN method of regionalization for the catchments in Serbia.

The right hand side charts in Fig. 19, Fig. 20 and Fig. 21 shed light on $ABSPE_{T,i}$ volume for $T = 100$ years in the large river basins. Taking the $ABSPE_{T,i}$ range of 0–10%

as a good result in regional flood quantile assessment, the best performing method in both the Una river basin and catchments in Serbia is BASIN, while it is WARD method for the Bosna river basin.

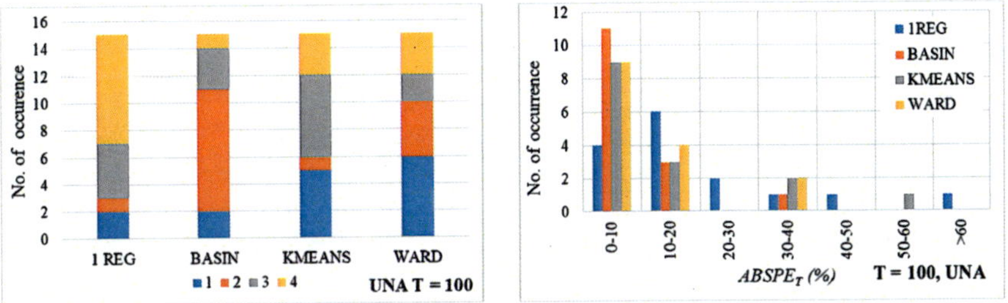

Fig. 19. The Una river basin: Left - Frequency of ranks for $Q_{100,\ REG,i}$ per regionalization method (1. best, 4. worst). Right – Frequency of $ABSPE_{100,i}$ in 10% intervals for $Q_{100,\ REG,i}$ assessed by each regionalization method.

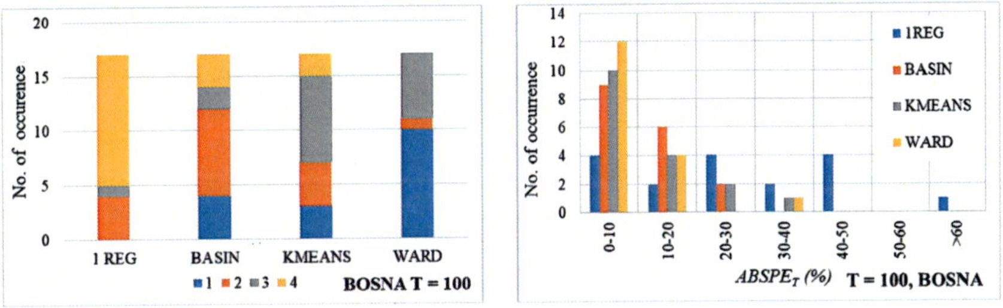

Fig. 20. The Bosna river basin: Left - Frequency of ranks for $Q_{100,\ REG,i}$ per regionalization method (1. best, 4. worst). Right – Frequency of $ABSPE_{100,i}$ in 10% intervals for $Q_{100,\ REG,i}$ assessed by each regionalization method.

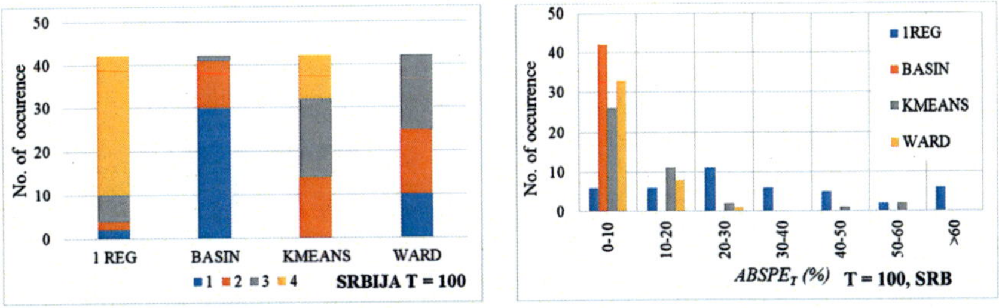

Fig. 21. The catchments in Serbia: Left - Frequency of ranks for $Q_{100,\ REG,i}$ per regionalization method (1. best, 4. worst). Right – Frequency of $ABSPE_{100,i}$ in 10% intervals for $Q_{100,\ REG,i}$ assessed by each regionalization method.

The proposed $RMSE_{T,RM}$ measure (Eq. 7) compares an absolute percent bias, or $ABSPE_{T,RM}$ to its mean value for quantiles estimated by a single regionalization method RM for one return period T. It is an attempt to provide a single value for regionalization

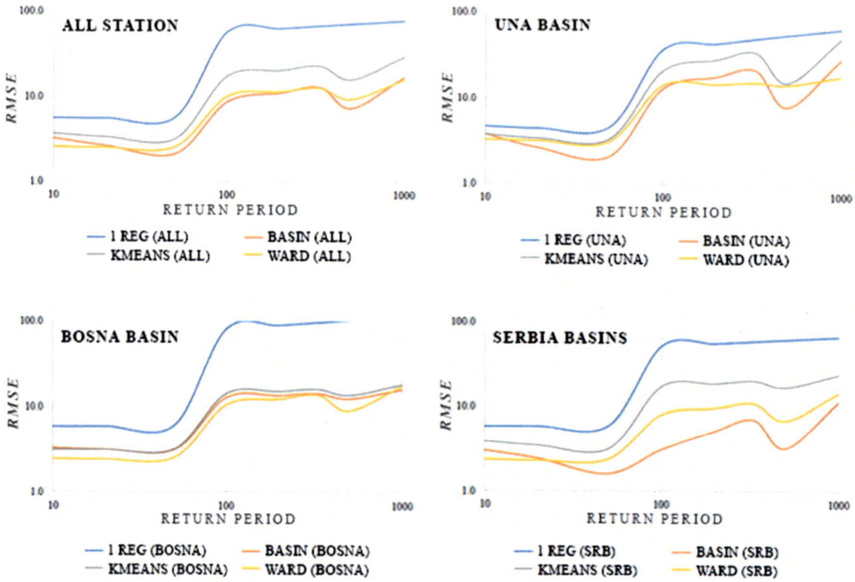

Fig. 22. $RMSE_{T,RM}$ change or $RMSE$ according to the analyzed return periods, taking into account stations (top left), basin Una (top right), basin Bosna (bottom left) and stations in Serbia (bottom right)

method evaluation, as shown in Fig. 22 across return periods for all stations and three main basins. In general, BASIN and WARD method perform better than KMEANS and 1REG. The $RMSE_{T,RM}$ values for presenting diagrams in Fig. 22, and ranking of $RMSE_{T,RM}$ are shown in Table 7.

Table 7. $RMSE_{T,RM}$ value and resulting rank of regionalization methods disaggregated by basin (Una, Bosna, Serbia)

	ALL STATION				UNA				BOSNA				SERBIA			
T	REG	BASIN	KMEANS	WARD	REG	BASIN	KMEANS	WARD	REG	BASIN	KMEANS	WARD	REG	BASIN	KMEANS	WARD
1000	79.2	17.1	29.8	16.0	63.0	28.0	48.6	17.9	111.9	16.0	18.4	17.4	67.5	11.5	24.0	14.6
500	72.1	7.3	16.0	9.3	54.0	7.9	15.2	14.3	102.2	12.4	13.7	9.0	62.4	3.3	17.1	6.9
200	63.6	11.1	20.5	11.5	43.7	18.0	28.4	14.7	89.8	13.6	15.2	12.3	56.4	5.2	19.0	9.7
100	57.9	8.8	17.6	10.2	37.3	12.9	21.0	14.2	81.0	12.8	14.3	10.5	52.5	3.1	17.5	8.1
50	5.8	2.2	3.3	2.6	4.7	2.1	3.4	3.2	6.1	3.2	3.2	2.6	6.0	1.6	3.2	2.4
20	5.6	2.7	3.4	2.6	4.5	2.7	3.5	3.3	5.9	3.2	3.2	2.4	5.9	2.5	3.5	2.3
10	5.7	3.3	3.8	2.6	4.8	3.9	3.9	3.4	5.9	3.3	3.2	2.5	5.9	3.1	3.9	2.4

Key: 1.0 1st, 2.0 2nd, 3.0 3rd, 4.0 4th

All extreme deviations for the BASIN regionalization method for all return periods originate from the BiH catchments, except for $T = 1000$ where the largest underestimation is at station S21 (not shown here). This station has a rather short flow record (16 years) in the studied period, because it has been relocated and its flow record is usually concatenated with the one of its predecessor, which was not done here. In Fig. 3 it can be seen this is the case with complementing stations S15 and S42, but treated separately, and also with S2 (its 'pair' is not included in the research). The data gaps in all these cases are described through perception thresholds in the FFA. However, the lower perception threshold bound in the research is set higher than recommended in [41], possibly causing higher flood quantile estimates in such stations. In the same light, out of 17 stations in the Bosna river basin, in 6 the data gaps are present, while in the Una river basin, out of 15 stations only one is with data gap. The flood quantiles in the Una river basin are predominantly underestimated, while in the Bosna stations they are overestimated by the regionalization methods.

Unlike the stations in the Bosna river basin where BASIN regionalization method does not give good results, for the catchments in Serbia this method is best according to all measures: $PE_{T,i}$ (Fig. 17), $PE_{T,i}$ (Fig. 21) and $RMSE_{T,RM}$ (Fig. 22, Table 7). This was already expected according to the coefficient of determination for reference flood quantiles (Fig. 14).

Focusing on the assessment of regionalization methods for quantile estimation in BiH, the measures $RMSE_{T,RM}$ and mean $ABSPE_{T,i}$ were recalculated to include stations from the Una and Bosna river basins only, and examine the two basins measures separately. This excludes the impact of regionalization results from the catchments in Serbia. REG1 and KMEANS still give the poorest results (except for $T = 10$ and 20 years in the Bosna river basin, when $RMSE_{T,RM}$ comes close to BASIN). In the Bosna river basin for all return periods, WARD gives the best results except for the 1000-year return period (Table 7, Fig. 22). For the 1000 years return period, $RMSE_{T,RM}$ of the BASIN and WARD methods are quite close. According to Fig. 17, several outliers contribute to increasing $RMSE_{T,RM}$.

In the Una river basin, the advantage of WARD over BASIN method is for return periods of 1000 and 200 years with significantly lower $RMSE_{T,RM}$, while for 500 years the results are more favorable according to BASIN. For other return periods according to $RMSE_{T,RM}$ measure, the results of BASIN and WARD are quite similar. Analyzing the frequency of $ABSPE_{T,i}$ for T = 100 years (Fig. 19) it can be seen that WARD in the Una basin has the highest frequency in the class 0–10%. The same was observed for all considered return periods (not shown here). For T = 500 years (not shown here), the frequency in class 10–20% is significantly higher for BASIN, and in addition to it, for 2 stations WARD gives $ABSPE_{T,i}$ in the interval 30–40%, which is the reason for worse $RMSE_{T,RM}$.

It can be concluded that the inclusion of stations from BiH mainly undermines the results in the catchments in Serbia (cases WARD and KMEANS) (Fig. 14, Fig. 15, Fig. 16, Fig. 17, Fig. 22), while the reverse is not valid. By including stations from Serbia, the results in the Bosna river basin are better compared to, for example, BASIN method, which takes into account only the catchments from the same large river basin. In the Una river basin for some return periods, the inclusion of stations from Serbia through WARD regionalization, improves the results, while for the rest, BASIN method is mostly preferred. However, according to the $RMSE_{T,RM}$ this advantage is not significant.

4 Conclusion

The results of this research supported expectation about the potential for flood quantile data transfer from catchments in Serbia to catchments in BiH by regionalization.

In general, Ward's clustering method performed better in the Bosna river basin where including stations from Serbia was beneficial, while in the Una river basin that benefit is not significant because acceptable results are obtained by flood quantile spatial interpolation from the stations within the basin. All results from the Una river basin should be treated with caution due to its karstic features, that contribute to uncertainty in regional analysis as early in the process as in the morphological attributes assessment.

The performance of regionalization methods can be anticipated from the determination coefficient plot across the return periods, obtained from the reference flood quantiles in the created region/cluster centers. For the methods and related regions where this coefficient is above 0.5, better results are achieved when spatially interpolating flood quantile data obtained by regression model with parameters assessed in the jack-knife procedure.

Besides the scree plot, in the number of clusters selection, additional consideration should be given to cluster compactness. For instance, a silhouette width index [48] seems a suitable candidate for assessing not only the fit of individual elements while clustering, but for the quality of clusters and the entire regionalization method. Such a measure coupled with the one resembling *RMSE* used in this research, would improve understanding of individual method efficiency.

Although the attributes for the distance matrix were selected according to correlation coefficient and t-test from the set of several morphological and geological attributes, all of the selected attributes are the catchment descriptors (drainage area, mean slope, mean altitude). It is felt including some of the climatological descriptors in catchments of the considered size (small and medium) would improve regionalization capability.

In an attempt to provide as many stations as possible for the analysis, some stations with poor input peak flow data were used. The data issues include inadequate record length and numerous detected PILFs in the processing period 1961–1990. Although FFA was conducted by the Bulletin 17C procedure capable of managing such a situation with input data, due to the short processing period, and too high lower perception threshold boundary, flood quantile estimates were burdened with uncertainty. Such flood quantiles contributed to forming ill-posed regions, where stations with solid input data resulted in flawed regional flood quantile estimates. In the next research phase with longer peak flow records processing period, poor input data stations will be eliminated.

Acknowledgement. The authors are grateful to the Federal Hydrometeorological Service of Bosnia and Herzegovina and the Republic Hydrometeorological Service of Serbia for the data provided for this research.

References

1. EU: Directive 2007/60/EC of the European Parliament and of the Council on the assessment and management of flood risks (2007). https://eur-lex.europa.eu/legal-content/EN/TXT/?uri=CELEX:32007L0060. Accessed 28 Feb 2021
2. IPF5 Consortium: TA2015030 R0 IPA/WB12-BIH-ENV-04C1, Flood Hazard and Flood Risk Maps Project in Bosnia and Herzegovina (2020). http://heis.ba/en/projects/flood-hazard-and-flood-risk-maps-project-in-bosnia-and-herzegovina-fhrm. Accessed 28 Feb 2021

3. UNDP: Technology transfer for climate resilient flood management in Vrbas River Basin (2021). https://www.ba.undp.org/content/bosnia_and_herzegovina/en/home/climate-and-disaster-resilience/GoALWaSH1.html. Accessed 28 Feb 2021
4. COWI A/S (Denmark): Flood Hazard and Risk Mapping, Component 2 of Serbia National Disaster Risk Management Plan (NDRMP) (2020)
5. Službeni Glasnik RS br 13/2017: Pravilnik o utvrđivanju metodologije za izradu karte ugroženosti i karte rizika od poplava (2017)
6. Viglione, A., Merz, R., Bloschl, G.: On the role of the runoff coefficient in the mapping of rainfall to flood return periods. Hydrol. Earth Syst. Sci. **13**, 577–593 (2009)
7. DWA-Regelwerk: Merkblatt DWA-M 552 Ermittlung von Hochwasserwahrscheinlichkeiten Hennef: Deutsche Vereinigung für Wasserwirtschaft, Abwasser und Abfall e.V (2012)
8. Wang, Q.J.: A Bayesian joint probability approach for flood record augmentation. Water Resour. Res. **37**, 1707–1712 (2001)
9. Vukmirović, V.: Analiza verovatnoće pojave hidroloških veličina. Naučna knjiga, Beograd (1990)
10. Afouda, A.: *FRIEND,* Flow regimes from international experimental and network data: Projects H-5-5 and 1.1: third report 1994–1997. Paris: UNESCO (1997)
11. Castellarin, A., et al.: Review of applied-statistical methods for flood-frequency analysis in Europe. NERC/Cenre for Ecology & Hydrology (2012)
12. England Jr., J.F., et al.: Bulletin 17C Guidelines for Determining Flood Flow Frequency. Chapter 5 of Section B, Surface Water, Book 4, Hydrologic Analysis and Interpretation. USGS (2018)
13. Institute of Hydrology: Flood Estimation Handbook. Wallingford, U.K. (1999)
14. Kjeldsen, T.R., Jones, D.A., Bayliss, A.C.: Improving the FEH statistical procedures for flood frequency estimation. Bristol: Environment Agency (2008)
15. Set of rules-design and construction: СП 33-101-2003 Определение основных расчетных гидрологических характеристик. Moskva: Государственный комитет Российской Федерации (2003)
16. Rao, A.R., Srinivas, V.V.: Regionalization of Watershads. An Approach Based on Cluster Analysis. Springer, Netherlands (2008). https://doi.org/10.1007/978-1-4020-6852-2
17. Donald, H.B.: Evaluation of regional flood frequency analysis with a region of influence approach. Water Resour. Res. **26**(10), 2257–2265 (1990)
18. Hosking, J.R.M., Wallis, J.R.: Regional Frequency Analysis an Approach Based on L-Moments. Cambridge University Press, Cambridge (1997)
19. Sharghi, E., Nourani, V., Soleimani, S., Sadikoglu, F.: Application of different clustering approaches to hydroclimatological catchment regionalization in mountainous regions, a case study in Utah State. J. Mountain Sci. **15**(3), 461–484 (2018). https://doi.org/10.1007/s11629-017-4454-4
20. Isik, S., Singh, V.: Hydrologic regionalization of watersheds in Turkey. J. Hydrol. Eng. **13**, 824–834 (2008)
21. Aytaç, E.: Unsupervised learning approach in defining the similarity of catchments: Hydrological response unit based k-means clustering, a demonstration on Western Black Sea Region of Turkey. Int. Soil Water Conserv. Res. **8**, 321–331 (2020)
22. Schaefer, M.G.: Regional analyses of precipitation annual maxima in Washington State. Water Resour. Res. **26**, 119–131 (1990)
23. Gingras, D., Adamowski, K., Pilon, P.J.: Regional flood equations for the provinces of Ontario and Quebec. Water Resourc. Bull. **30**, 55–67 (1994)
24. Cuanne, C.: Methods and merits of regional flood frequency analysis. J. Hydrol. **100**, 269–290 (1988)
25. Gingras, D., Adamowski, K.: Homogeneous region delineation based on annual flood generation mechanism. Hydrol. Sci. **38**, 103–121 (1993)

26. Merz, R., Blöschl, G.: A process typology of regional floods. Water Resour. Res. (2003)
27. Merz, R., Blöschl, G.: Flood frequency regionalisation-spatial proximity vs. catchment attributes. J. Hydrol. **302**, 283–306 (2005)
28. Merz, R., Blöschl, G.: Process controls on the statistical flood moments - a data based analysis. Hydrol. Process **23**, 675–696 (2009)
29. Burn, D.H.: An appraisal of the "region of influence" approach to flood requency analysis. Hydrol. Sci. J. **35**, 149–165 (1990)
30. Zrinji, Z., Burn, D.H.: Flood frequency analysis for ungauged sites using a region of influence approach. J. Hydrol. **153**, 1–21 (1994)
31. Nathan, R.J., McMahon, T.A.: Identification of homogeneous regions for the purpose of regionalization. J. Hydrol. 217–238 (1990)
32. Yang, X., Magnusson, J., Rizzi, J., Xu, C.: 'Runoff prediction in ungauged catchments in Norway: comparasion od regionalization approaches. Hydrol. Res. 49, 487–505 (2018)
33. Popović, V., Andrejević, D.: Velike vode u SR Srbiji. Vodoprivreda (1971)
34. Janković, D., Malošević, D.: Anvelope specifičnog oticaja velikih voda za teritoriju SR Srbije. Vodoprivreda, pp. 37–41 (1989)
35. Hrelja, H.: Definiranje nekih elemenata hidrološkog režima metodom regionalizacije. Vodoprivreda **37**(1–3), 21–34 (2005)
36. Imširović, Š.: Regionlana analiza karakteristika velikih voda na teritoriji Srbije i Bosne i Hercegovine u periodu 1961–1990. Diplomski rad, Građevinski fakultet u Sarajevu (2020)
37. Blagojević, B.: Razvoj modela za prostornu interpolaciju hidroloških vremnskih serija na neizučenim profilima. Doktorska teza, Univerzitet u Nišu, Građevinski fakultet (2011)
38. QGIS Development Team: QGIS Geographic Information System. Open Source Geospatial Foundation Project (2019). http://qgis.osgeo.org. Accessed 12 May 2019
39. U.S. Army Corps of Engineers, Hydrology Center: Statistical Software Package (HEC-SSP) version 2.1 (2016). https://www.hec.usace.army.mil/software/hec-ssp/download.aspx. Accessed 12 May 2019
40. Rallison, E.R., et al.: Bulletin 17B Guidelines for Determining Flood Flow Frequency. USGS (1981)
41. Blagojević, B., Mihailović, V., Đokić, N.: Ocena kvantila velikih voda iz nepotpunih nizova osmatranja na većim slivovima u Srbiji primenom pragova percepcije. Vodoprivreda, pp. 259–272 (2020)
42. AFIT Data Science Lab R Programming Guide . K-means Cluster Analysis. https://afit-r.github.io/kmeans_clustering. Accessed 26 Feb 2021
43. Alto, V.: Unsupervised Learning: K-means vs Hierarchical Clustering (2019). https://towardsdatascience.com/unsupervised-learning-k-means-vs-hierarchical-clustering-5fe2da7c9554. Accessed 19 Feb 2021
44. Sarhadi, A., Modarres, R.: Flood seasonality-based regionalization methods: a data-based comparasion. Hydrol. Process **25**, 3613–3624 (2011)
45. Kwon, Y., Lee, H., Lee, H.: Implication of the cluster analysis using greenhouse gas emissions of Asian countries to climate change mitigation. Mitig. Adapt. Strat. Glob. Change **23**(8), 1225–1249 (2018). https://doi.org/10.1007/s11027-018-9782-3
46. Rodda, J., Robinson, M.: Progress in modern hydrology: past, present and future (2015)
47. Mc Cuen, R.H., Smith, E.: Origin of flood. ASCE (2008)
48. Lengyel, A., Botta-Dukát, Z.: Silhouette width using generalized mean—a flexible method for assessing clustering efficiency. Ecol. Evol. (2019). https://doi.org/10.1002/ece3.5774. Accessed 19 Feb 2021

Research Financing Models for Construction of Underground Garages in Sarajevo

Suada Sulejmanović(✉), Žanesa Ljevo, Ammar Šarić, and Mirza Pozder

Faculty of Civil Engineering, University of Sarajevo, Sarajevo, Bosnia and Herzegovina

Abstract. One of the burning issues of almost every urban environment is parking space. Due to the lack of free land surface in densely populated areas, underground garages' construction is most often resorted to. However, as the adopted regulatory plan limits this space and the network of roads that should accept all the newly created traffic, the number of parking spaces usually does not justify underground garages' financial profitability. Due to the limited budget of local government levels, it is crucial to find the optimal solution for constructing this type of facility that will surely benefit the community and society. Moreover, we can observe this issue from an environmental point of view. The benefit can also be seen by reducing noise and air pollution due to the shorter time that vehicles are looking for a parking space. The paper presents two different ways of financing the construction of three underground garages in Sarajevo. A comparative analysis was performed using the traditional method of financing and public-private partnership. An example of achieving profitability of two financially unprofitable garages in a joint project with one financially profitable garage is presented.

Keywords: Underground garages · Parking space · Financial profitability · Public-private partnership

1 Introduction

Lack of parking space is one of the constant problems in urban areas. From the aspect of the space in which we live, there is a conflict between balancing supply and demand. On the other hand, there is a need to limit parking spaces in some urban regions according to a humane city's criteria as the most significant value we strive for in our work and activities.

Traffic connection and inclusion in the traffic network of planned garages are limiting factors that are often neglected even in the phase of adopting the regulatory plan. As the spatial requirements of stationary traffic are much greater than the objective possibilities, underground garages' construction is the option for resolving the parking needs [1]. Planned garages often do not have enough parking spaces to ensure cost-effectiveness and self-sustainability through toll parking spaces. The only remaining option is to increase the number of parking spaces by increasing underground floors. However, already with the "−2" floor, the construction costs double so that profitability is unattainable in such a way.

N. Ademović et al. (Eds.): IAT 2021, LNNS 316, pp. 457–468, 2022.
https://doi.org/10.1007/978-3-030-90055-7_36

Therefore, it is necessary to find a way to finance such facilities that certainly have their economic profitability and benefit to society, reducing travel time, reducing pollution and noise, opening ancillary activities, and many other benefits.

The paper gives an example of a cost-effectiveness analysis of the construction of three underground garages in Sarajevo as one joint project in Novo Sarajevo's municipality.

2 Locations of Planned Underground Garages

All three planned garages are located near the central city highway in a densely populated part of Sarajevo, the municipality of Novo Sarajevo (Fig. 1).

Figure 1 shows the subject garages' spatial positions, marked with A, B, and C.

Fig. 1. The position of the underground garages in municipality "Novo Sarajevo"

The garage in position A is located next to the Public Institution "Institute for Emergency Medical Aid of Sarajevo Canton" and behind the Federal Institute for Pension and Disability Insurance building. The entrance to the parking lot is directly from Kolodvorska Street. There are public parking areas free of charge, private parking areas, and one unmarked parking area with many vehicles. A large number of improperly parked cars were also recorded in the field (Fig. 2).

Fig. 2. Unmarked parking space at parking location A

A parking space marked as B is located between Envera Šehovića and Ložionička streets, and directly behind the "Hoše komerc" supermarket. This zone is bustling due to a large number of residential units and business facilities. The entrance to the parking lot is from a side street that is connected to Envera Šehovića Street. The area of the existing parking lot is 1692 m^2, with a total of 72 parking spaces (Fig. 3).

Fig. 3. Existing condition of parking lot B

The parking lot between the residential buildings and the Robot shopping center is marked with the letter C. There are two entrances in the parking lot, both from Aziza Šaćirbegović Street (Fig. 4) [2].

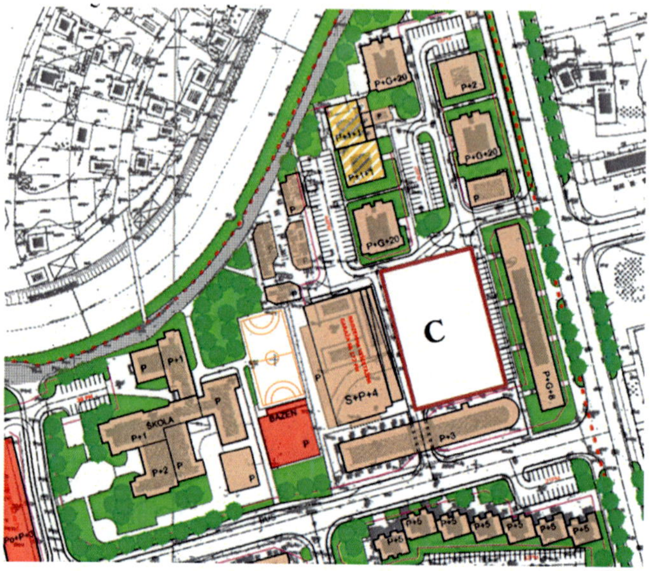

Fig. 4. Areas currently used for parking at the location C

3 Supply and Demand for Parking Spaces

To obtain relevant and quality data on the current number of available parking spaces, field recording and counting of user demand were performed. The conducted research included surveying car drivers and counting vehicles at the parking lot entrance, and counting parked vehicles [3].

By analyzing the collected data, the demand for parking spaces was obtained. The required number of parking spaces at location "A" is 129, at location "B" 149, while at location "C" 414 parking spaces are missing (Table 1).

Table 1. Costs of building underground garages (PPP)

Garage	Supply (PS)	Demand (PS)	Required number of parking spaces (PS)
A	349	478	129
B	72	221	149
C	320	734	414

After the identified parking demands, conceptual solutions of underground garages were designed at the analyzed locations, foreseen by the existing regulation plan [2, 4].

Since the construction price doubles with the increase in the number of underground floors, the conceptual solutions were designed with one underground floor and the retention of the existing parking area [5].

Designed garages may not fully meet demand, but they can mostly solve the problem.

4 Estimating of Investment and Operating Project Costs

A cost estimate lists all the costs involved in completing a project successfully, from implementation to completion (project duration). However, a cost estimate is more than a direct cost list: it also details each expense's assumptions.

Cost estimates are crucial to effective project management because, during the design and definition process, teams are expected to generate a reasonably detailed and accurate estimate. During the planning process, estimates are modified for precision, as project stakeholders and sponsors can ask for revisions until they are willing to approve a budget. The accuracy of estimates is systematically elevated after this early point.

Cost estimation is an evolving process, and revisions to estimates are expected to ensure consistency in the project's implementation [6].

4.1 Investment Costs

Investment costs are divided into several basic groups:

1. Construction costs,
2. Costs of design and preparation of technical documentation,
3. Construction supervision costs,
4. Costs of obtaining permits and approvals, and utility connections,
5. Costs of compensation for the right to build (in the case of PPP financing methods)

The cost of construction costs was determined based on the current average cost of construction of 1 m^2 of similar facilities in Sarajevo. This price includes exploration, preparation, earthworks, concrete works, protection of the construction pit, equipment, and traffic signals.

The cost of designing and revising technical documentation is determined as a percentage of the cost of construction. This price includes the design of all underground garage phases: architectural, construction, geotechnical, traffic, mechanical and electrical.

Other investment costs include the costs of supervision during construction, obtaining various permits and approvals (urban, construction, technical acceptance), the cost of utilities (electricity, water), and the cost of fees for the right to build paid to the landowner (in this case to the Municipality of Novo Sarajevo). The Municipality of Novo Sarajevo determines the amount of this fee, and it is 74.21 km/m^2 at locations "A" and "B", while at location "C" it is 61.85 km/m^2. In the case of the traditional method of financing (municipality), this fee is exempt. The following table provides an overview of construction costs and an overview of total investment costs in the case of PPP financing methods (Table 2).

Table 2. Investment costs of building underground garages (PPP)

Garage	Useful surface (m^2)	Construction costs (km)	The other investment costs (km)	Total (km)
A	1,560	1,310,400	375,025	1,685,425
B	1,710	1,436,400	391,854	1,828,254
C	2,000	1,680,000	399,670	2,079,670
			SUM:	**5,593,349**

4.2 Operating Costs

Operating costs include:

1. Staff costs,
2. Utility costs,
3. Maintenance costs,
4. Annuity costs, i.e., land use, and
5. Loan installment.

The following table provides a detailed overview of operating costs for the initial year (Table 3). Concerning historical data, the costs are projected for 20 years project period, while the loan installment lasts eight years.

Table 3. The operating costs for the initial year

Number	Type of costs	Costs for the initial year (km)		
		A	B	C
1	Staff costs	27,453	27,453	27,453
2	Utility costs	5,616	6,156	7,200
3	Maintenance costs	3,744	4,104	4,800
4	Annuity costs	221,842	240,642	273,735
	Sum	**258,655**	**278,355**	**313,188**

4.3 Total Revenue from Parking Fees

Total revenues from parking fees are shown in Table 4. Given that the project task focused on residential users, 50% of parking spaces for this type of users were planned in the initial analysis, and the remaining 50% is commercial. For residential users, the monthly ticket price is 70 km (average underground and aboveground parking space). In comparison, other users pay an hourly fee in the amount of 1.0 km as the average hourly price (day/night).

Table 4. Total revenues from parking fees

Garage	Parking capacity (PS)	Parking occupancy (%)	Resident/other beneficiary (%)		Coefficient of change	Billing system	Billing hours	Number of billing days	Per month (km)	Total per month (km)	Total per year (km)
A	132	100	Resid benef	50	-	70.00 KM/month	-	-	4,620	**46,422**	**557,061**
		70	Other benef	50	1.45	1.00 KM/hour	24	26	41,801		
B	88	100	Resid benef	50	-	70.00 KM/month	-	-	3,080	**30,948**	**371,374**
		70	Other benef	50	1.45	1.00 KM/hour	24	26	27,867		
C	70	100	Resid benef	50	-	70.00 KM/month	-	-	2,450	**25,076**	**300,915**
		70	Other benef	50	1.48	1.00 KM/hour	24	26	22,626		

5 Models of Project Financing

The financial evaluation assumes the project stands on its own and has to finance itself. Because of that, it is necessary to add loan costs at the current interest rate to projected costs.

The paper discusses two funding models: The Traditional Model (TM) and the Public-Private Partnership (PPP).

The Municipality would invest in the case of the traditional financing model, and the sources of financing are credit funds in the amount of 100%. The total amounts are reduced by the amount of compensation for the right to build because the Municipality is the landowner. The following table shows the sources of funding in the case of the traditional financing model (Table 5).

Table 5. Sources of funding in the case of the traditional financing model (TM)

Source of funds	Amount (km)	Percentage (%)
Own funds	0	0.00%
Credit funds	5,226,982	100.00%
Total	5,226,982	100.00%

The total required funds according to the traditional model for all three garages amount to 5,226,982 km. The interest rate for loans with the Development Bank of over 2 million km (variable rate, in km, June 2015. http://www.rbfbih.ba/bih/index.php) was 5.369%. The loan is calculated for ten years, including a "grace" period of two years with the principal's accrual of interest.

The calculation took into account the costs of loan processing of 1.5%.

5,306,581–1.5%*5,306,581 = 5,226,982 km.

Table 6 shows the calculation of credit liabilities in the case of the traditional financing model.

Table 6. Calculation of credit obligations of the garages according to the traditional financing model

Loan amount (km)	5,306,581
Interest rate	5.369%
Repayment years	10
Monthly installment (km)	77,845
Total interest (km)	2,166,542
Total loan cost (km)	7,473,124

In the PPP financing model, the investor will finance this project partly on his own and partly with credit funds obtained from commercial banks (Table 7) [7].

Table 7. Sources of funding in the case of the PPP financing model

Source of funds	Amount (km)	Percentage (%)
Own funds	1,678,004	30.00%
Credit funds	3,915,344	70.00%
Total	5,593,348	100.00%

The interest rate for commercial loans of over km 2 million (variable rate, in km, June 2015) was 6.369%.

The loan is calculated for ten years, including a grace period of two years with the principal's accrual of interest. The calculation took into account the costs of loan processing of 1.5%. Table 8 shows the calculation of credit liabilities in the case of PPP financing models.

Possible security instruments:

- bills of exchange of the company and personal bills of exchange of the owner,
- cash deposit,
- lien on real estate, movable property, stocks, and securities,
- third party guarantees (bills of exchange, equity and debt securities, cession, assignment, debt assumption, pledge of property and rights, and other instruments based on which the Bank may collect receivables through the market or court mechanism),
- insurance policy, and
- inflows under concluded contracts from quality clients.

Table 8. Calculation of credit obligations of the garages according to the PPP financing model

Loan amount (km)	3,974,969
Interest rate	6.369%
Repayment years	10
Monthly installment (km)	61,352
Total interest (km)	1,914,786
Total loan cost (km)	5,889,756

5.1 Project Financial Profitability

A significant element of any investment project is its financial analysis, based on which a decision is made on the profitability of investing in that project.

The financial analysis is performed by estimating cash flows and calculating net return indicators, such as financial net present value (FNPV) and financial internal rate of return (FIRR), which are performed by analyzing the return on investment cost and return on the national capital [8].

To summarise the overall relationship between the relative costs and benefits of a planned project, a benefit-cost ratio (BCR) is used. If a project has a BCR greater than 1.0, a firm and its investors are supposed to provide a positive net present value to the project.

It also estimated the payback period. The length of time needed for an investment's net cash receipts to completely cover the initial costs.

Fundamental assumptions that define the input data for calculation:

- the economic life of the project is 20 years,
- all amounts are in convertible marks (km),
- current prices are taken into account,
- inflation is not taken into account (insignificant),
- the recommended financial discount rate is 12%,
- VAT is included in the budget because it is an eligible cost to the Investor, i.e., VAT is not refundable to the Investor.

It should be noted that the financial analysis was conducted for each garage separately. The results showed that the underground garage at location A is financially viable, while the garages at locations B and C are unprofitable.

By establishing a joint project with a financially viable garage, two financially unprofitable garages' profitability has been achieved. For such a joint project, a financial analysis was performed, and the results are given below.

The following Fig. 5 shows the financial parameters for the traditional and concessionary form of PPP financing if the FBiH Law on Real Rights does not apply [9].

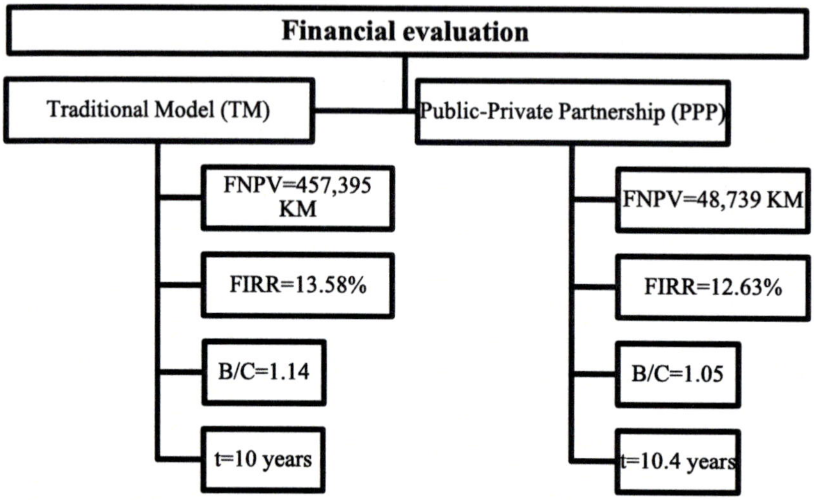

Fig. 5. Project financial evaluation parameters

Although all parameters are favorable, the analysis showed that the project is not financially sustainable for both financing models and that the payback time is unfavorable. Therefore, further analysis was based on finding a solution that would achieve financial sustainability.

At the request of the Investor, the model of public-private partnership according to the Law on Real Rights of the FBiH [10] was considered as a method of financing.

The settings of the analysis according to this Law are:

- Article 299, paragraph 2. The holder of the construction right is obliged to pay the landowner a monthly fee for the land in the amount of the average rent for such land unless otherwise specified.
- Article 312, Paragraph 3. The owner is obliged to give the person whose construction right has ceased to pay as much compensation for the building as his real estate in circulation is more valuable with that building than without it.
- The resident ratio in capacity reduced to 40% (A and B).

After the financial analysis, applying the mentioned Law, the following financial parameters were obtained for the case of PPP.

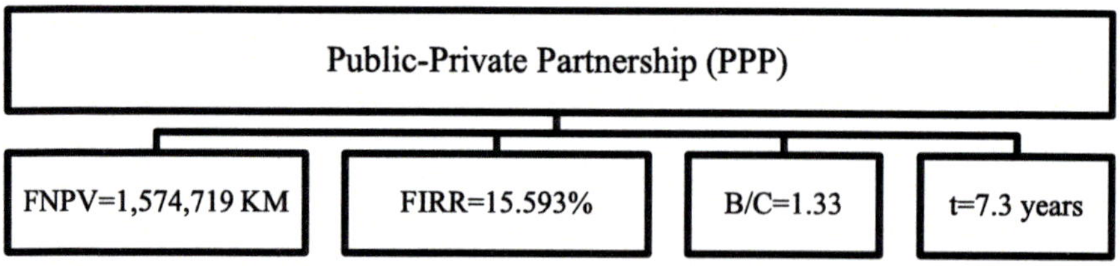

Fig. 6. Financial indicators according to the PPP model and the law on real rights of FBiH

As can be seen from Fig. 6, a solution that has been financially justified and sustainable has been reached.

6 Conclusion

One of the burning problems in urban areas is the lack of parking space. Therefore, it is increasingly resorting to the construction of underground garages. The crampedness of the built space and the already existing traffic infrastructure still dictate the size of underground garages, which often do not meet demand. However, their construction would greatly help partially solve the need for parking. These facilities are most commonly financially unprofitable and unsustainable. At the same time, there is indisputable economic efficiency, i.e., the gain to society.

The paper explains combining financially viable with financially unprofitable underground garages in one joint, financially viable project. Two ways of financing the project were compared: traditional and PPP. The public-private partnership has proven to be the only solution that is both financially profitable and sustainable. By applying a public-private partnership, the problem of insufficient budget funds is solved. The project's possible risks are shared between the partners, whereby the conditions under which this type of project is entered can be clearly defined. This approach also solves the problem of parking of residential users, which is also dictated by the Investor's conditions. The private sector's involvement significantly increases efficiency, and there are significant prospects that the project will be completed within a realistic time and budget.

References

1. Džebo, S.: Application of Sensitivity Analysis for Investment Decision in Building of Underground Garage Croatia, pp. 611–617. Department of Transportation, Faculty of Civil Engineering, University of Zagreb, Šibenik (2016)
2. Sarajevo Canton Development Planning Institute: Regulation plan Hrasno II (2021). https://zpr.ks.gov.ba/sites/zpr.ks.gov.ba/files/Regulacioni%20plan%20Hrasno%20II%20-%20izmjene%20i%20dopune.pdf. Accessed 20 May 2021
3. Brčić, D., Šoštarić, M.: Parkiranje i garaže. Zagreb: Fakultet prometnih znanosti Sveučilišta u Zagrebu (2012)
4. Sarajevo Canton Development Planning Institute: Regulation plan Center Novo Sarajevo (2021). https://zpr.ks.gov.ba/sites/zpr.ks.gov.ba/files/Regulacioni%20plan%20Centar%20Novo%20Sarajevo%20-%20izmjene%20i%20dopune.pdf. Accessed 25 May 2021
5. Banja Luka, I.G.: Feasibility study of the public investment project of underground garages in the settlements of Malta, Hrasno and Željeznička (2015)
6. Smartsheet: Ultimate Guide to Project Cost Estimating. https://www.smartsheet.com/ultimate-guide-project-cost-estimating. Accessed 25 May 2021
7. EIB: The Guide to Guidance – How to Prepare, Procure and Deliver PPP Projects. s.l.:s.n (2011)
8. European Commission: Directorate-General for Regional and Urban policy. Guide to Cost-Benefit Analysis of Investment Projects, Economic appraisal tool for Cohesion Policy 2014-2020. s.l.:s.n (2014)

9. Local development improvement project (Projekat unapređenja lokalnog razvoja - GOLD): Vodič za primjenu instituta prava građenja u Federaciji Bosne i Hercegovine, s.l.: USAID-Sida (2015). https://cupdf.com/document/vodic-za-primjenu-instituta-prava-gradjenja-u-fbih.html. Accessed 20 May 2021
10. Official Gazette of the FBiH: Law on Real Rights of the FBiH (2015). http://www.fbihvlada.gov.ba/bosanski/zakoni/2013/zakoni/26bos.html. Accessed: 25 May 2021

Test Operation of the MBR Pilot Plant for Leachate Treatment at the Sarajevo Sanitary Landfill

Amra Serdarević[1] (✉), Teresa Garstenauer[2], and Bernhard Mayr[2]

[1] Faculty of Civil Engineering, University of Sarajevo, Sarajevo, Bosnia and Herzegovina
amra.serdarevic@gf.unsa.ba
[2] EnviCare® Engineering GmbH, Graz, Austria
office@envicare.at

Abstract. Regional sanitary landfills and waste disposal centers are the strategic commitment in regard to municipal waste management in Bosnia and Herzegovina. In order to meet the environmental conditions, the obligation for each landfill is the treatment of leachate and disposal of landfill gas. Leachate from sanitary landfills is highly polluted wastewater, which changes its composition and quantity over time. Selection of an adequate leachate treatment technology is the first step in addressing the environmental impacts of leachate. At the Sarajevo Sanitary Landfill, in the period 2008–2011, a plant for the treatment of leachate started operation based on Membrane-Bio-Reactor technology (MBR) as a chosen technology. Due to a number of circumstances and malfunctions, the plant operation was stopped already soon after commissioning. The current situation at the Smiljevići Landfill is such that the landfill does not meet all of the technical requirements necessary to limit the negative environmental impact from the site. Leachate management continued to be one of the major issues at the landfill site since the leachate treatment plant (LTP) is out of function. There is a significant impact of leachate to the recipient (Lepenica stream). In 2016, the Government of Kanton Sarajevo (KS) started a five-year project for the reconstruction of the sanitary landfill and all associated facilities. One of the planned tasks is the reconstruction of the leachate treatment plant. For this purpose, a pilot plant was installed in 2018 to determine the composition of leachate and also to test a state of the art MBR - RO technology. The commissioning and related effects of the pilot plant for the treatment of the specific Smiljevići leachate in the period April–July 2018 are presented in this paper.

Keywords: MBR – MembraneBioReactor · Landfill leachate · Sludge · Pilot plant · Chemical characterization

1 Introduction

Regional sanitary landfill Smiljevići in Sarajevo Canton is the largest facility of this type in Bosnia and Herzegovina (BiH), with a yearly disposal amount of approximately 180,000 tons of waste. The landfill is located in the north of Sarajevo, and close to the

N. Ademović et al. (Eds.): IAT 2021, LNNS 316, pp. 469–483, 2022.
https://doi.org/10.1007/978-3-030-90055-7_37

settlements. The waste is disposed at the location since 1962 and until 1997 the site has been used as a simple dumpsite. Since 1997 operational and technical measures are introduced. Specific activities and construction have been carried out like multi-barrier bottom seal, landfill gas collection and utilization systems, a leachate collection system and a leachate treatment plant (LTP). These were the initial efforts to transform the site into the sanitary landfill complying with the state of the art [1].

Fig.1. Bird view of the sanitary landfill Smiljevici - Sarajevo Canton, BiH

The landfill is operated since several decades and the total area which was used to dump mainly household waste is approx. 411,000 m^2. The bigger part (approx. 338,000 m^2) is already planted or wooded, while the open section has approximately 73,000 m^2. The yellow line shown on Fig. 1 indicates the overall size of the landfill area and the red line indicates the open area for waste dumping [14].

One of the major issues at the Smiljevići sanitary landfill is a large amount of leachate generated from 250 up to 500 m^3/day. The average leachate discharge measured at the site in the period March 2017 – September 2018 was 3.42 l/s, 300 m^3/day. In order to be able to estimate the amount of leachate formed as the result of the yearly rainfall an infiltration coefficient of 10% for the covered area and 60% for the open area was assumed. The annual average precipitation is 961 mm.

This leachate contains significant amounts of ammonia (820 mg/L) and also chemical oxygen demand (COD 2,200 mg/L) and a treatment concept has to be found in order to comply with the discharge limits for the public sewer system.

The reason for leachate generation is the large open area of the landfill and areas covered with a temporary cover of approximately 0.5 m thick soil/inert material layer on which grass and small vegetation is grown (Fig. 1) [2].

In the year 2008 an unsuccessful attempt was undertaken to clean the leachate by installing a MembraneBioReactor (MBR) treatment based on non-permanent hydrophilic PE hollow fibre membranes operated in a submerged mode (Fig. 2, Fig. 3), in order to fulfil the environmental requirements [3]. MBR with submerged membrane modules was installed at the site and was put into operation in 2011. Unfortunately, after a very short period of time, due to the mechanical damages and improper configuration of the PE hollow-fibre membrane unit, plant operation had to be stopped.

Fig.2. MBR plant in operation (2008–2011)

Fig. 3. Membrane installation in the basin

This MBR plant failed due to improper design and execution and thus the leachate is still discharged into the environment ("Lepenicki potok") without any proper treatment causing harm to the environment due to the high concentrations of pollutants (COD, ammonia, sulphides, silicon, mercury, chlorides) (Fig. 4, Fig. 5) [4]. The problem of leachate treatment and discharge is still an unsolved issue at the Sarajevo sanitary landfill. Activities on the reconstruction of the treatment plant and monitoring of the leachate are undergoing.

The Government of Sarajevo Canton, Ministry of physical planning, construction and environmental protection of Sarajevo Canton has invited several companies to submit a concept and present possible solutions for leachate treatment.

Some techniques which are cutting of the edge and which could be applicable for the Sarajevo sanitary landfill were presented in January 2017 to the representatives of authorities, local experts and engineers from this field. The proposed solutions for the treatment of the leachate from Sarajevo sanitary landfill are based on biological treatment and membrane separation technique. The amount of leachate is up to 500 m^3/day, and the quality is typical for old landfills, i.e. methanogenic leachate [5].

The proposed technologies are based on the state of the art in this domain. Two-stages reverse osmosis (RO) and MBR in combination with ultra-filtration, nanofiltrations and activated carbon adsorption were considered as possible solutions. The activities on the

Fig. 4. Leachate drainage pipes in the collection sump

Fig. 5. Temporary discharge to the Lepenicki potok – open stream

comprehensive leachate monitoring were set up on the site in 2018 and were lasting for a period of one year. The second step was a pilot test on site [2, 6].

2 Material and Methods

A pilot test on a semi-technical scale is, from a procedural point of view, a very important instrument for limiting both technical and entrepreneurial risks. Therefore, a pilot test for a membrane bioreactor was conducted between 24thApril and 17th August 2018 by the company EnviCare® Engineering GmbH at the site of the landfill.

This paper presents the findings from the pilot test and further steps in activities regarding LTP reconstruction [4].

2.1 Task and Objectives of the Pilot Test

The main objective from the pilot test phase are results and findings which should be incorporated into the concept of the large-scale plant and allow an exact dimensioning of the essential plant components. The entrepreneurial and technical risk can therefore be correspondingly reduced.

The following aspects were considered and investigated during the test period [2]:

- Definition of the type and mesh width of the pre-screening
- Biodegradability of the COD and denitrification capacity
- Nitrification rate
- Requirement of an external carbon source
- Dry substance in the MBR, consistency and dewatering behaviour of the sludge
- Air and oxygen requirements for bioprocess and membrane operation
- Influence of hardness forming agents or metal salts on the throughput, fouling and scaling aspects in membrane operation
- Flux and permeability values of the membrane used
- Temperature influence on biology and membrane operation
- Power supply

- Amount of sludge produced
- Emission concentrations in the filtrate
- Foam formation and foam control

In order to assess the above-mentioned aspects, the following analyses were carried out in inflow, biology and filtrate:

- On-line measurements in the pilot-plant: temperature, conductivity, various throughputs and level, oxygen concentrations, pressures
- Off-line measurements in the operational laboratory at the landfill site: pH, COD, NH_4-N, NO_3-N, NO_2-N, PO_4-P, dry substance (DS), filterability
- Off-line measurements in authorized external laboratory in Austria: pH, conductivity, COD, NH_4-N, NO_3-N, NO_2-N, N_{tot}, TOC, Cl, Ca, Si, Ba, Sr)
- Microscopic examination of the activated sludge.

2.2 Brief Description of the MBR Pilot Plant

The pilot plant is designed as a container system, including a sedimentation tank, a denitrification and nitrification tank, a filtrate collecting tank, tanks for chemical cleaning agents, the necessary pumps, aggregates, fittings and an electrical control cabinet together with PLC control and data loggers (Fig. 6).

The membrane bioreactor system was fed by means of an immersion-controlled submersible pump, which feeds the effluent from the leachate buffer basin. A cone shaped inlet area serves as settling basin for the primary sludge (Fig. 7). This settled sludge can be discharged via an immersion pump back to the leachate basin.

Through a fine sieve with a mesh width of 0.5 mm the wastewater reaches the denitrification zone (4 m^3), which is an un-aerated volume where a stirrer ensures a homogeneous mixing and nutrient supply of the organisms. Alternatively, a screen with a mesh width of 1.5 mm can be installed. The sieve is cleaned by means of a controlled air flush and the resulting abrasion of the solid particles.

Fig. 6. Containerized pilot plant

Fig. 7. Settling area with fine mesh screen

The nitrification tank is constructed with a volume of 8 m^3, and therefore the total volume of the MBR was 12 m^3. The membrane module with a filter area of 200 m^2 was lifted into the basin via a maintenance crane (Fig. 8, Fig. 9).

Fig. 8. Membrane-Bioreactor (MBR): nitrification (left) and anoxic denitrification (right)

Fig. 9. Membrane module

Integrated measuring devices record the following parameters: temperature, conductivity and oxygen concentration. The level of the basin is controlled by two magnetic switches. A vacuum hose pump with a frequency converter (throughput adjustable from 250–2.000 L/h) is placed on the filtrate side of the submerged hollow fibre membrane module (Fig. 10, Fig. 11).

The recirculation pump immersed in the aeration tank also simultaneously serves to remove the biological excess sludge (time-controlled).

A flow measurement is installed in the feed line to the filtrate tank. The filtrate sample valve is also installed in this pipe. The total throughput of the system is recorded via a counter with magnetic contact.

Fig. 10. Suction pump and measuring devices

Fig. 11. Control panels and flow chart

Various cleaning strategies may be necessary to maintain the filtration performance. This ranges from regular filtration pauses and backwashes through chemical cleanings in the nitrification tank (clean in place – CIP) to external cleaning steps (clean on air) [7].

The pilot system is therefore equipped with a backflushing function, which allows cleaning solution and filtrate to be driven back against the normal filtration direction. A flushing pump with a wind boiler also allows this flushing to be performed intermittently.

For external cleaning or maintenance, the modules are lifted from the basin by means of a crane. The oxygen input into the aerated nitrification section is optionally via built-in pipe aerators and/or via the aeration pipes integrated in the modules.

Since the pilot plant is built in a container it can be easily transported and used for different applications in wastewater treatment, like food and beverage production, liquid waste treatment, galvanic industry and surface water treatment of solid waste handling sites, just to name a few [8].

In this individual case of application the capacity of the pilot plant was in the range of 3–10 m^3/day depending on the COD and ammonium load.

3 Performance and Results of the Pilot Test

The pilot plant was installed in April 2018 under the supervision by a team from EnviCare® and was in operation until the end of August 2018.

The experiment was carried out in extremely good cooperation between experts from EnviCare® (Graz) and the team of experts from Sarajevo (from the Faculty of Civil Engineering, Public Utility Company RAD and the Institute for Spatial Planning of Canton Sarajevo) [2].

3.1 COD Load and Inflow Characteristics

At the beginning, about 2.5 m^3 of sludge from the communal wastewater plant of Sarajevo were filtered and pumped into the MBR basin. The remaining volume was filled with rainfall water from a basin.

Figure 12 shows the DS-content and the COD load including the external carbon source (sugar, acetic acid). The COD concentration in the inflow was more or less stable throughout the experiment. Therefore, the desired COD load was easy to adjust. The COD load was increased following the biomass growth. Several peaks in the COD load are due to the addition of the external carbon source. Before day 89 sugar was added and during the last 20 days acetic acid was used as an external carbon source. Including the external carbon source, the planned COD load was exceeded during the last days of the pilot operation.

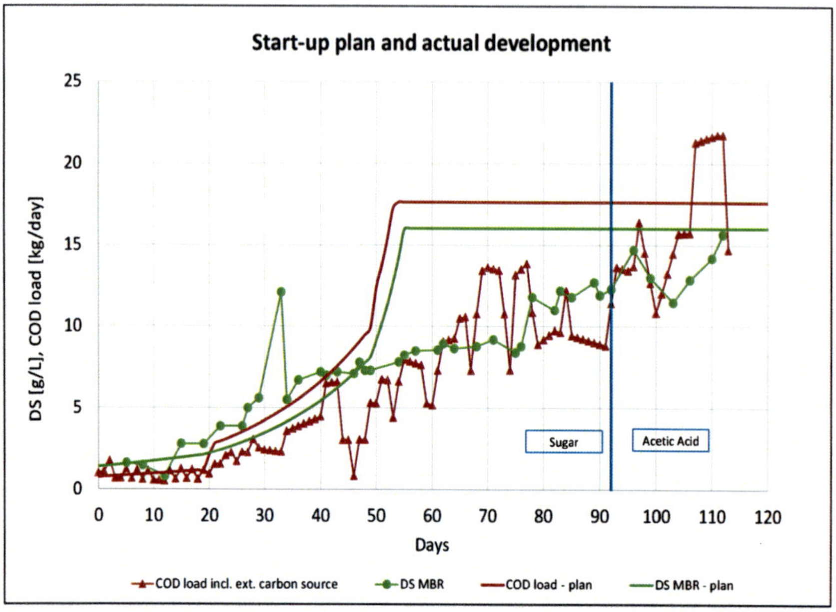

Fig. 12. Start-up plan and development during the test period

Concentration of the DS in the biological reactor was planned to reach a value of 16 g/L, but the biomass was not growing as fast as expected. The maximum measured DS at the end was 15.7g/L. It has to be considered that the salt content of the filtrate is very high (about 8.2 mS/cm) and contributes with 6.6 g/L (DS value in the filtrate) to the total DS. Therefore, the goal of 16 g/L was only reached by the end of the pilot phase (after four months of experiment).

Summarizing, the growth of the biomass with this kind of leachate is very slow. This has to be considered for the start-up of a large-scale plant. Additionally, the COD load has to be adjusted to the existing active biomass.

This slow growth and the tendency of the sludge to form very small and freely flowing flocs makes clear that a conventional biological treatment based on sedimentation will not be able to be used for cleaning this leachate.

In general, the inflow shows stable properties (Fig. 13). The slight decrease in the amount is related to less rainfall. Conductivity is indirectly linked to the amount of leachate as can be seen from Fig. 13.

The same slight increase in concentrations over time can be observed for all inflow parameters. Due to the stable inflow properties over time, it is quite simple to adjust the loads on the treatment plant and to operate the plant stable.

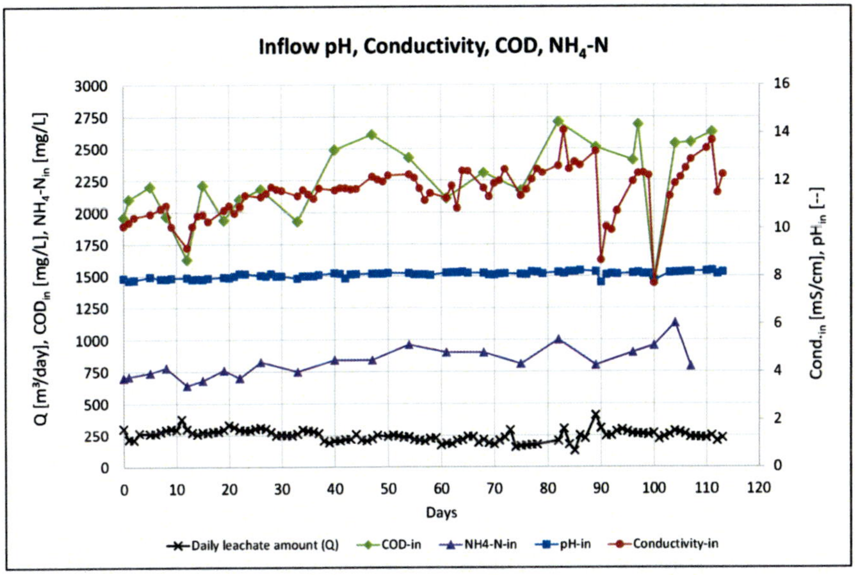

Fig.13. Inflow properties

3.2 pH and Ammonia in the Biology

During the first 20 days of the pilot phase the pH was at 8,9 and the ammonium oxidation was inhibited. With decreasing pH-values the ammonia degradation started to increase from day 20. A high pH and high ammonia concentration as well as low pH in combination with high nitrite concentration can be toxic for the biomass. The temperature should always be kept below 35 °C [9, 10]. These parameters must therefore be monitored closely in a large-scale plant.

3.3 Nitrogen Elimination and External Carbon Source

Nitrification and denitrification are the most critical issues of this particular biological leachate treatment. On the one hand ammonia in the inflow is quite high and on the other hand, the bacteria for nitrogen degradation are growing very slowly. Additionally, the denitrifying bacteria need an easily degradable carbon source in the inflow. Carbon compounds from an older landfill are typically not easily degradable. Therefore, the addition of an external carbon source (e.g. acetic acid) is necessary to reduce nitrate or nitrite [9].

In the first phase of the pilot test sugar was added as an additional carbon source in order to build up biomass and to reduce nitrate/nitrite. Sugar was used instead of acetic acid, because the later was not available at this time. Unfortunately, the reduction of nitrate/nitrite with sugar didn't yield the expected results. Bacteria are able to degrade sugar, but obviously it took too long for a successful denitrification. And as an unwanted side-effect it is possible that the sugar did not degrade completely and together with persistent organic compounds from the inflow (humic substances) a bulky, "fat", rubberlike sludge developed over time and deposited on the membrane surface (Fig. 14). This means the elastic aggregations were not introduced with the leachate through the sieve but were formed during the biological process. The organic matter was hindering the filtration performance significantly.

It was necessary to perform an extensive cleaning and adapt the membrane module, because the high and steadily rising filtration pressure required a quick reaction (Fig. 15). As a lesson learned it is now obvious that sugar cannot be used as a carbon source [11, 12].

Fig. 14. Membrane module before cleaning

Fig. 15. Membrane fibres after cleaning (June)

In the last phase of the pilot test (July–August) acetic acid (80%) was used as an external carbon source. Acetic acid is very fast and easily degradable by denitrifying bacteria. Before the dosage of acetic acid, all dissolved oxygen was used for the degradation of carbon compounds and nitrification. With the beginning of the successful denitrification (around day 93) the microorganisms started to use the nitrate-oxygen for degradation of the acetic acid and other easily degradable carbon compounds. Therefore, less dissolved oxygen was consumed and the content increased again up to 3 mg/L.

To sum up, for a successful denitrification anoxic conditions in the denitrification basin are essential [13]. The blower capacity and the recirculation flow rate have to be adjusted accordingly. Online measurement probes can be covered with sludge and therefore regular cleaning and manual reference measurements are necessary.

Table 1. Sludge characteristics

Date	**14.06.**	**25.06.**	**12.07.**
	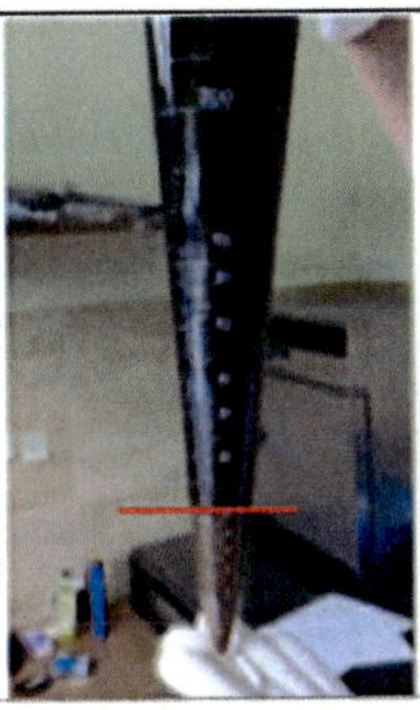	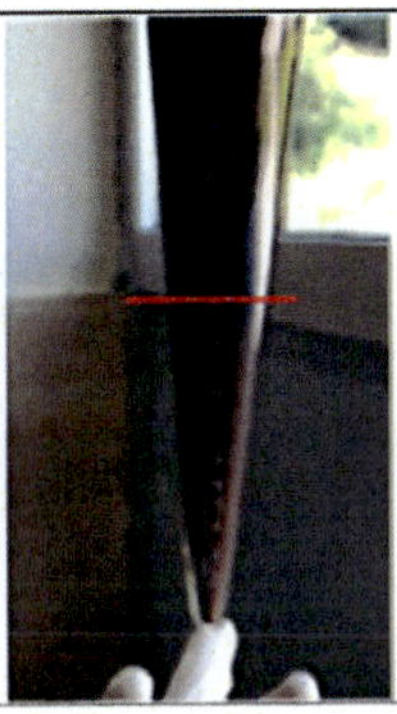	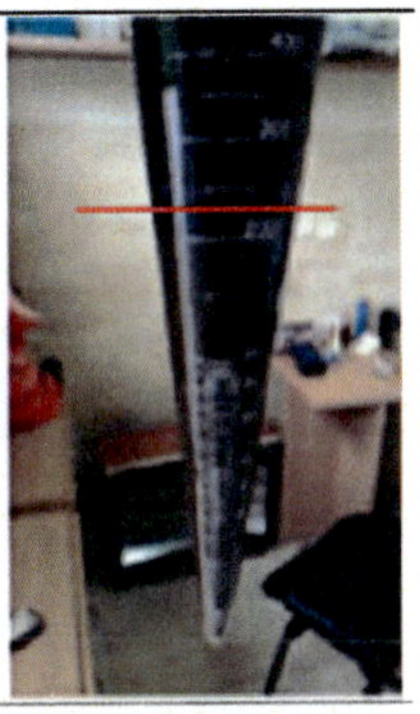
Sludge Volume [mL/L]	12	40	200
DS [g/L]	7.3	8.6	11.8

3.4 Sludge Characteristic

The following pictures in Table 1 show the settled sludge in the Imhoff funnel. An increase over time in the amount of sludge can clearly be observed. And the end of the pilot test the DS-content was 15.7 g/L (see Fig. 12).

Although the sludge was growing slowly, a steady growth was clearly proven throughout the pilot test. The leachate is not toxic to the special kind of microbiology which was cultivated in the membrane confined basin of the MBR system.

The pictures (Fig. 16) show sludge flocs, consisting of various not moving bacteria, a few higher microorganisms (single-celled eukaryotes) and large amounts of colonies of Vorticella convallaria (a species of ciliates).

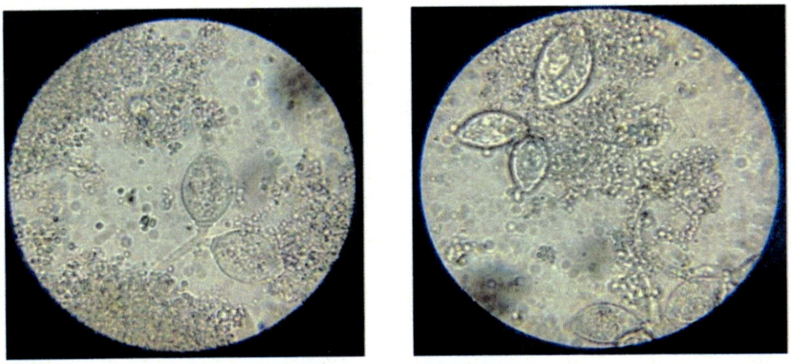

Fig.16. MBR sludge flocs – Microscope (June 2018)

4 Conclusions of the Pilot Test

4.1 Characteristic Values

The properties of the leachate are presented in Table 2.

Table 2. Leachate characteristics (relevant for biological treatment)

Leachate	Mean	Lower range	Upper range	Units
Flow	2.9	1.5	4.7	L/s
	10.3	5.3	17.1	m^3/h
	248	128	410	m^3/day
pH	8.1	7.7	8.2	-
Electrical conductivity	11,523	7,700	14,130	μS/cm
COD (excl. Carbon Source)	2,244	1,480	2,710	mg/L
Ammonia NH_4-N	823	640	1,130	mg/L
Total Nitrogen TN	968	755	1,235	mg/L
Ratio TN/NH_4-N	1.18	1.09	1.18	-
Phosphorus PO_4-P	7.8	1.2	14.5	mg/L

The membrane used for the pilot test shows the following characteristics (Table 3):

Table 3. Membrane characteristics used for pilot test

Membrane		
Material	PVDF	
Pore size	0.05	μm
Area	200	m^2
Flux	>10	L/(m^2.h)

Table 4 contains the most important characteristic values for wastewater treatment retrieved from the pilot test. The values are based on the period 27.07. (day 93) to 15.08.2018 (day 112).

4.2 Final Findings of the Pilot Test

The pilot plant based on MBR technology was tested in the period April–August 2018 on the Smiljevići sanitary landfill, and the most important conclusions are presented below:

Table 4. Characteristic values pilot test

Biology		
MBR-Volume	12	m^3
Dry Substance (DS) in MBR	15.7	g/L
DS in Filtrate	6.6	g/L
Air Flow	80	Nm^3/h
COD Out	1.537	mg/L
COD reduction	56%	
Ammonia NH_4-N Out	137	mg/L
Ammonia NH4-N reduction	85%	
Hydraulic Retention Time HRT	2.6	D
Nutrients		
ext. Carbon Source	1.092	mg COD/L
Fertilizer	23	mg PO_4-P/L

- Bacteria are adapting to the leachate slowly but steadily and they are growing much slower than in communal waste water treatment plants. This has to be taken into consideration during the start-up of a large-scale plant.
- COD load has to be adjusted to the existing active biomass.
- Due to the stable inflow properties over time, it is quite simple to adjust the loads on the treatment plant and to operate the plant consistently.
- pH, ammonia, nitrite and nitrate are very important key figures for the operation of a biological leachate treatment system and should always be monitored very closely to avoid a toxic environment for the biomass.
- The temperature should always be kept below 35 °C. Therefore, an external cooling system will be necessary for a large-scale plant.
- During the pilot test a nitrification degree of > 80% was achieved. The denitrification performance during the last phase of the pilot test was at around 46% related on the inflow of ammonia. For COD, a degradation of 45% was achieved during the last phase.
- For a successful denitrification anoxic conditions in the denitrification basin are crucial. The blower capacity and the recirculation rate have to be adjusted accordingly. Online measurement probes can be covered with sludge and therefore regular cleaning and manual reference measurements are advised.
- Sludge volume and dry substance as indicators for biomass in the MBR were increasing throughout time. The DS at the end of the pilot test was 15.7 g/L. In microbiological analysis bacteria in typical sludge flocs and some higher species (e.g. Vorticella convallaria) could be observed.
- Since the phosphorus concentration in the leachate is very low, an external phosphorus source has to be added. The sufficient supply is measured by the concentration in the outflow, which should not be lower than 10 mg PO_4-P/L. Therefore, a dosage of

30 mg PO_4-P/L in the inflow proofed to be necessary. Additionally, the incorporation of phosphorus indicates the growth of biomass.
- For a large-scale plant the membrane area must be chosen with care and it has to be recommended to take a conservative approach with sufficient membrane area.

These findings present a good basis for the design of large-scale plant. However, during scale-up the results have to be evaluated very carefully and a comprehensive concept for the overall leachate management has to be developed. Although the MBR removes more than 50% of the organic load as well as around 70% of nitrogen, it will have only a limited removal degree of approx. 30–50% on certain heavy metals while chloride will remain widely unaffected. It is, therefore, unavoidable to install a secondary treatment step in order to meet the stipulated effluent limits for COD, BOD_5, total nitrogen, nitrate and chloride.

Based on preliminary analysis, a 1 stage RO system is the best solution to meet the required discharge limits (particularly if insisted on discharge into a waterbody). According to the available analytical data, scaling in the RO system caused by the elements Ba, Si, Sr and Ca could become an operational problem, especially by using spiral wound membranes with high recovery rates.

The installation of a reverse osmosis system with disc tube (DT) design as post-treatment after the MBR step is recommended, because the DT design is far less sensitive to blocking/scaling induced by these elements. The given inflow and effluent parameters strictly request the use of membrane processes as a secondary treatment step. These processes are inevitably linked to a production of a concentrate which has to be disposed. It is a proven fact that this concentrate can be re-infiltrated into the landfill body for decades, if the following preconditions are fulfilled:

- Minimized amount (in the present case <40 m^3/day),
- Controlled injection into the landfill body,
- Professional maintenance and operation.

The pilot plant operation in the period April–August 2018 and analysed related impacts enable the selection of an appropriate technology for the design of an LTP. It is expected that this can contribute to find a sustainable solution regarding the leachate problem at the Smiljevići landfill.

References

1. Serdarevic, A.: Environmental damage assessment, waste management and overview of the current situation in Bosnia and Herzegovina. In: Ibrahimbegovic, A., Zlatar, M. (eds.) Damage Assessment and Reconstruction after War or Natural Disaster, pp. 357–381. Springer, Dordrecht (2009). https://doi.org/10.1007/978-90-481-2386-5_14
2. Mayr, B., Garstenauer, T.: Final Test Report and Scale-up MBR-RO for landfill Smiljevici – Sarajevo, EnviCare® Engineering GmbH (2018)
3. Serdarević, A., Tuhtar, D., Milašinović, Z., Babić, F.: Membrane filtration for the landfill Leachate treatment- Case study of the pilot plant in Sarajevo. In: Eleventh International Waste Management and Landfill Symposium (Proceedings), Cagliary, Italy, p. 8, 1–5 October 2007

4. Serdarevic, A.: Landfill Leachate management—Control and treatment. In: Hadžikadić, M., Avdaković, S. (eds.) IAT 2017. LNNS, vol. 28, pp. 618–632. Springer, Cham (2018). https://doi.org/10.1007/978-3-319-71321-2_54
5. Ehrig, H.-J.: Leachate quality. In: Christensen, T.H., Cossu, R., Stegmann, R. (eds.) Sanitary Landfilling: Process, Technology and Environmental Impact, pp. 213–230. Academic Press, London (1989)
6. Serdarević, A., Kralj, M.P.: Čiščenje Izcednih Voda Odlagališča Komunalnih Odpadkov - Izzivi Pri Iskanju Optimalne Rešitve. Ekolist **14**, 12–17 (2017)
7. Mayr, B.: Membranbioreaktortechnik in Österreich - Aktuelle Projekte. In: Rautenbach, R., et al.(eds.) Begleitbuch zur 3. Aachener Tagung Siedlungswasserwirtschaft und Verfahrenstechnik, Klenkes Druck, vol. 2, pp. 8–9, Aachen/Deutschland (2000)
8. Mayr, B., Garstenauer, T.: Membrane bioreactor as polishing step in the treatment of galvanic wastewater. In: FILTECH Conference Program, Cologne, Germany, vol. 10, pp. 22–24 (2019)
9. Anthonisen, A.C., Loehr, R.C., Prakasam, T.B.S., Srinath, E.G.: Inhibition of nitrification by ammonia and nitrous acid. J. Water Pollut. Control Fed. **48**, 835–852 (1976)
10. Mayr, B.: Hochleistungsbiologie und Membrantrenntechnik zur Abwasserreinigung: Fallstudie Deponiesickerwasser. Österr. Wasserwirtschaft **46**(7/8), 195 (1994)
11. Kjeldsen, P., Barlaz, M.A., Rooker, A.P., Baun, A., Ledin, A., Christensen, T.H.: Present and long term composition of MSW landfill leachate: a review. Crit. Rev. Environ. Sci. Technol. **32**(4), 297–336 (2002)
12. Kruempelbeck, I., Ehrig, J.G.: Long-term behaviour of municipal solid waste landfills in Germany. In: Proceedings of the Seventh International Waste Management and Landfill Symposium, 4–8 October. Cagliari, Italy (1999)
13. Mayr, B., Novak, S., Horvat, P., Gaisch, F., Narodoslawsky, M., Moser, A.: Purification of Landfill leachates by means of combined biological and membrane separation treatment. Chem. Biochem. Eng. Q. **9**(1), 39–45 (1995)
14. Google maps: Bird view of the sanitary landfill in Sarajevo Canton. https://www.google.com/maps/@43.8665583,18.3386654,905m/data=!3m1!1e3. Accessed 15 Feb 2021

Geodesy and Geoinformatics

GIS Application for the Promotion and Development of Ecotourism in Albanian's National Parks

Sonila Papathimiu[1](✉), Esmeralda Laci[2], and Sabri Laci[1]

[1] Faculty of History and Philology, University of Tirana, Tirana, Albania
sonila.papathimiu@fhf.edu.al

[2] Faculty of Humanities and Law, University of Vlora, Vlora, Albania

Abstract. Protected Areas management and planning is very complex and should be done using the right information and tools. Especially, when planning for Tourism development in these areas, should be taken in consideration every information about the natural values of the area and of any kind of possible conflicts between the local people and the legal framework, which can be gathered and analyzed with PPGIS or Public Preferences surveys and questionnaires. GIS application also offers a wide range of possibilities to create different thematic maps and spatial analyzes or predictive maps, which can help both managers and the other users of the area.

In this study, we used questionnaires to elicite and analyze Public preferences for conservation policies and land or water use inside and outside Protected Areas (PA). We also identified the potentials of the study area for the development of Ecotourism, which is one of the best sustainable forms of Tourism. The questionnaire was distributed randomly mostly with social media and the points for the attraction map were collected in the field by GPS.

The calculation of the means and median of local communities' preferences suggest that they are not very aware of the legal framework of the Protected Areas in Albania, especially for the Category II (National Park) which is our main focus. These preferences are more in favor of restricted activities and development inside PAs, which leads to the need for the promotion and development of Ecotourism in them. We used GIS application to create a map for the main Touristic attractions (which later can be easily adapted for Ecotourism) and analyzed their accessibility to the main roads. These two maps are just an illustration of the varieties of maps that can be created in GIS and how managers and other users of the Protected Areas can profit to get useful information from them.

Keywords: GIS application · Public preferences · Protected areas · Ecotourism · Planning and management

N. Ademović et al. (Eds.): IAT 2021, LNNS 316, pp. 487–501, 2022.
https://doi.org/10.1007/978-3-030-90055-7_38

1 Introduction

GIS has become nowadays a powerful tool and methodology for spatial analyses and planning. Its importance and wide use in different fields has been recognized by many researchers [1]. GIS is an indispensable tool for acquiring and analyzing objects and their associated attributes distributed geo spatially and thus contributing to the improvement of decision making. Other authors emphasize the application of GIS to enhance the efficient planning and development of Tourism and leading in its success [2].

The planning and management of every form of alternative and sustainable Tourism, especially Ecotourism is recommended to use GIS and if it is possible also PPGIS. GIS nowdays offers solutions for the complex problems in decision making for the development of tourism by offering many possibilities to elaborate, analyze and visualize geospatial data [3].

Despite all these qualities and possibilities that GIS offers for the tourism and ecotourism, in Albania, its use has begun only after the 1990s and still nowadays is not as much used as it should be. Although, in many strategic and management plans of Albania or some municipalities there are maps designed in GIS, there are no maps designed with PPGIS and there is also a lack of analytical and predictive maps, which are essential for a successful planning and management of the territory and economic activities [4]. The use of GIS and PPGIS is very helpful for the planning and management process for the protected areas (PA), because with it can be collected and analyzed data of large areas to identify any possible conflict for land use or to map the relationships between public preferences and governance [5, 6].

In this paper we aim to study the state of GIS use in Tourism development and planning in Albania and want to promote its use for Ecotourism and other sustainable forms of Tourism. Ecotourism, for a long time now has been widely practiced in many countries, but unfortunately this is not the case of Albania, although most of its National Parks (NP) have the potential to develop ecotourism and other sustainable tourism activities.

In this study we also aim to emphasize and promote the use of some important tools and calculations which are very important for the planning and management of Ecotourism and Tourism such as: tourist's management, inventory of the resources etc.

The management of tourists has been identified as one of the main risks for the natural protected areas, especially for the areas with many visitors per day. In most of the Strategies for Development approved in Albania for the whole country or for each district, the raise of the number of tourists visiting Albania every year has been mostly emphasized as an indicator of successful management of tourism [7] and it is also one of the main objectives for the development of Tourism to 2023 [8]. What is missing in these strategies, among others, is the calculation of the sustainable number of tourists or the sustainable carrying capacity of the protected areas, especially of National Parks. In the sustainable tourism planning this is a key element and should be included in all the strategies and management plans.

Another key element in planning for sustainable tourism, especially ecotourism is also the inventory of the resources and zoning, which help the managers and local community for a better protection and sustainable use of the protected area.

In this paper we have applied some of the above-mentioned key elements and methods in Divjaka-Karavasta National Park, focusing on the use of public preferences, GIS application for mapping and analyzing, as a demonstration or a first attempt for a better management and planning of sustainable activities inside National parks and especially for the promotion of GIS use for planning of Ecotourism and other sustainable tourism activities.

2 Study Area and Its Potential to Develop Ecotourism

Albania is located in the southwest of Balkan Peninsula and although it is a small country it is very rich in natural resources and biodiversity. Its relief changes naturally from lowland in the west to hilly and mountainous in the middle and eastern border. The protected areas are spread in the four regions of the country and according to the classification by IUCN, it has 799 protected areas covering almost 7% of the country's total surface, where the largest is Category II covering 3% of them and represented by 15 National Parks.

The selection of the study area for this paper was made taking into consideration the definitions of Ecotourism by "The International Ecotourism Society" (TIES) in 1991[1] and of IUCN (now World Conservation Union) in 1996[2], which both emphasize the presence of natural areas, relatively undisturbed as the primary resource for ecotourism development.

National Park Divjaka-Karavasta is situated in the western lowland region which has the most advantageous geographical location and conditions comparing with other regions. This National Park is also known for its coastal wetland ecosystems of Karavasta, the first Ramsar area in Albania [9].

The potentials of the NP Divjaka-Karavasta for practicing Ecotourism are:

First, its relative location is very advantageous, being only 75 km from Tirana, almost equidistant from two main seaports and important cities Durrës and Vlorë, only 50 km in distance from Berati (UNESCO site) and the archaeological site of Apollonia and Bylis and many other religious and cultural heritages.

Its subtropical Mediterranean climate, with mild winters and hot dry summers allows the development of ecotourism almost all year long, especially in the periods: *end of March - beginning of June* and *beginning of September - end of November.* In the proposed

[1] "*Ecotourism* is responsible travel *to natural areas* that conserves the environment and sustains the wellbeing of local people.".

[2] "*Ecotourism* is environmentally responsible travel and visitation to relatively undisturbed *natural areas* in order to *enjoy and appreciate nature* (and any *accompanying cultural features* -both past and present) that promotes conservations, has low negative visitor impact, and provides for beneficially active socio-economic involvement for local populations".

periods the air temperatures are not very high (in average18 °C–23 °C), the number of sunny days and the average sunny hours are very high (approximately 150 days/year with 2682 h/year) and rain or other forms of precipitation are rare. *The relief* in the region is mostly flat, which makes it very easy walking long distances in the forests and other parts of the wetlands. Also, the hilly ranges along the coast, which where terraced during the communist period[3] to be planted with fruit trees (peaches, figs, pears, citrus, and plums), olives and vineyards are very interesting to visit and observe the changes they have experienced in the last decades. *The soils* in the villages around these wetlands are very good and supportive for the production of many vegetables and cereals (potatoes, carrots, different kind of beans, peas, corn, wheat, etc.). The local agriculture production is very appreciated from the ecotourists, and they tend to buy and taste different kind of traditional foods in the areas they visit. The development of ecotourism in these wetlands will encourage the farmers to improve the quality of their products and of the food served in the restaurants and guest houses.

Second, NP Divjaka-Karavasta is very interesting and attracts many visitors, because it has many habitats that provide shelter to a large variety of species of animals and plants. The pine forest of Divjaka has great values too and is the best habitat for many species of animals, reptiles, birds and plants. Karavasta wetland ecosystem has been always classified and referred as the largest and most important one in Albania (almost 43 km^2). *"Karavasta lagoon and Divjaka pine forest will become increasingly important for ecotourism development"* [10].

Karavasta wetland ecosystem is rich with almost 12 species of mammals, 16 species of reptiles and over 210 species of bird. This site is very important for the presence in it of Dalmatian Pelican, which represent 5% of the world total population of this specie [11]. The small island in the north-western part of the lagoon is the nesting area to this migratory bird and this is the most western part where they appear. Lately (after the moratorium of the hunting of birds which continues from 2016 flamingos are seen in the lagoon. The lagoon is rich in different kind of fish, crabs, eels etc. The restaurants in Divjaka beach and city are furnished with fresh fish almost every day. The eel of Divjaka is very demanded in the market and from the restaurants. This NP is a perfect spot to visit from the ecotourists in search for wide natural and rich in biodiversity areas.

Third, the local population is well known as hard workers and of land cultivation, good fishermen, very good in livestock and poultry (especially in raising hens and turkeys) etc. Hospitability is one of the main characteristics of this population, which is very helpful for the development of tourism in general and particularly for ecotourism. The main agriculture products of the area are potatoes, watermelon, melons, carrots, broccoli, leeks, cabbage etc. Divjaka hills are planted with fruit trees, olives and vineyards. Traditional foods with these fresh local products are served in most of the restaurants near the beach and it can attract the ecotourists who want to experience as much as possible from the place they visit.

[3] The first action of terracing the hills in Albania started in 1966, in Jonufër, Vlorë and then also in Saranda (source: http://www.shekulli.com.al/web/).

Karavasta wetland possesses all the necessary resources for development of eco-tourism: it is a very wide and beautiful natural area, has a very favorable relative and absolute geographical position, provides habitat for a large number and variety of species and plants and has a young population that can be trained and orientated to practice eco-tourism. *Ecotourism* is the right choice not only for Karavasta, but for every Albanian wetland, whenever they are located.

3 Methodology and Sampling

Public preferences for Tourism planning and management in Albanian Protected Areas and the need for Ecotourism development

One of the main issues when dealing with Protected Areas planning and management for Tourism development is finding the best solutions to have a win-win situation. The local population and the community living and working near these areas are even more interested to be able to profit from them. In this paper, before analyzing the Ecotourism development and the necessity to practice it in the Albanian Protected Areas, with focus on the National Parks and especially on the National Park of Divjaka Karavasta, we wanted to have an overview and determine on the Albanian public preferences for the protected areas management and tourism development. This helped us to better understand the attitude of local community toward the land use in protected areas and to define if they are ready to be involved and support the development of Ecotourism and every other sustainable economic activity inside and around them.

This is very important for this study as one of the main components of successful and strong sustainable Ecotourism is the involvement of local community and creating jobs related with it such as: eco tour guides, eco lodges, eco restaurants with traditional and local food, camping, etc.

The maps were created and analyzed in ArcGIS pro (ArcGIS pro-2.6.0, Advanced). Position data for main tourist attractions were obtained from field measurements using GPS for GIS data collection (Garmin GPSMAP 76S). The coordinate system used for position measurement and maps creation is WGS 1984. For the creation of thematic and analytic maps was used the basemap offered online World Topographic map and World Hillshade. The same basemaps were used for the vectorization of the roads according to the category to be used in the analytical map of the accessibility of the objects (main touristic attractions) from the main roads.

Sampling

To understand better people's preferences for the use of Protected Areas resources in Albania and especially on the activities that they would prefer or oppose in them we wanted to perform a public preferences survey analyzes with a random house sampling, especially in the study area. Due to the restrictions for Covid-19, the sampling implemented in this study was random and the questionnaire was prepared in Google forms and was distributed via social Medias, LMS-s and personal contacts of the authors, ex-students living and working near the interested areas etc. In this study we chose to use public preferences survey and analyze to give examples on how this method can be used and to stress on its usefulness for the planning of the activities in the protected areas in accordance with what people favor or oppose.

We had respondents from different parts of Albania with a dominance of women (69%) and age group 15–29 (73.4%). There was a high percentage of the respondents (77.4%) claiming that they were living near a protected area, where 65% of them being not more than 35 km far. The geographical distribution of the respondents was wide and most of them lived near a National Park (74.3% of them declaring to live near a PA) which is the focus of this study, especially the NP Divjaka-Karavasta from which we had 13% of the respondents near a NP.

Our interest on the public preferences for land use inside and outside of the protected areas was designed to compare them with the legal framework of the restrictions inside and outside protected areas in Albania. Through the public preferences survey we also analyzed and determined whether the collective preferences are more toward sustainable activities and land use or if they tend to be interested in fast economic profits without any consent on the environmental aspect. If the residents living near protected areas are aware of their value and the state policy, we expect them to be more opposing (mean 1.5) than favoring (mean higher than 1.5) land and water development activities inside the protected areas such as: grazing, fishing, and hunting and favoring these activities outside them. The same preferences we expect also from the respondents living near (a range from 1–25 km) a National Park, as the Albanian law on protected areas does not favor too much these activities inside the NPs.

They also should have greater opposition to the use of motorized vehicles inside protected areas and opening of new roads than outside them. Also, we expect a big difference in the preferences of the people living in a range of no more than 24 km from a protected area, always focusing on the respondents near the National Parks. The respondents should have the same opposition and attitude or preferences toward predator control and other activities which by law are treated similar in both areas (inside and outside a protected areas).

In order to have another view and perspective on Tourism sector in Albania and to identify some of the difficulties and problems of this economic sector that could be solved by using GIS application, we had some deep interviews with touristic guides and distributed a structured interview to the Association of Albanian tour guides. From this group we expected to be oppose any form of unsustainable tourism and the need for ecotourism practices in National Parks as well the enhancement of GIS application, especially for creating touristic and ecotouristic maps.

4 Results and Discussion

Data gathered from the questionnaire were analyzed in SPSS (IBM SPSS Statistics 20 app) to compare and analyze if the public preferences are in favor of sustainable activities and in line with the legal framework for Protected Areas and the Strategic Plans or not. We used Analyze/Compare Means/Means with dependent list of activities inside and outside the protected areas and independent list with the variables of Gender, Distance from the PA and Category of PA, taken separately. For some other analyzes and calculations we used also Analyze/Compare Means/Paired-Samples T Test. The dependent list was filled with variables according to the type of activity such as: land and water use, predators' control, development use etc. These groups of variables were also compared with each other adding the variable of distance from the protected areas. The three variables: gender, distance and category influenced the public preferences, but for some variables the changes were not significant (Table 1). Both men and females are strongly opposed the fishing and hunting inside the PA, but they are slightly in favor of grazing. According to the legal framework for protected areas in Albania [10], these activities should be in moderation inside protected areas, especially inside National Parks. Hunting has the lowest Mean and Median values which is in line with the moratorium of hunting in Albanian from 2016 and the legal framework.

Table 1. Comparing means and median preferences by gender for consumptive use activities inside and outside the protected areas

Report							
Gender		Grazing **Inside** PA	Grazing Outside PA	Fishing **Inside** PA	Fishing Outside PA	Hunting **Inside** PA	Hunting Outside PA
Female	Mean	2.19	2.98	1.77	2.77	.92	1.34
	Median	2.00	3.00	1.00	3.00	.00	1.00
Male	Mean	2.30	2.84	1.89	2.29	.91	1.47
	Median	3.00	3.00	1.00	2.00	.00	1.00
Total	Mean	2.23	2.93	1.82	2.59	.92	1.39
	Median	2.00	3.00	1.00	3.00	.00	1.00

The respondents were also aware and opposed the building of luxury hotels and resorts inside the protected areas, while favoring the building of small vacation houses, inns and sustainable camping areas which is also in line with the Albanian legislation [13], but they were also very favorable to the use of motorized vehicles and opening of new roads both inside and outside the protected areas.

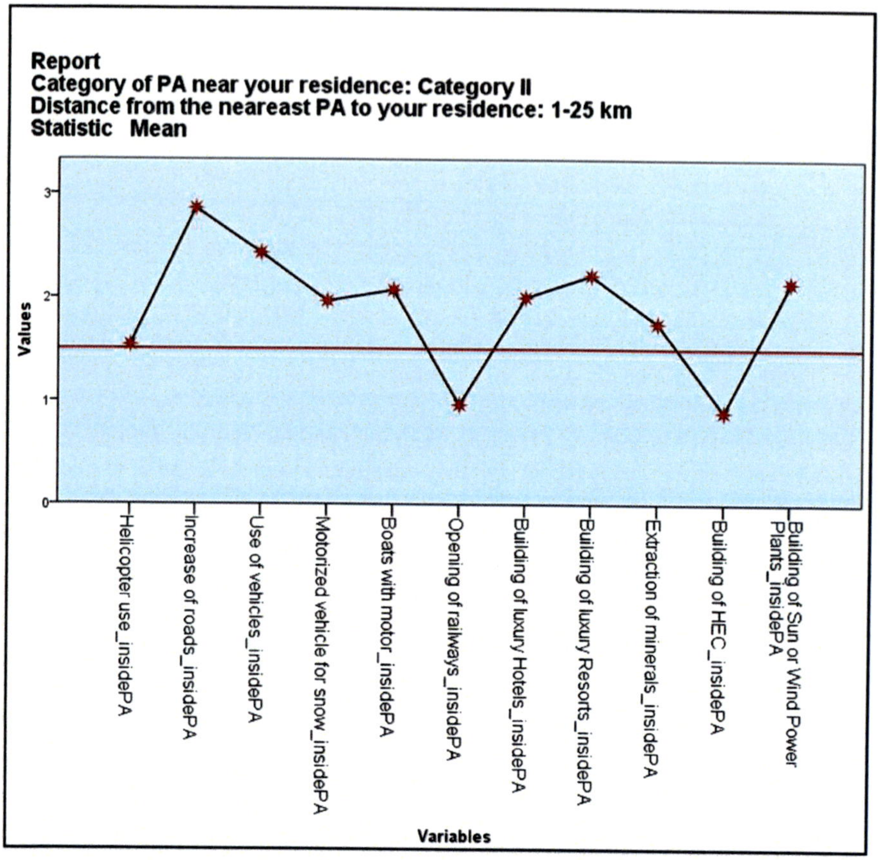

Fig. 1. Public preferences (people living no more than 25 km from a Category II PA) for the development and motorized vehicles used inside Protected Areas.

The odds that residents living near a protected area (1–25 km), especially the ones living near a National Park (Category II) were in oppose of the activities which are not allowed inside them, were significantly low and did not differ much from the total preferences, which should not be the case (Fig. 1, 2, 3, 4).

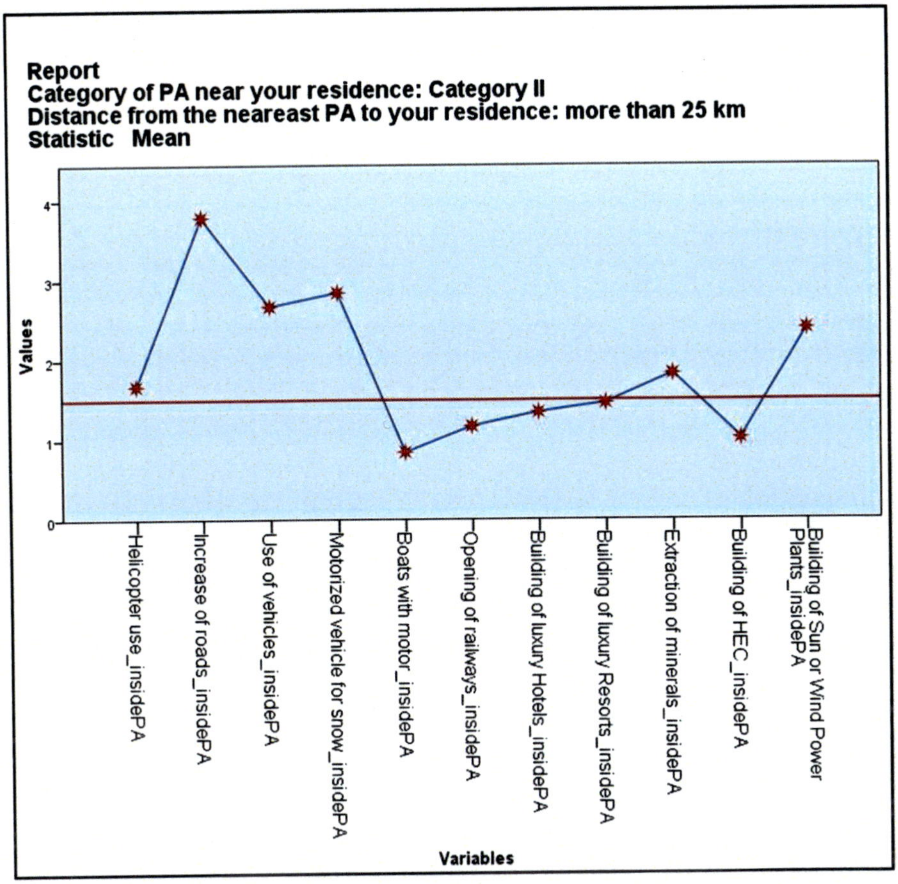

Fig. 2. Public preferences (people living more than 25 km from a Category II PA) for the development and motorized vehicles used inside Protected Areas.

The deep interviews with the tour guides and the results from the structured interviews were as we expected related to the necessity of promoting more sustainable type of Tourism and especially Ecotourism. Almost all interviewers, despite their working experience stressed out the lack of usable maps for tourism, which occasionally make it very difficult to offer the right service to the tourists.

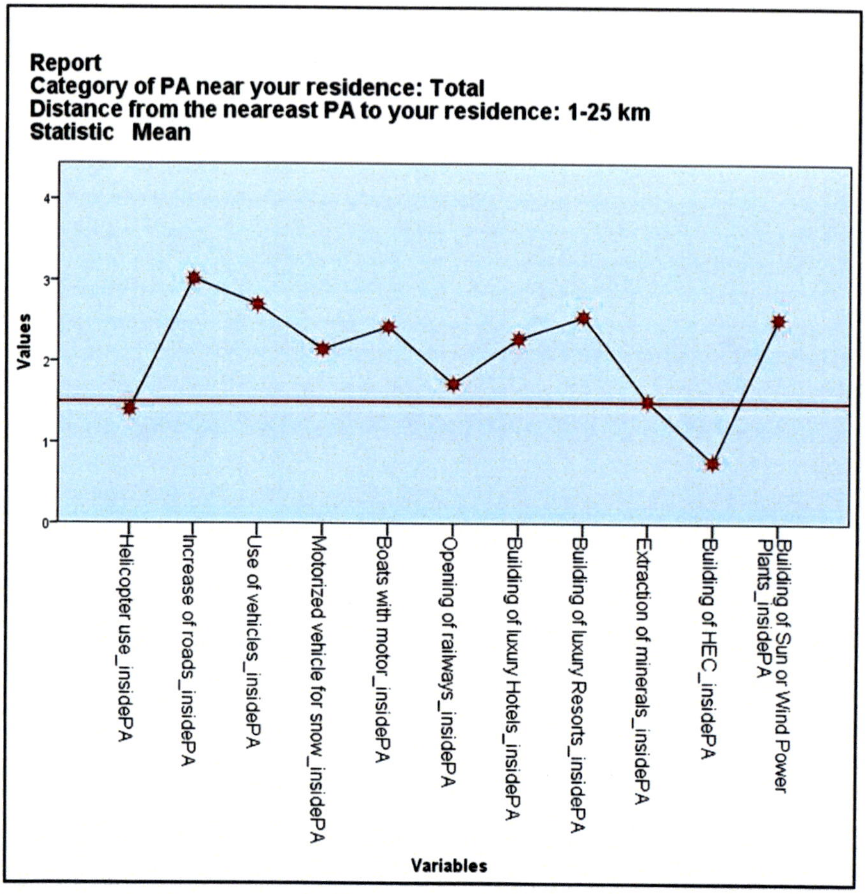

Fig. 3. Public preferences (people living no more than 25 km from a PA) for the development and motorized vehicles used inside Protected Areas.

In this paper we used ArcGIS pro to create and analyze a map for the promotion of Ecotourism. The maps created in GIS are very useful not only for the management and planning of Ecotoursim, but they can serve also for the tour guides and other tourists. In a GIS map you can add and update the information very easily. GIS application offers many possibilities and calculations for both managers and tourists of the Protected Areas. To create the map of the Main attractions in Divjaka Karavasta National Park we collected positioning data in the field, selecting only the attractions and services which are generally offered for Ecotourism and other sustainable Tourism. We focused on the

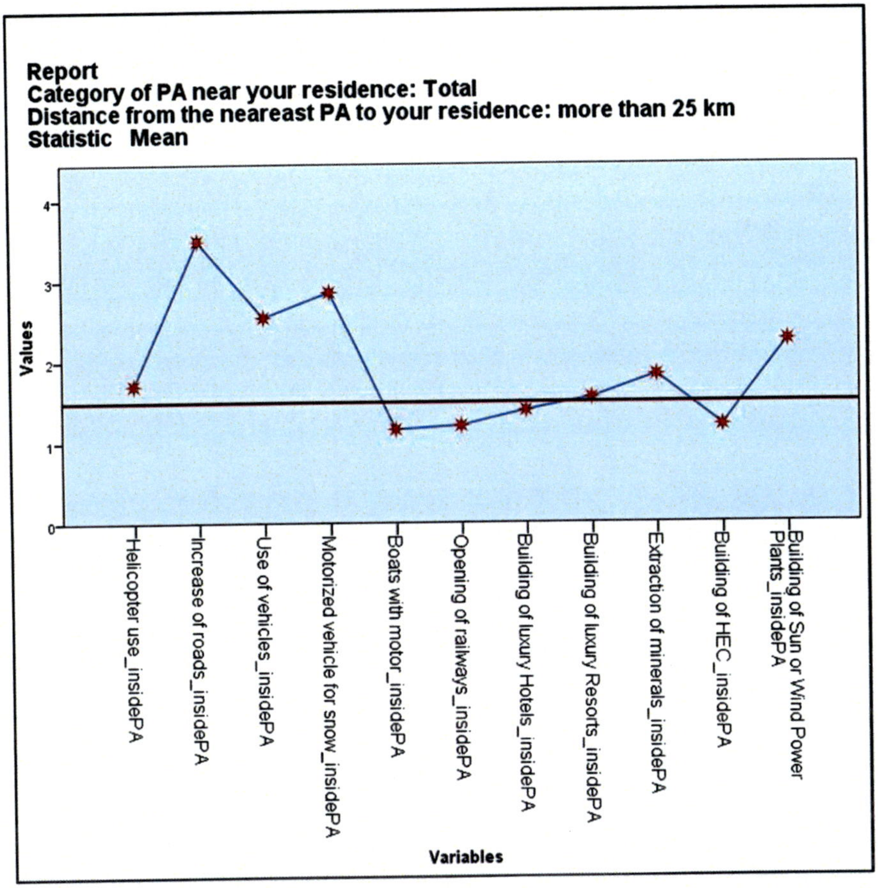

Fig. 4. Public preferences (people living more than 25 km from a PA) for the development and motorized vehicles used inside Protected Areas.

positioning data of the main archaeological sites near the park, old churches, guesthouses (which later can easily be adapted to the eco lodges), restaurants offering traditional food from the local productions etc. (Map 1). To prove that our study area not only offers many attractions sites, but most of them have a good accessibility from the main roads we performed the Near Analyze Tool (Map 2), which confirmed our expectations. From this analyze it resulted that from 57 attractions in the area, more than 50% (31 points) were in a distance not more than 1 km from a Main Road.

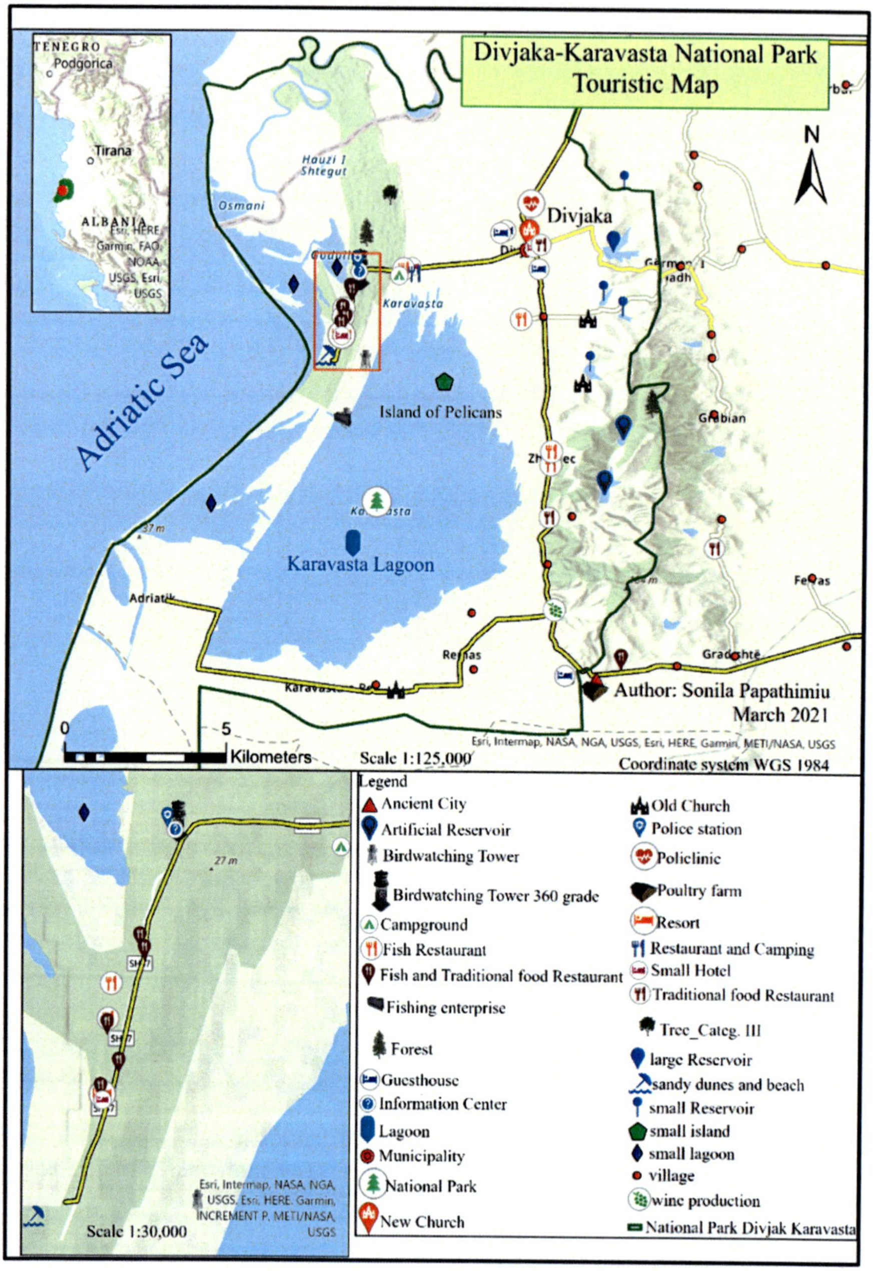

Map. 1. Touristic map of Divjaka Karavasta National Park

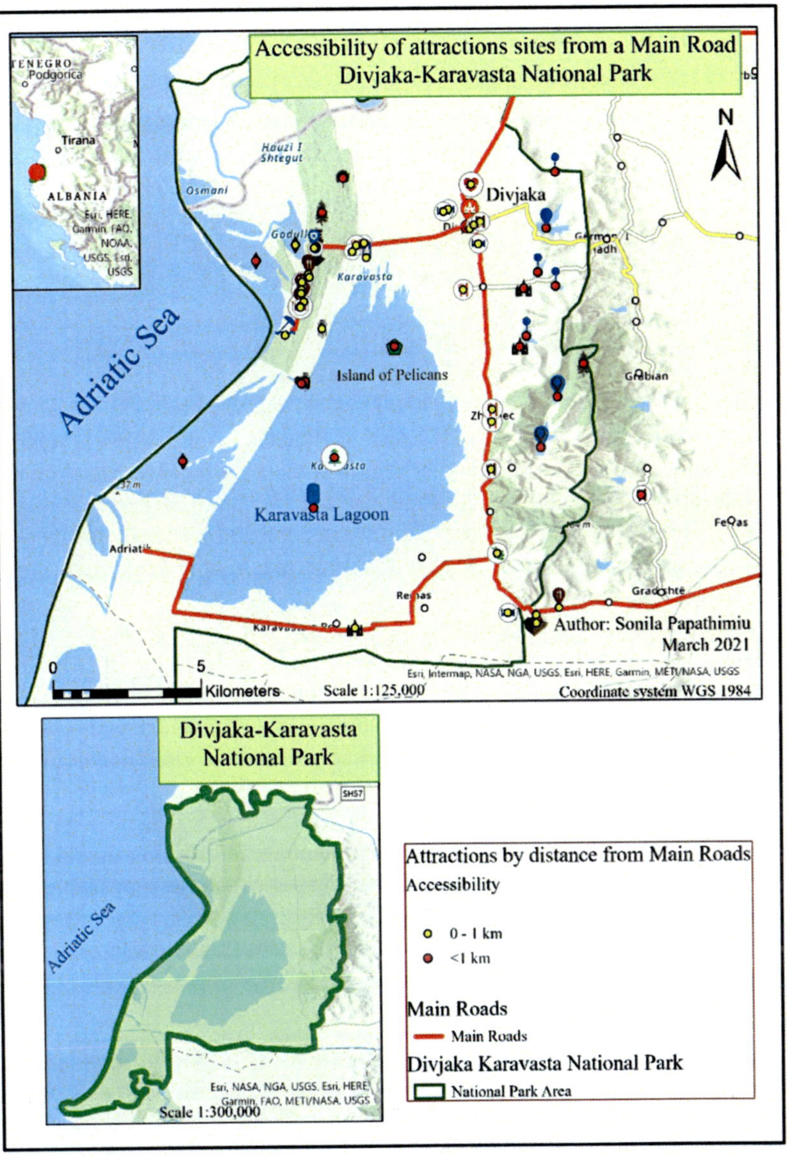

Map. 2. Accessibility of attraction sites from the Main Roads in Divjaka Karavasta National Park

5 Conclusion

This study is focused and brings contribution in two important issues, first the necessity to widen the use of GIS and Public Preferences analyzes for the management and planning of Tourism and second to demonstrate that the Protected Areas of Albania need to promote and develop more Ecotourism and other forms of sustainable tourism.

From our analyzes we found that people preferences, despite if they were living near of far from a Protected Area were not always in alignment with the conservation policy in Albania. Local people living not more than 14 km from a protected area of Category II have almost the same values of preferences (*from 0 not favor to 5 totally favor*) about development and motorized use, with people living more than 14 km away, and are not in alignment with the Albanian Law [10] and the strategies for the development of coastal areas [12] and especially of the Divjaka municipality [13], which restricts any type of intensive use inside this Category of Protected Areas. Their attitude on the preferences were also confusing in some cases as sometimes they were in align with the Law and the strategic development of the areas and in other preferences it was totally the contrary. This suggests that the population in general and especially the local people near protected areas are not very aware of their values and how they can be conserved and used in a sustainable way.

Another contribution in this paper is the promotion of the Ecotourism as a sustainable form of Tourism which will bring both natural and economic profits for the local population [14, 15]. Through the analyze of the potentials for developing Ecotourism and the creation of thematic and analytical maps in GIS we want to demonstrate that not only for this study area, but for most of the Albanian National Park, Ecotourism can be the best choice.

Our suggestion is that the local population and managers dealing with Protected Areas should be informed regularly about their potentials and training should be offered for the National Parks staffs to know and use GIS application for sustainable planning and management of every activity allowed inside it.

References

1. Kushwaha, A., Chatterjee, D., Mandal, P.: Potentials of GIS in Heritage and Tourism (2011)
2. Switzerland, Y.F., Poulicos, P.: GIS contribution for the evaluation and planning of tourism: a sustainable tourism perspective. In: Proceedings of the International Leisure and Tourism Symposium, ESADE, P. Singh Barcelona (2003). http://www.gipsynoise.gr/HellasGI/KoinonikesPraktikes/papers/Farsaki.pdf
3. Singh, P.: Role of geographical information systems in tourism decision making process: a review. Inf. Technol. Tourism **15**(2), 131–179 (2015). https://doi.org/10.1007/s40558-015-0025-0
4. Engen, S., Runge, C., Brown, G., Fauchald, P., Nilsen, L., Hausner, V.: Assessing local acceptance of protected area management using public participation GIS (PPGIS). J. Nat. Conserv. **43**, 27–34 (2010). https://doi.org/10.1016/j.jnc.2017.12.002
5. Karimi, A., Tulloch, A.I.T., Brown, G., Hockings, M.: Understanding the effects of different social data on selecting priority conservation areas. Conserv. Biol. **31**, 1–20 (2017). https://doi.org/10.1111/cobi.12947

6. Karimi, A., Brown, G.: Land Use Policy Assessing multiple approaches for modelling land-use conflict potential from participatory mapping data. Land Use Policy **67**(February), 253–267 (2017)
7. Plani i integruar ndersektorial: Bregdeti/RSH. Ministria e Zhvillimit Urban, AKPT, Tirane: Pegi (2016). https://doi.org/10.1016/j.landusepol.2017.06.004
8. Strategjia Kombëtare për Zhvillimin e Qëndrueshëm të Turizmit 2019–2023, Fletorja zyrtare 2019, Nr. 90. VENDIM Nr. 413, 19 June 2019
9. Ramsar site from October 31st, 1995 (www.ramsar.org) The List of Wetlands of International Importance Published 2 February 2021
10. Brew, D.S.: Geomorphology of the Albanian Adriatic coast: a study of short-and long-term changes at karavasta lagoon and their implications for coastal management. Geography 88(2), 88–98. JSTOR (2003). www.jstor.org/stable/40573827
11. Papathimiu, S.: The socio-economic and environmental development of Divjaka municipality and Qender commune (comparative analyze), Dissertation Thesis, January 2011
12. LIGJ Nr. 81/2017 PËR ZONAT E MBROJTURA [Albanian Law for Protected Areas, 2017]
13. Relacion i Planit te Pergjithshem Vendor te Bashkise Divjake, Miratuar me Vendim te Keshillit Kombetar te Territorit nr. 9, 08 February 2017 [Report of the General Local Plan of the Municipality of Divjake, Approved by the National Council of the Territory no. 9, 08 February 2017]
14. de Costa, T.O., de Assis, L.R., Calijuri, M.L., Assemany, P.P., Lima, G.S.: Defining priority zones for conservation and ecotourism in a protected area. Revista Árvore **40**(5), 769–779. https://doi.org/10.1590/0100-67622016000500001
15. Ceballos-Lascurain, H.: Tourism, Ecotourism, and Protected Areas: The State of Nature-based Tourism around the World and Guidelines for its Development, p. 315. IUCN-World Conservation Union, Gland (1996)

Simultaneous Use of Terrestrial Laser Scanning and Close-Range Photogrammetry for Documentation of Cultural and Historical Heritage Monuments

Admir Mulahusić[1](✉), Nedim Tuno[1], Jusuf Topoljak[1], Muamer Đidelija[1], and Enes Grabus[2]

[1] Faculty of Civil Engineering, Department of Geodesy and Geoinformatics, University of Sarajevo, Sarajevo, Bosnia and Herzegovina
{admir_mulahusic,nedim_tuno,jusuf.topoljak, muamer.djidelija}@gf.unsa.ba

[2] Tehnička škola Zenica, Zenica, Bosnia and Herzegovina
enes.grabus@tscze.ba

Abstract. Terrestrial laser scanning (TLS) and close-range photogrammetry (CRP) are of great importance in the preservation of cultural and historical heritage. Using both methods, it is possible to faithfully create 3D models of cultural and historical monuments. In this paper, the procedure of making a 3D model of an object using two mentioned methods is presented. Pix4Dmapper software was used to process close-range photogrammetry data, while Topcon ImageMaster Pro software was used to process point clouds from terrestrial laser scanning. A comparison of close-range photogrammetry and terrestrial laser scanning results is presented. The paper describes the geometric documentation of the replica monument "Zgošćanski stećak" (stećak - monumental medieval tombstone), which begins with the collection of data, followed with processing, and concludes with the presentation of the final results.

Keywords: Photogrammetry · Terrestrial laser scanning · Camera · Cultural and historical heritage

1 Introduction

The cultural heritage (cultural and historical heritage, national heritage, or just heritage) refers to the inheritance of physical artefacts and immaterial attributes of a group or society which constitutes the heritage of past generations and is carefully preserved in the present to be left in inheritance for the benefit of future generations. That kind of heritage, that has been historically preserved is often unique and irreplaceable and the responsibility for its protection lies with the next generation [1]. Photogrammetry is traditionally the most appropriate method for the collecting data about different

N. Ademović et al. (Eds.): IAT 2021, LNNS 316, pp. 502–513, 2022.
https://doi.org/10.1007/978-3-030-90055-7_39

objects. Photogrammetry has been defined by the American Society for Photogrammetry and Remote Sensing (ASPRS) as the art, science, and technology of obtaining reliable information about physical objects and the environment through processes of recording, measuring and interpreting photographic images and patterns of recorded radiant electromagnetic energy and other phenomena [2]. The laser scanning technology is based on measurements with the laser application. With the involvement of Terrestrial laser scanning, it is able to register huge amounts of data to be widely explored in many applications [3].

2 Basic Information About Stećci and the Processed Medieval Tombstone

Stećci (plural of Stećak) is the name for monumental medieval tombstones, that lie scattered across Bosnia and Herzegovina, and the border parts of Croatia, Montenegro and Serbia. An estimated 60 000 are found within the borders of modern Bosnia and Herzegovina and the rest of 10 000 are found in what are today Croatia (4 400), Montenegro (3 500), and Serbia (2 100), at more than 3 300 odd sites with over 90% in poor condition. Appearing in the mid-12th century, with the first phase in the 13th century, the tombstones reached their peak in the 14th and 15th century, before disappearing during the Ottoman conquest of Bosnia and Herzegovina in the very early 16th century [4]. Stećci were inscribed as a UNESCO World Heritage Site in 2016. It includes a selection of 4000 stećci at 28 necropolises (out of a total of approximately 70 000 stećci at 3 300 sites) on the territory of four states: Bosnia and Herzegovina, Republic of Croatia, Republic of Montenegro and Republic of Serbia. Although immersed in medieval European culture, the historical context and specific regional space where we find them, with traces of earlier influences (prehistoric, ancient and early medieval), stećci, by several aspects, remain a unique phenomenon in the medieval European artistic and archaeological heritage [5]. Most famous and the most decorated stećak is the one from Zgošća near Kakanj, from 15th century. It is considered that it belonged to Stjepan II Kotromanić, but there is no valid evidence. Now it is located in the garden of the National Museum of Bosnia and Herzegovina in Sarajevo. Also, there is a replica of this stećak in Sarajevo, in front of the Parliamentary Assembly of Bosnia and Herzegovina [6]. Replica of the mentioned tombstone from Zgošća (Fig. 1) was surveyed by using methods of terrestrial laser scanning and close-range photogrammetry.

Stećak from Zgošča is richly decorated with numerous ornaments and it is the most beautifully stećak in Bosnia and Herzegovina. Because of his originality, it was an inspiration to many archaeologists, museologists, historians and other experts, who carefully analyzed his motives.

Fig. 1. Replica of "Zgošćanski stećak" in front of the Parliamentary Assembly of Bosnia and Herzegovina [7]

3 Used Imaging Station and Digital Camera

The most important part of the equipment used for laser scanning was Topcon Imaging Station 3, a total station with the auto-scanning capabilities of a laser scanner [8]. Topcon Imaging Station 3 was accompanied with appropriate software solution Topcon ImageMaster Pro. Topcon ImageMaster Pro software has revolutionized capturing images in the field, whether used with an off-the-shelf digital camera, or in conjunction with the Topcon Imaging Station 3. Technical characteristic of Topcon Imaging station 3 are given in Table 1.

The Canon EOS 500D digital camera was used to obtain a 3D model of the stećak by terrestrial photogrammetry. The resulting digital photos were processed in Pix4Dmapper software. Technical characteristic of Canon EOS 500D are given in Table 2.

4 Fieldwork and Data Processing

For the purpose of Terrestrial laser scanning and Close-range terrestrial photogrammetry it was necessary to stabilize the geodetic network around the object (stećak) itself, and thus to establish a local network. The scan itself was performed in the local coordinate system, and the results of the 3D model were obtained in the mentioned one. After the stabilization of geodetic network points and determination of their coordinates in the local coordinate system, it was necessary to determine their coordinates in the state coordinate system. The points of the geodetic network are placed quite close to the scanning object (approximately 5 m from the scanning object). A well distributed set of control points (well defined details of tombstone decorative motifs and inscriptions)

Table 1. Technical characteristic of Topcon Imaging Station 3

Angle Measurement	
Method	Absolute Reading
Minimum Reading	1"/5" (0.1 / 0.5 mgon)
Tilt Correction	Dual Axis
Compensating Range	±6' (1.8 m)
Distance Measurement	
Range	
1 prism	9,840' (3,000 m)
3 prisms	13,120' (4,000 m)
9 prisms	16,400' (5,000 m)
Non-Prism Mode	4.9' - 820' (1.5 m – 250 m)
Non-Prism Long Mode	16.4' - 6,500' (5.0 m - 2,000 m)
Precision	
Fine	±(2 mm + 2 ppm x *D*)
Non-Prism Mode	± 3 mm
Non-Prism Long Mode Precision	±(10 mm + 10 ppm x *D*)
Imaging	
Cameras	(2) 1.3 mp
Image speed	1-10 FPS
Scanning Max	20 pts/sec
User Interface	
Operating System	Microsoft Windows® CE.NET 4.2
Processor Type	Intel PXA255
Processor Speed	400Mhz
Display Panel	Full Color Touch-screen

Table 2. Technical characteristic of the Canon EOS 500D digital camera

Image Sensor	
Type	22.3 x 14.9 mm CMOS
Effective Pixels	Approx. 15.1 M
Total Pixels	Approx. 15.5 M
Aspect Ratio	3:2
Low-Pass Filter	Built-in/Fixed with Self Cleaning Sensor Unit
Sensor Cleaning	EOS integrated cleaning system
Colour Filter Type	Primary Colour
Image Processor	
Type	DIGIC 4
Lens	
Lens Mount	EF/EF-S
Focal Length	Equivalent to 1.6x the focal length of the lens
Focusing	
Type	TTL-CT-SIR with a CMOS sensor
AF System/ Points	9 AF points (f/5.6 cross type at centre, extra sensitivity at f/2.8)
AF Working Range	EV -0.5 -18 (at 23°C & ISO100)
AF Modes	AI Focus One Shot AI Servo
AF Point Selection	Automatic selection, Manual selection
Selected AF Point Display	Superimposed in viewfinder and indicated on LCD monitor
Predictive AF	up to 10m
Manual Focus	Selected on lens, default in Live View Mode
Exposure Control	
Metering Range	EV 1-20 (at 23°C with 50mm f/1.4 lens ISO100)

(*continued*)

Table 2. (*continued*)

ISO Sensitivity	AUTO (100-1600), 100, 200, 400, 800, 1600, 3200 Expandable to 6400 + H (approx. 12800) in 1-stop increments
White Balance	
Type	Auto white balance with the imaging sensor.
Settings	AWB, Daylight, Shade, Cloudy, Tungsten, White Fluorescent light, Flash, Custom. White balance compensation: 1. Blue/Amber +/-9 2. Magenta/ Green +/-9
Shooting	
Modes	Auto, Portrait, Landscape, Close-up, Sports, Night Portrait, No Flash, Movie, Program AE , Shutter priority AE, Aperture priority AE, Manual, A-DEP
Picture Styles	Standard, Portrait, Landscape, Neutral, Faithful, Monochrome, User Defined (x3)
Colour Space	sRGB and Adobe RGB
Image Processing	Highlight Tone Priority, Auto Lighting Optimizer (Basic and Creative modes), Long exposure noise reduction, High ISO speed noise reduction, Peripheral illumination correction.
File Type	
Still Image Type	JPEG: Fine, Normal (Exif 2.21 [Exif Print] compliant) / Design rule for Camera File system (2.0), RAW: RAW (14bit, Canon original RAW 2nd edition), Digital Print Order Format [DPOF] Version 1.1 compliant.
RAW+JPEG Simultaneous Recording	Yes (RAW + Large JPEG only)
Image Size	JPEG: (L) 4752x3168, (M) 3088x2056, (S) 2256x1504 RAW: (RAW) 4752x3168

was determined on the object, with the aim of merging and georeferencing point clouds in the State Coordinate System of Bosnia and Herzegovina. The amount of overlap between each scan was about 20%, which was enough since the Topcon IS instrument makes it possible to acquire directly georeferenced point clouds of a scene. The object was scanned by grid with point spacing 4 mm × 4 mm. The issue of noise and outliers removal from the point cloud was the most demanding part of the TLS data processing.

Figure 2 shows the final results of scanning data collected using the Topcon Imaging station 3. Since this instrument is not intended exclusively for laser scanning, the scanning process took quite a long time (six days). The complete processing procedure is done in the software Topcon ImageMaster Pro.

Fig. 2. Final result of Laser Scanning

To create a 3D model in the Pix4Dmapper software, terrestrial photogrammetry data was collected with a Canon EOS 500D digital camera. A total of 75 photos were taken. Figure 3 shows the final results of close-range photogrammetry process. This process took only one day. The complete processing procedure is done in the software Pix4Dmapper.

Fig. 3. Final result of Close-range photogrammetry

5 Analysis of the Results

Twelve check points were set up on the object. Their layout is visible in the Figs. 4, 5 and 6. Coordinates of these points were determined in the State Coordinate System by the method of terrestrial positioning, based on measured horizontal directions, zenith angles and slope distances.

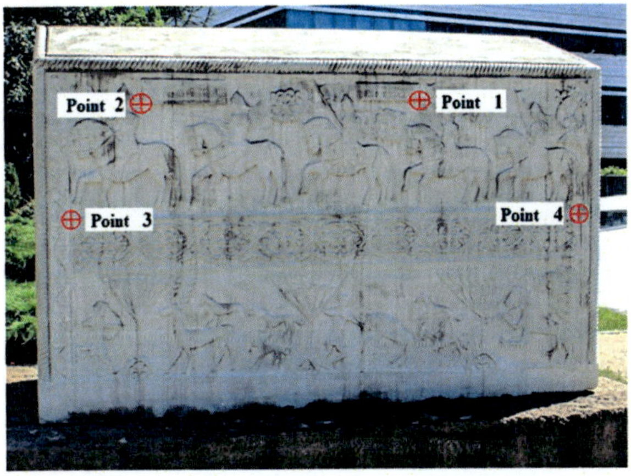

Fig. 4. Check points (north side of the object)

Fig. 5. Check points (west side of the object)

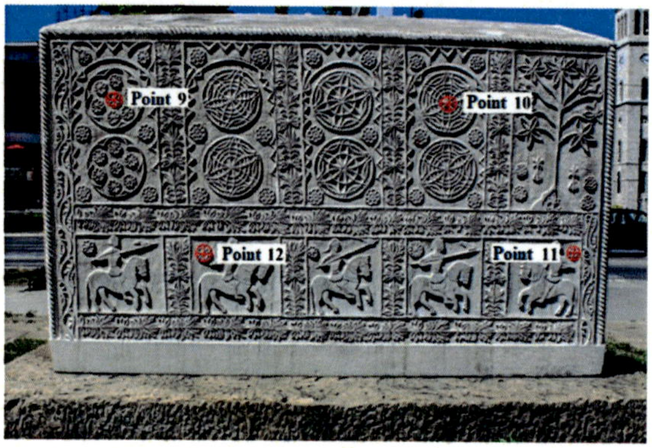

Fig. 6. Check points (south side of the object)

Table 3 gives an overview of differences between three-dimensional coordinates of points measured tachymetrically and corresponding points obtained from a 3D model by software Topcon ImageMaster Pro and Pix4Dmapper. Graphical overview of the coordinate differences is shown in Fig. 7. The obtained results show that 3D model obtained by laser scanning has significantly better geometric quality compared to model obtained by photogrammetry. This is also confirmed by the analysis of of Δx, Δy and Δh, according to the methodology described in [9] and [10], which revealed that the remaining systematic close-range photogrammetry errors are present in the coordinate differences. The maximum error in the photogrammetric point cloud (check point 2) may be explained by possible differences in illumination considering the moment of image acquisition or measurement noise that is often present around the object's edges. Since the point 2 is a corner point, formed by the intersection of two sharp edges, identifying and defining such a point as a vertex could produce a positional error.

Table 3. The coordinate differences of check points in the state coordinate system

Control point	Topcon ImageMaster Pro			Pix4Dmapper		
	Δx [mm]	Δy [mm]	Δh [mm]	Δx [mm]	Δy [mm]	Δh [mm]
1	0	0	3	4	25	0
2	−5	−1	4	−73	−60	−8
3	−3	4	−4	−19	37	−16
4	−4	−4	3	14	33	−8
5	−6	−3	0	−27	−11	−11
6	3	0	7	−9	9	−2
7	4	0	−5	−24	0	−4
8	−1	−5	0	−17	6	7
9	3	−3	2	28	−7	−1
10	−2	0	3	10	−15	2
11	1	4	−5	−1	−9	5
12	−4	10	−2	13	−1	6
Minimum	−6	−5	−5	−73	−60	−16
Mean	−1	0	1	−8	1	−3
Maximum	4	10	7	28	37	7
Range	10	15	12	101	97	23
Standard deviation	3	4	4	27	26	7

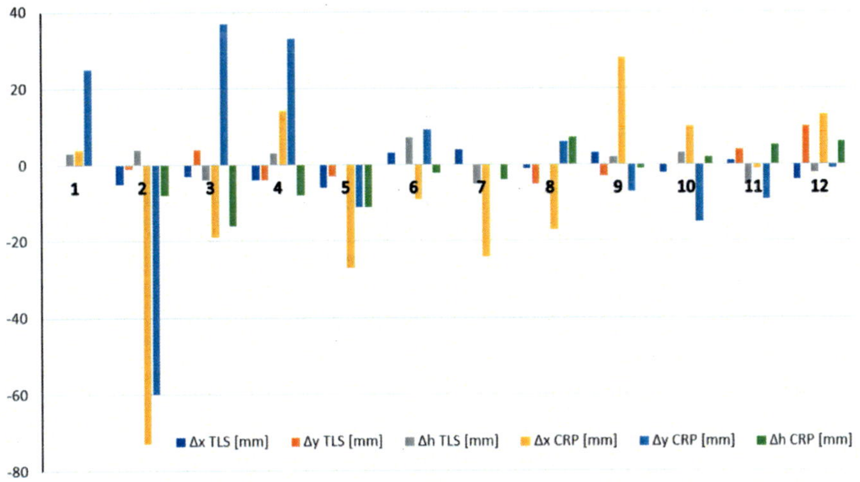

Fig. 7. Graphical interpretation of coordinate differences

6 Conclusion

The primary objective of this paper is to present the characteristics of laser scanning and close-range photogrammetry technologies in the documentation of cultural and historical heritage objects.

Generally speaking, the digital recording of monuments is a complex process that depends highly on the nature of subject as well as on applied surveying methods and processing. However, the creation, publication, and interaction of high-resolution 3D textured models remain rather challenging and difficult tasks. Such models, with a certain level of accuracy, are clearly invaluable tools for applications involving the restoration, maintenance, and monitoring of monuments [11]. Method of close-range photogrammetry is a technique for accurately measure objects directly from photographs or digital images captured with a camera at close range. Photogrammetry techniques allow you to convert images of an object into a 3D model. Using a digital camera with known characteristic (lens focal length, image size and number of pixels), we need a minimum of two pictures of an object to create 3D model [12]. Laser measuring concepts are applied in many technical fields by different instruments and methods. Generally, the meaning of non-contacting measuring an object is to execute processes in computer environment using the data obtained as cloud of points. Also, the meaning of 3D scanning process is to obtain the real coordinates of an object as three-dimensional [13]. The measurement principle of Laser instrument is to calculate the time taken between transmitting and receiving movements of distance signal when laser pulse is sent to object [14–16]. Both used methods have their advantages and disadvantages. The authors recommend sharing both methodologies and creating 3D models based on combined data from terrestrial laser scanning and close-range photogrammetry.

Acknowledgements. This research was supported by the Ministry of Science, Higher Education and Youth of Sarajevo Canton, under contract number 27-02-11-4375-5/21.

References

1. Mulahusić, A., Tuno, N., Gajski, D., Topoljak, J.: Comparison and analysis of results of 3D modelling of complex cultural and historical objects using different types of terrestrial laser scanner. Surv. Rev. **52**(371), 107–114 (2020). https://doi.org/10.1080/00396265.2018.1528758
2. Mulahusić, A., Tuno, N., Topoljak, J., Čengić, F., Kurtović, S.-Y.: Collecting geospatial data using unmanned aerial photogrammetric system. In: Avdaković, S., Mujčić, A., Mujezinović, A., Uzunović, T., Volić, I. (eds.) IAT 2019. LNNS, vol. 83, pp. 563–575. Springer, Cham (2020). https://doi.org/10.1007/978-3-030-24986-1_45
3. Burdziakowski, P., Tysiac, P.: Combined close range photogrammetry and terrestrial laser scanning for ship hull modelling. Geosciences **9**(5), 242 (2019). https://doi.org/10.3390/geosciences9050242
4. Wikipedia: Stećak (2021). https://en.wikipedia.org/wiki/Ste%C4%87ak#cite_note-Unesco-1
5. Unesco: Examination of nominations of cultural properties to the World Heritage List (2021). https://whc.unesco.org/en/decisions/6802

6. Admir Krilašević web site on Stećci (2016). https://admirkrilasevic.wixsite.com/stecci-tombstones/blank
7. Sarajevo Construction (2021). http://www.sa-c.net/images/stories/zgoscanski_stecak.jpg
8. Tuno, N., Mulahusić, A., Savšek, S., Kogoj, D.: Pet generacij integriranih elektronskih tahimetrov [Five generations of integrated electronic tachymeters]. Geodetski vestnik **63**(1), 41–56 (2019). https://doi.org/10.15292/geodetski-vestnik.2019.01.41-56
9. Tuno, N., Mulahusić, A., Kogoj, D.: Vrednotenje modelov transformacije geodetskih načrtov [Evaluation of transformation models for geodetic plans]. Geodetski Vestnik **59**(4), 736–751 (2015). https://doi.org/10.15292/geodetski-vestnik.2015.04.736-751
10. Tuno, N., Mulahusić, A., Kogoj, D.: Improving the positional accuracy of digital cadastral maps through optimal geometric transformation. J. Surv. Eng. **143**(3), 05017002 (2017). https://doi.org/10.1061/(ASCE)SU.1943-5428.0000217
11. Mulahusić, A., Tuno, N., Topoljak, J., Balić, Dž.: Izdelava 3D-modela kompleksnega kulturno zgodovinskega spomenika z uporabo digitalne fotogrametricne postaje [Using an imaging station in making a 3D model of a complex cultural and historical heritage object]. Geodetski vestnik **60**(1), 28–41 (2016). https://doi.org/10.15292/geodetski-vestnik.2016.01.28-41
12. Yakar, M., Yilmaz, H.M., Mutluoglu, O.: Close range photogrammetry and robotic total station in volume calculation. Int. J. Phys. Sci. **5**(2), 86–96 (2010)
13. Kraus, K.: Photogrammetry, Advanced Methods and Applications, vol. II, 4th edn. Dümmler, Bonn
14. Lichti, D.D., Stewart, M.P., Tsakiri, M., Snow, A.J.: Calibration and testing of a terrestrial laser scanner. Int. Arch. Photogrammetry Remote Sens. Part B5 **XXXIII**, 485–492 (2000)
15. Tuno, N., Mulahusić, A., Kogoj, D.: Od Reg Elta do Spatial Station: Štiri desetletja elektronskih tahimetrov Zeiss (Trimble) [From reg elta to spatial station: four decades of Zeiss electronic tacheometers (Trimble)]. Geodetski vestnik **56**(3), 415–426 (2012). https://doi.org/10.15292/geodetski-vestnik.2012.03.415-426
16. Tuno, N., Mulahusić, A., Marjetič, A., Kogoj, D.: Pregled razvoja elektronskih tahimetrov Leica Geosystems [Overview of development of electronic tachymeters Leica Geosystems]. Geodetski vestnik **54**(4), 643–660 (2010)

Development of Levelling Staff Scale Calibration Method by Integrating a CCD Camera

Sergej Baričević[1(✉)], Đuro Barković[1], Mladen Zrinjski[1], and Tomislav Staroveški[2]

[1] Faculty of Geodesy, University of Zagreb, Zagreb, Croatia
sergej.baricevic@geof.unizg.hr
[2] Faculty of Mechanical Engineering and Naval Architecture, University of Zagreb, Zagreb, Croatia

Abstract. High accuracy geometric levelling is the most precise existing classical geodetic surveying method. Considering the progress of levels, reduction of measuring uncertainty or increase in precision is no longer possible to achieve only by using better-quality levels. Nowadays, research is focused on the detection and modeling of the impact of systematic errors of measuring equipment. One influential systematic error is the imperfection of levelling staffs and levelling staff scales. Systematic error influence of levelling staff scale is modeled by calibrating it on a comparator. In this paper, a development of a levelling staff scale calibration method by integrating a charge-coupled device (CCD) camera to the existing comparator at the Laboratory for Measurements and Measuring Technique at the Faculty of Geodesy at the University of Zagreb is described. Results of initial tests are given with plans for research and testing.

Keywords: High accuracy levelling · Levelling staff · Systematic errors · Calibration · Comparator

1 Introduction

Geometric levelling, and especially high accuracy levelling, represents the most precise existing classical geodetic surveying method. Although one of the oldest, and despite the high level of development of other land surveying methods, even today precise geometric levelling has remained inviolable in terms of precise height difference determination. It is a method with up to one-hundredth of a millimeter precision with very small measurement uncertainty. Thus, Croatian *Ordinance on the Manner of Performing Basic Geodetic Works* [1] stipulates that when establishing a network of high accuracy levelling, the standard deviation may be up to 1.5 mm per $\sqrt{\text{km}}$, or for random errors 0.29 mm per $\sqrt{\text{km}}$, for systematic errors 0.86 mm per $\sqrt{\text{km}}$, and for measuring height differences 0.91 mm per $\sqrt{\text{km}}$. With the development of technology, instrumentation has been brought to a level when the reduction of measuring uncertainty or increase in precision is no longer possible to achieve only by using better-quality instrumentation. Therefore, research is focused on the detection and modeling of the impact of systematic

N. Ademović et al. (Eds.): IAT 2021, LNNS 316, pp. 514–521, 2022.
https://doi.org/10.1007/978-3-030-90055-7_40

errors to achieve the stipulated criteria of measurement uncertainty [2]. If the previous statement is applied to a high accuracy levelling, provided that the level is rectified [3], to achieve prescribed set criteria of measurement uncertainty, it is necessary to minimize the impact of systematic errors of measuring equipment, that is levelling staffs, on the height differences determination. This is done by determining the model of the influence of systematic errors of each individual leveling staff on the levelling measurements by calibrating the leveling staff scales. Complete elimination of the influence of systematic errors on measurements is not possible because the model itself determined by calibration of the leveling staff has a certain measurement uncertainty. In this paper, the development of levelling staff scale calibration method by integrating a CCD camera to the existing comparator at the Laboratory for Measurements and Measuring Technique at the Faculty of Geodesy at the University of Zagreb is presented.

2 Levelling Staff Influences and Errors

For the needs of high precision levelling, levelling staffs made in one piece are used. International standard *ISO 12858-1: Optics and optical instruments - Ancillary devices for geodetic instruments – Part 1: Invar levelling staffs* [4] recommends framework standards for the manufacture of precision leveling staffs to achieve the prescribed measurement uncertainties in the implementation of high accuracy levelling. The bodies of the staffs used to be made of wood, while today are most often made of aluminum alloy, which greatly reduces the impact due to the thermal deformation of the staff body. The scale graduation lines are not applied to the staff body but are located on a narrow invar strip that is attached to the bottom of the staff, while at the top of the staff invar strip is tensed by a spring with a force of 196.133 N [5]. The scales used in optical leveling are double (left and right) with a constant shift between, with a centimeter or half-centimeter distance between two graduation lines, while the coded graduation for digital levels is single-sided. Although the aluminum staff body can be significantly less deformed due to changes in temperature and humidity compared to wood, the deformation still exists. Due design solution in the form of attaching the invar strip at only two points, the invar strip is independent of the staff body and does not follow its deformation. Invar (nickel and steel alloy) has a much lower coefficient of thermal expansion than aluminum, so the scale is more stable and does not have as much magnitude of change as it would have if applied directly to the staff body [6, 7]. Part of the systematic influences caused by leveling staffs is related to invar strips and scale graduation on them [2, 5, 8, 9]. The error that occurs when the levelling staff footplate does not coincide with the imaginary zero graduation line, that is, when the invar strip is not attached at the correct distance from the bottom of the staff is called the levelling staff zero error and it is of a constant value for the individual levelling staff. Although the graduation is applied with a high precision laser on the invar strip, there is still a possibility of a deviation of the position of the graduation lines from its nominal values. It is called the error of applying the graduation lines on the invar strip or shorter, graduation error. The graduation error may be such that the graduation along the entire length of the invar strip is incorrectly applied, and the whole scale is shorter or longer than the nominal value, resulting in a levelling staff scale error. Graduation lines also can be incorrectly applied individually and/or at intervals. Although the invar has a very small coefficient of thermal expansion, it is not negligible, and a temperature change shortens or stretches the invar strip, and consequently the scale. This error is called the error due to the change in the length of the leveling staff, or just coefficient expansion error. The graduation error and the coefficient expansion error cannot be separated into two different systematic influences but are represented by the systematic influence of the leveling staff scale which depends on the temperature at which the graduation lines are read. All these errors can be tested and calibrated using a device for calibration of geodetic linear scales, a comparator. By using a comparator, the nominal values of graduation lines are compared to measured values of the same graduation lines. From the measured differences, the regression line is modeled. From the obtained model, the coefficient of levelling staff scale and zero error value are determined.

3 First Linear Comparator at LMMT

The first linear comparator (Fig. 1) at LMMT was developed and constructed as a part of professor Đuro Barković's doctoral dissertation in 2002 [3, 10, 11]. As seen in Fig. 1, the main parts of the comparator are iron stand (1) and bed (2) with clamping screws for levelling staff (9) in the horizontal position. On top of the bed are two parallel guide rails (7) for a trolley (3) that is equipped with a microscope (4), a lock (5), a fine adjustment screw (6), and a light source (10). A trolley is connected to a scanning carriage in a sealed linear encoder (8). By using a microscope, an individual graduation line can be manually aimed, and the trolley position is then obtained from the linear encoder that is connected to a computer with a cable (11). Software for data recording and processing was developed in Microsoft Visual Basic [10].

Fig. 1. First comparator at the LMMT [10].

Three-meter levelling staffs for optical leveling have a double graduation scale, as described in Sect. 1. There are usually 296 graduation lines on both sides. When determining the position of graduation lines, a position of both edges of each graduation line is obtained three times to avoid blunders. Then, the center line position is computed from both edges, and it should correspond to the nominal value of the graduation line. Hence, each graduation line must be aimed six times, which makes for one scale side 1,776 times, and for a double-sided scale 3,552 times. That amount of manual work is time-consuming and subjective to the operator's performance which directly influences the quality of carried out calibration.

4 Development of a New Calibration Method

To avoid the subjective influence of the human operator and speed up the calibration process, and consequently increasing the overall accuracy, a new levelling staff scale calibration method by integrating a CCD camera to the existing comparator has been developed at the LMMT. The comparator has been updated in a way that the existing trolley has been substituted with a servo-motorized trolley (Fig. 2) with a mounted CCD camera (1) and telecentric lens (2). Telecentric lens is used because it minimizes distortion of the picture (maximum distortion 0.51%) [12] and depth of field is sufficient for the possible unevenness of the invar strip.

Fig. 2. Servo-motorized trolley with CCD camera and telecentric lens.

For the control of the whole comparator system, an industrial computer running on Linux operating system was set up. Comparator software was and still is, developed in several open-source programming languages. The software consists of several modules, of which some can be operated through a graphical user interface (GUI) (Fig. 3), while others are operating in the background. The software enables controlling the position of the trolley and changing the parameters of the camera.

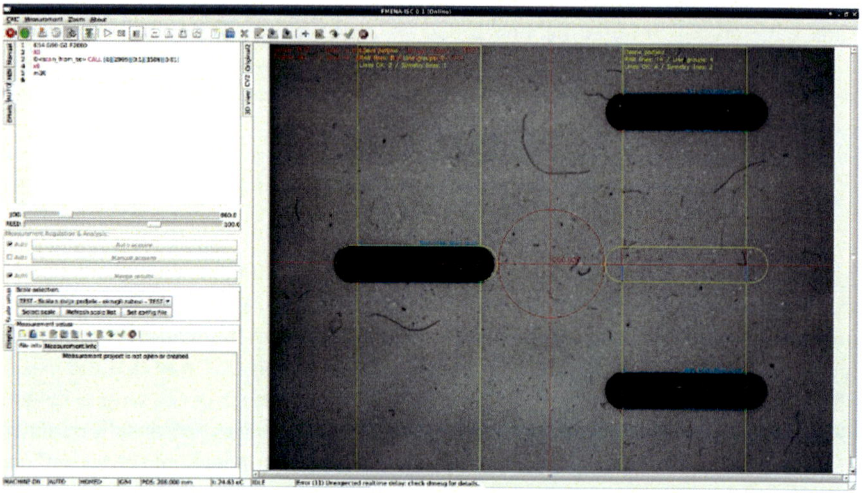

Fig. 3. Graphical user interface of the comparator software.

The position of the trolley is determined by the linear encoder to which the trolley is connected to a scanning carriage. When at a certain position, the camera is taking a picture and processing it. It is scanning for and detecting graduation line edges on the levelling staff invar strip. Considering different graduation patterns regarding levelling technology and manufacturer brands, different graduation types to detect can be chosen from the drop-down menu to match calibrated levelling staff. The calibration itself can be performed in several ways, manually, semi-automatically, and automatically. Usually, calibration is performed in automatic mode where the operator needs to input parameters from which the most important is the movement step of the trolley. In every step camera takes the picture, the software detects all the edges and writes the position of every edge in the picture. Also, the temperature from the thermometer that is mounted on the comparator is saved with the detected position of the edge. The smallest step can be 0.01 mm, and in that case, every edge will be detected around 1,460 times, or the biggest 8 mm when every edge will be detected only two times. When scanning in smaller step value, there was a loss of data. To troubleshoot that happening, a computer with a solid-state drive (SSD) was set as a remote data storage. That was the final hardware update, and the comparator system is now assembled as shown in Fig. 4.

After the scanning is finished, raw data is processed. The position of every edge is averaged and from pairs of edges of each graduation line, symmetry lines are computed. Further, the difference between the value of the measured graduation line and its nominal value is computed (Δ). From those differences, by using linear regression, levelling staff scale coefficient (e.g., $m_0 = -18.778$ ppm) is computed with zero error of the levelling

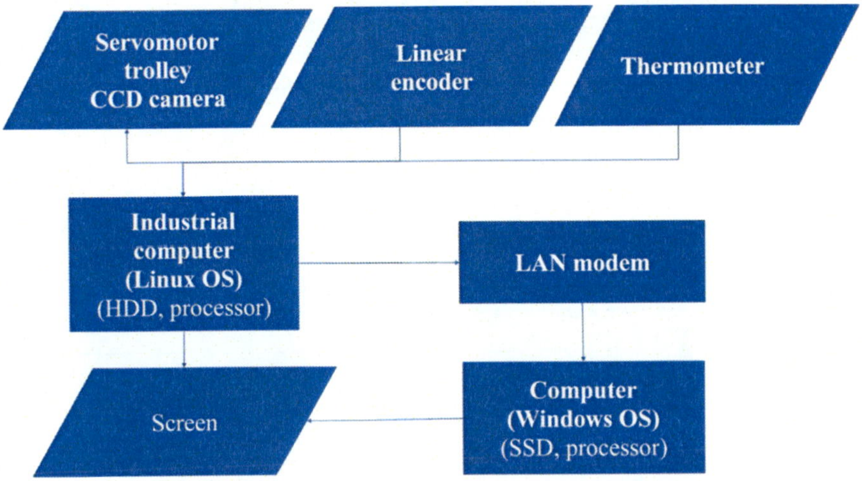

Fig. 4. Schematic representation of the updated comparator.

staff (e.g., 98.079 μm) (Fig. 5). The software automatically creates a calibration report in a PDF document.

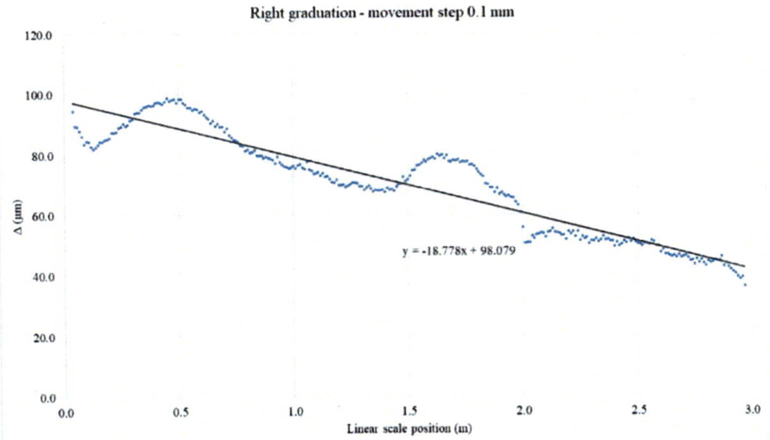

Fig. 5. Computed differences of graduation lines position with regression line.

5 Testing the System and the Method

After the comparator hardware was updated, preliminary hardware system tests were carried out. As mentioned in the previous section, there was a problem with data loss, and that was fixed by adding a computer with an SSD disk into the comparator system. After the hardware system was not showing any bugs, initial testing of the method was carried out. Firstly, one levelling staff was fixed into the comparator and scanned with six different movement steps of the trolley (Table 1). For trolley movement steps of 0.1 mm, 0.2 mm, 0.5 mm, 1.0 mm, and 2.0 mm results were consistent. At the step of 5.0 mm, there was a significant difference in obtained levelling staff scale coefficient from the other. That was logical because, with that step value, every edge is detected only three times.

Table 1. Results of levelling staff calibration with different trolley movement step.

Step [mm]	m_0 [ppm]	Standard deviation (k = 1.96) [ppm]
0.1	−20.88	0.774
0.2	−21.87	0.820
0.5	−21.61	0.750
1.0	−20.61	0.768
2.0	−20.66	0.771
5.0	−18.80	0.829

After that, the same levelling staff was scanned with the same trolley movement step of 0.5 mm (Table 2). Results were relatively consistent, but slight differences were noticeable. Further tests are recommended.

Table 2. Results of levelling staff calibration with same trolley movement step (0.5 mm).

Step [mm]	m_0 [ppm]	Standard deviation (k = 1.96) [ppm]
0.5	−22.54	0.800
0.5	−23.41	0.805
0.5	−22.04	0.809
0.5	−21.52	0.832

6 Conclusions and Plans

High accuracy geometric levelling is the most precise existing classical geodetic surveying method. Considering the progress of levels, reduction of measuring uncertainty or increase in precision is no longer possible to achieve only by using better-quality levels. Nowadays, research is focused on the detection and modeling of the impact of systematic errors of measuring equipment. One influential systematic error is the imperfection of levelling staffs and levelling staff scales. Systematic error influence of levelling staff scale is modeled by calibrating it on a comparator.

At the Laboratory for Measurements and Measuring Technique at the Faculty of Geodesy at the University of Zagreb existing linear scale comparator has been updated with a servo-motorized trolley with a CCD camera and a telecentric lens. A new calibration method for using the updated comparator is under development. Initial results are consistent and satisfactory.

In the future, further research will be carried out. The uncertainty budget of the comparator system must be modeled. Trolley movement step for the optimal cost (time) – benefit (accuracy) must be determined.

After the research and setting hardware and software parameters, the comparator should be fully operational and have overall accuracy under 5 µm/m of the levelling staff calibrated.

References

1. Republic of Croatia Parliament: Ordinance on the Manner of Performing Basic Geodetic Works – Official Gazette no. 15/2020, Zagreb, Croatia (2020)
2. Rožić, N.: Random and systematic errors investigation in the geometric levelling. Dissertation, Faculty of Geodesy, University of Zagreb, Zagreb (1995)
3. Zrinjski, M., Barković, Đ., Razumović, I.: Automation of testing the precision of levels and calibration of invar levelling staves. Geodetski List **64**(87) (4), 279–296 (2010)
4. International Organization for Standardization: ISO 12858-1: Optics and optical instruments – Ancillary devices for geodetic instruments – Part 1: Invar levelling staffs. ISO, Geneva (2014)
5. Macarol, S.: *Practical Geodesy*, 4th edn, Školska knjiga, Zagreb (1985)
6. Barković, Đ., Zrinjski, M., Nervo, M.: Determination of temperature extension coefficient of invar levelling staves. In: International Multidisciplinary Scientific GeoConference SGEM*(ed.)* Conference Proceedings, Volume II – 11th International Multidisciplinary Scientific GeoConference SGEM 2011, Albena, 20–25 June, pp. 247–254 (2011)
7. Benčić, D., Solarić, N.: *Measuring Instruments and Systems in Geodesy and Geoinformatics*, Školska knjiga, Zagreb (2008)
8. Feil, L.: Contribution to the consideration of the accuracy assessment model in the high accuracy levelling. Dissertation, Faculty of Geodesy, University of Zagreb, Zagreb (1984)
9. Rüeger, J.M., Brunner, F.K.: On system calibration and type testing of digital levels. Z. f. Vermessungswesen **125**, 120–130 (2000)
10. Barković, Đ.: Comparison of levelling staves by means of sealed linear encoder. Dissertation, Faculty of Geodesy, University of Zagreb, Zagreb (2002)
11. Barković, Đ., Zrinjski, M., Debeljak, T.: Levelling staves comparator on the basis of incremental measuring system. In: International Scientific Conference SGEM (ed.) Conference Proceeding, Volume II – 8th International Scientific Conference SGEM 2008. Albena, 16–20 June, pp. 509–516 (2008)
12. Sill Optics: Telecentric lens – list of products. https://www.silloptics.de/produkte/bildverarbeitung/telezentrische-objektive/produkttabelle. Accessed 12 Mar 2021

Rapid Assessment of the Verticality of Structural Objects with a Circular Base

Adis Hamzić[1](✉), Dina Kamber Hamzić[2], and Zikrija Avdagić[3]

[1] Hydro Power Plants On the Neretva River, Public Enterprise of Electric Utility of Bosnia and Herzegovina, Sarajevo, Bosnia and Herzegovina

[2] Faculty of Science, University of Sarajevo, Sarajevo, Bosnia and Herzegovina

[3] Faculty of Electrical Engineering, University of Sarajevo, Sarajevo, Bosnia and Herzegovina

Abstract. Structural health monitoring represents an important task that involves various measurements and methods. Due to its importance, it uses the best available equipment and the most advanced algorithms to give precise and reliable information about an object's health. For tall objects such as chimneys, towers, and masts the main criterion of stability is verticality deviation (inclination). However, sometimes it is necessary to get a rapid assessment of an object's stability and in that case, accuracy and reliability are sacrificed to save time. In this paper, an approach for rapid assessment of verticality of structural objects with a circular base is presented. Geodetic methods were used to measure points in two levels on the perimeter of the object. Three algebraic methods for circle fitting were used to determine the center coordinates and radius of the object in the two observed cross-sections. The final step was to determine the spatial vector which represents the object's axis and compare it to the theoretical vertical. The proposed approach is fast, simple, and inexpensive in terms of equipment, human resources, and software necessary for data processing but at the same time, it is less accurate and reliable compared to the state of the art approaches for object verticality assessment.

Keywords: Structural health monitoring · Verticality deviation · Inclination · Algebraic circle fitting · Analytical geometry

1 Introduction

Geodetic methods play a very important role in the structural health monitoring (SHM) of infrastructure objects. Usually, these measurements include angular measurements, distance measurements, and height differences by using precise total stations and precise levels. In modern times photogrammetry, laser scanning, and global positioning system (GPS) measurements also participate in SHM tasks. The lack of precision of these modern methods compared to classical methods is compensated with automatization and mass data acquisition. In order to derive as many benefits from all these geodetic methods, it is usually the case that multiple geodetic methods are combined when SHM is conducted. Additionally, to get the whole picture of an object's behavior, physical

N. Ademović et al. (Eds.): IAT 2021, LNNS 316, pp. 522–531, 2022.
https://doi.org/10.1007/978-3-030-90055-7_41

methods of monitoring (deformeters, clinometers, pendulums, etc.) are used together with geodetic measurements.

The inclination of an object is especially dangerous for objects for which the ratio of height and width is high. In cases when the inclination is significant the structural statics of the object may be impaired. The main reasons which cause the inclination of the object can be classified into three groups: subsoil properties, foundations, and anthropogenic factors [1]. Neither the aforementioned reasons for the object's inclination nor the methods for rectification of inclined objects are in the focus of this research – the focus is only on rapid, simple, and inexpensive assessment of the inclination of the objects.

Many researchers and practicians tackled the issue of determining the inclination of the tall structural objects with a circular base and a brief overview of some researches, which were available to the authors, is presented in the following text. Kregar *et al.* [2] used precise total station and laser scanner measurements to create a precise 3D model of an observed object. Measured points were fit to a cylinder by using the least-squares adjustment to determine parameters of the cylinder which represents a monitored object (industrial chimney). The authors concluded that both methods could be equally used for the inclination monitoring since neither method showed superior results compared to the other.

A method to measure the inclination of tall chimneys is presented in the research conducted by Barazzetti, Previtali, and Roncoroni [3]. The authors performed high-density points measurement of the object by the laser scanner, then horizontal slices derived from the point cloud were used to determine the center of the observed object in different levels. Manual and automatic methods were presented in this paper and the authors concluded that it is possible to give a precise estimation of the inclination by using both proposed methods.

The circular regression analysis and 3D linear regression were used by Zrinjski *et al.* [4] to determine the inclination of the tall industrial chimney. One of the main strengths of the proposed approach is that for all calculated parameters there is accuracy quantification obtained by the consistent utilization of the error propagation law.

Markiewicz, Zawieska, and Podlasiak [5] developed an original software for 3D point cloud processing obtained from multiple sources (terrestrial laser scanning, terrestrial photographs, and digital images). Developed software was able to verify the verticality of the object and assess the quality of utilized data.

Harshit, Jain, and Mishra [6] used an unmanned aerial vehicle (UAV) and Lidar for modeling and inclination assessment of the very tall industrial chimney (275 m). The results showed that both methods could be used for selected purposes but Lidar had better relative accuracy compared to UAV.

Another research of the tall industrial chimney stability was conducted by Muszynski and Milczarek [7]. The authors used laser scanning measurements conducted in poor conditions (night, snowfall, low air temperature) to test the ability of this method for the determination of the object's inclination. In their conclusion, the authors highlighted that inclination was calculated with satisfactory accuracy and that the main advantage of this approach is that detailed data regarding geometry of the entire object is obtained during the observations.

As seen from [2–7] the most common object of the research are industrial chimneys while favorite measuring methods are laser scanning [2, 3, 5, 7] and classical tachymetry measurements by using total stations [2–4]. The usual approach in the aforementioned studies was to determine coordinates of the circle center based on the observations, and later central axis of the object was calculated.

The problem of fitting circles to observed points plays an integral part in the proposed approach for inclination assessment. There are two major groups of proposed solutions for the circle fitting problem: geometric fit and algebraic fit. Among geometric fitting methods, the most popular are Gauss-Newton method with the Levenberg-Marquardt correction, Landau algorithm, and Späth algorithm, while among algebraic fitting methods the most commonly used are Kåsa's method, Pratt's method, and Taubin's method.

Circle fitting approaches that minimize the geometric distances from the circle to the data points are called "geometric fits" while approaches where the geometric distances are replaced by some other simple algebraic formulas are called "algebraic fits". Although there are many other approaches to the circle fitting problem described in literature, most of them are either quite slow or can be reduced to one of the algebraic fits [8].

According to Chernov [9]:

- Geometric fitting is more accurate than algebraic fitting but it requires an initial guess and it is at least 5 to 10 times slower than any algebraic fit.
- Kåsa fit is practically identical to the geometric fit when the data are sampled along the full circle but it is biased toward smaller circles when an incomplete arc is observed.
- Pratt circle fit has all the advantages of the Kåsa method and additionally performs well even when measurements are situated on the small arc.
- Taubin fit is slightly simpler compared to the Pratt fit, hence it is a bit less expensive computationally, also Taubin fit (unlike Pratt) can be generalized to ellipses and other algebraic curves.
- On the smaller arcs Taubin's characteristics are closer to the Pratt fit, and on larger arcs, to the geometric fit.

It is important to note that in the most practical applications (when data is distributed regularly) any of these methods would provide good results. But, taking into account all the above-mentioned advantages and disadvantages of these methods, Kåsa's, Pratt's, and Taubin's methods were chosen for circle fitting in this research.

The precise total station was used to obtain necessary data and MATLAB 2018b software was used to calculate the coordinates and radius of the circles by three commonly used algebraic methods for circle fitting. From calculated coordinates of the centers of the circles, the spatial vector which represents the central axis of the tower was determined. The final step was to calculate the inclination angle from the theoretical vector of verticality and calculated vector (central axis) of the wind turbine tower in Podveležje wind farm.

2 Data and Methods

The Podveležje wind farm is built near the town of Mostar and consists of 15 wind turbines which are installed on 92.5 m high steel conical towers. The measurements were conducted from an arbitrarily set geodetic network in a local coordinate system. Sokkia NET05AX precise total station (1 mm + 1 ppm reflectorless distance measurement accuracy, 0.5″ angle measurement accuracy) was used to observe points of interest on one of the wind turbine towers.

The measurements were conducted in two vertical levels from three measurement stations (Fig. 1):

- lower level (ground level) - the junction of terrain and tower, and
- upper level - the junction of the second and third segment of the wind turbine tower.

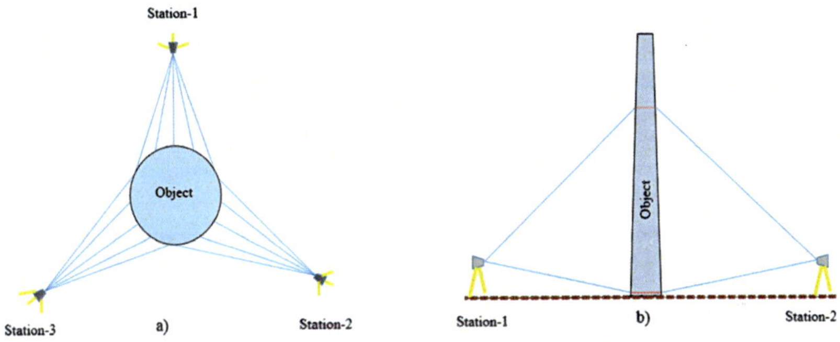

Fig. 1. Observation organization: a) horizontal plane; b) vertical plane

The local coordinate system is defined as follows:

- the origin of the coordinate system is in Station 1,
- local N-axis passes through the centers of Station 1 and Station 2, the positive direction is towards Station 2,
- local E-axis is perpendicular to the local N-axis in Station 1 and the positive direction of the E-axis is to the right, and
- H-axis is height.

When performing observations, points representing a circle of the object need to have a similar height which is especially important for objects shaped as a conical frustum. The easiest way to fulfill this condition is to follow a visible horizontal line on the object, for example, a conjunction of the two segments. Additionally, it is advised that observed points are distributed along the full circle or at least along the large arc, arcs smaller than 90° should be avoided. The distribution of observations is presented in Fig. 2, and the results of the conducted observations are presented in Table 1.

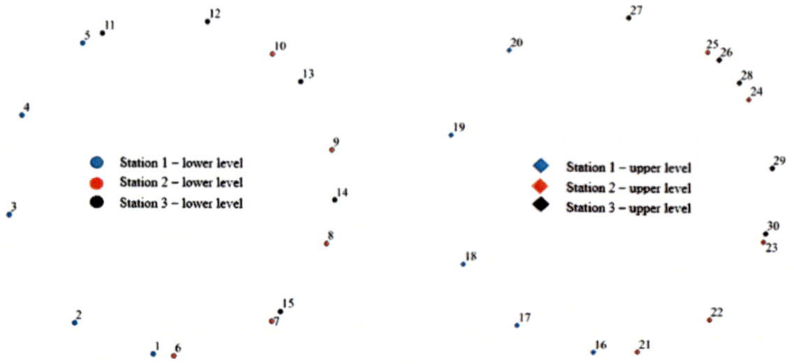

Fig. 2. Distribution of observations

As shown in Fig. 2, observations are well distributed in both levels and observed points cover a full circle. Since wind farms are built on open terrain, it was possible to choose good positions for measurement stations. Note that due to field conditions, favorable distribution of observations is not always possible, which is especially the case in crowded urban areas. To simulate unfavorable measurement conditions some of the measurements are intentionally removed during data processing. The results of these experiments are presented in Sect. 3.

Table 1. Coordinates of the observed points on the wind turbine tower

Point	N (m)	E (m)	H (m)	Point	N (m)	E (m)	H (m)
1	1073.977	1016.890	995.910	16	1073.968	1016.876	1032.918
2	1072.945	1017.276	995.925	17	1072.989	1017.213	1032.913
3	1072.098	1018.618	995.915	18	1072.287	1017.975	1032.906
4	1072.259	1019.865	995.911	19	1072.125	1019.594	1032.897
5	1073.047	1020.770	995.914	20	1072.866	1020.658	1032.901
6	1074.241	1016.866	995.920	21	1074.543	1016.886	1032.910
7	1075.496	1017.308	995.920	22	1075.461	1017.299	1032.908
8	1076.194	1018.280	995.920	23	1076.158	1018.278	1032.911
9	1076.257	1019.451	995.923	24	1075.959	1020.066	1032.915
10	1075.492	1020.645	995.918	25	1075.422	1020.651	1032.914
11	1073.301	1020.898	995.917	26	1075.565	1020.555	1032.915
12	1074.662	1021.048	995.923	27	1074.412	1021.077	1032.904
13	1075.853	1020.299	995.925	28	1075.831	1020.268	1032.911
14	1076.298	1018.821	995.923	29	1076.255	1019.205	1032.902
15	1075.609	1017.431	995.916	30	1076.179	1018.382	1032.904

Kåsa's approach for circle fitting [10] is probably the most obvious approach to deal with this problem. Kåsa proposed to minimize the function f_K to solve circle fitting problem, where

$$f_K(a, b, R) = \sum_{i=1}^{n}\left[(x_i - a)^2 + (y_i - b)^2 - R^2\right]^2 \tag{1}$$

In Eq. (1), (a,b) denotes coordinates of the center of the circle, R denotes the radius of the circle, (x_i, y_i) denotes plane coordinates of the point observed on the circle and n is the number of observed points.

Since Kåsa fit gives biased estimates when incomplete circular arcs are observed, Pratt [11] proposed to minimize another function:

$$f_P(a, b, R) = \frac{1}{4R^2} f_K(a, b, R) \tag{2}$$

Taubin [12] proposed a somewhat different approach for the circle fitting by minimizing the function f:

$$f_T(a, b, R) = \frac{\sum\left[(x_i - a)^2 + (y_i - b)^2 - R^2\right]^2}{4n^{-1}\sum\left[(x_i - a)^2 + (y_i - b)^2\right]} \tag{3}$$

For a more detailed explanation of the above-mentioned circle fitting methods, see [8–12]. By using these methods, coordinates of the center of the tower are determined in two observed levels.

Equation (4) is a well-known way to calculate an angle θ between the two vectors a $= (x_1, y_1, z_1)$ and $b = (x_2, y_2, z_2)$ in the 3D space:

$$\theta = \arccos\left(\frac{x_1x_2 + y_1y_2 + z_1z_2}{\sqrt{x_1^2 + y_1^2 + z_1^2} * \sqrt{x_2^2 + y_2^2 + z_2^2}}\right) \tag{4}$$

Point $O(N_O, E_O, H_O)$ is the observed center of the lower circle, and $P(N_P, E_P, H_P)$ is the observed center of the upper circle of the wind turbine tower. Point $R(N_R = N_O, E_R = E_O, H_R = H_P)$ is a point on the theoretical vertical in point O. Note that it is not necessary that $H_R = H_P$ but only $H_R > H_O$, but we choose $H_R = H_P$ for visualization purposes (Fig. 3).

Vector $\overrightarrow{OR}$ defines the theoretical vertical in the point O, while vector $\overrightarrow{OP}$ defines the observed axis of the object. By using Eq. (4) and applying calculated coordinates and notation of our local coordinates system, an equation for determining the inclination angle of the observed object has the following form:

$$\theta = \arccos\left(\frac{H_P - H_O}{\sqrt{(N_P - N_O)^2 + (E_P - E_O)^2 + (H_P - H_O)^2}}\right) \tag{5}$$

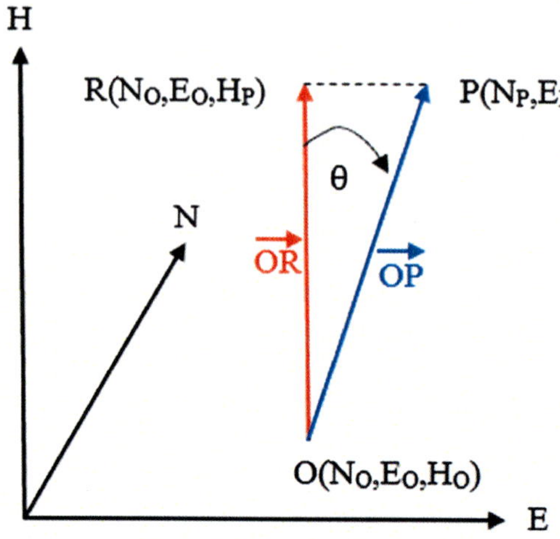

Fig. 3. The angle between the theoretical vertical of the object and the observed axis of the object

3 Results and Discussion

The data and methods presented in Sect. 2 were used to determine the coordinates of the circle center in the horizontal plane, circle radius, and inclination angle of the object's central axis. In Eq. (5) H_O and H_P were calculated as the mean value of the observed heights in the two measured levels. Additionally, the influence of the observed point distribution was investigated by using observations from different measuring stations. The obtained results are demonstrated in Tables 2, 3, and 4.

Table 2. Kåsa fit (N, E, R), inclination angle (θ), and deflection at the top of the tower

Station	Level	N (m)	E (m)	R (m)	Inclination angle ("-")	Deflection at the top (m)
1, 2, 3	Lower	1074.1863	1018.9852	2.1086	8–53.6	0.239
	Upper	1074.0907	1018.9806	2.0282		

When the observed points were evenly distributed along a full circle, Kåsa's method showed results comparable to Pratt's and Taubin's methods, but when observed points did not cover full circle (measurements performed from only one or two measuring stations) the results did not have any practical value.

By comparing the results from Tables 2, 3, and 4 it can be concluded that Pratt's and Taubin's methods show very similar results in every combination of observed points. Kåsa's method performed significantly worse compared to the other used methods. Pratt's and Taubin's methods performed well even when there is a small number of observations distributed along arcs of approximately 120°.

Strengths and weaknesses of the proposed approach are compared to the state of the art approaches [2–7] and some notable properties are presented in Table 5. It is important

Table 3. Pratt fit (N, E, R), inclination angle (θ), and deflection at the top of the tower

Station	Level	N (m)	E (m)	R (m)	Inclination angle (“-”)	Deflection at the top (m)
1, 2, 3	Lower	1074.1892	1018.9852	2.1168	3–34.5	0.096
	Upper	1074.1525	1018.9736	2.1163		
1, 2	Lower	1074.1899	1018.9851	2.1185	3–42.2	0.100
	Upper	1074.1525	1018.9713	2.1153		
1, 3	Lower	1074.1863	1018.9890	2.1144	3–22.7	0.091
	Upper	1074.1510	1018.9804	2.1152		
2, 3	Lower	1074.1929	1018.9826	2.1149	4–34.6	0.123
	Upper	1074.1456	1018.9689	2.1226		
1	Lower	1074.1738	1018.9881	2.1076	3–1.6	0.081
	Upper	1074.1427	1018.9783	2.1094		
2	Lower	1074.2022	1018.9746	2.1097	3–57.6	0.107
	Upper	1074.1619	1018.9606	2.1091		
3	Lower	1074.2002	1018.9948	2.1050	3–52.1	0.104
	Upper	1074.1598	1018.9849	2.1072		

Table 4. Taubin fit (N, E, R), inclination angle (θ), and deflection at the top of the tower

Station	Level	N (m)	E (m)	R (m)	Inclination angle (“-”)	Deflection at the top (m)
1, 2, 3	Lower	1074.1892	1018.9852	2.1168	3–34.5	0.096
	Upper	1074.1525	1018.9736	2.1162		
1, 2	Lower	1074.1899	1018.9851	2.1185	3–42.2	0.100
	Upper	1074.1525	1018.9713	2.1153		
1, 3	Lower	1074.1863	1018.9890	2.1144	3–22.7	0.091
	Upper	1074.1510	1018.9804	2.1152		
2, 3	Lower	1074.1929	1018.9826	2.1149	4–34.6	0.123
	Upper	1074.1456	1018.9689	2.1225		
1	Lower	1074.1738	1018.9881	2.1076	3–1.6	0.081
	Upper	1074.1427	1018.9783	2.1094		
2	Lower	1074.2022	1018.9746	2.1097	3–57.6	0.107
	Upper	1074.1619	1018.9606	2.1091		
3	Lower	1074.2002	1018.9948	2.1050	3–52.1	0.104
	Upper	1074.1598	1018.9849	2.1072		

to note that the proposed approach aims for simplicity and fast results and that is what distinguishes it from other approaches presented in the introductory chapter. The proposed approach is not intended as a means of regular monitoring of the object's verticality but only for rapid assessment of verticality. For the regular structural health monitoring (which includes monitoring of horizontal displacements, vertical displacements, and verticality assessment) methods, and instruments presented in [2–7] should be used. If instruments with the ability of reflectorless distance measurement are not available, some solutions rely on angular measurements and can be used for these purposes, e.g. see [13].

Table 5. A comparison between the proposed approach and the state of the art approaches for object verticality assessment

Characteristic	Proposed approach	State of the art
Time	From observations to results in less than 60 minutes	Time-consuming - usually takes a few days to perform observations and process the data
Complexity	Simple and easy to perform even for non-experts	Requires professional expertise for observations and data processing
Cost	Inexpensive in terms of required equipment and human resources	Expensive - requires specialized equipment and a team of experts
Software	Does not require specialized software, every spreadsheet application can be used for data processing (e.g.: Microsoft Excel, Google sheets, Open Office Calc)	Specialized applications for processing of geodetic observations and 3D modeling is needed
Modeling	Observed object simplified to cylinder or conical frustum	Precise 3D model of the observed object
Displacements	Cannot detect horizontal or vertical displacements	Can detect object movements in every direction
Stable geodetic network	Not required	Mandatory
Accuracy assessment	Limited options	Available
Reliability	Unknown	Reliable

4 Conclusion

In this paper, a novel approach for rapid assessment of the verticality of the structural objects with a circular base is introduced. This approach uses well-known surveying techniques and mathematical methods to provide a simple, inexpensive, and fast assessment of the object's verticality.

The data was acquired by using geodetic measurements on the circumference of the object in the two levels. Three algebraic circle fitting methods (Kåsa, Pratt, and Taubin) are applied to determine coordinates of the circle center which were later used to calculate inclination angle through analytical geometry. More precisely, the spatial vector which depicts the object's central axis is determined from the circle center's coordinates in the two levels and then an angle between this vector and theoretical vertical is determined.

The results of this research showed that Pratt's and Taubin's method for circle fitting performed better compared to Kåsa's method which should be avoided for these purposes unless there is a large number of points evenly distributed along the full circle.

The main disadvantage of the proposed approach is that there are limited options for accuracy assessment and unknown reliability since obtained results were not compared to other methods which are tested in practice and are known to be accurate and reliable. Comparing the proposed approach to other well-known methods for object verticality assessment will be part of future researches.

References

1. Kijanka, M., Kowalska, M.: Inclined buildings–Some reasons and solutions. IOP Conf. Ser. Mater. Sci. Eng. **245**(2), 022052 (2017)
2. Kregar, K., Ambrožič, T., Kogoj, D., Marjetič, A.: Determining the inclination of tall chimneys using the TPS and TLS approach. Measurement **75**, 354–363 (2015)
3. Barazzetti, L., Previtali, M., Roncoroni, F.: The use of terrestrial laser scanning techniques to evaluate industrial masonry chimney verticality. Int. Arch. Photogrammetry Remote Sens. Spatial Inf. Sci. **XLII-2**, 173 (2019)
4. Zrinjski, M., Tupek, A., Barković, Đ., Polović, A., Novosel, T., Vidoš, M.: Determination and analysis of chimney inclination. INGEO SIG **2020**, 155 (2020)
5. Markiewicz, J., Zawieska, D., Podlasiak, P.: Comparing multi-source photogrammetric data during the examination of verticality in a Monumental tower. The Int. Arch. Photogrammetry Remote Sens. Spatial Inf. Sci. **42**, 475 (2017)
6. Harshit, S.L., Jain, K., Mishra, V.: Analysis of survey approach using UAV images and lidar for a Chimney study. J. Indian Soc. Remote Sens. **49**(3), 1–6 (2020)
7. Muszynski, Z., Milczarek, W.: Application of terrestrial laser scanning to study the geometry of slender objects. IOP Conf. Ser. Earth Environ. Sci. **95**(4), 042069 (2017)
8. Al-Sharadqah, A., Chernov, N.: Error analysis for circle fitting algorithms. Electron. J. Stat. **3**, 886–911 (2009)
9. Chernov, N.: Circular and Linear Regression: Fitting Circles and Lines by Least Squares. CRC Press, Boca Raton (2010)
10. Kåsa, I.: A circle fitting procedure and its error analysis. IEEE Trans. Instrum. Meas. **1**, 8–14 (1976)
11. Pratt, V.: Direct least-squares fitting of algebraic surfaces. ACM SIGGRAPH Comput. Graph. **21**(4), 145–152 (1987)
12. Taubin, G.: Estimation of planar curves, surfaces, and nonplanar space curves defined by implicit equations with applications to edge and range image segmentation. IEEE Comput. Archit. Lett. **13**(11), 1115–1138 (1991)
13. Cristian, O.N.U., Asachi, G.: Geodesy and algorithm method for determining the geometrical axes inclination of tall buildings, with circular sections. J. Geodesy Cadastre RevCAD **11**, 179–186 (2011)

Physical, Geodetic Methods and Automatic Monitoring System

Esad Vrce[1](✉) and Mirnes Bojić[2]

[1] Faculty of Civil Engineering, University of Sarajevo, Sarajevo, Bosnia and Herzegovina
esad.vrce@gf.unsa.ba
[2] BNpro doo, Sarajevo, Bosnia and Herzegovina

Abstract. Fundamental changes in the geodetic deformation analyses methodology happens in the recent years. It is desirable to use several different methods of deformation analyses to improve conventional models. The paper gives a brief overview of the current state of physical and geodetic methods as well as the latest achievements in the analysis of deformations using an automatic monitoring systems. Some of the important automatic observation systems that are popular and frequently used are described. Automatic observation systems support geodetic and non-geodetic sensors used in the determination of the deformations. The implementation of monitoring at the Salakovac dam has been shown as a positive example. All types of geodetic measurements are previously processed and analysed in the GOCA software, so the results are subsequently entered into the ARGUS software for monitoring the condition of the object. The final results represent the displacements of the points on and nearby the dam. The example shows the vertical displacements of the dam crown on the upstream side and the vertical displacements on the plateau in front of the machine room.

Keywords: Deformation analysis · Physical and geodetic methods · Automatic monitoring

1 Introduction

For the needs of monitoring large infrastructure facilities, which directly or indirectly endanger people and material goods, displacements and deformations are periodically measured by various methods. Although deformation analysis began to be applied in the 1920s, the theory of deformation analysis has become popular in the last thirty years or so. Despite the extensive literature on measurement methods and techniques, rigorous analysis becomes feasible only with the advent of modern computers capable of processing large amounts of data.

Displacement and deformation monitoring (observation or monitoring) implies a set of operations whose ultimate goal is to determine the condition of the object as a whole or in its individual parts, as well as the condition of the terrain around the object. In general, methods for monitoring displacements and deformations are divided into non-geodetic and geodetic [1–7]. Non-geodetic methods are methods of measuring local and

N. Ademović et al. (Eds.): IAT 2021, LNNS 316, pp. 532–548, 2022.
https://doi.org/10.1007/978-3-030-90055-7_42

relative displacements on an object, while geodetic methods are mainly used to measure absolute displacements. The choice of measurement methods is made depending on the type of object as well as the conditions of the object being monitored. It is often necessary to combine several different measurement methods in order to obtain the most reliable result.

As automation, digitalization and remote control are not only more popular in the world today, but also more purposeful and useful, so many geodetic institutions dealing with the issue of monitoring have started to develop a system of automatic monitoring of displacements and deformations [8–12]. The main feature and goal of the automatic monitoring system is to combine the results of several different measuring sensors (non-geodetic and geodetic), process the results, and to provide data on changes in a certain period of displacement and deformation monitoring as the final result. Also, all automatic monitoring systems strive to give the results of movement on objects in real time, and possible alarms in case of significant and critical changes in objects. This paper will describe some of the significant automatic surveillance systems that are popular and frequently used today.

2 Methods of Monitoring Displacement and Deformation

Monitoring of displacements and deformations of different types of buildings and structures is performed in order to ensure the safety of these buildings for the environment and people, then for maintenance (for safe and efficient operation) and finally for scientific research. As the range of objects on which displacement and deformation are monitored is very wide, so the range of conditions and phenomena that are monitored on these objects is also very wide and diverse. The phenomena that are monitored depending on the kind and type of objects are: climatological, hydrological, hydrogeological and seismic phenomena. The conditions that are monitored are: the state of absolute and relative (horizontal and vertical) displacements of points on the surface and inside the building, the state of the terrain around the building, the state of physical and mechanical characteristics of installed materials and the state of the building load. Measurements of all the above states and phenomena can generally be divided into two groups of methods for monitoring displacements and deformations:

- physical methods, non-geodetic methods for monitoring displacements and deformations, and
- geodetic methods for monitoring displacements and deformations.

2.1 Physical Non-geodetic Methods

The purpose of physical methods for determining displacements and deformations is to determine mainly the relative displacements of points on an object, using various specialized measuring instruments. Physical methods also have some advantages over geodetic methods. Deformations within a deformable body and relative motions of different layers of terrain or deformations within a rock mass can only be determined by physical methods. Physical methods are based on the measurement of mechanical or electrical quantities. They are applied locally, which also has advantages, because they do not require visual inspection between points. With these methods, some other parameters can be determined, using measured physical quantities, such as load, voltages, groundwater pressure, groundwater level and the like. The advantage of these methods is the high accuracy of the achieved measurement results. Of the many methods, let us mention some here.

Water Level Measurement. Since the water level in the accumulation in hydrotechnical structures and the groundwater level in hydrotechnical and other objects can have a significant impact on displacements and deformations of objects, it is necessary to measure water levels in accumulation and/or groundwater levels. According to the frequency of measurements, we distinguish [1]: continuous (constant monitoring of water levels) and discrete (occasional monitoring of water levels).

Measurement of Water Levels in Accumulation. Water level measurement in the accumulation is performed at each accumulation, regardless of the purpose and type of accumulation. Data on the measurement of the water level in the accumulation together with the data of the topographic survey of the bottom of the reservoir provide information on the available volume of water in the accumulation, as well as data on the state of sediments in the accumulation. This information is crucial both for optimal use and management of the accumulation and for determining the pressure on the dam at different water levels. Devices that can measure the water level in accumulations are: water meter bar, limnograph and ultrasonic level meter.

Measurement of Groundwater Levels. Measuring the level of groundwater is a very important element in determining the condition of an object whose displacements and deformations are monitored and can give many answers when it comes to the safety and stability of objects, especially hydraulic objects. Boreholes called piezometers and wells are used to measure groundwater levels. Observing the groundwater level in a certain area provides insight into the state of groundwater levels as well as the direction of their movement. Various manual and electronic automatic methods are used to measure groundwater levels: "whistle", manometer or pressure gauge with membrane and electronic automatic groundwater level measurement.

Meteorological Measurements. Meteorological conditions are another factor that can significantly affect the state of stability of buildings. In addition, meteorological conditions can significantly affect the geodetic methods for monitoring displacements and deformations.

Pendulums. To monitor changes in the geometric shape of the object as a deformable body, or the rotation of the object as a whole, can serve a certain reference line that passes through a fixed point of the object. It is most often used in determining the displacements and deformations of dams. For the realization of this method, it is necessary for the reference line to adopt a vertical direction. In the existence of such a line, the horizontal displacements of the dam points in relation to that line can be determined, ie the relative horizontal displacements of these points.

Coordinate Pendulum. The coordinate pendulum belongs to a group of simple but very safe methods for controlling the displacement of points on the body and crown of the dam, in relation to the foundations [2]. A special instrument, a coordinate meter, measures the relative displacements of the vertical Invar thread, in relation to the point of the instrument. The displacement is determined with respect to the two components, usually in the direction of the watercourse and perpendicular to this direction.

Inverted Pendulum. An inverted pendulum is a device that measures the relative horizontal displacement of one point of a structure relative to a point at some lower level (usually in the rock below the foundation). The vertical, which serves as a reference direction, starts from the lower point which is considered fixed. The coordinate system is established by means of a coordinate device as with an ordinary coordinate pendulum.

Inclinometer. Due to the occurrence of relative horizontal displacements of individual layers of terrain or backfilled material, there is a corresponding tilting, ie bending of the originally vertically installed flexible inclinometer tube (pipe column). The device for measuring inclination is called an inclinometer. Measurements of the inclinometer tube inclination are performed using an inclinometer probe which is hung on a cord (chain, steel ribbon or cable) and lowered into the tube and in several places, in height, measures the inclination of the probe vertically. The measurement is performed electrically with remote reading on the ground surface. The accuracy of measuring with an inclinometer device is usually around 0.001 rad (2'–10').

Clinometer. Local measurement of the angle of rotation of the structure in one vertical plane is reduced to determining the change of angle (which a certain length related to the structure at a certain place makes with the horizontal or vertical direction). A clinometer is an instrument used to accurately measure angular changes in inclination. It is based on a very sensitive spirit level, the movements of which are controlled by a micrometric screw. It is mounted on firmly built-in fixed bases, single or in the form of chains. Measurements are performed at accessible points of the dam [2]. The number and arrangement of measuring points for measuring local angular deformations depends on the size and type of dam, on the quality of the substrate and the facts observed during the research works, design, construction and during its use. The sensitivity of the measurement or the smallest scale is different for various instruments and ranges from 1" to 10".

Extensometers. Instruments for measuring dilatations are called extensometers. These are instruments that are installed at different points of the dam and measure the local deformation in the direction in which they are placed. Their advantage is that deformations can be measured by remote transmission at inaccessible points, and measurements are performed with great precision and can be centralized [2]. Concrete dilatations are measured on the surface of the dam and inside it. Instruments for measuring dilatations on the surface are placed on the outer surfaces of the finished structure, and those on the inside are placed in the concrete during the concreting of the dam. The number and location of measuring points for each construction is different. It depends on the specifics of each construction.

Deformeter. For different concrete structures, it is often necessary to measure the opening or closing of joints. The joints are measured on the surface or inside the concrete mass. The instrument used to measure the joints is a deformeter and is a modified version of the extensometer, and has a larger measuring range and lower sensitivity. The deformeter is used to measure deformations and relative displacements [2], and usually has two separate metal permanently installed bearings, between which the change in distance is measured. Measurements at inaccessible points are performed with electric instruments, while measurements at accessible points can be performed with mechanical instruments. The number of measuring points on the structure depends on the number of joints and recorded cracks during the investigation, construction and operation of the structure. The measurement sensitivity of these instruments ranges from 0.01 mm to 0.02 mm, and the measurement range is up to 10 mm.

2.2 Geodetic Methods

Geodetic methods for monitoring displacements and deformations include determining the spatial positions of control points on the structure and monitoring the change in the positions of control points over time. Checkpoints in geodetic measurement methods are placed on the body of the structure, and on the space around the structure to determine the magnitude of displacements and deformations of the structure and the terrain around it, and to determine the interaction of the structure on the ground and vice versa.

Geodetic methods for monitoring displacements and deformations are performed before the start of construction work to determine the zero condition, during construction to determine changes caused by the construction process and during exploatation and use of the structure to determine the current state of the structure. Measurements are performed periodically at certain time intervals depending on the kind and type of object, purpose, method of use, and in different conditions of use (different weather conditions, different load conditions, etc.).

Geodetic methods for monitoring displacements and deformations distinguish between those methods used to determine horizontal displacements and deformations and those for determining vertical displacements and deformations, as well as methods for determining relative and absolute displacements and deformations. Geodetic methods for tracking displacements and deformations are diverse. Some of the methods, in the geodetic professional literature (triangulation, trilateration, geometric and trigonometric leveling) are described in detail, here are just a few:

Alining is a method that determines the horizontal relative radial displacements of a structure, quite quickly and easily. The methods used in practice can be classified according to the method of establishing a reference line on [3]: mechanical methods (by which a stretched wire establishes a reference line), optical method (by which sight or laser beam indicates a reference line) and diffraction method where the reference line is created by projecting a pattern of diffraction slits.

In order to observe the relative horizontal displacements by the method of alining, in agreement with the designer and the user of the structure, the points at which the measurement is performed are determined. The arrangement of points should be where the measurement with the coordinate height is envisaged (non-geodetic method). These two measurements serve each other as controls. Immediately after the completion of construction works on the structure, it is necessary to perform zero measurement (no load) by the method of alining. During the test load of the structure, and during sudden changes in load, it is necessary to perform more frequent measurements. The alining method is also used during regular exploitation.

GNSS/GPS (Global Navigation Satellite System/Global Positioning System). Many institutes and research centers have developed systems for monitoring the displacement and deformation of bridges (dams and other structures) based on GNSS technology [4, 5]. Today, two basic architectures of monitoring systems can be recognized through GNSS technology. The first is based on a network of fixed sensors and the second uses mobile sensors Most conventional monitoring systems use a network of fixed sensors that send data to a single control center where data is processed.

This architecture is completely conducive to GNSS. Each sensor knot consists of a GNSS receiver, a microcontroller and a radio. The receiver monitors the satellites, collects the data and sends it by radio to the central data processing unit. The microcontroller controls the operation of the receiver and the radio connection. The control center contains a radio computer with an appropriate program for controlling the operation of the system, data processing, communication, management, data verification and analysis of displacements and deformations. When it comes to the system that is installed on the object, the expected accuracy of calculating individual positions is better than 1 cm. With long-term monitoring of displacements and deformations, an accuracy of up to 1 mm can be achieved [4, 5].

Photogrammetric methods have so far been considered insufficiently accurate to determine displacement and deformation. Photogrammetry methods, ie digital photogrammetry, are a suitable tool for automatic measurement of positional (horizontal and spatial) displacements and deformations, such as cracks [6]. Given the relatively low cost of equipment for this type of measurement, and the processing speed and overall economy, this has become a very interesting method for monitoring displacements and deformations on large structures.

Digital photogrammetry techniques present a fast, economical and versatile tool for spatial deformation measurements, and the potential accuracy (2–3 mm) obtained by applying new techniques and new photographic equipment simply imposes this solution as acceptable, especially when a large number of points must be measured (for local analysis of shearing cracks, fissures, dilatations, etc.). Classical photogrammetric methods of terrestrial photogrammetry and aerial photogrammetry enable the detection of changes in the horizontal and spatial sense, they can also represent permanent documentation on the condition of the observed object on the day of photogrammetric measurements.

Laser Scanning. Terrestrial Laser Scanning (TLS) provides great potential in displacement and deformation projects. In the case of significant buildings, information can be collected on remote points of the facility and places with limited accessibility [7]. TLS can directly provide high-density spatial data of the entire area, not only at discrete points, as well as additional information. Although the accuracy of laser scanner measurement at a certain point is less than the accuracy of classical geodetic methods (triangulation, trilateration and leveling), laser scanning received significant attention mainly due to the high spatial resolution of the obtained data.

In order to obtain measurements that meet the accuracy requirements for use in monitoring displacements and deformations, it is necessary to perform a quality and reliable calibration of measuring equipment, which is a complicated and demanding procedure, which is not fully standardized, causing an additional difficulty.

With the laser scanning method, the simplest way to monitor deformations is to use measuring points that are signaled. A large number of stamps are placed on the object and by repeating the scan in each epoch, we can get shifts by subsequent processing. The advantage of laser scanning in tracking displacements and deformations is that it is possible to measure hard-to-reach or completely inaccessible places.

3 Automatic Monitoring

Given that today's society, in every respect, strives for digitalization and automation, there is no difference when it comes to monitoring displacements and deformations. World-renowned experts in this field strive to integrate multiple sensors, and automate the entire displacement and deformation monitoring system to ensure that any adverse conditions that may occur during the operation of the facility (especially during the test load) are detected as soon as possible.

Automation of displacement and deformation monitoring systems implies integration and automation of measurements with all available non-geodetic and geodetic systems, automation of displacement and deformation detection systems and establishment of an early notification system (alarm) about possible displacements and deformations. Where it is not possible to place geodetic instruments, then non-geodetic sensors are placed which are then combined. Non-geodetic sensors in principle determine the relative displacements, and give us other important information significant for the interpretation of geodetic results. A very interesting system is a combination of GNSS and pseudolith, as well as locata sensors that serve as a supplement to GNSS sensors.

In any case, in order to establish a system for automatic monitoring, a serious and professional approach to the analysis of the facility itself is needed. Then there is the economic aspect, which is often a limiting factor in the establishment of such systems. The advantages of an automatic monitoring system are manifold. The most significant advantage is the availability of real-time data and automatic warning, which due to the safety of people living in the area near the monitored facility is the most important aspect. Automatic monitoring systems have already been developed worldwide. Let us mention some, in alphabetical order.

3.1 GOCA

Multi-sensor system GOCA (GNSS/LPS/LS Online Control and Alarm System) real-time application of GNSS/GPS sensors, LPS - ground sensors (total stations, levellers and hydrostatic leveling instruments) and LS - local sensors (coordinate peaks, inclinometers, clinometers, extensometers, deformeters, etc.) for monitoring and analysis of deformations. A system designed to monitor natural hazards, geotechnical structures, building structures, buildings and dams [8].

The multi-sensor system GOCA, consists of numerous global and local sensors, which provide the vector of the state of displacement, velocity and acceleration of points on the object (Fig. 1). In addition, local sensors (pendants, inclinometers, clinometers, deformeters, extensometers, etc.) are used in integral deformation analysis, and it is possible to estimate additional physical parameters of objects.

The concept of deformation analysis implemented in the GOCA software package is a classic deformation analysis, which was developed to work in a fully automatic online mode. Straightening and deformation analysis in the GOCA system are designed in three steps:

Step one, initialization, equalization, is based on the theory of least squares, free adjustment of GPS/GNSS baseline networks and LPS data in relation to the user-defined initial, zero measurement era.

Step two, is based on the above GPS/GNSS and LPS data and works completely automatically during online monitoring: constrained adjustment of GPS/GNSS baselines and LPS data (lengths, zenith distances, directions, altitude differences), with constrained reference points. Any measurement epoch can be processed in post-processing mode.

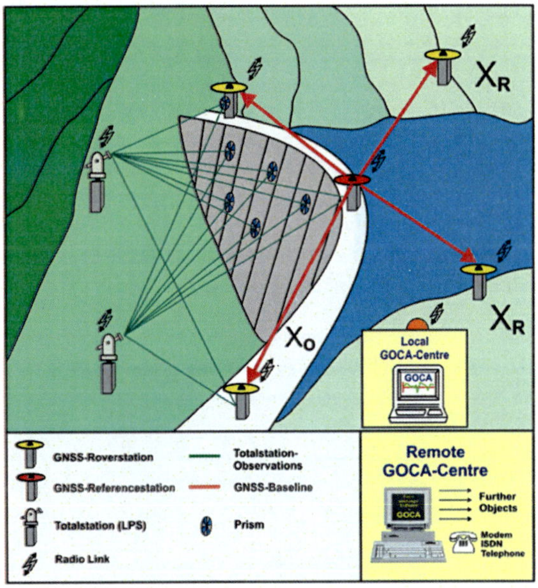

Fig. 1. GOCA system sensors [8]

Step three, the deformation analysis itself, implemented in the software, deals with the estimation of the parameters of various deformation functions. Parameter estimation refers to the results of the second step and there are different parameter estimation methods that the user of the GOCA software can choose for himself. In online mode, there are several different methods for estimating and defining deformation functions, including Kalman filtering, based on which displacements, velocities, and accelerations of points on an object are estimated.

Deformation functions and parameters can be limited to critical values or to significant changes, so that an automatic alarm can sound when certain parameters exceed the limit (critical) values. Alarm notification methods can be various: e-mail, SMS, fax, etc. A special feature of GOCA deformation analysis software consists in the automatic procedure of statistically strict testing of the congruence of horizontal and vertical displacement components.

The GOCA system has been successfully implemented in numerous monitoring projects of dams, landslides, bridges, tunnels, mines, etc. In addition to the GOCA system, the same faculty has developed the MONIKA (Deformation integrity monitoring for GNSS-positioning services by the Karlsruhe approach) monitoring system using GNSS networks (Fig. 2).

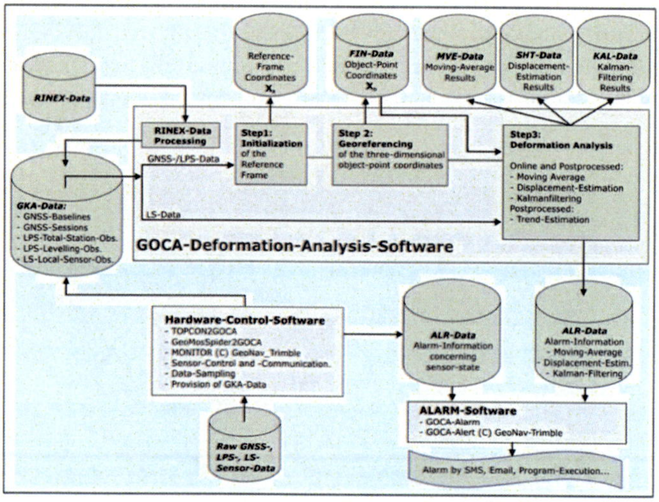

Fig. 2. GOCA system working principle [8]

3.2 Leica GeoMoS

Leica GeoMoS (Leica Geodetic Monitoring System) is an automatic deformation monitoring system developed by Leica. Leica GeoMoS is a solution that provides the necessary flexibility to deformation monitoring projects and that adapts to the specific needs of each user [9]. It is a suitable solution for continuous or periodic monitoring projects and offers a complete solution with appropriate accuracy and reliability.

Leica GeoMoS supports connection to any monitoring sensor: geodetic, geotechnical, meteorological, environmental or any other data collection. Accurate and reliable data collection is ensured through data validation, filtering and automatic re-measurement. There is a possibility of activating and controlling the limit or critical values of parameters. Responsible persons can be informed in a timely manner about exceedances of the limit values using various messaging options (e-mail, SMS, alert). Tracking technology has also been developed that provides photo-based information for documentation, review and detection from a remote location. Using different modules, the results can be accessed anytime and anywhere (Fig. 3).

Fig. 3. Leica GeoMoS system working principle [9]

3.3 Trimble 4D Control

Trimble 4D Control software is a fully customizable displacement and deformation monitoring solution developed by Trimble. The options of this software range from simple post-processing tracking operations, to complete real-time tracking systems that combine optical instruments, GNSS receivers and geotechnical sensors [10].

One of the key elements of the system is the Trimble 4D control software. Trimble 4D Control system supports and enables the integration of GNSS, LPS and other various sensors (seismographic, meteorological, geotechnical, etc.). It is possible to continuously monitor the facilities 24 h a day. Like all more advanced software, this software has an alarm system that warns of possible point movements. Trimble has developed its own sensors for installation on monitored objects. One such sensor is the Trimble Kestrel System seismic sensor, Model SG160-09, which represents the integration of a GNSS receiver (200 Hz) and a triple axis accelerometer.

Real-time monitoring involves software that manages total stations and/or GNSS receivers, and collects and stores data from all sensors, performs displacement analysis, and generates reports and alerts. The monitoring module provides the data needed to determine the speed, direction and magnitude of the displacement. Trimble 4D Control provides detailed data estimation and detects moving points, and also has developed options for detecting random or systematic measurement errors.

Trimble 4D Control supports a variety of functions to provide a wide range of options for different monitoring requirements. All the options and capabilities of this system are combined into various specific modules for specific monitoring operations such as instrument controls, data collection, data processing and data analysis.

Within this software, it is possible to manually add measurements for certain sensors or instruments, if the need arises. Within this system of automatic monitoring, a methodology for alarming and notifying about cyclical as well as about sudden and extraordinary phenomena has been developed (Fig. 4).

3.4 GIMS

GIMS (Geodetic Integrated Monitoring System) is a commercial low cost system based on EGNSS (European GNSS), Copernicus SAR (Synthetic Aperture Radar) and other in-situ sensors, such as inertial units of measurement, for the purpose of determining displacements and deformations with a focus on landslides and subsidence of the terrain. The system can determine displacements and deformations with great precision (of the

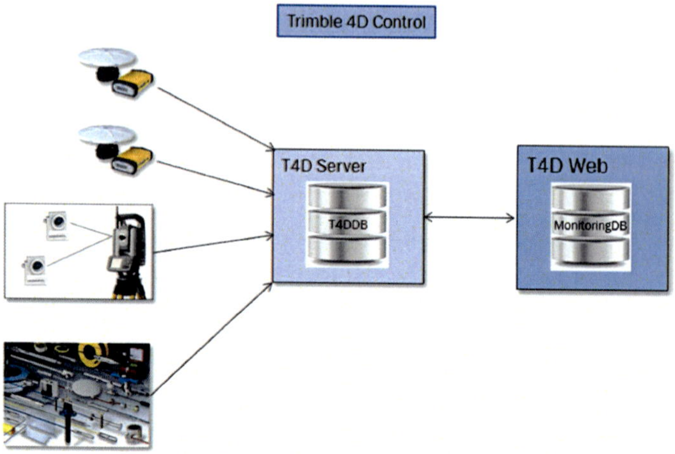

Fig. 4. Trimble 4D Control system working principle [10]

order of millimeters) even on a daily basis. The integration of accelerometers on site, in the event of sudden movements, can provide real-time warning [11].

Three different measurement techniques (EGNSS, SAR and accelerometers are complementary in time and space) can be integrated to gain a better understanding of the processes being monitored and to complete the knowledge of the deformation phenomenon. The ultimate goal of the GIMS project is to have detailed and timely knowledge about the geophysical behavior of parts of the Earth's surface and its impact on buildings, in order to reduce the impact on the population and material goods, and to plan better their maintenance.

3.5 Low-Cost Monitoring System

In addition to the above systems, there are a number of other systems developed within various scientific and professional projects. In addition, it should be noted that today a lot of work is being done on the development of the so-called low-cost (low-budget) deformation monitoring systems. The company Yetitmoves from Italy has thus developed a system for monitoring facilities (landslides, bridges, dams,...) that uses GNSS L1 sensors and which can compete with professional instruments currently on the market in terms of reliability and efficiency [12].

Such sensors are especially suitable for objects that are threatened with collapse, because the classical approach to measuring such objects would cause great risks for people, and on the other hand, the use of expensive geodetic systems is not economically justified.

4 Implementation of the Monitoring System at the Salakovac Dam

Technical monitoring is actively carried out at dams located under the management of the Hydroelectric Power Plant on the Neretva river, ie under the management of the Public

Company Elektroprivreda Bosne i Hercegovine (Electric Power Industry of Bosnia and Herzegovina). Jablanica, Grabovica and Salakovac dams are a very good example of automated dam monitoring.

The dam of HPP (Hydropower plant) Salakovac was built in 1982. It is a concrete gravity dam. The dam consists of 17 blocks. The basic technical characteristics of the dam are [13]:

- Dam type	gravity
- Construction height	70.00 m
- Geodetic height	52.00 m
- Length in the crown of the dam	230.50 m
- Elevation of the dam crown	127.00 m
- Elevation of max. accumulation levels	124.70 m
- Elevation of normal working level	123.00 m

The observation of natural and artificial structures of the area and the dam of HPP Salakovac since its construction was based on two technical observation projects. The modernization project for technical monitoring of dams of HPP Jablanica, HPP Grabovica and HPP Salakovac, which was made in 2009, envisaged the reconstruction, modernization and automation of the observation system on these dams.

The realization of this project, ie the modernization of the observation system was performed in the period from 2010 to 2012. Since 2012, the observation of displacements and deformations has been performed within the reconstructed and modernized system of technical monitoring, and according to the Rulebook and Program of technical monitoring of dams and terrains in the area of dams and accumulation by physical and geodetic methods.

Physical methods in the monitoring system at the HPP Salakovac are (some of the methods) [13]:

- water level in the reservoir (2 measuring points),
- bottom water level in drainage channels (1 measuring point),
- air temperatures in the dam area and in the dam galleries (5 measuring points),
- water temperatures in the accumulation (3 measuring points),
- precipitation in the dam area (1 measuring point),
- concrete temperature of the dam (33 measuring points),
- buoys in the dam (19 measuring points),
- flows in the dam (4 measuring points),
- groundwater level in the dam area (11 measuring points),
- dam displacement measured by coordinate peaks and coordinate meters (14 places),
- dam subsidence measured by hydrostatic leveling in the control gallery (12 places),
- rotations of measuring points in the vertical plane of the dam body by the clinometer method (36 places),
- operation of block joints measured by deformeters and dilatometers (76 places),
- stability of the dam sides measured by the method of rock extensometers (19 places),
- seismic observation (3 places) (Fig. 5).

Fig. 5. Appearance of HPP Salakovac dam, profile points [13]

Depending on the type and method of measurement, each of these methods has its own dynamics of observation and measurement. Namely, some measurements are performed continuously, and some periodically. Also, some measurements are done automatically and some manually. Regardless of the dynamics of the observation and whether the measurements are performed automatically or manually, all measurements are entered into the ARGUS (Audit Record Generation and Utilization System) object monitoring software. Those measurements that are performed automatically, they are automatically downloaded to the specified software, while manual measurements are manually entered into the ARGUS software and the condition of the dam is monitored.

Geodetic methods also play a significant role in the observation system of the Salakovac HPP dam, where three measurement methods were used, namely [13]:

- alining method (8 measuring points),
- triangulation and trilateration - microtriangulation network, (Fig. 6, 13 columns and 16 profile points), and
- geometric leveling (Fig. 7, 142 measuring points).

Depending on the position of the measuring points and depending on the type of measurement method, the dynamics of measurement at individual measuring points is also different. Namely, measurements at some measuring points are performed periodically twice a year, while measurements at some measuring points are performed periodically every month. Regardless of the dynamics of observations, all measurements by geodetic methods are entered into the software for monitoring the condition of the ARGUS object, but are previously processed and analyzed in the GOCA software.

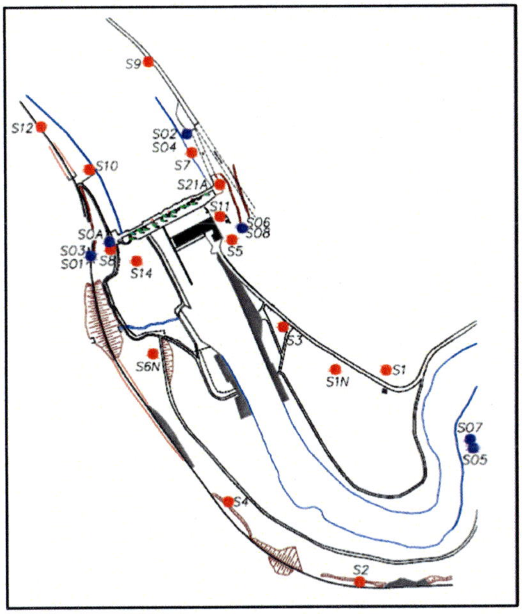

Fig. 6. Microtriangulation network [13]

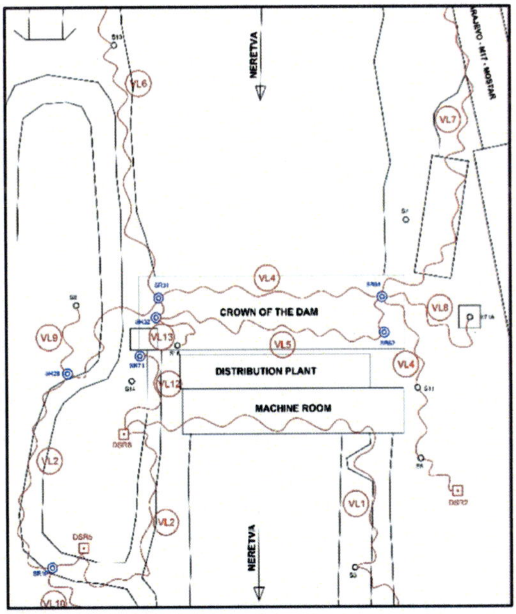

Fig. 7. Leveling network [13]

5 Results and Discussion

System of technical monitoring of HPP Salakovac dam, before the modernization and automatization was very scarce, partially abandoned and in some measure outdated.

As it can be visible from the above mentioned, the approach used during modernization and automatization process is multisensory automatic and manual measurement and monitoring of the dam state, integrated through the system of automatic monitoring. System of automatic monitoring is realized using ARGUS software, as well as some additional software solutions (Fig. 8 and 9).

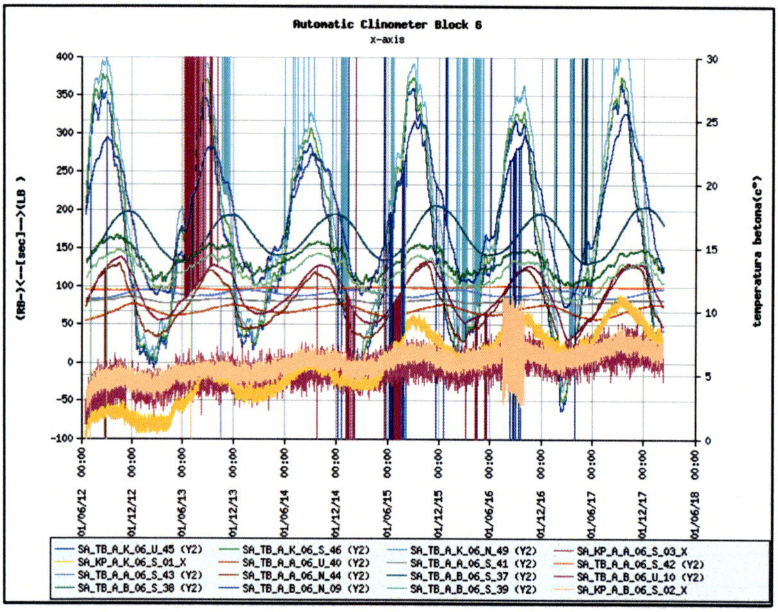

Fig. 8. Example report [13]

The biggest advantage of the established systems, in relation to the previous one, is incomparable larger number of sensors monitoring the dam state. Besides, the largest number of installed sensors automatically conducts measurements and delivers them to central monitoring system, when the measurements are integrated with the data of rest of the sensors. Also, large number of sensors and methods serve as a control to each other, which ensures better reliability of the monitoring system.

At the end, after integrating all the measurement methods, physical and geodetic, it is possible to inspect the measurement results, measurement history, and to make wide spectre of reports and graphical views in ARGUS software, all in order to gain a better insight in dam state. System of automatic monitoring, in case of exceeding previously defined limit values, automatically sends a warning on measurement sites, and alarms the system controllers.

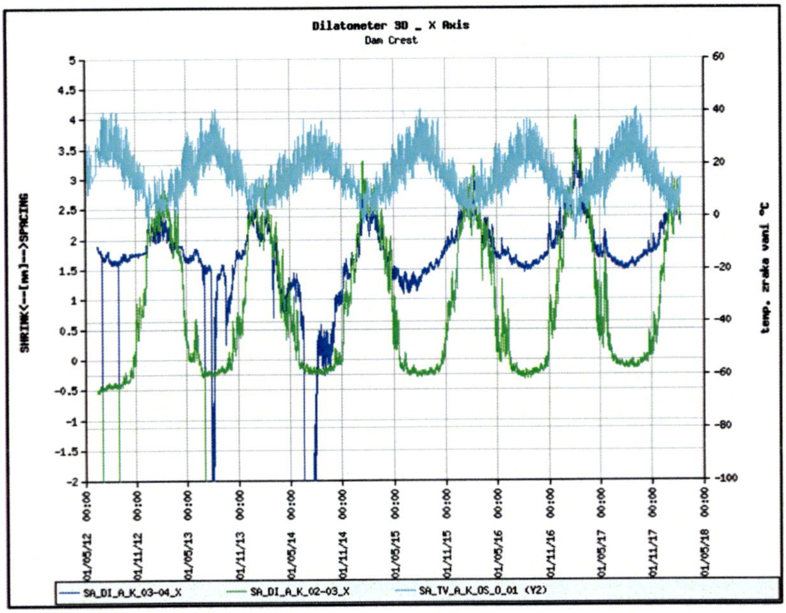

Fig. 9. Example report [13]

6 Conclusion

The most important requirement during construction as well as during the exploitation of capital construction and geotechnical facilities is safety. The safety of these facilities is reflected in their stability, which is periodically tested and determined by various methods of technical observation.

Deformation determination methods can be divided into non-geodetic and geodetic methods. Non-geodetic measurements are mainly relative and local measurements of displacements and changes (rotation and motion) of an object or terrain around an object. Geodetic measurement methods are absolute methods of measuring displacements and changes in the building and the terrain around the building.

When monitoring displacements and deformations, it is desirable to combine several measurement methods, all in order to obtain as many indicators of the behavior of the object and the terrain around the object. It is also desirable and somewhat necessary to bring different measurement methods to overlap in order for one to serve the other as a control.

Multisensor monitoring, with the application of telecommunication technologies and automation of measurement processes, sending and storing measurement results, as well as the development of a system solution that will automatically process measurement results from all sensors have resulted in automatic monitoring systems.

Systems for automatic tracking of displacements and deformations today operate on the principle of giving results in real or near real time via the web remotely, which allows insight into the state of the object anytime and anywhere.

References

1. Gjetvaj, G.: Eksperimentalna hidraulika. University of Zagreb, Faculty of civil engineering, Zagreb (2013). https://www.grad.unizg.hr/predmet/ekshid. Accessed 01 June 2020
2. Tehnička enciklopedija: Brane - brane, osmatranje (2020). https://tehnika.lzmk.hr/tehnickaenciklopedija/brane_osmatranje.pdf. Accessed 01 June 2020
3. Webarchiv: Deformation Measurement and Alignment Instrumentation (2006). http://webarchiv.ethz.ch/geometh-data/student/eg1/2006/02/Monitoring%20instruments.pdf. Accessed 01 June 2020
4. Roberts, G.W., Brown, C.J., Tang, X., Ogundipe, O.: Using satellites to monitor Severn Bridge structure. In: Proceedings of the Institution of Civil Engineers – Bridge Engineering (2015). https://doi.org/10.1680/bren.14.00008. Accessed 01 June 2020
5. Hyzak, M., Leach, M., Duff, K.: Practical application of GPS to bridge deformation monitoring. In: Proceedings of the 64th Permanent Committee Meeting and Symposium of International Federation of Surveyors (FIG). Washington, DC (2008)
6. Hampel, U., Maas, H.-G.: Application of digital photogrammetry for measuring deformation and cracks during load tests in civil engineering material testing. Institute of Photogrammetry and Remote Sensing, Dresden (2003). https://tu-dresden.de/bu/umwelt/geo/ipf/photogrammetrie/ressourcen/dateien/forschung/publikationen/pubdocs/2003/2003_Hampel_Maas_Opt3D2003.pdf?lang=en. Accessed 01 June 2020
7. Lovas, T.: Terrestrial Laserscanning in Deformation Measurements of Structures. International Archives of Photogrammetry and Remote Sensing, vol. 37 (2008). https://www.researchgate.net/publication/228910661_Terrestrial_Laserscanning_in_Deformation_Measurements_of_Structures. Accessed 01 Jun 2020
8. Karlsruhe, H.: GOCA - Research Project (2020). http://goca.info. Accessed 01 June 2020
9. Leica-geosystems: Leica GeoMoS Monitoring Solution (2020). https://leica-geosystems.com/products/total-stations/software/leica-geomos. Accessed 21 Sept 2020
10. Trimble Survey Division: Trimble Monitoring (2020). https://www.trimble.com/infrastructure/trimble-4d-control.aspx. Accessed: 07 Sept 2020
11. European Global Navigation SatelliteSystems Agency: GIMS Geodetic Integrated Monitoring System (2020). https://www.gsa.europa.eu/geodetic-integrated-monitoring-system. Accessed 29 June 2020
12. Yetitmoves: Leica GeoMoS Monitoring Solution (2020). http://www.yetitmoves.it. Accessed 21 Sept 2020
13. JP Elekreoprivreda dd Sarajevo: Elaborat o tehničkom osmatranju brane HE Salakovac geodetskim metodama. Subsidiaries HPPs on Neretva. Jablanica (2018)

Metre Convention Is a Root International System of Units

Džanina Omićević[(✉)], Dževad Krdžalić, and Esad Vrce

Faculty of Civil Engineering, Department of Geodesy, University of Sarajevo, Sarajevo, Bosnia and Herzegovina
dzanina.omicevic@gf.unsa.ba

Abstract. Metrology, the science of measurement, is part of the essential but largely hidden infrastructure of the modern world. We need it for high-technology manufacturing, human health and safety, the protection of the environment, global climate studies, information transfer, and the basic science that underpins all these. Nowadays, the need for highly accurate measurements exceeds the fields of physical sciences and engineering. The International System of Units, the SI (Système International d'unités), provides the internationally agreed means by which we make such measurements. At the meeting of the General Conference on Weights and Measures (CGPM), held in Paris on November 16, 2018 a new and revised SI, adapted to the 21st century, was approved. It has been 145 years since the Convention du Meter was formally signed for 17 countries on May 20, 1875. This treaty was concluded with the oldest international treaties that are still in force today, and form the basis of today's metrology and, above all, a globally coherent system of measures. Every year on May 20 is the anniversary of the Meter Convention, it marks World Measurement Day and is used as an opportunity to inform the public about the key roles of measurement and consistency of measurement in the modern world.

Keywords: Meter Convention · Metrology · International system of units - SI · Fundamental natural constants

1 Introduction

In 1875 in Paris, representatives of seventeen countries signed the Metre Convention: an agreement designed to standardize measurements. The International Bureau of Weights and Measures (BIPM) was created and formalized as part of this agreement. The primary role of BIPM is to standardize and coordinate systems of units around the world, and the importance of BIPM in the world of science and global trade has grown exponentially in the last 145 years. Every year on May 20, the anniversary of the Meter Convention is marked, declared World Metrology Day and used as an opportunity to inform the public about the key roles of measurement and measurement consistency in the modern world. The theme for World Metrology Day 2021 is "Measurement for Health". This theme was chosen to create awareness of the important role measurement plays in health, and thus in the wellbeing of every one of us.

N. Ademović et al. (Eds.): IAT 2021, LNNS 316, pp. 549–557, 2022.
https://doi.org/10.1007/978-3-030-90055-7_43

Fig. 1. Evolution of the systems of units of measurement [1]

Thanks to a global system of universally understandable measurements and international standards certified by competent organizations, consumers have access to everything from computer parts to efficient pharmaceutical products. World Metrology Day is an opportunity to proclaim the importance of measurement in everyday life and recognize the rigorous and hard work of surveyors around the world to ensure people's safety, facilitate the quality of consumer clothing and support global trade.

The way in which basic units are determined has changed greatly over time. Figure 1 shows the evolution of the system of units of measurement through the history of human civilization. Through constant progress and development, a practical system of units was created that corresponded to all member states of the Meter Convention, today is known as the International System (SI), which was adopted and defined in 1960 at the 11th General Conference on Weights and Measures (CGPM).

The International System of Units is not static; on the contrary, it evolves according to the growing demands of measurement requirements and the field of knowledge and reflects the best measuring practices of each period of history [1].

2 A Brief History of Metrology

Units of measurement have been essential in every civilization throughout history. Since the dawn of humanity, man has necessarily devised various basic measures to help himself in daily life. We assume that during the Paleolithic period (around 2,500,000 to around 9000 BC), homo sapiens was able to survive the seasons. Early weights were simple stones, an empty shell was used to measure capacity, and body parts, such as finger length and width, elbow and body span, were suitable for most length measurements. Time was probably measured by periods of the sun, moon, and other heavenly bodies [2].

The need for standardized units of measurement increased significantly as companies became involved in broad trade. It soon became necessary to pass laws regulating transactions and the accuracy of standards, as the need to combat fraud increased. Subsequently, the establishment of official weights and measures became an essential element in ensuring fair trade and, no less important, in ensuring payment to the ruling class of the right amounts of taxes [2].

The death penalty faced those who forgot or neglected their duty to calibrate the standard unit of length at each full moon. Such was the peril courted by the royal site architects responsible for building the temples and pyramids of the Pharaohs in ancient Egypt, 3000 years BC. The first royal cubit was defined as the length of the forearm from the elbow to the tip of the extended middle finger of the ruling Pharaoh, plus the width of his hand. The original measurement was transferred to and carved in black granite. The workers at the building sites were given copies in granite or wood and it was the responsibility of the architects to maintain them [3].

Unfortunately, it must be pointed out that there are very few preserved artifacts that testify to that time. Probably the fire, which occurred in 642, in the legendary library in Alexandria, the cosmopolitan center of science at the time, irreversibly took away the facts and knowledge gathered by thinkers such as Aristotle, Plato, Archimedes, Eratosthenes, Heron, and Claudio Ptolemy who based their knowledge on the knowledge and experiences of mathematics/geometry, mapping, and geodetic measurements of Ancient Egypt [4].

The European measuring system in the Middle Ages, until the adoption of the meter system, was based on Greek and Roman measuring systems that had anthropological units (finger, fist, pedal, etc.). Hollow measures (objects) were used for the preparation, which were simultaneously filled with water and served as the basis of various weights. That is why until today, units of measurement have been called "measures and weights" (kilograms and weight).

The main evolutions in the field of metrology during the period from the 17th to the 21st century was the development of scientific discoveries, the intensification of exchange and the establishment of international institutions. Table 1 provides an overview of these events summarized in 14 steps.

The South Slavs brought the natural system of the unit in the inhabited areas, and supplemented or replaced it with the units of the peoples with whom they came into contact. Anthropological units of hair, thumb, pedal, elbow, step, blade, grip, etc. are used for length. Anthropological or natural units are used for "quantity": wagons, jugs, eyes, etc. These units changed over time and most disappeared from use only after World

War II, but land units, due to the inertia of the cadaster, to some extent remained today [6].

Table 1. An overview of important events in the field of metrology [5]

Date	Event
1670	Proposition for a new length unit based on the terrestrial meridian
1799	Creation of the decimal metric system. Two platinum standards, representing the meter and the kilogram manufactured
1832	Austrian mathematician Gauss strongly promotes the application of the metric system, together with the second defined in astronomy, as a coherent system of units for the physical sciences; first measurements of the Earth's magnetic field take place
1860	Maxwell and Thomson formulate the requirement for a coherent system of units with base units and derived units
1880	Approval by International Electrotechnical Commission (IEC) of a mutually coherent set of practical units. Among them were the Ohm for electrical resistance, the Volt for electromotive force, and the Ampere for electric current
1875	Signing of the Metre Convention, which created the BIPM, established the CGPM and the CIPM, and adopted the MKS system
1889	The first conference of CGPM takes place
1901	The so-called rationalized proposal of Giorgi, for a single coherent four-dimensional system, by adding to the three base units a fourth unit, of an electrical nature such as the Ampere or the Ohm, and rewriting the equations occurring in electromagnetism
1939	Adoption of a four-dimensional system based on the meter, kilogram, second, and Ampere, and the MKSA system, a proposal approved by the CIPM in 1946
1954	Introduction of the Ampere, the Kelvin, and the Candela as base units, respectively, for electric current, thermodynamic temperature, and luminous intensity
1960	The name International System of Units, with the abbreviation SI, is given to the system
1971	Introduction of the last SI base unit: the mole, as the base unit for amount of substance, bringing the total number of base units to seven
1999	Signature of the CIPM-MRA (Mutual Recognition Agreement), for international recognition of national measurement standards
2018	New definition adopted concerning four base units on 7

3 International Institution for Metrology

Today's context of metrology is reflected in a set of closely related international institutions:

- General Conference for Weights and Measures (CGPM),
- International Committee for Weights and Measures (CIPM); and
- Bureau International for Weights and Measures (BIPM).

These institutions have received authority to act in matters of world metrology from the Convention of the Metre in Fig. 2 (a diplomatic treaty between 51 nations initially, but today approved by almost all nations).

This particularly concerns the demand for measurement standards of ever increasing accuracy, range, diversity, and the need to demonstrate equivalence between national measurement standards.

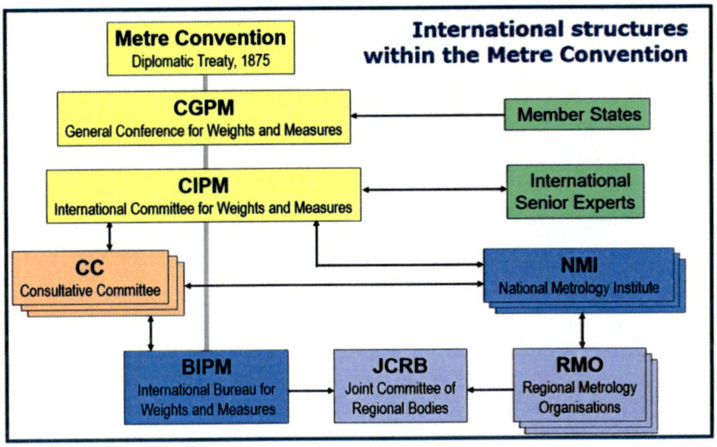

Fig. 2. International structure within the Metre Convention [7]

The National Metrology Institutes (NMIs), such as IMBIH (Institute of Metrology of Bosnia and Herzegovina) constitute the local relays of the international institutions. The global organization is completed by World Regional Institutes (RMOs), according to the map in Fig. 3.

The regional institutes are less known than NMIs; however, their role is important. They have responsibilities:

- To facilitate traceability to primary realisations of the SI;
- To coordinate comparisons of national measurement standards;
- To make mutual reviews of technical competencies and quality systems;
- To cooperate in metrology research and development;
- To operate joint training and consultation; and
- To share technical capabilities and facilities.

Examples of RMOs are:

- EURAMET (European Association of National Metrology Institute);

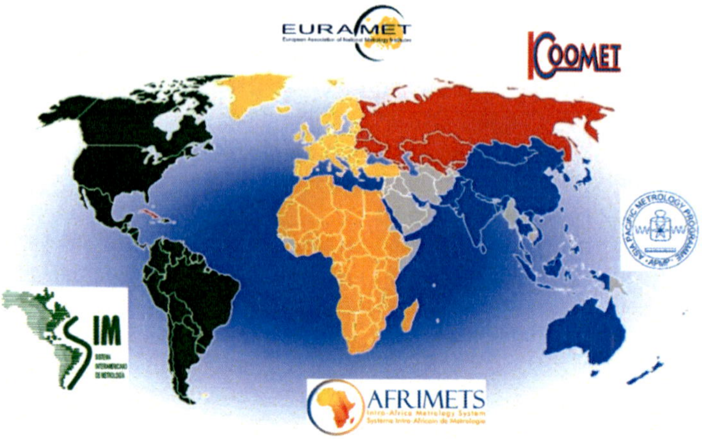

Fig. 3. World Regional Institutes (RMOs) [5]

- AFRIMET (Intra-Africa Metrology System supported by the Technical Cooperation of Physikalisch-Technische Bundesanstalt (PTB));
- COOMET (Euro-Asian Cooperation of National Metrological Institutions);
- APMP (Asia Pacific Metrology Programme)

IMBIH is an associate member:

- BIPM (International Bureau of Weights and Measures),
- OIML (International Organization of Legal Metrology),
- WELMEC (European Cooperation in Legal Metrology),
- COOMET (Euro-Asian Cooperation of National Metrological Institutions)
- GULFMET (Gulf Association for Metrology), and
- SMIIC (The Standards and Metrology Institute for Islamic Countries).

Among other tasks, the international institutions have issued a set of fundamental documents: the VIM (International Metrology Vocabulary) [8], the GUM (Guide to the expression of Uncertainty in Measurement) [9], and defined the rules for mutual recognition between NMIS, for national measurement standards, and for calibration and measurement certificates (CIPM Mutual Recognition Arrangement 1999).

4 Redefinition of the International System of Units - SI

The ever-increasing achievements of science and technology have been made possible by the application of the science of measurement. The average civilized citizen is unaware that a large number of precise measurements are needed to make everyday life possible.

The desire to increase accuracy has just led to another of the greatest scientific successes of recent years - the redefinition of units in the International System - SI. In November 2018, the General Conference on Weights and Measures (CGPM) approved the upgrade of the International System of Units, i.e., redefinition in the SI system in Fig. 4, which is an internationally agreed basis for expressing measurements at all levels of accuracy and in all areas of science, technology and human action. The four units used to measure mass (kilogram), electric current (Ampere), temperature (Kelvin) and substance quantity (moll) has been redefined with the help of fundamental natural constants (Planck constant - h, elemental charge - e, Boltzmann constant - k, Avogadro's number - NA).

The revision of the SI system establishes a set of seven fundamental natural constants, which will be known as the "defining constants of the SI system". By defining and expressing units of measurement through fundamental natural constants, they become measurable and repeatable anywhere and at any time, which is the basis of metrology.

Thus, Max Planck's vision of 120 years ago [10], became a reality: "with the help of fundamental constants we have the possibility of establishing units of length, time, mass, and temperature, which necessarily retain their significance for all cultures, even unearthly and nonhuman ones."

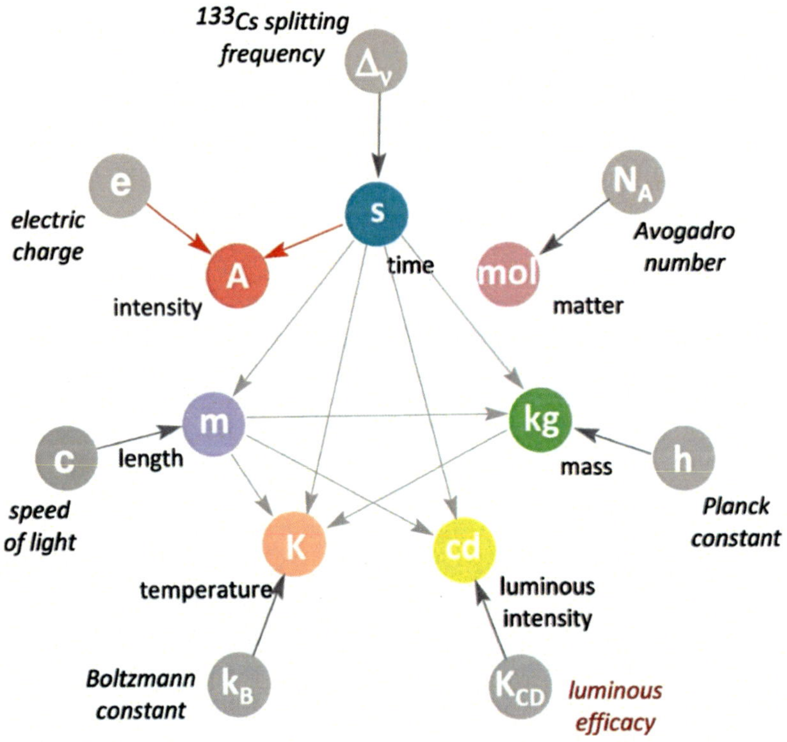

Fig. 4. New redefining SI with the help of seven fundamental natural constants [5]

The numerical values of the seven defining constants have zero uncertainty. They are summarized in Table 2. Definitions based on defining constants are called explicit-constant definitions; they are based on the fundamental constants of nature, in contrast to explicit-unit definitions which are based on particular experimental procedures.

Table 2. The seven defining constants of the SI, and the seven corresponding symbols, numerical values, and units

Defining constant	Symbol	Numerical value	Unit
Hyperfine transition frequency of cesium	$\Delta\nu_{Cs}$	9 192 631 770	Hz
Speed of light in vacuum	c	299 792 458	$m\ s^{-1}$
Planck constant	h	$6.626\ 070\ 15 \times 10^{-34}$	J s
Elementary charge	e	$1.602\ 176\ 634 \times 10^{-19}$	C
Boltzmann constant	k	$1.380\ 649 \times 10^{-23}$	$J\ K^{-1}$
Avogadro constant	N_A	$6.022\ 140\ 76 \times 10^{23}$	mol^{-1}
Luminous efficacy	K_{cd}	683	$lm\ W^{-1}$

5 Conclusion

People have had to go through a significant number of obstacles throughout history to reach the first internationally recognized units of measurement. In parallel with the development of technology, there was a need for a unique measuring system to be used around the world. By signing the Meter Convention 145 years ago, it enabled the use of a single measuring system. Before the introduction of a single measurement system, chaos reigned, each of the states had its own measurement system and did not want to give it up. It is only with the introduction of the metric system that, over time, the common language is accepted and the advantages of such a practical system are realized. The international system of units is not static and has evolved since its establishment.

As the world evolved, so did the need for increasingly universal and more accurate standards of basic physical quantities. Therefore, the units of measurement must be redefined from time to time. Units of measurement have an unbreakable connection with natural constants. This connection has always existed and is especially emphasized in the latest change in the international system of units of measurement. With the last change of the international system of units of measurement, the last standard, the diameter of kilograms, was abolished. The decision to abolish the weight kilogram was made at the 26th session of the International Bureau of Weights and Measures, held in Paris on November 16, 2018. The decision came into force on May 20, 2019. Following this decision, whole system of units of measurement is based on seven fixed values of natural constants. Units of measurement that were basic and had their own definitions, based on standards, now have new definitions based on the fixed values of the seven natural constants.

Although the development of metrology began as a support to people's economic activities, in the first place exchanges and trade, the level of metrological performance achieved far exceeds the real needs in this field; further progress is now being sought to facilitate scientific progress. Today, an even higher level of precision is needed for certain domains of science such as space, astrophysics, medical care, etc. all of these domains actually contribute to people's well-being.

It is encouraging to note that metrology is one of the few areas of efficient and sincere cooperation between world nations. Almost all nations recognize the metric system (with the remaining exception of the United States); more than 100 laboratories around the world have signed a document called the CIPM-MRA agreement.

In the end, it should be said that there has always been resistance to change. It is very difficult to understand, comprehend and accept new discoveries, to give up everyday way of thinking and habits. This resistance to change can be explained by the maintenance of SI compatible units (angle degrees, litre, electron-volt, etc.) and even SI non-compatible units (carat in jewellery, faraday in chemistry, bar in meteorology, horsepower in mechanics, calorie for food, and all the Anglo-Saxon units).

It is unlikely that we will express the distance in seconds, which is completely logical: 1 m corresponds to 3.335 ns. On the other hand, we accepted the light year as a distance.

References

1. Reyes-Ponce, Y., Regla, A.: Reasons for, concerns. Paris: OIML (2018). https://www.oiml.org/en/publications/bulletin/pdf/oiml_bulletin_jan_2018.pdf Accessed 15 Feb 2021
2. Gyllenbok, J.: Encyclopedia of Historical Metrology, Weights, and Measures. Vol. 1. Cham, Birkhaüser (2018)
3. Howarth, P., Redgrave, F.: Mjeriteljstvo Ukratko, 3rd edn. IMBIH, Sarajevo (2009)
4. Mulić, M., Vrce, E., Omićević, D.ž, Đonlagić, E.: Geodezija od Mesopotanije do globalnog geodetskog opažačkog sistema. Geodetski glasnik **46**, 132–168 (2017)
5. Fanton, J.: A brief history of metrology: past, present, and future. Int. J. Metrol. Qual. Eng. **10**, 5 (2019). https://doi.org/10.1051/ijmqe/2019005
6. Džemić, Z.: *Mjeriteljski sistem u BIH i organizacija Instituta za Mjeriteljstvo BiH*. In stitut za mjeriteljstvo BiH (IMBiH) (2018). https://www.met.gov.ba/. Accessed 15 Feb 2021
7. Schmid, W.: Harmonised Measurements and their International Recognition. Seminar Metrology & Conformity Assessment, Brussels, 1-2 March 2010, Euramet (2010)
8. JCGM: International Vocabulary of Metrology: Basic and General Concepts and Associated Term. VIM BIPM, Paris (2008)
9. JCGM: Guide to the Expression of Uncertainty in Measurement. BIPM, Paris (2008)
10. Max Planck,: Ueber irreversible Strahlungsvorgänge. Ann. Phys. **306**(1), 69–122 (1900). https://doi.org/10.1002/andp.19003060105

Accuracy Analysis of Third High Accuracy Leveling Measurements in Federation of B&H

Alma Tabaković[1](✉), Mirnes Bojić[1], and Ervin Redžepagić[2]

[1] BNpro d.o.o. Sarajevo, Sarajevo, Bosnia and Herzegovina
[2] Federal Administration for Geodetic and Real Property Affairs, Sarajevo, Bosnia and Herzegovina

Abstract. In this article authors tried to give an insight of conducted measurements performed in Third High Accuracy Leveling project in Federation of Bosnia and Herzegovina. As a measure of accuracy, reference mean square error of measured height differences, leveling lines and leveling loops are given, together with loop closures. These are the first calculated data of accuracy criteria given for the Third High Accuracy Leveling project in Federation of Bosnia and Herzegovina.

Keywords: High accuracy leveling · Leveling lines · Leveling loops · Accuracy criteria · Reference mean square errors

1 Introduction

Geometric leveling is considered to be the most precise measurement technique in geodesy. It is the basis for determining vertical control networks. Depending of the accuracy achieved and used leveling procedures, national geometric leveling can be distinguished in orders. First-order leveling is the most precise procedure, while the second, third and fourth-order leveling serve to densify the first-order network. Due to time variations, crustal movements and different geodynamic processes, bench marks change their position (in vertical sense). That is the reason why it is very important to relevel national vertical control networks from time to time. Preferable time interval between two observations differs from author to author, but it is in range of 10 to every 20 to 30 years.

On the territory of Bosnia and Herzegovina, until now, only two high accuracy leveling projects have been implemented. First high accuracy leveling (NVT 1) was performed in period from 1945 to 1953. This leveling network was relied on 1875 vertical datum, and the heights determined then are still in use today [1]. From 1970 to 1973 as a part of Yugoslavian network, activities on second high accuracy leveling (NVT 2) project were performed. Network comprised of 13 leveling loops, with total length of cca 1967 km. It is important to stress out that these results were never put in the official use (due to disintegration of Former Yugoslavian Republic – FYR). Also, analysis of the results obtained in NVT 1 and NVT2 showed that heights obtained in NVT1 were higher than ones obtained in NVT2 for about 30 cm. While including the

N. Ademović et al. (Eds.): IAT 2021, LNNS 316, pp. 558–572, 2022.
https://doi.org/10.1007/978-3-030-90055-7_44

neighboring Croatia in United European Leveling Network - UELN, 17 nodal points on the territory of Bosnia and Herzegovina were recalculated [1]. In addition to that, in Bosnia and Herzegovina, two precise leveling project were conducted, one during Austro-Hungarian monarchy, and other in 1929.

2 Third High Accuracy Leveling in Federation of B&H

Almost 40 years later, activities on third high accuracy leveling (NVT 3) were taken into account. Detailed technical specification for NVT 3 was announced in the beginning of 2017. Specification was written by prof.dr.sc. Asim Bilajbegović and prof.dr.sc. Oleg Odalović.

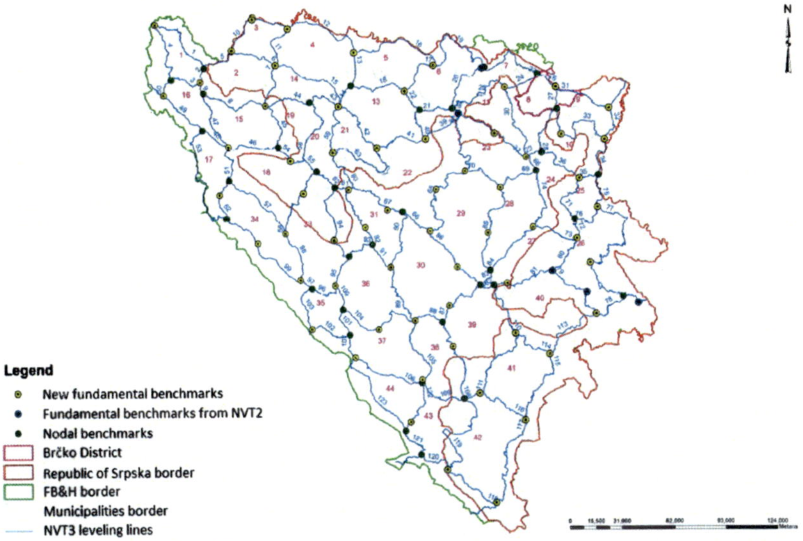

Fig. 1. NVT 3 leveling network of Bosnia and Herzegovina [2]

As shown in Fig. 1 leveling network of Federation of Bosnia and Herzegovina is a part of leveling network of Bosnia and Herzegovina, which also includes networks of Republic of Srpska and Brčko District. NVT 3 network comprises of 44 leveling loops covering the whole territory of B&H, and in total 123 leveling lines. Total length of the leveling lines in the leveling network is 4485 km, with the average length of a leveling line of 36.5 km [2].

From above mentioned, in total 2692 bench marks are positioned on the territory of FB&H. These bench marks are positioned along the 55 leveling lines lying in FB&H, and part of the lines that connect neighboring entity and neighboring Croatia. There are in total 35 leveling loops creating the network of FB&H, from which 13 of them are entirely in FB&H, and with average loop length of 178 km.

If observing bench mark types comprising this network, NVT 3 network is consisted of 1 normal bench mark (positioned in Maglaj), 33 fundamental bench marks, 92 spatial reference network points, 972 vertical and 1594 horizontal bench marks. From 33 fundamental bench marks, 15 of them are taken from NVT 2 leveling network [3].

Leveling measurements of NVT 3 leveling network were conducted in period from August 2019 until September 2020. In this time period 55 leveling lines and 9 parts of leveling lines were measured, in total length of cca 2100 km. Not all terrain works are finished, and is expected that the leveling measurements will continue in 2021. Planned measurements refer to parts of leveling lines that are only partially in territory of FB&H. As all the leveling loops on FB&H territory have been leveled, authors found no reasons against to make accuracy assessment of conducted measurements.

3 Accuracy Assessment of Conducted Leveling Measurements

In geodetic works, measurement process, besides the measurement of a quantity itself, includes the assessment of accuracy with which the quantity is obtained. Assessment accuracy is needed in order to determine whether certain measured result is reliable or not, i.e. whether it is useable [4]. Since the assessment is never conducted based on one measured value, quantity is always measured more than once. There are a few accuracy assessment parameters used in geodetic practice. One parameter is a disagreement between measured and true value, or between two measured values of the same quantity. Still, most commonly used accuracy assessment parameter would be mean error. Sometimes used are also average and probable errors [4].

For leveling measurements, accuracy assessment could be divided in two groups: a priori assessment (prior to adjustment) and a posteriori accuracy assessment (after the leveling network adjustment) [5].

In this paper a priori assessment of measurements has been determined. Further, the assessment is based on the assumption that there are only random errors and no systematic errors present in the measurements. All raw measurements have been corrected for index error and scale factor of the leveling rods.

Following the mentioned, accuracy assessment of leveling measurements is based on:

- Discrepancies of double measured height differences
- Discrepancies of double measured leveling lines
- Discrepancies of leveling loop closures

For every mentioned quantity corresponding value of reference mean square error can be calculated. Reference mean square error of double leveling is calculated as [6]:

$$m'_{Ri} = \pm\frac{1}{2}\sqrt{\frac{1}{n_R}\left[\frac{\rho\rho}{R}\right]}$$

where:

m'_{Ri} is reference mean square error of double leveling height difference obtained in mm,

n_R is total number of height differences in leveling line.
$\rho\rho$ is a square of double measured discrepancies (difference between forward and backward measurement), expressed in square mm.
R is the length of leveling line, expressed in km.

4 Accuracy Assessment of Double Measured Height Differences

As specified in detailed technical specification [5] and in accordance with law on the increase of errors, standard deviation s of an individual measurement forward or backward is given with the following expression: $s = \frac{\rho}{\sqrt{2}}$. This expression will not be used as an accuracy assessment of measured height differences for obvious reasons (length between two neighboring benchmarks is at least 1 km, and length of view for digital levels is restricted to 22.5 m ± 0.5 m [6], which further points to more standpoints for every height difference), and reference mean square error is given as an accuracy parameter.

Considering the large amount of data, for analysis purposes, in this article accuracy assessment of height differences in leveling line L50 is shown in Table 1. For every measured height difference tolerance was calculated using expression [5]:

$$Z = 0.5R + 1.5\sqrt{R}$$

Table 1. Accuracy assessment of height differences in leveling line L50

From	To	n_R	R in km	Tolerance Z in mm	ρ in mm	m'_{Ri}
L3R027	L50R001	8	0.31	1.00	−0.34	0.11
L50R001	L50R002	22	1.27	2.32	0.78	0.07
L50R002	L50R003	23	1.25	2.30	−0.82	0.08
L50R003	L50R004	25	1.16	2.19	1.19	0.11
L50R004	L50R005	22	0.92	1.90	−0.38	0.04
L50R005	L50R006	17	0.85	1.81	−0.64	0.08
L50R006	L50R007	21	1.03	2.04	−0.01	0.00
L50R007	L50R008	16	0.86	1.82	0.78	0.11
L50R008	L50R009	13	0.69	1.60	1.01	0.17
L50R009	L50R010	10	0.52	1.35	0.74	0.16
L50R010	L50R011	7	0.34	1.04	−0.02	0.01
L50R011	L50R012	11	0.59	1.44	−0.44	0.09
L50R012	L50R013	14	0.81	1.75	0.90	0.13
L50R013	L50R014	11	0.58	1.44	0.09	0.02
L50R014	L50R015	10	0.51	1.33	0.59	0.13
L50R015	L127R001	14	0.75	1.67	0.71	0.11

All leveling discrepancies (differences between measured height differences forward and backward) meet the tolerance condition, i.e. all discrepancies are smaller than the tolerance value Z.

The average reference mean square error for the 16 height differences shown in Table 1 is 0.09 $mm/\sqrt{km}$. The largest reference mean square error is calculated for height difference between benchmarks L50R008 and L50R009, while the smallest reference mean square error is calculated for the height difference between benchmarks L50R006 and L50R007.

5 Accuracy Assessment of Double Measured Leveling Lines

Just like previously mentioned for measured height differences, the same assessment procedure was done for measured leveling lines. Table 2 shows 55 leveling lines with their corresponding calculated reference mean square errors. Measured leveling lines are positioned through the entire territory of Federation of Bosnia and Herzegovina, which means that almost every line has a different geomorphological position. Nevertheless, from Table 2 it is visible that measured lines are different in terms of length and number of standpoints. Average leveling line length is 38 km, and the line length varies from 8 km to 84 km. In terms of number of standpoints, average number of standpoints is 40, and this number varies from 10 to 99, depending on a leveling line.

Table 2. Accuracy assessment of measured leveling lines

Line	From	To	n_R	R in km	Tolerance Z in mm	ρ in mm	m'_{Ri}
L1	L1R001	L1R061	60	60.30	41.80	20.30	0.17
L2	L1R061	L2R010	10	10.30	9.90	1.40	0.07
L3	L2R010	L3R027	27	22.10	18.10	−4.10	0.08
L4	L3R027	L1R001	53	47.10	33.90	13.30	0.13
L8	L9R011	L8R059	59	56.60	39.60	25.20	0.22
L9	L2R010	L9R011	11	8.70	8.70	4.40	0.23
L30	L30R001	L68R001	66	56.10	39.30	16.50	0.14
L45	L8R059	L45R038	38	35.20	26.50	12.90	0.18
L46	L45R038	L46R041	41	37.70	28.10	13.10	0.17
L47	L9R011	L46R041	48	45.30	32.80	−15.70	0.17
L48	L46R041	L48R023	23	20.70	17.20	10.20	0.23
L49	L48R023	L127R001	37	35.70	26.80	10.60	0.15
L50	L3R027	L127R001	16	12.40	11.50	4.10	0.15
L51	L46R041	L51R031	31	29.60	23.00	3.60	0.06
L52	L51R031	L52R034	34	32.30	24.70	5.60	0.08

(continued)

Table 2. (*continued*)

Line	From	To	n_R	R in km	Tolerance Z in mm	ρ in mm	m'_{Ri}
L53	L51R026	L48R023	53	56.60	39.60	1.40	0.01
L57	L51R031	L98R038	51	53.20	37.50	10.30	0.10
L64	L64R001	L70R001	50	47.10	33.80	10.60	0.11
L65	L70R001	L65R091	91	77.40	51.90	43.70	0.26
L66	L65R091	L66R029	29	22.90	18.60	6.90	0.13
L67	L67R001	L66R029	54	52.90	37.40	18.70	0.17
L68	L68R001	L69R001	16	12.40	11.50	0.40	0.02
L69	L69R001	L70R034	36	29.60	22.90	3.10	0.05
L70	L70R001	L70R034	33	31.50	24.10	5.70	0.09
L74	L69R001	L84R049	55	47.60	34.10	15.30	0.15
L83	L83R001	L82R014	19	16.60	14.40	7.60	0.21
L84	L83R001	L84R049	49	47.80	34.20	13.40	0.14
L85	L70R034	L83R001	99	83.80	55.60	3.70	0.02
L86	L82R016	L65R091	60	52.10	36.90	16.00	0.14
L87	L82R016	L87R054	34	45.40	32.80	18.30	0.23
L88	L87R054	L88R029	29	23.20	18.80	−1.00	0.02
L89	L88R029	L89R057	57	53.70	37.80	9.60	0.09
L90	L89R057	L66R029	49	47.10	33.80	−2.40	0.03
L91	L89R057	L91R021	21	19.30	16.20	1.50	0.04
L92	L67R001	L91R021	57	48.90	34.90	4.30	0.04
L93	L91R021	L95R001	28	24.60	19.70	14.20	0.27
L95	L95R001	L96R001	23	21.60	17.80	−8.00	0.18
L96	L96R001	L96R029	28	24.10	19.40	10.60	0.20
L97	L96R029	L97R011	11	8.10	8.30	0.40	0.02
L98	L97R011	L98R038	38	36.50	27.30	15.30	0.21
L99	L97R011	L52R034	68	68.40	46.60	11.90	0.09
L100	L96R001	L100R018	18	17.90	15.30	3.40	0.10
L101	L100R018	L101R019	19	17.80	15.30	−2.90	0.08
L102	L101R019	L102R028	28	28.60	22.30	9.50	0.17
L103	L96R029	L102R028	35	33.70	25.60	7.90	0.12
L104	L88R029	L100R018	67	65.20	44.70	12.20	0.09
L105	L88R029	L105R060	60	48.60	34.80	18.50	0.17
L106	L105R060	L106R055	55	52.00	36.80	−1.80	0.02

(*continued*)

Table 2. (*continued*)

Line	From	To	n_R	R in km	Tolerance Z in mm	ρ in mm	m'_{Ri}
L107	L101R019	L106R055	21	19.40	16.30	9.00	0.22
L109	L109R011	L105R060	32	30.40	23.50	5.10	0.08
L120	L121R032	L118R044	28	29.00	22.60	0.90	0.02
L121	L122R038	L121R032	32	33.60	25.50	0.70	0.01
L122	L105R060	L122R038	38	36.50	27.30	−0.80	0.01
L123	L106R055	L122R038	71	70.60	47.90	12.70	0.09
L125	L121R032	L125R026	26	25.20	20.10	−9.70	0.19

If observing reference mean square error, from Table 2 it is visible that the average value is 0.12 mm/$\sqrt{km}$. The value varies from 0.01 mm/$\sqrt{km}$ for leveling lines L53, L121, L122, to 0.27 mm/$\sqrt{km}$ for leveling line L93. Average length of a leveling line between two bench marks is 0.924 km, while the average number of standpoints is 23.

6 Accuracy Assessment of Leveling Loop Closures

Accuracy assessment of leveling loops is given with 3 indicators: discrepancy ω between two levelings (forward – backward), reference mean square error m_F calculated using the formula [6]:

$$m'_F = \pm\sqrt{\frac{1}{n_F}\left[\frac{\omega\omega}{F}\right]}$$

where:
m'_F is the reference mean square error of leveling loop closure,
n_F is total number of leveling lines in leveling loop.
$\omega\omega$ is a square of double measured discrepancies (difference between forward and backward loop closure), expressed in square mm.
F is the length of leveling loop, expressed in km.
and probable reference error obtained by expression [6]:

$$u'_F = \frac{2}{3}m'_F$$

Reference mean square error and probable reference error are calculated on two ways: considering the number of standpoints (m'_F and u'_F) and without considering the number of standpoints (m''_F and u''_F).

Tables 3–14 show the assessment accuracy of 13 leveling loops closed in measurements of NVT 3 performed so far.

Table 3. Accuracy assessment of NVT 3 leveling loop 1

Leveling line	No. of benchmarks	Length in km	Max Δh [m]	Min Δh [m]	∑Δh [m]
L1	61	60.34	70.07585	−58.70011	2.53938
L2	11	10.27	3.89431	−1.00662	13.45590
L3	28	22.15	5.60215	−2.39848	40.21956
L4	54	47.14	58.59215	−62.10550	−56.21897
∑Δh forward			∑Δh backward		
0.01133			0.01958		
Allowed discrepancy in mm			23.66		
ω [m]	m_F' [mm/$\sqrt{km}$]	m_F'' [mm/$\sqrt{km}$]		u_F' [mm/$\sqrt{km}$]	u_F'' [mm/$\sqrt{km}$]
−0.0082	0.35	0.70		0.23	0.46

Table 4. Accuracy assessment of NVT 3 leveling loop 15

Leveling line	No. of benchmarks	Length in km	Max Δh [m]	Min Δh [m]	∑Δh [m]
L8	60	56.62	57.28844	−70.76035	−154.95816
L45	39	35.18	57.00313	−52.25144	191.72931
L46	42	37.72	83.03854	−61.97794	308.73449
L47	49	45.31	52.55410	−31.55849	−345.49409
∑Δh forward			∑Δh backward		
0.02930			0.00621		
Allowed discrepancy in mm			26.44		
ω [m]	m_F' [mm/$\sqrt{km}$]	m_F'' [mm/$\sqrt{km}$]		u_F' [mm/$\sqrt{km}$]	u_F'' [mm/$\sqrt{km}$]
0.0231	0.87	1.75		0.58	1.16

Table 5. Accuracy assessment of NVT 3 leveling loop 16

Leveling line	No. of benchmarks	Length in km	Max Δh [m]	Min Δh [m]	∑Δh [m]
L9	12	8.65	57.93789	−16.22479	166.91153
L47	49	45.31	52.55410	−31.55849	345.49409
L48	24	20.68	41.39676	−21.37223	−15.01227
L49	38	35.74	63.67631	−56.03430	−431.22553
L50	17	12.42	7.27334	−1.35697	−25.94270

(*continued*)

Table 5. (*continued*)

Leveling line	No. of benchmarks	Length in km	Max Δh [m]	Min Δh [m]	∑Δh [m]
L3	28	22.15	5.60215	−2.39848	−40.21955
∑Δh forward			∑Δh backward		
0.01039			−0.00075		
Allowed discrepancy in mm			24.08		
ω [m]	m_F' [mm/√km]	m_F'' [mm/√km]		u_F' [mm/√km]	u_F'' [mm/√km]
0.0111	0.38	0.92		0.25	0.62

Table 6. Accuracy assessment of NVT 3 leveling loop 17

Leveling line	No. of benchmarks	Length in km	Max Δh [m]	Min Δh [m]	∑Δh [m]
L53	54	56.55	93.19557	−74.54110	1.81525
L48	24	20.68	41.39676	−21.37223	15.01227
L51 (dio)	27	25.22	70.68616	−62.78382	−16.82682
∑Δh forward			∑Δh backward		
0.00847			0.00708		
Allowed discrepancy in mm			20.24		
ω [m]	m_F' [mm/√km]	m_F'' [mm/√km]		u_F' [mm/√km]	u_F'' [mm/√km]
0.0014	0.08	0.14		0.05	0.09

Table 7. Accuracy assessment of NVT 3 leveling loop 28

Leveling line	No. of benchmarks	Length in km	Max Δh [m]	Min Δh [m]	∑Δh [m]
L84	50	47.76	50.60438	−73.52555	151.57870
L74	56	47.56	81.68324	−59.66326	−405.97107
L69	37	29.58	54.69635	−55.03306	69.12795
L85	100	83.83	71.25189	−85.31960	185.27254
∑Δh forward			∑Δh backward		
0.02584			0.00962		
Allowed discrepancy in mm			28.89		
ω [m]	m_F' [mm/√km]	m_F'' [mm/√km]		u_F' [mm/√km]	u_F'' [mm/√km]
0.0162	0.56	1.12		0.37	0.75

Table 8. Accuracy assessment of NVT 3 leveling loop 29

Leveling line	No. of benchmarks	Length in km	Max Δh [m]	Min Δh [m]	∑Δh [m]
L83	20	16.58	20.82391	−17.30629	30.17120
Part of L82	3	1.94			3.49147
L86	61	52.12	41.83437	−54.92464	−135.19548
L65	92	77.41	28.11485	−19.32751	−167.47415
L70	34	31.46	15.96549	−11.68992	83.73294
L85	100	83.83	71.25189	−85.31960	185.27254
∑Δh forward			∑Δh backward		
0.03683			0.03980		
Allowed discrepancy in mm			32.46		
ω [m]	m_F' [mm/√km]	m_F" [mm/√km]		u_F' [mm/√km]	u_F" [mm/√km]
−0.0030	0.07	0.18		0.05	0.12

Table 9. Accuracy assessment of NVT 3 leveling loop 30

Leveling line	No. of benchmarks	Length in km	Max Δh [m]	Min Δh [m]	∑Δh [m]
L87	35	45.39	61.26312	−85.13528	−224.95401
L88	30	23.20	27.24546	−48.37674	−90.56087
L89	58	53.68	66.89169	−104.27595	472.93166
L90	50	47.09	35.10097	−65.38120	−163.59334
L66	30	22.91	22.95568	−5.92929	−129.03317
L86	61	52.12	41.83437	−54.92464	135.19548
∑Δh forward			∑Δh backward		
−0.00659			0.02190		
Allowed discrepancy in mm			31.27		
ω [m]	m_F' [mm/√km]	m_F" [mm/√km]		u_F' [mm/√km]	u_F" [mm/√km]
−0.0285	0.74	1.82		0.50	1.21

Table 10. Accuracy assessment of NVT 3 leveling loop 31

Leveling line	No. of benchmarks	Length in km	Max Δh [m]	Min Δh [m]	∑Δh [m]
L67	55	52.94	78.03591	−77.60470	122.85332
L90	50	47.09	35.10097	−65.38120	163.59334
L91	22	19.25	5.34462	−10.69450	−93.87652
L92	58	48.87	38.51420	−22.95644	−192.56896
∑Δh forward			∑Δh backward		
0.01220			0.00984		
Allowed discrepancy in mm			25.93		
ω [m]	m_F' [mm/√km]	m_F" [mm/√km]		u_F' [mm/√km]	u_F" [mm/√km]
0.0024	0.09	0.18		0.06	0.12

Table 11. Accuracy assessment of NVT 3 leveling loop 34

Leveling line	No. of benchmarks	Length in km	Max Δh [m]	Min Δh [m]	∑Δh [m]
L99	69	68.44	42.84045	−40.79598	141.60377
L52	35	32.28	75.04322	−50.66133	−369.90806
L57	52	53.17	97.99793	−42.69082	423.61612
L98	39	36.53	62.61437	−55.77596	−195.29415
∑Δh forward			∑Δh backward		
0.03920			0.00384		
Allowed discrepancy in mm			27.60		
ω [m]	m_F' [mm/√km]	m_F" [mm/√km]		u_F' [mm/√km]	u_F" [mm/√km]
0.0354	1.28	2.56		0.85	1.71

Table 12. Accuracy assessment of NVT 3 leveling loop 35

Leveling line	No. of benchmarks	Length in km	Max Δh [m]	Min Δh [m]	∑Δh [m]
L101	20	17.85	8.77583	−13.34157	−2.73015
L102	29	28.62	67.87667	−68.65873	−149.12154
L103	36	33.71	44.06415	−39.78221	−6.97343

(*continued*)

Table 12. *(continued)*

Leveling line	No. of benchmarks	Length in km	Max Δh [m]	Min Δh [m]	∑Δh [m]
L96	29	24.11	54.86273	−58.86039	203.21325
L100	19	17.89	42.32666	−55.07929	−44.37681
∑Δh forward			∑Δh backward		
0.02565			0.00299		
Allowed discrepancy in mm			22.11		
ω [m]	m_F' [mm/√km]	m_F" [mm/√km]		u_F' [mm/√km]	u_F" [mm/√km]
0.0227	0.92	2.05		0.61	1.37

Table 13. Accuracy assessment of NVT 3 leveling loop 36

Leveling line	No. of benchmarks	Length in km	Max Δh [m]	Min Δh [m]	∑Δh [m]
L93	29	24.60	65.20601	−95.04145	599.64215
L95	24	21.58	16.62213	−45.75643	−232.57050
L100	19	17.89	42.32666	−55.07929	−44.37681
L104	68	65.17	96.90000	−60.06996	−701.76425
L89	58	53.68	66.89169	−104.27595	472.93166
L91	22	19.25	5.34462	−10.69450	−93.87652
∑Δh forward			∑Δh backward		
0.00219			0.03074		
Allowed discrepancy in mm			28.44		
ω [m]	m_F' [mm/√km]	m_F" [mm/√km]		u_F' [mm/√km]	u_F" [mm/√km]
−0.0286	0.82	2.01		0.55	1.34

Table 14. Accuracy assessment of NVT 3 leveling loop 37

Leveling line	No. of benchmarks	Length in km	Max Δh [m]	Min Δh [m]	∑Δh [m]
L105	61	48.64	37.31790	−41.00061	134.74942
L104	68	65.17	96.90000	−60.06996	701.76425
L101	20	17.85	8.77583	−13.34157	−2.73015

(continued)

Table 14. (*continued*)

Leveling line	No. of benchmarks	Length in km	Max Δh [m]	Min Δh [m]	∑Δh [m]
L107	22	19.44	29.07019	−52.13986	−241.81447
L106	56	51.99	66.66591	−43.78675	−591.98014
∑Δh forward			∑Δh backward		
0.00643			0.02860		
Allowed discrepancy in mm			28.50		
ω [m]	m_F' [mm/√km]	m_F'' [mm/√km]		u_F' [mm/√km]	u_F'' [mm/√km]
−0.0222	0.70	1.56		0.46	1.04

Table 15. Accuracy assessment of NVT 3 leveling loop 44

Leveling line	No. of benchmarks	Length in km	Max Δh [m]	Min Δh [m]	∑Δh [m]
L106	56	51.99	66.6659	−43.7868	591.98014
L123	72	70.57	70.7236	−75.4153	−643.88860
L122	39	36.55	10.2513	−13.2127	51.91465
∑Δh forward			∑Δh backward		
0.02184			0.00946		
Allowed discrepancy in mm			25.23		
ω [m]	m_F' [mm/√km]	m_F'' [mm/√km]		u_F' [mm/√km]	u_F'' [mm/√km]
0.0124	0.57	0.98		0.38	0.65

As visible from Tables 3–15, most of the loop closures are in the allowed tolerance, except for the loop 34, which has exceeded the allowed tolerance, while loops 35 and 36 are on the verge of exceeding.

Accordingly, the highest reference mean square error has been obtained for loop 34, even 1.28 mm/$\sqrt{km}$, while the average reference mean square error for closed loops is 0.57 mm/$\sqrt{km}$. Interesting thing is that the lines forming the loop 34 are all in the allowed tolerance, which may implicate on the presence of accumulated random errors due to variable topography of the loop path.

For comparison, Table 16 shows short analysis of leveling loop closures of NVT2 leveling measurements.

Table 16. Accuracy assessment of NVT 2 leveling loop closures

Leveling loop	ω [mm]	No. of leveling lines	Length [km]	m_F' [mm/√km]	m_F" [mm/√km]	u_F' [mm/√km]	u_F" [mm/√km]
8	7.14	6	442.4	0.14	0.34	0.09	0.23
9	−30.13	7	556.5	0.48	1.28	0.32	0.85
10	−77.07	5	429.5	1.66	3.72	1.11	2.48
11	−92.31	6	453.4	1.77	4.34	1.18	2.89
13	−4.48	5	583.2	0.08	0.19	0.06	0.12
14	6.72	6	615.3	0.11	0.27	0.07	0.18
15	−8.84	6	614	0.15	0.36	0.10	0.24

Total of 7 leveling loops where leveled on the territory of Bosnia and Herzegovina during the NVT 2 project. Average leveling loop length was 528 km, with average of 6 leveling lines constituting the loop. Overall reference mean square error and probable reference error look acceptable, except for the leveling loops 10 and 11. These loops have larger discrepancy than the allowed tolerance. Understanding why loops 10 and 11 have larger reference mean square errors compared to the other loops from NVT2 requires analysis leveling lines constituting the loops, which will not be covered in this paper.

7 Conclusion

Overall measurement accuracy is acceptable in terms of meeting the requirements of prescribed tolerance defined by technical specification for NVT 3. Still, analysis was performed with the presumption that all systematic errors were eliminated, and there are only random errors present in the measurements, which is not entirely true presumption. Due to lack of funds, leveling rods were only calibrated for index correction and scale factor, but coefficient of expansion was not determined. This however should be done as soon as the conditions are created, but should not wait for too long, since the adjustment of the control network needs to be performed.

While analyzing leveling lines, reference mean square error was within acceptable range, but very uneven, rating from 0.01 to 0.27 mm/$\sqrt{km}$. Adequately, the same was noticed in leveling loop closures. Length of leveling loops varies from loop to loop, ranging from 102 km in loop 17 to 263 km in loop 29. Since the mean square error and consequentially probable reference error depend on number of standpoints, length of leveling line/leveling loop and leveling discrepancy, variable length of leveling loops yields for variations in reference mean square error.

Some of the leveling lines, such as lines 34, 35 and 36 should be inspected again, and perhaps impact of systematic errors should be calculated.

When comparing obtained accuracy results of NVT3 measurements with the ones obtained for NVT2, following could be noted: average reference mean square error m_F'

calculated for NVT 2 leveling loops is slightly larger than the one calculated for NVT 3 leveling loops. This is not unusual, having in mind the more precise equipment used for NVT3 measurements. While the average reference mean square error for NVT3 leveling loops is 0.57 mm/$\sqrt{km}$, this value for NVT 2 leveling loops is 0.63 mm/$\sqrt{km}$. What could be noted is that the NVT 2 loops are larger than NVT3 loops. Length of NVT2 loops is consistent, between 400 and 600 km, while the NVT 3 loops length is much more variable and at least two times smaller than the length of NVT 2 loops.

Despite of the positive indicators that the accuracy assessment of measurements is showing, definite analysis should be performed and given after adjustment of the leveling network.

References

1. Krzyk, T.: Nivelmanske mreže viših redova i vertikalni datumi na području Bosne i Hercegovine. Geodetski glasnik – Glasilo Saveza udruženja građana geodetske struke u Bosni i Hercegovni **35**, 22–34 (2001)
2. Vasiljević, S., Vasić, D., Sarajlić, T.: Potreba i značaj uspostavljanja novog verikalnog referentnog okvira Republike Srpske. In: IV Kongres o katastru u BiH, Neum 29–31 October (2019)
3. Federal Geodetic Administration Osnovni geodetski radovi – Nivelman visoke tačnosti (NVT 3) – projektni zadatak, Tenderska dokumentacija na nabavku usluga Realiziranje nivelmanskih mjerenja. Nivelman visoke tačnosti 3 Bosne i Hercegovine". FGU: Sarajevo (2020)
4. Pašalić, S.: Račun izravnanja. Građevinski fakultet Sarajevo (1984)
5. Bilajbegović, A., Odalović, O.: Detaljne tehničke specifikacije nivelmana visoke tačnosti – NVT3. Dresden and Belgrade, Report Nr: BAL2–4–1 (2017)
6. Rezo, M. et al.: Analysis of accuracy leveling measurements and integral adjustment of precise leveling network (II. NVT). Geodetski list **69**(1), 1–25 (2015)

Computer Modelling and Simulations for Engineering Applications

Numerical Analysis of Steady State and Transient Motion of a Vibro-Impact System with Non-ideal Excitation with a Nonlinear Spring

Džanko Hajradinović[1](✉) and Miodrag Zuković[2]

[1] Department of Mechanics, Faculty of Mechanical Engineering, University of Sarajevo, Sarajevo, Bosnia and Herzegovina
hajradinovic@mef.unsa.ba

[2] Centre for Vibro-Acoustic Systems and Signal Processing, Faculty of Technical Sciences, University of Novi Sad, Novi Sad, Serbia

Abstract. This paper is studying the behaviour of a vibro-impact system with non-ideal excitation where the impacting oscillator is connected to a driving slider with a non-linear cubic spring. The paper is giving an insight in modelling and numerical analysis of steady state and transient motion of the system. The analysed system consists of an oscillating slider, a hard stop on one side of the oscillator, on the other side the oscillator is connected via a nonlinear spring to a slider. The driving slider is connected to a balanced disk with a link. The disk is attached to an electromotor shaft. The system is excited with a DC electromotor via the balanced disk. The electromotor characteristic is assumed to be linear and the mathematical model of it is shown in the paper. The impact model is inelastic. The equations of motions are derived using Lagrange's equations. The numerical analysis is based on the continuation method of a run up and close down simulation by changing the stall torque value. Results are shown in the form of frequency response diagram, average value of the excitation frequency versus the stall torque value diagram and oscillation amplitude versus the stall torque value diagram for the steady state solutions. Based on these results transient motion is also analysed and results are shown in the form of diagrams which are describing the displacement and excitation frequency change in function of time. The aim of this paper is to show how the system behaves if the installed electromotor is weak or strong one. Through this analysis impact and non/impact solutions are obtained. It is pointed out that for which range of the stall torque values the system will have impact solutions or non-impact solutions. In reference to different regions the goal is to keep the system from the impact region assuming that the machine should just oscillate. The part of the paper where transient solutions are analysed is showing how the system can start as an vibro-impact system and switch to non-impact motion.

Keywords: Vibro-impact systems · Non-ideal excitation · Nonlinear spring · Steady state solutions · Transient solutions · Numerical analysis

N. Ademović et al. (Eds.): IAT 2021, LNNS 316, pp. 575–590, 2022.
https://doi.org/10.1007/978-3-030-90055-7_45

1 Introduction

Excitation of a vibration system can be ideal where the excitation frequency is constant or they can be non-ideal where the excitation frequency is variable due to inertia and opposing forces or torques in the system. Both systems, with ideal or non-ideal excitation are nonlinear because of the impact motion and the nonlinear spring, but an additional nonlinearity relates to the system with non-ideal excitation because of the variable excitation and it has one degree of freedom more. By adding a stop in which the oscillating body can impact a different type of system is formed which is in this paper denoted as a vibro-impact system with non-ideal excitation. This type of system has two nonlinearities, one is related to the variable excitation frequency and the other is related to the impact. Introducing a nonlinear spring gives the system a third layer of nonlinearity and such a system is analysed in this paper. The possible motions of such system which can be analysed are steady state motions, transient motions and chaotic motions. In this paper steady state and transient motions are analysed with it's full complexity. Because of the three nonlinearities that are existing in this system and with the idea to analyse it without any approximations and simplifications the analysis is conducted using numerical simulations. Machines, such as hand-held percussion machines, pile driving machines, cutting and grinding machines, etc. are all working as a vibro-impact system.

The work presented in this paper is a continued research based on the papers [1, 2]. In paper [1] different systems are analysed, simple and complex ones, with a combination of analytical and numerical methods. The most complex system which was analysed is a vibro-impact system with non-ideal excitation with a linear spring. Steady state motion is analysed where the motion of the driving slider is approximated. Results presented in this paper are mainly shown as frequency response diagrams with basins of attraction. The main parameter which is changed is the excitation frequency control parameter. In this paper besides that the system is physicaly different, the changing parameter is the stall torque and transient motion will be analysed. Paper [2] gives an insight how the variation of the stall torque influences the behaviour on the steady state motion of a vibro-impact system with non-ideal excitation with a linear spring. The work in this paper is different by the spring that is used to connect the hitting oscillator with the slider and transient motions will be analysed as well. Papers [1, 2] were focused how to adjust the parameters so that the system works as an vibro-impact system, where in this paper the idea is to adjust the parameters so that the oscillator doesn't impact a wall or a body which is close to it and it is fixed.

With only few available publications [3–8] considering vibro-impact systems with the non-ideal excitation, there is an evident lack of literature on this particular subject. Analysis of a system with non-ideal excitation is analysed in paper [3] where clearance in the connecting joints exists. The clearance is imposing non-smoothens in the systems behaviour which is one way similar to the system analysed in this paper but also different because it doesn't cause a change in the velocity direction. Beside that difference in the mentioned paper chaotic motions is analysed. Study on electromechanical system with an impact element is given in [4] and [5]. Papers [4] and [5] are very similar where paper [5] is widen with additional results from paper [4] that add quantitative value to it. In [5], a non-impacting system where both types of excitation are acting on the system simultaneously, where a vibro-impact system was used as an attachment with the purpose to absorb energy from the system. In [6], a system with a non-ideal excitation with three degrees of freedom was investigated, where the impact was modelled as a linear stiff spring. The impacting bodies motion was not directly presented, it was related with the motion of the dribing body and presented as relative motion. Frequency-response diagrams of this relative displacement were found numerically, but no run up or close down simulations were carried out. In [7], a possible use of vibro-impact systems for chaos control was analysed for dynamical systems with a non-ideal excitation. The paper presented a deep study of chaotic motion in systems with non-ideal excitation and with the nonlinear characteristic of the spring, but the analysis of the vibro-impact system was related to the ideal excitation and the way it can be used to damp chaotic motion in a system with non-ideal excitation. In [8], a vibro-impact attachment was added to the system with the purpose to show how it would damp the chaotic motion of a system with non-ideal excitation. The characteristic of the spring in [8] is nonlinear. The difference between the studies [7, 8] and the study presented in this paper is related to the steady-state and transient motion of the impacting body which has a non-ideal excitation source.

Papers [9–13] give an insight of the behaviour of vibro-impact systems with ideal excitation where the importance to observe vibro-impact systems with non-ideal excitation is given through papers [1] and [2] where clear qualitative differences are shown. Paper [13] shows the use of an vibro-impact system as an dynamic absorber which is similar to papers mentioned earlier but very different from the research given in this paper.

A wide and through research of systems with non-ideal excitation is done and shown in papers [14–19]. The main difference between the system analysed in this paper and the research done in papers [14–19] is the missing impact possibility of the oscillator. The strong nonlinearity caused by the possible impact of the oscillator influences the results and the system behaviour both qualitative and quantitative.

This paper consists of five sections. In Sect. 2 is given the description of a physical model of the system under consideration with the relevant mathematical modelling of it. Section 3 presents the results obtained by numerical simulations using the run-up and close down simulation of the system's steady state motion. In Sect. 4 possible transient motions are shown and it is pointed out in which region related to the stall torque value such types of motions can occur. Paper summarised conclusions are given in Sect. 5.

2 Typological and Structural Survey of the Built-Up Area

Physical model of the system under consideration (Fig. 1a) consists of an oscillator C which is connected with a nonlinear spring to a driving slider on one side. On the other side of the oscillator is a barrier located. The driving slider is part of a mechanism which consists, beside the driving slider, of a lever AB, homogeneous disk with the radius $a = 0.004$ [m] and moment of inertia $J = 0.06$ $\left[\text{kg/m}^2\right]$. The centre of the disk is located on the electromotor shaft which is driving the whole system. Geometrical and mechanical parameters are indicated in Fig. 1a) and b). The lever AB is assumed to be have a negligible mass and the length is $\overline{AB} = l = 0.12$ m. The lever is connected to the driving disk at point A. The mass of the driving slider is $m_B = 0.2\ [\text{kg}]$ and of the oscillator $m_C = 0.35\ [\text{kg}]$; The free length of the spring is $l_0 = 0.36\ [\text{m}]$. The mass m_C can impact the wall on it right side under certain regimes. These regimes will be investigated in the further analysis in this paper. The driving torque of the electromotor shaft is assumed to be $T(\dot{\varphi})$ and it depends on the angular velocity of the shaft, or in this case as per the physical model the disk.

The system has two degrees of freedom with the generalized coordinates being the displacement of the oscillator x and the angle of rotation of the eccentric drive φ, as shown in Fig. 1 a). The equations of motion are formed based on Lagrange's equations of motion:

$$\frac{d}{dt}\frac{\partial E_K}{\partial \dot{x}} - \frac{\partial E_K}{\partial x} = -\frac{\partial \text{V}}{\partial x} + Q_x;\ \frac{d}{dt}\frac{\partial E_K}{\partial \dot{\varphi}} - \frac{\partial E_K}{\partial \varphi} = -\frac{\partial \text{V}}{\partial \varphi} + Q_\varphi \tag{1}$$

As per Eq. (1) the equations of motion can be expressed via the kinetic energy E_K, potential energy V, their corresponding derivatives and the generalized forces Q_x and Q_φ.

The potential energy of the system is calculated as following:

$$\text{V} = \frac{c}{2}(\Delta l)^2 + \frac{\beta}{4}(\Delta l)^4 \tag{2}$$

where $\Delta l = 2a$ is the spring deflection. As per Fig. 1 b) the spring deflection can be calculated as following:

$$\Delta l = x - u(\varphi) - l_0 \tag{3}$$

where $u(\varphi)$ represents the motion of slider B.

The motion of slider B and the corresponding derivatives are:

$$\begin{aligned}
u(\varphi) &= a\cos\varphi + l\sqrt{1 - \left(\frac{a}{l}\right)^2 \sin^2\varphi} = a\cos\varphi + l\sqrt{1 + \frac{1}{2}\left(\frac{a}{l}\right)^2(\cos 2\varphi - 1)} \\
u'(\varphi) &= -a\sin\varphi - \frac{l}{2}\left(\frac{a}{l}\right)^2 \frac{\sin 2\varphi}{\sqrt{1 + \frac{1}{2}\left(\frac{a}{l}\right)^2(\cos 2\varphi - 1)}} \\
u''(\varphi) &= -a\cos\varphi - l\left(\frac{a}{l}\right)^2 \frac{\cos 2\varphi}{\sqrt{1 + \frac{1}{2}\left(\frac{a}{l}\right)^2(\cos 2\varphi - 1)}} \\
&\quad - \frac{l}{4}\left(\frac{a}{l}\right)^4 \frac{\sin 2\varphi}{\left[1 + \frac{1}{2}\left(\frac{a}{l}\right)^2(\cos 2\varphi - 1)\right]^{3/2}}
\end{aligned} \tag{4}$$

From the mentioned the potential energy is fully described and it can be written in the following form:

$$\text{V} = \frac{c}{2}(x - u(\varphi) - l_0)^2 + \frac{\beta}{4}(x - u(\varphi) - l_0)^4 \tag{5}$$

The kinetic energy of the system is:

$$E_K = \frac{1}{2}J\dot{\varphi}^2 + \frac{1}{2}m_B\dot{x}_B^2 + \frac{1}{2}m_C\dot{x}^2 \tag{6}$$

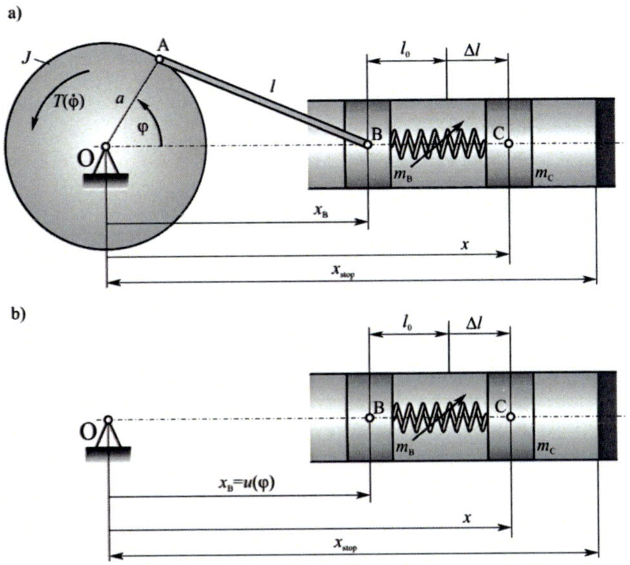

Fig. 1. Vibro-impact system: a) mechanical model of the system under consideration, b) geometry of the system and generalized coordinates

where the first derivative of the slider B motion can be written as follows:

$$\dot{x}_B = \frac{dx_B}{dt} = \frac{du(\varphi)}{dt} = \frac{d\varphi}{dt}\frac{du(\varphi)}{d\varphi} = \dot{\varphi}u'(\varphi)$$

based on which the kinetic energy becomes:

$$E_K = \frac{1}{2}J\dot{\varphi}^2 + \frac{1}{2}m_B\left(\dot{\varphi}u'(\varphi)\right)^2 + \frac{1}{2}m_C\dot{x}^2 \tag{7}$$

The generalized forces have the following form:

$$Q_x = -b\dot{x}, Q_\varphi = T(\dot{\varphi}) \tag{8}$$

The driving torque is assumed to be a linear function of the angular velocity of the eccentric drive:

$$T(\dot{\varphi}) = T_0\left(1 - \frac{\dot{\varphi}}{\Omega_0}\right) \tag{9}$$

Equation (9) is also called the characteristic of the electromotor. How the characteristic of the electromotor depend on the parameter T_0 is shown on Fig. 2. It is seen as the characteristic of the electromotor is steeper that the stall torque T_0 value is bigger and that the system behaves more similar to a system with ideal excitation. The variation of the stall torque as the control parameter will be in the range $0.028 \leq T_0 \leq 5$ [Nm].

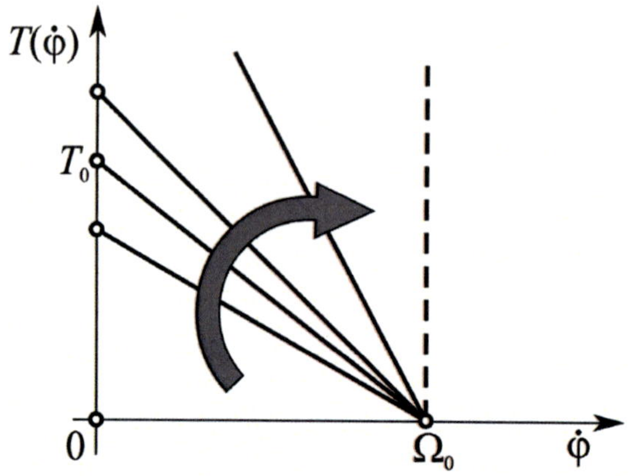

Fig. 2. Characteristics of driving torque as a function of the angular velocity $\dot{\varphi}$ for different values of the parameter κ

Summarizing Eqs. (9), (8), (7), (5) and (4) and their corresponding derivatives into Eq. (1), equations of motions are obtained:

$$\begin{gathered}\ddot{x} + 2\delta\dot{x} + \omega^2(x - u(\varphi) - l_0) + \gamma(x - u(\varphi) - l_0)^3 = \omega^2(l_0 + u(\varphi)) \\ \left(m_B u'(\varphi)^2 + J\right)\ddot{\varphi} + m_B\dot{\varphi}^2 u'(\varphi)u''(\varphi) \\ = T_0\left(1 - \frac{\dot{\varphi}}{\Omega_0}\right) + cu'(\varphi)(x - l_0 - u(\varphi)) + \beta u'(\varphi)(x - u(\varphi) - l_0)^3\end{gathered} \tag{10}$$

where:

$$\omega^2 = \frac{k_0}{m_C}, 2\delta = \frac{c}{m_C}, \gamma = \frac{\beta}{m_C} \tag{11}$$

where $k_0 = 9\pi^2$ [N/m], $c = 0.5$ [Ns/m], $\beta = 0.01$[]

As it can be noted in Fig. 3, the oscillator can hit the base, where the coordinate marked by x_{stop} represents the position of a fixed wall at which the oscillator can impact. The position of the impact wall can be obtained as:

$$x_{stop} = l + l_0 + \Delta \tag{12}$$

The initial position of the oscillator is also shown in Fig. 3 and it is the position when the oscillator is at distance Δ from the impact wall.

If the impact occurs periodically during the oscillating motion of the attachment, the system corresponds to a vibro-impact system. The impact model is assumed to be inelastic and it can be described as:

$$x(t^-) = x(t^+) = x_{stop}, \dot{x}(t^+) = -k\dot{x}(t^-) \tag{13}$$

The full mathematical model of the system is now described with Eqs. (10), (11), (12) and (13). The coefficient of restitution is $k = 0.25$.

3 Steady-State Motion Analysis

This chapter of the paper analyses steady state motion of the oscillator and the excitation frequency. For this analysis the method of continuation is used for the run up and close

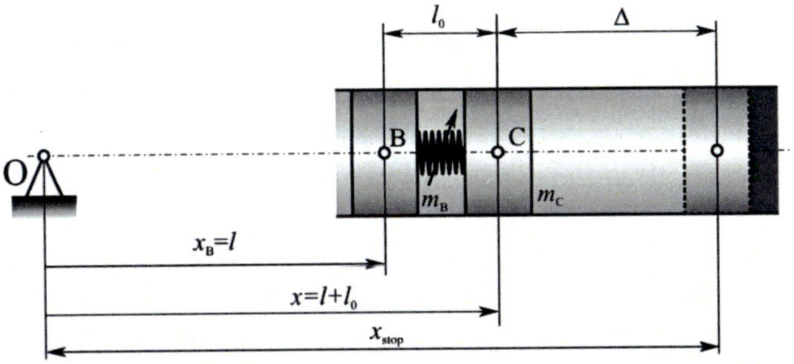

Fig. 3. Definition of the stop position x_{stop} and the gap Δ

down simulation. The varying parameter is the stall torque. Differential Eqs. (10) with the impact model (13) using the parameters are solved in Mathematica. During the run up and close down simulation for every value of the stall torque values for the maximum and minimum displacement of the oscillator is extracted and the average value of the excitation also. Combining the obtained values different diagrams are formed and presented in the paper.

Diagram which describes the change of the oscillator displacement in reference to the stall torque is given on Fig. 4.

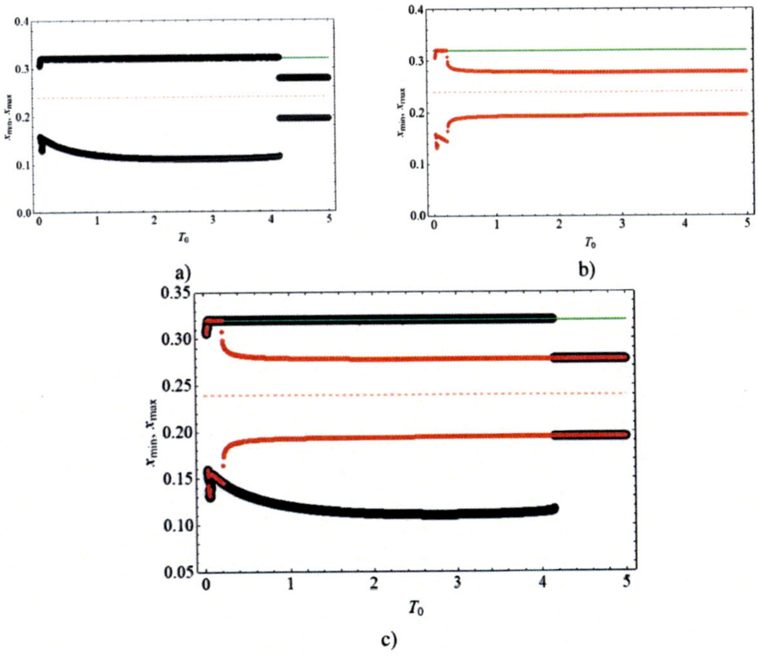

Fig. 4. Oscillator maximum and minimum displacement value in reference to the stall torque value: a) run up simulation; b) close down simulation; c) combined run up and close down simulation

Results of the run up and close down simulation are shown separately as the summary diagram of these two simulations is shown on Fig. 4 c) from these diagrams regions of impact and non-impact solutions can be distinguished in reference to the stall torque. The necessity to do both simulations, run up and close down, with the goal to find all solution can be seen from these diagrams where a big difference of results for each simulation can be seen. A wide range of the stall torque shows the possibility of impact and non-impact solutions. The main idea is to show for which values of the stall torque non-impact solutions are existing. Where the region of non-impact solutions is when the stall torque value is higher than 4.2 Nm. A short region can be distinguished from the diagram when only non-impact solutions exists and they are related to very small values of the stall torque. The biggest region is related to multiple solutions. The dependence on which solution can be obtained in this region is mainly related to the initial conditions of the system. On the run up diagram Fig. 4 a) the jump is clearly shown and only the change of the oscillator displacement is shown. For the close down simulation a very small jump exists and it is happening for a very small change of the stall torque.

Diagram which describes the change of the excitation frequency in reference to the stall torque is formed and shown on Fig. 5. In this figure the behaviour of the excitation frequency is in someway predicted by the diagram of the electromotor characteristic. By increasing the stall torque value at some point the system starts to behave identically to a system with ideal excitation. Both simulations, run up and close down, are conducted where the results are shown in the same way as in the previous two figures. Position of jumps during the simulations are easy to notice because they represent the positions where sudden change of the excitation frequency is happening during the increase (run up) or decrease (close down) of the stall torque value. The value of the excitation frequency in Fig. 5 is the average value in the steady state work regime of the system.

Frequency response diagram which, Fig. 6, is showing how the extreme values of the oscillator displacement in reference to the average value of the excitation is changing. Different regions in the frequency response diagram can be noticed. Region of non-impact solutions where the maximum value of the oscillator displacement is below the position of the impact wall. Region where impact solutions are existing, on the diagram this region is when the maximum displacement is equal to the position of the impact wall. On the summary diagram where the results from the run up and close down simulation are shown, Fig. 6 c), region with two solution can be distinguished where impact and non impact solutions can occur depending on the initial conditions of the system. During the run up and close down simulation oblique jumps are noted. These jumps are characterized with a sudden change in the excitation frequency and oscillation amplitude and they are illustrated on the diagram with pink arrows. The frequency response diagram is related to steady state solutions, which are extracted from each simulation for different values of the stall torque. From the frequency response diagram possible transient solution and regions where they could occur can be noticed. Transient solutions are related to regions where a change from non-impact to impact solutions or vice versa is happening. These solutions are analysed in the following chapter of the paper.

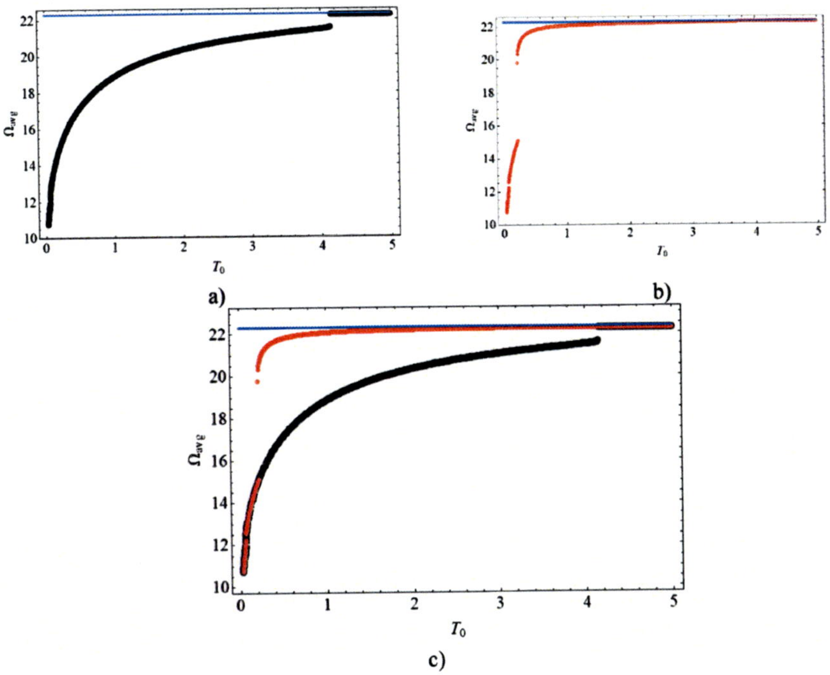

Fig. 5. Average values of the steady state excitation frequency value in reference to the stall torque value: a) run up simulation; b) close down simulation; c) combined run up and close down

4 Transient Motion Analysis

On diagrams shown in Fig. 4, Fig. 6, and Fig. 5 the results shown are related to the steady state motion where the excitation frequency is shown as it's average value in Fig. 4 and Fig. 4. From these diagrams some transient regions between impact and non-impact solutions are noted and a region with multiple solution's is noted. To understand what is happening in the transient regions and how the exact value of the excitation frequency is behaving it is necessary to analyse the full motion of the system from the start to the moment when the system is getting in the steady state work regime.

In the following figures diagrams of the oscillator displacement and the excitation frequency are shown for different values of the stall torque. The values of the stall torque which are chosen for the transient motion analysis are based on the analysis of the steady state motion and the specific regions which were noted and mentioned in the previous chapter.

On Fig. 7 and Fig. 8 diagrams are for small values of the stall torque. The oscillation amplitude increases where by reaching the steady state motion it doesn't hit the wall but it is very close. The excitation frequency increases also and when it reaches the steady state work regime it varies a little bit about a value that was used as the average value in the steady state analysis. On diagrams c) and d) of every following figure the full green line shows the values of the excitation frequency which is representing the value

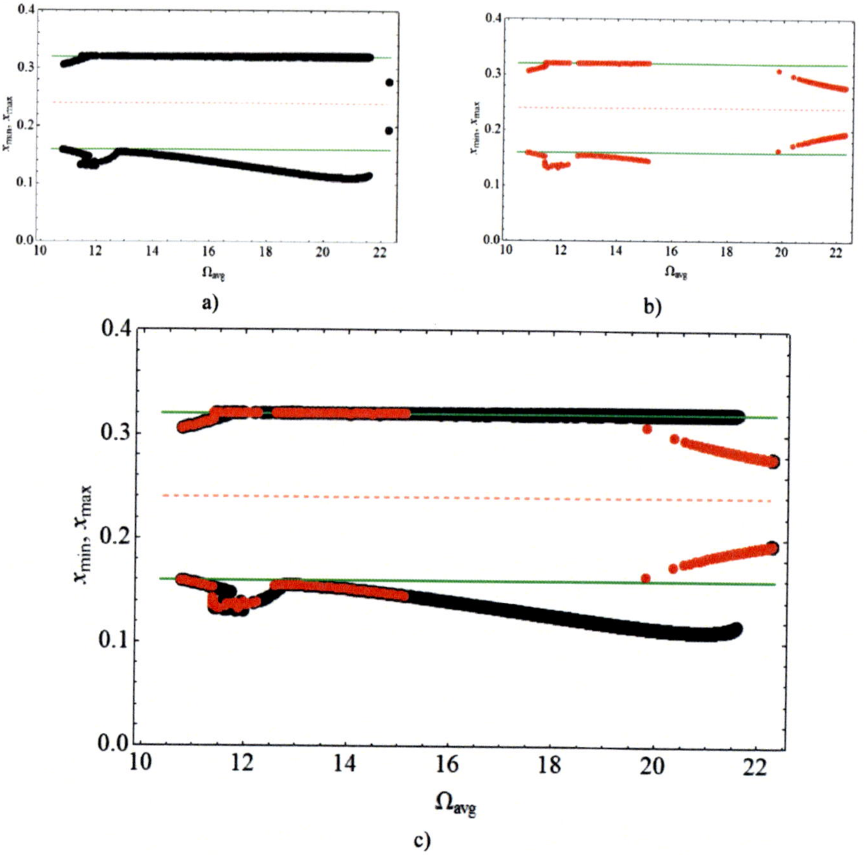

Fig. 6. Frequency response diagram: a) run up simulation; b) close down simulation; c) combined run up and close down

if the system would be with ideal excitation. The two figures represent the region with non-impact solutions as shown on Fig. 4, Figure, Fig. 6 and Fig. 5.

On Fig. 9 are shown results which are related to the region with multiple solutions. The displacement diagrams points out that the solutions are impact ones. From the earlier figures shown in the previous chapter they could be non-impact solutions and this depends only on the initial conditions. The excitation frequency diagrams show that for higher values of the stall torque the excitation frequency response is much faster and the variation about the average value in the steady state regime are bigger.

Figures Fig. 10, Fig. 11 and Fig. 12 show results for increased values of the stall torque. Three different values are analysed and they are related to the transient region during the run up simulation where the impact happens. As it can see the same time interval is analysed in all three cases. The first case, shown on Fig. 10, relates to impact solutions where the steady state regime last almost for the full interval. The change of the excitation frequency is stable when it reaches steady state work regime. For an slight increase of the stall torque, Fig. 11, the system behaves differently. The final solution

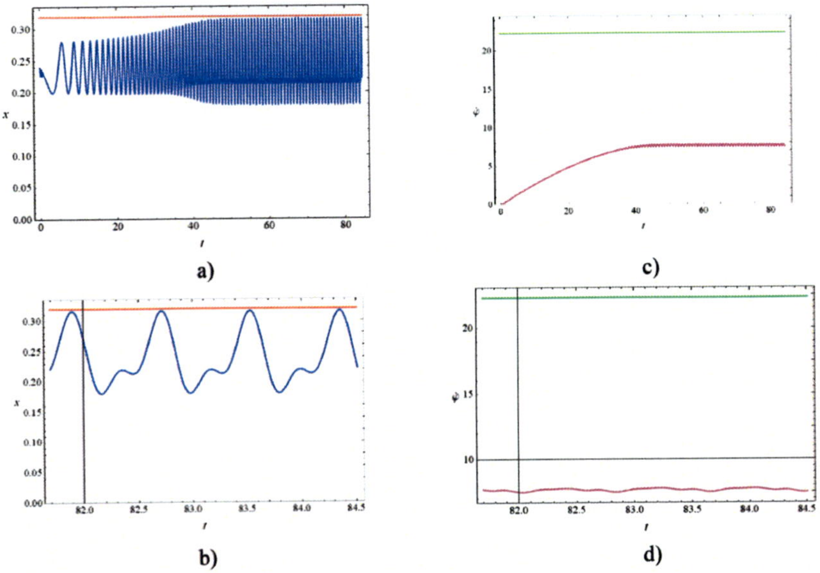

Fig. 7. Displacement diagram of the oscillator and excitation frequency diagram in function of time for the case $T_0 = 0.0180$ Nm: a) Full range $x - t$ diagrm, b) Last two period $x - t$ diagrm, c) Full range $\dot{\varphi} - t$ diagram; d) Last two $\dot{\varphi} - t$ diagram

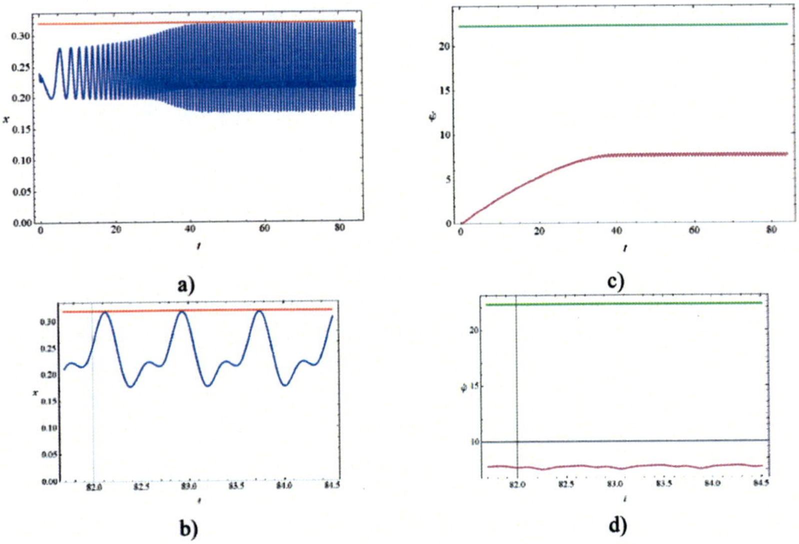

Fig. 8. Displacement diagram of the oscillator and excitation frequency diagram in function of time for the case $T_0 = 0.020$ Nm: a) Full range $x - t$ diagrm, b) Last two period $x - t$ diagrm, c) Full range $\dot{\varphi} - t$ diagram; d) Last two $\dot{\varphi} - t$ diagram;

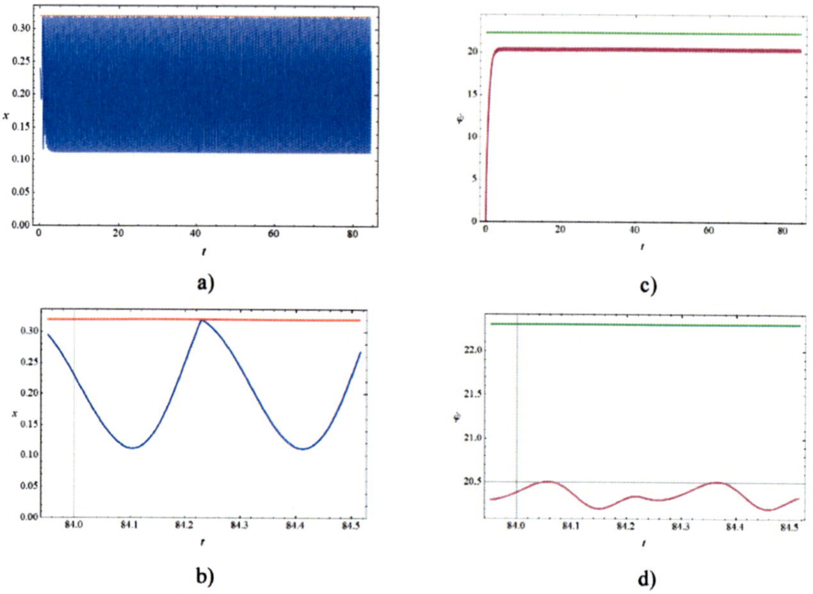

Fig. 9. Displacement diagram of the oscillator and excitation frequency diagram in function of time for the case $T_0 = 2.000$ Nm: a) Full range $x - t$ diagrm, b) Last two period $x - t$ diagrm, c) Full range $\dot{\varphi} - t$ diagram; d) Last two $\dot{\varphi} - t$ diagram;

in the steady state work regime is non-impact and it changes suddenly from an impact work regime. During that moment sudden change of the excitation frequency happens and the oscillation amplitude drops. This phenomenon is related to the jumps in the frequency response diagram. As explained in [1] the jumps are occurring so that the change happens from an unstable to a stable solution. As per the shown diagram the unstable solution is the impact solution and the stable one is the impact solution. By additional increase of the stall torque the jump happens earlier. Based on the steady state diagram the jump happens only ones. From this analysis it can be noticed that the jumps are happening after some time period and based on the analysis of the transient motion the time interval plays a big role in what solution will be obtained for the steady state motion. Based on this and the practical importance of this discovery leads us to analyse these kinds of systems from different viewpoints. One and the first one is related to the different regions which are based on the steady state motion analysis, second one is related to the transient motion analysis where the time interval in which the system operates can lead the system to behave as an impact or non-impact system. From the perspective if the system should work as an impact or non-impact system for a specific value of the stall torque as the two values on Fig. 11 and Fig. 12. If the machine stops to operate every 60 or 40 s for example, but if the machines would work longer they would operate just as oscillating machines.

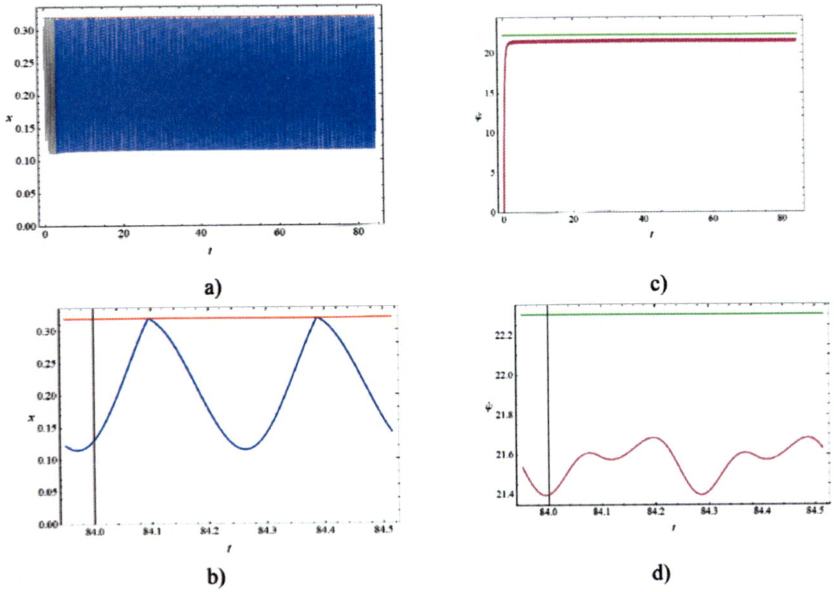

Fig. 10. Displacement diagram of the oscillator and excitation frequency diagram in function of time for the case $T_0 = 4.140$ Nm: a) Full range $x - t$ diagrm, b) Last two period $x - t$ diagrm, c) Full range $\dot{\varphi} - t$ diagram; d) Last two $\dot{\varphi} - t$ diagram;

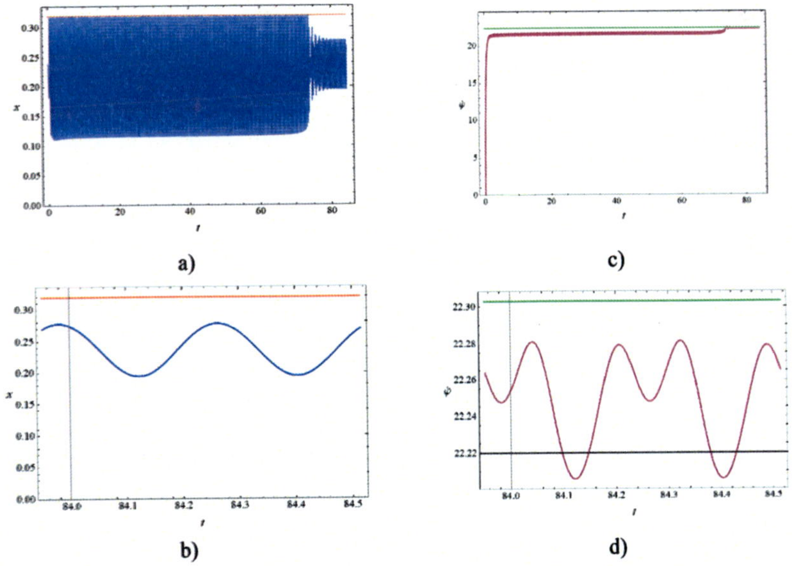

Fig. 11. Displacement diagram of the oscillator and excitation frequency diagram in function of time for the case $T_0 = 4.145$ Nm: a) Full range $x - t$ diagrm, b) Last two period $x - t$ diagrm, c) Full range $\dot{\varphi} - t$ diagram; d) Last two $\dot{\varphi} - t$ diagram;

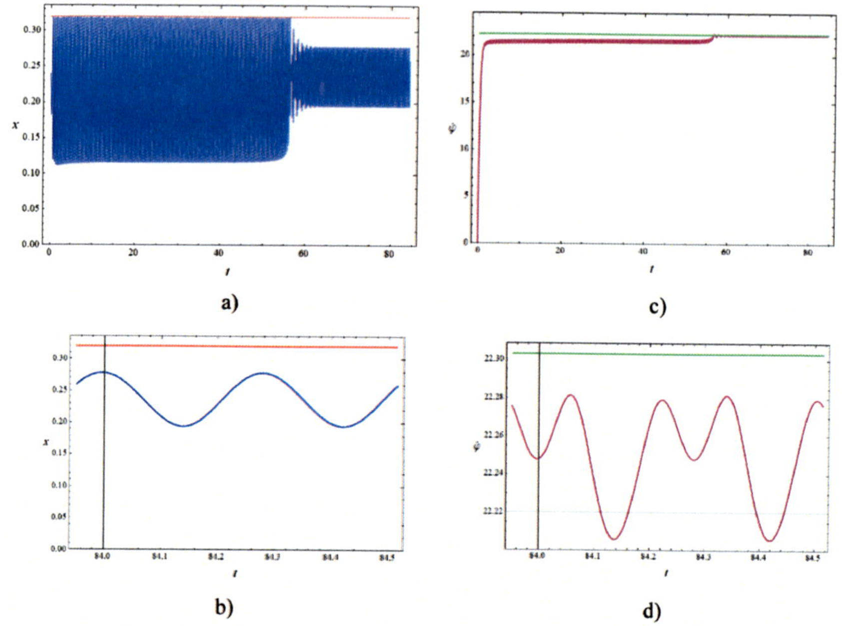

Fig. 12. Displacement diagram of the oscillator and excitation frequency diagram in function of time for the case $T_0 = 4.150$ Nm: a) Full range $x - t$ diagrm, b) Last two period $x - t$ diagrm, c) Full range $\dot{\varphi} - t$ diagram; d) Last two $\dot{\varphi} - t$ diagram;

5 Conclusions

A physical model of a vibro-impact system with non-ideal excitation is shown and analysis of such system with a nonlinear spring is considered. The physical model is explained and a mathematical model is formed and represented as a coupled system of nonlinear differential equations of second order without any approximations. The impact model is an inelastic model where the coefficient of restitution describes how much the oscillator speed is changing after every impact. The mathematical model is solved using numerical methods in Mathematica. Run up and close down simulation's are used to obtain all possible stable solutions. The varying parameter is the stall torque. Three different type of diagrams are obtained and shown in the chapter related to the steady state motion analysis. First diagram is the frequency response diagram which is shown, the second one presents how the displacement amplitude of the oscillator changes with the stall torque and the third diagram illustrates the change of the excitation frequency with varying the stall torque. From this analysis different regions are noticed, how the stall torque is influencing the oscillator motion. Regions of impact, non-impact and multiple solution are pointed out as positions where jumps are occurring. The jumps are pointing out that for some values of the stall torque and specific values of the initial conditions a sudden change in the excitation frequency and the oscillation amplitude could happen. These analysis is additionally widen by looking in the displacement and the excitation frequency diagram in reference to time. From these diagrams was noticed that a system can work as an impact system for a specific time interval and then change to an non-impact work regime. For different parameters it is noticed that these interval can be shorter or longer. The practical importance of this research is pointed out.

Future work related to the transient solution could be based on a wider spectre of the influencing parameters. Based on the method used in this analysis, the method of continuation, where different initial conditions are used the influence of them could have a big practical use in the sense how to adjust this kind of machines. This can be done using basin of attraction diagrams. Because of the complexity of the mathematical model analysis of unstable solution would be also interesting using the shooting method.

References

1. Zukovic, M., Hajradinovic, D., Kovacic, I.: On the dynamics of vibro-impact systems with ideal and non-ideal excitation. Meccanica **56**(2), 439–460 (2021). https://doi.org/10.1007/s11012-020-01280-5
2. Hajradinovic, D., Zukovic, M., Kovacic, I.: Influence of the stall torque on a vibro-impact system with non-ideal excitation. In: 2nd International E-Conference on Engineering, Technology and Management - ICETM 2020, place: Online (Via Video Conference), pp. 1–6 (2020)
3. Zukovic, M., Cveticanin, L.: Chaos in non-ideal mechanical system with clearance. J. Vib. Control **15**(8), 1229–1246 (2009)
4. Lampart, M., Zapomel, J.: Dynamics of the electromechanical system with impact element. J. Sound Vib. **332**, 701–713 (2013)
5. Lampart, M., Zapomel, J.: Dynamic properties of the electromechanical system damped by an impact element with soft stops. Int. J. Appl. Mech. **06**(02), 1450016 (2014)
6. Moraes, F.H., Pontes, B.R., Jr., Silveira, M., Balthazar, J.M.: Influence of ideal and non-ideal excitation sources on the dynamics of a nonlinear vibro-impact system. J. Theor. Appl. Mech. **51**, 763–774 (2013)
7. Souza, S.L.T., Caldas, I.L., Viana, R.L., Balthazar, J.M.: Control and chaos for vibro-impact and non-ideal oscillators. J. Theor. Appl. Mech. **46**, 641–664 (2008)
8. de Souza, S.L.T., Caldas, I.L., Viana, R.L., Balthazar, J.M., Brasil, R.M.L.R.F.: Impact dampers for controlling chaos in systems with limited power supply. J. Sound Vib. **279**, 955–967 (2005)
9. Babitsky, V.I.: Theory of Vibro-Impact Systems and Applications. Springer, Heidelberg (1998)
10. Marzbanrad, J., Shahsavar, M., Beyranvand, B.: Analysis of force and energy density transferred to barrier in a single degree of freedom vibro-impact system. J. Central South Univ. **24**(6), 1351–1359 (2017). https://doi.org/10.1007/s11771-017-3539-0
11. Raouf, A.I.: Vibro-Impact Dynamics - Modeling, Mapping and Applications. Springer, Heidelberg (2009)
12. Batako, A.D., Babitsky, V.I., Halliwell, N.A.: Modelling of vibro-impact penetration of self-exciting percussive-rotary drill bit. J. Sound Vib. **271**, 209–225 (2004)
13. Yuri, V.M., Reshetnikova, S.N.: Dynamical interaction of an elastic system and a vibro-impact absorber. Math. Probl. Eng. **2006**, 1–15 (2006)
14. Zukovic, M., Cveticanin, L., Maretic, R.: Dynamics of the cutting mechanism with flexible support and non-ideal forcing. Mech. Mach. Theory **58**, 1–12 (2012)
15. Cveticanin, L., Zukovic, M., Balthazar, J.M.: Dynamics of Mechanical Systems with Non-Ideal Excitation. Mathematical Engineering, Springer, Cham (2018)
16. Balthazar, J.M., Mook, D.T., Weber, H.I., Fenili, A., Belato, D., Felix, J.L.P.: An overview on non-ideal vibrations. Meccanica **38**, 613–621 (2003)

17. Cveticanin, L., Zukovic, M.: Motion of a motor-structure non-ideal system. Eur. J. Mech. A. Solids **53**, 229–240 (2015)
18. Cveticanin, L., Zukovic, M.: Non-ideal mechanical system with an oscillator with rational nonlinearity. J. Vib. Control **21**(11), 2149–2164 (2013)
19. Karthikeyan, M., Bisoi, A., Samantaray, A.K., Bhattacharyya, R.: Sommerfeld effect characterization in rotors with non-ideal drive from ideal drive response and power balance. Mech. Mach. Theory **91**, 269–288 (2015)

Numerical Modeling of Forced Convection of Nanoionic Liquid [C_4mpyrr] [NTf_2] with Al_2O_3 Particles

Amra Hasečić[1(✉)], Armin Hadžić[2], Siniša Bikić[3], and Ejub Džaferović[1]

[1] Faculty of Mechanical Engineering, University of Sarajevo, Sarajevo, Bosnia and Herzegovina
hasecic@mef.unsa.ba

[2] Faculty of Mechanical Engineering, University of Ljubljana, Ljubljana, Slovenia

[3] Faculty of Technical Sciences, University of Novi Sad, Novi Sad, Serbia

Abstract. The problem of forced heat transfer of a nanoionic liquid [C_4mpyrr] [NTf_2] with Al_2O_3 nanoparticles in a straight tube under the laminar flow regime and constant heat flux on the tube wall is numerically modeled and analyzed. The heat transfer characteristics were considered by analyzing the Nusselt number and heat transfer coefficient. Numerical simulation of three-dimensional and stationary problem was performed in the STAR-CCM + 2019.2 using the finite volume method. Thermophysical properties obtained experimentally were used to perform the numerical simulation as well as to calculate the characteristic quantities that describe the heat transfer. For the purpose of numerical simulations was used 3D numerical model. Numerical research was done with the following approximations: the effects of natural convection were neglected and fluid flow through a part of tube in the shape of a longitudinal wedge with angle of 5° was analyzed. The grid independence study was done, so the results presented are grid independent. The results obtained numerically are validated using the Shah's equation for ionic liquid. This paper is a follow up of the research published by Hadžić et al. [1] in 2020.

Keywords: Nanoionic fluid · Forced convection · Nanofluids · Heat transfer · CFD

1 Introduction

Nanofluids are a mixture of pure fluid and small particles up to 100 nm called nanoparticles. Conducted theoretical research shows that fluids with nanoparticles (nanofluids) have a higher thermal conductivity compared to basic fluids. In the recent years, ionic fluids have been increasingly used as basic fluids to form nanofluids [2–8]. Nanofluids in which particles are dispersed in ionic fluids are called nanoionic fluids. The basic ionic fluids that are most often used to obtain nanoionic fluids are: cations (pyrrolidinium, sulfonium), organic anions (axilsulfate, mathanesulfonate) and inorganic anions (Bis (trifluoromethylsulfonyl) imide, hexfluorophosphate). Ionic fluids are non-flammable and non-volatile under atmospheric conditions, and the suspension of nanoparticles in

N. Ademović et al. (Eds.): IAT 2021, LNNS 316, pp. 591–599, 2022.
https://doi.org/10.1007/978-3-030-90055-7_46

these fluids improves the thermophysical properties of ionic fluids [9]. In order to be able to compare the heat transfer characteristics of ionic and nanoionics liquid, in this paper we will perform the computational study of flow and heat transfer of ionic and nanoionic liquid in a tube in a forced convection regime.

Bondareva et al. [10] performed a numerical simulation of heat transfer within a heated paraffin-filled abyss, which was enhanced by nanoparticles. Hussein et al. [11] performed a numerical simulation of heat transfer and pressure drop characteristics of a hybrid nanofluid under a turbulent flow regime. Abdolbaqi et al. [12] performed a numerical study to analyse the improvement of heat transfer of nanofluids with a different volume concentration of nanoparticles in a turbulent flow regime through a flat channel with a constant state of thermal flux whereas Rostamani et al. [13] analysed the turbulent flow regime with different volume concentrations in a two-dimensional channel under constant heat flux conditions. A numerical study was conducted to investigate the convective heat transfer of Al2O3/water-ethylene glycol and CuO/water-ethylene glycol nanofluids in a circular tube with constant heat flux under a laminar flow regime [14]. Karima [14] focused their research on forced heat transfer and observed the changes that occur in the range of mass concentrations from 0% to 100% of nanoparticles in the ethylene glycol-water base fluid. Numerically, using single-phase and two-phase nanofluid models, Frad et al. [15] studied the convective heat transfer of nanofluids in the laminar mode in a circular tube at a constant wall temperature using the CFD approach. The results showed that the heat transfer coefficient increases significantly with increasing volumetric concentration of nanoparticles in the basic fluid. Sajjd et al. [16] performed a numerical analysis of forced heat transfer in a laminar flow regime of a mixture, obtained by suspending graphene oxide nanoparticles in a mixture of water and ethylene glycol, in a horizontal tube under constant heat flux conditions. Berberović and Bikić [17] performed modelling and analysis of the laminar flow of non-Newtonian fluid (ethylene glycol-Si3N4) in a horizontal tube during forced convection of heat transfer. Heat transfer was estimated by determining the Nusselt number and heat transfer coefficient for different volume fractions of Si3N4 nanoparticles at different nanofluid flow rates. A numerical study of the melting of n-octadecane enhanced by the addition of CuO nanoparticles within a horizontal cylindrical capsule was performed by Dhaidan et al. [18]. Mahdi et al. [19] performed a numerical simulation of heat transfer in organic PCM with improved nanoparticles and porous foam inside a heated tube. The analysis showed that nanoparticles contribute to the reduction of convection, so the effectiveness of nanoparticle addition depends on the porosity and increases with an increasing volume fraction of metal foam. Titan and others [20] performed a numerical simulation of the natural convection of N-butyl-N-methylpyrrolidinium bis {(trifluoromethyl) sulfonyl} imide ([C4mpyrr] [NTf2]) ionic fluid enhanced by alumina nanoparticles. The results of the research show that at a certain Rayleigh number nanoionic fluids have a lower Nusselt number compared to ionic fluids. Titan [21] performed a numerical study of the thermophysical properties and forced convection of an ionic fluid [C4mim] [NTf2] enhanced by alumina nanoparticles. The results of the research showed that at the laminar flow regime and at a higher concentration of nanoparticles, the nanoionic fluid has a longer hydrodynamic inlet length.

In this research the forced heat transfer of a ionic liquid [C4mpyrr] [NTf2] with Al_2O_3 nanoparticles in a straight tube under the laminar flow regime and constant heat flux on the tube wall is numerically modeled and analyzed. Numerical research was done with the following approximations: the effects of natural convection were neglected and fluid flow through a part of tube in the shape of a longitudinal wedge with angle of 5° was analyzed.

Numerical research can be based on two different approaches: modelling and simulation of single-phase nanofluid flow with unique properties that depend on the volume fraction of nanoparticles and modelling and simulation of two-phase flow with an explicit calculation of nanoparticle interaction with the basic fluid [22], here the single-phase approach is used where nanoionic fluid was treated as a Newtonian fluid.

2 Mathematical Model and Numerical Method

The mathematical model can be summarized in the following governing and constitutive equations:

Continuity equation

$$\frac{d}{dt}\int_{v}^{0} \rho dV + \int_{s}^{0} \rho v \bullet ds = 0 \tag{1}$$

where $\boldsymbol{\rho}$ is density, $\mathbf{v}$ is velocity vector

Momentum equation

$$\frac{d}{dt}\int_{v} \rho v dV + \int_{s} \rho v v \bullet ds = \int_{s} T \bullet ds + \int_{v} f_b dV \tag{2}$$

where $\mathbf{T}$ the Cauchy stress tensor and $\mathbf{f}_b$ is the resultant body force.

Thermal energy equation

$$\frac{d}{dt}\int_{v} \rho c_v T dV + \int_{s} \rho c_p T v \bullet ds = -\int_{s} q \bullet ds + \int_{v} (T : \text{grad} v) dV \tag{3}$$

where c_p is the specific heat, T is the temperature and $\mathbf{q}$ is the heat flux vector.

Stoke's law

$$T = 2\mu\dot{\mathbf{D}} - \frac{2}{3}\mu \text{div} v I - \rho I \tag{4}$$

where

$$\dot{\mathbf{D}} = \frac{1}{2}[\text{grad} v + (\text{grad} v)^T] \tag{5}$$

Is the rate of strain tensor, μ is the dynamic viscosity, p is the pressure and $\mathbf{I}$ is the unit tensor.

Fourier's law

$$q = -k\, \text{grad} T \tag{6}$$

where k is thermal conductivity.

Physical properties (density, thermal conductivity, viscosity and specific heat) of nanoionic liquid [C_4mpyrr] [NTf_2] with Al_2O_3 nanoparticles which is used in this study are experimentally investigated and reported in [22].

Numerical method employed for modeling the flow and heat transfer of ionic liquid was finite volume method. The methodology follows closely the one presented in [23] and [24].

3 Numerical Modeling of Forced Convection of Nanoionic Liquid

The steady-state flow of nanoionic fluid [C4mim] [NTf2] with Al_2O_3 nanoparticles through the horizontal straight tube of 1.75 m length and 0.0014 m diameter was analyzed with included convective heat transfer. The geometry of the problem is the same as in the [1]. The case study was analyzed for initial inlet temperatures of $T_{in1} = 293$ K and for three values of mass fraction of particles times 100 (wt%); 0.5, 1.0 and 2.5. The wall heat flux was constant $q = 13$ kWm^{-2}. At the outlet the pressure was set to 0 Pa. Every case was analyzed for two values of Reynold's number $Re = 100$ and $Re = 512$. For the purpose of set up inlet boundary conditions plug velocity profile was used. The heat transfer coefficient and Nusselt number were calculated using the equations given in [18]. The validation of the results along with the grid refinement was done for the ionic liquid [C4mim] [NTf2] and the results were compared with results obtained by using the Shah's equation. The results are presented in [1]. Here, the grid sensitivity study was also performed for 4 different values of cell sizes but for nanoionic liquid, where the cell size was systematically refined. The characteristics of the analyzed grids were given in the Table 1, while the used grid was shown on the Fig. 1. Characteristic cell size which is represented in the Table 1 is actually base size of cell.

Table 1. Grid characteristics

Grid	Characteristic cell size [mm]	Number of CVs
1	2.4	46688
2	1.2	198356
3	0.6	1353488
4	0.3	8493674

In the Fig. 2 the values of heat transfer coefficient along the pipe for $T_{in1} = 293$ K and $Re = 100$ for different grid sizes are shown.

It is seen from the Fig. 2 that the results are approaching each other as the grid refines and the grid independent result is obtained for grid 3 with 1353488 CVs, and cell size of 0.6 mm, therefore the grid 3 will be used in the following analyses.

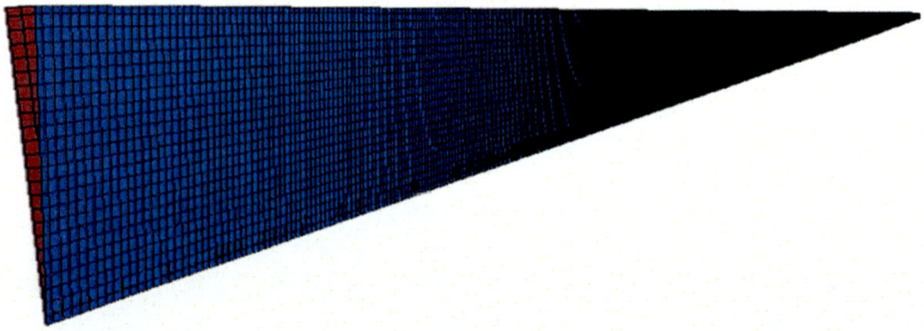

Fig. 1. The numerical grid

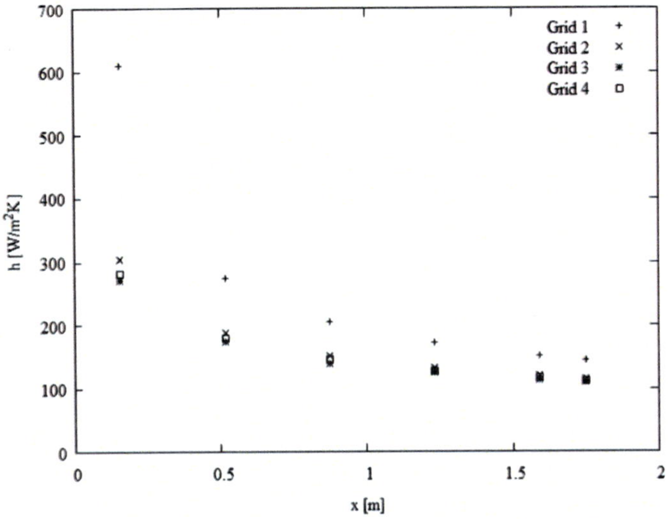

Fig. 2. Grid sensitivity study

4 Results and Discussion

The numerical results of heat transfer coefficient for ionic liquid and for nanoionic liquids with different wt% for Re = 100 are shown in the Fig. 3. It could be seen that the heat transfer coefficient increases as the wt% increases.

Figure 4 represents the values of Nusselt number along the pipe for different values of wt% as well as for ionic liquid. One can conclude that with the increase of the wt%, the Nusselt number increases. Also, the highest values of Nu number are at the entrance of the pipe.

Heat transfer coefficient and Nu number values along the pipe for different values of wt% and for Re = 512 are presented at Fig. 5 and 6. As for Re = 100, for Re = 512

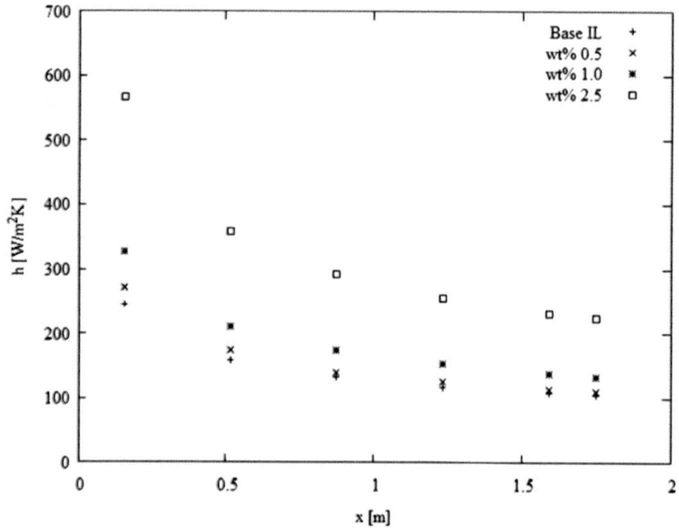

Fig. 3. Heat transfer coefficient along the pipe for ionic liquid (Base IL) and for different values of wt% for Re = 100

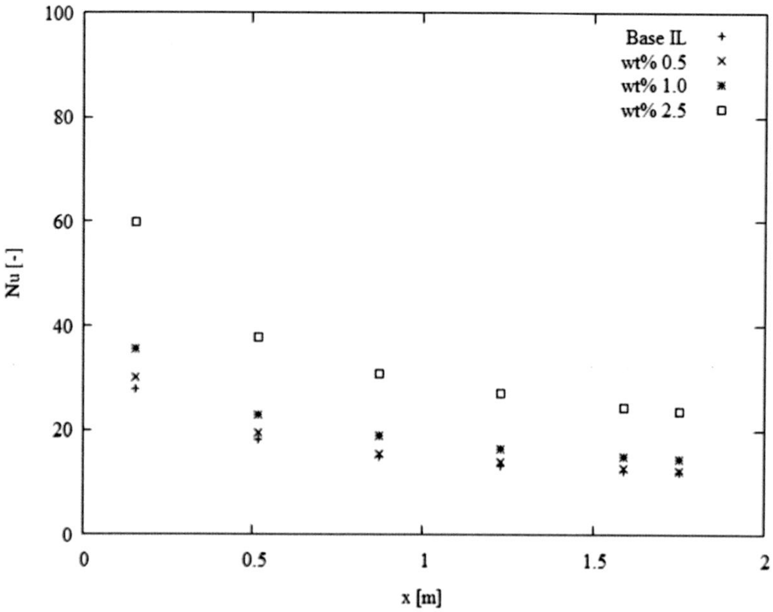

Fig. 4. Nusselt number values along the pipe ionic liquid (Base IL) and for different values of wt% for Re = 100

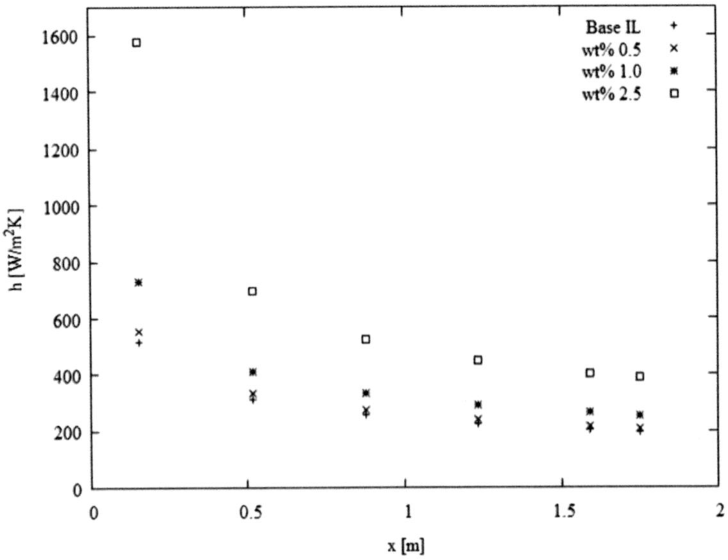

Fig. 5. Heat transfer coefficient along the pipe for ionic liquid (Base IL) and for different values of wt% for Re = 512

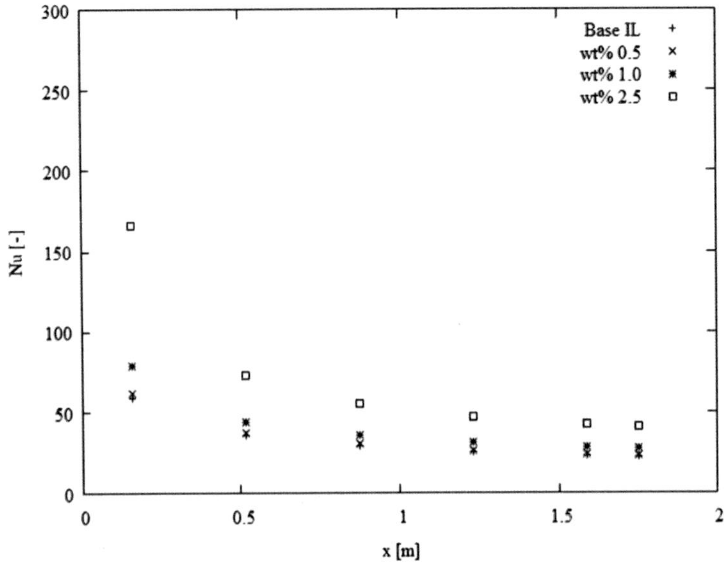

Fig. 6. Nusselt number values along the pipe ionic liquid (Base IL) and for different values of wt% for Re = 100

it can be seen that with the increase of wt% the values of both heat transfer coefficient and Nu number increses.

Also, it can be concluded that for higher values of Re number, higher values of both heat transfer coefficient and Nu number are observed.

5 Conclusion

This paper presents the numerical investigation of the forced heat transfer of a nanoionic liquid [C_4mpyrr][NTf_2] with Al_2O_3 nanoparticles through the straigth tube. The geometric and physical models were defined that adequately describe the problems, as well as the numerical grid on which the simulations were performed in the STAR-CCM + 2019.2 software. Numerical research was done with the following approximations: the effects of natural convection were neglected and fluid flow through a part of tube in the shape of a longitudinal wedge with angle of 5° was analyzed. In order to obtain a grid independent solutions the grid sensitivity study was performed. The presented analysis of the obtained results of the Nusselt number and the heat transfer coefficient of the ionic and nanoionic liquid shows the change of the considered parameters along the tube for different wt% and different Reynolds numbers. During the analysis of the numerical results of the Nusselt numbers and the heat transfer coefficient of the both ionic and nanoionic liquid, a decrease in the Nusselt numbers and the heat transfer coefficient with a decrease of a wt% was found. Also, it is concluded that the increase in a Reynolds number results with the increase in both heat transfer coefficient as well as for Nusselt number. The Nusselt number and the heat transfer coefficient of the ionic and nanoionic liquid decrease exponentially along the tube for any case under consideration.

References

1. Hadžić, A., Hasečić, A., Bikić, S., Džaferović, E.: Numerical Modeling of Forced Convection of Ionic Liquid [C_4mpyrr] [NTf_2]. In: ICNAAM (2020)
2. Choi, S.U.S.: Enhancing Thermal Conductivity of Fluids with Nanoparticles. International Mechanical Engineering Congress and Exposition, New York (1995)
3. Titan, C.P., Morshed, A.K.M.M., Khan, J.A.: (2013) Nanoparticle enhanced ionic liquids (NEILS) as working fluid for the next generation solar collector. Proc. Eng. **56**, 631–636 (2013)
4. Bhattacharjee, A., et al.: (2014) Thermophysical properties of sulfonium and ammonium-based ionic liquids. Fluid Phase Equilib **381**, 36–45 (2014)
5. Titan, P.C., Morshed, A.K.M.M., Fox, E.B., Visser, A.N., Bridges, N.J., Khan, J.A.: (2014) Thermal performance of ionic liquids for solar thermal applications. Exp. Therm. Fluid Sci. **59**, 88–95 (2014)
6. Titan, P.C., Morshed, A.K.M.M., Fox, E.B., Khan, J.A.: Effect of nanoparticle dispersion on thermophysical properties of ionic liquids for its potential application in solar collector. Proc. Eng. **90**, 643–648 (2014)
7. Titan, P.C., Morshed, A.K.M.M., Fox, E.B., Khan, J.A.: Thermal performance of Al_2O_3 nanoparticle enhanced Ionic Liquids (NEILs) for concentrated Solar power (CSP) applications. Int. J. Heat Mass Transf. **85**, 585–594 (2015)

8. Titan, P.C., Morshed, A.K.M.M., Fox, E.B., Khan, J.A.: Enhanced thermophysical properties of NEILs as heat transfer fluids for solar thermal applications. Appl. Therm. Eng. **110**, 1–9 (2017)
9. Chereches, E.I., Sharma, K.V., Minea, A.A.: A numerical approach in describing ionnanofluids behavior in laminar and turbulent flow. Cont. Mech Termodyn. **2018**(30), 657–666 (2018)
10. Bondareva, N.S., Buonomo, B., Manca, O., Sheremet, M.A.: Heat transfer inside cooling system based on phase change material with aluminia nanoparticles. Appl. Therm. Eng. **144**, 972–981 (2018)
11. Hussein, A.M., Noor, M.M., Kadirgama, K., Ramasamy, D.: Heat enhancement using hybrid nanoparticles in ethylene glycol through a horizontal heated tube. Int. J. Automim. Mech. Eng. **14**, 4183–4419 (2017)
12. Abdolbaqi, M.K., Azwadi, C.S., Mamat, R., Azmi, W.H., Naiafi, G.: Nanofluids heat transfer through straight channel under turbulent flow. Int. J. Autom. Mech. Eng. **11**(1), 2294–2305 (2015)
13. Rostamani, M., Hosseinizadeh, S.F., Gorji, M., Khodadadi, J.M.: Numerical study of turbulent forced convection flow of nanofluids in a long horizontal duct considering variable properties. Int. Commun Heat Mass Transf. **37**(10), 1426–1431 (2010)
14. Karima, B., Mahfoud, K.: Convective heat transfer of Al_2O_3 and CuO nanofluids using various mixtures of water-ethylene glycol as base fluids. Eng. Technol. Appl. Sci. Res. **7**(2), 1496–1503 (2017)
15. Frad, M.H., Esfahanv, M.N., Talaie, M.R.: Numerical study of convective heat transfer of nanofluids in a circular tube two-phase model versus single-phase model. Int. Commun Heat Mass Transf. **37**(1), 91–97 (2010)
16. Sajjad, M., Kamran, S.M., Shaukat, R., Zeinelabdeen, M.I.M.: Numerical investigation of laminar convective heat transfer of grapheme oxide/ethylene glycol-water nanofluids in a horizontal tube. Eng. Sci. Technol. Int. J. **21**(4), 727–735 (2018)
17. Bikić, S., Berberović, E.: Computational Study of Flow and Heat Transfer Characteristics of EG-Si3N4 Nanofluid in Laminar Flow in Pipe in Forced Convection Regime. Energies **13**(1), 74 (2019)
18. Dhaindan, N.S., Khodadadi, J.M., Al-Hattab, T.A., Al-Mashat, S.M.: Experimental and numerical study of contained melting of n-octadecane with CuO nanoparticle dispersion in horizontal cylindrical capsule subjected to a constant heat flux. Int. J. Heat Mass Transf. **67**, 523–534 (2013)
19. Mahdi, J.M., Nsofor, E.C.: Melting enhancement in triplex-tube latent heat energy storage system using nanoparticles-metal foam combination. Appl. Energy **191**, 22–34 (2017)
20. Titan, C.P., Morshed, A.K.M.M., Khan, J.M.: Numerical investigation of natural convection of nanoparticle enhanced ionic liquids (NEILs) in enclosure heated from below. In: AIP Conference Proceedings, vol. 1754, no. 1, p. 050019 (2016)
21. Titan, P.C.: Investigation of Thermal Performance of Nanoparticle Enhanced Ionic Liquids (NEILs) for Solar Collector Applications, PhD thesis, Scholar Commons, University of South Carolina (2014)
22. Minea, A.A., et al.: A benchmark study on the numerical approach in nanofluids simulation. Int. Commun. Heat Mass Transf. **108**, 104292 (2019)
23. Demirdžić, I., Muzaferija, S.: Numerical method for coupled fluid flow, heat transfer and stress analysis using unstructured moving meshes with cells of arbitrary topology. Comput. Methods Appl. Mech. Engrg **125**, 235–255 (1995)
24. STAR-CCM + User Manual, Siemens

Computational Prediction of Water Flow Rates Through a Simplified Tainter-Gate Model

Tarik Čorbo[1], Muris Torlak[1](✉), and Mensud Điđelija[2]

[1] Mechanical Engineering Faculty, University of Sarajevo, Vilsonovo Šetalište 9, 71000 Sarajevo, Bosnia-Herzegovina
torlak@mef.unsa.ba

[2] Mechanical Engineering Faculty, University Džemal Bijedić in Mostar, Sjeverni Logor B.B., 88104 Mostar, Bosnia-Herzegovina

Abstract. The water flow rates through a 2D model of a tainter-gate are calculated. The geometric *computer-aided design* (CAD) model is created as parametric one, allowing thus simple and efficient variations of the model dimensions, compatible with automatic update of the solution domains and generation of computational meshes. The problem of water flow is considered as fluid flow with free-surface, which is computed using a *computational-fluid dynamics* (CFD) method based on *Volume-of-Fluid* (VoF) approach. The flow rates and their corresponding discharge coefficients are calculated for a number of different values of the water level in reservoir, the gate opening size and the intake depth.

Keywords: Hydraulics · Hydropower plants · Free-surface flows · Computational fluid dynamics (CFD) · Parametric CAD

1 Introduction

Tainter gates are hydraulic structures which are, among others, frequently used at hydropower plant dams in order to maintain the water level in reservoirs and control the necessary water flow rates of the discharge over the dam spillway. Accurate control of flow rate is especially important for hydropower plants in cascade arrangement or for power plants in the densely populated areas. In case of too large flow rates expected, e.g. caused by sudden changes in weather conditions [1], significant threats and security risks are possible and may cause severe damages and high economic costs due to floodings downstream the reservoirs (e.g. erosion and damage of power-plant structural elements, submergence of power house, debris accumulation, flooding of infrastructure and settlements, environmental damages etc.).

In a number of cases, the water flow rate is not directly measured at the hydropower plant dams or gauging is not done properly, such as in case described in [2]. Instead, the flow rates are estimated from the characteristic curves of the gates, obtained in experimental model testings, showing the relation between reservoir water level, gate opening, and flow rate. Beside the results of hydraulic-model testings (usually achieved on reduced models, where the scale effects may play a certain role) [3], availability of

N. Ademović et al. (Eds.): IAT 2021, LNNS 316, pp. 600–609, 2022.
https://doi.org/10.1007/978-3-030-90055-7_47

calculated data obtained on detailed computational models in realistic scales, such as those created by methods of *computational fluid dynamics* (CFD) [4], would increase reliability and reduce uncertainty of discharge impact assessments.

The goals of this work are: (a) to examine applicability of the available mathematical models and the computational method, (b) to provide a set of discharge values for a typical model of a tainter gate for an orientation estimation of expected flow rate in real cases with similar conditions, (c) to examine applicability of the choice of boundary condition types, and (d) identify conditions and issues for further improvements.

2 Problem Description

Figure 1 shows a simplified 2D model of a tainter gate, which is considered in the present analysis. Influence of the following parametric values on the water flow rate notepaunder the gate is investigated: h_1 – the water level in reservoir, h_2 – the opening size beneath the gate, h_3 – the intake depth. The variables p_{tot}, p_{stat} and p_0 denote total pressure, static pressure and atmospheric pressure, respectively.

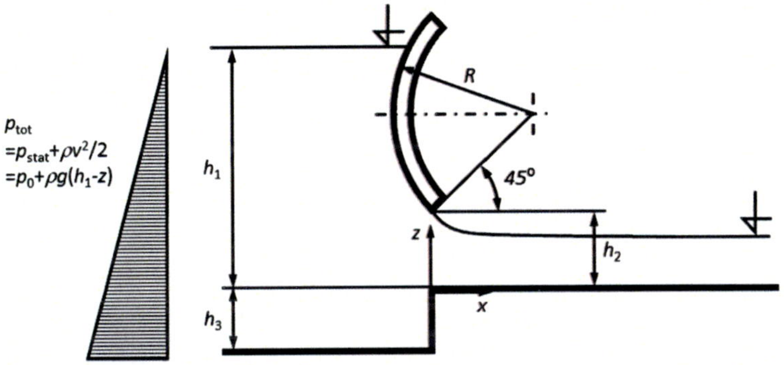

Fig. 1. Cross-section of a tainter gate with the basic dimensions used in calculations. Water flow direction is from left to right.

In the case of quiescent water in reservoir, static-pressure increases linearly with depth. In the present analysis, the domain boundary on the reservoir side (left side in Fig. 1) is relatively close to the gate. Hence, the total-pressure distribution along that domain boundary is adopted to be equal to the static-pressure distribution in the far, quiescent reservoir region and it is specified as constant boundary condition. On the downstream side, condition of zero-gradient of flow properties in downstream direction is assumed. The shape and position of the water free-surface is not known in advance, since it depends on the input geometric and boundary conditions. This implies that it cannot be used as a boundary condition, and it has to be obtained as a part of the solution.

3 Method and Computational-Model Setup

Solution of the problem is accomplished through an integrated workflow, which includes: parametric CAD geometric modelling, automatic computational-mesh generation, and numerical solution of the two-phase fluid flow (water-air) with free surface. The complete process is done using a single computer program, the commercial software *STAR-CCM+®* [5].

The geometric model is created as a parametric one. The dimensions of the geometric model are shown in Fig. 2. In the present study, however, only the following dimensions are varied: h_1 describing the reservoir water level and herewith the effect of the pressure on the inflow side (not shown in figure, since it is not a geometric dimension), h_2 describing the size of the gate opening, and herewith indirectly the resistance to the flow, and h_3 describing the intake depth, and herewith the flow resistance at the inflow side. All other dimensions are kept constant.

For each simulated case, the computational mesh is automatically generated using Cartesian cells (typically, square-shaped) in the core region, which are trimmed at the boundaries in order to appropriately describe the wall shapes. In the region around the expected water level in the reservoir, the mesh is refined in z-direction in order to capture the free surface motion appropriately. The mesh is extruded expanding toward the inflow side and toward the outflow side, yielding fewer mesh cells than in case of equal cell sizes, in order to reduce computational costs. At the walls, 5 to 20 prism layers are used in order to capture strong velocity variations in the boundary layers. A typical mesh, created for the case $h_1 = 5$ m, $h_2 = 2$ m, and $h_3 = 2$ m, containing about 240 000 cells, is shown in Fig. 3.

The Navier-Stokes equations describe the motion of a single-phase fluid, which can be given in integral form for a part of space considered as control volume as follows:

$$\frac{\partial}{\partial t} \int \rho \mathrm{d}V + \oint \rho \vec{v} \bullet \mathrm{d}\vec{S} = 0, \tag{1}$$

$$\frac{\partial}{\partial t} \int \rho \vec{v} \mathrm{d}V + \oint \rho \vec{v} \otimes \vec{v} \bullet \mathrm{d}\vec{S} = \oint \boldsymbol{\sigma} \bullet \mathrm{d}\vec{S} + \int \rho \vec{f_b} \, \mathrm{d}V, \tag{2}$$

where t is the time, ρ is the density, V is the volume of the considered control volume, $\vec{v}$ is the velocity vector, $\vec{S}$ is the control surface bounding the control volume, $\boldsymbol{\sigma}$ is the stress tensor, and $\vec{f_b}$ is the vector of body forces, e.g. taking gravity into account.

Depending on nature of the fluid response and the flow, as in the cases considered here (negligible volume change when exposed to external pressure, flow velocities significantly lower than the speed of sound), fluids is regarded as incompressible. Thus, the density can be considered constant. Consequently, the constitutive equation for incompressible fluids is reduced to:

$$\boldsymbol{\sigma} = 2\mu\dot{\boldsymbol{\varepsilon}} - p\mathbf{I}, \tag{3}$$

where p is the pressure, $\mathbf{I}$ is the unit tensor, μ is the dynamic viscosity, and $\dot{\boldsymbol{\varepsilon}}$ is the strain-rate tensor:

$$\dot{\boldsymbol{\varepsilon}} = \frac{1}{2}\left[\nabla\vec{v} + \left(\nabla\vec{v}\right)^{\mathrm{T}}\right] \tag{4}$$

Modeling of immiscible two-phase flows, such as water-air flows with free surface as they appear in the problem considered in this study, is typically designed combining the

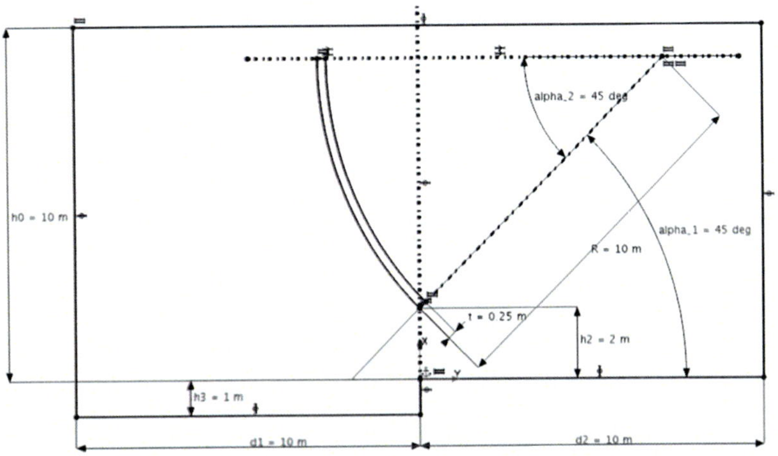

Fig. 2. Parametric geometric model used in the computations: the model is automatically updated upon the variation of the parameters h_1[1], h_2 and h_3.

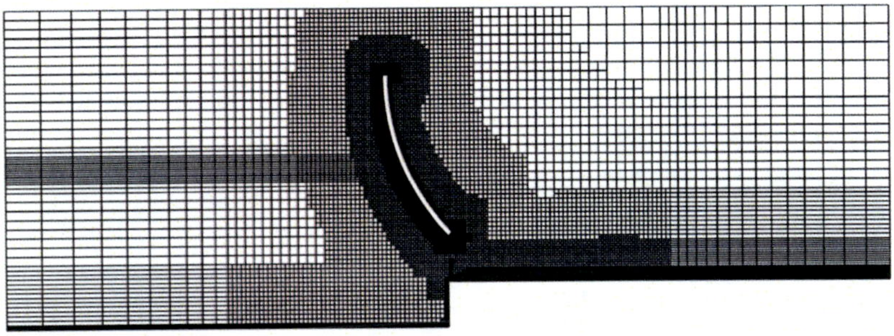

Fig. 3. The computational mesh for the case $h_1 = 5$ m, $h_2 = 2$ m, and $h_3 = 2$ m.

Eqs. (1–4) and interface capturing by virtue of the *Volume-of-Fluid* approach, such as in the works of [6,7]. One of the phases is considered as the transporting fluid (typically, the lighter, gas phase), while the other phase is considered as the transported one (typically, the heavier, liquid phase). The mathematical model includes, in addition to Eqs. (1–4), the scalar equation for the volume fraction c_l to describe distribution of the transported phase, and is written as:

$$\frac{\partial}{\partial t} \int c_l \mathrm{d}V + \oint c_l \vec{v} \bullet \mathrm{d}\vec{S} = 0 \qquad (5)$$

Distribution of the primary, transporting phase is obtained using the condition: $c_g + c_l = 1$. Turbulence effects can be captured by additional transport equations, depending on the turbulence model adopted. In this work, the variant of realizable k-ε model for all y^+ values [8,9] is used.

[1] The parameter h_1 is not shown in the figure, since it describes the position of the free surface, not a geometric dimension of the solution-domain boundaries.

The governing transport Eqs. (1–2), (5), and the turbulent transport equations are discretized using a *finite-volume method*, as described in [4, 5, 10]. Note that for incompressible fluids Eq. (1) reduces to divergence-free condition, which is not directly discretized. Instead, it is used in combination with the momentum Eq. (2) to derive the Poisson-type pressure-correction equation, which is discretized basically in the same way as the other transport equations.

Computational points, where values of the solution variables are calculated, are positioned at the geometric centers of the mesh cells. The volume and the surface integrals are approximated by the *midpoint rule*. For discretization of the convective terms the 2nd order accurate *linear-upwind scheme* is used. In Eq. (3), however, the convective term is discretized using the *high-resolution interface capturing scheme* HRIC [6,7] which is a non-linear blend of upwind and downwind cell face values, depending on the Courant number. Gradients at the cell centers are calculated as a blend of the approximation by virtue of Gauss method and the weighted least-squares approximation.

Essentially, the considered problem and the underlying mathematical model are time-dependent, requiring thus the simulated period of time to be discretized by dividing it into time steps, and the time-dependent terms to be discretized as well. In this work, anticipating the near-stationary flow, the 1st order-accurate implicit time integration scheme is used for that. The time-step size of 0.001 s is used at the start, advancing it to 0.02 s, having the flow developed and as the near-stationary state is approached.

The systems of linear algebraic equations obtained by discretization are solved using an *algebraic-multigrid* method.

4 Results

The main monitored value in this study is the non-dimensional discharge coefficient defined as:

$$c_D = \frac{\dot{V}}{w\delta\sqrt{2gh^*}}, \tag{6}$$

where is $\dot{V}$ the flow rate through the gate, w is the width of the gate (in 2D models it is equal to 1), δ is the water depth below the gate (in gate-obstructed flows, it is equal to h_2, otherwise $\delta < h_2$), g is the gravitational acceleration, and h^* is the elevation difference between the free surface in the reservoir and the centerline of the water stream below the gate: $h^* = h_1 - \delta/2$. In the cases of gate-obstructed flow, the water depth • is obtained from the computed free-surface position below the gate.

The water flow through the gate is simulated for a number of different combinations of h_1, h_2, and h_3.

Refinement of the computational mesh by halving the mesh cell size in cases with: (a) $h_1 = 3$ m, $h_2 = 3$ m, and $h_3 = 0$ m; (b) $h_1 = 5$ m, $h_2 = 2$ m, and $h_3 = 1$ m; (c) $h_1 = 7$ m, $h_2 = 1$ m, and $h_3 = 2$ m delivered the differences in flow rates about 1% or less, so that the same, original mesh template, as shown in Fig. 3, is considered to provide sufficient flow resolution and is used for all computations. Additionaly, for the cases with: (a) $h_1 = 7$ m, $h_2 = 1$ m, and $h_3 = 0$ m; and (b) $h_1 = 7$ m, $h_2 = 3$ m, and

$h_3 = 0$ m, the position of the left (upstream) and the right (downstream) boundary were moved further away from the region of interest, which yielded differences less than 1%. According to this, it was concluded that the locations of the inflow and outflow boundary do not disturb the region of interest.

The resulting discharge coefficients and the flow rates obtained on the original-mesh template are shown in Table 1 and Table 2, respectively.

Table 1. The calculated discharge coefficients [–].

	h_1 [m]								
	3			5			7		
	h_2 [m]			h_2 [m]			h_2 [m]		
h_3 [m]	1	2	3	1	2	3	1	2	3
0	0.698	0.666	0.630*	0.715	0.674	0.668	0.724	0.693	0.680
1	0.684	0.603	0.547*	0.709	0.674	0.630	0.719	0.692	0.664
2	0.677	0.587	0.525*	0.704	0.663	0.616	0.712	0.682	0.656

* Note that these cases represent unobstructed water flow with $\delta < h_2$.

Table 2. The calculated discharge per unit width [m^3/s / m].

	h_1 [m]								
	3			5			7		
	h_2 [m]			h_2 [m]			h_2 [m]		
h_3 [m]	1	2	3	1	2	3	1	2	3
0	4.887	8.339	8.46	6.723	11.945	16.608	8.177	15.038	21.207
1	4.789	7.55	8.065	6.664	11.944	15.652	8.114	15.021	20.702
2	4.74	7.355	7.818	6.613	11.743	15.325	8.046	14.808	20.457

In the considered range of parameter variations, the values of discharge coefficients vary significantly, from 0.525 to 0.724 (a change for about 38%). The variation of the flow rates is even more significant, from 4.74 m^3/s to 21.21 m^3/s (per unit width), or about 4.5 times. The variation of the discharge coefficient demonstrates clearly non-linear relation between the flow rate $\dot{V}$, the water level in the reservoir h_1, and the opening size h_2. At lower values of h_1 relation $\dot{V} - h_2$ is non-linear due to weak flow obstruction, while it is nearly linear for higher h_1 values.

From the discharge coefficient definition, Eq. (6), and the relation:

$$c_D{}^2 \rho g h^* = \frac{\rho v_2{}^2}{2}, \tag{7}$$

where v_2 is the depth-averaged water flow velocity below the gate, one can conclude that between about 27.5% (in case $h_1 = 3$ m, $h_2 = 3$ m, $h_3 = 2$ m) and about 52.5% (in case $h_1 = 7$ m, $h_2 = 1$ m, $h_3 = 0$ m) potential energy of water in reservoir (assuming quiescent condition there) is converted to kinetic energy below the gate. The remaining part of energy is assumed to be converted to the pressure below the gate, or dissipated.

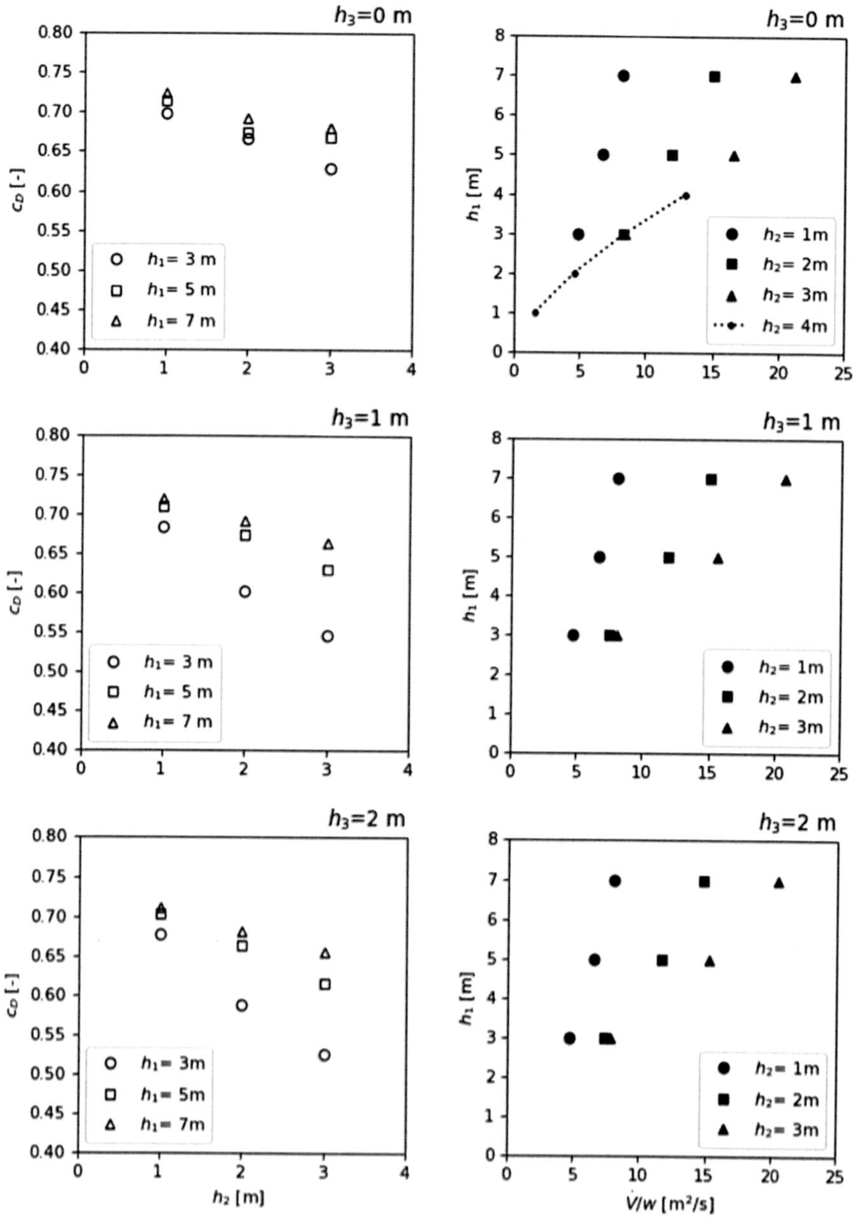

Fig. 4. Variation of the discharge coefficients (left) and the discharge per unit width (right) for different input-parameters values.

The trends of discharge coefficient variations and discharge per unit width are shown in diagrams given in Fig. 4.

Increase of the water level in reservoir (h_1) increases both the discharge coefficient and the flow rate, which is a consequence of the increased pressure level within the reservoir, while the outlet pressure remains at the same level.

Increase of the opening size (h_2) increases the flow rate at the constant value of h_1, as expected, since the shape-based flow resistance of the gate is getting smaller. Obviously, the flow rate through the gate varies non-linearly with the variation of the opening size h_2. On the other side, the discharge coefficient decreases.

Increase of the intake depth (h_3) reduces both the discharge coefficient and the flow rate. This is expected due to increased shape-based flow resistance ("minor" energy loss) at the intake. The change is however not very strong.

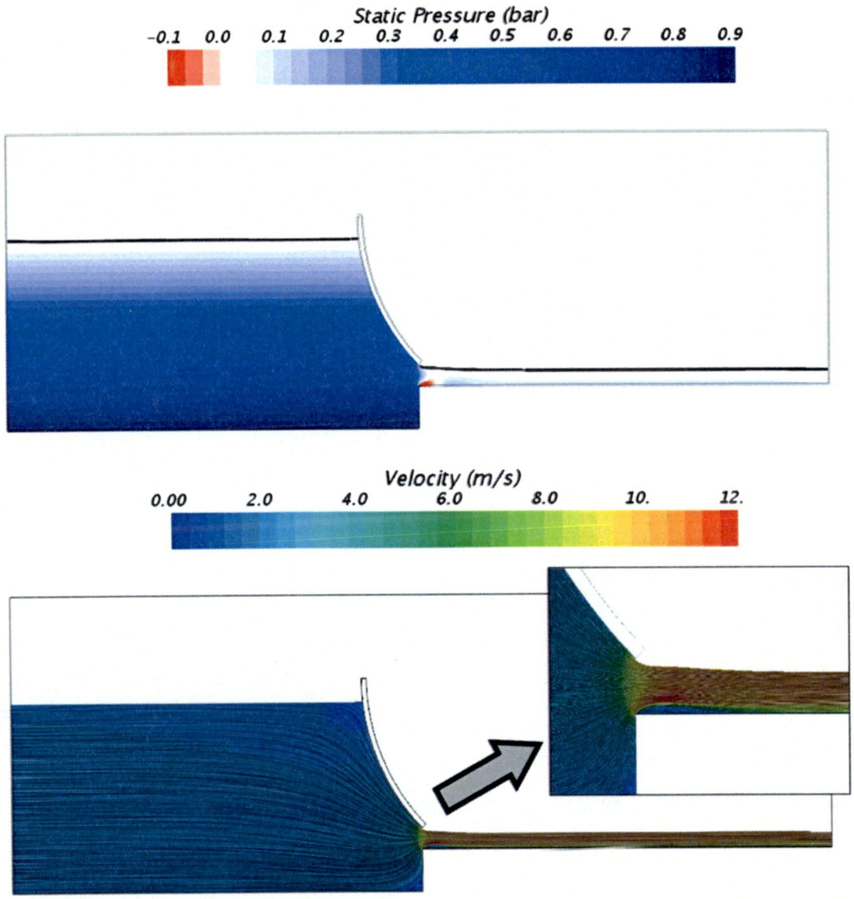

Fig. 5. Distribution of the static pressure and the velocity during water flow through the taintergate, the case with $h_1 = 7$ m, $h_2 = 1$ m, $h_3 = 2$ m.

Figure 5 shows the part of space occupied by water with the free surface, the static-pressure distribution, and the water flow velocity distribution, for the case with $h_1 = 7$ m, $h_2 = 1$ m, $h_3 = 2$ m.

The free surface in the reservoir seems to be very smooth and stable, and the flow toward the tainter gate is uniformly distributed with clearly seen nearly-hydrostatic pressure distribution. Due to significant cross-section area reduction determined by the tainter gate and the step-shaped bottom wall, the flow is redirected toward the gate opening, resulting in a considerable acceleration. Due to sharp edge of the step, immediately after the opening, a small recirculation region arises at the bottom wall, behind the upper edge of the step. In the core of this recirculation region, the pressure drops to a value below the atmospheric one (note negative static pressure values at this place), but still far enough above the saturation pressure, so that cavitation does not appear. The length of this low-pressure, recirculation region is comparable to the opening size. The elevation of the free surface downstream the gate, i.e. water depth, reduces to about $0.75h_2$, which is less than the critical depth $\delta_c = \left[\left(\dot{V}/w\right)^2/g\right]^{1/3} = 1.88$ m, revealing supercritical flow. After this, the flow keeps smooth, stable, nearly flat and almost horizontal free-surface until it reaches the outflow boundary (20 m downstream the gate). No hydraulic jump is detected in the tailwater within the solution domain used here. Consequently, the velocity field in the tailwater is quite uniform in the streamwise direction.

5 Conclusion

The problem is analyzed using one, single computer program for parametric description of the geometric model, automatic mesh generation, and the computation of the water flow with free-surface. The adopted mathematical model of the water free-surface flow and the applied computational-model setup, including total-pressure distribution as the boundary condition at the inlet side, turn out to be acceptable.

The flow rates and their corresponding discharge coefficients are calculated for a number of different values of the water level in reservoir, the gate opening size and the intake depth, and are given in Tables 1 and 2 and Fig. 4. These values can be used for an orientation estimation of expected flow rate in real cases with similar conditions. The trends of variation of the discharge coefficients and the flow rates seem to be plausible. In the considered range of parameter variations, the values of discharge coefficients vary significantly, from 0.525 to 0.724 (a change for about 38%). The variation of the flow rates is even more significant, from 4.74 m^3/s to 21.21 m^3/s (per unit width), or about 4.5 times.

Further improvements could be achieved by extending the model to three-dimensional one in order to investigate the side effects on the flow properties.

References

1. Wikipedia the free encyclopedia Oroville Dam – 2017 spillway failure (2021). https://en.wikipedia.org/wiki/Oroville_Dam. Accessed: 02 Mar 2021

2. Wikipedia the free encyclopedia taum Sauk hydroelectric power station – 2005 upper reservoir failure (2021). https://en.wikipedia.org/wiki/Taum_Sauk_Hydroelectric_Power_Station.Accessed 02 Mar 2021
3. Ettema, R.: Hydraulic Modeling: Concepts and Practice. In: ASCE Manuals and Reports on Engineering Practice, No. 97. ASCE (2000)
4. Ferziger, J.H., Perić M., Street, R.L.: Computational Methods for Fluid Dynamics. 4th ed. Springer, Berlin (2020)
5. Siemens Digital Industries Software: Simcenter STAR-CCM+ Documentation v. 2020.3. (2020)
6. Muzaferija, S., Perić, M.: Computation of free surface flows using interface-tracking and interface-capturing methods. In: Mahrenholtz, O., Markiewicz, M. (eds.) Nonlinear water wave interaction. Chap. 3. Southampton: Computational Mechanics Publications (1998)
7. Muzaferija, S., Perić, M., Sames, P., Schelin T.: A two-fluid Navier-Stokes solver to simulate water entry. In: Proceedings Twenty-Second Symposium on Naval Hydrodynamics (1998)
8. Shih, T.H., Liou, W.W., Shabbir, A., Yang, Z., Zhu, J.: A new k-ε eddy viscosity model for high reynolds number turbulent flows -- model development and validation. In: NASA TM 106721 (1994)
9. Rodi, W.: Experience with Two-Layer Models Combining the k-ε Model with a One-Equation Model Near the Wall. In: 29th Aerospace Sciences Meeting, Reno, NV, 7–10 January, pp. 91–0216. AIAA (1991)
10. Demirdžić, I., Muzaferija, S.: Numerical method for coupled fluid flow, heat transfer and stress analysis using unstructured moving meshes with cells of arbitrary topology. Comput. Meth. Appl. Mech. Engrg. **125**, 235–255 (1995)

Interpolating Rotations with Non-abelian Kuramoto Model on the 3-Sphere

Zinaid Kapić(✉) and Aladin Crnkić

Faculty of Technical Engineering, University of Bihać, Bihać, Bosnia and Herzegovina
zinaid.kapic@unbi.ba

Abstract. The paper presents a novel method for interpolating rotations based on the non-Abelian Kuramoto model on sphere $\mathbb{S}^3$. The algorithm, introduced in this paper, finds the shortest and most direct path between two rotations. We have discovered that it gives approximately the same results as a Spherical Linear Interpolation algorithm. Simulation results of our algorithm are visualized on $\mathbb{S}^2$ using Hopf fibration. In addition, in order to gain a better insight, we have provided one short video illustrating the rotation of an object between two positions.

Keywords: Rotations · Quaternion · Interpolation · Non-Abelian Kuramoto model

1 Introduction

Data in the form of 3-dimensional rotations are applied in many spheres of computer science, kinematics, robotics, computer vision, etc. The most common problems with 3-dimensional rotation data are its averaging and interpolation. The problem of rotation averaging has been extensively studied in the last decade by many authors [1, 2].

In this paper, we have studied rotation interpolation problem which is one of the fundamental problems in computer animation, robotics and structure from motion [3–5]. Due to the need for smooth interpolation between camera frames, it is often used in computer graphics [6]. It also finds application in animations where each joint of a body is usually represented by a rotation so animations in keyframe are created interpolating between these rotations [7]. Furthermore, it is worth mentioning that motion control is required in robotics, where interpolation is used for obstacle avoidance and object approach, such as 'pick and place' examples of robot movement [8].

There are several representations of the rotations such as rotational matrices, Euler angles, and quaternions [1]. Matrix representation of rotations is not a recommended way for interpolation purposes because simple linear combinations of the coefficients can sometimes result in non-positive orthogonal matrices, and therefore non-valid rotations. Euler angles are better than rotation matrices, but the main disadvantage of this representation is a Gimbal lock problem [7]. Quaternions arise as the most efficient way of rotation representation, especially for rotation interpolation.

N. Ademović et al. (Eds.): IAT 2021, LNNS 316, pp. 610–616, 2022.
https://doi.org/10.1007/978-3-030-90055-7_48

Quaternions represent a number system that expands complex numbers and they are defined as [9]:

$$q = a + b\mathbf{i} + c\mathbf{j} + d\mathbf{k},$$

where a, b, c and d are real numbers and $\mathbf{i}^2 = \mathbf{j}^2 = \mathbf{k}^2 = \mathbf{ijk} = -1$. A special group of quaternions is unit quaternions that correspond to a rotation and represent a point on unit sphere $\mathbb{S}^3$ in 4-dimensional space. In other words, if the Euclidean norm

$$\|q\| = \sqrt{a^2 + b^2 + c^2 + d^2}$$

is equal to 1, then a quaternion is called a unit quaternion [9].

Quaternion inverse is defined as $q^{-1} = \overline{q}/\|q\|^2$, where $\overline{q} = a - b\mathbf{i} - c\mathbf{j} - d\mathbf{k}$ is a conjugated quaternion. The equation $q^{-1}q = qq^{-1} = 1$ further defines that $q^{-1} = \overline{q}$ is valid for each unit quaternion. Quaternion popularity rose with aircraft simulations and computer graphics where quaternions were used for describing 3-dimensional rotations [9].

Throughout this paper we will limit ourselves on interpolation between two rotations. The most common and a standard way of rotation interpolation is a Spherical Linear Interpolation (SLERP) algorithm [10]. Using this method, interpolation from unit quaternion p to unit quaternion r with parameter $t \in [0, 1]$ is defined as [10]:

$$Slerp(p, r; t) = p\left(p^{-1}r\right)^t.$$

A novel method for rotation interpolation has been proposed in this paper. This method is based on generalizations of the famous Kuramoto model to higher dimensions. Kuramoto model represents the most significant model for studying collective behavior and self-organization in large populations of coupled oscillators [11]. One of the generalizations of this model has been introduced by Lohe and is known as the non-Abelian Kuramoto model [12]. Authors in the paper [13] used a variation of this model in quaternion form for studying low-dimensional dynamics. That model is defined as

$$q_i = q_i f q_i + w q_i + q_i u - \overline{f}, \;\; i = \overline{1, N}, \tag{1}$$

where $f = f(t, q_1, \ldots, q_N)$ is a quaternionic function that represents the coupling between particles [13]. This model is a non-Abelian Kuramoto model on the 3-sphere.

Different choices of coupling functions in the model (1) solve various problems in different scientific fields. For instance, this model is useful for coordination and consensus on groups $\mathbb{S}^3$ and $SO(3)$ [14], clustering of static and stream data [15, 16], and other applications in science and engineering [17].

The paper is organized as follows. In the next section, we will explain in more detail our algorithm for interpolating rotations. In Sect. 3 we present simulation results that illustrate our method and visualize them as curves on $\mathbb{S}^2$ using Hopf fibration. This section will also provide a video for visualization of object's rotation interpolation. Finally, in the last section we will present our conclusion, possible upgrades of this method, and further research.

2 Algorithm

In this section we will introduce our method for interpolating rotations using non-Abelian Kuramoto model on $\mathbb{S}^3$. Let us primarily suppose that the rotations p and r are represented by unit quaternions. Setting $N = 1$ and $u = w = 0$ in (1), we get following quaternion-valued ordinary differential equation.

$$q'(t) = q(t)fq(t) - \bar{f}, \tag{2}$$

where f is a quaternionic coupling function. Note that differential Eq. (2) preserves the unit sphere $\mathbb{S}^3$. This means that if the initial condition of Eq. (2) satisfies $q(0) \in \mathbb{S}^3$ then the solution to that equation satisfies $q(t) \in \mathbb{S}^3$ at any moment t. We can observe the last Eq. (2) as matrix ordinary differential equation on $SU(2)$ because the group of unit quaternions is isomorphic to the group of special unitary matrices $SU(2)$.

By substituting $f = -\frac{1}{2}\bar{r}$ in (2) and using initial condition $q(0) = p$, one obtains

$$q'(t) = -\tfrac{1}{2}(q(t)\bar{r}q(t) - r), \tag{3}$$

which is a dynamical model that will be used in our algorithm. Solution to this differential equation is an interpolating smooth curve $q(t) \in \mathbb{S}^3$ which connects the initial unit quaternion p and the final one r. In other words, the initial condition of model (3) is given by the initial unit quaternion ($q(0) = p$), and as time t increases, the solution curve $q(t)$ tends to the final unit quaternion ($r = \lim_{t\to\infty} q(t)$).

If we substitute $q(t) = q_1(t) + q_2(t)\mathbf{i} + q_3(t)\mathbf{j} + q_4(t)\mathbf{k}$, $p = p_1 + p_2\mathbf{i} + p_3\mathbf{j} + p_4\mathbf{k}$ and $r = r_1 + r_2\mathbf{i} + r_3\mathbf{j} + r_4\mathbf{k}$ in (4), we will obtain the following system of ordinary differential equations:

$$\begin{cases} q_1'(t) = r_1 - q_1(t)\cos\theta(t) \\ q_2'(t) = r_2 - q_2(t)\cos\theta(t) \\ q_3'(t) = r_3 - q_3(t)\cos\theta(t) \\ q_4'(t) = r_4 - q_4(t)\cos\theta(t) \end{cases}, \tag{4}$$

with initial conditions $q_1(0) = p_1, \ldots, q_4(0) = p_4$. This model is more convenient for simulating than (3). In system (4), $\cos\theta(t)$ is an inner product (dot-product) of unit quaternions $q(t)$ and r, or mathematically expressed as $\cos\theta(t) = q(t) \cdot r = q_1(t)r_1 + q_2(t)r_2 + q_3(t)r_3 + q_4(t)r_4$.

Furthermore, we will explain our method for interpolating between two rotations in more detail through the algorithm. We tend to refer to this algorithm as KLI (Kuramoto-Lohe Interpolation) algorithm.

The KLI algorithm is as follows:

```
Enter p, r
Choose tolerance ε, step δ, and define T=0
Solve (3) with q(0) = p
loop
    if ||r − q(T)||<ε then
            return q(t) for t ∈ [0, T]
    else
            return T=T+δ
    end if
end loop
```

Algorithm 1 KLI algorithm for interpolating between two rotations

3 Simulations

This section presents simulation results of our method. We will assume that the initial rotation (represented by unit quaternions) has value $p = 0 + 0\mathbf{i} + 0\mathbf{j} + 1\mathbf{k}$ and the final one $r = 0.5 + 0.5\mathbf{i} + 0.5\mathbf{j} + 0.5\mathbf{k}$. The goal is to find interpolating curve between p and r, which lies on the surface of the unit sphere $\mathbb{S}^3$ in the 4-dimensional space. Since such curve is problematic to visualize, we will use Hopf's fibration to describe it as a curve lying on a unit sphere $\mathbb{S}^2$ in the 3-dimensional space. In 1931, Heinz Hopf defined this mapping in the paper [18], and since then it has found applications in magnetic monopoles, rigid body mechanics, and quantum information theory. Hopf fibration is defined as a mapping $h : \mathbb{S}^3 \to \mathbb{S}^2$ [19]:

$$h(a, b, c, d) = \left(a^2 + b^2 - c^2 - d^2, 2(ad + bc), 2(bd - ac)\right).$$

Using the KLI algorithm with tolerance $\varepsilon = 10^{-5}$ and step $\delta = 0.01$, interpolating curve $q(t)$ reached the final unit quaternion at moment $t = 11.66$, as shown in Fig. 1a). From closer inspection of Fig. 1a) and Fig. 1b), we can see that KLI and SLERP algorithms give approximately the same results.

Table 1 presents quaternion values of interpolating curve $q(t)$ at different moments. We have visualized those values as thick points in Fig. 1a).

Figure 2 illustrates the evolution of interpolation by displaying rotating object in 3-dimensional space at the same moments provided in Table 1. For this purpose, we have used polar form of unit quaternion defined as.

$$q = \cos\left(\frac{\theta}{2}\right) + U sin\left(\frac{\theta}{2}\right),$$

where $U = q/q$ and $tg(\theta/2) = q/q_0$ [20]. In that way, the object's rotation can be specified by the unit vector U that defines an axis of rotation and the rotation angle θ

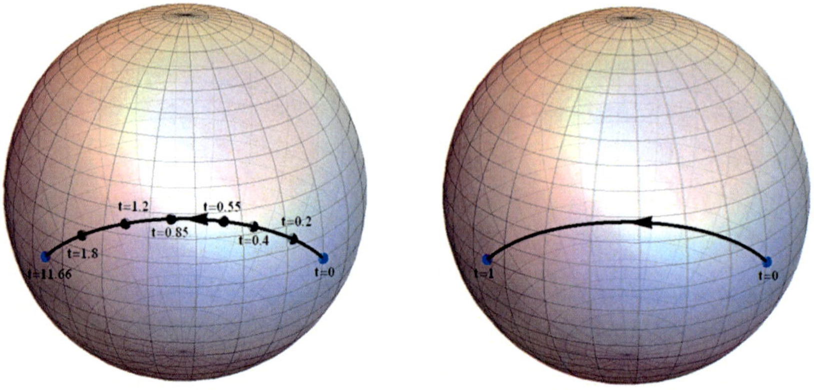

a) KLI algorithm a) SLERP algorithm

Fig. 1. Interpolation curves obtained with KLI and SLERP algorithms

Table 1. Quaternion values of q(t) at different moments t

t	q(t)
0	0.00 + 0.00i + 0.00j + 1k
0.2	0.09 + 0.09i + 0.09j + 0.98k
0.4	0.17 + 0.17i + 0.17j + 0.95k
0.55	0.23 + 0.23i + 0.23j + 0.92k
0.85	0.31 + 0.31i + 0.31j + 0.84k
1.2	0.37 + 0.37i + 0.37j + 0.76k
1.8	0.44 + 0.44i + 0.44j + 0.65k
11.66	0.49 + 0.49i + 0.49j + 0.5k

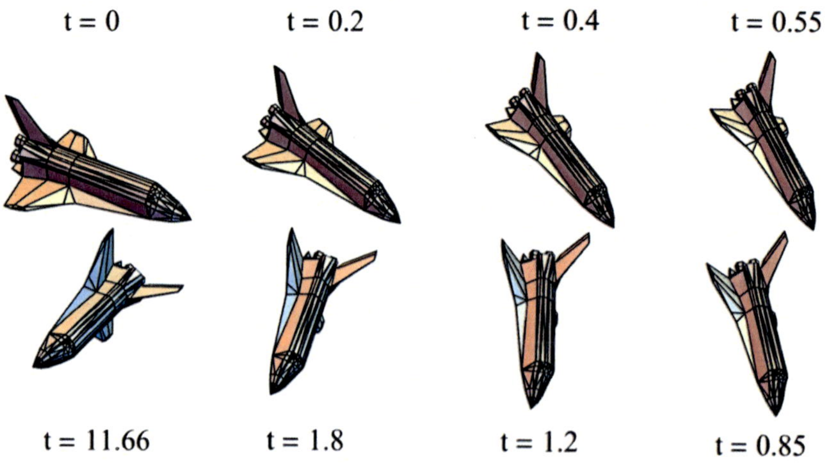

Fig. 2. Interpolation of object's rotation

around that axis. However, such representation can cause some peculiar effects, since two antipodal quaternions q and $-q$ correspond to the same rotation.

Furthermore, for the purpose of better illustration, we have provided one video demonstrating the continuous motion from p to r.[1]

4 Conclusion

The paper introduces a novel method for performing an interpolation between two rotations. This method is based on the so-called non-Abelian Kuramoto model. Simulation results of our algorithm are visualized on $\mathbb{S}^2$ using Hopf fibration and illustrated on a rotating object. Furthermore, to gain better insight, we have provided one short video demonstrating the continuous motion of that object.

The simulation results indicate that our algorithm gives approximately the same results as the SLERP algorithm finding the shortest and most direct path between two rotations. The algorithm proposed in this paper can be used, for instance, for various cases in robotics, robotic hand movement, motion control and object avoidance. For that purpose, it would be interesting to implement this algorithm in ROS (Robot Operating System). ROS represents a meta-operating system for writing robot software that offers users a way to quickly build, maintain and expand their robot [21]. In addition, it can find application in animations in computer graphics.

It is noticeable that this algorithm can be used only for interpolation between two rotations. Hence, we find it interesting to introduce a model that will interpolate multiple rotations. We will address this open question in further research.

Acknowledgements. The second author acknowledges the support of the COST action CA16228 "European Network for Game Theory".

References

1. Hartley, R., et al.: Rotation averaging. Int. J. Comput. Vis. **103**(3), 267–305 (2013). https://doi.org/10.1007/s11263-012-0601-0
2. Kapić, Z., et al.: A new dynamical model for solving rotation averaging problem. In: 2021 20th International Symposium INFOTEH-JAHORINA (INFOTEH), pp. 1–6. IEEE (2021). https://doi.org/10.1109/INFOTEH51037.2021.9400663
3. Kang, I.G., Park, F.C.: Cubic spline algorithms for orientation interpolation. Int. J. Numer. Meth. Eng. **46**(1), 45–64 (1999). https://doi.org/10.1002/(sici)1097-0207(19990910)46:1%3c45::aid-nme662%3e3.0.co;2-k
4. Jüttler, B.: Visualization of moving objects using dual quaternion curves. Comput. Graph. **18**(3), 315–326 (1994). https://doi.org/10.1016/0097-8493(94)90033-7
5. Park, F.C., Ravani, B.: Smooth invariant interpolation of rotations. ACM Trans. Graph. **16**(3), 277–295 (1997). https://doi.org/10.1145/256157.256160

[1] Supplementary video that demonstrates rotation interpolation can be found at https://www.youtube.com/watch?v=IkE_olufAuI.

6. Haarbach, A., Birdal, T., Ilic, S.: Survey of higher order rigid body motion interpolation methods for keyframe animation and continuous-time trajectory estimation. In: 2018 International Conference on 3D Vision (3DV), pp. 381–389. IEEE, Verona (2018). https://doi.org/10.1109/3DV.2018.00051
7. Dam, E., Koch, M., Lillholm, M.: Quaternions, Interpolation and Animation (2000)
8. Svejda, M., Cechura, T.: Interpolation method for robot trajectory planning. In: 2015 20th International Conference on Process Control (PC), pp. 406–411. IEEE, Strbske Pleso (2015). https://doi.org/10.1109/pc.2015.7169997
9. Günaşti, G.: Quaternions Algebra, their applications in Rotations and Beyond Quaternions (2012)
10. Shoemake, K.: Animating rotation with quaternion curves. In: Proceedings of the 12th Annual Conference on Computer Graphics and Interactive Techniques - SIGGRAPH 1985. ACM Press, New York, pp. 245–254 (1985). https://doi.org/10.1145/325334.325242
11. Kuramoto, Y.: Self-entrainment of a population of coupled non-linear oscillators. In: Araki, H. (ed.) International Symposium on Mathematical Problems in Theoretical Physics, pp. 420–422. Springer-Verlag, Berlin/Heidelberg (1975). https://doi.org/10.1007/BFb0013365
12. Lohe, M.A.: Non-Abelian Kuramoto models and synchronization. J. Phys. A: Math. Theor. **42**(39), 395101 (2009). https://doi.org/10.1088/1751-8113/42/39/395101
13. Jaćimović, V., Crnkić, A.: Low-dimensional dynamics in non-Abelian Kuramoto model on the 3-sphere. Chaos Interdiscip. J. Nonlinear Sci. **28**(8), 083105 (2018). https://doi.org/10.1063/1.5029485
14. Crnkić, A., et al.: Consensus and coordination on groups $SO(3)$ and S^3 over constant and state-dependent communication graphs. Automatika **62**(1), 76–83 (2020). https://doi.org/10.1080/00051144.2020.1863544
15. Crnkić, A., Jaćimović, V.: Data clustering based on quantum synchronization. Nat. Comput. **18**(4), 907–911 (2018). https://doi.org/10.1007/s11047-018-9720-z
16. Crnkić, A., et al.: Swarms on the 3-sphere for online clustering of multivariate time series and data streams. Futur. Gener. Comput. Syst. **112**, 11–17 (2020). https://doi.org/10.1016/j.future.2020.05.018
17. Ha, S.Y., Park, H.: A dynamical systems approach for the shape matching of polytopes along rigid-body motions. arXiv: Mathematical Physics arXiv:2006.11533 (2020)
18. Hopf, H.: Über die Abbildungen der dreidimensionalen Sphäre auf die Kugelfläche. Math. Ann. **104**, 637–665 (1931)
19. Lyons, D.W.: An elementary introduction to the hopf fibration. Math. Mag. **76**(2), 87–98 (2003). https://doi.org/10.1080/0025570x.2003.11953158
20. Sangwine, S.J., Bihan, N.L.: Quaternion polar representation with a complex modulus and complex argument inspired by the cayley-dickson form. Adv. Appl. Clifford Algebras **20**(1), 111–120 (2008). https://doi.org/10.1007/s00006-008-0128-1
21. Estefo, P., et al.: The robot operating system: package reuse and community dynamics. J. Syst. Softw. **151**, 226–242 (2019). https://doi.org/10.1016/j.jss.2019.02.024

Structural Stress and Strain Analysis Using a 3D Scanner

Elvedin Kljuno, Faruk Razić, Elmedin Mešić, and Alan Ćatović(✉)

Mechanical Engineering Faculty, University of Sarajevo, Sarajevo, Bosnia and Herzegovina
{kljuno,razic,mesic,catovic}@mef.unsa.ba

Abstract. The paper shows a way how a 3D scanner can be used in a combination with numerical simulations and corresponding analytical results to estimate deformations and stresses in a loaded structural element, such as beams, pipes, shafts, etc. To compare measurement results, a series of numerical simulations were performed for a test object, which was a short pipe that was loaded externally by a force obtained via a manual clamp, which was continuously measured via a force sensor. A general approach to measure strain via a 3D scanner at the object's surface is explained. However, the surface strain is not sufficient without an analytical consideration to obtain the information what is really occurring inside a loaded structural element, regarding deformations and stresses. Finally, the simulation and the analytical results are compared with the real measurements that were used to calculate strain tensor components, which are used to calculate stress tensor components using the generalized Hooke's law.

Keywords: Deformation · Stress · Strain · 3D scanner

1 Introduction

1.1 Digital Image Correlation

Optoelectronic systems with non-contact and non-invasive measuring features have highly evolved in last few years. The optic technique specifies displacement and strain features for the whole surface of considered specimen with or without movement of rigid body [1].

Digital Image Correlation (DIC) is an optoelectronic method for measuring strains, which was introduced in the 1980s [2]. With great potential of measuring surface strains by optical method and possibility of image capturing for the sample during deformation, this method surpasses limitations linked with attaching strain gauges (SG) to the sample or the measured area. DIC has many advantages such as measuring surface strains on samples with unorthodox shapes and various sizes with an easy procedure of sample preparation, where the accuracy of SGs is one order of magnitude higher than one of DIC.

DIC devices obtain information about displacement/deformation changes in two and three-dimensional systems, where principles of their work are founded on tight connections between mechanics of continuous media [3].

N. Ademović et al. (Eds.): IAT 2021, LNNS 316, pp. 617–634, 2022.
https://doi.org/10.1007/978-3-030-90055-7_49

GOM ATOS Core is a non-contact optical system, based on DIC working principle, for 3D measuring which studies and calculates deformations and dynamic state of measuring object. This system requires an input composed from a series of digital images, where images correspond to loading steps. Based on the change between images, full-field displacements and strains are obtained.

Deformations of the subjected structure are evaluated with computations of 3D coordinates for discrete points in a certain period of time, where initial and final positions of discrete points are compared. With such a method, software not only calculates displacements and strains, but also derivate features, such as speed and acceleration [4].

Surface of the object's undergoing deformation has a printed pattern, which is used to determine the displacement under bearing with time. This displacement is then used to calculate the surface strain on the object. The technique requires the use of images taken in sequence over a period of time. For 3D correlation the system requires two sets of images of the target taken from separate camera angles. A system calibration is required to determine the 3D space where the event occurs. This is then used to correlate the images for the determination of deflection and strain [5].

Coordinates can be acquired with stereo camera and different types of patterns in the object surface. Patterns can be random or have specific referent marks. Random pattern is created with the random application of the marks with usage of paint or sprays. With this pattern, the system can clearly relocate the pixels of the camera images (called *facets*). Conducted experiments mostly use random pattern, as it is more appropriate for the complex determinations of the strains and displacements. DIC also offers a method for capturing full-field deformation on the sample surface, regardless of their material, shape or size.

The DIC technology comes with limitations. The most significant limitation is dictated with direct line of sight for both cameras in order of obtaining paired images for a successful processing. In comparison to measuring techniques, such as SGs, the DIC technique has the necessity of calibration in accordance with the applicable procedure.

1.2 The Working Principle of DIC Devices

DIC's working principle is based on comparison between two digital images of the subjected surface, where images are obtained for undeformed and deformed state, also known as reference and solution state. With this comparison, DIC provides full-field displacements with sub-pixel accuracy. For easier tracking of the displacement, surface of the sample using a DIC system, is sprayed with random patterns. The surface is divided into facets of the pixels. Facets centers are called measuring points (Fig. 1a), where each facet has a unique grey-value distribution of pixels and location in space. With the method of image correlation, the facets are identified and matched along the image series. The best matching region of the two images, where the maximum correlation coefficient is found, is taken as the new position of the measurement point [6]. After determination of points coordinates for reference and deformed image, calculation of displacements in x, y and z-direction are performed along with difference between the deformed and the reference states, e.g. x-x*, y-y* or z-z*.

For 3D computation, two cameras are needed in order to calculate the coordinates in the deformed stage via triangulation. A triangular mesh is placed onto the analyzed region, with the nodes being the measurement points where the displacements were computed (Fig. 1b). The coordinates A, B and 0 correspond to the reference image, and they establish one triangular element.

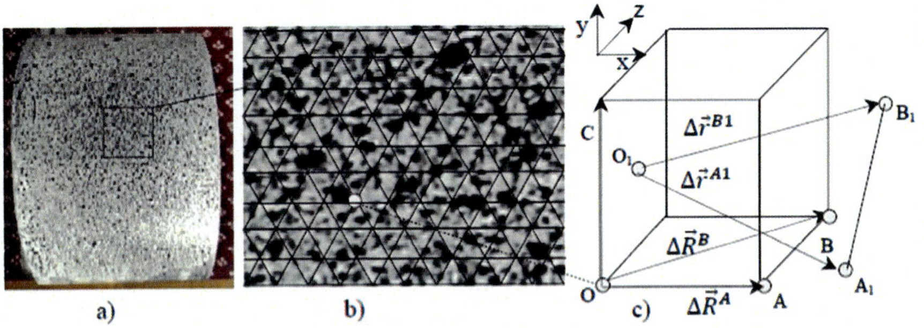

Fig.1. Surface area that is analyzed (a), Measurement point on the surface (b) and the relative position vectors before and after the deformation (c)

The deformation gradient is computed as [6]:

$$\left[\Delta\vec{r}^{A1}, \Delta\vec{r}^{B1}\right] = F \cdot \left[\Delta\vec{R}^{A}, \Delta\vec{R}^{B}\right], \tag{1}$$

where F is the deformation gradient (a matrix), $\Delta\vec{r}$ and $\Delta\vec{R}$ are the position of a measurement point in the deformed and reference position, respectively. The position of the measurement point in the deformed state is:

$$\Delta\vec{r} = \Delta\vec{R} + \Delta\vec{u}, \tag{2}$$

where $\Delta\vec{u}$ is the relative displacement. Using (1) and (2), the movement and deformation of an element (O, A, B) can be computed based on the movement of its points and can be described as following for point A:

$$\begin{bmatrix} a_x \\ a_y \end{bmatrix} = \begin{bmatrix} u_x \\ u_y \end{bmatrix} + \begin{bmatrix} F_{11} & F_{12} \\ F_{21} & F_{22} \end{bmatrix} \begin{bmatrix} A_x \\ A_y \end{bmatrix}. \tag{3}$$

Since the point distribution is based on triangles (Fig. 1b), one needs at least three points with known undeformed and deformed coordinates for computing the displacement and the deformation. The Green–Lagrange strain tensor can be computed as per the following equation:

$$\varepsilon = \frac{1}{2}\left(F^T \cdot F - I\right) \tag{4}$$

where I is the identity tensor. From the strain tensor, specific strains in x, y and z direction of an element can be obtained and assigned to the measurement point.

1.3 Software Simulations to Determine Stress and Strain

Every part or assembly that is in process of doing some work: lifting weight, punching, pressing, cutting, bending etc. is exposed to some stress. In order to provide information about that stress, deformation or displacement that is produced during an assembly or a part loading, while performing work, certain analysis needs to be done. These analyses will provide data that can clearly show what these elements needs to withstand in order to work properly. The best way to perform this analysis is testing. This testing assumes sample of the part which will be subjected to tensile test or some other test that will provide required data. Sample shape, dimensions and type of the sample piece will be defined by certain standard such as [7].

Beside testing, simulation is another way of performing analysis of the part or assembly that is exposed to stress such as pressure, force, torque etc. Simulation can provide data about stresses that are produced in part or assembly while in operation. This kind of analysis has some advantage over part testing in a way that it is repetitive, costs much less than real testing, it is not time consuming in a way that test piece needs to be made each time when test needs to be performed (test piece is usually destroyed after testing). Simulations cannot replace real testing but they can provide parameters that will be used in order to improve part design. During design process, part which is designed will always be subjected to real testing that will provide its final limits of load at which it can be subjected, working hours and other limits defined through project assignment.

Simulations sometimes provide very accurate stress analysis of a part or assembly which is in good agreement with real test results. But these results depend on type of load that is produced on piece, force applied, torque, fixtures, shape of the piece, process of deformation: is it stochastic process, dynamic or static process, and many other factors that will influence accuracy of the results and data that is gained through this process of analysis. In order to provide stress, deformation and displacement analysis of the part from this paper, FE simulation was performed using SolidWorks simulation add-in. John F. Moxnes et al. [8] performed simulation of the tensile test using SolidWorks FE simulation module which provided good agreement with LS-DYNA, and these two codes gave insignificant differences. Also, measured engineering stress vs measured strain, which was measured with extensometer, was in good agreement with simulation.

In order to perform correct simulation process, it is extremely important to provide information about the part which is about to be subjected to the simulation process. Part which was used in this paper was a pipe fitting piece made out of Grey Iron. Material properties where assigned from SolidWorks library for Cast Grey Iron. Some of the properties had to be changed (Modulus of Elasticity (calculated for the fitting material) and Poisson's ratio that was taken for material designated as EN-GJL-200. Modulus of elasticity for EN-GJL-200 varies from 80 to 110 GPa [9] and mechanical properties of EN-GJL-200 are defined in [10].

2 Test Problem Setup

The test case for the comparison with the results of the 3D scanning of deformations, displacements and strain components is chosen to be the statically indeterminate case internally as shown in Fig. 2.

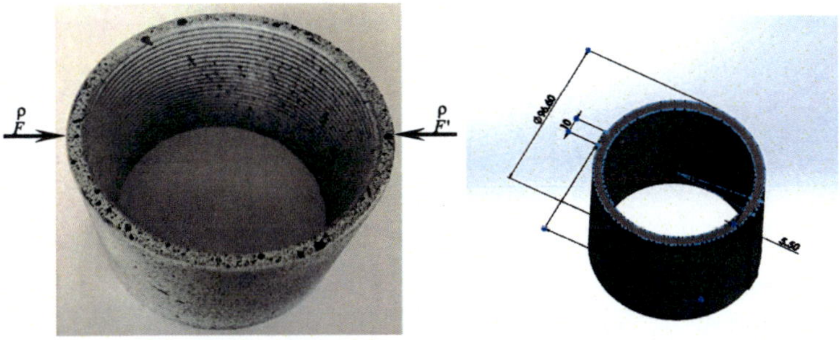

Fig. 2. Prepared test object - a pipe of length 70.5 mm with a threading

The loading of the short pipe is done by a clamp along with the force measuring, as shown in Fig. 3.

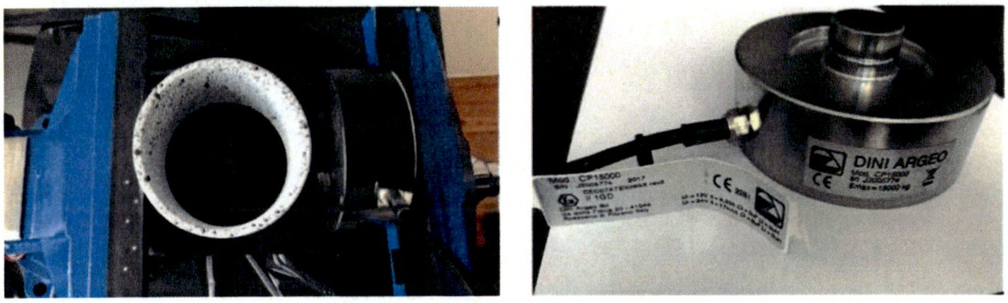

Fig. 3. Loading and deforming the test object using a manual clamp

Force sensor has the measuring range up to 150 kN and the max error of 0.05% and the output of 2 mV/V for the full measuring scale.

An amplifier circuitry is prepared to amplify the signal from the force sensor, using dual operational amplifier TLC272CP and resistors, as shown in Fig. 4. If a precise multimeter is available, then the amplifier circuitry is not necessary.

The piece of pipe had an internal threading such that for the analytical calculations the mean diameter of the threading was used, as well as for the numerical simulations. With this approximation, a relatively small error was introduced for the stiffness of the cylinder, which is bounded by the stiffness of the cylinder with thickness of 6 mm on one side and the thickness of 5 mm on the other side (minimum and maximum diameter of the threading).

3 Analytical Results for the Stress and Strain Tensor Components

Two main approximations were made in order to obtain analytical results. First, the inner threading is removed, and the inner cylinder is assumed to be at the middle diameter of the threading. Second, the material of the testing object was not known, such that several measurements had to be done in order to determine mechanical properties of

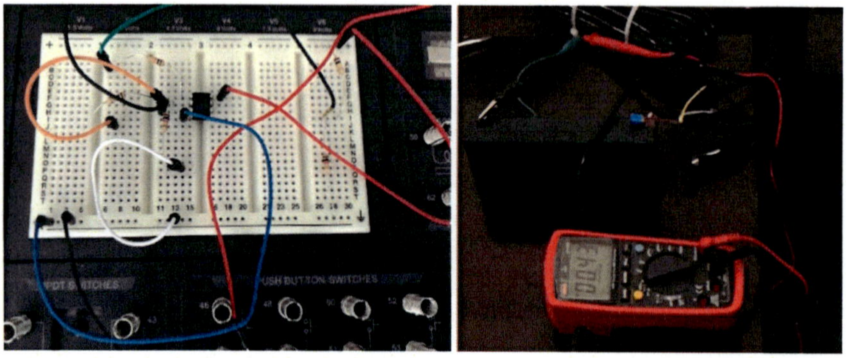

Fig. 4. Measuring the signal from the force sensor

the material that are needed for this comparison with measurement results on the real object. One of the mechanical properties that was determined along this process was the Young's modulus. The other one was measurement at what force there would be a measurable plastic deformation, such that the range of the loading was established to ensure the proportional range, i.e. that the Hooke's law applies in the analysis.

The generalized Hooke's law gives the relationship between the stress and strain tensor components. Given stress components, the strain tensor components can be calculated as [11]:

$$
\begin{aligned}
\epsilon_x &= \frac{1}{E}\left[\sigma_x - \nu\left(\sigma_y + \sigma_z\right)\right],\\
\epsilon_y &= \frac{1}{E}\left[\sigma_y - \nu(\sigma_x + \sigma_z)\right],\\
\epsilon_z &= \frac{1}{E}\left[\sigma_z - \nu\left(\sigma_x + \sigma_y\right)\right],\\
\gamma_{xy} &= \frac{\tau_{xy}}{G},\ \gamma_{yz} = \frac{\tau_{yz}}{G},\\
\gamma_{xz} &= \frac{\tau_{xz}}{G},
\end{aligned} \tag{5}
$$

where ν is the Poisson's ratio, E is the Young's modulus and G is the shear modulus which is related to the previous parameters as $G = \frac{E}{2(1+\nu)}$.

If a cylindrical coordinate system is used, which is more convenient considering the geometry of the object, then (5) becomes:

$$
\begin{aligned}
\epsilon_r &= \frac{1}{E}\left[\sigma_r - \nu\left(\sigma_\varphi + \sigma_z\right)\right],\\
\epsilon_\varphi &= \frac{1}{E}\left[\sigma_\varphi - \nu(\sigma_r + \sigma_z)\right],\\
\epsilon_z &= \frac{1}{E}\left[\sigma_z - \nu\left(\sigma_r + \sigma_\varphi\right)\right],\\
\gamma_{r\varphi} &= \frac{\tau_{r\varphi}}{G},\ \gamma_{rz} = \frac{\tau_{rz}}{G},
\end{aligned}
$$

$$\gamma_{\varphi z} = \frac{\tau_{\varphi z}}{G}, \tag{6}$$

where r is the radial coordinate, φ is the circular angle, and z is the axis in the direction of the simmetry axis of the pipe.

However, to calculate the stresses on the right hand-side of (5) or (6), the internal static indeterminacy has to be solved. The minimum deformational energy method (the Castigliano's method, see [11]) can be used to calculate the bending moments at the cross-sections where the load from the clamps is applied.

The clamp loads the pipe in the radial direction which causes the bending moment and corresponding normal stress due to the bending moment in the wall. An equivalent system, which is statically determined, is obtained by considering the bending moment to be an external load. Then, the additional conditions onto the elastic line are applied, as shown in Fig. 5.

Due to the symmetry, there is no normal force at cross-section D (otherwise, there would be the same force at A, which would not satisfy sum of forces in the direction perpendicular to the loading direction). From the moment balance equation, the following is obtained:

$$M_A = M_D. \tag{7}$$

Since the loading force F is known, the system is the first order statically indeterminacy. The total energy of the elastic deformation is given by [11, 12]:

$$W_d = 2\left[\int_0^s \frac{M_b^2}{2Er_0S_n}ds + \int_0^s \frac{F_n^2}{2EA}ds + \int_0^s \frac{M_bF_n}{EAr_0}ds + \frac{\kappa}{2GA}\int_0^s F_t^2 ds\right]. \tag{8}$$

Here the bending moment is $M_b = M_b(\varphi) = M_D - F_1 r_0\cos(\varphi)$, the normal force $F_n = F_n(\varphi) = -F_1\cos(\varphi)$, the transversal force $F_t = F_t(\varphi) = F_1\sin(\varphi)$, k is the coefficient dependent on the cross-section geometry $k = 1.2$ (see [11]) for a rectangular cross-section (shown as cross-section P-P in Fig. 5), $S_n = eA$ is the static moment of the cross-section with respect to the neutral layer, which is positioned at the distance e from the center point of the cross-section.

However, the expression (8) can be simplified if the diameter (mean) is much greater than thickness, which is the case for the test object. In this case, the curved elastic line can be considered as a straight line, such that (8) becomes:

$$W_d = \int_{-s}^{s} \frac{M_b^2}{2EI_z}ds + \int_{-s}^{s} \frac{F_n^2}{2EA}ds + \frac{\kappa}{2GA}\int_{-s}^{s} F_t^2 ds, \tag{9}$$

where $I_z = \frac{bh^3}{12}$ is the inertial moment of the cross-section for the central axis z. Using the Castigliano's method, the rotation of the cross-section D (see [13]) due to the load is:

$$\frac{\partial W_d}{\partial M_D} = \varphi_D = 0, \tag{10}$$

$$\frac{\partial W_d}{\partial M_D} = \frac{1}{EI_z}\int_{-s}^{s} M_b\frac{\partial M_b}{\partial M_D}ds + \frac{1}{EA}\int_{-s}^{s} F_n\frac{\partial F_n}{\partial M_D}ds + \frac{\kappa}{2GA}\int_{-s}^{s} F_t\frac{\partial F_t}{\partial M_D}ds. \tag{11}$$

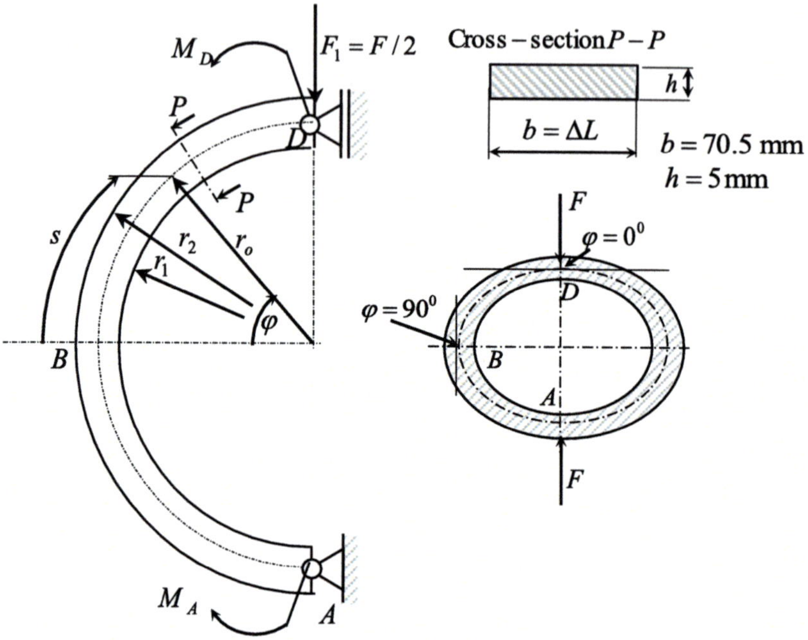

Fig. 5. Statically determined equivalent system of the pipe

The curved line coordinate s can be replaced by $s = r_0\varphi$, i.e. $ds = r_0 d\varphi$. The bending moment, normal force and transversal force functions are:

$$M_b = M_D - F_1 r_0 \cos(\varphi), \tag{12}$$

$$F_n = -F_1 \cos(\varphi), \tag{13}$$

$$F_t = F_1 \sin(\varphi), \tag{14}$$

along with their partial derivatives:

$$\frac{\partial M_b}{\partial M_D} = 1, \frac{\partial F_n}{\partial M_D} = 0 \text{ and } \frac{\partial F_t}{\partial M_D} = 0. \tag{15}$$

Using (12–15) in (11) and integrating over the interval $\left[-\frac{\pi}{2}, \frac{\pi}{2}\right]$, the following is obtained:

$$\int_{-\frac{\pi}{2}}^{\frac{\pi}{2}} (M_D - F_1 r_0 \cos(\varphi)) r_0 d\varphi = (M_D\varphi - F_1 r_0 \sin(\varphi))|_{-\frac{\pi}{2}}^{\frac{\pi}{2}} = M_D\pi - 2F_1 r_0 = 0, \tag{16}$$

$$M_D = \frac{2F_1 r_0}{\pi}. \tag{17}$$

In the similar way, using the Castigliano's theorem, the displacement of the center of cross-section D can be calculated as:

$$\delta_D = \frac{1}{EI_z}\int_{-\frac{\pi}{2}}^{\frac{\pi}{2}} M_b \frac{\partial M_b}{\partial F_1} r_0 d\varphi + \frac{1}{EA}\int_{-\frac{\pi}{2}}^{\frac{\pi}{2}} F_n \frac{\partial F_n}{\partial F_1} r_0 d\varphi + \frac{\kappa}{2GA}\int_{-\frac{\pi}{2}}^{\frac{\pi}{2}} F_t \frac{\partial F_t}{\partial F_1} r_0 d\varphi, \quad (18)$$

$$\delta_D = \frac{1}{EI_z}\int_{-\frac{\pi}{2}}^{\frac{\pi}{2}} M_b \frac{\partial M_b}{\partial F_1} r_0 d\varphi + \frac{1}{EA}\int_{-\frac{\pi}{2}}^{\frac{\pi}{2}} F_n \frac{\partial F_n}{\partial F_1} r_0 d\varphi + \frac{\kappa}{2GA}\int_{-\frac{\pi}{2}}^{\frac{\pi}{2}} F_t \frac{\partial F_t}{\partial F_1} r_0 d\varphi, \quad (19)$$

$$\begin{aligned} \delta_D = &\frac{1}{EI_z}\int_{-\frac{\pi}{2}}^{\frac{\pi}{2}} \left(\frac{2}{\pi}F_1 r_0 - F_1 r_0 \cos(\varphi)\right)\left(\frac{2}{\pi}r_0 - r_0\cos(\varphi)\right) r_0 d\varphi \\ &+ \frac{1}{EA}\int_{-\frac{\pi}{2}}^{\frac{\pi}{2}} F_1 \cos^2(\varphi) r_0 d\varphi + \frac{k}{GA}\int_{-\frac{\pi}{2}}^{\frac{\pi}{2}} F_1 \sin^2(\varphi) r_0 d\varphi, \end{aligned} \quad (20)$$

which finally results in the following:

$$\delta_D = \frac{2F_1 r_0}{EA}\left[\frac{A}{I_z r_0^2\left(-\frac{2}{\pi} + \frac{\pi}{4}\right)} + \frac{\pi}{4}(1 + k2(1+\nu))\right] \quad (21)$$

After substituting the measured force $2F_1 = F = 11,94\,\text{kN}$, the cross-section area $A = bh = 388\,\text{mm}^2$, the moment of inertia $I_z = \frac{bh^3}{12} = 977\,\text{mm}^4$, the mean radius $r_0 = 45.3\,\text{mm}$, the Poisson's ratio $\nu = 0.3$, the indirectly measured Young's modulus $E = 108\,\text{GPa}$ the following is obtained:

$$\delta_D = \frac{11,94\cdot 10^3\,\text{N}\cdot 45,3\,\text{mm}}{108\cdot 10^3\frac{\text{N}}{\text{mm}^3}\cdot 388\,\text{mm}^2}\left[388\frac{\text{mm}^2}{977\,\text{mm}^4}45,3^2\,\text{mm}^2\left(\frac{\pi}{4} - \frac{2}{\pi}\right) + \frac{\pi}{4}(1 + 2\cdot 1.2(1+0.3))\right] = 1.65\,\text{mm} \quad (22)$$

Since the material of the pipe was unknown at the beginning of the test, a small fraction of the pipe was cut and made of it a simple beam (by a plastic deformation). Then, the Young's modulus was measured in the way that the beam was loaded by a known force and the result was compared with the analytical result for the elastic line of the simple beam at the middle of the beam.

Since the displacement of the simple beam was measured on one hand and obtained analytically on the other hand, then an equation was formed:

$$E = \frac{F_{beam} L_{beam}^3}{48 I_{xbeam} \delta_{middle}} = 108\,\text{GPa}, \quad (23)$$

where: $F_{beam} = 2,7\cdot 10^3\,\text{N}$, $L_{beam} = 115,7\,\text{mm}$, $I_{xbeam} = 384\,\text{mm}^4$, $\delta_{middle} = 2.1\,\text{mm}$.

Since the strain ϵ_φ that will be used for the comparison with the results obtained via 3D scanner, here we will obtain the strain at the outer cylinder for $r = r_2$ at the middle, where $\varphi = 0$. To obtain this strain, the normal stress σ_φ has to be determined first, according to the generalized Hooke's law, shown earlier.

The normal stress σ_φ due to the bending moment includes the consideration of the curvature of the elastic line [12, 13]:

$$\sigma_\varphi^{M_b} = \frac{M_b(\varphi)}{S_n}\frac{\eta}{r}, \quad (24)$$

where $M_b(\varphi)$ is the bending moment (here considered positive if the top layer is stretched out, since the η coordinate is oriented in positive radial direction), η is the distance of the considered layer with respect to the neutral layer, $S_n = eA$ is the static moment of the cross-section with respect to the neutral layer and r is the radial distance of the considered layer (radial coordinate of the cylindrical coordinate system).

The normal stress in the circular direction due to the normal compressive force is:

$$\sigma_{\varphi}^{F_n} = \frac{F_n(\varphi)}{A}, \tag{25}$$

where $F_n = -F_1 \cos(\varphi)$ and $A = bh$.

Since $e = 0.06$ mm (the neutral layer position with respect to the central layer) is negligible relatively to h, then the regular analysis will be applied to calculate the normal stress due to the banding moment. The normal stress at the outer surface for $\varphi = 0^0$ is equal to:

$$\sigma_{\varphi}(r_2) = \frac{M_b(0)}{I_z}\frac{h}{2} + \frac{F_n(0)}{A} = 277\,\text{MPa} - 15\,\text{MPa} = 262\,\text{MPa}, \tag{26}$$

while the stress at the inner surface for the same cross-section is equal to:

$$\sigma_{\varphi}(r_1) = \frac{M_b(0)}{I_z}\left(-\frac{h}{2}\right) + \frac{F_n(0)}{A} = -277\,\text{MPa} - 15\,\text{MPa} = -292\,\text{MPa}. \tag{27}$$

Using these values for the circular normal stress and considering that other stress tensor elements in the generalized Hooke's law are negligible (the radial stress σ_r and axial stress σ_z), then the circular strain at the outer and the inner cylinder are:

$$\epsilon_{\varphi}\,|_{r=r_2}^{\varphi=0} \approx \frac{1}{E}\sigma_{\varphi}\,|_{r=r_2}^{\varphi=0} = 2.4 \cdot 10^{-3}, \tag{28}$$

$$\epsilon_{\varphi}\,|_{r=r_1}^{\varphi=0} \approx \frac{1}{E}\sigma_{\varphi}\,|_{r=r_1}^{\varphi=0} = -2.7 \cdot 10^{-3}. \tag{29}$$

These values will be used for comparison with the simulation results and with measured values using the 3D scanner and the testing setup.

4 Simulation Results Using Commercial Software

Simulations are performed in the commercial software SolidWorks, for an approximated case, where the inner threading is removed. The simulation results are performed assuming the linear range of the stresses, such that the Hooke's law applies, which was ensured by defining the range for the loading force. The Young's modulus was estimated via measurements to be about 108 GPa. However, this estimation introduces some error, which will eventually result in a certain discrepancy with the measurements using the testing equipment.

The loading and the numerical mesh are shown in Fig. 6.

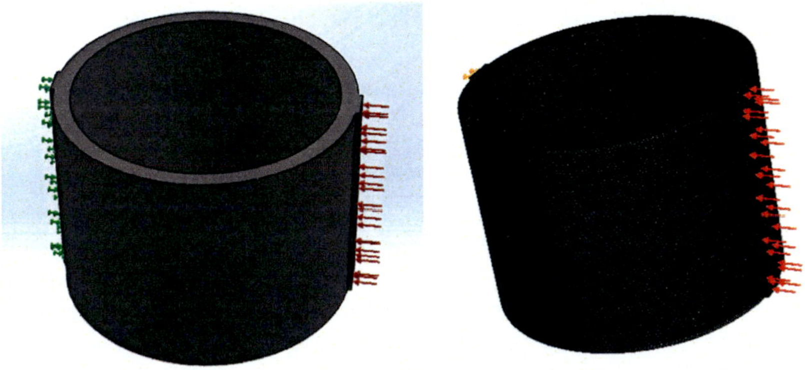

Fig. 6. The loading direction (left) and the numerical mesh (right)

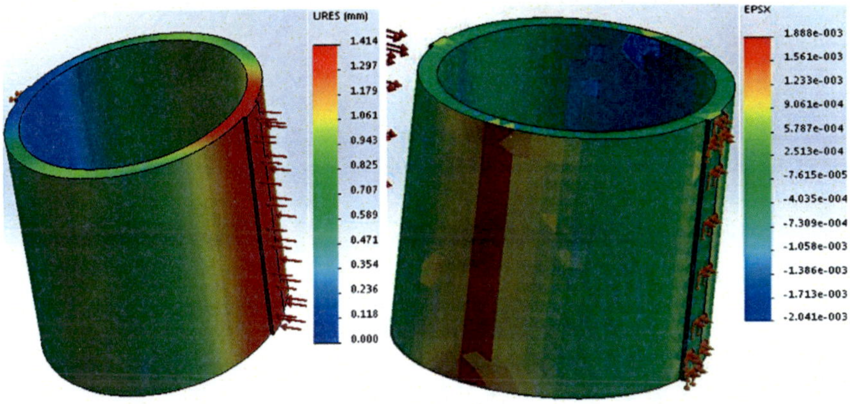

Fig. 7. The displacement in the loading direction (left) and the strain (right)

The radial displacement and strain, obtained through the numerical simulation, are shown in Fig. 7. For the comparison, the result for the critical cross-section (for the angle $\varphi = 0^0$) is of the particular importance.

If we compare the analytic results ($\delta_D = 1.65\,\text{mm}$) and the simulation results ($\delta_{D_{numerical}} = 1.41\,\text{mm}$), there is a relative error of about 15%. However, this relative error can be partially justified by the fact that there is a flat side added in the numerical model of the pipe, in the tangential plain where the load is applied as shown in Fig. 8. Since there is no such an additional part to contribute to the stiffness on the pipe in the analytical model, this is one of the reasons why the analytical model predicts slightly higher displacement. This flat part was necessary in the numerical model to be able to apply the force over a surface area instead of just applying it over the contact line of the cylinder, which would cause extremely high contact stresses that would dominate in the normal stress plots.

The comparison of the analytical results regarding the circular strain ϵ_φ can be done by comparing the circular strain at the critical cross-section, at the inner wall and the outer wall of the cylinder, as explained earlier. The circular strain results are given by (28) and (29), as $\epsilon_{\varphi}|_{r=r_2}^{\varphi=0} \approx 2.4 \cdot 10^{-3}$ and $\epsilon_{\varphi}|_{r=r_1}^{\varphi=0} \approx -2.7 \cdot 10^{-3}$.

On the other side, the results of the simulation for same points of the critical cross-section are about $1.9 \cdot 10^{-3}$ and $-2.04 \cdot 10^{-3}$, respectively. The relative error in the results is higher, but with the same reasoning for the higher strain predicted analytically, as in the case of the displacement.

Finally, the normal circular stress is shown in Fig. 8, along with the numerical mesh data. The stress distribution in the figure can be used for a comparison with circular stress for the critical cross-section, for which the coordinate stress σ_x is equal to the circular normal stress σ_{φ}. It is the critical cross-section, where the normal stress reaches the extreme absolute values (the red and the blue zone), for the angle $\varphi = 0^0$ (defined in the analytical results section).

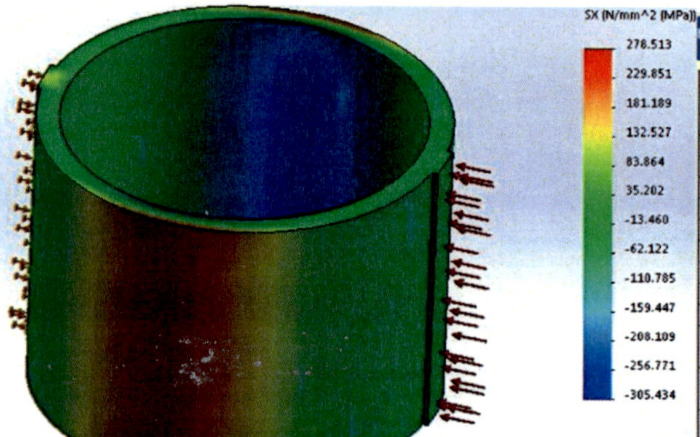

Mesh Details	
Study name	Static 4 (-Default-)
Mesh type	Solid Mesh
Mesher Used	Standard mesh
Automatic Transition	Off
Include Mesh Auto Loops	Off
Jacobian points	4 points
Element size	1 mm
Tolerance	0.05 mm
Mesh quality	High
Total nodes	1119898
Total elements	774414
Maximum Aspect Ratio	5.358
Percentage of elements with Aspect Ratio < 3	100
Percentage of elements with Aspect Ratio > 10	0
% of distorted elements (Jacobian)	0
Time to complete mesh(hh:mm:ss)	00:01:43
Computer name	BLUES

Fig. 8. Stress σ_x for the loading of 11.94 kN (left) and num. mesh data (right)

As shown earlier, the analytical results are $\sigma_{\varphi}^{an}(r_2) = 262\,\text{MPa}$, $\sigma_{\varphi}^{an}(r_1) = -292\,\text{MPa}$, and it can be seen in the figure that the corresponding numerical results are $\sigma_{\varphi}^{num}(r_2) = 278.5\,\text{MPa}$, $\sigma_{\varphi}^{num}(r_1) = -305.4\,\text{MPa}$. The relative errors are: 6% and −5%, respectively.

Several mesh refining steps were performed until the change in the results became insignificant.

5 Measuring the Strain and the Displacement via the 3D Scanner

The scanner that was used for these measurements is ATOS CORE 200 (GOM, GmbH). It working principle is based on the measurement of the relative displacement points that the system registers at the 3D object surface. This means that the surface of the measured object has to be prepared in the way that there is a fine raster of white and black dots evenly distributed all over the measured surface area of the object.

Figure 9 shows the prepared object area, that was sprayed first with a white color and then a fine spray of black non-reflective color is applied on to the surface.

Fig. 9. The prepared object surface for 3D scanning of deformations

Figure 10 shows the scanned strain at the critical cross-section after loading the object with 11.94 kN force externally via the manual clamp. For this deformation, on side of the clamp was fixed and the other yaw of the clamp had a displacement of about 2 mm.

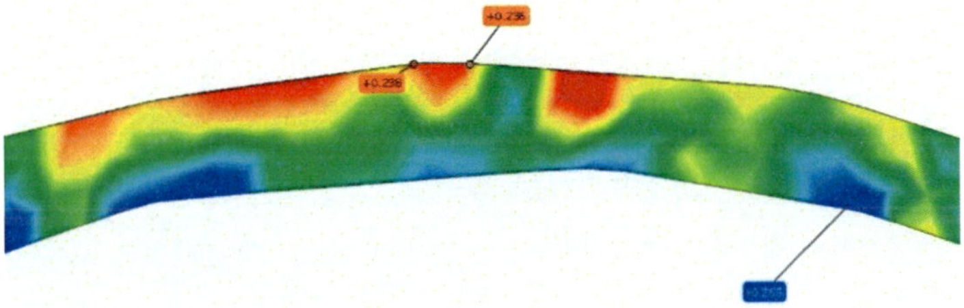

Fig. 10. The measured strain (shown in %) around the critical cross-section at the outer and the inner wall of the pipe (via GOM Correlate software)

How well the object is prepared for scanning depends on the number of points on the surface and their size, as well as contrast with respect to the background surface color. Figure 11 shows the registered point in the green squares that are monitored and used through the scanner system in order to obtain their relative displacement to finally obtain the strain.

Since one side of the clamp is fixed, then the displacement of the other side of the yaw represents the measure of the pipe deformation. This displacement of the left hand-side of the yaw is represented by the displacement $\delta_D = 1.64$ mm in the analytical calculation.

Figure 12 shows the displacement distribution in the cross-sectional plain.

The highest value of the displacement is at the contact of the force sensor with the cylindrical surface, which is 1.889 mm. Therefore, overshoots the analytical value for about 15%.

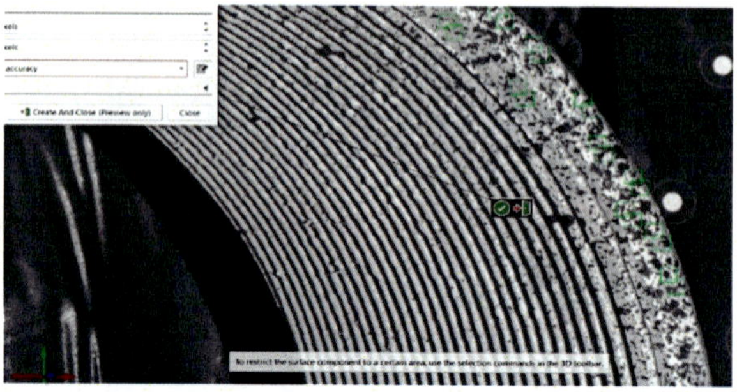

Fig. 11. Distinctive point detected automatically by the GOM Correlate software (the green squares) that are monitored through the deformation process to calculate the strain

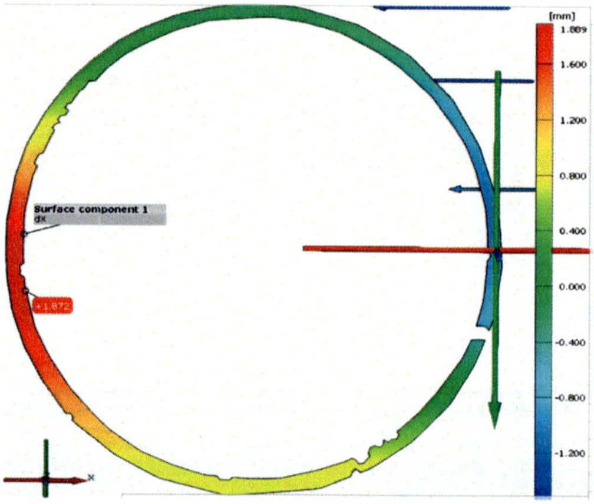

Fig. 12. Displacement distribution in the cross-sectional plain obtained using the 3D scanner

Besides the measurement of the surface strain in the plain perpendicular to the pipe axis of the symmetry, the strain was optically measured at the outer cylindrical surface. To measure the strain at the outer surface, the pipe was painted in white and then sprayed with black nonreflective color. Several trials were done to obtain the satisfactory patter quality.

Figure 13 shows the quality of the pattern regarding the reflectiveness, size and frequency of the black dots and the distance of the scanner with respect to the scanned surface. Since the surface is not flat, the focus cannot be obtained for the entire upper cylindrical surface.

The height was set such that the focus is at the surface about 5 mm below the highest line, such that most of the relevant surface area can be analyzed. Here the relevant area means the area where the highest normal strains and the normal stresses occur.

Surface area, which was scanned for normal strain components, was limited to the rectangular area in the figure around the highest line of the cylindrical surface. The right hand-side diagram shows the quality is mostly in the green area.

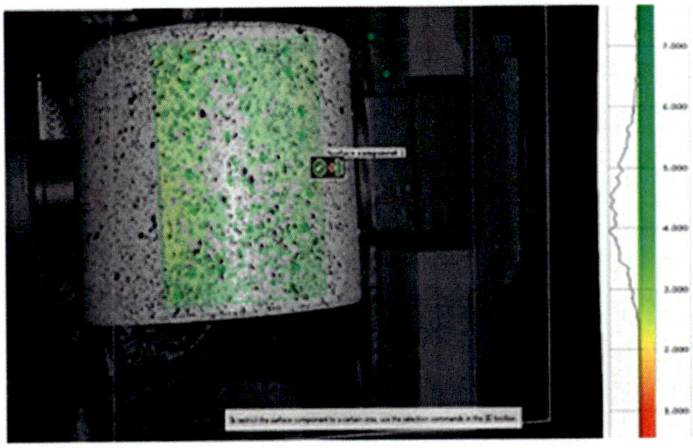

Fig. 13. The analysis of the pattern quality of the outer surface

Figure 14 shows the strain ϵ_x in the loading direction (in %). These values should correspond to the values at the outer layer in the analysis of the strain in the cross-sectional plain (shown in Fig. 10). The strain value near the highest line is not uniform, but it has some variation, depending on the pattern quality.

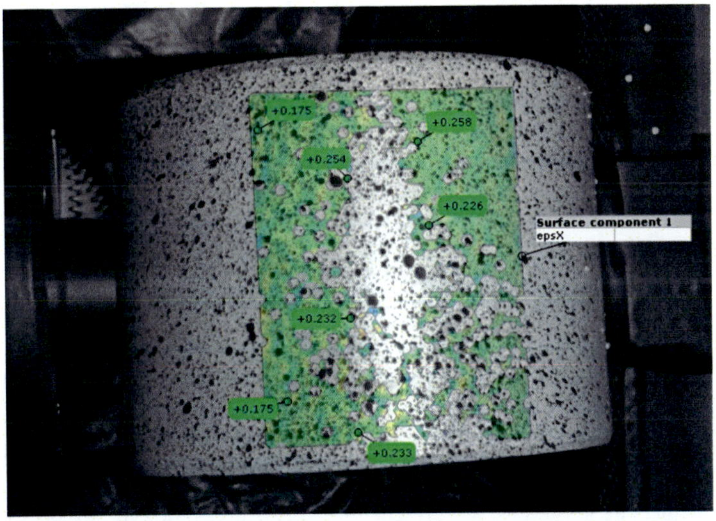

Fig. 14. Results using the 3D scanner - the deformation analysis of the outer surface of the pipe loaded with 11.94 kN

Displacement distribution at the outer cylindrical surface is shown in Fig. 15. Since the focus cannot be provided over the entire upper surface area, the area near the contact between the force sensor and the cylindrical surface is not registered automatically by the GOM Correlate software. The label is positioned at about the closest available point to the contact, where the displacement should have the highest value.

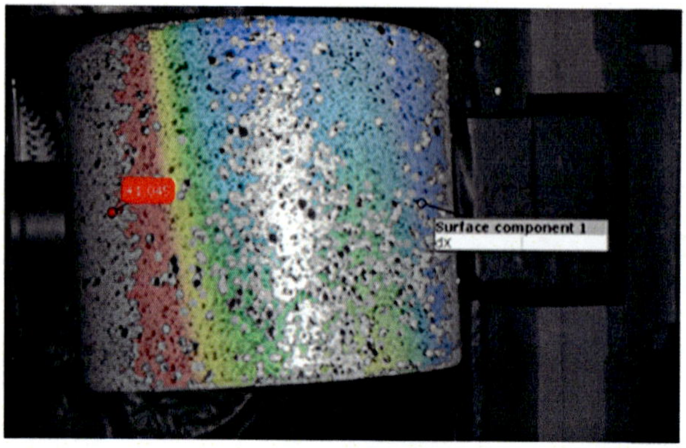

Fig. 15. Displacement distribution measured at the outer cylindrical surface

6 Results Analysis and Discussion

The normal strain was analyzed in two views: the cross-sectional plain and the side view at the cylindrical surface. The normal strain results in the cross-sectional plain show agreement with the analytical results. The measured strains in x direction (the loading direction) of the object at the inner and the outer wall at the critical cross-section are about +0.238% and −0.265%, respectively, which is $+2.38 \cdot 10^{-3}$ and $-2.65 \cdot 10^{-3}$, respectively.

If compared with the analytical results ($\epsilon_\varphi|_{r=r_2}^{\varphi=0} \approx 2.4 \cdot 10^{-3}$ and $\epsilon_\varphi|_{r=r_1}^{\varphi=0} \approx -2.7 \cdot 10^{-3}$), it can be seen that there is relatively good agreement, and relatively good agreement with the simulation results, as well. This comparison is valid for the cross-section where the x axis is in the circumferential direction (the cross-section denoted by $\varphi = 0^0$).

Using the generalized Hooke's law and by considering that there are no other normal stresses significant relatively to the hoop stress, the corresponding predicted stresses are:

$$\sigma_{\varphi max}^{predicted} = 110 \cdot 10^3\,\text{MPa} \cdot 0.00238 = 261.8\,\text{MPa}, \tag{30}$$

$$\sigma_{\varphi min}^{predicted} = -110 \cdot 10^3\,\text{MPa} \cdot 0.00265 = -291.5\,\text{MPa}, \tag{31}$$

which have relatively small errors comparing to the analytical results.

Considering the side view at the cylindrical surface, the highest value of strain recorded at the outer cylindrical surface is 0.258%, i.e., 0.00258, which overshoots the analytical value of 0.00238 by 0.0002. This result has the error about 8%. Based on the strain value, predicted using the scanner, the calculated stress is:

$$\sigma_{\varphi max}^{predicted} = 110 \cdot 10^3\,\text{MPa} \cdot 0.00258 = 283.8\,\text{MPa}, \tag{32}$$

where all other normal stresses are zero or negligible (the radial and axial stress) yielding the simple correlation via the Hooke's generalized law. This predicted stress has the error of 8%, with respect to the analytical value of 262 MPa (due to the linearity, it is the same error as in the prediction of the strain).

The displacement in the loading direction was measured via the scanner in two views, the cross-sectional plain and the plain of the side view. Since the recording is by only one camera, then there is no information about the depth. Hence, only 2D image is analyzed.

The highest displacement in the cross-sectional plain was 1.889 mm, which has an overshoot of 0.249 mm or 15% with respect to the analytical value ($\delta_D = 1.64\,\text{mm}$). The highest displacement in the side view cannot be considered since there was no way to register those points near the force sensor contact (that side of the yaw is moveable, the other side is fixed) due to the surface inclination. The measured deformation by the contact method on the clamp was 1.72 mm, which is in relatively good agreement with the analytical result for the displacement.

The numerical results for the stress (are $\sigma_\varphi^{num}(r_2) = 278.5\,\text{MPa}$, $\sigma_\varphi^{num}(r_1) = -305.4\,\text{MPa}$) have certain discrepancy with respect to the analytical results ($\sigma_\varphi^{an}(r_2) = 262\,\text{MPa}$, $\sigma_\varphi^{an}(r_1) = -292\,\text{MPa}$), with absolute values of relative errors are: 6% and 5%, respectively. These errors are partially produced by the way the boundary conditions were applied, since the force was represented by the uniform pressure over the side area to avoid applying a concentrated force at one cell element which would produce high stress locally.

7 Conclusion

A method was described and discussed how to combine measurements of 3D deformations of a loaded structural object along with an analytical approach to predict structural stress tensor components at critical points. Three results were compared, the analytical solution, the results from the numerical simulations and measured results for the strain components at the critical cross-sections of the loaded object.

In the way of obtaining the strain components analytically, in the same way the measured strains can be used to predict the stress tensor components, using the same analytical expressions. This means that the measured strain at specific areas of the loaded object can provide a way to obtain the overall loading of the analyzed structural element (in the case if the element cannot be disassembled and is a part of much larger complex structure), as well as to obtain the stress tensor components for the measured position and predict the stress distribution for other places (if the overall loading can be modelled analytically).

Further on, if an analytical solution is difficult or not possible to obtain, then the measured strains can be used as a check point for a numerical solution or as a particular condition that the numerical solution has to satisfy throughout an iterative procedure. This will be a subject for the further future work in the frame of the project.

References

1. Radosław, J., Krzysztof, S., Jarosław, D.: Application of the DIC technique to remote control of the hydraulic load system. Remote Sens. **12**(21), 1–24 (2020). https://doi.org/10.3390/rs12213667
2. Ackermann, F.: Digital image correlation: performance and potential application in photogrammetry. Photogram. Rec. **11**(64), 429–439 (1984)

3. Sciammarella, C.A., Sciammarella, F.M.: Digital image correlation (DIC). In: Experimental Mechanics of Solids, pp. 607–629. Wiley, New Jersey (2012)
4. Hild, F., Roux, S.: Digital image correlation: from displacement measurement to identification of elastic properties - a review. Strain **42**(2), 69–80 (2006)
5. Sutton, M.A., Orteu, J.J., Schreier, H.: Image Correlation for Shape, Motion and Deformation Measurements: Basic Concepts, Theory and Applications. Springer, New York (2009). https://doi.org/10.1007/978-0-387-78747-3
6. Amraish, N., Reisinger, A., Pahr, D.H.: Robust filtering options for higher-order strain fields generated by digital image correlation. Appl. Mechanics **1**(4), 174–192 (2020). https://doi.org/10.3390/applmech1040012
7. DIN Deutsches Institut fur Normung e. v., Berlin: Test pieces for the tensile testing of metallic materials - DIN 50 125. Beuth Verlag GmbH, Berlin (1992)
8. Moxnes, J.F., et al.: Experimental and numerical study of the fragmentation of expanding warhead casings by using different numerical codes and solution techniques. Defence Technol. **10**, 161–176 (2014)
9. vonroll-casting: Grey Iron (2017). https://www.vonroll-casting.ch/en/grey-cast-iron.html. Accessed 15 Mar 2021
10. DIN Deutches Institut fur Normung: Grey Cast Iron. Beuth Verlag GmbH, Berlin (1997)
11. Dolecek, V., et al.: Elastostatika. University of Bihac, Bihac (2004). (in Bosnian)
12. Timoshenko, S., Goodier, J.N.: Theory of Elasticity. McGraw-Hill, New York (1951)
13. Kljuno, E., Torlak, M.: Pipe stress analysis using an analytical and a finite-volume method. In: Avdaković, S., Volić, I., Mujčić, A., Uzunović, T., Mujezinović, A. (eds.) IAT 2020. LNNS, vol. 142, pp. 501–512. Springer, Cham (2021). https://doi.org/10.1007/978-3-030-54765-3_35

Design and Development of Compression and Torsion Springs Using CAD/CAE

Isad Šarić, Mirsad Čolić, Enis Muratović(✉), Muamer Delić, and Adis J. Muminović

Faculty of Mechanical Engineering, University of Sarajevo, Sarajevo, Bosnia and Herzegovina
muratovic@mef.unsa.ba

Abstract. In this paper we displayed the problematic of spring design, with usage of CAD/CAE (Computer Aided Design/Computer Aided Engineering) methods. Spring 3D (Three-dimensional) parametric modeling is performed in CATIA (Computer Aided Three-dimensional Interactive Application) software for purposes of achieving fast manipulation of spring geometry, so various spring variants can be made. Using the same software, FEM (Finite Element Method) analysis is used to determine maximum stresses and deflections. Utility of developed parametrized spring system is best reflected for cases of new spring design, where new geometric and FEM models are created with simple parameter application and update feature. This research resulted in spring prototyping with utilization of 3D printing techniques and spring manufacturing.

Keywords: Design · Spring · CAD/CAE · 3D modeling · FEM analyis · Parametrized system · Prototye

1 Introduction

Springs play very important role in modern industry. As one of the most commonly used elements in mechanical engineering and wide range of application, they are often used to react to the external loads, maintain a tension, absorb the shocks and reduce the vibrations of the system [1]. Mechanical spring is defined as an elastic body that has the primary function to deflect or distort under load, and return to its original shape when the load is removed [2]. As with other common mechanical components, the problem of spring design is often resolved using tables and charts containing pre-selected specifications and objectives. Even with computer assistance, designers are bound to oversimplify design procedures [3]. The parametric modeling enables creation of new concepts based on previous designs. Because these modeling tools offer interoperability in a parametric manner, no model re-creation is necessary, which is giving designer more time to explore a wider range of design alternatives [4]. The specification of spring quantitics is, inevitably, both variable and unknown. Though it has been refined over the years, and correction factors added as experience increased, the range of sizes, shapes and materials is so great that theory is stretched to the extremes. For that reason, it is not easy to correctly simulate realistic mechanical behavior of elastically deformable objects [5–7]. CAE software is a popular tool to analyse the robustness and performance of components

N. Ademović et al. (Eds.): IAT 2021, LNNS 316, pp. 635–644, 2022.
https://doi.org/10.1007/978-3-030-90055-7_50

and assemblies. The FEM analysis is a computing method implemented to find solutions to the boundary value problems in engineering [8, 9]. FEM method simulates a physical part of model's behavior by dividing the geometry into a number of elemets of standard shapes, applying loads and constrains, then calculating variables of interest [10, 11]. Based on that, much research effort needs to be expanded on the determination of spring parameters for proper mechanical simulation [12, 13]. Prototyping is related to physical and digital representation of a certain design, used for gaining an answer to a specific question or testing assumptions [14].

2 Paremetric 3D Modeling and FEM Analysis

Parametric modeling represents technology developed to utilize previous design to develop a new design and it enables quick and efficient product design phase [15]. Spring models, along with necessary design parameters are formed with CATIA software. Geometric parameters, for compression and torsion springs, are shown in Table 1.

Table 1. Compression and torsion spring design parameters

Compression springs parameters	Torsion springs parameters
Number of revolutions [-]	Number of revolutions [-]
Pitch value [mm]	Pitch value [mm]
Spring diameter [mm]	Spring diameter [mm]
Wire diameter [mm]	Wire diameter [mm]
	Length of free ends [mm]

As shown in Table 1, spring geometry, for both compression and torsion springs, can be described with a simillar parameters. Figure 1 shows compression spring (left) and torsion spring (right) with their predefined parameters located in the program tree.

New design solution is made by changing the values of parameters directly from program tree. Once parameters are set to the new values, an automatic update feature will create a new geometry. This way, various design solutions can be made in a short period of time. Figure 2 shows an example of variating geometry for compression and torsion springs.

With modification of specific parameters, different design solutions can be obtained. FEM models are created in the same software. Development of these models requires proper definition of material properties, adequate setting of the mesh, constrains, loads and boundary conditions. Material properties used for FEM simulation are shown in Table 2. Table 2 is considering Grade 1. patented and cold-drawn steel wire material with recommended values for spring permissible stresses [16].

Figure 3 shows meshed parts. Meshing is performed with parabolic finite elements, since they are more customized for curved surfaces. Mesh size and absolute sag, for both types of springs, are set to 0,2 mm and 0,1 mm, respectively.

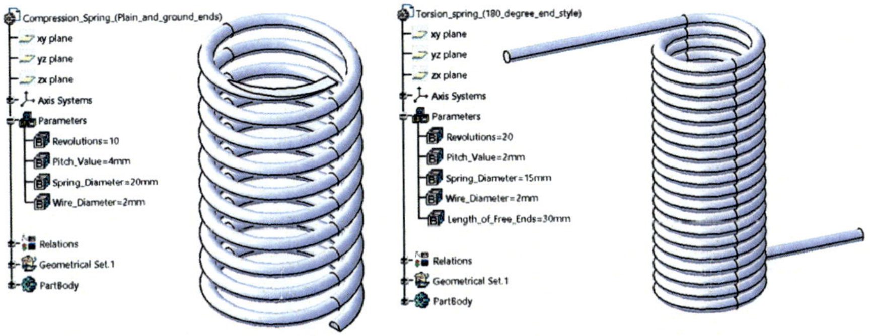

Fig. 1. Compression spring with specified parametric values (left), torsion spring with specified parametric values (right)

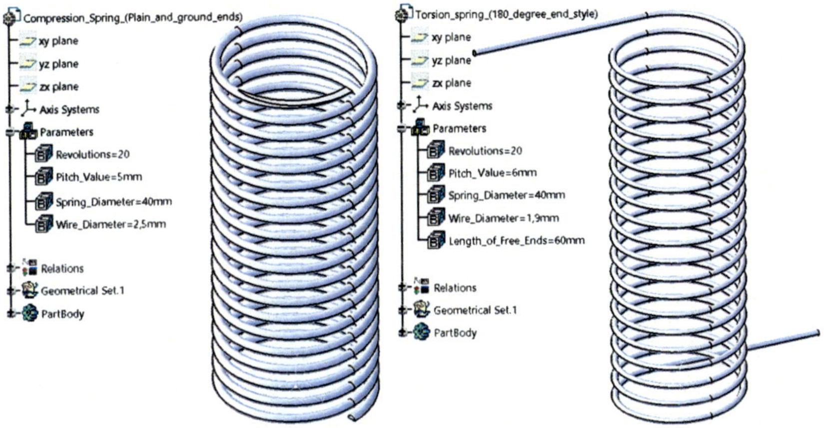

Fig. 2. Change of parameters for: compression spring (left) and torsion spring (right)

Table 2. Properties of the spring material and recommended permissible stresses

Size	Value
Ultimate tensile strength, S_{ut} [MPa]	1050
Permissible bending stress, σ [MPa]	315
Permissible shear stress, τ [MPa]	630
Modulus of elasticity, E [MPa]	210000
Modulus of rigidity [MPa]	81370

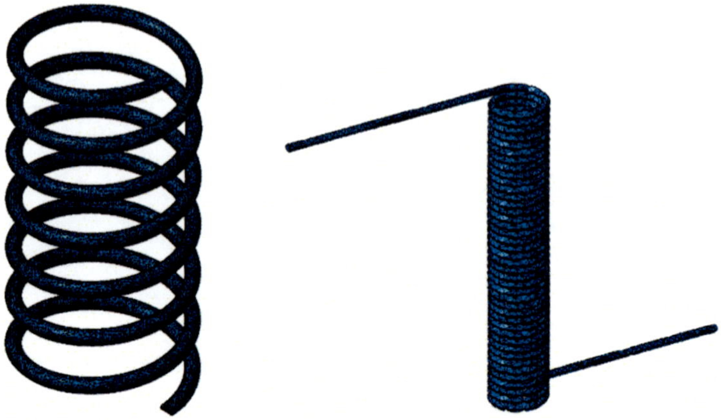

Fig. 3. Mesh settings applied on: compression spring model (left) and torsion spring model (right)

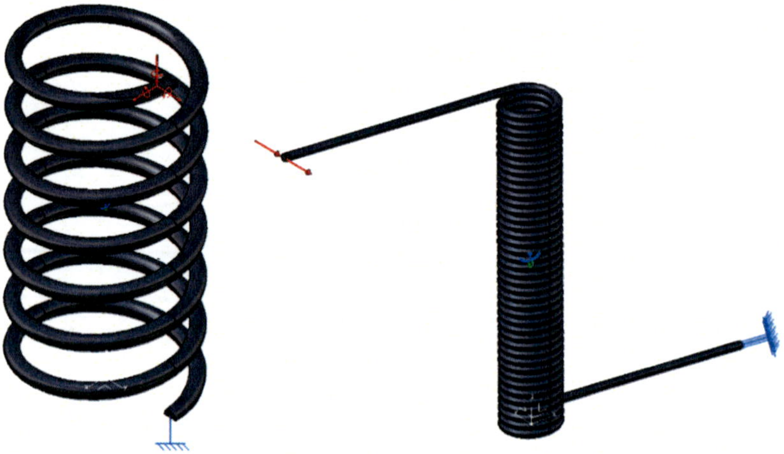

Fig. 4. Setting up loads, constrains and boundary conditions for: compression spring model (left) and torsion spring model (right)

Figure 4 (left) shows final setup of compression spring FEM model. Bottom surface of the spring is constrained with clamp. Also, user defined restrain is specified so that spring can only move in vertical direction. This way housing of the spring is simulated. Top coil of the spring is loaded with the distributed force. Figure 4 (right) shows final setup of torsion spring FEM model. One of the free ends is clamped (in this case bottom end), while end of the other is loaded with the distributive force (top end). Table 3 shows specific parameters used for FEM anlysis.

Figure 5 and Fig. 6 represent FEM analysis results for specific set of parameters described in Table 3. Figure 5 shows Von Mises stress distribution for compression and torsion springs. Compression spring was loaded with 60 [N] of distributed force. Maximum Von Mises stress according to colormap amounts 179 [MPa], but this stress

Table 3. Specific spring parameters used for FEM analysis

Parameter	Compression spring	Torsion spring
Revolutions []	6	41
Pitch value [mm]	16,7	2
Spring diameter [mm]	55	13,5
Wire diameter [mm]	5	2
Length of free ends [mm]	–	30

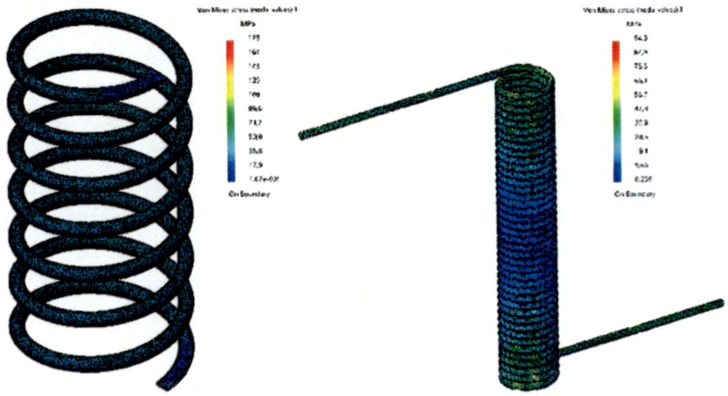

Fig. 5. Representation of equivalent (Von Mises) stress for: compression spring (left) and torsion spring (right)

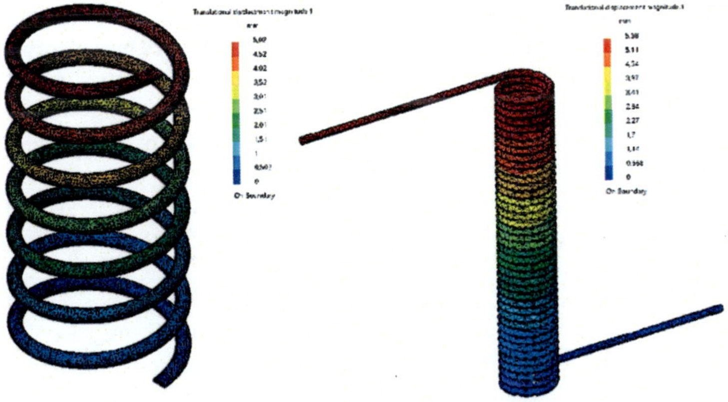

Fig. 6. Representation of deflection for: compression spring (left) and torsion spring (right)

is caused due to stress concentration on the "crossing zone" between plain end and the first coil. Maximum working stresses occure on the inner side of the spring coil, caused with uneaven distribution due to wire twist. It should be noted that although spring is under compression, the wire of the compression spring is not subjected to compressive stress. Torsional shear stresses are induced in a spring wire. Theese stresses are in range of 82–87 [MPa]. Torsion spring is loaded with 1,5 [N] distributed force i.e. moment equivalent of 45 [Nmm] due to length of free end. Such a small force is applied because of the spring dimensions, for which spring tangles easily during handling. Maximum Von Mises stress is located on the fixed end of the spring and it amounts 94,3 [MPa]. This stress is also prescribed to stress concentration. Maximum working stresses that occure in a spring wire are in range of 68–76 [MPa]. Also, it is noticeable that maximum stress components values locate in different places. This is caused with complex stress state, where bending moment, beside inducing bending stresses, tends to wind up the spring. Since Grade 1. patented and cold-drawn material was used, with specifications shown in Table 2, it is obvious that both of the springs will satisfy the strength criterium, compression spring will satisfy the permissible shear stress condition and torsion spring will satisfy the permissible bending stress condition [17]. Figure 6 shows deflections of compression and torsion springs. Maximum deflection of the compression spring is at the top coil and amounts 5,02 [mm], while maximum deflection of the torsion spring amounts 5,38 [mm] and it is located at the loaded end of the spring.

3 Prototyping

Prototypes are defined as pre-serial products, intended for testing and analyzing before serial manufacturing process. They represent product approximation considering one or more characteristics of interest. Prototypes are mostly used in a phase of design concept, where it is vital to know the level of functionality, that developed product has reached [18, 19]. 3D printing techniques and tools are traditionally focused on supporting design process. These methods open new design spaces for creating interactive and functional objects and devices on demand [20]. 3D printing of spring prototypes is realized with Ultimaker 3 3D printer shown on Fig. 7 (left), where 3D models of springs are prepared in the Cura software. Before preparation process, it is necessary to convert CATIA 3D models to STL (Standard Triangle Language) files which represent industry standard type for 3D printing. These file types use a series of triangles to represent the surfaces of a solid model.

Figure 7 (right) shows print setting dialog of the Cura software. For the considered models fine printing profile is used with layer thickness of 0,1 [mm]. Also, infill is set to 100% which means that spring will have solid interior. Since Ultimaker 3 has two extruders they can be configured in a way that one extruder operates simultaneously on main and supportive geometry, or one extruder can operate on the main while other operates on supportive geometry. Considered prototype printing considers usage of two extruders, where main and supportive geometry are printed in different colors. Printing of support is mandatory for models that have bridges, overhangs and other specific contours. Support material has a secondary constructive function in a 3D printing process in which additional material has a support role to the main object. For more stable start of the

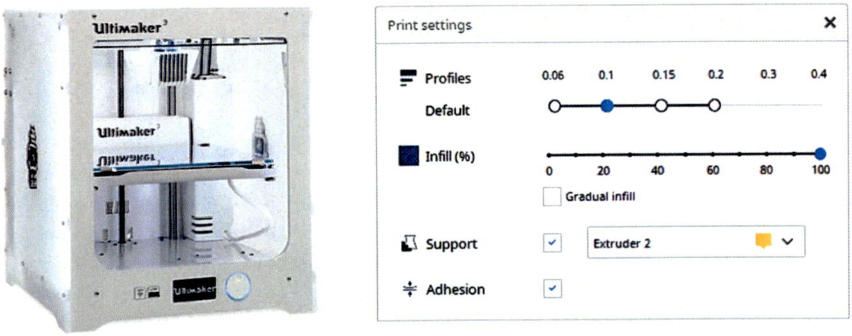

Fig. 7. Ultimaker 3D printer (left) and Cura software print setting dialog (right)

printing process adhesion option is checked. This way a flat area, which is easy to remove, will be added under the printing objects. Printing material used for prototype creation is PLA (Polyactic Acid). Once initial printing setup is complete, imported models can be positioned in the Cura software coordinate system which is correspondive to the actual printing area size of the real printer.

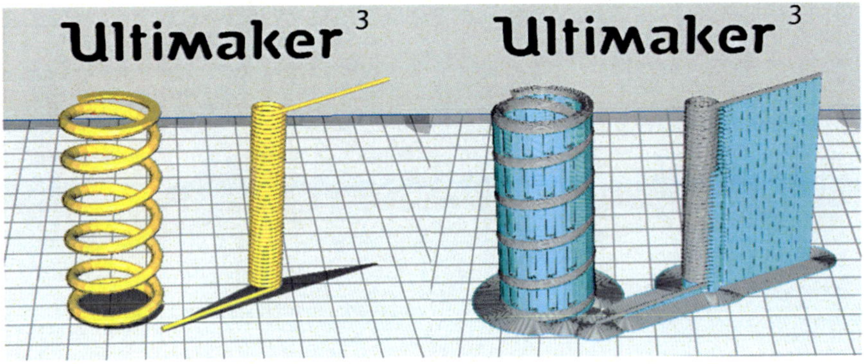

Fig. 8. Spring models positioned in the coordinate system (left) and preview of the printing process (right)

Figure 8 (left) shows spring models positioned in the centar of the virtual printing area. Figure 8 (right) shows the preview of the final prototype, where difference between main and supportive geometry is noticeable, due to different colors. Figure 9 shows the final printed prototypes for compression spring (Fig. 9 left) and torsion spring (Fig. 9 right). As shown on Fig. 9 there is a certain amount of support material that needs to be removed.

Once support material is removed, prototyping phase of a design process is completed. Next step is manufacturing of the springs. Figure 10 shows comparative view of prototyped springs (Fig. 10 left) and manufactured springs (Fig. 10 right).

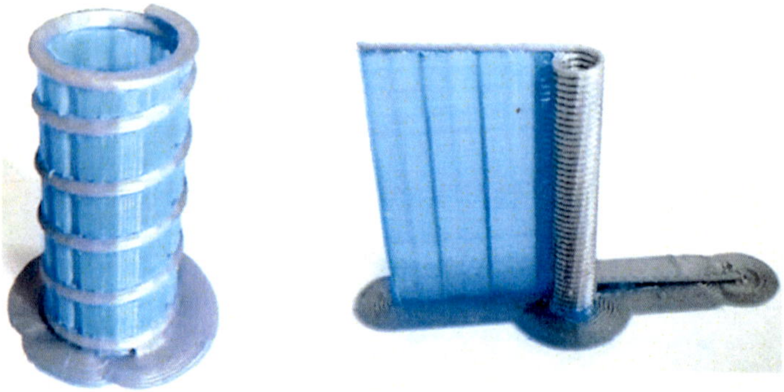

Fig. 9. Prototype of compression spring (left) and torsion spring (right)

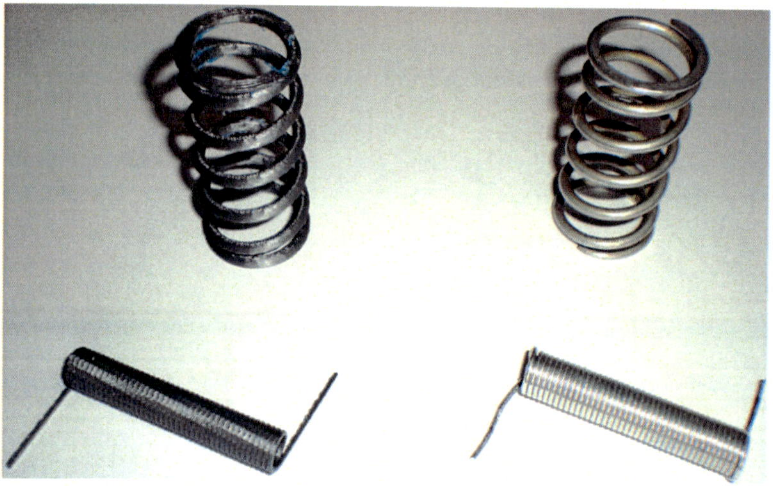

Fig. 10. Prototyped springs (left) and manufactured springs (right)

4 Conclusion

This paper demonstrates a design process supported with CAD/CAE tools. As shown, parametric modeling enables fast manipulation of created parameters which results in variety of design solutions. Considering the specific problematic of spring design, this paper shows one variant of parameter definition, where spring height is defined indirectly as product of pitch and number of revolutions. Paremeters could have been set differently where user would input total height of the spring and pitch, but this variant is limited to constant spring pitch definition. Parametrization method with current parameter definition, shown in research, was chosen because it can be easily upgraded to system where variable spring pitch can be defined.

FEM analysis gives clear insight into stress and deflection conditions which are dependant on current spring parameter definition and applied load. With a change of any parameter, FEM models are automatically updated, so new stress and deflection images are created. By knowing the characteristics of material, it can be established if current combination of spring parameters can withstand the applied load. This way, it is easy to conclude if spring design is safe or not.

With utilization of prototyping, future spring product can be examined, where flaws and defects, if there are any, can be detected. 3D printing technique, used in the research, is dependable on print settings conducted during the preparation of the certain spring model. Infill setting described in paper resulted in a more time-consuming prototyping (over 3 h). If final prototype is going to be used like a non-functional part, without being conducted to a moderate or significant force during examination, infill settings can be considerably decreased, which results in a shorter period of time needed for prototype manufacturing. Therefore, print settings can be adjusted with correspondence to prototype requirements.

References

1. Yang, C.J., Zhang, W.H., Ren, G.X.: Modeling and dynamics analysis of helical spring under compression using a curved beam element with consideration on contact between its coils. Meccanica **49**, 907–917 (2014)
2. Sarkate, T.S.: A finite element approach for analysis of a helical coil compression spring using CAE tools. Appl. Mech. Mater. **330**, 703–707 (2013)
3. Paredes, M., Sartor, M., Daidie, A.: Advanced assistance tool for optimal compression spring design. Eng. Comput. **21**, 140–150 (2005)
4. Ghionea, I.G., Devedžić, G., Ćuković, S.: Parametric modeling of surfaces using CATIA v5 environment. Appl. Mech. Mater. **760**, 93–98 (2015)
5. Fakhreddine, D., Mohamed, T., Said, A., Abderrazek, D., Mohamed, H.: Finite element method for the stress analysis of isotropic cylindrical helical spring. Eur. J. Mech. A/Solids **24**, 1068–1078 (2005)
6. Keller, S.G., Gordon, A.P.: Equivalent stress and strain distribution in helical compression springs subjected to bending. J. Strain Anal. Eng. Des. **4**(6), 405–415 (2011)
7. Ko, K.W., Kim, H.S., Bae, S.I., Kim, E.S., Lee, Y.S.: Determination of spring constant for simulating deformable object under compression. Key Eng. Mater. **417–418**, 369–372 (2021)
8. Zhang, P., et al.: Fatigue failure analysis and finite element assetment of the twins torsion spring. Eng. Failure Anal. **122**, 105–187 (2021)
9. Stpehen, C., Selvam, R., Suranjan, S.: A comparative study of steel and composite helical springs using finite element analysis. In: 2019 Advances in Science and Engineering Technology Internacional Conferences (ASET), University of Hertfordshire, De Havilland Campus, United Kingdom, 3–5 September (2019)
10. Fatola, B.O., Keogh, P., Hicks, B.: Modelling flat spring performance using fea. J. Phys. Conf. Ser. **189**, 1–9 (2009)
11. Werma, K.K., Aggarwall, M.L.: Finite element analysis and optimization of mono parabolic leaf spring using CAE software. Eng. Solid Mech. **3**(2), p85-92 (2015)
12. Rahat, A.M., Raiyan, M.F., Hossain, S.: Design and fabrication of a spring constant testing machine and determination of spring constant of a compression spting. Int. J. Eng. Res. **4**(10), 574–578 (2015)
13. Shaikh, Z.K.: Experimental study of helical compression spring. Int. J. Eng. Tech. Sci. Res. **4**(12), 970–975 (2017)
14. Jensen, M.B., Elverum, C.W., Steiner, M.: Eliciting Unknown unknowns with prototypes: introducing prototrials and prototrial-driven cultures. Des. Stud. **49**, 1–31 (2017)

15. Pandzic, A., Hodzic, D., Hajro, I.: Development of integrated intelligent CAD system for design of cylindrical helical compression valve springs. In: Proceedings of the 29th International DAAAM Symposium "Intelligent Manufacturing & Automation", Zadar, Croatia, 24–27 October (2018)
16. Bhandari, V.: Design of Machine Elements, 3rd edn. Tata McGraw-Hill, New Delhi (2010)
17. Mott, R.L.: Machine Elements in Mechanical Design, 4th edn. Pearson Education/Prentice Hall, Upper Saddle River, New Jersey (2004)
18. Saric, I., Muminovic, A.J.: Industrial Design, Faculty of Mechanical Engineering Sarajevo (2020)
19. Saric, I., Muminovic, A.J.: Product Development and Design, Faculty of Mechanical Engineering Sarajevo (2018)
20. He, L., Peng, H., Lin, M., Konjeti, R., Guimbretière, F., Froehlich, J.E.: Ondulé: designing and controlling 3D printable springs. In: Proceedings of UIST, vol. 19, pp. 1–12 (2019)

Computation of the Fluid Flow Around Octahedral Bodies for a Wide Reynolds Number Range

Tarik Čorbo, Almin Halač(✉), and Muris Torlak

Mechanical Engineering Faculty, University of Sarajevo, Vilsonovo šetalište 9, 71000 Sarajevo, Bosnia and Herzegovina
halac@mef.unsa.ba

Abstract. This paper presents results from a computational study of an incompressible flow around an octahedral body for Reynolds numbers ranging from Re = 100 to Re = 300 000. Different flow regimes are considered by solving the Navier-Stokes equations in time, imitating implicit large-eddy simulation (ILES) and using turbulence models based on Reynolds-averaged Navier-Stokes equations (RANS). The main aim of the paper is to calculate the dependence of the drag coefficient with respect to the Reynolds number. The results are compared with the experimental and empirical data from the literature. The best agreement is obtained by the ILES computations, while the steady-state turbulence models showed a certain discrepancy.

Keywords: Aerodynamics · Hydrodynamics · Bluff bodies · Non-spherical particles · Computational fluid dynamics (CFD)

1 Introduction

Transport of granular materials in the industry is in many cases done using pneumatic conveying. The design of such mechanical processes is, among others, reliant on the flow characteristics around the transported particles. The latter is strongly affected by the properties of the particles: size, shape, orientation, relative motion with respect to the transporting fluid, among others. Traditionally, the particles are for the sake of simplicity represented by spheres of appropriate size. In some cases, this idealization turns out to be inadequate. Instead, the particles can be represented by polyhedral bodies, as they much closer represent granular particles with sharp edges which promote flow separation. In this work we focus on the investigation of the octhaedral bodies. Accurate values of the drag coefficients and the behavior of the particles in the surrounding flow are significant for the optimized design of the transport processes.

This paper presents the investigation of incompressible fluid flow around a stationary octahedron in a wide range of Reynolds number using a computational method. In turbulent flow regimes, several turbulence models are tested to examine the most optimal approach in this application to be used for further research on different orientations of particles with respect to the incoming fluid stream.

N. Ademović et al. (Eds.): IAT 2021, LNNS 316, pp. 645–650, 2022.
https://doi.org/10.1007/978-3-030-90055-7_51

2 Problem Description

The solution domain is described by a 3D box set-up around the considered octahedron, see Fig. 1. It extends from –10d to 20d in the stream-wise direction. The square-shaped channel cross-section extends from –5d to 5d in the other direction. The octahedron is located 10d from the inlet, and is centered with respect to the cross-section. Here, d represents the diameter of the circumscribed sphere around the octahedron. In the computations presented here, the size of the octahedron is adopted so that its dimensions are: edge length $a = 7.07$ mm, and $d = 10$ mm.

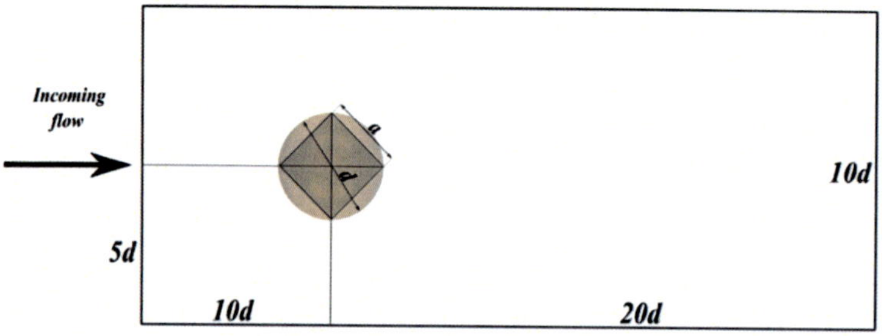

Fig. 1. The body and the solution domain with dimensions

The boundary conditions used are: no-slip wall for the surfaces of the bodies, slip walls for the surrounding solution domain, uniform velocity distribution for the inlet, and zero-gradient boundary condition for the outlet.

In order to calculate the Reynolds number for valid results comparisons, an appropriate reference length is needed. Usually, the equivalent-sphere diameter d_e is used and defined as the diameter of a sphere with the same volume as the tested non-spherical body. The equivalent-sphere diameter can be calculated as:

$$d_e = \left(\frac{V_{body}}{\frac{1}{6}\pi} \right)^{\frac{1}{3}}, \tag{1}$$

where V_{body} is the volume of the body. Using this diameter, the Reynolds number is given as:

$$Re_e = \frac{\rho v d_e}{\mu}, \tag{2}$$

Where ρ is the density of the fluid, v is its velocity, and μ is the dynamic viscosity of the fluid.

The relation used to calculate the drag coefficient is given as:

$$c_D = \frac{F_x}{\frac{\rho v^2}{2} A_{oct}}, \tag{3}$$

Where F_x is the drag force exerted on the object in the stream-wise direction, and A_{oct} is the area of the projected polyhedron surface onto a plane parallel to the inlet boundary.

3 Method and Computational-Model Setup

Simulations are done by using the program Simcenter STAR-CCM+ [1]. The simulation domain is divided into a finite number of computational cells. The mass conservation equation and momentum conservation equation are solved for each cell. The flow is considered as incompressible, and a constitutive equation for the Newtonian fluid is applied.

The mass conservation equation and momentum equation in their integral form can be written as [2]:

$$\frac{\partial}{\partial t}\int \rho dV + \oint \rho\vec{v}\cdot d\vec{S} = 0, \tag{4}$$

$$\frac{\partial}{\partial t}\int \rho\vec{v}dV + \oint \rho\vec{v}\otimes\vec{v}\cdot d\vec{S} = \oint \boldsymbol{\sigma}\cdot d\vec{S} + \int \rho\vec{f}_b dV, \tag{5}$$

where t is time, ρ is the density, $\vec{v}$ is the velocity vector, $\boldsymbol{\sigma}$ is the stress tensor, $\vec{f_b}$ is the vector of body forces, V is the control volume, bounded by the control surface S. Additionally, for the cases where the effects of turbulence is considered, the equations for the turbulence models are solved.

The equations are discretized using the *finite volume method*, where all solution variables are calculated at cell centres. The convective term is discretized by the linear upwind scheme, while gradients are calculated by the Gauss method. The simulations are performed using integration in time, where an implicit, second-order accurate scheme is applied for approximation of time-dependent properties. For the cases of low Reynolds numbers, the laminar flow solution is applied. For the larger Reynolds numbers, the laminar flow solution is applied, and the turbulence models: the k-ε model, k-ω *SST* turbulence model, and a Reynolds stress model (RSM). A detail description of the turbulence models can be found in [1].

One Cartesian computational mesh is generated for all cases. The mesh is trimmed around the octahedron to capture its shape properly. Furthermore, three wake refinement regions are introduced around and behind the body to achieve a trade-off of the computational resources needed while maintaining a high degree of solution accuracy. A total of 15 prism layers are added around the octahedron to capture the flow in the boundary layer accurately. The total number of cells is 1 700 125, where the biggest cell is of size 8 mm, while the smallest cell in the finest wake refinement is 0.25 mm, and the thickness of the first boundary layer cell is 0.0155 mm. For the RANS simulations, $y+$ wall treatment is used for all $y+$ values, while for the ILES simulations $y+$ is irrelevant since no turbulence model is used. The mesh is illustrated in Fig. 2.

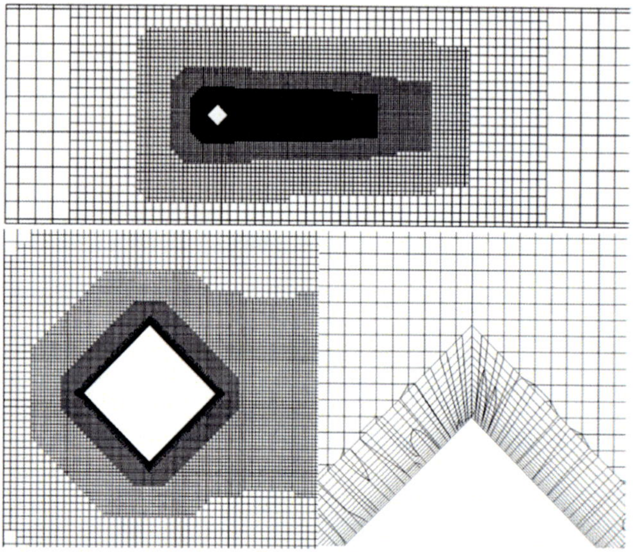

Fig. 2. The computational mesh, along with the used refinement zones and prism layers.

4 Results

The calculation of the drag coefficient described by Eq. (3) is the main focus of this study. Calculations are done for different Reynolds numbers extending from the laminar flow regime to the turbulent flow regime. All the Reynolds numbers calculations are performed by solving Navier-Stokes equations in time without a turbulence model, relying on the resolution of the mesh to capture the large-scale flow patterns even in strongly-turbulent regimes, while the fine-scale structure contributions are compensated by numerical-discretization effects. Herewith the simulations imitate *implicit large-eddy simulation* (ILES). Additionally, calculations with three typical, commonly used turbulence models based on Reynolds-averaged Navier-Stokes equations (RANS) are done for Reynolds numbers from 6828 onwards. The calculations are performed on a computer with 12 cores, with three simulations being done simultaneously, each taking four cores, with calculation time around 6 h per one calculation.

The validation of results is performed using experimental data from [3, 4], as well as empirical equations for octahedrons by: Ganser [5], Loth [6], Hölzer and Sommerfeld [7], Haider and Levenspiel [8]. This is illustrated in Fig. 3. Note that the correlation given by Leith [9] reduces to the same as Ganser [5] equation given for isometric particles.

According to Fig. 3, the best results are obviously obtained by solving Navier-Stokes equations in time without turbulence model. These results are in considerable agreement with the experimental results obtained by Pettyjohn and Christiansen [3], and Barker [4] for octahedral particles. They also agree well with the equations by Ganser [5] and Loth [6], both in the region where experimental results are available, and in the region that has not been experimentally tested.

On the other hand, the results obtained by using the RANS-based turbulence models differ significantly from the experimental results and the correlations, exhibiting considerably lower drag values. Furthermore, the trends of drag variation with Re-increase is different: the realizable k-ε model shows a slight decline of the drag coefficient, the k-ω SST model shows a slight increase, while the elliptic-blending RSM resembles the trend of the ILES-computations, yet suffering from a certain under-estimation. Among the tested RANS-models, the elliptic-blending RSM seems to be the most accurate one, as expected, since it is capable of modelling a certain number of eddies from the turbulent scale.

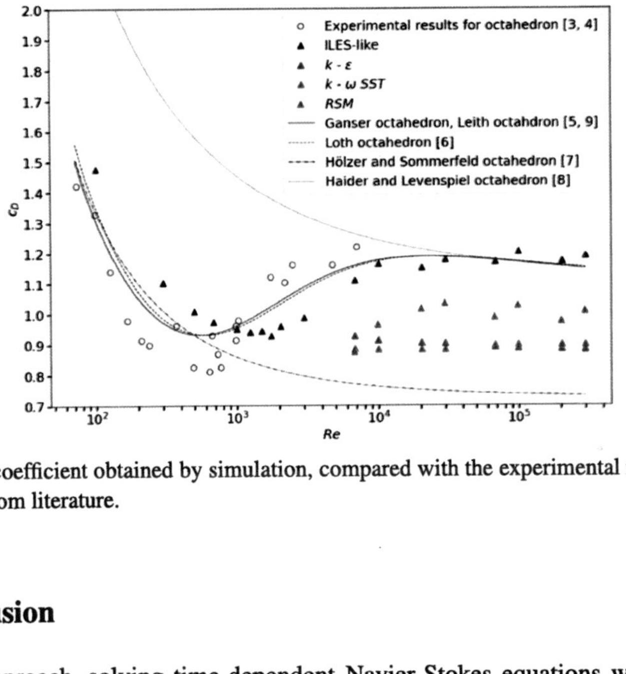

Fig. 3. Drag coefficient obtained by simulation, compared with the experimental results and the correlations from literature.

5 Conclusion

ILES-like approach, solving time-dependent Navier-Stokes equations without turbulence model, delivers most accurate drag coefficient computation, in a very wide range of Re numbers, from the laminar regime (Re = 100) to the strongly turbulent one (Re = 300 000). In the low Re range (laminar-to-transient), the calculated values of drag coefficient are slightly over-estimated, while in the turbulent regime, the calculated drag coefficient turns out to be slightly under-estimated. The results could also be improved using a finer mesh that would capture more of the turbulent fluctuations. The commonly used RANS-based turbulence models (realizable k-ε, k-ω SST and elliptic-blending RSM) yield a notable under-estimation, with different variation trends. This can be explained by the fact that all the tested RANS models give a time-averaged flow field that neglects the fine fluctuations caused by turbulence, significantly impacting the wake region and turning the drag coefficient values. These findings provide a basis for further investigations on different polyhedral shapes in different orientations.

References

1. Siemens Digital Industries Software: Simcenter STAR-CCM+ Documentation v. 2020.3 (2020)
2. Ferziger, J.H., Perić, M.: Computational Methods for Fluid Dynamics. Springer, Heidelberg (2013). https://doi.org/10.1007/978-3-642-56026-2
3. Pettyjohn, E.S., Christiansen, E.B.: Effect of particle shape on free-settling of isomeric particles. Chem. Eng. Progress **44**(2), 157–172 (1948)
4. Barker, H.: The effect of shape and density on the free settling rates of particles at high Reynolds Numbers, Ph.D. thesis, University of Utah (1951)
5. Ganser, G.H.: A rational approach to drag prediction of spherical and nonspherical particles. Powder Technol. **77**(2), 143–153 (1993)
6. Loth, E.: Drag of non-spherical solid particles of regular and irregular shape. J. Atmos. Sci. **28**(1), 110–115 (2008)
7. Hölzer, A., Sommerfeld, M.: New simple correlation formula for the drag coefficient of non-spherical particles. Powder Technol. **184**(3), 361–365 (2008)
8. Haider, A., Levenspiel, O.: Drag coefficient and terminal velocity of spherical and nonspherical particles. Powder Technol. **58**(1), 63–70 (1989)
9. Leith, D.: Drag on nonspherical objects. Aerosol Sci. Technol. **6**(2), 153–161 (1987)

Using OOMMF to Study the Effect of Microstructure on Magnetic Hysteresis Loops

Mehrija Hasičić[1](✉) and Aphrodite Ktena[2]

[1] Faculty of Engineering and Natural Sciences, International Burch University, Sarajevo, Bosnia and Herzegovina
mehrija.hasicic@ibu.edu.ba

[2] Energy Systems Laboratory, National and Kapodistrian University of Athens, Athens, Greece

Abstract. This paper presents a study of the effect of magnetic material parameters and microstructure on macroscopic magnetic properties with the open-source soft-ware for micromagnetic calculations OOMMF. Micromagnetic calculations are carried out using direct energy minimization. The effect of cell size, easy axis, exchange, and anisotropy on hysteresis loop parameters have been studied. Results show that the easy axis orientation has the greatest impact on coercivity, followed by cell size. In the case of randomized anisotropy, an increase in the exchange energy parameter resulted in slightly higher coercivity in comparison to the default value while the change in the anisotropy constant did not show any discernible effect on the material configuration studied.

Keywords: Micromagnetic calculations · Microscopic modeling · Magnetic properties · Hysteresis loop · Energy minimization · OOMMF

1 Introduction

Magnetic materials play a prominent role in modern technology. They are key components of motors, generators, and transformers [1]. The magnetic properties of materials used in the design of electrical machines, among others, are crucial to obtain the best characteristics and effectiveness of the machine. Numerous studies have been conducted in order to analyze and understand the effects of microstructure and microscopic parameters, such as grain size, magnetic anisotropy and exchange on macroscopic magnetic properties such as saturation M_s, permeability μ, coercivity H_c, et al. [2–8]. To investigate the material properties, characteristics, phenomena and/or relations between them, the following modeling approaches are used:

- macroscopic or phenomenological models among which are the Stoner-Wohlfarth model [9], the Jiles-Atherton model [10] and the Preisach model [6, 8, 11, 12].
- microscopic modeling which explores the magnetization reversal of ferromagnetic and ferrimagnetic materials at an intermediate scale between magnetic domain and crystal lattice [13].

N. Ademović et al. (Eds.): IAT 2021, LNNS 316, pp. 651–660, 2022.
https://doi.org/10.1007/978-3-030-90055-7_52

- atomistic modeling for investigating the influence of atomistic-scale phenomena on macroscopic quantities [3].

Relating material parameters and/or microstructural properties through different modeling approaches has been the research interest of many scientists [5, 7, 14–17]. With this work, we aim to contribute to this area through the investigation of the potential and limitations of the open-source Object Oriented Micromagnetic Framework (OOMMF), developed by NIST, in studying the effect of microstructure on the magnetic hysteresis loop. Hysteresis is a property associated with memory and energy storage capacity which is key to applications such as data storage devices or generators and motors respectively. On the other hand, it is a highly nonlinear effect not desirable in sensing applications. In the following, we propose that the microstructure is modeled through changes in the OOMMF source code, present the reasoning for choosing certain values of parameters and discuss their impact on the coercivity, the shape and the area of the loop, which is directly related to hysteresis losses, a figure of interest in materials such as electrical steel laminates used in transformers and electrical machines.

The paper is organized as follows. Section 2 presents the state of art on micromagnetic calculation in dynamic and quasi-static settings. Section 3 gives an insight on Object Oriented Micromagnetic Framework (OOMMF), materials dimensions and properties used in this study as well as an explanation for choosing certain parameters. In Sect. 4 results are presented and discussed through the parameters introduced in the previous section. In the last section, we summarize the research and draw conclusions based on the presented results.

2 Micromagnetic Calculations

Micromagnetism can be defined as a continuum approximation which allows the calculation of magnetization structures and magnetization reversal mechanisms assuming the magnetization to be a continuous function of position, and the derivation of relevant expressions for the contributions of exchange, magnetostatic, anisotropy, or other energy terms which come into play during a magnetization process [13].

When a field H_{eff} acts on magnetization M, the latter tends to align with the field in finite time, in the range of nanoseconds, through a precising motion which is described by the well-known Landau-Lifshitz-Gilbert (LLG) equation:

$$\frac{dM}{dt} = -|\overline{\gamma}|M \times H_{eff} - \frac{|\overline{\gamma}|\alpha}{M_s} M \times \left(M \times H_{eff}\right) \qquad (2.1)$$

where:

M is the magnetization i.e., the magnetic moment per unit volume;
H_{eff} is the effective magnetic field acting on M, which minimizes the material's free energy density (2.2) with respect to the orientation of M;
α is a dimensionless damping factor, assuming, for simplicity reasons, to be a constant between 0.1 and 1;

$|\overline{\gamma}|$ is the Landau and Lifshitz electron gyromagnetic ratio i.e., the ratio of the magnetic dipole moment to the mechanical angular momentum of the system expressed in radians per second per Tesla.

The LLG equation is used in all dynamic micromagnetic calculations. LLG calculations are tedious calculations requiring a lot of computing time and memory.

However, when the modeling of the hysteresis loop is desired, there is no need to solve the LLG since the hysteresis loop measurement is a quasi-static process. The applied field changes in time steps of several seconds, which means that the magnetization vector M is given the time to settle, through precession, to its equilibrium value before a new field is applied.

In hysteresis loop calculations, micromagnetic solvers that minimize the free (Gibbs) energy directly are more efficient and considered more suitable [18].

The total free (Gibbs) energy of a magnetic volume V may be broken down in energy terms, each one reflecting a different phenomenon, and is given by:

$$E_t(M) = \int (E_{ex} - E_{an} - E_H + E_m)dV \tag{2.2}$$

The first term is the exchange energy density E_{ex} (2.3) and reflects the nearest neighbor interactions whose strength depends on the exchange constant A_{ex}, expressed in Joules per meter.

$$E_{ex} = \frac{A_{ex}}{(M_s)^2}(\nabla M)^2 \tag{2.3}$$

The magnetocrystalline anisotropy energy density E_{an} (2.4) reflects the tendency of M to align towards the easy axis, or axes, of the material and, in the case of uniaxial anisotropy, it is determined by the (first) anisotropy constant K_1, expressed in Joules per cubic meter.

$$E_{an} = \frac{K_1}{M_s}(u_c \cdot M)^2 \tag{2.4}$$

Where u_c is a unit vector corresponding to the easy axis direction.
The effect of the external applied field H_{ext} is given by the Zeeman energy term:

$$E_H = \mu_0 M \cdot H_{ext} \tag{2.5}$$

The last term corresponds to the magnetostatic energy which tends to demagnetize the material and opposes the effect of the applied field. It reflects the effect of the long-range interactions, and it depends on the existing magnetic configuration:

$$E_m = \frac{\mu_0}{2} \int H_d^2(M)dV \tag{2.6}$$

where H_d represents the field acting on an elementary magnetic volume; it represents the torque exerted on the elementary magnetic volume by the magnetic moments in the rest of the material. This term is computationally the most challenging to calculate and a variety of formulations and approaches have been proposed. Some involve the use of vector potential $\boldsymbol{A}$ and the cominimization of (2.2) with respect to both $\boldsymbol{A}$ and $\boldsymbol{M}$ [13, 19].

Depending on the system and the effect being studied, it is also possible to include other energy terms such as the magnetoelastic energy term where the coupling between mechanical and magnetic properties is explicitly considered.

The magnetization saturation M_s, the exchange constant A_{ex} and the magnetocrystalline anisotropy constant K_1 are temperature dependent parameters but in quasi-static calculations such as that of a hysteresis loop, they are assumed to be constants.

In the results shown below, the total energy of a magnetic volume in an applied magnetic field is assumed to be the algebraic sum of the exchange energy, the magnetocrystalline anisotropy energy, the Zeeman energy, and the demagnetizing energy, as shown in (2.2). The minimization of (2.2) yields the magnetization direction for each elementary volume in the material for which the material has reached quasi-static equilibrium under the effect of a given vector applied field.

3 OOMMF Calculations

The Object Oriented Micromagnetic Framework (OOMMF) is a widely-available software for micromagnetic modeling that can produce programs with good performance as well as extensibility [20]. C++ based OOMMF core interfaces using Tcl/Tk to write a setup for performing the simulation. It can be modified at three levels: a) the top level where new modules are developed by using individual programs which interact using defined protocols across network sockets; b) at the second level, modifications of the Tcl/Tk script where scripts can be introduced and executed at run time; c) at the lowest level, it is possible to change the C++ source.

OOMMF has been used in various studies on micro scale modeling [2, 4, 21–24].

In this study we have focused on investigating the effects of A_{ex} and K_1 on hysteresis loops for different material microstructures by intervening at the second level in order to generate the desired structural and magnetic configurations and therefore study the effect of microstructure on magnetic parameters.

In the experiments shown here, we have modeled a material of size $1008 \times 108 \times 15$ [nm] with a cell size of 3 nm^3 and observed the effects of easy axis orientation and distribution, anisotropy, and exchange constants. We have also examined the effect of the cell size by comparing the standard problem with a material of size $1000 \times 100 \times 15$ [nm] with a cell size of 5 nm^3. Variations in dimensions of mentioned materials are due to the change in cell size. Each one of the three dimensions has to be a multiple of the cell size, therefore any change in material dimensions is modeled as a result of a change in cell size.

The choice of the cell size value should be based on two characteristic exchange lengths: the magnetocrystalline exchange length $l_{ex,K}$ for hard materials and magnetostatic exchange length $l_{ex,A}$ for soft materials, given in (3.1) and (3.2) respectively.

$$l_{ex,K} = \sqrt{\frac{A_{ex}}{K_u}} \tag{3.1}$$

$$l_{ex,A} = \sqrt{\frac{2A_{ex}}{\mu_0 M_s^2}} \tag{3.2}$$

According to (3.2), assuming a soft magnetic material with $K_1 = 48\frac{kJ}{m^3}, A_{ex} = 21\frac{pJ}{m}, M_s = 1700\frac{kA}{m}$, the maximum cell size should be 3.4 nm^3. Because the sample resolution should not be coarser than that, a cell size of 3 nm^3 is chosen.

For the results shown in the remaining of this paper, parameter values defined in the paragraph above correspond to the base case scenario and are called "default" values for convenience.

In the next section, we show results for a sample with these parameters modeled as a collection of identical cells with random magnetization orientations, similar to an amorphous or nanocrystalline material. Each arrow represents the magnetization direction of one cell. Each cell has the same magnetization value. The initial magnetic configuration of the case studied is shown in Fig. 1.

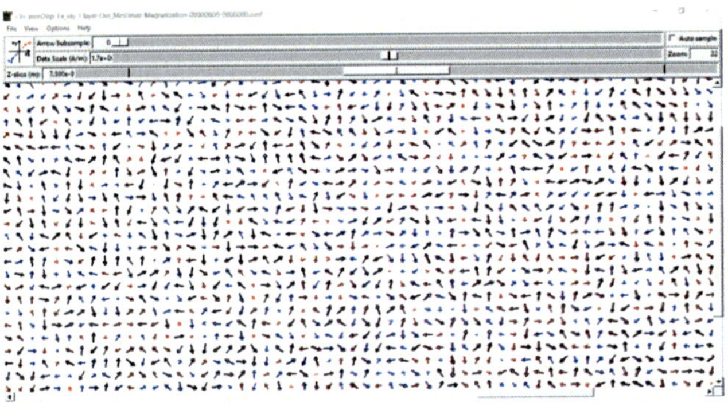

Fig. 1. Initial magnetic configuration of a sample with random magnetization directions, where each arrow represents the magnetization direction of one cell and each cell has the same magnetization magnitude

4 Results and Discussion

Results of the simulations are presented through the plots of hysteresis loops and discussed through the change in area of the loop. We varied the cell size, anisotropy constant and exchange with respect to the base case scenario. The external field H is assumed to be always applied along the x direction, i.e. along the length of the configuration shown in Fig. 1.

When a bigger cell size of 5 nm^3 is used, a slightly wider hysteresis loop with larger coercivity H_c is obtained, as shown in Fig. 2. Another aspect we investigated is the effect of initial easy axis orientation, therefore we compared the initial configuration shown in Fig. 1, with that of a material having the easy axis uniformly lying along the x direction or along 45° in the xy plane. The results obtained show (Fig. 3) that the randomly oriented material (Fig. 1) has the smallest H_c, while the biggest H_c is observed when the easy

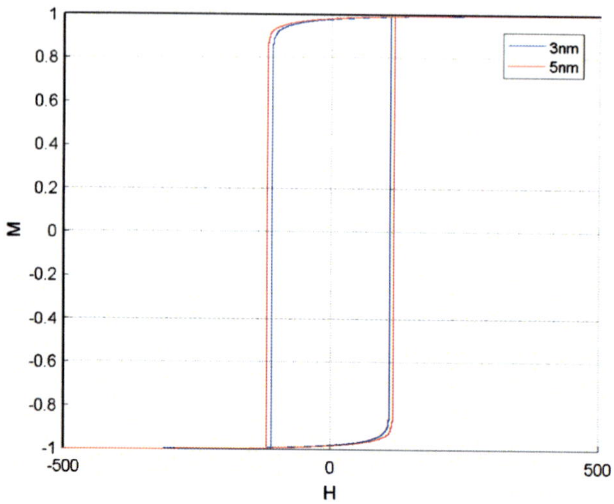

Fig. 2. The effect of cell size on the hysteresis loop with all other coefficients constant

axis is in the x direction and parallel to the direction of the applied field *H*, which is in line with experimental observations and other theoretical results.

Aiming for the narrowest loop which corresponds to lower hysteresis losses we continued our work on the material with 3 nm^3 cell size with random initial anisotropy directions and investigated the effect of exchange and anisotropy, as presented in Fig. 4 and Fig. 5, respectively.

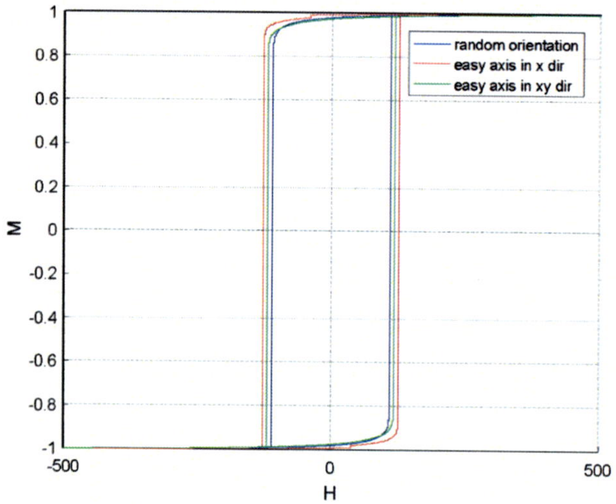

Fig. 3. The effect of the easy axis direction on hysteresis loop with all other coefficients constant

Given that the magnetostatic exchange length is directly proportional to A_{ex} and that it is already defined for our material we decided on minimum value for A_{ex} that can be

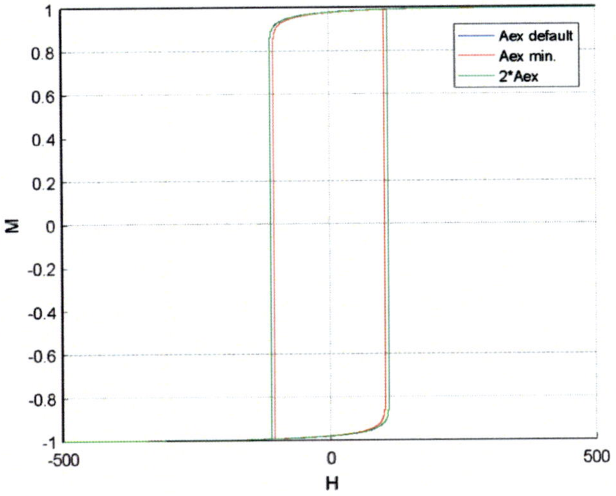

Fig. 4. The effect of the exchange constant A_{ex} on the hysteresis loop with all other coefficients constant

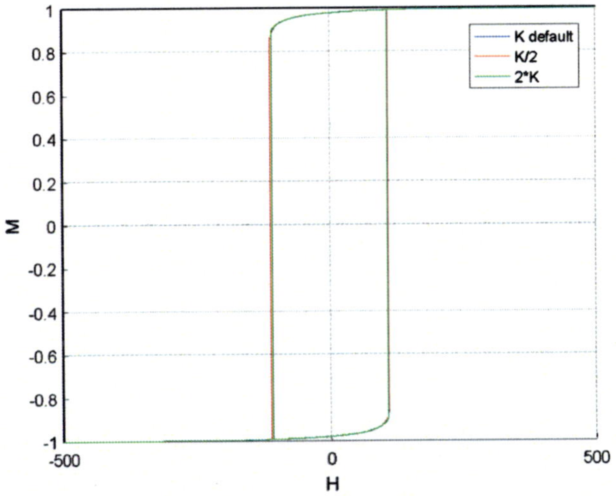

Fig. 5. The effect of the anisotropy constant K_1 on the hysteresis loop with all other coefficients constant

used to inspect the effect of decrease in exchange energy. As for trying a higher value, we doubled the default value since there is no constraint in choosing a smaller cell size than the value calculated through Eqs. (3.1) and (3.2).

To study the anisotropy effect, we used half the default value as lower value and twice the default as higher value. As shown in Fig. 4, H_c increases with the exchange energy. However, only a slight change is observed when the exchange is set equal to the minimum value, while for a similar decrease in anisotropy, the change in hysteresis loop is not discernible, Fig. 5.

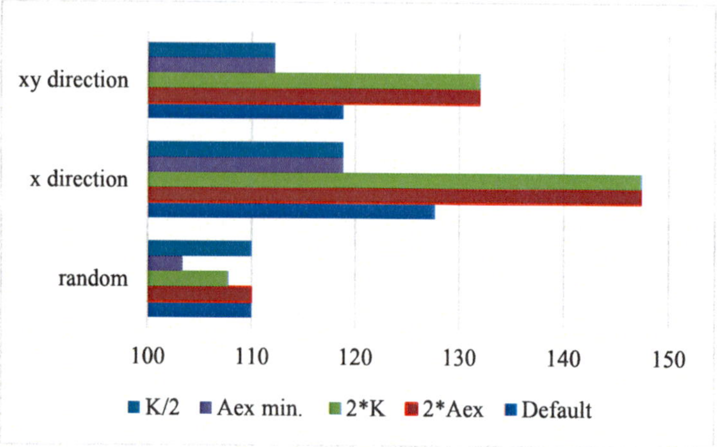

Fig. 6. The effect of the easy axis directions, anisotropy and exchange constants on the coercivity of the material

The results are summarized in Fig. 6. We observe that the anisotropy profile rather than the values of the microscopic parameters A_{ex} and K_1 has an effect on the hysteresis loop width. For the case where the material is uniformly magnetized and has a uniform uniaxial anisotropy profile, increasing or decreasing A_{ex} and K_1 leads to an increase or decrease in H_c. However, in the case of a material with random easy axes, the results are different: decreasing K_1 did not affect H_c but increasing K_1 slightly decreased H_c. As for A_{ex}, we notice that decreasing the exchange decreased H_c but increasing it had no effect on H_c.

The above cases and results are indicative of the potential OOMMF calculations have to offer in the study of the effect of microstructure on macroscopic magnetic properties. Changing the cell size, material size, material structure and microscopic parameters, various microstructural configurations can be obtained and through the hysteresis loop calculation based on the free energy minimization their effect on coercivity, permeability or the overall shape of the loop and the hysteresis losses can be modeled.

5 Conclusion

In this paper we have presented the results of our micromagnetic study using the open-source software OOMMF on the effects of different microscopic parameters on the magnetic hysteresis loop. The software has been modified at the intermediate level. Easy axis orientations, exchange and anisotropy parameters have been varied to observe the changes in coercivity i.e., the width of the loop which is related to the hysteresis losses. The study has shown that the easy axis direction affects the coercivity the most. Materials with random anisotropy profiles resulted in the narrowest hysteresis loop.

Increasing or decreasing anisotropy and exchange constants results in respective changes in coercivity when the material has a uniform anisotropy profile along a given axis. In the case of randomized anisotropy, increasing the anisotropy decreases the coercivity while decreasing the anisotropy and increasing the exchange did not have any significant effect.

Materials with random initial magnetization and anisotropy will be further studied in future work by adding impurities and voids while the effect of stress on the hysteresis loop will be emulated through changes in the anisotropy profile as well in order to model stress – related hysteresis losses.

References

1. Szymczak, H.: Magnetic materials and applications. In: Encyclopedia of Condensed Matter Physics, pp. 204–211. Elsevier (2005)
2. Zou, J., et al.: A novel analytical model for hysteresis loops of exchange-coupled hard-soft magnets. J. Rare Earths **37**(10), 1034–1039 (2019). https://doi.org/10.1016/j.jre.2019.03.007
3. Poluektov, M., Eriksson, O., Kreiss, G.: Coupling atomistic and continuum modelling of magnetism. Comput. Methods Appl. Mech. Eng. **329**, 219–253 (2018). https://doi.org/10.1016/j.cma.2017.10.010
4. Kim, H., You, C.Y.: Embedded object-oriented micromagnetic frame (OOMMF) for more flexible micromagnetic simulations. J. Magn. **21**(4), 491–495 (2016). https://doi.org/10.4283/JMAG.2016.21.4.491
5. Dupré, L., Bertotti, G., Basso, V., Fiorillo, F., Melkebeek, J.: Generalisation of the dynamic Preisach model toward grain oriented Fe–Si alloys. Physica B Condens. Matter **275**(1–3), 202–206 (2000). https://doi.org/10.1016/S0921-4526(99)00767-X
6. Ktena, A., Fotiadis, D.I., Spanos, P.D., Berger, A., Massalas, C.V.: Identification of 1D and 2D Preisach models for ferromagnets and shape memory alloys. Int. J. Eng. Sci. **40**(20), 2235–2247 (2002). https://doi.org/10.1016/S0020-7225(02)00116-7
7. Dupré, L.R., Bán, G., Von Rauch, M., Melkebeek, J.A.A.: Relation between the microstructural properties of electrical steels and the Preisach modelling. J. Magn. Magn. Mater. **195**(1), 233–249 (1999). https://doi.org/10.1016/S0304-8853(98)01158-5
8. Ktena, A.: Residual stresses and vector hysteresis modeling. Physica B Condens. Matter **486**, 29–33 (2016). https://doi.org/10.1016/j.physb.2015.12.008
9. Stoner, E.C., Wohlfarth, E.P.: A mechanism of magnetic hysteresis in heterogeneous alloys. Phil. Trans. R. Soc. Lond. A **240**(826), 599–642 (1948). https://doi.org/10.1098/rsta.1948.0007
10. Jiles, D.C., Atherton, D.L.: Theory of ferromagnetic hysteresis (invited). J. Appl. Phys. **55**(6), 2115–2120 (1984). https://doi.org/10.1063/1.333582
11. Preisach, F.: Uber die magnetische Nachwirkung. Z. Physik. **94**(5–6), 277–302 (1935). https://doi.org/10.1007/BF01349418
12. Ktena, A., Fotiadis, D.I., Berger, A., Massalas, C.V.: Hysteresis modeling of Gd films and AFC thin-film recording media. Physica B Condens. Matter **343**(1), 101–106 (2004). https://doi.org/10.1016/j.physb.2003.08.034
13. Fidler, J., Schrefl, T.: Micromagnetic modelling - the current state of the art. J. Phys. D Appl. Phys. **33**(15), R135–R156 (2000). https://doi.org/10.1088/0022-3727/33/15/201
14. Dupré, L., Sablik, M., Van Keer, R., Melkebeek, J.: Modelling of microstructural effects on magnetic hysteresis properties. J. Phys. D-Appl. Phys. **35**(17), 2086 (2002)

15. Dupré, L.R., Van Keer, R., Melkebeek, J.A.A.: Identification of the relation between the material parameters in the Preisach model and in the Jiles-Atherton hysteresis model. J. Appl. Phys. **85**(8), 4376–4378 (1999). https://doi.org/10.1063/1.369789
16. Van de Wiele, B., Dupré, L., Olyslager, F.: Memory properties in a 3D micromagnetic model for ferromagnetic samples. Physica B Condens. Matter **403**(2), 342–345 (2008). https://doi.org/10.1016/j.physb.2007.08.045
17. Sablik, M.J.: Modeling the effect of grain size and dislocation density on hysteretic magnetic properties in steels. J. Appl. Phys. **89**(10), 5610–5613 (2001). https://doi.org/10.1063/1.1359167
18. Exl, L., Bance, S., Reichel, F., Schrefl, T., Stimming, H.P., Mauser, N.J.: LaBonte's method revisited: an effective steepest descent method for micromagnetic energy minimization. J. Appl. Phys. **115**(17), 17D118 (2014). https://doi.org/10.1063/1.4862839
19. Berkov, D.V., Ramstöck, K., Hubert, A.: Solving micromagnetic problems. Towards an optimal numerical method. Phys. Stat. Sol. (a) **137**(1), 207–225 (1993). https://doi.org/10.1002/pssa.2211370118
20. Donahue, M.J., Porter, D.G.: OOMMF Userd's Guide, Version 1.0. National Institute of Standards and Technology, Interagency report NISTIR 6376 (1999)
21. Liao, Z.-M., Lu, Y., Zhang, H.-Z., Yu, D.-P.: Hysteresis magnetoresistance and micromagnetic modeling of Ni microbelts. J. Magn. Magn. Mater. **322**(15), 2231–2234 (2010). https://doi.org/10.1016/j.jmmm.2010.02.016
22. Muñoz-Sandoval, E., Torres-Heredia, J.J., López-Urías, F.: Micromagnetic simulations of hysteresis loops in ferromagnetic Reuleaux's triangles. J. Appl. Phys. **97**(10), 10E318 (2005). https://doi.org/10.1063/1.1858113
23. Musuroi, C., Volmer, M.: OOMMF modelling of magnetization dynamics in micrometer sized structures for sensing applications (2018)
24. Mahalingam, S.S., Manikandan, B.V., Arockiaraj, S.: Review – micromagnetic simulation using OOMMF and experimental investigations on nano composite magnets. J. Phys. Conf. Ser. **1172**, 012070 (2019). https://doi.org/10.1088/1742-6596/1172/1/012070

Trends and Innovation in Geographic Information Systems

Envisioning Sarajevo as Pandemic-Resilient City: A Case Study on Public Spaces

Emina Zejnilović[1(✉)], Erna Husukić[1], Jasmin Taletović[2], and Mirza Ponjavić[1]

[1] Faculty of Engineering and Natural Sciences, International Burch University, Sarajevo, Bosnia and Herzegovina
emina.zejnilovic@ibu.edu.ba

[2] Department of GIS, Institute of Development Planning of the Sarajevo Canton, Sarajevo, Bosnia and Herzegovina

Abstract. The influence of the urban environment on health has been recognized as a determining factor in creating contemporary urban models that favor health and well-being of city inhabitants. Today, as the spread of COVID-19 pandemic is accelerating, it is evident that the risks to health in urban environments need to be readdressed, as the cities that cater to contemporary lifestyle needs, demonstrate immense liability to the virus spread. Imposed change in living dynamics, the growing demand for the use of public spaces, and the relationship between public spaces and disease transmission, are only some of the most vigorously debated topics of today, raising numerous questions about the future of the city. Within this academic framework, the paper examines existing inventory of public spaces in Sarajevo aiming to address the possibility of the city to transit towards a wider and more homogeneous supply of public space. Through cartographic representation the research produces maps - an atlas of the main categories of public spaces such as parks, gardens, plazas and pedestrian areas. The results provide an insight into the assessment of vulnerability and risk in Sarajevo in terms of public areas availability and propose specific spatial interventions that could provide a more adequate response for the changing social behavior during and after COVID-19 pandemic.

Keywords: Urban health · Public spaces · Spatial mapping · GIS · Covid-19 · Sarajevo

1 Introduction

The danger of pandemics and their connection with densely populated cities is not a new phenomenon unique to our time, but is in fact, as old as the concept of the formally established special pattern of human settlements [1, 2]. This does not imply that urban areas are susceptible to calamities by nature, but the accelerated urbanization, contemporary movement patterns, and population density create fertile ground for environmental disasters and other hazards, significantly increasing city vulnerability [3].

N. Ademović et al. (Eds.): IAT 2021, LNNS 316, pp. 663–675, 2022.
https://doi.org/10.1007/978-3-030-90055-7_53

Under the influence of the ongoing pandemic, the UN Habitat is calling for comprehensive, rapid city-wide profiling and assessments that could help in preparations of a response plan [4]. In particular, public spaces are defined as an important asset in a time of present crisis, as promoting health through space by directing and changing behavior is likely to impact large proportions of population and create long lasting effects. World Health Organization (WHO) advocates for shifting the focus from understanding how the risks can influence the health of city inhabitants, to how well-planned cities can produce benefits for health [5].

Although adequate amount of public space differs significantly from place to place, it is stated that an average of 40% of the surface area of European cities is made up of urban green infrastructure, with around 18.2 m of publicly accessible green space per inhabitant, while 44% of Europe's urban population lives within 300 m of a public park [6].

Aside from the amount of public space available, the pandemic has put into the focus issues of uneven distribution of public and accessibility to shared spaces such as parks, gardens and playgrounds within a 15-min walk from home. Furthermore, it has significantly affected change in consolidated social behaviors, particularly in urban areas, demonstrating that cities around the world are poorly prepared to confront this mega challenge. The demand for spatial response in the creation of a new dynamics of living, raises numerous questions:

- How is the current COVID-19 public health crisis changing our cities?
- What will be the impact of the virus on the future of the design, use and perception of public spaces?
- How might we use planning and design strategies to improve resilience in the face of future pandemics?
- What are the possible solutions, in terms of safe, easily reachable public areas that could contribute to the living quality?

Using a research model performed by Systematica on the city of Milan [7], this study presents a survey of Sarajevo's public space in its four municipalities, aiming to assess the current state in terms of availability and accessibility, finally creating an atlas of the main categories of public spaces, quantifying emerging needs and outlining targeted responses to these new collective requests. Overlapping the information obtained from spatial analyzes with demographic examines of population, provide information on current 'pressures' on the existing public spaces in the locations of four municipalities in Sarajevo. A 'scan' of the selected areas shows the current state of public spaces and based on that, the research proposes possibilities for adapting the urban environment that could provide a more adequate response for the changing social behavior during and after COVID-19 pandemic.

2 Health and Urban Planning

Health and urban planning are natural allies, [8] due to the strong interconnectedness between public health and space. In its broadest sense urban health refers to the study of the health of urban populations. It includes two principal aspects: the description of the health of urban populations, both as a whole and as particular subgroups within cities, and an understanding of how characteristics of cities themselves affect the health of urban populations [9]. According to the World Health organization (WHO) 'being a Healthy City depends not on current health infrastructure, rather upon a commitment to improve a city's environs and a willingness to forge the necessary connections in political, economic, and social arenas' [10]. The links have evolved over time from the period of hygiene applied to urban planning in the 19th century, functional urban planning up to the 1970–80s, and finally the current period of "sustainable" urban planning. There has been a growing academic and legislative body on the topic aiming to contribute to the systematic influence of urban planning on the environment and health, underlining its role in the primary prevention, and the importance of a holistic approach in designing health-oriented cities [11–15].

These initiatives have made great innovations but have also encountered many challenges [16]. The risk of these actions having isolated short-term impact; the insufficient support from the local governments in establishing and implementing long-term policies; the lack of a more all-inclusive approach as the topic to often focuses primarily on conserving environmental resources and controlling the risk of the inevitable urban transformation on health; the challenge of finding the compatible solutions among different city models in regards to their urban, socio-cultural, economic and political context; are only some of the issues the urban health initiatives have faced. Thus, the complex theme of public health from urban perspective is an ambitious, continuous, large scale endeavor, particularly at present global COVID-19 pandemic situation.

It is argued that a very few phenomena throughout human history have shaped our societies and cultures the way outbreaks of infectious diseases have [17]. Over centuries, pandemic outbursts have caused major social, political, religious and economic changes, wiped out entire populations, but reciprocally they have also set the stage for innovations and advances in science and medicine, public health and city development [18]. From the II century Antonine Plague in the Roman Empire, the deadliest 14th century Black Death, to Victorian era Cholera and the early 20th century Spanish fever, [19] infectious diseases spread demanded healthier urban metabolism and a safer socio-natural landscape [20]. Sanitation, inadequate water supply, as well as the need for more light and air, over time transformed the unhealthy cities in terms of infrastructure, but also set direction for health oriented urban planning. Such spatial interventions make strong reference to the fact that physical and social environments play major roles in the health of communities, and that urban planning can and does serve as a form of primary prevention and contributor to health outcomes [16].

Surprisingly, the relationship between the established health policies and urban planning became particularly fragile in the late 20th century, while planning policies facilitated urban trends that have negative effects on public health [8]. Additional contemporary risk factor for health and well-being, one that is dramatically influencing city vulnerability particularly when talking about the risk of the infectious diseases spreading, are also contributed by projections that 55% of the world's population will live in urban areas by 2050, an increase of 68% [21]. It is estimated that population migration from rural to urban areas, combined with total world population growth, could increase the urban population by 2.5 billion by 2050. The UN World Urbanization Prospect from 2018 also estimates that by 2030, the world could have 43 so-called mega-cities (with more than 10 million inhabitants), which is 12 more cities than the current 31, mostly in developing countries. This denotes that as cities are creating environments that provide a number of socio-economic opportunities, they are correspondingly becoming increasingly risky places to live. Urban mechanisms for controlling and influencing health in systematic way are in fact contributing to the city liability.

Consequently, the connection between contagious disease outbreaks and urban settlements has never been as hazardous as it is now, under current conditions of intensified urbanization and globalization. Rising global connections and interactions are progressively becoming a driving force behind pandemics. Effects of extensity, intensity, velocity and impact propensity, as elements that define globalization [22] establish fertile ground for accelerated traffic of pathogens, contributing to the ongoing global health crises. Increased interconnection between cities as the patterns of flow transcend localities, and the overall intensification in the speed with which information, and people travel, have major impact on city health risks. This allows for a seemingly distant event to quickly have unexpected, emergent, large scale impact at local levels across the globe. Therefore, it is no surprise that it took just a few months for COVID-19 virus to reach Italy and the rest of the world from China, while it took three years for Bubonic plague (Black Death) that originated in China in 1334, to arrive to Europe in 1347, following the Silk Road, whereas the rest of the world was affected over the course of 5 years.

The current 'microbe vs. man' crises of the COVID-19 pandemics, puts to the forefront the issues of networked, fluid nature of urban connectivity, health policies and urban planning, but also questions the role of spatially fixed sites on local scale. It is now evident that the ongoing pandemics is not a temporary condition, but one that will require a long-term change in consolidated social behavior, that should be supported through adequate spatial interventions and use of space aimed at health protection. There is a high probability that the impact will be variable not only at the level of different cities, but also at the level of different parts within cities, as well as that cities may experience multiple peaks in virus spread [23]. The reason for this is that cities in their morphology do not have a universal form or structure, but are formed on various social, political, biological and economic locations and scales.

Cities across the world are rapidly working on assessing spatial risks and their mitigation, through the establishment of a more efficient system of urban resilience to infectious diseases. Such activates aim to increase quality of living and create safer outdoor environment, by building urban withstand through zoning and alterations in land use. Considering social distancing as the pillar of today's living dynamics, under the influence of the spread of the COVID-19 virus, many cities are already reacting with specific spatial interventions. Milan, Bogota, Mexico and Vienna are working on the configuration of the streets through the widening of the sidewalks, as well as Boston, London, Portland and Vancouver, in order to provide more space for cyclists and pedestrians, while maintaining the prescribed distance. Squares and parks are being redesigned with clear signs related to safe distance. The possibility of introducing a model of Barcelona's 'superblock' for redirecting traffic in order to get larger pedestrian zones within residential areas is being considered by many cities across the globe. Furthermore, research shows that under the influence of the pandemic, there could be decentralization and changes in the typology and distribution of green spaces in urban areas, as a quick response to the need to increase smaller green zones or parks within residential areas. In many cities, analyzes of the available potentials of open spaces are being done, ie. activation of unused green areas, brownfields, and roofs of buildings. Issues of the future of large public spaces within the urban fabric, and the need for a new typology, all related to the cultural and social specifics of cities, are discussed [24].

Though, as acknowledged in the WHO's 2016 Global Report on Urban Health, not every city can do 'extreme makeovers for health' but 'every city can take steps in direction of healthier planning.' Thus, the COVID-19 virus pandemics calls for examination of the reciprocal links between the society, urban planning, public space, and public health, on local scale. Once again, urban spaces have a key role to play in promoting and implementing the new living dynamics, and in helping suppress the spread of infection.

3 Epidemiological Characterization of the Space Related to the Case Study

Spatial-temporal spread of COVID-19 infection, related to the location of the study, is characterized by the dynamics of spread in Bosnia and Herzegovina (BiH), the magnitude of the disease in individual administrative units until July 23, 2020 (for initial wave of spread), and the current time, and the degree of disease in densely populated urban areas [25].

The Rt diagram (Fig. 1) shows the dynamics of the spread of COVID-19 with an overview of daily registered cases of infection during the reporting period, from the beginning of the epidemic to July 23, 2020 [26]. In addition, the spatial distribution of infection by administrative units (municipalities) in Bosnia and Herzegovina is shown for the same period using a choropleth map (Fig. 2).

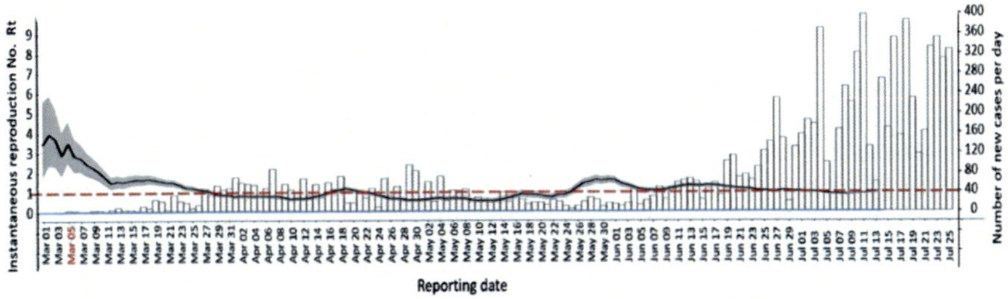

Fig. 1. Rt values and daily number of new infected cases in BiH (March 5–July 23, 2020)

There is a noticeable increase in the number of COVID-19 confirmed cases in larger urban areas such as Prijedor, Banja Luka, Doboj, Bijeljina, Tuzla, Zenica, Sarajevo and Mostar [27].

After this period, this trend continues as seen on the map with the spatial distribution of the total number of infected per 100,000 inhabitants by January 23, 2021 (Fig. 3).

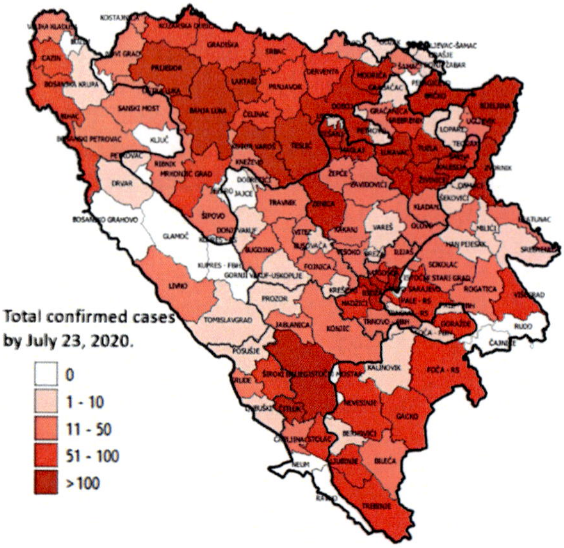

Fig. 2. Map of spatial distribution of total confirmed cases by July 23, 2020 in BiH

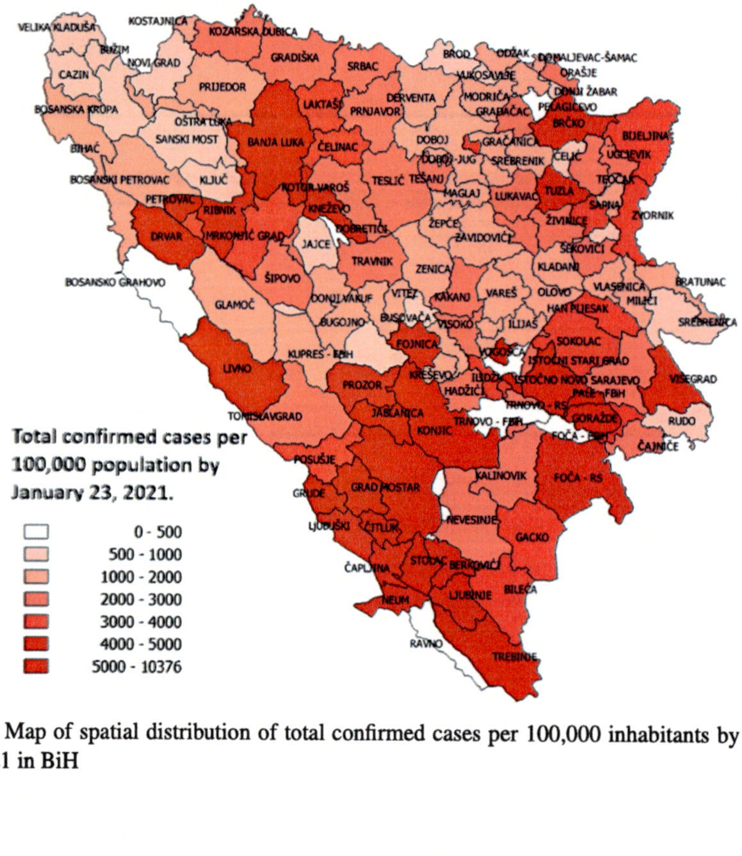

Fig. 3. Map of spatial distribution of total confirmed cases per 100,000 inhabitants by January 23, 2021 in BiH

In general, it can be estimated that throughout the period administrative centers were more exposed to the consequences of the epidemic, and that the incidence rate was higher in densely populated urban areas.

The higher incidence rate is particularly characteristic of administrative centers with higher population density, such as Banja Luka and Sarajevo. This is especially emphasized for the municipalities of the city of Sarajevo (Fig. 4) in Canton Sarajevo, whose areas was selected as the focus of this study.

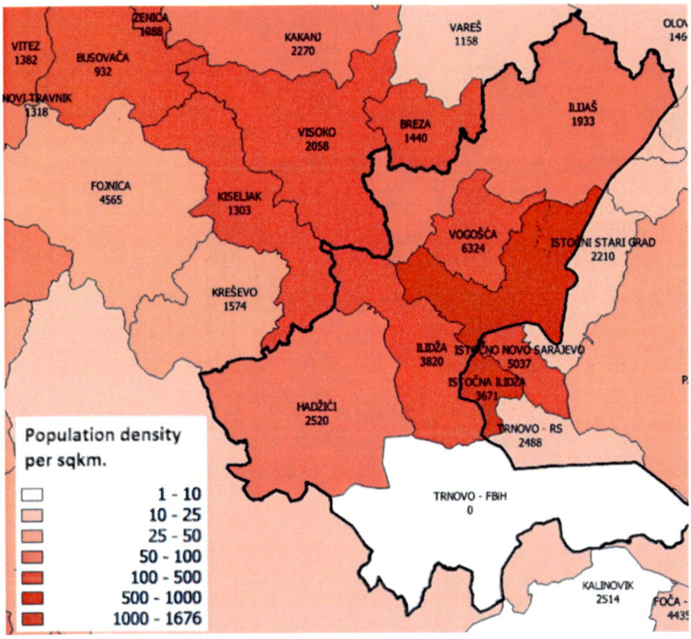

Fig. 4. Population density map in Sarajevo Canton with the incidence rate until January 23, 2021.

4 Overview of Open Public Spaces – Case Study Sarajevo

Sarajevo is a linear city positioned in about 10 km long valley, occupying area of 1.276.9 km^2 with population of 414.721 inhabitants [28]. Accelerated massive urbanization, urban sprawl, inefficient use of land and failure to construct intended area of the park-forest planned in the 'Urban Plan of Sarajevo, 1986–2015', are the consequence of the weaknesses in cities physical and institutional infrastructure.

Out of 1543.66 ha or 29.8 m^2 per inhabitant planned for the park-forest area, only 2.45 ha were realized, which is only 0.08 m^2 per inhabitant. Total planned area of parks for the same period is amounted to 392.6 ha or 7.58 m^2 per capita. Currently in the urban area of Sarajevo, there is 36.18 ha of park area, or approximately 1.2 m^2 per inhabitant. Total square greenery in urban area of Sarajevo covers roughly 6.14 ha, which amounts for 0.20 m^2 per inhabitant [29, 30].

This is far below the planned norms of green areas present in European cities, which range from a minimum of 15–25 m^2 per capita, or even the recommendation given by the World Health Organization which is 9 m^2 of green open space per capita [31]. Thus, substantial number of people in Sarajevo live in areas with minimal public recreational space, which forces people to move to other areas to seek out green space, causing spatial pressure and often overcrowding local parks in the city, and picnic areas on the local mountains. The relationship between the population and the extent and distribution of the public greenery and areas for general use, varies between municipalities (Fig. 5, Table 1).

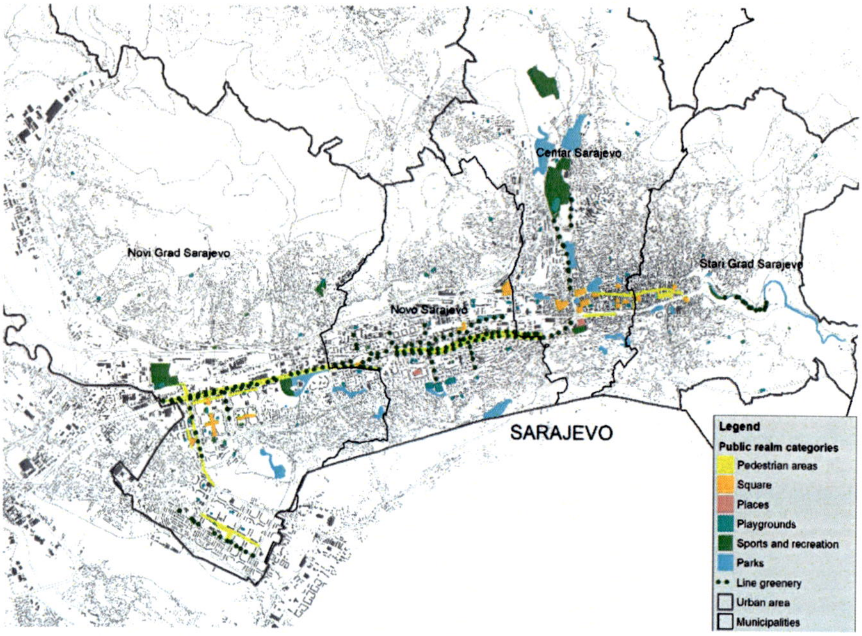

Fig. 5. Overview of open public spaces in Sarajevo (Source: Institute for Planning of the Sarajevo Canton)

All analyzed categories of urban greenery are present in municipality Center, the administrative core of the city, with public green areas available per capita equal to an average supply of 8.4 m^2. In municipalities Stari Grad, the historical center of the city, the area of public spaces per person is drastically reduced in comparison to municipality Center to an average of 0.4 m^2 per person. Novo Sarajevo, the municipality with largest density, has an average of only 0.8 m^2 per capita available, while in municipality Novi Grad, the largest residential and most populated area of the city, the distribution of open public spaces is only 0.2 m^2 per inhabitant. It must be noted that this analysis does not include the group of greenery adjacent to collective housing facilities, which sums to an average of 10, 85 m^2 per capita in the city of Sarajevo [28].

Table 1. Overview of the state of urban greenery according to user groups and municipalities, (Source: Institute of Development Planning of the Sarajevo Canton)

Municipalities of Sarajevo City	Park (P)/(ha) Square (S)/(ha) Park forest (PF) i Special Use Greenery (SUG)/(ha) Linear greenery (LG)/(m) Archeological park (AP)/(ha)				
	S	P	PF - SUG	LG	AP
Stari Grad	1,45	–	–	899	0,25
Centar	3,02	15,8	30,95	2493,67	0,13
Novo Sarajevo	0,79	4,9		8640,36	–
Novi Grad	–	–	2,45	6218,93	–

It is argued that parks and recreational greenery play the greatest role among all urban green areas in Sarajevo. Based on their size and proximity to residential areas, the proximity maps are presented in geoinformation system (GIS) (Figs. 6 and 7), identifying urban areas whose residents live less than 5 min away on foot from a park area. This distance is also known as the 'pedestrian shed' which is considered to be the distance people are willing to walk before opting to drive, and is represented by a radius of 400 m [32]. This analysis also includes a more generous walking distance of 10-min walk threshold, which is also applied in urban planning policies, and is represented by a radius of 800 m.

Such mapping of the city identifies that the supply of parks in Sarajevo is also unevenly distributed across the city, leaving some areas of the city uncovered. It indicates that the parks are predominantly located within the valley of the city, while the hillsides are not treated in that respect, with an exception of the locations in the northern section of municipality Center – 'the green transversal', location of Olympic Hall Zetra and Kosevo Stadium, constructed for the 1984 Olympic Games (Fig. 6).

Presented analysis shows that the diffusion of park areas across municipality Center, covers most of its space within a 10-min walking distance, and results in 2.6 m^2 supply per inhabitant. However, all other municipalities have less then 2 m^2 park area available for their inhabitants, leaving large areas disclosed by the 10-min radius. The deficiency is largest in municipality Novo Sarajevo, with less than 0.1 m^2 per capita, while the least coverage is in municipality Novi Grad, with the large residential areas distant from the nearest parks and their supply reduced to 0.2 m^2 per capita, on average. Although, there are significant number of areas of greenery adjacent to collective housing facilities in municipality Novi Grad, they are mostly unused as they do not provide activity content for the users. (Fig. 7) This indicates that based on their size and proximity to residential areas, the possibility for crowding risks in this municipality are significant.

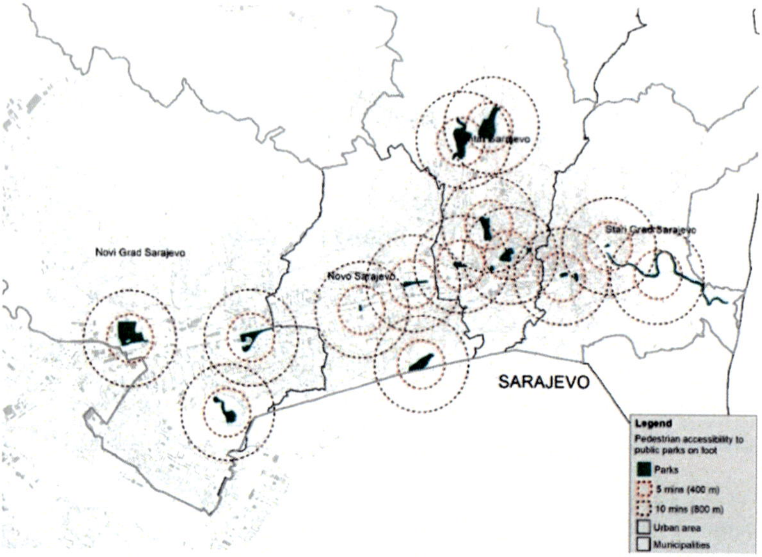

Fig. 6. Proximity map: pedestrian accessibility to large public parks in Sarajevo, 5- and 10-min accessibility (Source: Institute of Development Planning of the Sarajevo Canton)

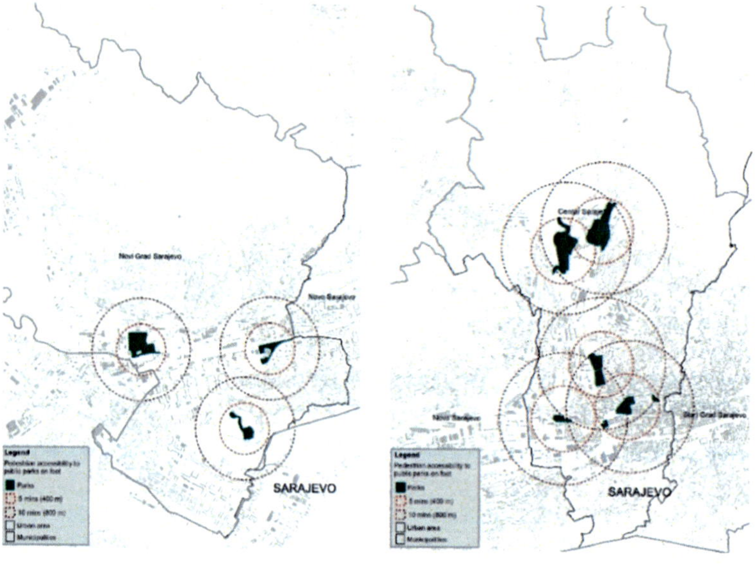

Fig. 7. Proximity map for Municipalities Center and Novi Grad: pedestrian accessibility to large public parks in Sarajevo, 5- and 10-min accessibility (Source: Institute of Development Planning of the Sarajevo Canton)

5 Conclusion

The ongoing pandemics has revealed fragility of human existence. Instantaneously, millions of people, particularly urban residents across the globe became dangerously vulnerable, as the contemporary city was and still is unable to provide the needed response to the present public health needs and legislation. In fact, it has exposed all existing city flaws and has put to the forefront issues of contingency planning and resilience strategies, which present an immense challenge even to the cities in developed countries. However, the current situation should be used to take a proactive role in creating and implementing necessary recommendations and build resilience enabling cities to create healthier urban environments.

Analysis of public space availability and its spatial distribution throughout the city of Sarajevo, confirms that the supply of public spaces in relation to the residing population is insufficient to provide safe outdoor environment under current conditions, indicating that Sarajevo is immensely vulnerable. In the context of the ongoing pandemic, and practices recommended and implemented in European cities, the paper emphasizes the importance of access to open-air gathering spaces for physical and mental health of the citizens, and proposes how to improve special availability.

An increase in public areas must be of high priority, one that would allow increase in the number of citizens who have access to parks and playground within 15 min from their residence. Ideally, it should be a value that would bring the areas concerned closer to the current public space provision norms in the European cities (15–25 m^2 per capita). Furthermore, the study states following recommendations:

1. De-densification of Sarajevo could be initiated by prioritizing pedestrian infrastructure and spaces, over massive, dense construction, as to allow more usable space with proper social distancing in the event of the current or future health crises. Using the model of some European cities, such as Barcelona or Milan cities roads network should be reconsidered, aiming to increase pedestrian public space, as this can play a crucial role in virus control. Research suggests that redesign should be directed towards potentially closing certain areas to vehicle traffic, at least temporarily during the pandemic spread, or occasionally during weekends to ensure the enhanced provision of secure public spaces where needed.
2. City as well as private developers should be encouraged to reactivate existing outdoor space, particularly the green zones adjacent to collective living, by adding activity content, by promoting and maintain these public areas, as they can provide significant relief of the current pressure put on the parks.
3. Expansion of outdoor sidewalks and bike lanes, as this has become one of the most exploited activities during the pandemics. The enlargement of streets for active mobility is an impactful measure to make physical distancing possible on pavements.
4. Activating and increasing amount of picnic areas, and areas for sports and recreation on the nearby Olympic Mountains, as their proximity to the city allows them to be an important repository to the urban greenery.

It is recognized that the implementation of these ideas may seem difficult of even utopian, considering that such a shift asks for proactive coordinated efforts of public agencies on all levels. However, the seriousness of the current state of open public spaces in Sarajevo, and the urban health challenge, calls for urgent development of resilience strategies, more comprehensive plans, regulations, and design changes that would respond faster and more effectively to health crises, and in the long run offer improved living quality.

References

1. Mark, J.J.: Plague in the Ancient & Medieval World (2020). https://www.ancient.eu/article/1528/plague-in-the-ancient--medieval-world/#references. Accessed 15 Dec 2020
2. Grant, J.L.: What cities can learn from lockdown about planning for life after the coronavirus pandemic (2020). https://theconversation.com/what-cities-can-learn-from-lockdown-about-planning-for-life-after-the-coronavirus-pandemic-136699. Accessed 30 June 2020
3. UN: Cities and Urban Vulnerability. UNDP (2007)
4. UN-Habitat: Guidance on COVID 19 Public Space (2020). https://unhabitat.org/sites/default/files/2020/06/un-habitat_guidance_on_covid-19_and_public_space.pdf. Accessed 28 Jan 2021
5. D'Onofrio, R., Trusiani, E.: For the "human" development of cities in an era of climate change. In: D'Onofrio, R., Trusiani, E. (eds.) Urban Planning for Healthy European Cities. SG, pp. 3–7. Springer, Cham (2018). https://doi.org/10.1007/978-3-319-71144-7_1
6. European Commission: The future of cities (2021). https://urban.jrc.ec.europa.eu/thefutureofcities/space-and-the-city#the-chapter. Accessed 01 Feb 2020
7. Systematica: Archdaily (2020). https://www.archdaily.com/941723/systematica-releases-first-assessment-on-milan-public-realm-green-areas-and-gathering-places. Accessed 28 June 2020
8. Burton, H., Grant, M., Mitcham, C., Tsourou, C.: Healthy urban planning in European cities. Health Promot. Int. **24**, i91–i99 (2009)
9. Galea, S., Vlahov, D. (eds.): Handbook of Urban Health: Populations, Methods, and Practice. Springer, Boston (2005). https://doi.org/10.1007/b104167
10. WHO: Ottawa Charter for Health Promotion. WHO, Ottawa (1986)
11. de Leeuw, E., Simos, J. (eds.): Healthy Cities: The Theory, Policy, and Practice of Value-Based Urban Planning. Springer, New York (2017). https://doi.org/10.1007/978-1-4939-6694-3
12. WHO: Health 2020 a European policy framework and strategy for the 21st century. WHO Regional Office for Europe, Kopenhagen (2013)
13. HUP: Hup Initiatives, 20 June 2020. https://www.hupinitiatives.com/about/. Accessed 13 Nov 2020, Accessed 01 Feb 2021
14. United Nations: The Sustainable Development Goals Report. United Nations (2020)
15. United Cities and Local Governments: The European Charter for the Safeguarding of Human Rights in the City. UCLG (2012)
16. Duhl, L., Sanchez, A.: Healthy cities and the city planning process. WHO, Copenhagen (1999)
17. Huremović, D.: Brief history of pandemics (pandemics throughout history). In: Huremović, D. (ed.) Psychiatry of Pandemics, pp. 7–35. Springer, Cham (2019). https://doi.org/10.1007/978-3-030-15346-5_2
18. Chotiner, I.: How Pandemics Change History (2020). https://www.newyorker.com/news/q-and-a/how-pandemics-change-history. Accessed 15 July 2020
19. LePan, N.: A visual history of pandemics (2020). https://www.weforum.org/agenda/2020/03/a-visual-history-of-pandemics. Accessed 28 Dec 2020
20. Swyngedouw, E., Heynen, N.: Urban political ecology, justice and the politics of scale. Antipode **35**(5), 898–918 (2003)

21. Nations, U. (2019). World Urbanization Prospects The 2018 Revision. 2019: United Nations.
22. Held, D., McGrew, A., Goldblatt, D., Perraton, J.: Rethinking globalization. In: Held, D., McGrew, A. (eds.) The Global Transformations Reader: An Introduction to the Globalization Debate, pp. 67–74. Blackwell, Oxford (2002)
23. Kissler, S.M., Tedijanto, C., Goldstein, E., Grad, Y.H., Lipsitch, M.: Projecting the transmission dynamics of SARS-CoV-2 through the postpandemic period. Science **368**(6493), 860–868 (2020)
24. Honey-Rosés, J., et al.: The impact of COVID-19 on public space: a review of the emerging questions. OSF Preprints, pp. 1–20 (2020)
25. Ponjavic, M., et al.: Spatio-temporal data visualization for monitoring of control measures in the prevention of the spread of COVID-19 in Bosnia and Herzegovina. Med. Glas. **17**(2), 265–274 (2020)
26. Hukic, M., et al.: SARS-CoV-2 virus outbreak and the emergency public health measures in Bosnia and Herzegovina: January–July 2020. Bosn. J. Basic Med. Sci. **21**(1), 111–116 (2021)
27. Hukic, M., et al.: COVID-19 pandemic in Bosnia and Herzegovina: March–June 2020. Academy of Sciences and Arts of Bosnia and Herzegovina, Sarajevo (2020). https://publications.anubih.ba/handle/123456789/710. Accessed 30 Nov 2020
28. Kremić, E.: Kanton Sarajevo u brojkama. Sarajevo: Federalni zavod za statistiku (2018). http://fzs.ba/wp-content/uploads/2018/07/Kanton-9-Kanton-Sarajevo.pdf
29. Canton Sarajevo, Institute for Planning: Urban plan for urban zone of Sarajevo - Analysus and evaluation of possible development. Institute for Planning of the Sarajevo Canton, Sarajevo (2019)
30. Zavod za planiranje: Urbanistički plan urbanog područja Sarajevo, Postojeće stanje i problemi u prostoru i uređenju prostora. Zavod za planiranje razvoja Kantona Sarajevo, Sarajevo (2019)
31. WHO: Urban Planinning and Human Health in the European City. WHO (2009)
32. Morphocode: The 5 minute walk. https://morphocode.com/the-5-minute-walk/. Accessed 5 Jan 2021

Crowdsourcing Application in the Development of a Dynamic Noise Map

Vesna Poslončec-Petrić(✉), Iva Cibilić, and Stanislav Frangeš

Faculty of Geodesy, University of Zagreb, Zagreb, Croatia
vesna.poslonceC@geof.hr

Abstract. One of the most important components of smart cities are smart citizens, i.e. people who actively participate in the functioning of the city. Current communication technologies and smart devices are components that make this possible in near real-time. Crowdsourcing is becoming increasingly present in cartography and geovisualization, and the availability of smart devices with many sensors gives it new significance. In addition to well-known cartographic services such as Open Street Map and Google Maps, this paper describes the creation of a dynamic noise map in which students of the Faculty of Geodesy participate in the teaching of Geovisualization. It is described how the use of free applications and programs and community engagement can provide high quality and informative cartographic representations, and the active participation of the community in collecting spatial data raises their awareness in creating and maintaining smart cities.

Keywords: Crowdsourcing · Smart city · Dynamic noise map · Geovisualization

1 Introduction

According to UN data published in 2014 [1], in the 1950s only 29% of the world's population lived in cities. Today, that percentage rises to about 50% of the population and is projected to increase to as much as 70% of the population by 2050. For Europe, this percentage is even higher and population growth in large cities is expected to increase by up to 80%. Such a sharp increase in the urban population creates challenges in solving the infrastructural problems of cities such as energy, energy efficiency, public lighting, urban water supply and sewerage, transport, and of course, the impact on the environment, and the answer to these challenges gives the concept of smart cities.

The increase in the number of inhabitants in cities has encouraged the further development of information and communication technologies and enabled the development of new concepts of city management with the aim of their sustainable development. A smart city is a place where traditional networks and services are more efficient thanks to the use of digital and telecommunications technologies, all to improve the quality of life of its citizens and achieve the progress of the business sector.

N. Ademović et al. (Eds.): IAT 2021, LNNS 316, pp. 676–683, 2022.
https://doi.org/10.1007/978-3-030-90055-7_54

Closely related to the concept of smart cities is the concept of smart people: citizens who actively participate in the implementation of smart cities, either through participation in decision-making processes or more specifically, in the collection of spatial information and data (crowdsourcing).

Voluntary collection of information about space or the situation in space through citizens represents a significant step forward in the design of management systems and enables the elimination of certain weaknesses of the concept of smart cities [2].

2 Crowdsourcing

Crowdsourcing was defined in 2005 as a combination of the words crowd and outsourcing. It was created by Jeff Howe and Mark Robinson, editors of Wired magazine, to describe how companies used the Internet to "outsource work to the crowd." In other words, instead of using their resources, companies publish the problem online to help Internet users solve it.

Crowdsourcing is an outsourcing process that is traditionally performed by employees and transferred to an undefined group of people in the form of an open call [3].

The emergence and prevalence of mobile telephony have taken crowdsourcing to a new level. Today, most people own smartphones that are available at very affordable prices and come with many built-in sensors such as camera, GPS receiver, microphone, proximity sensor, compass, accelerometer, gyroscope, etc. Built-in sensors allow you to get a variety of information quickly and easily, and with the development of technology, this information will only become more precise and diverse [4].

Although crowdsourcing was first described in 2005, the method itself goes much further back in history. So, Wikipedia cites some of the ancient examples. In France, in the period before the French Revolution (around 1750), citizens were invited to present their proposals for solving certain social problems, and the best solution was rewarded. In the 18th century, the British Parliament used mass data collection to find solutions to the problems in the field of science at the time. Napoleon also used the power of the crowd in search of the most efficient way to deliver food to military troops on distant battlefields.

With the advent of the Internet, there is a rapid development of crowdsourcing that is becoming simpler and more accessible to the general public and its most important advantages include cost reduction, speed, quality, flexibility, upgradeability and diversity of ideas. Yet we can say that its biggest benefit is the ability to get quality results from a large number of people, allowing companies to choose the best of what is offered by various exhibitors, instead of the classic approach of choosing the best from a single source. Also, it can be said that mass data collection enables the development of ideas that would not survive in traditional business due to insufficient support or capital, all in record time [5].

3 Crowdsourcing as a Basis for Cartographic Representations

The popularity of crowdsourcing has grown steadily across the spatial domain, including cartography, spatial data collection, and Earth observation [6]. One of the most prominent and widespread examples of mass spatial data collection is the Open Street Map (OSM) [7]. OSM is a collaborative project to create a free, editable map of the world. The geodata that forms the basis of the map is considered the primary result of the project. The creation and growth of OSM is motivated by limitations in the use or availability of cartographic data in the world and the emergence of cheap portable satellite navigation devices. It was created by Steve Coast in the UK in 2004, inspired by the success of Wikipedia and the predominance of non-available data. Since then, OSM has grown to over two million registered users [8, 9]. Currently, the most popular mapping application is Google Maps [10]. It is a web mapping service developed by Google. Launched in February 2005, Google collects spatial data from a wide range of sources, including road sensors, user contributions through Map Maker and local transportation departments, among others. Google Map Maker was discontinued in March 2017, but contributions to Google Maps through mass sources have not [11].

The view of the city center of Zagreb in the vicinity of the Faculty of Geodesy on OSM (Fig. 1) and Google Maps (Fig. 2) shows significant differences between these two cartographic services. While Google Maps is primarily developed for navigation, OSM focuses on positionally accurate and detailed mapping. From a cartographic point of view, the OSM view is more readable and clearer than Google Maps, and we conclude that crowdsourcing, as a mass data collection method, can produce up-to-date and high-quality cartographic representations of the same quality as commercial ones.

In addition to these cartographic services, crowdsourcing is the basis for some of the thematic maps. The noise map is a thematic map showing the existing and projected levels of noise immissions at all places within the observed area, depending on one or all specific noise sources. The noise map provides insight into noise management problems and provides a clear, unambiguous and easy-to-read picture of these problems, enabling more efficient spatial planning, noise protection planning of existing areas and implementation of "acoustic zone zoning" in accordance with legally permitted noise levels. The production of official noise maps and their maintenance by standard methods is a very demanding and expensive task in the competence of the local government.

The project "Interoperable Smart City services through an Open Platform for urban Ecosystem" (i-SCOPE) [12] shows how to develop and test technology for smart city services. One of the project-developed services is noise monitoring by real-time mapping using an application on a mobile phone.

Noise monitoring is enabled by the active participation of citizens who measure noise via a mobile application and thus contribute to the creation of a dynamic noise map. It is created by mapping the recorded noise levels in real-time, involving citizens who take on the role of sensors by measuring the noise level with an application on their mobile phone.

The data collected by citizens are processed and used to create a noise map available on the i-Scope web platform. The data collected relate to the total noise level present at the time of collection. In this way, it is possible to get a very detailed view of the noise level in the city, and the amount of data that citizens can easily collect gives a very accurate view of the situation and pollution in a large, busy and urban city.

Fig. 1. Part of the city of Zagreb in the vicinity of the Faculty of Geodesy at OSM

Fig. 2. Part of the city of Zagreb in the vicinity of the Faculty of Geodesy at Google Maps

4 Application of Crowdsourcing in the Creation of a Dynamic Noise Map in the Course Geovisualization

A dynamic noise map is a map that shows real-time noise levels over different periods. Students of the geodesy and geoinformatics at the Faculty of Geodesy, University of Zagreb, as part of Geovizualization classes, encouraged by the i-SCOPE project, are actively involved in creating a dynamic noise map and since 2016 regularly collect real-time noise data, process them and analyze. The aim of the course Geovisualization is [13] to acquaint students with the basic theoretical settings related to geovisualization and the acquisition of practical knowledge needed to create various cartographic representations through available technology. The volunteer data collection project for smartphone sensors provided additional motivation for the active involvement of students in teaching.

The process of creating a voluntary noise map includes data collection (spatial, temporal component and noise intensity), data processing (sorting and analysis of collected data) and visualization (classification and production of a thematic map) (Fig. 3). Data collection was performed with the Noise Tube application. Noise Tube is a mobile application that measures the noise level in a certain place at a certain time [14]. Since the noise levels in decibels dB (A) and coordinates (WGS84) are available for each data, it proved to be very good and reliable for this task. Each smartphone model has to be calibrated by the NoiseTube team to decrease the microphone's error in order to produce a relatively accurate data set. The accuracy is estimated for around 5dB. Data processing and final visualization were performed in QGIS or ArcGIS Online programs. Input datasets are in the form of CSV files. First step is data validation; data with unreliable coordinates and noise values are eliminated (ie. Data caused by an unforeseen source of noise).

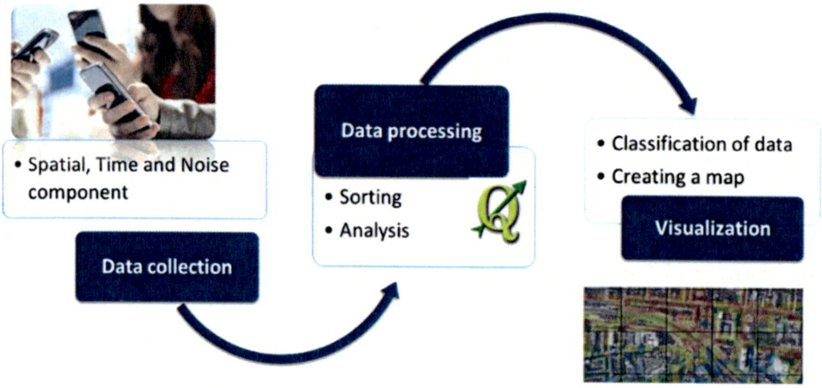

Fig. 3. The process of making a volunteer noise map [15]

Geometry is described by polygonal features of the vector grid. In this exercise, students practice the tasks of linking attribute tables with geometry using operations and expressions to manipulate tabular data. Recommended methods of interpolation are IDW interpolation, TIN interpolation and method of Spatial Join. To get a better and simpler representation of the obtained data, arithmetic mean for each polygon was calculated. Students perform data analysis based on the obtained data. The motivation of students for this task is extremely high. The project has been performed as a practical exercise since the academic year 2016/2017 and during the past 4 years, an extremely large amount of data on noise in the center of Zagreb has been collected, which are additionally processed and analyzed through bachelor and master theses (Fig. 4) [16].

Given a large amount of data, special attention is paid to data analysis. Using data gathered this way, it is possible to analyze noise levels in different locations getting information about the main source of the noise presenting the whole aspect through

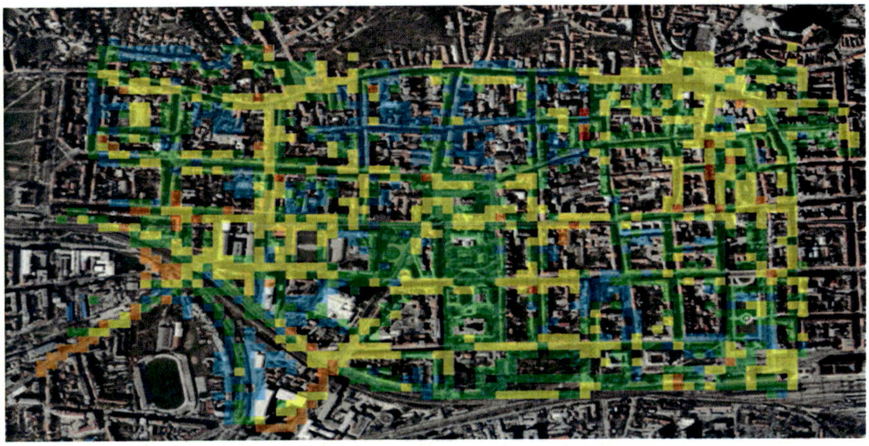

Fig. 4. Interpolation and classification of noise data [16]

different thematic maps. One of the analyses required of students is the change of the noise levels during the day (Fig. 5) or to compare the noise levels on weekdays and weekends, etc. Data is classified in five categories: First category includes noise values lower than 30 dB (blue color), green displays values between 31 and 40 dB, yellow class displays values from 41 to 50 dB, Orange includes levels from 51 to 60 dB, and red shows highest noise values, more than 60 dB.

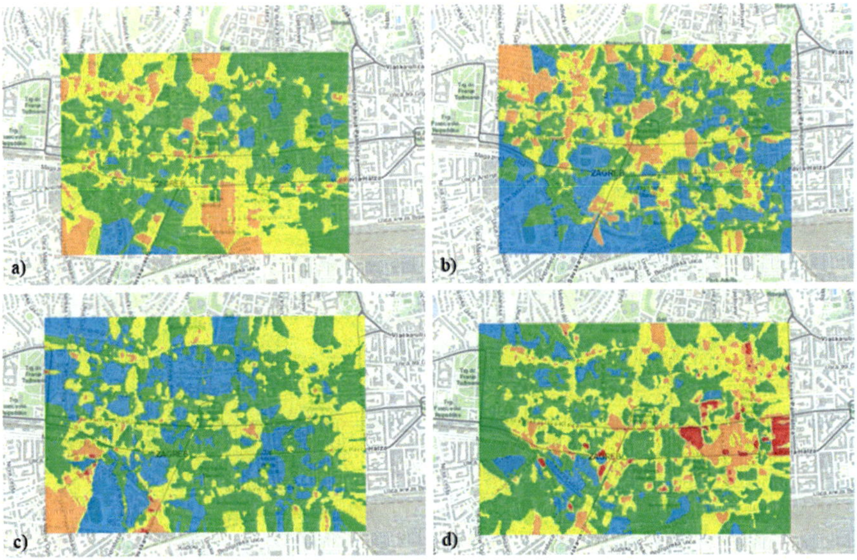

Fig. 5. Noise change over time a) 9–10 h, b) 10–11, c) 11–12 and d) 12–13 h [16]

Comprehensive information on spatial and temporal distribution of noise is crucial to address the issue by relevant authorities and experts. Although a noise map made in this way cannot replace official noise maps, it has many advantages. The greatest value is that it gives a realistic and complete picture of noise in the environment regarding all noise sources.

In accordance with the program of the course Geovisualization, this exercise covers several learning outcomes and after completing the course students have the knowledge: to make different geovisualizations and use geovisualization tools and interpret geospatial data and phenomena with their visualization and mutual communication using cartographic representations. check the effectiveness of geovisualization.

5 Conclusion

Crowdsourcing has proven to be a successful method in creating quality information and content, not only in cartography but also in other disciplines. Although the concept has been around for a long time, it experienced its rise at the beginning of the 21st century with the development of data collection sensors.

In addition to numerous financial advantages, there is a more important aspect of this method: raising citizens' awareness of the problems that occur in their environment. Smart cities cannot be smart if citizens are not active - their participation in creating, understanding and manipulating the data obtained and maintaining smart cities is essential.

Students of the Faculty of Geodesy through practical exercises in Geovisualization participate in the voluntary creation of noise maps. By mastering the knowledge of cartography and geoinformatics, data processing and visualization procedures, their awareness in creating and maintaining smart cities is raised. Smart citizens are not only those who participate in data collection, but also those who implement it.

The given examples show that crowdsourcing will be an increasingly important source of data collection.

References

1. UN: World Urbanization Prospects (2014). https://population.un.org/wup/. Accessed 26 Feb 2021
2. Schuurman, D., Baccarne, B., De Marez, L., Mechant, P.: Smart Ideas for smart cities: investigating crowdsourcing for generating and selecting ideas for ICT innovation in a city context. J. Theor. Appl. Electron. Commer. Res. **7**(3), 49–62 (2012)
3. Howe, J.: Crowdsourcing: Why the power of the crowd is driving the future of business (2006). http://www.crowdsourcing.com. Accessed 29 June 2015
4. Poslončec-Petrić, V., Vuković, V., Frangeš, S., Bačić, Ž.: Voluntary noise mapping for smart city. In: First International Conference on Smart Data and Smart Cities, Split, Croatia, 7–9 September 2016 (2016)
5. Crowdsourcing. https://hr.wikipedia.org/wiki/Nabava_iz_mno%C5%A1tva. Accessed 15 Sept 2020

6. See, L., Mooney, P., Foody, G., Bastin, L., Comber, A., Estima, J., et al.: Crowdsourcing, citizen science or volunteered geographic information? The current state of crowdsourced geographic information. ISPRS Int. J. Geo-Inf. **5**(5), 55 (2016). https://www.mdpi.com/137996
7. Open Street Map. https://www.openstreetmap.org. Accessed 26 Feb 2021
8. Open Street Map. https://en.wikipedia.org/wiki/OpenStreetMap. Accessed 26 Feb 2021
9. Neis, P., Zipf, A.: Analyzing the contributor activity of a volunteered geographic information project—the case of OpenStreetMap. ISPRS Int. J. Geo-Inf. **1**(2), 146–165 (2012)
10. Google Maps. https://cloud.google.com/maps-platform/. Accessed 26 Feb 2021
11. Google Maps. https://www.cnet.com/how-to/use-google-maps-to-see-how-fast-youre-driving/. Accessed 26 Feb 2021
12. i-Scope. https://iscopeproject.net/. Accessed 26 Feb 2021
13. Opis predmeta Geovizualizacija. https://www.isvu.hr/javno/hr/vu7/nasprog/2019/pred50405.shtml. Accessed 26 Feb 2021
14. i-Scope. https://iscopeproject.net/noisetube. Accessed 26 Feb 2021
15. Poslončec-Petrić, V.: Instructions for making a dynamic noise map, teaching materials. Faculty of Geodesy, University of Zagreb, Zagreb (2020)
16. Vrljičak, M.: Dynamic noise map of the center of Zagreb. Master thesis, Faculty of Geodesy, University of Zagreb, Zagreb (2020)

A Modern Base Map and 3D City Model Production - A Case Study "City of Sarajevo"

Jasmin Taletović[1] (✉) and Slobodanka Ključanin[2]

[1] Department of GIS, Institute for Planning of Canton Sarajevo, Sarajevo, Bosnia and Herzegovina

[2] Technical Faculty, University of Bihac, Bihać, Bosnia and Herzegovina

Abstract. In 2018 the Institute for planning of Canton Sarajevo and Federal Administration for Geodetic and Property-Legal Affairs (FGA) realised the project "The Base Map Revision in the scale 1:5000". The project requires updating the existing topographic bases at a scale of 1:5000, as well as creating a 3D City model for the urban area of Sarajevo Canton based on new photogrammetric images, i.e., digital orthophoto (DOF). One of the main purposes of the project was to provide the basis for urban policy decision making for the city of Sarajevo. The project has been realized in accordance with the Topographic Data Model of the Federation of Bosnia and Herzegovina.

The paper provides a description of the process of creating the Base Map the Federation of Bosnia and Herzegovina including data sources and documentation, data structure, processing and presentation of data, and deliverables.

Keywords: Base map production · Topographic database · Topographic Information System · 3D city model

1 Introduction

By establishing rules and procedures for the formation and establishment of a basic topographic database, FGA works to modernize its systems, update and improve data availability, encouraging economic development [1]. On the other hand, for the purpose of making feasible urban plans, it is necessary to use up-to-date topographic maps of larger scales 1:5000 (so-called basic topographic maps), and bearing in mind that these topographic maps have not been updated and they are outdated, new rules and procedures are needed. In response to this question, and at the request of the Institute for Development Planning of Sarajevo Canton, a project task entitled "Updating the basic topographic map 1:5000 urban areas of Sarajevo Canton" followed, which, in addition to making a topographic map scale 1:5000, required creating a database of the height of buildings and other structures. To create a project task ToR applied the guidelines from the Strategy for the development of official cartography of the Federation of Bosnia

N. Ademović et al. (Eds.): IAT 2021, LNNS 316, pp. 684–693, 2022.
https://doi.org/10.1007/978-3-030-90055-7_55

and Herzegovina, the rules of the existing system of topographic information and other professional experience of the author. The importance of this project was manifested in the application of new technology in the field of urban planning, as well as the application of topographic information system (based on which the database was formed) which facilitates the integration of different types of data and their sharing.

2 Input Data and Documentation

Production and updating of topographic cartographic products is in the competence of the Federal Administration for Geodetic and Property Affairs (FGA). Since the current topographic and cartographic products are older than 30 years, the Institute for planning of Canton Sarajevo requested their updating due to creating a new urban plan. The Institute for planning of Canton Sarajevo and the FGA funded this project. The result of the project is the establishment of a topographic database from which topographic maps in the scale of 1:5000 and thematic maps on the same or smaller scale were printed. According to the ToR, the implementation of the project is divided into three phases:

1. Aerial photogrammetric surveying and digital orthophoto,
2. Acquisition of data on the height of buildings and other structures,
3. Creating a basic topographic database at a scale of 1:5000.

Aerial photogrammetric surveying and digital orthophoto (Phase 1) included numerous preliminary works (preparation flight plan, orientation and control points, geodetic measurements, etc.) which were approved by investors. Each partial and/or complete delivery of the product was accompanied by an appropriate ToR. In this phase of the project implementation, 438 aerial photographs were taken, from which are made 76 DOFs (digital orthophotos). DOFs are indexed according to the nomenclature for topographic maps at a scale of 1:5000 (BTM5).

Collecting data on the height of buildings and other structures (Phase 2) implied adding data on the height (h) to the existing data on the position (x, y) of each individual building. The technical report submitted to the investor after the completion of the work contained a list of coordinates of points of buildings (x, y, h) (obtained by one of the geodetic methods of data collection) and other data required by the ToR.

Creating a basic topographic database at a scale of 1:5000 (Phase 3) demanded the proper database creation are prescribed by the TIS (Topographic Information System) and the ToR. BTM5 data are organized thematic data sets according to the topographic model:

a. Administrative boundaries
b. Buildings
c. Hydrography
d. Transport network

e. Land cover
f. Land use
g. Elevation
h. Utility lines and appliances
i. Toponyms
j. Other facilities.

ToR prescribed the form of the report the contractor for the purpose of verifying the quality and quantity of data (metadata about the completeness of delivered data, the data sources, conformity of data with the topographical model, about the completeness of thematic data sets and the topological data consistency) [2].

3 Data Structure

The data sets on cadastral parcels, buildings and other facilities, transport networks, land use, geographical names, administrative boundary, elevation, vegetation, utility network, and geodetic points were taken according to the Methodology and procedures for establishing and maintaining the basic topographic database [3]. After data collection, selection and cartographic generalization were performed. In addition, data were collected on striking objects that can be used for spatial orientation. Furthermore, in order to maintain the Basic Topographic Database (BTD), it is necessary to continuously implement:

a. Changes in the geometry of existing buildings, caused by regular and extraordinary maintenance, submitted by their owners upon completion of works,
b. Geodetic sketch of newly constructed or newly constructed parts of the building, submitted by their owners upon completion of works,
c. Data collected for the purposes of creating a real estate cadastre database (RECD), and
d. Data collected by vectorization of objects from digital orthophoto [3, 4].

The data sets that are not part of the RECD are used to meet the conditions of integrity and accuracy of BTD. The BTD update timeline is closely related to the availability of new data sets [3].

The BTD consists of 11 packages. Each packet has a number of object classes and a list of values. Each object class consists of object class names, definitions, descriptions, stereotypes, and a certain number of attributes and relationships (Fig. 1).

The value lists give you the option of selecting already predetermined values for a given attribute (Fig. 2).

1.2.4.1.15 AbstractBuilding

Class «featureType» in package 'BuildingsBase'

-- Name --
Abstract building

-- Definition --
Abstract spatial object type grouping the common semantic properties of the spatial object types Building and BuildingPart.

AbstractBuilding
Version 1.0 Phase 1.0 Proposed
karjoh created on 9/17/2010. Last modified 5/24/2015
Extends AbstractConstruction

OUTGOING STRUCTURAL RELATIONSHIPS

Generalization from «featureType» AbstractBuilding to «featureType» AbstractConstruction
[Direction is 'Source -> Destination'.]

INCOMING STRUCTURAL RELATIONSHIPS

Generalization from «featureType» BuildingPart to «featureType» AbstractBuilding
[Direction is 'Source -> Destination'.]

Generalization from «featureType» Building to «featureType» AbstractBuilding
[Direction is 'Source -> Destination'.]

ATTRIBUTES

buildingNature : BuildingNatureValue Public
Multiplicity: ([0..*], Allow duplicates: 0, Is ordered: False)

-- Name --
Building nature

-- Definition --
Characteristic of the building that makes it generally of interest for mappings applications. The characteristic may be related to the physical aspect and/or to the function of the building.

-- Description --
This attribute focuses on the physical aspect of the building; however, this physical aspect is often expressed as a function (e.g.

Fig. 1. An example of the object class description for AbstractBuilding (see: [5])

1.2.4.1.24 BuildingNatureValue

Class «codeList» in package 'BuildingsBase'

-- Name --
Building nature value

-- Definition --
Values indicating the nature of a building.

-- Description --
NOTE 1: This code list does not aim to be exhaustive as the attribute buildingNature addresses only noticeable buildings.
NOTE 2: The values included in this code list address mainly (but not only) two international use cases: air flights where buildings may be obstacles and marine navigation where buildings may be landmarks.
NOTE 3: This code list should only be applied for buildings, even if it may be applicable to other constructions (for example, not all dams are buildings).

Fig. 2. An example of the list value description - BuildingNatureValue [5]

4 Data Processing and Presentation

The region for which the map was made has an area of about 300 km^2 and is covered by 76 sheets of map at a scale of 1:5000. The region is divided into two zones: the northern which covers the urban area of Sarajevo, Hadzici, Ilijas and the southern which covers the urban area of Trnovo. The input data for forming a digital map at a scale of 1:5000 are data collected and stored in the ESRI personal geodatabase (Microsoft Access database). ArcGIS was chosen for the environment in which the topographic maps were formed because the Investor has the experience and necessary resources when it comes to ESRI software. The process of making a topographic map went through the following phases: defining the description of the topographic map, creating 38 Arcgis map files containing 76 maps (in each arcgis map file there are two topographic maps), created the appropriate symbols and mapped the appropriate thematic contents, integration of map contents was performed, cartographic material was controlled.

Administrative boundaries were provided by investors are official data and the contractor has no right to correct them (boundaries of municipalities, cantons, cadastral municipalities, local communities, censuses, and statistical circles). The transport network, in this case, consists of the road and railway network. It is completely digitized according to the instructions defined in the INSPIRE specification and Terms of Reference. Digitization of the transport network and the width of the road infrastructure was done in the Auto-CAD environment and was based on orthophoto plans. The transport networks classification was performed based on the parameters defined by the TIS. The street names along with the street code are taken from the official gazette of the Sarajevo Canton from 2014. The watercourses were checked by their location and height. The banks of watercourses have been digitized using DOF. When creating InspireID for watercourses, the Law on Waters ("Official Gazette of the Federation of BiH" No. 70/06) and the Law on Waters (Sarajevo Canton) were used. The land cover was taken from the CORINE Land Cover project, which was completed in 2011 [5], and the update was done by comparing its content with the content of the new DOF. Geographical names were taken from the existing Basic Topographic Map (BTM), from geodetic plans (in cases where there were no appropriate names on BTM), as well as from data obtained by investors (e.g., names of buildings, bridges, etc.). The Elevation data integrated into the database was taken from investors or obtained from data collected during the development of the Digital Terrain Model (DTM).

All necessary topological controls and corrections of new data (made in GIS software) were performed, for example, deleting duplicate data, deleting small objects that occurred as a result of errors during digitization, the extension of lines, deletion of double vertices, etc. Arranged data files were imported into a database according to TIS. Isohypses were generated from DTM data developed in the form of a 2.5 m grid in the first phase of the project (Fig. 4). Elevation points showing the vertical representation of the terrain on topographic maps (intersections, bridges, terrains without isohypses, etc.) were collected by stereo restitution. The data source for defining the basis of the buildings was the real estate cadastre database. This data was compared with DOF and supplemented with new building features. Visualisation of BTM5 in digital and analog format (Fig. 3) was performed after data collection and their cartographic processing.

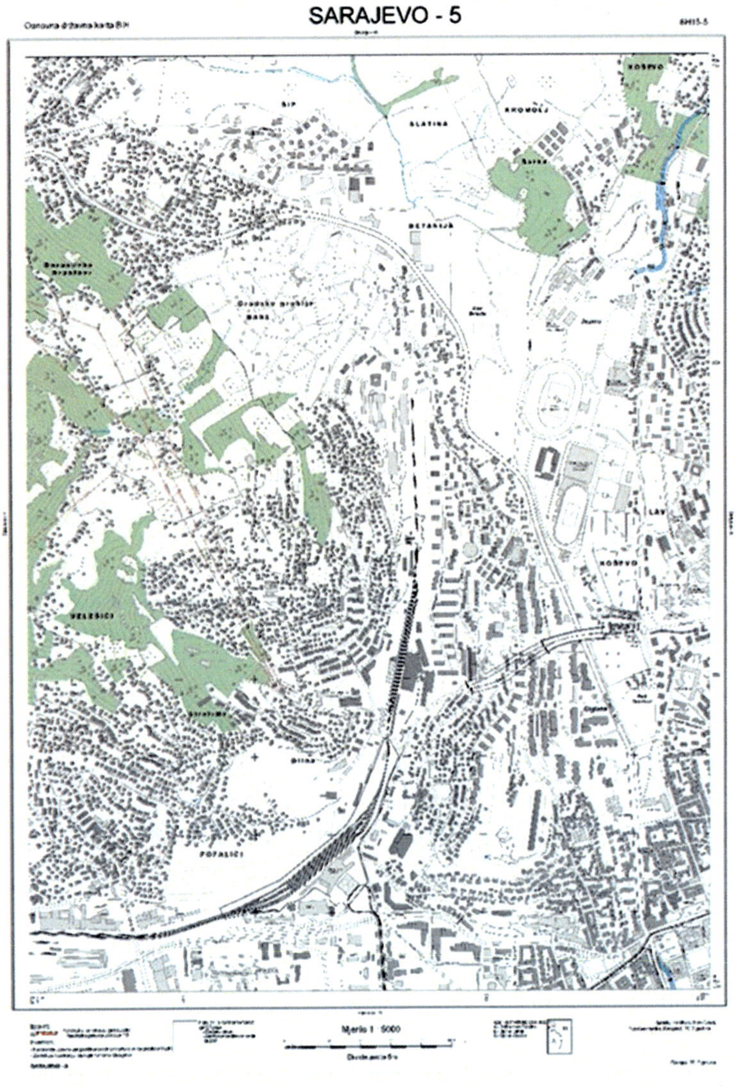

Fig. 3. Topographic map scale 1:5000 (Source: Institute for planning of Canton Sarajevo)

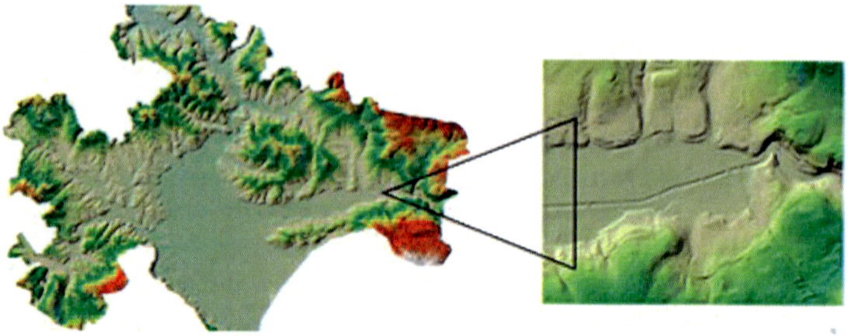

Fig. 4. DTM resolution of 2.5 m in the urban area of Sarajevo [6]

5 3D City Model Data Processing and Presentation

To meet the demand for efficient visualization of 3D city models, the new Open Geospatial Consortium (OGC) has created CityGML that defines a 3D model of the city in five different levels of detail (LOD) (Fig. 5). CityGML is an open data model for storage and exchange of 3D city models [7].

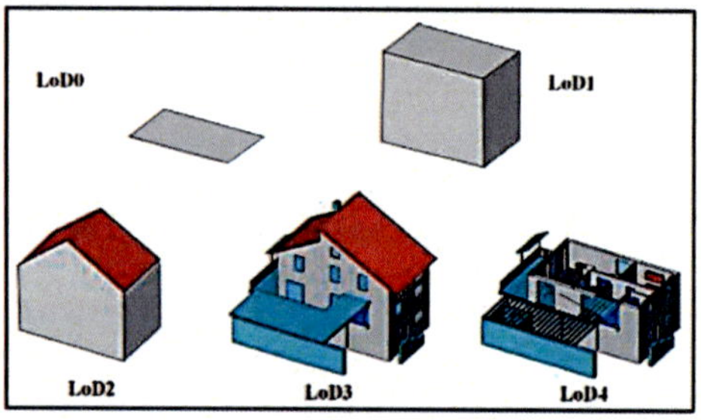

Fig. 5. Five different LoDs defined in CityGML [8]

Objects with their spatial position (base) and their height are defined. 3D geometry of objects, for creating a 3D City Model, obtained by raising the contour of the object to a height and elongated by the height of the object, thus obtaining an object of LoD1 definition. In LoD1, the building model was represented in a block model of buildings (Fig. 5, Fig. 6, Fig. 7). This geometry is placed in the spatial database and then attributes are added to it. Geometry can be viewed in applications that support 3D (e.g., in ArcScene directly). Orthophotos were used to obtain point clouds to determine the highest point of the object and DTM was used to determine the height of the point of the object on the ground. This presentation of buildings is sufficient for spatial analysis at the level of the urban plan (e.g., population density assessment, calculation of Construction Gross Area (CGA), simulation and modelling of ventilated urban corridors, creating a Smart Digital City, etc.).

Fig. 6. LoD1 for Sarajevo city (Source: Institute for planning of Canton Sarajevo)

Fig. 7. 3D City Model of Sarajevo (Source: Institute for planning of Canton Sarajevo)

6 Conclusion

This paper demonstrated the process of collecting, processing, and visualizing spatial data according to the Topographic Information System of Federation of Bosnia and Herzegovina. The paper also presents the methodology for updating topographic maps at a scale of 1:5000. Through the presented project "The Base Map Revision in the scale 1:5000", seventy six topographic maps were updated at a scale of 1:5000, a 3D City model was created for the urban area of Sarajevo Canton based on new digital orthophoto using ESRI ArcGIS software. The results were presented in a case study of the urban area of the City of Sarajevo.

Updating topographic and cartographic products is important for the preparation of spatial and urban plans, but it is extremely important to apply new technologies in the collection, processing, distribution and storage of spatial data. The application of new technologies can enable the formation of a database compliant with INSPIRE specifications and other international standards, which ultimately makes it easier to connect with other institutions responsible for specific sets of spatial data. TIS database data can be used multiple times for different needs (eg. visualization of different thematic maps, performing spatial analyses, etc.).

Given the TIS defined the modelling of three-dimensional data (currently this option is not applied), the next step is to increase the level of detail of objects currently in the spatial database from LoD1 to LoD2.

Available data on the boundaries of cantons, municipalities and cadastral municipalities were used to develop the project. The content of the Base Map should be expanded with boundaries and information for local communities, census units and statistical circles.

The daily need for up-to-date spatial data in a fast and easy way established TIS database provides the possibility of exchanging data with other institutions through the national infrastructure of spatial data or web services (WMS, WFS). Institutions dealing with spatial planning are one example of how they can increase their productivity at work by using web services.

This project was not included in the implementation of the address system in the sense that it should support the extension of the document that serves as the basis for the realization of basic topographic database: "TISFBIH - Model Report, Topogrfski information system of the Federation of Bosnia and Herzegovina" and implement of existing information contained in other institutions.

Basic topographic database is a unique database at the level of Federation of Bosnia and Herzegovina. Thise database represents a methodological and technological progress that will make this unique database as crucial for updating topographic maps in the Federation of Bodne and Herzegovina.

References

1. Kljucanin, S., Posloncec-Petric, V., Ponjavic, M., Karabegovic, A., Landek, I.: Strategija razvoja sluzbene kartografije Federacije Bosne i Hercegovine. Ekspertiza. Federalna uprava za geodetske i imovinsko-pravne poslove, Sarajevo - Geometrika d.o.o., Grude, Bosna i Hercegovina (2014)
2. Azuriranje osnovne topografske karte 1:5000 urbanih podrucja Kantona Sarajevo. Zavod za planiranje razvoja Kantona Sarajevo, Novembar 2016. https://zpr.ks.gov.ba/sites/zpr.ks.gov.ba/files/7.1_projektni_zadatak_otk5_up_ks_24112016.pdf. Accessed 10 Dec 2019
3. Modrinic, Z.: Metodologija i procedure za potrebe uspostavljanja i odrzavanja temeljne topografske baze podataka. Ekspertiza. Federalna uprava za geodetske i imovinsko-pravne poslove, Sarajevo - Geometrika d.o.o., Grude, Bosna i Hercegovina (2016)
4. Kljucanin, S.: The new topographic information system and establishing the basic topographic database of the Federation of Bosnia and Herzegovina. In: Hadžikadi, M., Avdakovi, S. (eds.) Advanced Technologies, Systems, and Applications II. Part of the Lecture Notes in Networks and Systems Book Series. IAT 2017. LNNS, vol. 28, pp. 701–710. Springer, Cham (2018). https://doi.org/10.1007/978-3-319-71321-2_61
5. Vojnikovic, S., Taletovic, J., Balic, B., Ljusa, M., Zurovec, O., Custovic, H.: Estimation of forest area in the Sarajevo Canton following the 4th level of the CORINE Land Cover nomenclature. Journal Naše Šume **12**(32/33), 12–22 (2013). Sarajevo
6. Taletovic, J.: Updating the basic topographic map of 1:5000 scale in urban areas of the Sarajevo Canton, 2nd Conference of Western Balkan Geodetic Forum 19–21 October 2017, University of Džemal Bijedić Mostar, Bosnia and Herzegovina

7. Mao, B.: Visualisation and Generalisation of 3D City Model. Stocholm: Royal Institute of Technology (KTH) (2011)
8. Akmalia, R., Setan, H., Majid, Z., Suwardhi, D., Chong, A.: TLS for generating multi-LOD of 3D building model. In: IOP Conference Series Earth and Environmental Science, January 2014

Using Satellite Data for Assessing the Land Use and Land Cover Change in Bosnia and Herzegovina

Melisa Ljuša[1](✉), Hamid Čustović[1], Jasmin Taletović[2], Mirza Ponjavić[3], and Almir Karabegović[4]

[1] Faculty of Agriculture and Food Sciences, University of Sarajevo, Zmaja od Bosne 8, 71000 Sarajevo, Bosnia and Herzegovina
[2] Department of GIS, Institute for Planning of Canton Sarajevo, Branilaca Sarajeva 26, 71000 Sarajevo, Bosnia and Herzegovina
[3] International Burch University, Francuske revolucije bb, 71210 Ilidza, Bosnia and Herzegovina
[4] Faculty of Electrical Engineering, University of Sarajevo, Zmaja od Bosne, 71000 Sarajevo, Bosnia and Herzegovina

Abstract. Coordination of Information on the Environment Land Cover (CORINE, CLC) provides information on the biophysical characteristics of European land cover. The CLC has become the primary spatial data source on land for Bosnia and Herzegovina (B&H). B&H became part of the CLC in 1998, after which databases were produced for the years 2000, 2006, 2012, and 2018. In updating the CLC database, the standard procedure as well as CLC nomenclature were used. The main source of remote sensing data was high-quality satellite imagery covering different seasons. In CLC2006 the single date Landsat TM was replaced by two satellite images IRS and SPOT-4. In CLC2012 IRS/ResourceSat and SPOT-4 and SPOT-5 were used. Sentinel-2 satellite imagery - the 1st European satellite dedicated to land monitoring was used in CLC2018. This paper examines the land use and land cover real changes that occurred in B&H in the periods 2000–2006, 2006–2012, and 2012–2018 and aims to present the importance of having high-quality satellite data for understanding land cover and its dynamics. In addition, difficulties encountered and challenges in understanding land cover classes are discussed.

Keywords: CORINE CLC · Land cover · Changes · Satellite imagery · Remote sensing

1 Introduction

The Corine Land Cover inventory is a Pan-European land use and land cover mapping program which based on the interpretation of satellite imagery provides maps of land cover and land cover change on a six-year basis at the national scale. The CLC databases aim to provide an inventory of the Earth's surface features for managing the environment.

N. Ademović et al. (Eds.): IAT 2021, LNNS 316, pp. 694–708, 2022.
https://doi.org/10.1007/978-3-030-90055-7_56

The first CLC project was implemented in most of the EU countries, as well as in the 13 partner countries in Central and Eastern Europe [1]. Based upon lessons learned from the first CLC inventory, a list of user needs, and the available satellite images, the second CLC inventory was implemented. The overall aim was to produce an updated CLC database (CLC2000) and the database of land cover changes between the first CLC inventory and 2000 [1]. The same approach was applied in other CLC projects. After 2000, the update cycle of CLC has become six years.

The first CLC project in Bosnia and Herzegovina (B&H) started in 1998 and was completed in 2000. During this CORINE project, two main data sets were produced: the CORINE Land Cover data for B&H and two change detection studies. Three more CLC projects have been implemented in 2006, 2012, and 2018 in which land cover and land cover change on a six-year basis are being produced.

The first CLC project was successfully implemented by the Institute of Geodesy of Bosnia and Herzegovina. CLC2006, CLC2012, and CLC2018 were implemented by the Faculty of Agriculture and Food Sciences University of Sarajevo.

CLC projects used different satellite imagery in producing the data. Remote sensing is currently the most reliable tool for monitoring varied spectrally sensitive changes of the Earth [2]. Remote sensing via satellite imagery is an excellent tool to study land use/changes because images can cover large geographic extents and have high temporal coverage [3].

2 Data and Methodology

In updating the CLC database, the Updated CLC illustrated nomenclature guidelines [4] were used. The standard CLC nomenclature was applied. This nomenclature is grouped in a 3-level hierarchical classification system and has 44 classes at the third and most detailed level. The five main (1-level) categories are: 1) artificial surfaces, 2) agricultural areas 3) forests and semi-natural areas, 4) wetlands and 5) water bodies.

General technical characteristics of the CLC mapping, as well as the mapping method of visual interpretation of multispectral high-resolution satellite imagery, were applied. The general rules in the production of real land cover changes applied in the mapping process were to identify changes: larger than 5 ha; wider than 100 m, occurred in six years (e.g. between 2012 and 2018); and detectable on satellite images regardless of their position.

The main working steps were: 1) revision and correction of previous CLC database where every mistake found were corrected, and 2) updating of CLC database that included mapping of CLC changes in a period of six years 2000–2006; 2006–2012; 2012–2018 [5–7].

The CLC-Change database is produced directly with the method of computer-aided photo interpretation. The photo interpretation includes a visual comparison of satellite images, detection of real land cover changes visible on the images, and delineation of change polygons based on polygon boundaries. All changes >5 ha were mapped and not only those that are associated with existing polygons. In the case of Bosnia and Herzegovina, the most common are polygons of a real change which represents a change between land cover classes that occurred in reality [5–7].

Production of the CLC database will be presented on the CLC2018 example. Production of CLC2018 database was synthesized in a GIS process as:

$$\text{CLC2018} = \text{CLC2012}_{\text{revised}} + \text{CLC-Changes}_{2012,2018}$$

Where:

- CLC2018 means CLC database for 2018 (25 ha min. mapping unit, 100 m min. mapped width, standard level-3 classes)
- $\text{CLC2012}_{\text{revised}}$ means the corrected/revised CLC database for 2012 (25 ha min. mapping unit, 100 m minimum mapped width, standard level-3 classes)
- $\text{CLC-Changes}_{2012,\,2018}$ means CLC changes between 2012 and 2018 (5 ha min. mapping unit, 100 m minimum mapped width, two attributes, each according to standard level-3 nomenclature)

+ means a GIS process, including automatic generalization and some actions of a photo interpreter.

Verification of the databases consisted of thematic verification and checking topological, geometric, and attribute specifications. Verification was carried out in two levels, internal and external.

When it comes to the accuracy of data, two European validation studies have shown that the achieved accuracy is above the specified minimum (85%) for CLC, as well as for CLC changes [8].

Satellite Data

The type of satellite imagery used in CLC production was various. The quality and resolution of imagery have improved over time. Key characteristics of satellite imagery used in CLC projects are presented below. The major disadvantages of remote sensing include: the inability of many sensors to obtain data through cloud cover, distinct phenomena can be confused if they look the same to the sensor, the resolution of the satellite imagery, and price of very high-resolution satellite imagery [3].

In the CLC2000, Landsat imagery from 1998 at the scale of 1:100000 was used. Landsat Thematic Mapper (TM) imagery has seven spectral channels as presented in Table 1.

Table 1. Landsat 5 TM images: spectral channels and application possibilities [9]

Channel	Spectral bands in μm	Colour and main characteristics
1	0.45–0.52	Blue. Deep penetrating in water dividing of vegetation
2	0.52–0.60	Green. Maximal reflection of Vegetation
3	0.63–0.69	Red. Sensitive on chlorophyll absorption
4	0.76–0.90	NIR (Near of infrared). A strong reflection of vegetation; Larger contrast between land and aquatic surfaces and vegetation and bare spaces

(continued)

Table 1. *(continued)*

Channel	Spectral bands in μm	Colour and main characteristics
5	1.55–1.75	MIR (Middle infrared). A larger difference between different type of soils and types of stones, and content water in soil and vegetation
6	2.08–2.35	MIR (Middle infrared). A larger difference between different type of soils and types of stones, and content water in soil and vegetation, and geological mapping
7	10.40–12.50	TIR (Thermal infrared). Registering of energy emission (thermal) from lands and aquatic surfaces

Five Landsat 5 scenes were available for B&H. All images were acquired in the year 1998, in July, except one image that was acquired in August.

In CLC2006 French SPOT-4 and SPOT-5 and Indian IRS P6 high-resolution satellite imagery acquired in 2005 and 2006 were used. SPOT-4 imagery has five spectral channels as presented in Table 2.

Table 2. SPOT-4 images [10]

Channel	Spectral bands in μm	Color and main characteristics
1	0.50–0.59	Green. Maximal reflection of Vegetation
2	0.61–0.68	Red. Sensitive on chlorophyll absorption
3	0.78–0.89	NIR (Near of infrared). A strong reflection of vegetation; Larger contrast between land and aquatic surfaces and vegetation and bare spaces
4	1.58–1.75	MIR (Middle infrared). A larger difference between different types of soils and types of stones, and content water in soil and vegetation
5	0.61–0.68	Panchromatic channel

Overview of the main parameters of IMAGE2012 satellite imagery used to derive CLC2012 IRS Resourcesat 1, 2, SPOT-4 and SPOT-5 is presented in Table 3.

Satellite data dominantly used in CLC2018 was Sentinel-2.

Sentinel-2 is a multispectral high-resolution imaging mission for global land monitoring to provide different imagery such as imagery on vegetation, soil and water cover etc. Sentinel-2 also has high revisit capability to provide enhanced continuity of data provided by SPOT-5 and Landsat-7. Overview of the main parameters of Sentinel-2 imagery is presented in Table 4.

Table 3. Imagery used for B&H covered the period from June to October of 2011 and 2012 [4].

	IRS Resourcesat 1,2 (Indian)	SPOT-4 and SPOT-5 (French)
Swath width (km)	141	60–80 (depending on looking angle)
No. of bands	4	4
Bands	Green, red, NIR, SWIR	Green, red, NIR, SWIR
Ground sampling Distance (m)	23.5	20 and 10
Bit depth	7	8
Delivered resolution (m)	20	20

Table 4. Overview of the main parameters of Sentinel-2 imagery [9]

	Sentinel-2 Multispectral Imager (MSI)
Swath width (km)	290
Number of bands	13 (altogether)
	4 in VIS
	6 in NIR
	3 in SWIR
Ground sampling distance (m)	10 bands 2,3,4 (VIS) and band 8 (NIR)
	20 bands 5,6,7,8a (NIR) and bands 11,12 (SWIR)
	60 band 1 (VIS), band 9 (NIR) and band 10 (SWIR)
Bit depth (recording)	12
Repeat cycle at the Equator (days)	10 (with 1 satellite) 5 (with 2 satellites)
Data access	Free, full and open access
Delivered resolution (m)	10/20 /60 (depending on band)

Ancillary Data

The purpose of ancillary data is to provide support to photo-interpretation of satellite images.

The main ancillary data to support the CLC2006 were topographic maps and aerial images, both covering the whole territory of the country. Topographic maps at the scale of 1:25,000 were produced in 1976, while 60 updated topographic maps at the same scale were produced in 2004. Black & white aerial images used in the project were the most valuable source of detailed information. Those images at the scale of 1:5,000 were produced in 2003.

The main ancillary data to support the CLC2012 were orthophotos, covering the whole territory of the country. Those images at the scale of 1:5,000 were produced in 2008 and 2012.

In CLC2018, orthophoto images from 2012 covering the whole country were used as well as orthophoto images from 2017 for Sarajevo Canton. CLC for several municipalities (e.g. Tuzla, Gradacac, Gracanica, Srebrenik) and cantons (e.g. Sarajevo Canton) at the scale of 1:10000 were essential in identifying and confirming the identification of certain land cover features or change.

CLC for municipal or cantonal level is being produced within the Land Use Studies prepared by the Faculty of Agriculture and Food Sciences University of Sarajevo. The main objective of the Land Use Study is to prepare a thematic database and map of land use value, as well as to examine and analyze the area of a municipality or canton using the GIS system and remote sensing sources of information (satellite and ortho-photo images) [11].

3 Results and Discussion

CLC 2018 in B&H is characterized by 35 out of 44 classes of land cover (Table 5, Fig. 1). The analysis of the CLC database revealed that semi-natural and forest areas dominate in the country with 64.4% of coverage, followed by 33.1% of agricultural areas. Land classified as artificial areas occupied about 1.7%, while about 0.8% was classified as wetlands (0.1%) or water bodies (0.7%).

Table 5. The overall composition of CLC classes (2018)

Class		Class description	Area (ha)	% in category	% of the total area
Urban fabric	111	Continuous urban fabric	205	0.2	1.7
	112	Discontinuous urban fabric	64,742	74.6	
Industrial, commercial and transport units	121	Industrial or commercial units and public facilities	7,511	8.7	
	122	Road and rail networks and associated land	1,092	1.3	
	124	Airports	1,131	1.3	
Mine, dump and construction sites	131	Mineral extraction sites	8,704	10.0	
	132	Dump sites	1,379	1.6	
	133	Construction sites	683	0.8	
Artificial non-agricultural vegetated areas	141	Green urban areas	26	0.03	
	142	Sport and leisure facilities	1,337	1.5	

(continued)

Table 5. (*continued*)

Class		Class description	Area (ha)	% in category	% of the total area
Total of Artificial surfaces			**86,810**	**100.0**	
Arable land	211	Non-irrigated arable land	176,381	10.4	33.1
	212	Permanently irrigated arable land	2,252	0.1	
Permanent crops	221	Vineyards	3,022	0.2	
	222	Fruit tree and berry plantations	7,995	0.5	
Pastures	231	Pastures, meadows and other permanent grasslands under agricultural use	324,163	19.1	
Heterogeneous agricultural areas	241	Annual crops associated with permanent crops	52	0.003	
	242	Complex cultivation patterns	740,637	43.6	
	243	Land principally occupied by agriculture, with significant areas of natural vegetation	442947	26.1	
Total of Agricultural areas			**1,697,449**	**100.0**	
Forests	311	Broad-leaved forest	1,670,346	50.7	64.4
	312	Coniferous forest	270,256	8.2	
	313	Mixed forest	432,781	13.1	
Shrubs and/or herbaceous vegetation associations	321	Natural grassland	288,273	8.7	
	322	Moors and heathland	20,843	0.6	
	323	Sclerophyllous vegetation	223,155	6.8	
	324	Transitional woodland/shrub	282,503	8.6	
Open spaces with little or no vegetation	331	Beaches, dunes, and sand plains	93	0.003	
	332	Bare rock	5,893	0.2	
	333	Sparsely vegetated areas	99,864	3.0	

(*continued*)

Table 5. (*continued*)

Class		Class description	Area (ha)	% in category	% of the total area
	334	Burnt areas	2,267	0.1	
Total of Forests and semi-natural areas			**3,296,273**	**100.0**	
Wetlands	411	Inland marshes	5,272	88.3	0.1
	412	Peatbogs	699	11.7	
Total of Wetlands			**5,971**	**100.0**	
Water bodies	511	Water courses	13,918	40.5	0.7
	512	Water bodies	20,445	59.5	
Total of Water bodies			**34,363**	**100.0**	
Total			**5,120,867**		**100**

Source: EEA, Faculty of Agriculture and Food Sciences, University of Sarajevo

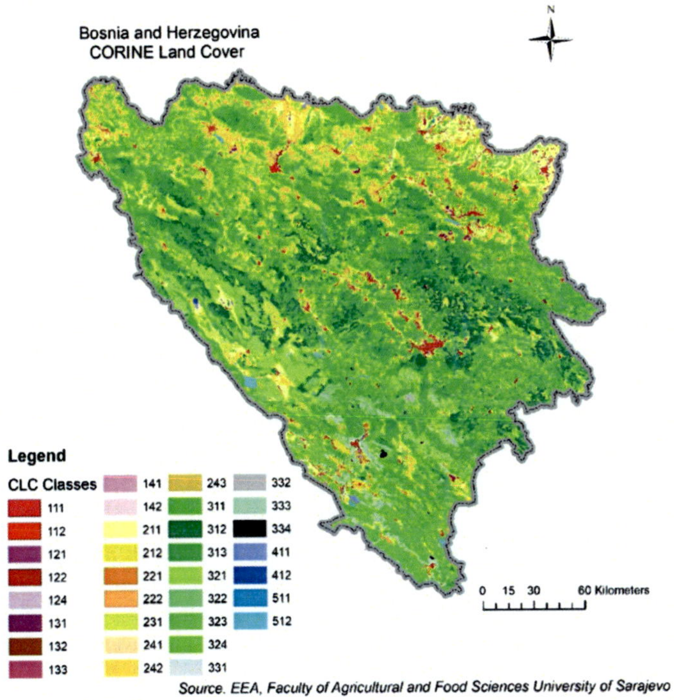

Fig. 1. CORINE Land Cover in Bosnia and Herzegovina

CORINE data shows that in the 2000–2006 period in B&H, land cover changes occurred on a total area of 47,685 ha. In the 2006–2012 period, land cover changes occurred on a total area of 19,702 ha while in the 2012–2018 period, those changes occurred on a total area of 24,223 ha.

In the period 2000–2018, the increase of artificial surfaces amounted to 14,075 ha (Fig. 2). Within this category, the urban fabric has the biggest increase of 6,402 ha as a result of population migration, urbanization, and post-war reconstruction of the country. Industrial, commercial and transport units category had also increased and this change was recorded in the amount of 1,868 ha in the period 2000–2018. On the other hand, a continuous increase of mine, dump and construction sites was evident in the period observed (5,733 ha in total).

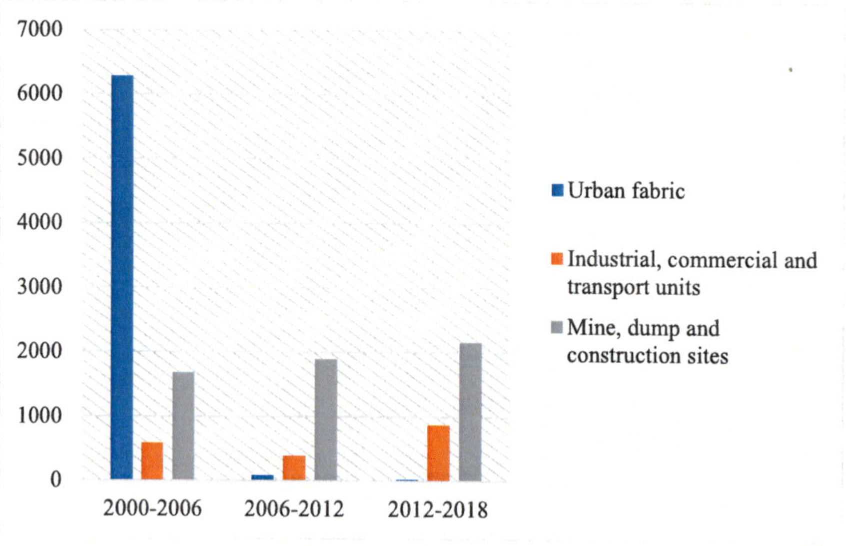

Fig. 2. Increase of artificial surfaces

The increase of agricultural areas amounted to 3,135 ha in the 2000–2018 period (e.g. broad-leaved forest) converted to land principally occupied by agriculture, with significant areas of natural vegetation class). Trends of increase in the three observed periods (2000–2006, 2006–2012, 2012–2018) are different. The increase of agriculture areas was 552 ha (2000–2006), 1,548 ha (2006–2012), and 1,035 ha (2012–2018). The arable land category had the biggest increase of 625 ha in the 2006–2012 period, followed by pastures and heterogeneous agricultural areas, by 200 ha and 550 ha respectively. Within heterogeneous agricultural areas, the biggest increase was recorded in complex cultivation patterns. Permanent crops were increased continuously: 53 ha (2000–2006), 174 ha (2006–2012) and 686 ha (2012–2018).

As a result of agriculture intensification in certain areas of B&H, a total of 6,238 ha transitioned from one agricultural class into another one (e.g. pasture) converted to complex cultivation patterns). This kind of change is more visible in the period 2000–2006 which could be linked with the return of population in rural areas and donor support to agriculture at that time. The transition of classes within agriculture areas decreased over time, so in the period 2012–2018 those changes were recorded on the surface of 421 ha only.

Larger areas of agricultural land are being used for non-agricultural purposes (Fig. 3). In a twenty years' period, a total of 10,400 ha of agricultural areas had converted into artificial surfaces. Again, the biggest changes were recorded in the 2000–2006 period, when 7,266 ha agricultural areas had converted into artificial surfaces, of which 81% refers to conversion into discontinuous urban fabric class. Obviously, this is a direct consequence of urbanization due to the migration of the population from rural to urban areas. This kind of agricultural area losses have decreased after 2006, and the recorded changes amounted to 1,202 ha in 2006–2012, and 1,933 ha in 2012–2018.

When it comes to the conversion of agricultural areas into forests and semi-natural areas, it amounted to 3,213 ha in the period 2000–2018. It mainly referred to conversion into transitional woodland/shrub.

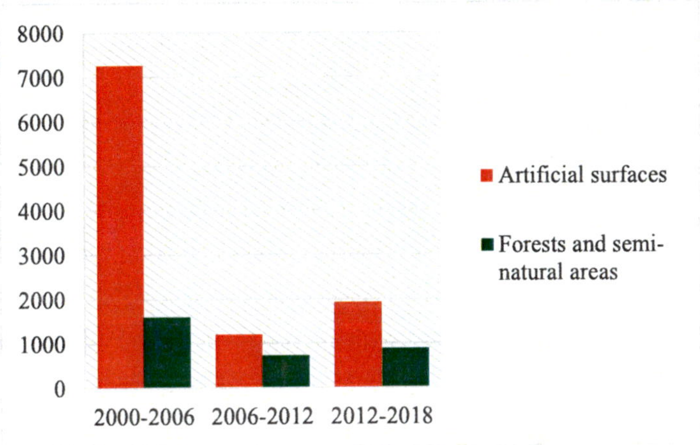

Fig. 3. Increase of artificial surfaces and forests and semi-natural areas on account of agricultural areas

In the 2000–2018 period, changes of forests and semi-natural areas category amounted to 68,487 ha. The most significant changes were the conversion of forests into transitional woodland/shrub in the amount of 23,882 ha (Fig. 4), mainly broad-leaved forest (19,870 ha). On the other hand, a total of 18,715 ha under transitional woodland/shrub converted to forests, out of which 14,886 ha of transitional woodland/shrub converted to broad-leaved forest (Fig. 5).

In the 2000–2018 period, the increase of the wetland category amounted to 48 ha. The water bodies category increased for 853 ha.

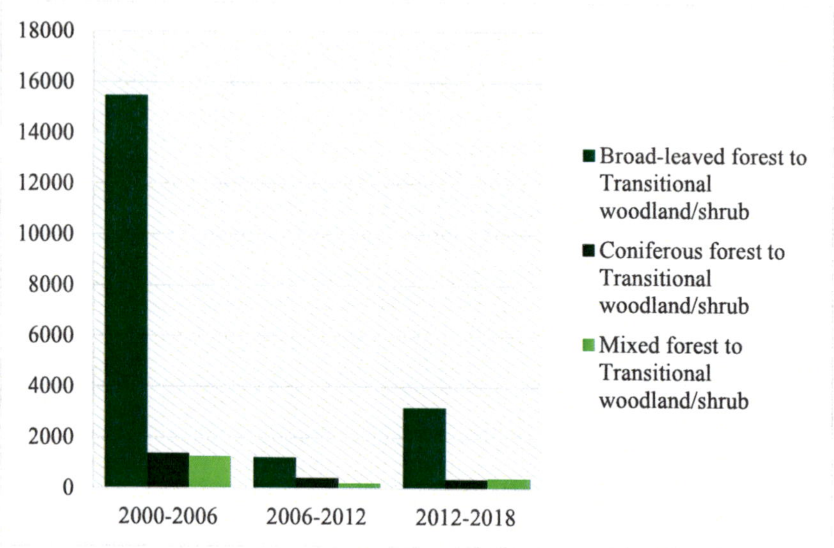

Fig. 4. Changes in Forests - Forests to Transitional woodland/shrub

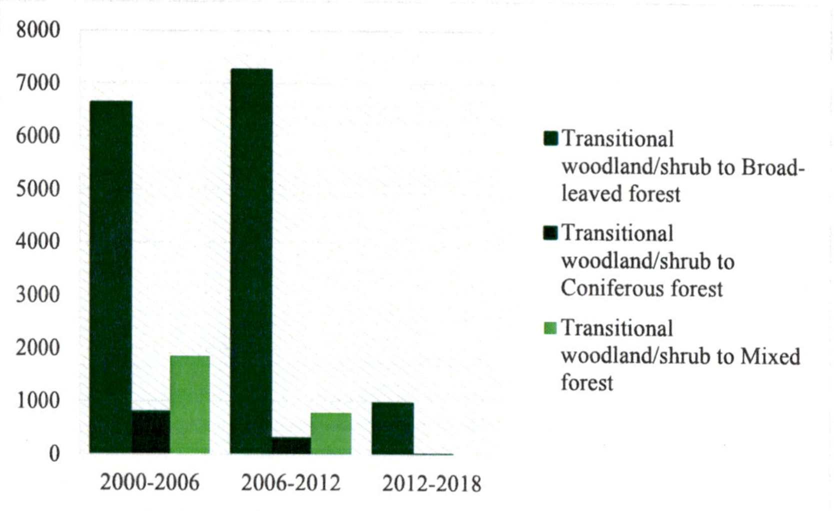

Fig. 5. Changes in Forests - Transitional woodland/shrub to Forests

Difficulties Encountered and Solutions Found

Certain shortcomings were noticed during the CLC revision process thanks to the high quality of satellite imagery. Different satellite imagery for the same area used in CLC production is presented below (Fig. 6). There is an obvious difference in the quality and better resolution of the imagery from 2018 compared to previous years.

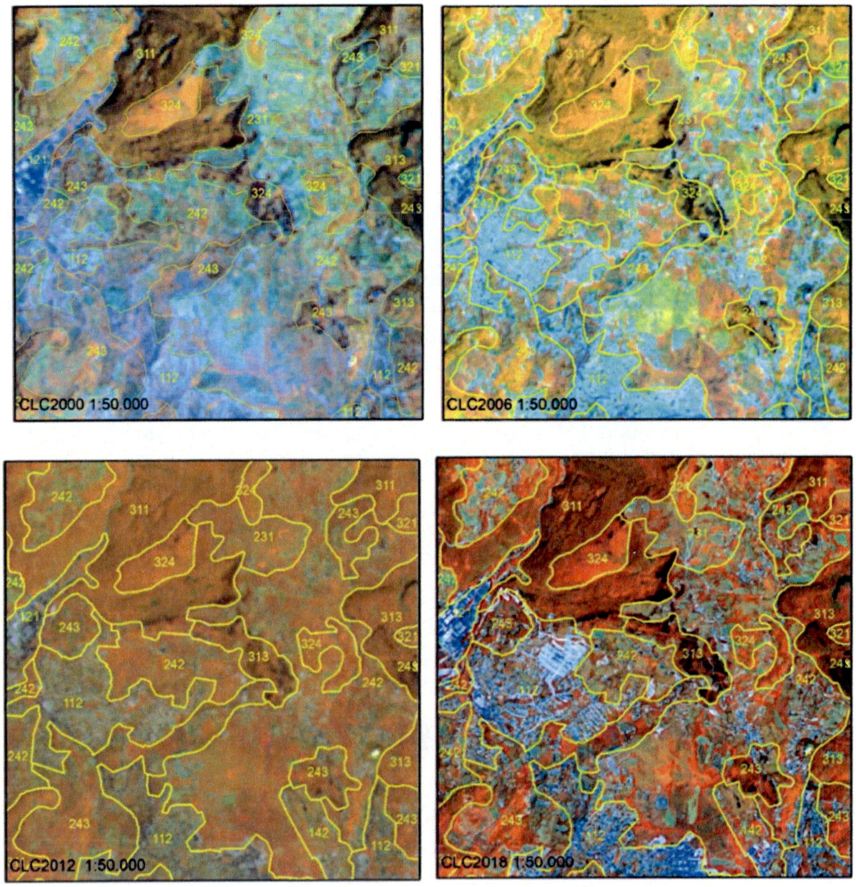

Fig. 6. Satellite imagery from different period of time where the difference in quality and resolution is clearly visible

In the CLC2000 revision, a class moors and heathland in the sub-Mediterranean region was wrongly interpreted. Following the description of this class, it appeared that vegetation in the moors and heathland class never appeared in the sub-Mediterranean region. This class, in the case of B&H, is typical for mountain areas.

In the CLC2006 revision, pastures in some parts of the Herzegovina region were wrongly interpreted as natural grassland.

In the CLC2012 revision, class sclerophyllous vegetation was wrongly interpreted as transitional woodland shrub and in some cases as broad-leaved forest. This was a systematic mistake identified in the southern part of the country, in the Herzegovina region. In several cases, class fruit trees and berry plantations were frequently incorrectly registered and mixed with class transitional woodland to shrub.

In general, commonly identified problems in CLC2012 and CLC2018 were over-estimation of the class land principally occupied by agriculture, with significant areas of natural vegetation. In fact, there was an evident need for better delineation of certain classes within agricultural areas as it was the case with the class land principally occupied by agriculture with significant areas of natural vegetation.

In terms of difficulties in interpretation of some classes in previous projects especially when it comes to agriculture, it should be noted that fieldwork was not possible during the CLC2000 project due to the post-war circumstances, political challenges as well as the existence of (unmapped) minefields. In addition, there was a lack of ancillary data in CLC2000. No digital data from the pre-war period, e.g. CLC1990 was available, which would give a basis for comparing land cover and changes. To note is that land cover changes which have occurred until 2000 were not recorded, particularly changes regarding conversion to succession in abandoned areas [12].

High-resolution images from 2017/18 were extremely useful in the identification of certain classes within agricultural areas that were not visible previously due to small parcel size and low resolution of images (e.g. separation of fruit tree and berry plantations in agricultural areas). In addition to this, one of the reasons for having overestimated class land principally occupied by agriculture is that the post-war period was characterized by abandoned land. Many agricultural fields are not used anymore, so those areas are converted into a certain type of natural vegetation.

High-quality images as well as ancillary data, used especially in CLC2018, enabled a better understanding and mapping of changes, especially those on agricultural areas, given the dynamics of activities and changes in agriculture. However, identification of abandoned land is still a challenge in B&H. When it comes to mapping abandoned cropland using remote sensing technology, it is still challenging as it is difficult to distinguish abandoned cropland from fallowed land [13].

It should be noted that the revision of the CLC implied a more precise delineation of all classes. This especially refers to the built-up areas that were more difficult to notice in previous CLC due to low quality of imagery (Fig. 7).

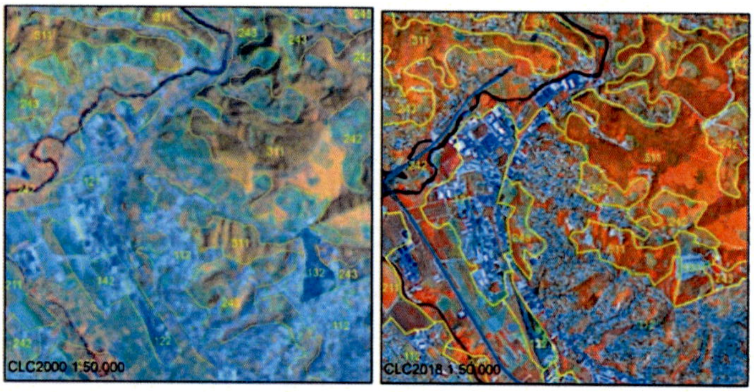

Fig. 7. Quality of satellite imagery on the example of built-up areas

4 Conclusion

Different sets of satellite imagery were used in four CLC projects in B&H, namely: Landsat 5 TM images, SPOT-4, SPOT-5, Indian IRS P6, IRS Resourcesat 1, 2, Sentinel-2. Comparing the imagery clearly shows the development in the quality and resolution of the imagery. For instance, Sentinel-2, SPOT-4, and SPOT-5 have a resolution of 10 m while IRS Resourcesat has a resolution of 20 m. Sentinel-2 has 13 bands, IRS Resourcesat, SPOT-4, and SPOT-5 have 4 bands each.

Satellite coverage of B&H for different seasons has also evolved. Thus, in CLC2000, all imagery was from July 1998, which certainly affected the quality of interpretation, since ancillary data were very limited at that time. Lack of ancillary data was also a constraint in understanding land cover classes and those changes.

In CLC2018, the seasonal coverage was excellent and the imagery covered seasons from April to October 2018, and partly 2017, which enabled quality review of the previous CLC, and better class delineation, especially within agricultural areas.

The type and quality of ancillary data have improved over time. Satellite imagery was accompanied by various ancillary data such as orthophoto images, CLC for several municipalities/cantons at the scale of 1: 10000. Those data were necessary to support the mapping of certain land cover features or land cover changes.

CLC 2018 resulted in the identification of 35 out of 44 classes of land cover. Semi-natural and forest areas dominate in the country with 64.4% of coverage, followed by 33.1% of agricultural areas, 1.7% of artificial areas, 0.1% of wetlands and 0.7% of water bodies.

Total real changes in land cover identified in the period 2000–2018 amount to 91,610 ha.

CLC data shows that trends in observed periods (2000–2006, 2006–2012, 2012–2018) are different. In the 2000–2006 period in B&H, land cover changes occurred on a total area of 47,685 ha; in the 2006–2012 period, land cover changes occurred on a total area of 19,702 ha; in the 2012–2018 period, those changes occurred on a total area of 24,223 ha.

In the period 2000–2018, the increase of artificial surfaces amounted to 14,075 ha. In total, 10,400 ha of agricultural areas had converted into artificial surfaces in the same period. Changes of forests and semi-natural areas category amounted to 68,487 ha. The most significant changes were the conversion of forests into transitional woodland/shrub (23,882 ha). A total of 18,715 ha under transitional woodland/shrub converted to forests.

CORINE Land Cover is a unique database at the level of B&H. The above analysis provides a short overview of the key changes both in B&H's land typology and technological advancements which made this unique database crucial in measuring land cover and all the changes in the past 20 years.

Acknowledgement. CLC data were produced with funding by the European Union. We would like to express special thanks to BiH CLC team: Hamid Custovic-Team leader, Melisa Ljusa, Jasmin Taletovic, Sead Vojnikovic, Besim Balic and Fahrudin Djuzo.

References

1. Büttner, G., Maucha, G., Kosztra, B.: European Validation of Land Cover Changes in CLC2006 Project (2011). http://earsel.org/symposia/2011-symposium-Prague/Proceedings/PDF/Land%20Use%20Land%20Cover/39%20ok25-a2421-buttner.pdf. Accessed 15 Mar 2021
2. Akhtar, A., Sultan Bhat, M., Maheen, M.: Using Landsat satellite data for assessing the land use and land cover change in Kashmir valley. GeoJournal (2020) **85**,1529–1543 (2019). https://link.springer.com/article/10.1007/s10708-019-10037-x. Accessed 15 Mar 2021
3. Fonji, S.F., Taff, G.N.: Using satellite data to monitor land-use land-cover change in North-eastern Latvia. SpringerPlus **3**, 61 (2014). https://springerplus.springeropen.com/articles/10.1186/2193-1801-3-61. Accessed 15 Mar 2021
4. Kosztra, B., Büttner, G., Hazeu G., Arnold S.: Updated CLC illustrated nomenclature guidelines (2017). https://land.copernicus.eu/user-corner/technical-library/corine-land-cover-nomenclature-guidelines/docs/pdf/CLC2018_Nomenclature_illustrated_guide_20190510.pdf. Accessed 17 Mar 2021
5. Custovic, H., Ljusa, M., Taletovic, J.: B&H CLC Report. Faculty of Agriculture and Food Sciences University of Sarajevo (2006)
6. Custovic, H., Ljusa, M., Taletovic, J.: B&H CLC Report. Faculty of Agriculture and Food Sciences University of Sarajevo (2012)
7. Custovic, H., Ljusa, M., Taletovic, J.: B&H CLC Report. Faculty of Agriculture and Food Sciences University of Sarajevo (2018)
8. Büttner, G.: CORINE land cover and land cover change products. In: Manakos, I., Braun, M. (eds.) Land Use and Land Cover Mapping in Europe. RSDIP, vol. 18, pp. 55–74. Springer, Dordrecht (2014). https://doi.org/10.1007/978-94-007-7969-3_5
9. Olujic, M.: Kartiranje zemljišta i vegetacije područja Istre prema Corine programu. - Korištenje tla i vode u funkciji održivog razvoja i zaštite okoliša. ANU BiH Knjiga CIX, Odjeljenje prirodnih i matematičkih nauka Knjiga 16. Sarajevo
10. European Environment Agency: CLC2006 technical guidelines. Technical report No 17/2007. EEA, Copenhagen (2007)
11. Čustović, H., Ljuša, M., Schlingloff, S. (eds.): Sustainable land management – approaches and practices in Bosnia and Herzegovina. University of Sarajevo (2020)
12. Ljusa, M., Custovic, H.: Agricultural land use and land losses in Bosnia and Herzegovina in the period 1961–2018. Agric. Forestry **65**(1), 147–156 (2019). Podgorica (2019)
13. Wei, S.: Mapping cropland abandonment in mountainous areas using an annual land-use trajectory approach. Sustainability **11**, 5951 (2019). https://www.mdpi.com/2071-1050/11/21/5951. Accessed 17 Mar 2021

Application of GIS Technologies in Sustainable Land Management at Local Level

Hamid Čustović[1](✉), Melisa Ljuša[1], Jasmin Taletović[2], Mirza Ponjavić[3], and Almir Karabegović[4]

[1] Faculty of Agriculture and Food Sciences, University of Sarajevo, Zmaja od Bosne 8, 71000 Sarajevo, Bosnia and Herzegovina
[2] Department of GIS, Institute for Planning of Canton Sarajevo, Branilaca Sarajeva 26, 71000 Sarajevo, Bosnia and Herzegovina
[3] International Burch University, Francuske revolucije bb, 71210 Ilidza, Bosnia and Herzegovina
[4] Faculty of Electrical Engineering, University of Sarajevo, Zmaja od Bosne, 71000 Sarajevo, Bosnia and Herzegovina

Abstract. Food production is a primary role of land. The importance of soil, as one of land's essential components in socio-economic planning should be seen through the fact that soil is a limited and hardly renewable natural resource or a non-renewable resource for a human lifetime. The usage of land can be multi-functional and its role in natural ecosystems and life on Earth is not an easy one to understand. It is a dynamic system that performs numerous functions which are essentially divided into environmental and technical ones. In addition, land provides services for ecosystems and human activities. Bosnia and Herzegovina (B&H) has scarce amounts of good quality agricultural land. The top three soil bonity categories account for only 14.2%. Land use and development are directly influenced by socio-economic trends and sectoral policies. In addition to this and in more general terms the land is also influenced by the attitude that the society holds towards it, as one of the key resources. Sustainable land management has a very important agro-ecological and socio-economic dimension. This paper aims to present the importance, characteristics as well as the use of the different approaches in evaluating and sustainable management of land at the local level. Application and importance of GIS as a tool will be presented through several research studies, namely: soil data of Popovo polje, bonity data of Mostar City, land cover and land use changes of Tuzla City, and application of agroecological zoning of Srebrenik municipality.

Keywords: GIS · Land evaluation · Bonity · Land cover/land changes · Agro-ecological zoning

1 Introduction

"Land is a delineable area of the earth's terrestrial surface, encompassing all attributes of the biosphere immediately above or below this surface including those of the near-surface climate the soil and terrain forms, the surface hydrology (including shallow

N. Ademović et al. (Eds.): IAT 2021, LNNS 316, pp. 709–718, 2022.
https://doi.org/10.1007/978-3-030-90055-7_57

lakes, rivers, marshes, and swamps), the near-surface sedimentary layers and associated groundwater reserve, the plant and animal populations, the human settlement pattern and physical results of past and present human activity (terracing, water storage or drainage structures, roads, buildings, etc.)." [1] For determining the productivity and sustainability of agro-ecosystems, the interaction between the components of land resources is vital components of land resources is vital [2].

Soil is a basic and hardly renewable natural resource, which with water, air, and organisms constitutes the ecosystem. Due to its unique characteristics, the land is a key factor of sustainability, possibility, and quality of life in social development [3]. Ensuring sufficient food supply and the functioning of terrestrial ecosystems are two major concerns when it comes to land [4].

Agricultural land is used for agricultural production. One of the most important tasks at all levels of government is to ensure efficient use and protection of agricultural land [5]. According to the Law on Agricultural Land in the Federation of Bosnia and Herzegovina "*agricultural land as a natural resource and property of general interest to the Federation and Bosnia and Herzegovina enjoys special protection, is used for agricultural production, and cannot be used for other purposes, except in the cases and under the conditions laid down in this Law"*[6].

The latest statistical data [7] shows that B&H has a total of 1,843,832 ha of agricultural land. "Top-quality land (the three first soil bonity categories) covers an area of only 14.2%".To add, the four first soil bonity categories cover about 31%, or slightly less than 1/3 of the total territory of B&H [8].

A permanent reduction in the existing agricultural land areas is evident in B&H. The process of conversion of quality agricultural land to non-agricultural could be significantly influenced if legal permissions were respected and the awareness of society raised.

Today, sustainable soil and land management, in general, is not feasible without the application of GIS technologies, especially at the local level. The data from GIS can be used for various purposes, especially for soil/land management and monitoring, monitoring and development of agriculture as well as rural and urban areas, monitoring of changes in the environment (changes in land cover), production of thematic maps etc.

2 Data and Methodology

This paper presents the data and results of the research conducted by the Faculty of Agriculture and Food Science at the University of Sarajevo and the Institute for Hydrotechnics in Sarajevo, B&H. The research for Mostar region was conducted in cooperation with the Institute, while the research in the cities of Tuzla, Srebrenik and the municipality of Ravno has been conducted by the Faculty of Agriculture and Food Science in Sarajevo. The research activities were conducted during implementation of the land capability map projects and the preparation of soil databases based on pedological research and other data collection in all locations. These instruments are basis for spatial planning, protection and monitoring of soil/land, as well as agriculture and rural development.

Several approaches in producing and modeling of data applied in the research will be presented in this paper, namely:

- soil data of Popovo polje,
- land capability map of Mostar City,
- land cover and land use changes of Tuzla City,
- and application of agroecological zoning of Srebrenik municipality.

Management of soil data (field and laboratory, M 1:10000) and usage of the data for monitoring and planning purposes will be presented on the example of Popovo polje. Valorization of soil quality and its production capacity will be presented on Mostar City. This analysis is a part of the land capability map which is a basic tool for sustainable use and planning of land resources at the level of municipality or canton.

The main objective of the land capability map is to prepare a thematic database and map of land use value at the scale of M: 1:10000 as well as to analyze the area of a municipality or canton using the GIS system and remote sensing sources of information. The purpose is to establish the balances of land use, identify certain forms of land damage, perform valorization through soil quality rating, and analyze the pedological characteristics of certain soil types (field and laboratory analysis) [5, 9, 10].

Soils are classified into eight bonity classes. Bonity class I represent the best quality soils, and class VIII the lowest quality ones.

In the process of soil categorization (rating), the following important properties were considered:

a) Type of soil and its properties: morphological (depth, compaction, skeletalness, plasticity, etc.); physical (texture, structure, permeability, etc.); chemical (pH, humus content, the content of $CaCO_3$ etc.);
b) General characteristics of the terrain: relief features, exposure, inclination, the position of the plot, rockiness, etc.; hydrological conditions (flooding, drainage, irrigation, etc.).

The land quality evaluation is used to determine the level of its suitability for agricultural production based on the scoring of natural conditions for crop production (soil, climate, relief), in our example for Mostar City. The land quality evaluation is performed by awarding points to the land rating properties. The highest-quality land of the I rating category is awarded 100 points. The lowest quality land of the VIII rating category is awarded 10 points. Land quality rating is calculated using the following formula [11].

$$B = \sqrt{\text{Tx x Kx x Rx x}\left(1 - \frac{\sum \% \ \textit{special} \quad \textit{effects}}{100}\right)}$$

B-bonity score; T-soil properties score; K-climate properties score; R-relief properties score.

The application of GIS and orthophoto imagery in the analysis of land cover and its changes will be presented in the example of Tuzla City. Land cover/land use map for 2012 has been prepared at a scale of 1:10000. The analysis of land cover changes (>0.5 ha) was performed for the period 2008–2012. A unique IV level of CORINE nomenclature prepared for B&H by [12] was applied in this research.

The general rules in the production of land cover applied in the mapping were:

- polygons larger or equal to 1 ha (smaller polygons are also included due to their significance such as orchards, succession, afforestation)
- wide of polygons is >10 m,
- all polygons are visible on orthophoto imagery.

Agroecological zoning (AEZ) is a suitable methodology to assess the suitability of land for agricultural production by analyzing different layers This methodological approach will be presented in the example of Srebrenik Municipality.

The AEZ is used to determine specific constraints for particular crops to be grown in certain conditions of climate, soil and terrain, but also in certain (assumed) conditions of investment in agriculture and management [13].

The particular parameters used in the definition focus attention on the climatic and edaphic requirements of crops and on the management systems under which the crops are grown [14].

In our example, taking into account certain soil parameters (pH, texture, humus), AEZ divides areas into regions or zones that are homogeneous concerning soil characteristics and which are also uniform in terms of opportunities for growing crops. Each zone has a similar combination of constraints and potentials for land use and serves as a basis to effectively use the potentials and comparative advantages of individual production areas for growing crops that are suitable for the given conditions.

Suitability levels are defined as follows: S1-very suitable; S2-suitable; S3-moderately suitable; N-non suitable. In principle, S1 has no or has slight constraints in terms of soil parameters. S2 has very small constraints, S3 has a medium or one larger constraint, while N has very high constraints.

Through the examples of GIS applications at the local level, the possibilities for sustainable planning and management of land resources at the local level will be pointed out.

3 Results and Discussion

Soil Data – Study of Popovo Polje

Popovo Polje (*field*) is part of a great karst depression, which stretches in the direction of Dinarides (northwest-southeast), roughly in line with the coastline, from Hutovo to Hum in Herzegovina. The field occupies a territory of approximately 4500 hectares of fertile land on altitude from 227 to 251 m [15].

Field and lab research was carried out on the territory of Popovo Polje in six sectors: Hutovo, Turkovići, Velja Međa, Orašje, Most and Dračevo-Grmljani-Sedlari (Fig. 1 - soil samples in Dracevo-Grmljani-Sedlari sectors). Field and lab research served to determine pH reactions in H_2O and M-KCl, content of total and physiologically active lime, content of humus, accessible phosphor and potassium, texture properties and structure of soil.

Field research included the collection of average samples of soils from depths 0–30 cm and 30–60 cm. A total of 366 locations or 732 soil samples were analyzed.

All locations of collected soil samples were recorded with GPS. The.shp file for each sector represents the locations of collected samples which are linked with the data obtained through lab research exclusively for the given sector.

Soil data prepared and presented in GIS could serve as a concept for monitoring soil conditions as well as for future agricultural development in this area. Based on this data for Popovo polje, the level of soil suitability for the cultivation of individual cultures is given in the form of a model, with taking into account optimal requirements by different crops, and limitations or advantages of these soils in relation to individual culture. The soil fertility and modern systems of plant nutrition, with special attention to the condition of the land in Popovo Polje, could be developed based on those data.

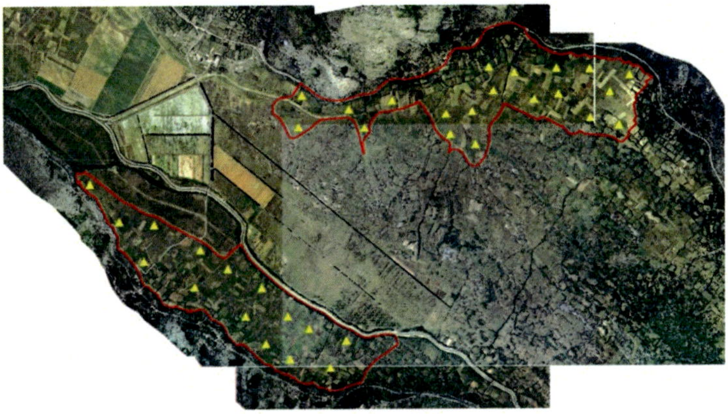

Fig. 1. Soil samples in Dracevo-Grmljani-Sedlari sectors

Soil Bonity – Study of Mostar City

In Mostar City, soil bonity categories that range from II to VIII were registered (Fig. 2) [16]. Bonity classes II, III and IVa are suitable for cultivation, IVb is less suitable, while V, VI, VII and VIII bonity classes have big constraints and they are unsuitable for cultivation.

Land suitable for cultivation (II, III, IV bonity classes) occupies 4,146.23 ha which is 3.56% of the total area of the Mostar City.

Category II bonity class occupies 7.03 ha which is 0.01% of the total city area. This bonity class includes soils that has minor limitations which reduces the number of suitable crops and also requires application of certain agrotechnical measures. These soils require more careful tillage, including soil conservation, preventing deterioration of their condition, and improving the air-water ratio in the soil. Limitations are small and can be easily removed in practice.

Category III bonity class occupies 837.97 ha which is 0.72% of the total city area. Soil has some limitations and requires the application of more complex agrotechnical measures compared to the previous bonity class, especially if they are used for intensive production. Soil of this bonity class can be used for crops and vegetable production as well as some other cultures (orchards, meadows).

Category IV bonity class occupies 3,301.22 ha or 2.83% of the total city area. Soil has significant limitations and requests certain control measures as well as land reclamation measures. Cultivation of soil is difficult and it requests the application of sustainable land management practices.

Soils unsuitable for cultivation are soils of pastures, meadows as in some cases plough fields. These are V, VI, VII and VIII bonity classes. Today, due to advanced technology, especially mechanized cultivation and irrigation, these soils can be used for the production of grapevine, olives and immortelle.

Category V bonity class occupies 6,444.22 ha or 5.53% of the total city area. Soil of this bonity class may or may not be exposed to erosion, but has certain limitations for cultivation. This soil can be exceptionally used as arable land, mainly as meadows, pastures and orchards. Category VI bonity class occupies 16,164.65 ha or 13.87% of the total city area. Due to its limitations, this soil category can only be used as meadows, pastures and forests.

Category VII bonity class occupies 30,253.72 ha or 25.96% of the total city area. Soil of this bonity class has significant limitations for cultivation and other types of use that cannot be removed.

Category VIII bonity class occupies 1,772.49 ha (1.52%). This soil cannot be used for bio–reproduction. Instead, it can be used for the development of infrastructure, mineral exploitation etc.

Spatial distribution of soil bonity classes, its potential and limitations for specific land use certainly facilitate any planning processes and investment.

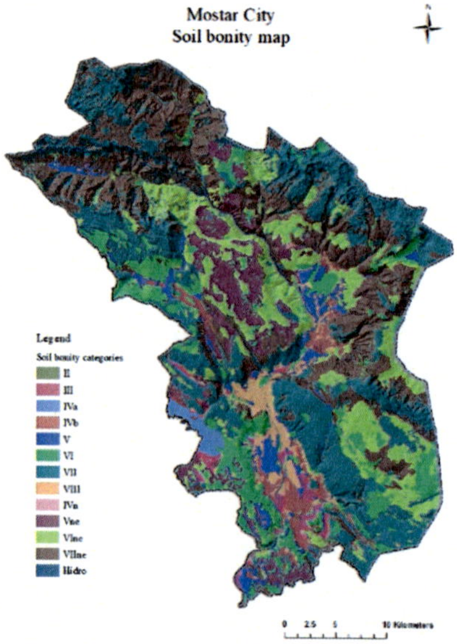

Fig. 2. Soil bonity map for Mostar City

Land Cover and Land Use Changes – Study of Tuzla City

On the basis of orthophoto imagery, an analysis of the existing use value of the land

and spatial dispersion of agricultural, forest and barren land, as well as water surfaces, was made for City of Tuzla (Fig. 3) [9]. In other words, a land cover/use map at a scale of 1:10000 for 2012 was produced. CORINE IV level nomenclature has been applied. Orthophoto imagery was the basis for the detection of all land cover changes >0.5 ha that occurred in the period 2008–2012.

The analysis of the land cover in Tuzla City revealed that semi-natural and forest areas dominate with 15,382.2 ha or 64.4% of coverage, followed by agricultural areas that occupy 9,327 or 31.5%. Land classified as artificial areas (built up and degraded) occupies about 4,842.5 or 16.4%, while around 40.7 ha or 0.1% was classified as water bodies. Following IV level of CORINE nomenclature, the dominant artificial surfaces category is the family houses and gardens class which cover 3,594 ha or 74% of total artificial surfaces. In the category of the agricultural area, complex cultivation patterns classes with and without scattered houses are the dominant classes that cover 7,357 ha or 79% of total agricultural areas. Pastures without natural trees and shrubs (<15%) were a significant land cover class in Tuzla City. This class covers 1,270 ha or 13.6% of total agricultural areas. Arable land with parcels bigger than 5 ha covers only 2.9% of agricultural areas. In the semi-natural and forest category, the dominant class is broad-leaved forest (88% of this category). Transitional wood-land/shrub as specific and very important land cover class was detected on the surface of 1,711 ha, on 418 locations distributed on the territory of Tuzla.

Land cover changes >0.5 ha that occurred in the period 2008–2012 were identified in 42 different locations covering 51.7 ha. The biggest changes are evident in the increasing orchards (22 ha) and dumpsites (21.5 ha).

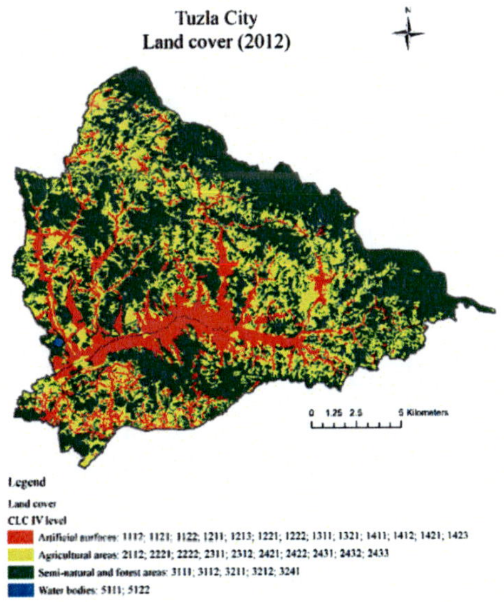

Fig. 3. Land cover for Tuzla City

Agroecological Zoning –Study of Srebrenik City

AEZ methodology focuses on the requirements of crops specifically in relation to the climate (e.g. sum of temperature, length of growing period, precipitation etc.), soil (pH, depth, texture etc.), relief (e.g. slope, exposition), and compares natural conditions of certain areas with crops requirements to determine land suitability and potential constraints of land for crops growth.

AEZ approach requests climate, soil and terrain data in GIS. Numbers of thematic layers are being used in AEZ modeling. By combining different layers and comparing them with crop requirements, we could easily get spatial information about the potential and constraints of certain areas.

For Srebrenik Municipality, soil pH thematic map and suitability map as a combination of soil ph, texture and humus content will be presented for corn. Soil pH in Srebrenik Municipality ranges from 4.3 (very acid) to 8.1 (very alkaline), while the average content of humus is from 0.73 (very low) to 8.4% (very high). The optimal pH for corn is 6.5–7.2 and humus content >3%.

A set of thematic layers and suitability maps for 14 different crops were produced for Srebrenik Municipality (Fig. 4 - suitability for land for growing corn) [10]. This is a very important output as thematic maps give us a clear picture of soil conditions as a basis for applying agrotechnical measures where necessary and for investment opportunities. Suitability maps should serve as key inputs in the development of agriculture and rural policy as well as for investment planning.

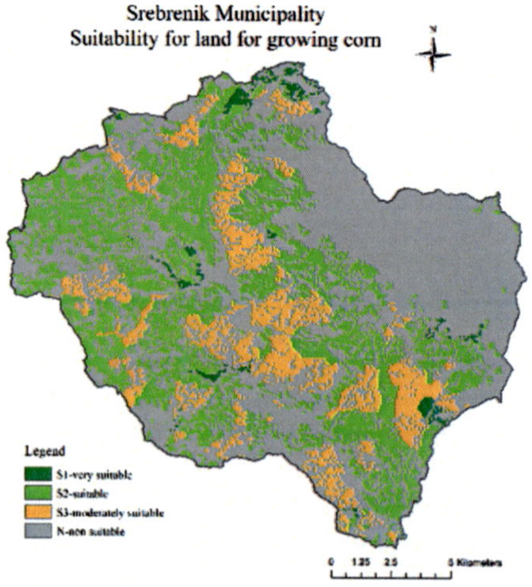

Fig. 4. Suitability for land for growing corn for Srebrenik municipality

4 Conclusion

Sustainable land management at the local level request a good database for planning the approach on how land/soil will be used in the future. This can only be assured by having databases in GIS and applying different methodological approaches as presented in this paper.

All data treated by the land capability map should be part of a single information system in the area of spatial planning of cantons and municipalities [17]. Additionally, it is very important to have the appropriate and single methodology for map development applied [5].

Land cover/use map is a basis for:

- determination and delineation of productive agricultural and forest land from nonproductive land (land use for other purposes);
- displaying soil quality through bonity classes;
- delineation and presentation of agro zones of agricultural land.

GIS database provides the spatial distribution of land, which allows observation and planning of land for various purposes based on its status and environmental conditions [18].

Spatial data facilitate decision-making on what land can be excluded from their primary function and used for other purposes such as the development of infrastructure, urbanization, etc.

The combination of GIS and remote sensing gives us the possibility to monitor the status of land cover and its dynamics of changes, as presented for Tuzla City. This information would be a crucial input for land/soil protection and its use in a sustainable manner. This could also serve as a basis for the evaluation of different policies at the local level such as spatial planning, development policy, agriculture/rural development policy, etc. [19].

AEZ approach results in a set of thematic and suitability maps. This assessment facilitates the definition and implementation of agrotechnical measures aimed at mitigating or eliminating certain limitations and phenomena. This approach also supports planning and economical evaluation of measures that are not technically and economically justified. Suitability maps, as presented in practice, are a very useful tool for sustainable planning and investments in agricultural production.

To have a sustainable land policy at the local level, it is necessary to introduce and develop a land information system. This information system should integrate land resources data, soil quality data, land cover/use data, bonity data, etc., which will be further used as a basis for land monitoring activities.

References

1. FAO: Planning for Sustainable Use of Land Resources: Towards a New Approach. Land and Water Bulletin 2. Rome (1995)
2. FAO: Land & Water ((2021)). http://www.fao.org/land-water/land/en/. Accessed 17 Mar 2021
3. Várallyay, G.: Soil, as a multifunctional natural resource. Columella – J. Agric. Environ. Sci. **2**(1) (2015)
4. Vogel, H.J., et al.: A systemic approach for modeling soil functions. Soil **4**, 83–92 (2018). https://soil.copernicus.org/articles/4/83/2018/. Accessed 18 Mar 2021

5. Čustović, H., Ljuša, M., Schlingloff, S.: Sustainable land management – approaches and practices in Bosnia and Herzegovina. University of Sarajevo (2020)
6. The Law on Agricultural Land in the Federation of Bosnia and Herzegovina. FBiH Official Gazette, no. 52/98
7. Agency for Statistics of Bosnia and Herzegovina-BHAS. https://bhas.gov.ba/. Accessed 10 June 2018
8. Jones, R.J.A., Houšková, B., Bullock, P., Montanarella, L. (eds.): Soil Resources of Europe, second edition. European Soil Bureau Research Report No.9, EUR 20559 EN. (Custovic H.: Section 2. Country report: An overview of general Land and Soil Water Conditions in Bosnia and Herzegovina) (2005)
9. Custovic, H., Ljusa, M.: Studija upotrebne vrijednosti zemljišta za područje Grada Tuzla. Poljoprivredno-prehrambeni fakultet Univerziteta u Sarajevu (2017):
10. Custovic, H., Ljusa, M.: Studija upotrebne vrijednosti zemljišta za područje općine Srebrenik. Poljoprivredno-prehrambeni fakultet Univerziteta u Sarajevu (2018)
11. Ljuša, M., Čustović, H., Cero, M.: Land capability study and map in function of land protection, spatial planning and agro-ecological zoning. Works Faculty Forest. Univ. Sarajevo **1**(1), 191–203 (2016)
12. Custovic, H., et al.: Priručnik za CORINE nomenklaturu IV nivo. Sarajevo (2012)
13. Biancalani, R., Predić, T., Leko, M., Bukalo, E., Ljuša, M.: Tip iskorištavanja zemljišta. FAO, Sarajevo (2004)
14. FAO: Agroecological zoning. FAO Soils Bulletin 73. Rome (1996)
15. Custovic, H.: Pedological Study of Popovo Polje. Faculty of Agriculture and Food Sciences University of Sarajevo (2007)
16. Custovic, H., et al.: Studija upotrebe zemljišta Grada Mostara. Institut za hidrotehniku Građevinskog fakulteta Sarajevo (2010)
17. Karabegovic, A., Ponjavic, M., Ferhatbegovic, E., Karabegovic, E.: Spatial data and processes integration in local governance of Bosnia and Herzegovina. In: MIPRO 41st International Convention, IEEE Conference in DE-DS - Digital Economy and Digital Society. Opatija, Croatia. May, 2018 (2018). https://doi.org/10.23919/MIPRO.2018
18. Ponjavic, M., Karabegovic, A., Custovic, H., Hivziefendic, J.: Upgrading the capability of an online biomass atlas by developing a functional extension based on spatial interaction model. Int. J. Sustain. Energy (2020). https://doi.org/10.1080/14786451.2020.1845169
19. Karabegovic, A., Ponjavic, M.: Integration and interoperability of spatial data in spatial decision support system environment. In: MIPRO IEEE Croatia Conference. Opatija, Croatia. May, 2010 (2010)

Sustainable Urban Development: Designing Smart, Inclusive and Resilient Cities

Greening in the Margins: Children's Perception on the Sustainability of Urban Gardening in Informal Settlements in San Jose del Monte City, Philippines

Aireen Grace Andal(✉)

Centre for Global Urbanism, Ural Federal University, Yekaterinburg, Russia
aandal@urfu.ru

Abstract. This paper aims to examine the views of urban poor children on the sustainability of urban gardening in mid-tier cities. The focus is on children engaged in urban micro-gardening in an informal settlement in San Jose del Monte (SJDM) City, Philippines. Informal settlements in SJDM comprise households under poverty that use vacant lands for micro-gardening as part of their general urban survival. Through remote qualitative interviews with 9–12 years old children (n = 25), this article analyzes children's opinions on the ecological benefits of their micro-gardens and the potentials they see in urban farming. Responses show contrasting sentiments. On the one hand, children perceive micro-gardens as helpful to the city's ecological sustainability and well-being. They consider urban gardens as a major household survival strategy against food insecurity. But on the other hand, these children have little hopes that urban policymakers will notice their contributions towards the city's sustainability. They do not expect significant improvement occurring in their micro-gardens and they fear to lose their micro-gardens due to threats of eviction. Such views of the children put into question the city's green governance and urban garden's capacity to address equity issues on sustainable food security in cities. This paper proposes that mid-tier cities have the potential to accommodate the vision of sustainable development through urban farming designs. However, this effort could only gain traction by having appropriate city policy acknowledgement of poor households as contributors to urban sustainability. The discussions raised in this work speak more broadly to the field of urban studies to recognize marginalized children's voices and their contributions to sustainable development.

Keywords: Urban farming · Food security · Sustainable development · Children · Mid-tier cities

N. Ademović et al. (Eds.): IAT 2021, LNNS 316, pp. 721–734, 2022.
https://doi.org/10.1007/978-3-030-90055-7_58

1 Introduction

Mid-tier[1] cities are gaining traction due to the renewed interest to diversify urbanity [1]. Generally, mid-tier cities are secondary or tertiary capitals of the country or even the lesser known business hubs, which are endorsed as alternative business and residential settlements in response to the overcrowded and expensive global cities [2]. These "non-global" cities are of particular interest to the visions of Sustainable Development Goals (SDG) because of their rich urban biotopes and wild nature for cultivating a more environment-friendly urbanization and alternative to unsustainable food production and consumerism [3]. In particular, Urban Agriculture (UA) or urban farming is promoted in non-major cities for its contribution to sustainable development in cities for diversifying urban land use, contributing to the urban ecosystem balancing the temperature and reducing urban heat islands [4, 5]. More importantly, UA has long been internationally acclaimed for its potential to improve food production in the city by decreasing transport distances and the overall level of pollutant emissions [6, 7]. Urban farming thus potentially benefits mid-tier cities in a vast array of sustainable development goals such as food justice and urban resilience [8].

In view of the urban farming potential in mid-tier cities, this paper considers how such efforts to the greening of mid-tier cities extend to address the interests and needs of children, especially those living under poverty. Children have mutual relationships with more-than-humans such as animals and plants. Yet despite the evidence that mid-tier cities are promising for urban growth, many of the mid-tier urban designs have focused on modelling these cities under global cities, presumably with the expectations of having the same economic impact in these big cities [9, 10]. This has taken its toll on children in mid-tier cities, where the young population bear the consequences of food insecurity and ecologically-extractive economies in mid-tier cities [11]. As such, this paper aims to shed light on how children in mid-tier cities themselves perceive environmental sustainability of their city. This paper then discusses a shift in framework from globalizing mid-tier cities to local urban gardening as an entry-point for mid-tier urban design toward children's well-being.

Specifically, this work examines the context of micro-gardens in an urban informal settlement in San Jose del Monte (SJDM) City, Philippines. San Jose del Monte City is experiencing a rapid population growth as observed in the increasing population from a majority of the barangays in the City, with an expected increase in the city's population by roughly 1.9 million people within the next 30 years [12]. The increasing density of children in San Jose del Monte City and its unprecedented levels of urbanization pose great challenges for SJDM City's coherent planning of sustainable development. Young dependents (0–14 years old) in this city are around 151,786. The barangay of interest is Francisco Homes-Mulawin, with about 1,979 residents. Informal settlements in SJDM comprise poor neighborhoods that use vacant lands for micro-gardening to secure food supply during the pandemic lockdown and for their general urban survival.

[1] It is to be noted that there is no single definition of 'mid-tier cities', which is usually determined by national and regional classifications. The OECD defines cities with more than 100,000 residents as 'medium-sized'. (See Dijkstra and Poelman, 2012) Dijkstra, L. and Poelman, H. (2012). Cities in Europe: the new OECD-EC definition. Regional focus 1: 1–13.

2 Literature Review

It is well established that there are disparities in food security between lower- and higher-income households [13]. In the Philippines, many low-income residents who live in informal settlements struggle to make ends meet for their families. The burdens of the high price of food weigh heavily on children [14]. To combat food insecurity, households under poverty engage in urban farming on the outskirts of cities. Aside from feeding the entire household, urban farming contributes to the broader sustainable development goals such as recycling of urban waste and improving air quality [4–6]. The interest in mid-tier cities enters the discussion because such cities are still filled with interim spaces and untouched areas for gardening compared to global cities. Previous literature points to the rich presence of green "interim spaces" and other spontaneous manifestations of nature in mid-tier urban slum neighborhoods, which residents use for urban farming [15]. In such poor urban neighborhoods, urban gardens provide children with environmental benefits through green space and a community gathering place that serves the collective empowerment of vulnerable populations [16, 17]. Community gardens are spaces where both social and environmental justice can be practiced and sustained [18, 19]. Specifically, micro-gardens help in increasing awareness of environmental stewardship and reducing anxiety over an unstable food system [20]. Micro-gardening is especially important for poor households because it can be easily managed by the entire household– children, women, men, the elderly and the disabled [21]. Aside from planting crops in vacant lots, the urban poor can plant vegetables in a container filled with soil or any substitute from natural materials, such as coconut fibre, rice husks, and peanut shells. Moreover, micro-gardens reduce ambient air temperature especially in tropical countries and help control rainfall runoff and flooding [22].

3 Methodology

Through remote unstructured qualitative interviews with children (n = 25) in poor neighborhoods, this article analyzes children's opinions on urban gardening. The respondents of this study are children from informal settling areas, wherein houses are usually makeshift from recycled materials. The houses are located in a secluded hill away from the main road where they are not visible to the general public. Using a convenience sampling, the main criteria for participation are that the children of 9–12 years old lived within the research site of informal settling areas in San Jose del Monte City, Bulacan, Philippines and that they and their parents give consent to participate in the study. With the help of a local organization serving the children's neighborhood, an online meeting interview took place. Parents of participants were given information about the purpose of the study and information about the voluntary nature of participation. The children's names are documented but not mentioned in the results in order to protect their anonymity. Children were asked about their views on the ecological benefits of their micro-gardening practices, state of their neighbourhood and the possibility of improving their micro-gardens. To maintain a comfortable environment for the children, they were interviewed in groups of three or four peers—7 triads and 1 quartet [23]. Interviews were evaluated using qualitative content analysis. The time period of investigation is during the Modified General Community Quarantine (MGCQ) which refers to "the transition phase between GCQ [General Community Quarantine] and the New Normal.... limiting movement and transportation, the regulation of operating industries, and the presence of uniformed personnel to enforce community quarantine protocols" [24]. Under the MGCQ condition, children are still highly encouraged to remain in their residences at all times. Indoor and outdoor physical activities are allowed provided that citizens follow minimum public health standards of wearing masks and keep physical distancing protocols.

Paying attention to children in slums is important to diversify the voices recognized in urbanity. The children from the urban poor community in this study experience substandard living conditions such as lack of access to running water, clean toilets, and safe housing materials against disasters. They also face the threat of evictions and displacements [25]. Yet there are opportunities for making ends meet through urban farming, in which households utilize adjacent vacant lots to plant various vegetables for their consumption. Children's exposure to the natural environment is helpful to understand what children think about their own food security and micro-gardens. It is important to consider children's views on sharing a free space which they transform with other vulnerable populations altogether. Besides, urban farming will only be sustainable if it is accepted and perceived positively by those living in close proximity, including children.

4 Results

4.1 City Context

The vision of informal urban farming in San Jose del Monte City only tangentially meets with the urban priorities of the city. While SJDM aims for an eco-friendly environment to boost its tourism, the city's focus centers on preparing to be a hub for industrial activities and transportation projects, as it was proclaimed as a "highly-urbanized city" by the country's president [26]. For instance, digitalization is a major priority of the city, being one of the new locations of *Digital Cities of 2025* declared by the Information Technology and Business Process Association of the Philippines and the Department of Information and Communications Technology (DICT) [27]. The city's mayor has signed a Memorandum of Agreement (MOA) with industrial partners for the installation of free and high-quality connectivity that are accessible in selected public areas in the city [28]. The city is also undergoing infrastructural development as part of the Manila Metro Rail Transit (MRT) Line 7 in the Philippines, a 22.8 km-long elevated rapid transit line [29]. With such priorities, slum areas are driven at the periphery of discussions. Moreover, the growing economic activities dominating the spaces of both the residential and public areas of the city mean less land allocated for agricultural activities.

4.2 Participants and Their Small-Scale Urban Farms

The households of children in the sample build micro-gardens in their neighborhood located at the periphery of San Jose del Monte City. The total land area of the slum neighborhood is nearly one hectare (see blue polygon in Fig. 1) but they only engage in small-scale informal farming. Since households only engage in small-scale urban farming, they only have smaller plots (50–101 m^2) of cultivating vegetables for home consumption. Some households find vacant lots where they share farming inputs such as seeds from already existing plants nearby their houses.

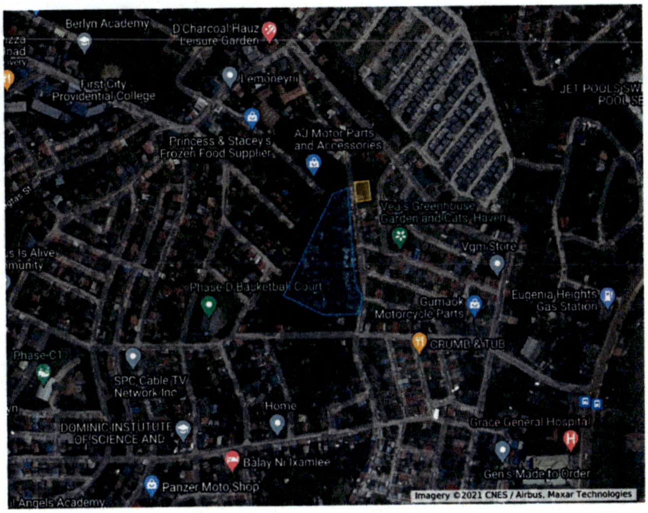

Fig. 1. Blue polygon: area coverage of Children participants' slum in San Jose del Monte City

In terms of children's characteristics, 64% were female, the average age was 11 years old. Almost all the interviewees have fairly large families, with an average household size of five persons. Participants either live with both parents or a parent and step-parent. Two children live only with their mother. Many parents are construction workers, retail workers, workshop helpers, street vendors or offer laundry or cleaning services for well-off families. Some are self-employed in the informal sector, selling candies, rags, or newspapers on the street or in markets.

They usually plant edibles such as eggplant, chilly and various green leafy vegetables. They also plant these vegetables in pots near their houses (see Fig. 2). Most of the children actively participate in urban farming alongside their parents and siblings. These children postponed going to school altogether when the coronavirus pandemic quarantine restricted social gatherings. They cannot afford devices and regular internet services for distant learning.

4.3 Children's Opinions

For children in San Jose del Monte City, their micro-gardens can play a major role in enhancing not only household but also community food security. The children admitted that before environmental concerns, they see urban farming as serving themselves first rather than the sustainability needs of the city. For these children, economically disadvantaged households struggle to get by, especially during the coronavirus pandemic.

Fig. 2. Urban gardening in poor neighborhoods, San Jose del Monte City. Photo by: Ken Abryl Eleazar Salanio and Erlinda Andal

They face challenges in meeting their nutritional needs for developing a stronger immune system. As such, they primarily consider urban gardens as a major household survival strategy against food insecurity. The children also mention positive health benefits from their micro-gardens since they can exercise and at the same time eat nutritious food. However, for them, urban farming is not a leisure activity but rather "work" that they need to do. All children recognize that they have unmet economic needs and almost all say they would like to work to help their families survive daily. Majority of the children say they combine household chores and urban farming, spending time in their micro-gardens either before or after finishing household chores. A few commented that they could not imagine having an abundance of food stocked in their house. What is more, for these children, not being hungry is good enough. Child A (10), for instance, says that: "I cannot imagine having extra food" and her family needs to have fewer wants to survive. In the same vein, Child B (11) mentions that getting by everyday is "already a blessing" and desiring for more is too much.

The children participants also conceptualize urban farming as a collective activity for adults and children alike. They talk about the relationships they create while farming together with their neighbors. Micro-gardens, for them, is a way of developing, maintaining, or strengthening their social ties among vulnerable people like them. What Child C (9) likes most about urban farming is that he can play while working. Child D (10) is upset that another child's string beans are longer and look healthier. Child E (11) says that a "good day" means "to show others some chillies and eggplants and share it with neighbors". In general, these children have similar views of their marginalized neighbourhood as a cohesive community. They see the farming efforts of their neighbourhoods as a positive action, especially those willing to help those who struggle in growing vegetables. However, it is also inevitable to have conflicts sometimes. For instance, Child F (12) says about "children who criticize your yields or call you out for neglecting your vegetables". When asked how conflict in gardening affects their relationship with each other in the neighborhood, most respondents express that they could resolve the conflict very easily and they would rather not say anything to make it a big deal. Overall, however, children's motivation to engage in urban farming can also be found in the collective spirit they experience in their farming neighborhood.

Moreover, these children perceive micro-gardens as helpful to the city's ecological sustainability and to their well-being. These children know that their micro-gardens, broadly speaking, benefit their vicinity by contributing to oxygen production. They also expressed that they help to reduce waste by recycling organic waste and waste water as part of their micro-gardens Despite many challenges such as heavy rainfall or too much heat, most children voiced their commitment toward continuing the existing micro-gardens in their neighborhood. Several interviewees mention the sense of accomplishment they feel when bringing home their harvest. Some children also say that they are impressed by learning more about agriculture through listening to adults. Child G (11) claims: "I did not care much about the environment and vegetation. But when I started helping out in the urban garden, I was so impressed by the things I learned such as discerning the health of the vegetables by just looking at colors". They also say that urban farming teaches them practical skills like plowing, crop rotation, harvesting as well as moral values like self-sufficiency and frugality. They also claim that urban farming allows them to appreciate the value of food: Child H (11) says "I know where food comes from, how much effort, time and emotions it took." There are only 2 interviewees who do not express liking urban gardening. They are also the only ones who prefer playing basketball or walking in the streets rather than urban farming.

4.4 Children's Expectations

In general, the children participants in this study have a vague view of their micro-gardens' future status. The children rarely think that authorities or policymakers such as their city mayor have their best interests in mind. Children have little hope that urban policymakers will notice their contributions towards the city's sustainability. Mostly, they believe that authorities are only concerned about "beautifying" the city, building new subdivisions and attracting new business hubs. Child I (11) says: "Those who did not live in poverty will not understand [food insecurity]". Likewise, Child J (11) exclaims: "They always have food in their houses; they also have refrigerators to store food; we grew up in different situations". In general, children in SJDM City's slums do not have high expectations that their food security is the concern of the city. They are aware that they contribute to clean air and general ecological well-being but they are on their own and their realities will never be recognized. They view their efforts to the greening of the city as a "happy accident". As Child K (10) puts it, "from the eyes of other people, what we do is illegal so they do not see our needs. In our neighborhood, we just want to eat. We just need to keep our small vegetable gardens to live". City policies, in their eyes, are restrictive and do not help them to make ends meet.

When asked about significant changes they wish to happen in their settlements and micro-gardens, these children almost unanimously say "not much". Several of them state that if any change takes place, it largely depends on volunteer groups that help them and not from the city government. These children do not feel confident that they are visible enough to be noticed by the city government, or to obtain any assistance in having proper settlements or improving their micro-gardens. They say that many are aware of their deprivation in the slums, such as issues on water and sanitation, health and shelter, and food and nutrition. Some private individuals sometimes give aid but mostly, non-government organizations help them. The city government has not been active in addressing their needs. Interestingly, several children mention that they are aware of their need for some help but they refuse to claim that they are helpless. They believe that they were just born as poor so it is more difficult for them to fare with those born out of initial endowments. The way to go forward, for these children, is just getting by with low expectations from both themselves and the city authorities to avoid disappointments from high hopes.

Children also express their fear of losing their micro-gardens due to threats of eviction. Children participants face the threat not only of the reduction of spaces for vegetation but also, and more importantly, the loss of their settlement due to being converted into building land. They are aware that they might be driven off their occupied lands by urban dwellers or figures of authority. The children know that their micro-gardens are tolerated by the public authorities because there are no construction interests where they live so far. However, they anticipate that their micro-gardens can disappear in favour of the construction projects necessary for the installation of new houses. If they are not evicted, children still fear that their settlements might be sites of industrial waste from constructions nearby, affecting the quality of soil for their food production. The children know that they are less powerful than the well-off urban dwellers and that they are in positions of dependence. As such, they measure their value by how well they manage not to be hungry. They value their micro-garden because they develop relationships and sense of dignity and pride out of what they do.

Moreover, these children receive unwelcome comments from urban dwellers. Child L (10) mentions: "some old ladies shouted at us for using a space we do not own to plant our food; they told us we are taking advantage of a space we do not deserve. They yelled that we should not stay in our city". Likewise, Child M (9) reembers: "a man told me on the street once that we are 'illegal' then threatened me that he would call the police if I did not stop". In other words, children experience threats and fear because of using a portion of land they do not own. For instance, the children mention that they are financially inferior to real estate developers and thus unable to compete against evictions. Indeed, they share countless stories of being called out or humiliated for their micro-gardens. While no one mentioned about being in danger while attending to their micro-gardens, some did mention getting their yields damaged because other urban dwellers let their dogs step on the micro-gardens. Finally, these children are also doubtful that urban farming can help feed urban poor populations because they can only grow small quantities.

5 Discussion

5.1 Implications

The interview from children in SJDM City's slums suggest that the context of food subsistence in slum-led urban agriculture comes from a place of lack of access to food and not from a higher goal of environmental preservation or connecting with nature [30]. The issue of food security has been a long-standing concern for many developing countries including the Philippines. To survive the urban life, households from informal settling areas engage in urban agriculture, which is aligned with the principles of sustainable development goals. In contrast to the search for an alternative to globalised food markets and consumerist lifestyles [31], urban farming in the slums of San Jose del Monte City takes place in the context of food deprivation and economic hardship. The poor urban neighborhood produces their own food by their own efforts because they cannot afford food. Moreover, since SJDM City is experiencing a high influx of newcomers, slum dwellers turn to have an antagonistic relationship with commercial real estate investors who are interested in residential construction projects, which ultimately threaten micro-gardens [32, 33].

The views of the children in this study implies challenges to San Jose del Monte City's preparedness for green governance and to urban garden's capacity to address equity issues on sustainable food security in cities. The wider implication of urban farming in slums is that urban agriculture's benefits may not apply in equal measure to all urban farmers. While Urban Agriculture is promoted in cities, this does not necessarily mean an inclusive vision of sustainable development in mid-tier cities [34]. In SJDM City, there is a strong rhetoric of ownership/non-ownership binary in urban spaces, which has implications on the agency of children who feel threatened as they occupy a portion of vacant land to feed themselves. Hence, while these children believe in the promise of environment-friendly ways for food production, they do not consider themselves deserving enough to be relevant agents in pursuing ecological feats for a collective cause. These children do not see themselves as powerful participants or contributors for sustainable development. However, the children interviewed in this paper can still offer insights and counter-narratives to such deprivations. For instance, they do not judge actions and practices as ecologically beneficial or harmful based on ownership/non-ownership. Rather, it depends on the capacity of each urban dweller to be self-accountable for the environment.

5.2 Key Areas for Child-Friendly Urban Sustainable Development

Mid-tier cities such as San Jose del Monte have a potential to accommodate the vision of sustainable development through urban farming. However, this potential will remain only as a potential if cities fail to recognize the efforts of poor neighborhoods in micro-gardening. In order for San Jose del Monte City to make a real contribution to sustainable development, it is important for urban planning to consider the issues children raised in the interview. This includes the following notable insights for policy action and further studies:

First, it is important for sustainable development efforts in cities to align urban resources according to food security and nutrition needs of children, especially those with disadvantaged socio-economic status. The children interviewed in this study grasp urban farming as a pragmatic and consequential part of their everyday survival. As such, urban planning is encouraged to focus on initiating child-friendly urban gardening programs—be it horticulture, greensites, or agriculture urbanism. For example, child-focused urban agriculture educational programs can enrich the child-environment relationship through extending these programs to green spaces [35, 36]. Children can be provided a child-friendly system of streets, greenways, and spaces for growing their food. Urban planning can also secure green infrastructure systems that reinforce children to engage in urban farming such as provision of child-appropriate weight for farming materials and cheap inputs from the provincial or municipal agriculturist office (P/MAO).

Second, there is a need for longitudinal research on success in urban farming with specific focus on developments in children's views and participation. Participatory appraisals in urban farming are instructive to examine the sustainability of micro-gardens in informal settlements in a life-cycle context. Moreover, participatory research acknowledges children's special capacity to discern and express the needs of their fellow disadvantaged peers. As Aitken puts it, "children see things in environments that we may have forgotten to see, let alone understand" [37]. Children develop ecological competence growing up and unique spatial experience as they go around the urban spaces [38–40]. For instance, children can participate in collaborative community-led planning activities to demonstrate the importance of their unique ecosystem for biodiversity conservation and physical protection of the environment [41, 42].

Third, it is crucial that policy research create or design frameworks that are sensitive to children's diverse backgrounds in urban slums. For instance, children might have mobility needs that prevent them from engaging in urban farming [43]. This potentially neglects children who are under poverty, which potentially leads to excluding them from participation in sustainable development goals of the city. Much work is to be done on how children's experiences in mid-tier city slums and how their lives are shaped by their background such as religion, race, gender, and/or ethnicity. Urban policies, budgets and programmes need to be sensitive to and proactively look for these disparities in order to ensure inclusivity of the city and to address "recognition gaps" or the deprivations felt about one's own worth [44]. Such information on cross-cutting deprivations can assist city government, urban planners and development agencies to prioritize and fulfil the needs of the most deprived and vulnerable children not only in urban farming but also in terms of their general well-being. Finally, this also includes research that examines children's view regarding their adjustments on opportunities from urban farming such as increase in socio-economic status or incentives for alternative activities mitigating living standards.

6 Conclusion

Urban farming has allowed the most deprived populations to eat well as they face the ever-increasing cost of living in a growing city. It has also facilitated the survival of migrants from their rural exodus. However, being able to get by through urban farming is insufficient in view of having a more inclusive city for vulnerable populations. Interviews from this study reveal that building sustainable cities also runs the risk of neglecting vulnerable populations, in particular poor children. The pursuit of sustainable development in cities is futile without addressing the needs of vulnerable populations. As such, this paper proposes a shift in framework from globalizing mid-tier cities to an inclusive local urban gardening as an entry-point for mid-tier urban design. Micro-gardening as a point of departure for urban design of mid-tier cities offers a pragmatic and future-oriented response for the irreversible climate consequences global cities have done. A departure from "urban competition" [45] to a focus on the sustainability of vulnerable populations does not only function to represent diversity between cities but also to bolster the generation of "new geographies of theory" [46].

In moving forward, a child-oriented approach to urban sustainability offers an insightful perspective because "children are a kind of indicator species. If we can build a successful city for children, we will have a successful city for all people" [47]. A continued commitment to monitor children's conditions in the city's marginal terrains can provide a nuanced and multifaceted view of what the term "sustainability" really means. Food security and ecological sustainability through urban micro-gardens are beneficial to children not only for food security but also for their social well-being and sense of dignity. This effort could be supported by appropriate city policy incentives for poor households that contribute to the greening of the city.

This research hopes to contribute to the literature on sustainable development and food governance by raising some urban issues from the children's perspective toward achieving social justice goals in urban agriculture. On a much broader scale, the findings of this work offer several insights to food and agriculture policy making endeavors. Such policies have the potential to influence emerging trends in urban neighborhoods and communities, stimulate the growth of child-friendly urban farming and further assess the ecological implications of urban agriculture. Such results also provide directions into how urban farming is perceived by the young population, thereby promoting a future-oriented urban agriculture. Finally, the discussions raised in this work speak more broadly to the field of urban studies to acknowledge marginalized children's voices and their contributions to sustainable development.

References

1. Cardoso, R., Meijers, E.: Secondary yet metropolitan? The challenges of metropolitan integration for second-tier cities. Plan. Theory Pract. **18**(4), 616–635 (2017). https://doi.org/10.1080/14649357.2017.1371789
2. Song, S., Gambrill, J., Wood, W.: The rise of second-tier cities. Gensler. https://www.gensler.com/research-insight/blog/the-rise-of-second-tier-cities. Accessed 10 Jan 2021
3. Gaast, K., Leeuwen, E., Wertheim-Heck, S.: City-region food systems and second tier cities: from garden cities to garden regions. Sustainability **12**(6), 2532 (2020). https://doi.org/10.3390/su12062532
4. Programme, U.N.D. (ed.): Urban Agriculture. Food, Jobs and Sustainable Cities. United Nations, New York (1996)
5. Pearson, L.J., Pearson, L., Pearson, C.J.: Sustainable urban agriculture: stocktake and opportunities. Int. J. Agric. Sustain. **8**, 7–19 (2010). https://doi.org/10.3763/ijas.2009.0468
6. Santo, R., Palmer, A., Kim, B.: Vacant lots to vibrant plots. A Review of the Benefits and Limitations of Urban Agriculture. Johns Hopkins Center for a Livable Future (2016)

7. Artmann, M., Sartison, K.: The role of urban agriculture as a nature-based solution: a review for developing a systemic assessment framework. Sustainability **10**(6), 1937 (2018). https://doi.org/10.3390/su10061937
8. Horst, M., McClintock, N., Hoey, L.: The intersection of planning, urban agriculture, and food justice: a review of the literature. J. Am. Plann. Assoc. **83**(3), 277–295 (2017). https://doi.org/10.1080/01944363.2017.1322914
9. Robinson, J.: Ordinary Cities: Between Modernity and Development. Routledge, London (2006)
10. Roy, A.: Slumdog cities: rethinking subaltern cities. Int. J. Urban Reg. Res. **35**, 223–238 (2011)
11. UNICEF and UN Habitat: Analysis of Multiple Deprivations in Secondary Cities in Sub-Saharan Africa. https://www.unicef.org/esa/reports/analysis-multiple-deprivations-secondary-cities-sub-saharan-africa. Accessed 15 Feb 2021
12. Republic of the Philippines: City of San Jose del Monte. https://csjdm.gov.ph/socioeconomic-and-physical-profile/demography/. Accessed 10 Jan 2021
13. Asian Development Bank: Macro analysis of covid-19 threats to food security and livelihoods in Asia and the pacific. World Food Programme. https://reliefweb.int/sites/reliefweb.int/files/resources/WFP-0000118195.pdf. Accessed: 05 Feb 2021
14. Briones, R., Antonio, E., Habito, C., Porio, E., Songco, D.: Food security and nutrition in the Philippines. World Food Programme. https://docs.wfp.org/api/documents/WFP-0000015508/download/. Accessed 20 Feb 2021
15. Gottlieb, R., Fisher, A.: Community food security and environmental justice: searching for a common discourse. Agric. Hum. Values **13**(3), 23–32 (1996). https://doi.org/10.1007/BF01538224
16. Levkoe, C.Z.: Learning democracy through food justice movements. Agric. Hum. Values **23**, 89–98 (2006). https://doi.org/10.1007/s10460-005-5871-5
17. Smit, J., Bailkey, M.: Urban agriculture and the building of communities. In: van Veenhuizen, R. (ed.) Cities Farming for the Future. Urban Agriculture for Sustainable Cities, pp. 145–170. RUAF Foundation, IDRC and IIRR (2006)
18. Colding, J., Barthel, S.: The potential of 'urban green commons' in the resilience building of cities. Ecol. Econ. **86**, 156–166 (2013). https://doi.org/10.1016/j.ecolecon.2012.10.016
19. Schmelzkopf, K.: Urban community gardens as contested space. Geogr. Rev. **85**, 364–381 (1995). https://doi.org/10.2307/215279
20. Appolloni, E., et al.: 'From microgarden technologies to vertical farms: innovative growing solutions for multifunctional urban agriculture. Acta Horticult. (2020). https://doi.org/10.17660/ActaHortic.2020.1298.10
21. Food and Agriculture Organization Micro Gardens: With micro-gardens, urban poor "grow their own."' http://www.fao.org/ag/agp/greenercities/en/microgardens/index.html. Accessed 5 Feb 2021
22. Peters, C.J., Bills, N.L., Wilkins, J.L., Fick, G.W.: Foodshed analysis and its relevance to sustainability. Renewable Agric. Food Syst. **24**(1), 1–7 (2008)
23. Corsaro, W.: The Sociology of Childhood. SAGE, London (2015)
24. Republic of the Philippines, Inter-agency Task Force for the Management of Emerging Infectious Diseases: Omnibus Guidelines on the Implementation of Community Quarantine in the Philippines (with Amendments as of October 08, 2020) (2020). https://www.officialgazette.gov.ph/downloads/2020/10oct/20201008-IATF-Omnibus-Guidelines-RRD.pdf. Accessed 20 Jan 2021
25. Elsey, H., et al.: Public health risks in urban slums: findings of the qualitative "healthy kitchens healthy cities" study in Kathmandu, Nepal. PLoS ONE **11**(9), e0163798 (2016). https://doi.org/10.1371/journal.pone.0163798

26. Ramos, C.M.: Duterte proclaims San Jose del Monte as a highly-urbanized city – Bulacan solon. Philippine Daily Inquirer 6 December (2020) https://newsinfo.inquirer.net/1368691/duterte-proclaims-san-jose-del-monte-as-a-highly-urbanized-city-bulacan-solon
27. Campos, O.: BPO group names 25 digital cities in new normal. Manila Standard 11 July (2020). https://manilastandard.net/mobile/article/328349
28. Globe Philippines: San Jose Del Monte Becomes Newest GoWiFi City in Bulacan' Globe 21 February (2020). https://www.globe.com.ph/about-us/newsroom/corporate/san-jose-del-monte-bulacan-newest-gowifi-city.html#gref
29. Railway Technology: Manila Metro Rail Transit (MRT) Line 7 (n.d.). https://www.railway-technology.com/projects/manila-metro-rail-transit-mrt-line-7/
30. van der Jagt, A.P.N., et al.: Cultivating nature-based solutions: the governance of communal urban gardens in the European Union. Environ. Res. **159**, 264–275 (2017). https://doi.org/10.1016/j.envres.2017.08.013
31. Calvet-Mir, L., March, H., Nordt, H., Pourias, J., Čakovská, B.: Motivations behind urban gardening – 'Here I feel alive.' In: Bell, S., et al. (eds.) Urban Allotment Gardens in Europe, pp. 320–341. Earthscan from Routledge, New York (2016)
32. Mougeot, L.J.: Urban agriculture and the millennium development goals. In: Mougeot, L.J. (ed.) AGROPOLIS. The Social, Political and Environmental Dimensions of Urban Agriculture. Earthscan, London (2005)
33. McClintock, C., Miewald, E., McCann, N.: The politics of urban agriculture: sustainability, governance, and contestation. In: Jonas, A.E., Miller, B., Ward, K., Wilson, D. (eds.) The Routledge Handbook on Spaces of Urban Politics, pp. 1–15. Routledge, London (2016)
34. Purcell, M., Tyman, S.K.: Cultivating food as a right to the city. Local Environ. **20**(10), 1132–1147 (2015). https://doi.org/10.1080/13549839.2014.903236
35. De Zeeuw, H.: Cities, climate change and urban agriculture. Urban Agric. Mag. **25**, 39–42 (2011)
36. Roggema, R. (ed.): Sustainable Urban Agriculture and Food Planning. Routledge, New York (2016)
37. Aitken, S.: Geographies of Young People: The Morally Contested Spaces of Identity. Routledge, London (2001)
38. Chawla, L., Heft, H.: Children's competence and the ecology of communities: a functional approach to the evaluation of participation. J. Environ. Psychol. **22**(1–2), 201–216 (2002)
39. Karsten, L.: It all used to be better? Different generations on continuity and change in urban children's daily use of space. Child Geogr. **3**(3), 275–290 (2005). https://doi.org/10.1080/14733280500352912
40. Freeman, C., Lingam, G.I., Burnett, G.: Children's changing urban lives: a comparative New Zealand-Pacific perspective. J. Urban Des. **20**(4), 507–525 (2015)
41. Urban Gardens Program: Child Friendly School Gardens: A Handbook for OVC Care in School Garden Settings. DA, Bethesda (2012a)
42. Urban Gardens Program: Discussion Group Participants Handbook, 1st edn. DAI, Bethesda (2012)
43. Schwartz, N., Buliung, R., Wilson, K.: Disability and food access and insecurity: a scoping review of the literature. Health Place **57**, 107–121 (2019). https://doi.org/10.1016/j.healthplace.2019.03.011
44. Lamont, M.: Addressing recognition gaps: Destigmatization and the reduction of inequality. Am. Sociol. Rev. 83, 419–444 (2018). https://doi.org/10.1177/0003122418773775
45. Smith, R.: The ordinary city trap. Environ. Plan A **45**, 2290–2304 (2013)

46. Roy, A.: The 21st-century metropolis: new geographies of theory. Reg. Stud. **43**, 819–883 (2009)
47. Peñalosa, E., Ives, S.: The politics of happiness: land and people. Albuquerque Model City Council Curriculum (2004). https://www.commoncause.org/new-mexico/wp-content/uploads/sites/19/2018/07/abq-model-city-council-lesson-2-penalosa.pdf. Accessed 10 Feb 2021

Designing the Future Residential Buildings with Low Environmental Impact - Case Study Buildings in Bosnia and Herzegovina

Renata Androšević[1,2](✉) and Damir Androšević[3]

[1] GSA Laboratory – Geometry-Structure-Architecture, National Graduate School of Architecture Paris-Malaquais, Paris-Est University, 14 Rue Bonaparte, 75006 Paris, France

[2] Faculty of Architecture, University of Sarajevo, Patriotske lige 30, 71000 Sarajevo, Bosnia and Herzegovina

[3] Doctoral School SIMPPE, Laboratory ERPI, Université de Lorraine, 8 Rue Bastien Lepage BP647, 54010 Nancy Cedex, France

Abstract. Quick development in technology in the last decade affected many industries and should be the main force that will help the construction industry to develop the methods and processes that will decrease the environmental impact of the built structures. Computational technology could be helpful in precise early building design and also in the decision making process. Predicting possible building scenarios in the first stage of design has a lot of benefits, such as finding better solutions for the environment and creating a sustainable green building for the future. In any case, we must agree on the right protocol and methods to achieve that goal. Residential buildings are the most important, as these buildings are the most numerous buildings in every country. The proposed model of a sustainable building is the model where the building and its parts have low environmental impact. The representative residential buildings in Bosnia and Herzegovina will be the case study to develop some conclusions and recommendations for future buildings. These buildings and their parts will be analyzed from the environmental impact /construction waste standpoint. Existing buildings will be analyzed and conclusions will contribute to properly make the next generation of buildings, which will be environmentally friendly for another century and future generations. Further research to find the method and protocols to lower the construction in the design phase is developed, based on this first analysis.

Keywords: Sustainable and circular building · Environmental impact · Construction waste · Building life-cycle · Building technology · Future building design

1 Introduction

Construction industry has a huge role in environmental impact, especially in terms of the construction waste. It produces more than a third of a total waste in the world [1]. This article will try to explain the principles and suggest methods to lower the construction

N. Ademović et al. (Eds.): IAT 2021, LNNS 316, pp. 735–743, 2022.
https://doi.org/10.1007/978-3-030-90055-7_59

waste that residential buildings make during and after its first lifecycle. The conclusions will be drawn after analyzing residential buildings in Bosnia and Herzegovina, as these buildings are the dominant type of buildings.

Even in XXI century, residential buildings in BiH are still made using the same technology and almost the same materials as in the XX century. This is similar in other countries, as the construction industry continues to use the same construction technology and almost the same materials for this type of buildings. Building industry stakeholders are testing new and different technologies and materials mostly on office and other public buildings, but residential buildings mainly stay the same. There are individual examples of the residential buildings made using new construction innovation, but just as isolated case study. Only the shaping design of residential buildings has some variations. Traditional way of building is not the solution for future buildings, because the problem of the building waste is increasing. Many hazardous materials are not in use anymore, but there is still a really big problem with construction demolition waste after a building finishes its lifecycle. There are many strategies being developed and recommended by the European Commission in order to solve the problem of waste [2], but in the meantime, new buildings are being built in the same way. The European Commission prefers avoiding waste, and then comes other type of recovery [3]. There must be another way to design and build to avoid this kind of waste at the very beginning of the project. Some suggestions will be presented below.

2 Residential Buildings in Bosnia and Herzegovina

The periodization of typology of residential buildings in Bosnia and Herzegovina was defined in each historical period by Arnautovic-Aksic [4]. The buildings built in the period from 1992 to 2014 are in focus for this research. These buildings are going to last and will be used by present and future generations. Residential buildings in Bosnia and Hercegovina at the beginning of this century were made with the same technology and almost the same materials as in the XX century, in accordance with valid regulations at the time of the building [4]. The situation is similar in other developing countries where sustainability and energy efficiency protocols in force don't follow the latest discoveries in this area. According to the energy efficiency requirements, situation in the building industry is developing at the beginning of XXI century due to increase of awareness about importance of construction of energy efficient buildings [4].

Another problem occurs when residential buildings change their function and should be transformed or, in most cases destroyed and make room for other buildings with different function.

In Netherlands, for example, many buildings are changing its function every 5 to maximum 20 years and if a building is not designed for transformation, it is going to be destroyed and create construction waste. A new building is constructed in the same place, to satisfy and fulfil the new purpose. If this new function is not predicted, a new building will have the same destiny as the previous one.

In BiH settlements it is not a widespread practice to change the function of a residential building and these buildings keep their purpose for almost a century or more. The ground floor is usually for public use as a store, market or office and interior changes are happening very often, but the upper residential floors usually keep their purpose. The same is true for the individual dwellings and collective housing.

Most individual houses have garage/storage or store at the ground floor, family living unit at the first floor (or ground floor) and one or two additional floors corresponding to the number of the children in the family. Each child (male child mostly) should get one whole floor with a flat to accommodate his own future family. These houses are built mostly with reinforced concrete structure and bricks or prefabricated concrete blocks between structural elements. Investor's/owner's presumption is to build a house which will last for the next generation or maybe two generations. This idea has roots from the period when only rich families could have a house which could accommodate 2 or 3 generations with luxury of space. Each generation had to have its own flat. But the situation has changed. Children don't stay at the same house with their parents, and they have their own flat or a house with their new family. This became a practice during and after the war in BiH. Many young people have gone abroad and settled in another country. Some of them have found a job and stayed in another town after the war. Also, new generation has a different concept of living, which presumes life away from parents. These big houses are staying half-empty and in less than 50 years they will be construction waste. But, the mindset of a big house for two or three generations remains and these houses are still built in the same way even in XXI century.

It is really difficult to change the mindset of people and it is a long process and cannot be solved by persuading people to make smaller houses. We can try to educate people about construction waste and environmental impact, but a better solution would be to think about technology possibilities which could lower the waste no matter what scenario will happen in the future.

Instead of preparing the strategy for lower construction demolition waste after the first life-cycle, stakeholders can prevent the waste at the very beginning, while proposing the design and technology solution which will be acceptable for investors and for the environment.

3 Individual Residential Buildings' Problems and Possible Solutions

When we are considering high sustainability for building practice, it is clear that the building industry should follow these main recommendations to lower the environmental impact of built structures:

- lower landscape, water and nature pollution;
- lower land use, in terms of quantity of the construction waste;
- use only construction materials that are not hazardous;
- decrease global warming and
- establish high energy efficiency standards.

As previously mentioned, residential buildings in BiH at the beginning of XXI century are built with less hazardous materials (considering enhanced regulations) which means lower pollution. Higher energy efficiency standards are satisfied in some cases, as the investors, who are building these residential buildings, started installing thermal insulation of greater thickness than required [4]. Some reusable energy sources are considered for heating these buildings, but at a large scale. This segment has to be endorsed more by government and non-government projects.

When we speak about social impact, in terms of health, users' comfort, safety and participation, there are some important things to mention. As previously mentioned, individual houses are imagined and made to be big and comfortable for the next generation. There is the root of the problem. The house is usually too big and has one or even two additional floors which are not in use. This space is foreseen for the new family that the child should establish in 15 or 20 years. Building a structure with additional floors requires more financial assets and the house remains unfinished in many cases, usually without final layers of the façade. The thermal insulation and final layer of plaster is missing for years. In a typical overview of 10 houses in the neighborhood, 3 to 4 houses remain unfinished on the outside, because of the lack of funds. The photo below is showing unfinished houses without façades, which have had the same look for last 20 years and will remain unfinished and have the same look for another decade at least (Fig. 1). Comfort of the current user is not satisfied in a house like this, as energy efficiency is critically low. House is cold in the winter, even if the energy use is high and it is also hot in the summer, due to lack of thermal insulation.

Fig. 1. Photo that shows typical look of the houses in a Sarajevo suburb. (photo by author)

There is also economic impact, where, even if enormous funds are used for heating in the winter, the comfort of user is still low. Usually, user doesn't have enough money, and thus the entire family stays in one room which has heating. In rare cases, one whole floor is heated. This is visible in the night, where a three floor house has only a few windows with light. This means that only one room is warm and whole family stays there. Houses like this could be called "one-eyed house" and, unfortunately, it is typical example of a house in BiH.

There must be a way to avoid these mistakes in the future. An investor/user of a house can be educated and informed about possible scenarios, but it is really difficult to change the mind of the people. Customs endure for decades and centuries, and the idea of a big house will still be in many minds. This building practice also means that the unfinished/unprotected building structure will last shorter and will become construction waste sooner.

If mind changing is a long-term process, or even impracticable solution, then there must be a technological solution. Maybe economic view of costs (capital and maintenance), preserved value of investment, efficiency of use could be considered and demonstrated.

In any case, technology of the residential buildings has to be adapted according to new discoveries in the area of sustainability. Sustainable buildings should allow flexibility, so that they can satisfy any scenario that happens. Many researchers compared energy and structural concepts of renewable (wood) and non-renewable materials (masonry, concrete, steel) in the context of low-energy dwellings and assign better results to renewable materials [5–7]. Some researchers are pointing out the significance of "design for deconstruction" in building industry [8]. Guy et al. [8] found that wood is the preferable material when building transformation takes place. He also concluded that the best results are accomplished when using interstitial connecting devices. All these researchers can help us to find a better solution for buildings in the future.

Even though wood is available and affordable material in BiH, the stakeholders usually don't recommend it or use it for building a house. The structure of residential buildings in BiH is built with reinforced concrete and external walls are built with bricks with external layer of thermal insulation and plaster (if a house is finished). Layer of thermal insulation (if it exists) is usually thick enough to satisfy the existing energy efficiency standards, but materials used for this purpose are not really environmentally friendly (polystyrene is used in most cases).

A solid concrete house is not suitable for transformation or any change and when inhabited, it becomes a construction waste very soon. Unfortunately, investors usually don't consider renewable materials for building their dream houses for the next generations, but there has to be some compromise.

One solution which comes to mind is a hybrid structure concept, which consists of:

- the ground floor (made traditionally with reinforced concrete),
- the next one or two floors (made with prefabricated wooden systems designed for easy assembly and disassembly).

Schematic visual of the proposed hybrid house structure where ground floor is made traditionally (without possibility of transformation) with solid reinforced concrete structure and bricks and first and second floor are made with wooden prefabricated structure systems and with the wooden prefabricated roof on top (Fig. 2).

In this case, the possibility of disassembling the upper floors is an option if a grown-up child doesn't want to live in the same house with his parents. The upper wooden floors keep their value and can be dismantled and sold separately.

This concept can be acceptable in terms of many aspects:

- Mind-set of the investor will be satisfied,
- More affordable solution,
- More sustainable design,

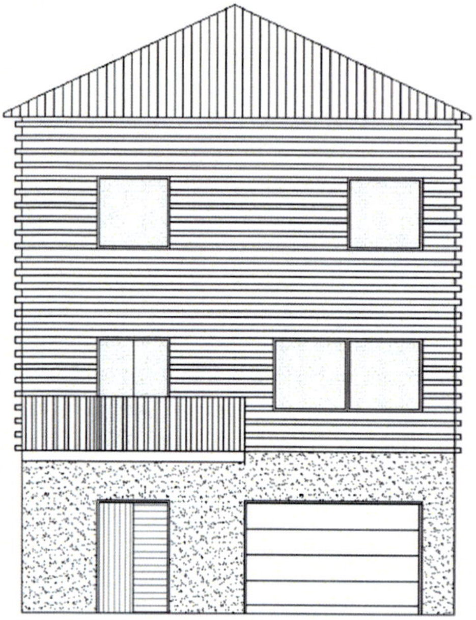

Fig. 2. Schematic visual of the proposed hybrid house structure

- Quicker assembly and move in,
- Investment value can be returned…

4 Residential Building External Walls in Bosnia and Herzegovina

Considering the longevity of the residential buildings and their parts, enclosure system life expectance is approximately 20 years [9]. The external building skin [10] is exposed to different weather conditions and human/animal impact. These final layers of a building (roof and walls) are often changed, repaired, adapted and upgraded more frequently, in a shorter time period. The wall is made of a lot of different materials which are connected in a way that it is very hard to divide them. In any case of reparation, transformation or upgrade, the existing materials become construction waste. It is very difficult and, in most cases economically unjustified, to perform the process of recycling. Recycling in this case becomes down-cycling. This requires different approach to the design of the external wall. While struggling with the construction demolition waste, stakeholders should find new ways to avoid it when designing new buildings, in this case especially external walls, which require special attention. The possible scenarios should be predicted in the early phase of the building design [11]. The building practice in BiH continues to use the same building technology in this century, so we must react very soon to implement new technologies, which include a new way of sustainable thinking, especially for the building parts which have frequent changes.

When we are speaking about energy efficiency, there are regulations in each country. Even though the regulations vary in their form and expected values from one country to another, the principle is clear - the energy losses and energy use should be minimal, and the energy produced from renewable energy source is preferred. Many studies show, and it is known by now, that the biggest energy losses are through the walls and they amount to approximately 35% (Fig. 3).

External walls have the main role to preserve the energy and keep internal comfort for the user.

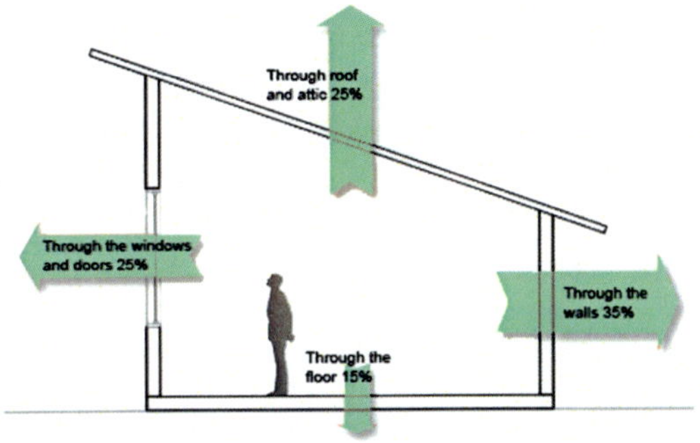

Fig. 3. Schematic visual of the average energy losses considering different parts of the house [12].

There are 4 main types of existing representative external walls of residential buildings made in BiH at the end of XX and beginning of the XXI century, according to study made in 2016 [4]. All representative types are suggested for energy efficiency improvement with additional 10 or 20 cm thermal insulation layer with 1 cm façade plaster. There are some projects that cover part of the costs to upgrade some collective houses' envelopes. But single-family and terraced houses which account for as much as 97.36% of the entire stock in BiH [4], are still losing energy for heating during winter and using a lot of energy for cooling the house during hot summers.

Eventually, this project showed investors that they can upgrade the existing envelope and some new investors are putting additional thermal insulation on the new buildings [4]. This solves energy consumption, but we should consider how long these solutions will last and what kind of materials should be used to minimize total environmental impact. Usually, these residential buildings in BiH will last at least next 50 years in the state they are now, with occasional envelope surface reparation. The mentioned study for energy efficiency has to be continued and upgraded with the research of environmental impact caused by different choices of insulation materials. Also, the next study has to be about the predictions of the frequencies of the future envelope reparations.

Among the above types, there is no wooden wall type, which should be the preferred one. Wood is available and wood as material is sustainable (obtained from nature, requires minimal energy to process, stores Co_2 in itself), such houses have quick installation time and time to move-in, construction is possible at any time of the year, and wooden structure is behaves well in fire and earthquake [13]. The wood is a natural and environmentally friendly material, which can last very long with proper maintenance. Wood has excellent thermal characteristics and provides natural thermoregulation without adding another material. It is flexible and has great bearing capacity but lower specific weight than many other building materials. There were 3 types of traditional wooden walls that satisfied the energy efficiency [14], but all these walls were forgotten in development of the building technology.

Some producers saw the benefit of the wood and started production of wooden walls for prefabricated energy efficiency houses. Unfortunately, these houses are mostly made for export, as other European countries are more interested than investors in BiH. The only flaw that these prefabricated houses have, is that they are not easy to disassemble and cannot be transformed once made. The study of reuse potential of this type of wall has been done and it showed very low results [15]. This problem can be solved using appropriate connections between materials and planning ahead for material connections in order to make future disassembly possible [15].

5 Conclusion and Recommendations

Following the positive examples in Europe and around the world, the technology of the residential buildings should be changed in developing countries as well.

For a Bosnian house, the promising solution is a “hybrid house”, as the concept of the big house which accommodates next generations is embedded in the minds of investors. This means that the building technology should be adapted so that, at least, the upper half of an individual house should be transformable.

The producers of the energy efficient wooden houses should also consider easy disassembly and fully apply the principles of circular building construction in XXI century. Wooden houses ensure healthy living, with many benefits and low environmental impact. For this research, the important feature is that a wooden house can be built in order to be disassembled and assembled again, which means reuse and longevity. If this hybrid concept becomes “real time action”, then wooden parts won’t be losing their value during time and will not become the waste after the first life cycle of a house.

References

1. Revoltella, D., Kolev, A., et al.: Investment Report 2017/2018 From Recovery to Sustainable Growth. Published by the European Investment Bank. https://www.eib.org/attachments/efs/economic_investment_report_2017_en.pdf. Accessed: 25 Feb 2021
2. European Commission: EU Construction and Demolition Waste Protocol and Guidelines (2018). http://ec.europa.eu/growth/content/eu-construction-and-demolition-waste-protocol-0_en. Accessed 25 Feb 2021
3. European Commission: Final Implementation Report for Directive 2008/98/EC on Waste (2018). https://ec.europa.eu/environment/archives/waste/reporting/pdf/Final_Implementation_Report_2013_2015_WFD.pdf. Accessed 15 Feb 2021
4. Arnautović-Aksić, D., et al.: Typology of Residential Buildings in Bosnia and Herzegovina. GIZ, Sarajevo (2016), ISBN:978-9958-691-50-8
5. Cole, R.J., Kernan, P.C.: Life-cycle energy use in office buildings. Build. Environ. **31**(4), 307–317 (1996). https://doi.org/10.1016/0360-1323(96)00017-0
6. Gerilla, G.P., Teknomo, K., Hokao, K.: An environmental assessment of wood and steel reinforced concrete housing construction. Build. Environ. **42**(7), 2778–2784 (2007). https://doi.org/10.1016/j.buildenv.2006.07.021

7. Mithraratne, N., Vale, B.: Life cycle analysis model for New Zealand houses. Build. Environ. **39**(4), 483–492 (2004). https://doi.org/10.1016/j.buildenv.2003.09.008
8. Guy, B., Shell, S., Sherick, H.: Design for deconstruction and materials reuse. Proc. CIB Task Group **39**(4), 189–209. http://citeseerx.ist.psu.edu/viewdoc/download?doi=10.1.1.482.8512&rep=rep1&type=pdf. Accessed 27 Feb 2021
9. Durmisevic, E.: Transformable Building Structures: Design for Disassembly as a Way to Introduce Sustainable Engineering to Building Design & Construction. Delft University of Technology, Delft (2006)
10. Brand, S.: How Buildings Learn: What Happens After They're Built, Viking, New York (1994)
11. Androsevic, R., Brocato, M., Bijedic, Dz., Rustempasic, N.: Life cycle assessment as a tool for low waste building systems, case study building external wall. IOP Conf. Ser. Mater. Sci. Eng. **603**, 042067 (2019). https://doi.org/10.1088/1757-899X/603/4/042067
12. Souza, E.: Learn How to Avoid Energy Loss in Your Buildings (2019). https://www.archdaily.com/915546/learn-how-to-avoid-the-main-sources-of-energy-loss-in-your-home?ad_medium=gallery. Accessed 27 Feb 2021
13. Husetić, A., Beganović, E.: Prednosti izgradnje drvenih porodičnih kuća u BiH u odnosu na klasičnu gradnju. In: 11th International Scientific Conference on Production Engineering Development and Modernization of Production, Sarajevo, 4–7 Octobre http://rim.tfb.ba/common/ZBORNIK2017.pdf. Accessed 27 Feb 2021
14. Hadrović, A.: Arhitektonsko kulturno-historijsko nasljeđe i energetska efikasnost. Faculty of Architecture Sarajevo, Sarajevo (2017)
15. Androsevic, R., Durmisevic, E., Brocato, M.: Measuring reuse potential and waste creation of wooden facades. In: SBE19 Brussels BAMB-CIRCPATH, IOP Conference Series Earth Environmental Science, vol. 225, p. 012017. https://doi.org/10.1088/1755-1315/225/1/012017

Sustainable Strategies in Vernacular Architecture Practices in Sarajevo: The Architectural Ensemble of Alija Djerzelez House as a Case Study

Maida Halilović[1](✉) and Adna Berković[2]

[1] Faculty of Engineering and Natural Sciences, Architecture department, International University of Sarajevo, Sarajevo, Bosnia and Herzegovina

[2] Faculty of Architecture, Konya Technical University, Selçuklu/Konya, Turkey

Abstract. Contemporary architecture and the construction sector are responsible for a significant increase in energy demand and its effects on environmental issues. Vernacular building principles, if adopted and adapted, are promising in addressing this problem using sustainable design approaches. This paper is based on an investigation of sustainability principles applied in vernacular Bosnian house and vernacular neighborhoods in Sarajevo. It uses primary sources such as in-situ analysis and aims to investigate sustainable building strategies in vernacular building design and the possibilities of adopting them in contemporary architecture. Particularly, the paper is focused on the Sarajevo region and the architectural ensemble of Alija Djerzelez house, as it is one of the most representative examples of vernacular residential architecture from the 17th century. It was found that some sustainability principles recognized in the example of Alija Djerzelez house and the vernacular neighborhoods, if improved and implemented in contemporary architecture, could become part of the solution concerning climate change. Moreover, vernacular Bosnian houses, if optimized for the needs of the 21st century, can provide outstanding climate comfort which is a crucial component of contemporary architectural design.

Keywords: Sustainability · Vernacular architecture · Passive design features · Local materials · Alija Djerzelez house

1 Introduction

Increased awareness of the potential of sustainable dwellings' positive effect on environmental issues brings vernacular architectural solutions to the forefront to understand how to apply sustainable principles to newly designed dwellings. Vernacular architecture offers essential solutions for building sustainability considering that it has compelling characteristics that act in response to sustainable architecture. The exhibition appointed as 'Architecture without architects' was performed in 1964 in New York under the direction of Bernard Rudofsky, and drew the attention of researchers to the architectural field

N. Ademović et al. (Eds.): IAT 2021, LNNS 316, pp. 744–762, 2022.
https://doi.org/10.1007/978-3-030-90055-7_60

[1]. It was Rudofsky who first mentioned the term vernacular in an architectural context [2]. Vernacular architecture, as stated in the Charter on the built vernacular heritage [3], is housing that is continuously adapting to the needs of society, created as a creative response to limitations of the surrounding. It consumes local goods, shapes the traditions, expresses the culture, and addresses everyday needs in social environment. Aside from reflecting local customs and creative processes in craftsmanship and creating identity [4], vernacular architecture contributed to establishing and implementing sustainable practices in built heritage. These practices responded to the local climate by using local materials and establishing a set of rules, usually unwritten, for obtaining pleasant living conditions. In time, due to globalization, modernization, increased urbanization and use of cheaper materials, vernacular sustainable practices faded away. Vernacular architecture exhibits features that can be improved over time as needs and conditions change. While preserving local characteristics, it provides balance between past and future [5].

In this research, sustainable-vernacular features are demonstrated on the example of both vernacular residential neighborhoods and one representative house in Sarajevo, Bosnia and Herzegovina to reveal lessons that can be learned from the past and used to acquire more sustainable approaches in modern-building design. Bosnia and Herzegovina became part of the Ottoman Empire in the 14th century and remained under Ottoman rule until 1878 when the Congress of Berlin awarded Austria-Hungary the right to occupation. Since the Ottoman rule over Bosnia and Herzegovina lasted for over four hundred years, the features found in vernacular Sarajevo residential dwellings and neighborhoods depict a modification of the Ottoman houses from the region of Northern Greece [6]. Vernacular residential urban layout in Sarajevo is in harmony with topography, where residential neighborhoods are dislocated from public areas (usually located on steep hills), which is common for Ottoman settlements. Environment-compatible residential architecture is established by using local materials, respecting climate, and meeting residents' needs.

In the last decades, post war renewal of Sarajevo attracted an interest for significant investments, dominated by free market capitalism [7] in which new developments overpower official planning regulations [8]. As a result, unsustainable urban planning and dwelling units majorly contribute to light, sound, and air pollution in the city. Contemporary architecture is built out of context and does not follow the community values once established, while authentic vernacular Sarajevo's urban spaces are disappearing, or have lost their identity.

The purpose of this study is to document, rediscover and determine the sustainable building strategies in vernacular settlements with a goal to examine principles that can be integrated into modern-day building techniques. Particularly, this study investigates sustainability of the architectural ensemble Alija Djerzelez house in Sarajevo. The factors that affected the selection of this topic are brief investigation of sustainability of vernacular neighborhoods and individual houses in Sarajevo and the lack of literature, especially in the example of this house. Furthermore, the architectural ensemble Alija Djerzelez house is among several houses remaining from the 17th century that are preserved in their authentic shape, and as such is a representative building.

2 Literature Review

In present day, as the world is in constant search for new ways of reducing energy demand and urges for the use of sustainable energy, reconsidering good practices from the past serves as a possible solution. Since approximately 40% of the global energy is consumed in the construction sector [9], vernacular architecture became one of the hot topics discussed and researched worldwide [10–15]. This is particularly as a result of demands to respond to climate change, global pollution, and to find solutions on the decrease of non-renewable energy consumption in the buildings [16, 17]. The annual number of studies on vernacular architecture has increased since 2007 [18]. Understanding deficiency of energy sources aims attention at energy efficiency and preservation of natural deposits. For that reason, vernacular architectural practices can provide solutions in environment-friendly, sustainable green architecture, as many researchers agree that vernacular architecture is a form of sustainable architecture.

Some studies focused on the particular features of sustainability in vernacular architecture, while others discussed the impact of vernacular dwellings in the complete urban layout. Possible integration of different architectural features, for example building orientation, shading, use of vegetation and courtyards into contemporary architecture, was proposed by many of them. Galán-Marín et al. emphasized the importance of courtyard in vernacular house design. As the study proved the positive effect on the thermal characteristics, it further proposed enhancement and integration of courtyards into contemporary architecture [19]. Another study that involved a survey based on 300 houses with different typologies in Palestine, concluded that the private courtyards represent the main element to improve in order to have housing sustainability, while outdoor spaces do not meet the inhabitants' needs [20].

Good thermal performances of vernacular houses were also widely discussed in terms of climate. For instance, Webber and Yannas argued that climate is important to consider in terms of building design whereas passive principles for cooling can be implemented into contemporary dwelling units relying mostly on fossil fuels. Another study compared vernacular and contemporary cities at urban scale in order to determine vernacular architecture principles that can contribute to improve sustainability of new urban areas, as the study compares city layouts, skylines, public spaces and infrastructure [21].

Others discussed the differences between vernacular and contemporary cities of at an urban scale and have presented a comparative study between vernacular and contemporary urban areas, in order to allow architects and other stakeholders to identify the principles of the vernacular architecture that can be used in improving the sustainability of new urban areas [22]. The area of sustainability of vernacular urban units was also discussed in the sociocultural frame. The study focused on the city of Nablus examined the impact of sociocultural values and their implementation as a possibility to improve new housing design sustainability [23]. In order to create more sustainable environment, these values must be used with design approaches of contemporary buildings.

Although the sole implementation of vernacular building techniques is nowadays difficult to obtain, as these usually do not comply with the nowadays building standards and regulations [18], the benefits of vernacular architecture in terms of lessons or good practices that can contribute sustainability in a positive manner are valuable. However, many of these studies concluded that it is not possible to copy or use vernacular features in stereotyped manner in contemporary architecture, and that there is still much to learn from the knowledge implanted in the vernacular dwelling units.

This study continues the examination of the sustainability principles in vernacular buildings, however, as most of the studies mentioned above focus either on urban scale or a single house as a case study, this study focuses on investigation of vernacular architecture sustainability principles in the urban core of Sarajevo both on the level of neighborhoods and on detail examination of a case study. Additionally, this research contributes to further development of analysis of Bosnian vernacular houses, as in the case of Bosnia this topic is still very briefly discussed and usually does not focus on detail analysis of the case studies.

3 Methodology

This research analyses sustainable architectural principles in vernacular Bosnian residential neighborhoods located in Sarajevo, Bosnia and Herzegovina. The selected area of the study are vernacular neighborhoods dating from the Ottoman period and the architectural ensemble of Alija Djerzelez house as the case study.

The following methodology has been used to write this research:

- Analytical-historical method where analyses were conducted to define the role of sustainability in traditional vernacular architecture in neighbourhood scale. Therefore, analyses of vernacular architecture were evaluated regarding sustainable design principles such as: urban layout, building form and orientation, use of the local materials, natural wind induced ventilation, walls thickness, and integration of greenery into building in the matter of sustainability.
- Secondary method of the case study was conducted to demonstrate vernacular sustainable principles in micro-scale and was obtained first-hand, from the recordings of site analysis and observations. The selected case study is the house of Alija Djerzelez, a 17th-century residential house from the Ottoman period in Sarajevo. The selected case study provided a detailed examination and precise description of the sustainable design strategies.

4 Vernacular Architecture and Sustainability Strategies in Sarajevo Neighborhoods

Sustainability integrated in vernacular architecture, based on traditional building construction, building technologies and materials used, was a necessity. With technological advancements and sustainability available nowadays, looking back for vernacular principles may seem sentimental, non-advanced or backward [13]. Sarajevo's urban morphology follows diverse history of the city generated under Ottoman, Austro-Hungarian, and socialist rules. Traditional, thereby vernacular Bosnian neighborhoods started emerging on the building traditions of Ottoman Turks who adopted and developed the vernacular

Seljuk building techniques of ancient Anatolia. The concept of the Ottoman neighborhood, *mahala*, refers to the particular residential physical part of a city that has distinguished social characteristics. The fundamental elements of the neighborhood, *mahala*, are: mosque, masjid, dervish lodge, coffee house, grocery, carpenter, barber, and several repairmen; that make neighborhood itself sustainable Fig. 1. The following chapters investigate and discuss positive aspects of vernacular architecture in Sarajevo *mahalas*, while recognize the possible problems and shortcomings of their implementation nowadays. However, even if some are hardly applicable today, all positive aspects are important to consider as a valuable lesson of cumulative knowledge.

Disrespect to sustainable (and vernacular) design principles in modern day housing in Sarajevo results in city (neighborhoods) losing sustainable living conditions and comfort that cannot be provided in such neighborhoods Fig. 2. The concept of vernacular neighborhoods reveals neighborhood unit which is as well part of our cultural heritage that could be reconsidered for urbanization and adapted to today's conditions.

Fig. 1. Streets of vernacular Sarajevo neighborhoods [24]

To evaluate sustainable practices of Sarajevo traditional neighborhoods, principles of urban layout were examined, including form and orientation, introversion, using local materials, natural wind induced ventilation, courtyards, openings and integration of greenery into building complex design as all of above mentioned are considered as sustainable design strategies. Vernacular architecture has been developed with multiple approaches to solve climatic restraints, such as conserving resources and energy, producing almost zero waste, providing recycling etc. [25]. Sarajevan vernacular architecture has been studied in the light of these approaches.

Modern-day construction is mostly driven by economic features and traditional layouts are being replaced with high rise buildings, and sole implementation of vernacular features in contemporary design is neither always possible nor recommended. However, considering the benefits of vernacular features such as orientation, building form, openings, etc., should be implemented by architects, as all these can contribute to designing optimal living conditions.

Urban Layout

The fundamental vernacular principles of neighborhood forms are as follows: they abide

Fig. 2. Analogy of Sarajevo vernacular neighborhood compared to contemporary neighborhood (Source: Authors)

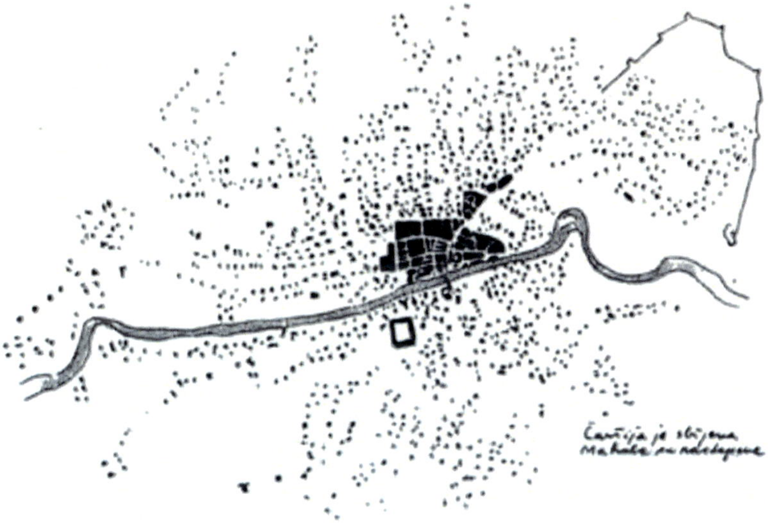

Fig. 3. Display of Sarajevo during Ottoman period with compact center-carsija and dispersed organic neighborhoods-mahala [24]

by 'the neighbor law' meaning that every residential unit is built in a way that doesn't obstruct the view of the neighbor's residential unit; the importance to the aesthetics is emphasized; it is built in 'human scale'. Moreover, Sarajevo's traditional neighborhood has organic structure that takes nature as reference. Figure 3 represents urban layout of Sarajevo in Ottoman period where residential neighborhoods were built in harmony with topography and formed organic structures which provided comfort to its residents. As many neighborhoods in Sarajevo built after Ottoman period do not resemble the same organic features, the practices of vernacular heritage should be considered into future developments, while the sustainability of un-sustainable modern-day inorganic structures can be enhanced on a lower scale by using individual architectural elements on houses, such as shadings, courtyards, or vegetation. Furthermore, as construction nowadays follows the regulation and urban plans and cannot be solely based on the 'unwritten rules' previously applied to form organic neighborhoods, however, the practices of vernacular architecture should be considered in the process of planning.

Building Form and Orientation

Traditional Bosnian house design was based on a few principles: dwellings were built in accordance to human scale, ground-floor was surrounded by courtyard walls, therefore protected from outside views, upper floors were lightweight and used for main living, porches and verandas were integrated into dwellings and had function of providing privacy to the tenants, transitional spaces formed spatial continuation from interior to exterior, ass visible in Fig. 4 [6]. Furthermore, houses were oriented towards south with the longest side, in order to perceive sunlight during day. This feature was particularly important during winter mode. However, building orientation and form, even if facing sun, could not be enough to prevent heat loss if the materials used were not proper. Nevertheless, in the process of modern-day construction, all these features should be taken into consideration, as these are valuable lessons usually very easy to achieve with many positive impacts in terms of thermal properties of the buildings.

Introversion

Closeness and compaction of houses inside of one neighborhood was due to hilly terrain. Although houses were densely distributed, courtyards were divided by fences, as more intimate spaces for private family atmosphere. Due to the customs in local communities, the houses that were aligned on the sides of narrow streets had high walls as fences to protect the family's privacy from the streets. The courtyards served for transferring part of house functions to 'residential courtyards' [27]. The effect of the different heights of the enclosure walls was examined. The results demonstrate that, by increasing the height of the walls, the noontime temperature decreases and the nighttime air temperature is higher than in the courtyards with lower walls.

Bosnian traditional houses are introverted units surrounded by courtyard walls forming organic streets of residential neighbourhood Fig. 5. The needs of the families for privacy nowadays may differ from the previous times, as residents may not be willing to enclose their courtyards. However, introversion had many positive impacts regarding control of the exposure to sun during the day. As the exposure to sun, if designed carefully, can contribute to enhancement of thermal characteristics during winter and summer mode, this vernacular feature, as valuable as it is, can be reconsidered and redesigned in terms of architecture to accommodate aesthetics values of today.

Using Local Materials

Sustainable construction was established on the usage of natural resources, such as adobe, stone and wood, in such a way to fulfill economic, environmental and cultural needs. The use of these materials in vernacular architecture is based on trial and error and evolved on experience of community. Furthermore, sustainable architecture is ensuring that the materials used are not disregarding the housing needs of the generations to come [28]. Integral part of sustainability in vernacular houses are also passive design features in heating and cooling. Renewable energy and energy efficiency are considered to be crucial aspects of sustainable energy policy [29]. Furthermore, vernacular houses are regarded as forerunners of sustainable environments [30].

Vernacular houses were built using locally available materials, using local manpower, with low amount of construction manufacturing energy and implementing ecological technologies. Moreover, vernacular houses, due to locally available materials used for construction, have low impact of transport, and do not increase pollution. Stone, wood, adobe, vernacular materials used in Sarajevo, are biodegradable or reusable, with low environmental impact. Comparison of these materials with concrete, the most common building material used nowadays in building construction in Bosnia and Herzegovina, proves poor thermal characteristics of the latter, which consequently contributes to environmental problems.

Materials with low diffusivity factor that absorb the heat and release it back to the environment in the winter mode can contribute to a slower process of cooling the interior due to retardation of heat transfer. In comparison to brick and wood, limestone has slightly higher diffusivity coefficient, but is highly better absorbent of heat than concrete. In the summer mode, when outdoor temperatures are very high, limestone enables fast diffusion of the heat to the inside, keeping interiors cooler [13]. When compared with concrete, adobe has lower thermal conductivity but higher heat capacity,

Fig. 4. Traditional Bosnian house forms, Sarajevo [26]

Fig. 5. The introversion of Sarajevo traditional neighborhoods (Source: Authors)

which grants it better thermal stability [28]. Furthermore, adobe has much lower and much less Global Warming Potential. Together with low diffusivity, high thermal inertia contributes to energy efficiency of limestone, which can be classified as good absorbent of heat that will not rapidly change temperature, making it ideal building material for both summer and winter mode. Another mean to obtain the thermal comfort inside of the house was through the thickness of the walls. To achieve structural stability, exterior walls were thick. This contributed to designing a building able to withstand long hours under sun, without transferring heat to the interior. As this study acknowledges the positive impact of vernacular materials in building technology, it also recognizes that some of the features regarding materials, may not be easy to obtain in contemporary architecture. For example, the walls built in vernacular houses were wider than today's norms, and can nowadays be expensive to purchase. Similarly, vernacular principles in this area included special treatment of the wood during its preparation for a building material, and these contributed to the overall durability of houses [27]. These techniques were provided by local carpenters and used traditional tools, but the cumulative knowledge on this faded away as it was not preserved and traditional building techniques may be very hard and expensive to acquire. When it comes to usage of adobe, although it is considered environmentally friendly and has many advantages, it also has shortcomings, considerable mostly in winter mode. Due to high U-value and permeability, it takes more energy to heat up the space, while the adobe itself is a subject to dampness. Moreover, due to the higher humidity rate in the interior, heat loss between interior and exterior space is also higher. For these reasons, in order to use adobe in modern-day construction, upgrades and advancements are required [31]. Vernacular housing units also used wooden elements. Wood, as natural element, was subjected to the decay and fungi. The disadvantages to wood in building construction further include vulnerability to water, fire and termites. In time, this could endanger the structure of the complete house.

Furthermore, the overall impact of the vernacular materials needs to be reconsidered and surely its use needs to be enhanced using contemporary techniques easily available.

Natural Wind Induced Ventilation

One of the most practical solutions to reduce energy demands in houses is by implementing passive ventilation solutions [9]. Wind ventilation is an efficient design strategy for the natural cooling system. In determining movements of the air in the interior, building orientation and position of ventilation openings play a crucial role [32]. As vernacular houses have sustainability elements as their integral part, they implemented practices of passive ventilation systems. In vernacular houses, windows were sometimes covered by wooden lattice elements. In addition to protecting the privacy of its tenants, these were designed in such a way to allow unobstructed air flow. To further enhance the movements of the air in the interior, on the upper parts of the walls there were holes for pushing the hot air from interior back to the streets. The open and semi open spaces in vernacular houses further contributed to the wind flow. This feature of vernacular architecture is widely examined in the studies worldwide, as one of the passive design cooling strategies. However, some research studies argue that the actual knowledge of the natural ventilation and its passive design features is incomplete, mostly as a result of complex design parameters, such as form of the roof, materials used, opening size, shape, and angle, etc. [33]. Furthermore, in contemporary design, these vernacular features can be additionally enhanced using innovative solar or wind panels that can be implemented into design.

Openings

In vernacular architecture, every building element was designed in order to contribute to the comfort of the occupants. Therefore, the role of the openings was to balance the air movements and limit the sunlight in order to prevent overheating or unnecessary cooling. In order to prevent discomfort during summer, openings needed to reduce heat gain in the house. The influence of openings in the aspect of sustainability was examined in studies [34, 35]. In order to obtain pleasant conditions, number, size and orientation of the openings was carefully considered. Furthermore, windows were protected by sun shading elements, usually designed of wood in form of lattice, which did not obstruct the circulation of the wind.

Courtyards and Integration of Greenery into Design

Houses inside neighborhood were closely linked to each other but courtyards were divided by fences, creating more intimate spaces and private family atmosphere. Inner courtyards, due to pleasant weather conditions served for part of house functions transferred to 'residential courtyards' [27]. It is being considered that these spaces represented the liveliest space of the house, or the heart [36]. Furthermore, the courtyard addressed the problem of overheating, as integration of water and greenery inside of it contributed to passive cooling design. Courtyards were important to reduce sound pollution. As a void area, courtyards played a great role in preventing sounds transmission in the context of neighborhood.

Greenery was an important aspect in obtaining better thermal comfort in the house. Aside of satisfying aesthetics and providing privacy from the street or neighbors, proper use of greenery in the courtyard could allow or block the direct sunlight into the house, and affect the interior temperature. High trees were usually planted on the south or west side, to protect the house from overheating during summer time. As the leaves on the trees are already fallen off in winter, obtaining sunlight, which is important in reaching optimal temperature inside, is not obstructed. Another useful element of greenery was cooling down the air before it reached the building by increasing air moisture through vegetative processes or evapotranspiration [36].

5 Case Study – Alija Djerzelez House Contextualization and Application of Passive Systems by Using Sustainable Architectural Strategies

The architectural ensemble of Alija Djerzelez house is located in Sarajevo in the cities northern part, in the neighborhood called Vrbanjusa. The ensemble is surrounded by Sagrdzije Street to the east and Alija Djerzelez Street to the north. The access to the ensemble is from the eastern side. The Alija Djerzelez house itself dates from the 17th century and depicts one of the most representative pexamples of residential architecture in Sarajevo dating from the Ottoman period Fig. 6.

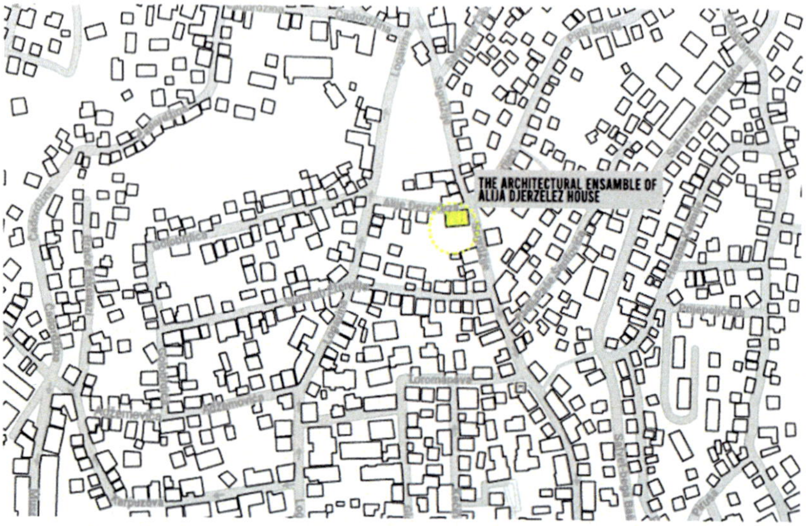

Fig. 6. The position of the architectural ensemble of Alija Djerzelez house located in organic formed traditional neighborhood (Source: Google Maps, processed by authors)

Spatial concept of the house demonstrates visible division in horizontal and vertical sense. In horizontal sense house is divided in men's - public area, and women's – semi-public and private area. In vertical sense house is divided in winter quarter located on the ground floor, and summer quarters assumed for living on the upper floor. The overall area of the architectural ensemble is 350 m^2. The residential area layout is presented in Table 1. In year 2002 architectural ensemble of the Alija Djerzelez house was listed as Category I monument and is on the Provisional List of National Monuments of Bosnia and Herzegovina (Commission to Preserve National Monuments). The findings on sustainability practices are presented in Table 2.

The analysis of sustainable architectural strategies applied in the architectural ensemble of Alija Djerzelez house in Sarajevo demonstrated that vernacular architectural practices reveal passive design principles. These strategies exploited terrain, solar and wind energy, and local materials to the extent that they can be applied to respond to contemporary needs as well as to decrease environmental impact.

Table 1. Architectural drawings of the architectural ensemble of Alija Djerzelez house [24]; processed by authors

Ground floor plan

1- Gate
2- Men's courtyard
3- Space under divanhana
4- Stairs
5- Storeroom
6- Room
7- Small room
8- Kitchen
9- Storeroom
10- Storeroom
11- Anteroom
12- Women's courtyard
13- Summer kitchen
14- Privy
15- Janitor
16- Tool shed
17- Small gate between courtyard and garden
18- Exit gate
19- Garden
20- Street

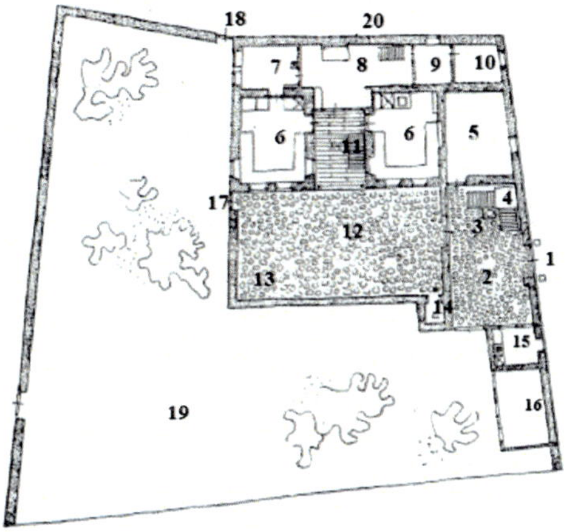

First floor plan

1- Men's divanhana
2- Kamarija (Dais in divanhana)
3- Cardak (Room)
4- Cardacic (Small room)
5- Dressing room
6- Women's divanhana
7- Kamarija (Dais in divanhana)
8- Cardak (Room)
9- Two-storey kitchen

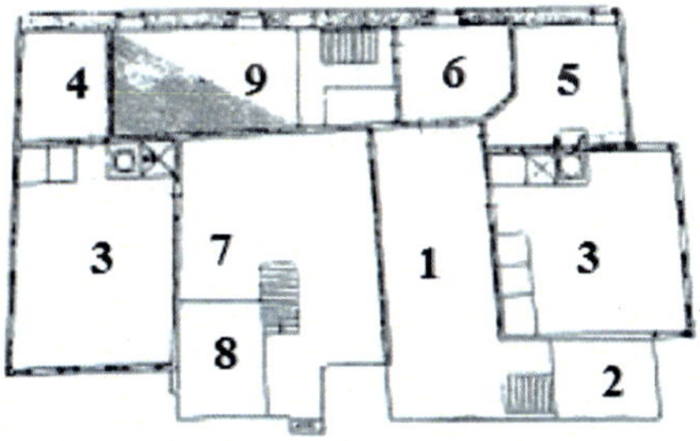

(*continued*)

Table 1. (*continued*)

Cross section	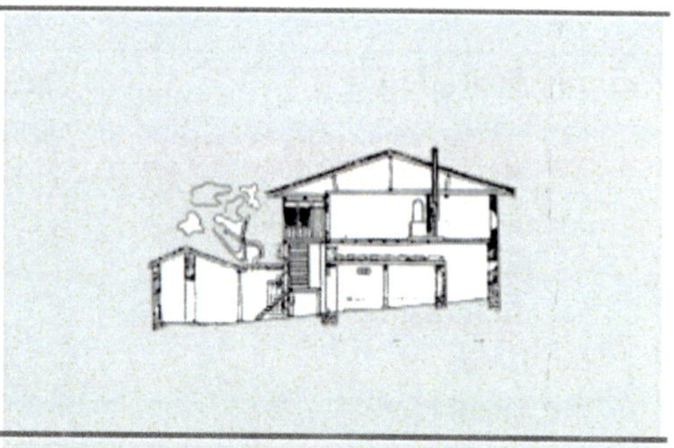
Longitudinal section	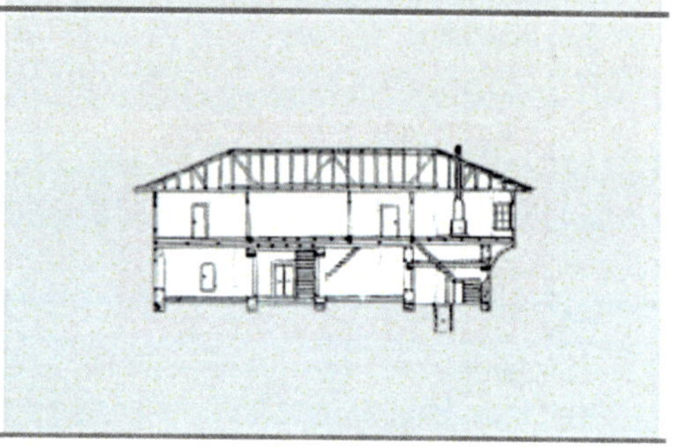

Table 2. The descriptive analysis of sustainable architectural strategies applied in the architectural ensemble of Alija Djerzelez

Building form and orientation The ensemble is built in 'human scale', in balance with topography and is one-storey residential dwelling with introverted building form typical to the architecture from Ottoman period in Sarajevo. The orientation of Alija Djerzelez house depended on the slope of the hill where the house is located. As the hill is facing south, the house is situated towards the south. In this way house gets appropriate insolation, especially during the winter. The roof is pitched which makes indoor air circulation smoother. Although positive impacts of building form and orientation are wide, in terms of enhancing thermal properties, surely, these sustainability features alone are not sufficient.	

(*continued*)

Table 2. (*continued*)

Introversion The ensemble of Alija Djerzelez house is a typical Ottoman courtyard house which makes all the surfaces introverted and oriented to a climate-controlled courtyard.	
Using local materials The choice of building materials depends on geomorphology and diversity of local materials. In Sarajevo, main building materials are limestone, wood, and adobe. Therefore, building materials in the ensemble of Alija Djerzelez house were: - Quarry stone used for foundations - Cut limestone for the walls in plinth area and decorations in the house - Adobe with horizontal reinforcements in form of timber tie beams for the massive ground-floor walls (60 cm thick) and for courtyard walls - Timber structure with brick filling (so called bondruk system) for upper floor - Close set of half-logs filled with mixture of clay and lime to make it fire resistant for floor structures - Timber for pillars and roof structure - Hollow tiles (ceramide) for roof covering and courtyard wall covering. Although sustainable and easily approachable, local materials had negative impacts as well, that left visible marks on the house itself. Adobe that was used in construction, due to higher U-value, contributed to creation of damp in the interior. Moreover, due to higher amount of humidity in the interior, heat loss amount is always higher. Generally, in order to protect wooden elements from the decay and fungi, traditionally smoke was used. However, as this measure of protection was not always effective, there are visible traces of mold and fungi in the wooden elements.	

(*continued*)

Table 2. (*continued*)

Natural wind induced ventilation Natural ventilation is accomplished with combination of openings (windows and ventilation apertures) that are placed high in all living spaces of the house. There are two ways of ventilating the house: - Cross ventilation that means opening windows all over the house simultaneously during the summertime - Opening window and ventilation aperture placed usually on sofa wall Additionally, lightweight walls on first floor made with 'bondruk system' provided fast cooling over the nights. Shading of first floor is provided by roof eaves. Ground floor as it is drawn inside gets shading from the first floors projections.	
Openings Most of the house openings are situated in east and south facing parts in order to gain more sun power. Openings facing north are minimized both in number and size; the reason for this is reducing heath losses. Openings on the ground floor are narrow in order to provide privacy to the tenants so the insolation levels in ground floor are low, therefore ground floor is used as storage and during the winter.	
Courtyards and integration of greenery into building Climate-controlled courtyard was established by introducing greenery and providing enough shading/projections. This resulted in reducing heat gains during the summer months, as well as minimizing the wind influence to the house. Greenery increased air moisture and cooled the air streams which resulted in controlled micro climate around the house.	

6 Discussion

In the study the architectural ensemble of Alija Djerzelez house as well as traditional Sarajevo neighborhoods were analyzed based on different criteria regarding sustainable design principles that were obtained from various studies. According to the analyses, traditional Sarajevo neighborhoods have organic street texture, are positive in terms of orientation and prevailing winds, are compatible with topography and the urban forms are in human scale which reveals that they were built according to sustainable strategies. Although these traditional neighborhoods have positive aspects, it must be recognized that the social and economic conditions that produced these urban structures no longer exist today. The design of architectural ensemble of Alija Djerzelez house provides thermal comfort in the interior spaces and courtyards. The form of the ensemble is in human scale with both courtyard and garden; it provides natural ventilation along with greenery and water elements and reveals that it is built in accordance to sustainable design strategies.

The analysis of the design of architectural ensemble of Alija Djerzelez house showed that building used the best in terms of orientation; it faces south side in order to receive sunlight during all day. The thick walls of stone in the ground floor were containing the heat during day, making the interior pleasantly cool. During the night, however, the absorbed heat was released back to the interior, making nights warmer. Together with climate control courtyard that protected the building from overheating during the day by natural ventilation and vegetation that cooled the air around the house, it provided conditions for pleasant living conditions. Vegetation in this case was mostly placed from the west side, preventing the overheating. As the leaves fall during winter, in this period the sunlight was not further prevented from entering the house, making the interior warmer. Traditionally, one of the most important reasons for creation of the courtyards was transferring of some of the house functions outside, which is no longer relevant.

Even with all benefits from the above-mentioned vernacular features, negative aspects in terms of thermal properties still occurred. As the upper floor of the house was built using wooden structural elements and filling of adobe, in the winter mode this contributed to creation of heat losses. Adobe, although environmentally friendly material, has a high U-value, that contributed to creation of damp in the interior. Furthermore, wood used in the structure of the building, was subjected to the decay and fungi. Traditionally, this was prevented by using smoke that would protect the wood and also consolidate structural elements; however, as this measure of protection was not always completely effective, there are visible traces of mold and fungi in the wooden elements. This study emphasizes the need to further address the negative aspects mentioned here prior to implementation of vernacular features into contemporary design.

7 Conclusion

The purpose of this study was to document, rediscover and determine the sustainable building strategies in vernacular settlements and to explore principles that can be integrated into modern-day building techniques. Aside of investigation vernacular features of the neighborhoods, this study particularly focused on the architectural ensemble Alija

Djerzelez house in Sarajevo. It examined the elements of sustainability used in vernacular house by evaluating following sustainable design strategies: urban layout, building form and orientation, introversion, usage of local materials, natural wind induced ventilation, courtyards, openings and integration of greenery into building complex design. This study found that sustainability principles recognized in the example of Alija Djerzelez house, if adapted, improved and implemented in modern-day architecture, could become part of the solution concerning climate change. Furthermore, this research contributed to analysis of Bosnian vernacular houses, as this topic is not discussed enough and does not focus on detail analysis of the case studies.

This study suggests that positive aspects of vernacular architecture, such as sustainability features, can contribute to reducing energy demands and designing environmentally-friendly neighborhoods. For example, the traditional materials such as stone, adobe and wood could be used more in contemporary architecture, whereas passive design solutions, such as proper use of vegetation, once used in vernacular architecture could be integrated and enhanced in buildings in Sarajevo.

In addition, lessons in vernacular architecture can contribute to the development of a more sustainable urban layout, however, as in the case of Sarajevo, the lessons from cumulative knowledge and practices of sustainable vernacular architecture have yet to be reused and incorporated into new urban development plans. On the other hand, architects and researchers need to be aware that implementation of here investigated sustainable design principles are difficult to obtain or copy in contemporary designs, as these cannot be stereotyped, rather a deeper understanding of the lessons from vernacular practices will be needed for implementation into modern—day architecture. Moreover, this leaves space to researchers and architects to further examine and develop in order to build more sustainable environments.

References

1. Rudofsky, B.: Architecture Without Architects: A Short Introduction to Non-pedigreed Architecture. University of New Mexico Press (1987)
2. Berkovic, A., Canan, F.: The influence of climate and culture in the formation of vernacular settlements in Počitelj, Bosnia and Herzegovina and Safranbolu, Turkey. bab J. FSMVU Faculty Architect. Des. **1**(2), 260–277 (2020)
3. ICOMOS: Charter on the built vernacular heritage. ICOMOS (1999)
4. Benslimane, N., Biara, R.W.: The urban sustainable structure of the vernacular city and its modern transformation: a case study of the popular architecture in the Saharian Region. Energy Procedia**157**, 1241–1252 (2019)
5. Correia, M., Carlos, G., Rocha, S.: Vernacular Heritage and Earthen Architecture, 1st edn. Routledge (2014)
6. Pecar, M.: Bosnian Dwelling Tradition: Continuity and Transformation in the Reconstruction of Sarajevo. Traditional Dwellings and Settlements Review **11**(1), 49–55 (1999)
7. Tufek-Memisevic, T.: Architecture of consumption case study Sarajevo. Konsumpcja i Rozwój **3**(8), 73–86 (2014)
8. Pobric, A., Robinson, G.M.: Recent urban development and gentrification in post-Dayton Sarajevo, Bosnia and Herzegovina. Cities **89**, 281–295 (2019)
9. Duraković, B.: PCM-Based Building Envelope Systems - Innovative Energy Solutions for Passive Design. Springer (2020)

10. Gil-Piqueras, T., Rodríguez-Navarro, P.: Tradition and sustainability in vernacular architecture of Southeast Morocco. Sustainability **13**(2), 1–18 (2021)
11. Qtaishat, Y., Emmitt, S., Adeyeye, K.: Exploring the socio-cultural sustainability of old and new housing: two cases from Jordan. Sustain. Cities Soc. **61**, 102250 (2020)
12. Mazraeh, H.M., Pazhouhanfar, M.: Effects of vernacular architecture structure on urban sustainability case study: Qeshm Island, Iran. Front. Architect. Res. **7**(1), 11–24 (2018)
13. Halilovic, M.: Vernacular architecture sustainability principles: a case study of Bosnian stone houses in Idbar village. Period. Eng. Nat. Sci. **8**(4), 2564–2574 (2020)
14. Örnek, E.: Sustainable characteristics of the vernacular house and its impact to the building physics. Herit. Sustain. Dev. **1**(1), 14–20 (2019)
15. Sharifi, A., Murayama, A.: Changes in the traditional urban form and the social sustainability of contemporary cities: a case study of Iranian cities. Habitat Int. **38**, 126–134 (2013)
16. Basyigit, C., Alkayis, M.H., Kartli, M.I.: Environmental effects of utilization of sustainable building materials. Herit. Sustain. Dev. **3**(1), 64–70 (2021)
17. Baykal, F.Y.: The polluter should... pay? Herit. Sustain. Dev. **3**(1), 71–77 (2021)
18. Nguyen, A.T., Truong, N.S.H., Rockwood, D., Le, A.D.T.: Studies on sustainable features of vernacular architecture in different regions across the world: a comprehensive synthesis and evaluation. Front. Architect. Res. **8**(4), 535–548 (2019)
19. Galán-Marín, C., López-Cabeza, V., Rivera-Gómez, C., Rojas-Fernández, J.: On the influence of shade in improving thermal comfort in courtyards. In: Proceedings, vol. 2, p. 1390 (2018)
20. Hussein, M., Barlet, A., Semidor, C.: Socio-Environmental Dimensions of Private Outdoor Spaces in Contemporary Palestinian Housing. Open House Int. **35**, 67–76 (2010)
21. Weber, W., Yannas, S.: Lessons from Vernacular Architecture. Routledge, Abingdon-on-Thames (2013)
22. Al Tawayha, F., Bragança, L., Mateus, R.: Ecology and environment, an urban scale comparative study between contemporary and vernacular city of Nablus. In: Proceedings of the SBE16 Brazil & Portugal—Sustainable Urban Communities towards a Nearly Zero Impact Built Environment, Vitória, Brazil, 7–9 September (2016)
23. Tawayha, F., Bragança, L., Mateus, R.: Contribution of the vernacular architecture to the sustainability: a comparative study between the contemporary areas and the old quarter of a Meditteranean City. Sustainability **11**(3), 896 (2019)
24. Grabrijan, D., Neidhardt, J.: Arhitektura Bosne i Put U Suvremeno. Ljubljana: Državna založba Slovenije (1957)
25. Uyanik, I., Özkan, O., Mihçiokur, H.: Waste management in a university campus. Sustain Eng. Innov. **3**(1), 49–53 (2021)
26. Aga Khan Visual Archive ML: Ottoman Houses of Sarajevo. Vernacular housing of Sarajevo (Variant) (1983). https://archnet.org/sites/5284/media_contents/9238. Accessed Jan 2021
27. Kadić, M.: Starinska seoska kuća u Bosni i Hercegovini. Sarajevo: Veselin Masleša (1967)
28. Revuelta-Acosta, J., Garcia-Diaz, A., Soto-Zarazua, G., Rico-Garcia, E.: Adobe as a sustainable material: a thermal performance. J. Appl. Sci. **10**(19), 2211–2216 (2010)
29. Prindle, B., Eldridge, M., Eckhardt, M., Frederick, A.: The Twin Pillars of Sustainable Energy: Synergies between Energy Efficiency and Renewable Energy Technology and Policy. American, Washington, DC (2007)
30. Halıcıoğlu, F.: Analysis of vernacular architecture in terms of sustainable considerations: the case of Şirince Village in Western Turkey. Alam Cipta **5**(2), 39–54 (2012)
31. Obafemi, A., Kurt, S.: Environmental impacts of adobe as a building material: the north cyprus traditional building case. Case Stud. Constr. Mater. **4**, 32–41 (2016)
32. Wahab, I.A., Ismail, L.H.: Natural ventilation performance of Kedah vernacular house. Int. J. Sustain. Constr. Eng. Technol. **5**(2), 53–65 (2014)

33. Moosavi, L., Mahyuddin, N., Ghafar, N.A., Ismail, M.A.: Thermal performance of atria: an overview of natural ventilation effective designs. Renew. Sustain. Energy Rev. Iss. **34**, 654–670 (2014)
34. Fernandes, J., Mateus, R., Bragança, L., Correia da Silva, J.: Portuguese vernacular architect the contribution of vernacular materials and design approaches for sustainable construction. Archit. Sci. Rev. **58**, 324–336 (2015)
35. Meier, I., et al.: The vernacular and the environment towards a comprehensive research methodology. Eindhoven, The Netherlands, s.n., pp. 719–724, 9–22 September 2004
36. Tawayha, F.A., Braganca, L., Mateus, R.: Contribution of the vernacular architecture to the sustainability: a comparative study between the contemporary areas and the old quarter of a Mediterranean City. Sustainability **11**(896), 1–20 (2019)

Introducing the Healthy Urban Environment Courses in Bosnia and Herzegovina

Samir Lemeš[1](✉), Senaida Halilović Terzić[2], and Maja Roso Popovac[3]

[1] Polytechnic Faculty, University of Zenica, Zenica, Bosnia and Herzegovina
samir.lemes@unze.ba
[2] Faculty of Architecture, University of Sarajevo, Sarajevo, Bosnia and Herzegovina
[3] Faculty of Civil Engineering, University "Džemal Bijedić" Mostar, Mostar, Bosnia and Herzegovina

Abstract. This paper describes the results of the international Erasmus + project "HURBE: Healthy URBan Environment Developing Higher Education in Architecture and Construction in Bosnia and Herzegovina". The project consortium developed the curricula in Higher Education Institutions dealing with the field of healthy urban environment in the Faculties of Architecture and Construction in Bosnia and Herzegovina. The project involves six universities from Bosnia and Herzegovina, Italy, Croatia and Bulgaria. The main results include modernisation of the curriculum by developing new and innovative courses and installing new laboratories.

Keywords: Health · Urban planning · Environmental protection · Curriculum development

1 Introduction

The concept of the healthy urban environment was described in Zagreb declaration [1] as "A healthy city offers a physical and built environment that supports health, recreation and well-being, safety, social interaction, easy mobility, a sense of pride and cultural identity and is accessible to the needs of all its citizens". This declaration set out the principles for an inclusive, supportive, and sustainable approach based on a balanced ratio between economic development and urban life's environmental and social aspects.

Giofrè and Đukanović in [2] described the evolution of health and its relationship with the urban environment and design, from urban planning and architectural point of view. They pointed out that all urban interventions should bear in mind the human comfort, health and inclusion (see Fig. 1).

This initial idea led to an Erasmus + National Joint Project that started in November 2018, and it will last by November 2021. The project aims to develop the curricula in Higher Education Institutions in Healthy URBan Environment (hence the acronym HURBE) in the Faculties of Architecture and Construction in Bosnia and Herzegovina. The project involves six universities: "Džemal Bijedić" University of Mostar; University of Sarajevo, University of Zenica; Sapienza University of Rome, Italy (coordinator);

N. Ademović et al. (Eds.): IAT 2021, LNNS 316, pp. 763–770, 2022.
https://doi.org/10.1007/978-3-030-90055-7_61

the University of Zagreb, Faculty of Architecture, Croatia; University of Architecture, Civil Engineering and Geodesy, Bulgaria; and five associate partners from Italy, Bulgaria, Croatia and B&H. The project activities focused on modernising curriculum by developing new and innovative courses and installing new laboratory facilities [3].

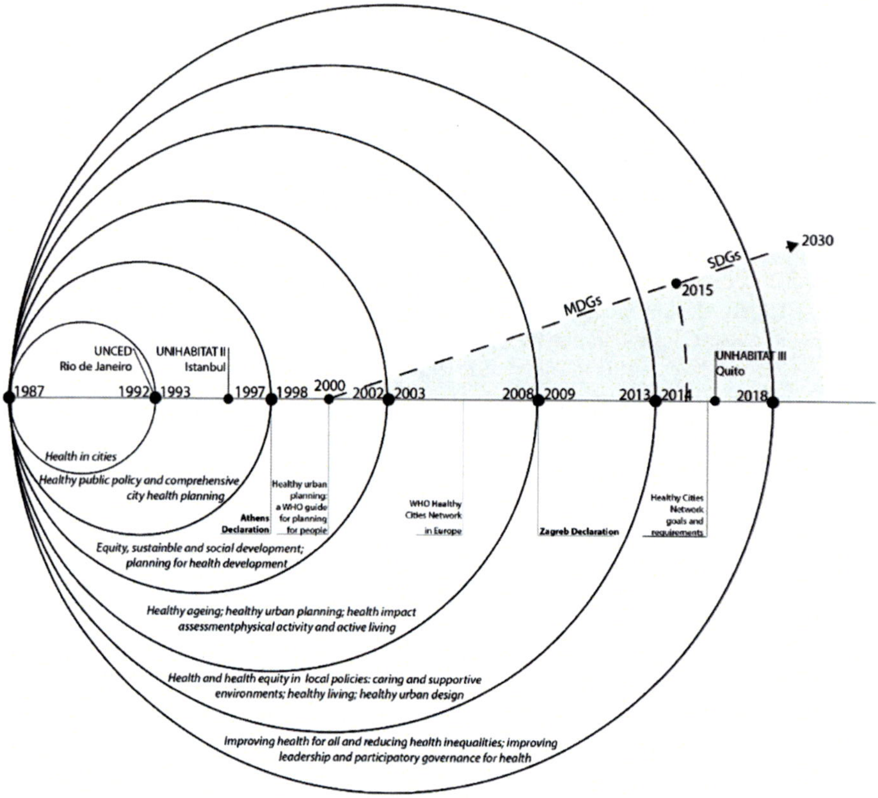

Fig. 1. Evolution of Healthy City concept [2]

2 Healthy Urban Environment

According to the UN, the world cities, occupying only 3% of the Earth's land, consume between 60% and 80% of energy, generate 70% of human-induced greenhouse gas emissions, produce 70% of global waste and make 70% of the world economy. More than half of the world's population lives in urban areas, and this proportion is expected to increase to 68% by 2050 [4]. These facts introduce new challenges, primarily affecting the environment, health and quality of life in general.

To achieve a healthy urban environment, the World Health Organization (WHO) initiated and supported several global projects and movements. Some of them are "Healthy Cities 1986", "Shanghai Consensus on Healthy Cities 2016", "The UN Agenda for Sustainable Development 2015" [5], encouraging governments and institutions to develop commitment, partnership-based planning and capacity-building projects.

According to WHO, a healthy city depends on a supportive environment, sanitation, hygiene, and healthcare to achieve a good quality of life, which relies more on improving the urban environment than on existing appropriate health infrastructure. This commitment largely depends on available resources and technologies usually unknown to the decision-makers. There lies a role of engineers, architects, and urban planners, who shape the future cities. They need to acquire the necessary knowledge, awareness, expertise, and instruments to deliver appropriate solutions. They should also maintain a continuous dialogue with local decision-makers, academics and actors of urban transformation.

Urban planners have to integrate health considerations into planning processes and establish the necessary capacity and institutional commitment. Urban designers should promote designs that meet all citizens' expectations for safety, accessibility, comfort, and active living. Both architects and construction engineers should use environmentally friendly materials and technologies with the most negligible environmental impact, providing that the built environment offers a good quality of life. Landscape architects and civil engineers can design and make cities accessible for everyone, including people with disabilities.

All these professions have to interact and cooperate, making this topic multidisciplinary and interdisciplinary. Only a synergy of different occupations and expertise can achieve the desired goals.

This project actualizes strategies of urban regeneration to improve the built environment, promote physical activities and the use of urban public space, encourage healthy behaviour, prevent diseases and promote healthier living. This project's consortium aims to create a comprehensive educational offer that prepares experts who act directly on city transformation.

3 Project Work Packages

The project started by analysing how healthy urban environment topics were covered in the existing curricula in Architecture and Civil Engineering Faculties in Bosnia and Herzegovina (Fig. 2). The analysis revealed a need for curriculum innovation by introducing new, interdisciplinary courses at three partner universities. The project integrated innovative educational approaches by focusing on decisive content in the disciplinary sectors of Architecture, Construction, Urban Planning, Landscape Architecture, Technology of Architecture and Civil Engineering.

Following the initial self-evaluation of the existing curricula, the project partners conducted surveys with students and different stakeholders, including urban-planning and academic institutions, policymakers, construction and architectural companies, and non-governmental organisations. They all expressed the need to introduce new courses providing students with the knowledge and skills required to implement environmental protection, healthy living, sustainability, and urban spaces' resilience.

The project included establishing the new laboratories at all three partner universities in Mostar, Sarajevo and Zenica. The equipment included measurement devices for different environmental monitoring purposes, such as noise meters, EMF and ionised radiation meters, air quality stations, drones, thermal imaging cameras, computer software and hardware.

Fig. 2. Project kick-off meeting in Rome, February 2019.

The vital part of the project was design and creation of Massive Open Online Course (MOOC), which will be available publicly, and students will be able to earn 1 ECTS if they attend it. This aspect of the course became an added-value as the pandemic conditions in 2020 transformed most of the teaching and other activities online.

The HURBE project is divided into seven Work Packages (WP) [3]:

- WP1: Startup: needs on degree programmes and credit transfer,
- WP2: Training of Trainers on Healthy Urban Environment (HURBE),
- WP3: Permanent laboratories,
- WP4: Educational modules on Healthy Urban Environment,
- WP5: Quality Assurance, monitoring service and sustainability,
- WP6: Dissemination and exploitation, and
- WP7: Management of project activities.

Within the Work Package 1, the project partners analysed the study programmes in three partner Universities in B&H to facilitate credit transfer and course accreditation and determine how the HURBE topic is dealt with in the teaching courses the Consortium Faculties.

Work Package 2 covered the Training of Trainers on HURBE, to increase local trainers' competences (professors, assistants, PhD students) in the HURBE topic.

Three permanent laboratories were established within the Work Package 3. The laboratory equipment has been chosen to support the implementation of innovative practices for the teaching methods and the development of new services - for HEI and other institutions - in the Bosnian Universities. This will create a sustainable funding resource by offering services to external clients after the project lifetime.

The result of the Work Package 4 is a set of educational modules on HURBE. The education modules were prepared according to the International Quality Standard for HEIs. The original project proposal planned to introduce one 8 ECTS module at each university, but the study programs' existing curriculum could not accept such a course. The project partners agreed to establish two study courses at each university. Some topics will be shared between the universities, thus increasing teacher mobility and cooperation between HEIs. Other outcomes of this Work Package are two MOOC modules, which are worth 1 ECTS.

Work Packages 5, 6 and 7 are regular parts of each Erasmus + project, aimed to ensure that all project activities are carried out with the highest possible standards and that the project objectives are reached efficiently.

4 Project Results

The courses were designed according to the analysis of existing curricula, student surveys, and stakeholders' opinions. Project partners invited stakeholders from the private sector (architectural studios, construction companies, individual experts), public sector (utility companies, institutions, government) and civil society (non-governmental organizations, academia). Figure 3 illustrates which topics the stakeholders identified as necessary for the new courses. The results revealed that most stakeholders are aware that spatial planning, design and construction are greatly influenced by the investors' interests, introducing the need to cope with ethical issues and corruption. Another important observation was made, pointing out that graduates need to improve their ability to use theoretical knowledge with practical skills, by introducing obligatory internship for final year students.

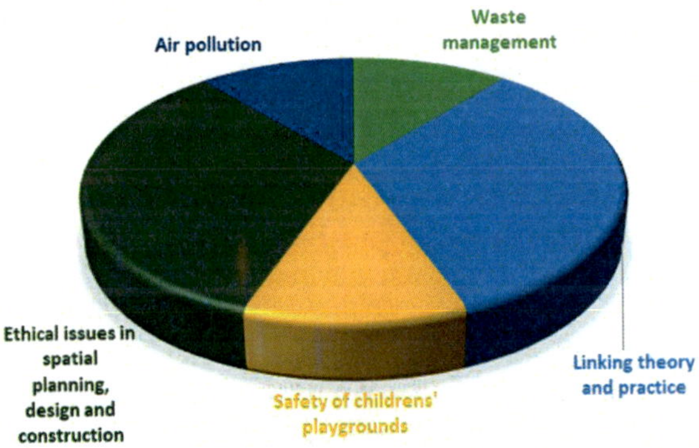

Fig. 3. Stakeholder survey results

Another survey performed at the beginning of the project collected data from 216 students and revealed interesting facts (Fig. 4). Only 8% of students had some educational experience abroad during university course (internship, study, exchange). 56% of students have never heard of UN Sustainable Development Goals (SDGs). 81% could not identify air pollutants. 63% of students think that HURBE is a multidisciplinary topic, covering healthcare, economy, politics, planning, design, architecture and engineering. 90% of students considered HURBE an essential topic for their future profession. Most of the students expressed their will to learn about HURBE through practice/internship and in laboratories, and only one third would like to participate in student exchange [6].

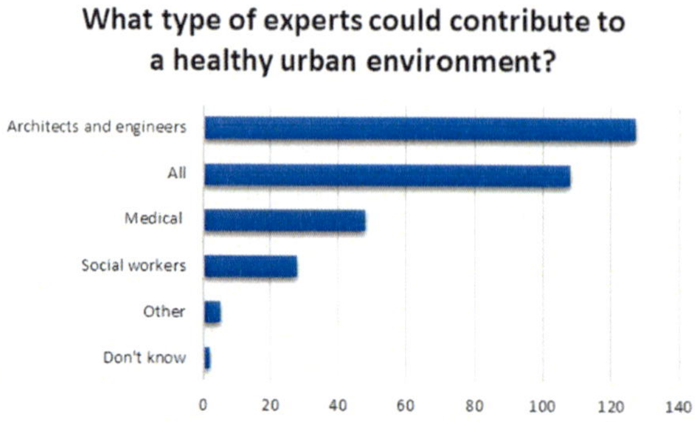

Fig. 4. Student survey showed that students are aware of interdisciplinarity of the topic

The most satisfying and inspirative result for the project team was that students started to ask whether they can do their final diploma papers related to this topic by the end of the semester. That proved the course was exciting and considered useful by the students, despite online classes' limitations.

The key project results are new university courses developed for students of architecture and civil engineering in Bosnia and Herzegovina. The project web page contains materials collected and prepared through project activities and used in the new courses [7].

The new courses at the University of Sarajevo are elective courses at the master level. The winter semester course HURBE 1 is a 3 ECTS course covering overview and background of health, architecture and environment. The summer semester course is a 6 ECTS elective course about City Conversation - Healthy Cities and the Built Environment, delivered in the form of Design studio/Project.

The new elective courses at the University "Džemal Bijedić" are at the master level. The winter semester course HURBE 1 is a 5 ECTS course aimed to raise the students' awareness of the healthy and sustainable environment to empower them to be future engineers and managers and promoters of importance of the healthy urban (and rural) environment. The summer semester course is a 5 ECTS elective course covering topics about sustainable design of open spaces, barrier-free city, urban mobility, sustainable building design, passive houses, green buildings, local natural materials, reversible building design, and circular economy.

The new elective courses at the University of Zenica are at the final year of the bachelor level. The winter semester course HURBE 1 is a 5 ECTS course aimed to raise students' awareness about environmental impacts on human health and increase the importance of sustainable and resilient city infrastructure. Figure 5 shows how students performed measurements and presented their results. The summer semester course (HURBE 2) is a 5 ECTS elective course which will enable students to understand healthy city planning through methods, strategies and case studies that highlight successful elements for urban health equity; to design and build sustainable and resilient buildings and open spaces; to effectively organise mixed urban/industrial use of space, and to design sustainable and green urban mobility with special attention for persons with disabilities.

Fig. 5. Students used the new laboratory equipment to measure outdoor noise, EMF and radiation levels and then presented results online and in the laboratory

Another project result is a Massive Open Online Course (MOOC), divided into two modules. The aim of the first module, "Healthy Urban Environment: Spatial Planning and Architecture" is to teach what spatial planners and architect can do to mitigate a healthy urban environment. The second module "Healthy urban environment engineering" familiarises attendants with the engineering solutions for the healthy urban environment, from renewable energy, recycling, and new materials to smart cities and similar IT solutions.

The final event of the project is an international conference "Making healthy cities for people. Education, research, practice in planning, architecture and engineering. HURBE2021" to be held in Sarajevo, in October 2021. The conference aims to share all the results of the HURBE project and interdisciplinary vision studies, plans, projects, and experiences for making Healthy City. The conference covers three thematic areas: (1) Healthy spatial planning, (2) Healthy urban and architectural design, and (3) Healthy engineering.

The project web page (https://hurbe-project.eu/) contains all other project results not mentioned here, valuable materials and presentations that could be used as a role model for other universities if they would like to introduce this important topic into their curriculum. The project partners and stakeholders intend to continue with activities, expanding the network of organisations and institutions related to the healthy urban environment.

The project confirmed the importance of innovative teaching methodologies integrated with the use of practical experience and laboratories. This method is replicable for curriculum development in other higher education institutions in Bosnia and Herzegovina and any other country. It highlighted the importance of using research as a fundamental base for designing and teaching academic courses [8].

5 Conclusions and Lessons Learned

The consortium realised the project under specific conditions, as the peak of activities was interrupted by pandemic, preventing any mobility. However, all project partners showed a high level of responsibility and adopted quickly to new conditions. Project partners learned that stronger cooperation between academic institutions, business sector and decision makers is essential to achieve sustainability.

The project demonstrated that interdisciplinary courses are increasingly demanded, in order to answer the challenges imposed by the changing global environment and rapid urbanisation. Achieving resilient and sustainable cities requires experts from different areas to cooperate and act in a synergy, engaging all available technology to cope with modern challenges, such as climate change, environmental degradation, aging population, social inequality and massive migration from rural to urban areas.

Finally, the project could be marked as successful, having valuable results and outcomes. Long-term impact – changed awareness of students and acquired skills could help Bosnia and Herzegovina protect the environment and cope with rapid urbanisation and climate change.

Acknowledgement. The authors would like to express their gratitude to all project participants, students, teaching and administrative staff who made this possible. This project has been funded with support from the European Commission within the Erasmus+ Programme, 598503-EPP-1–2018-1-IT-EPPKA2-CBHE-JP.

The coordinators of each partner university are Francesca Giofrè (lead coordinator), Sapienza University of Rome, Faculty of Architecture, Department Architecture and Design; Vesna Mikić, University of Zagreb, Faculty of Architecture; Milena Tasheva-Petrova, University of Architecture, Civil Engineering and Geodesy; Maja Roso Popovac, "Džemal Bijedić" University of Mostar; Senaida Halilović Terzić, University of Sarajevo; and Samir Lemeš, University of Zenica.

References

1. World Health Organization: Zagreb Declaration for Healthy Cities: Health and health equity in all local policies. WHO Regional Office for Europe, Copenhagen Ø, Denmark (2009). https://www.euro.who.int/__data/assets/pdf_file/0015/101076/E92343.pdf. Accessed 9 Feb 2021
2. Giofrè, F., Đukanović, Z.: Healthy urban environment and design: the outdoor spaces. In: Keeping up with Technologies in the Context of Urban and Rural Synergy: Book of Conference Proceedings/4th International Academic Conference "Places and Technologies 2017", Sarajevo, 8–9 June 2017, pp. 155–168 (2017)
3. HURBE Project: Project overview (2019). https://hurbe-project.eu/. Accessed 9 Feb 2021
4. UN: 68% of the world population projected to live in urban areas by 2050 (2018). https://www.un.org/development/desa/en/news/population/2018-revision-of-world-urbanization-prospects.html. Accessed 9 May 2021
5. World Health Organization: Health Promotion: Healthy cities (2016). https://www.who.int/healthpromotion/healthy-cities/en/. Accessed 9 May 2021
6. HURBE Project: WP1 T1.5 The Bosnian students' awareness survey results (2019). https://hurbe-project.eu/Content/documents/160620201215_WP1. T1.5 TheBosnian students' awareness.pdf. Accessed 11 Feb 2021
7. HURBE Project: Project Results (2020). https://hurbe-project.eu/ProjectResults. Accessed 11 Feb 2021
8. Giofrè, F., El Edeisy, M.: Between research and teaching: identifying new compe-tencies for healthy cities. In: EAAEARCC International Conference & 2nd VIBRArch, pp. 1969–1970 (2021). https://doi.org/10.4995/EAAE-ARCC-IC-2020.2020

Integration of Solar Photovoltaic Power Plant Along National Highway Corridors

Emina Ušanović, Maja Muftić Dedović(✉), Aida Zugor, Nejra Viteškić, Lamija Zametica, Esad Đakovac, Behadem Muminović, Neila Milišić, and Samir Avdaković

Faculty of Electrical Engineering, University of Sarajevo, Sarajevo, Bosnia and Herzegovina
maja.muftic-dedovic@etf.unsa.ba

Abstract. In recent years, the use of photovoltaic systems in traffic has become more frequent. The field of application within the field of transport refers to signaling such as high poles, navigation, railway signals, traffic signs, information boards, as well as in terms of public lighting and safety lighting. The paper will provide a detailed review of the literature regarding the applied renewable solar energy and all applicable technologies for highway corridors. Also in this paper, the installation of a solar photovoltaic power plant for meeting the energy demand of highway corridors and tunnels is analysed. The production of electricity from these PV power plants is used to power highway corridors own consumption, such as tunnels and lighting today, and tomorrow for mass use to charge electric vehicles. In this paper, a techno-economic analysis of a solar photovoltaic power plant with an installed capacity of 1 MW in the village Tarčin, next to the A1 highway, is performed. From the obtained results, it can be concluded that the investment in solar power plant represents a cost-effective techno-economic solution for reducing electricity costs. The other benefits of using solar photovoltaic power plant along national highway corridors are in emission reduction and better transport planning. Integrating PV system into national grids can reduce transmission and distribution line losses, increase grid resilience, lower generation costs, and reduce requirements to invest in new utility generation capacity.

Keywords: Solar photovoltaic power · Techno-economic analysis · Solar highways · Renewable energy · Payback period

1 Introduction

Climate change and the enormous air pollution population have witnessed in the last few decades on a global scale have caused a drastic change of course in world energy policy [1]. Electricity producers who use conventional sources (coal, oil, nuclear energy) such as thermal power plants and nuclear power plants, through the energy plans of developed countries are gradually abolished, and fast-growing technologies based on renewable energy sources (wind farms, solar power plants, geothermal power plants, power plants on biomass, etc.). The world is facing a constant increase in energy needs,

N. Ademović et al. (Eds.): IAT 2021, LNNS 316, pp. 771–786, 2022.
https://doi.org/10.1007/978-3-030-90055-7_62

but also a gradual decrease in fossil fuel stocks, which has further encouraged the mass construction of plants using renewable energy sources. The price of MWh produced from wind and solar power plants has been declining rapidly in recent years, and today it can be established that it is lower than MWh of electricity produced in thermal power plants [2]. Bosnia and Herzegovina is a country rich in natural resources and there is significant potential for the use of wind and solar energy. Renewable energy sources participate in the total production up to 40% per year, while the share of other technologies is much smaller. So far, 180 photovoltaic power plants in Bosnia and Herzegovina are built, with an installed capacity of just over 18 MW [3]. Solar power plants participate in the total capacity of electric power plants with only 0.37%. Such a small percentage is largely the result of complicated procedures and other barriers that prevent significant capitalization of renewable energy sources. Solar energy has by far the greatest potential, given that the amount of solar radiation that reaches the Earth in one year is as much as 10,000 times greater than the actual needs. With the constant reduction of production costs of photovoltaic systems, in areas with not so much solar radiation, with the correct choice and dimensioning of the system, adequate profitability can be achieved.

2 Literature Overview

In this paper, the construction of solar highways as a smart, safe and efficient investment in renewable energy projects is considered. As solar highways correspond to the production category, solar energy is directly to the local electricity grid sent. The generated electricity for a variety of applications, such as road and tunnel lighting, tollbooth operation is used, and over time, electric cars could use the energy as well. So the energy will actually be generated where it is needed, providing a big step towards an energy neutral system. Typical spaces to convert solar energy into electricity are resorts, unused land near highways and roofs of buildings.

Tunnel lighting design presents multiple challenges related to safety, economy and environmental protection. Due to its special features, the tunnel requires careful lighting design to get the driver's eyes used to the sudden change in lighting as soon as possible when entering the tunnel from a bright environment. Therefore, the lighting level at the entrance to the tunnel (threshold zone) must be high enough, especially during the day. This means that a large number of luminaires, which consume electricity and represent 80% of the tunnel's consumption, are needed. To reduce the electricity consumption of the tunnel, scientists have begun to consider various strategies to reduce the required level of illumination of the threshold zone.

The paper [4] presents for the first time a strategy based on the installation of vertical, black solar panels, whose reflection factor is very small, at the entrance to the tunnel. In this way, a reduction in the required level of illumination in the threshold zone is achieved.

Reducing the required level of illumination at the entrance to the tunnel can be achieved by using double-sided solar panels in front of the entrance to the tunnel, as well as by using a "solar roof". Reducing the electricity consumption of the tunnel, without reducing the required level of illumination of the threshold zone, can be achieved by using solar fiber optic lighting systems. These panels have solar cells placed on both sides. The side facing the sun also gets its energy mainly from the direct component of the sun's radiation. However, the side facing away from the sun receives energy through the reflected component of the sun's radiation. In [5] uses vertically placed double-sided panels on the highway, so that during the first half of the day, one side of the panel is exposed to direct solar radiation and the opposite side during the second half of the day.

Transparent solar panels whose main purpose is to use solar energy along the entire surface of the highway are tested in Germany in terms of their cost-effectiveness and efficiency. The paper [6, 7] states that the biggest challenges of this type of project are reflected in ensuring the safety of road users. The benefits are manifold, with much more research needed to make the application of these panels in highways fully cost-effective [8]. The solar fiber optic lighting system consists of a collector that collects sunlight and sends it with fiber optic cables to the light source - the lamp [9].

Photovoltaic barriers (PVNBS) are a combination of noise protection systems and photovoltaic (PV) systems. Noise barriers are physical barriers designed to reduce noise levels between noise sources and sensitive receptors, such as hospitals, schools, and residential areas. The PVNBS literature, most of which is several years old, generally agrees that there is great potential for using existing and planned new noise barriers to produce solar energy. According to the information gathered, noise barriers are designed to produce energy without compromising their ability to reduce noise and in some cases can improve their performance. PVNBS may require either the retrofitting of existing noise barriers with PV modules (i.e., solar panels) or the integration of PV modules into the design of new noise barriers. In both cases, the protective insulation serves as a substructure for the PV modules. High-end built-in designs that provide additional area to the existing noise barrier structure are currently the most common PVNB approach [10]. Where noise absorption is required, an integrated cassette or zigzag design that allows a combination of sound reflection (outside PV glass) and sound absorption (via absorption material in non-PV areas) is applied. The disadvantage of cassette configuration or zigzag from an energy perspective is that the shape of the barrier encourages shading of the solar panels.

The latest generation of photovoltaic panels, thanks to recent technological innovations, can withstand heavy vehicle loads due to its resistant structure (hyper-resistant). The use of this type of solar panels could transform traditional asphalt roads into huge energy generators [11]. The panels consist of several elements: LED lighting to create lines and signposts without the use of paint, heating elements to prevent the accumulation of snow, ice, and microprocessors, which makes them intelligent. This allows the panels to communicate with the central control station. The panels are made of specially formed tempered glass that can withstand the weight of a vehicle with an asphalt-like surface [11]. Solar corridors could provide the so-called smart roads of the future by integrating weight sensors into strips and panels to prevent road hazards. When a person, object or animal crosses the road, sensors placed on the boards are activated by illuminating that part of the road and thus reduce the risk of accidents. In addition, roads as screens for sending messages, notifications and warnings to drivers are used [11].

Highway areas have physical and topographical features that complement solar energy production, such as well-maintained vegetation, easier access to a solar plant along the road and minimal presence of trees or other objects that can obscure sunlight [12].

The construction of solar panels next to highways, in addition to the installation of solar panels in noise barriers, represents a great potential for the conversion of solar energy into electricity with little investment, high space utilization and high cost-effectiveness [13].

3 Development Plans of Bosnia and Herzegovina

Bosnia and Herzegovina has significant potential for the installation of photovoltaic systems, especially the southern regions with many hours of sunshine per year and a significant amount of total radiation received during one year. In order to ensure optimal use of solar energy, it is necessary to define the technical characteristics of the facility on which the photovoltaic modules are placed. The roof or some other surface suitable for the construction of a solar power plant should be directed towards the south, southeast or southwest at a certain angle in order to ensure maximum electricity production (the angle depends on the geographical location of the facility). The arrangement of photovoltaic modules should be such that they do not overshadow each other, and the entire surface on which the power plant is planned to be built must not be in the shadow of the roof structure, chimney, ventilation openings, buildings or trees in the vicinity. The lifespan of solar power plants is 30 years, and the manufacturers guarantee the efficiency of the equipment for up to 25 years. One of the main characteristics of photovoltaic systems is the high degree of efficiency of the equipment during its lifetime [14].

As the production of electricity from renewable sources has been encouraged, thus greenhouse gas emissions have been reduced, various models for subsidizing photovoltaic plants (feed-in premiums, feed-in tariffs, trading in green certificates, etc.) have been developed in the world and in Bosnia and Herzegovina. However, with the sharp decline in production prices, photovoltaic systems have become fully competitive with conventional energy sources, so that any form of subsidy already frames some countries. In Bosnia and Herzegovina, the principle of subsidizing photovoltaic systems is based on a feed-in tariff that changes and adjusts every 18 months, and the number of current applications for the status of privileged and qualified electricity producers is significantly higher than available dynamic quotas for production from renewable energy sources. The permitting process is quite complex and is defined by laws, regulations and ordinances at the entity, district and cantonal levels. In order to build on the construction of photovoltaic power plants, apart from the long and complicated administrative procedure, it is necessary to wait several years. However, the downward trend in prices opens up the possibility of building solar power plants that are cost-effective even without achieving a special status and the right to purchase electricity at guaranteed prices.

The cost of investing in photovoltaic systems generally are divided into [15]:

- Costs of investing in photovoltaic modules;
- Investment costs in DC / AC inverters;
- Costs of investing in voltage regulators and charging batteries, accumulators (in the case of photovoltaic power plant models with electricity storage);

- Costs of investing in other equipment;
- Costs of consulting and design services;
- Costs of installation of equipment and connection to the electricity network;
- Equipment operation and maintenance costs.

The best indicator of the potential success of the project is the techno-economic analysis of its profitability, with information on cash flow and net financial and economic benefits during the period of plant operation. One of the best methods is the payback period method, which contains projections of annual revenues and expenditures during the use of a photovoltaic power plant.

The cost-effectiveness of all power generation technologies, and thus photovoltaic systems, is determined by [16]:

- Savings and/or revenues from the use of the system;
- Initial investment costs;
- Operating costs;
- Maintenance and servicing costs;
- Equipment dismantling costs at the end of the plant's service life.

Within the techno-economic analysis, the environmental aspect of the plant is also determined, ie. what impact does the newly built plant have on reducing carbon dioxide emissions.

4 Photovoltaics Market

The solar market in the world tends to grow rapidly. In the middle of the past decade, the annual growth rate exceeded 50%, and in 2008 about 100%, with about 3,000 MWp of total power of PV devices produced annually, which corresponds to a market value of over 5 billion US dollars per year. However, the global economic crisis has affected the market for PV devices in the past two years, significantly slowing down growth. Many PV companies simply disappeared, while others lowered the prices of manufactured PV devices just to survive [17].

The price of solar PV devices on the world market is falling at an average annual rate of about 5%. It is predicted that in the next few years, prices will fall at a higher rate, thanks to constant technological innovations and increasing competition in the market, and some analysts suggest an annual level of falling prices for PV devices of 20% and 30%.

This trend of constant development of technology and constant cheapening of equipment for PV energy, and on the other hand the growth of prices of traditional fossil fuels and electricity in all countries around the world, gives the right to expect equalization of prices of 1kWh of energy produced from conventional sources and directly from solar energy before 2030.

In the Federation of Bosnia and Herzegovina, the market for PV devices is poorly developed and too expensive, the production of PV devices does not exist, investments in research in this area are negligible, PV systems are only sporadic cases based on personal enthusiasm. Significant factors that determine the overall state and perspective of solar energy development are the global energy policy of the world community, as well as the state policy of development and investment in energy and renewable energy sources, and especially in solar energy [18].

If science and technology fully master the use of solar energy and its economically viable conversion into electricity, humanity and planet Earth will not feel the effects of global pollution and fossil fuel shortages in the future. Solar-to-electricity photovoltaic converters are probably the fastest and cleanest way to solve these problems.

5 Techno-Economic Analysis of PV System

The costs of electricity production are depending on the type of power plant, so there are differences in production costs between conventional and unconventional power plants that affect the profitability of electricity production. The most general division of production costs are investment costs, labor and maintenance costs, and decommissioning costs. This chapter is organized as follow: firstly in Sect. 5.1 is described a study case, location of PV plant and weather conditions of PV integration sites. In this Sect. 5.2 is presented calculation of possible electricity production from the future solar power plant. In the Sect. 5.3 cost-effectiveness of the project is observed. Additionally, in the Sect. 5.4 Helioscope online software are briefly introduced.

5.1 Case Study

A good location is essential when it comes to installing solar panels. This study focuses on a village Tarčin in the municipality of Hadžići. It is located around 30 km west of Sarajevo in the Sarajevo Canton within the Federation of Bosnia and Herzegovina. The location of solar power plant is next to the part of the Corridor Vc, A1 motorway between latitudes 43°47′51″N, 43°47′24″N and longitudes 18°05′17″E, 18°04′49″E, with an area of 15,000 km^2. The altitude of the case study from the sea level is 659 m. The average temperature of the area is 8,5 °C, The warmest month is July with average temperatures between 20 °C and 22 °C. The annual distribution of clouds shows that the cloudiest part of the year is winter, while in the summer half of the year, the clouds are below 50%, which can explain the long duration of sunshine in this area. The average solar radiation in the selected case study is 1,300 $kWhm^{-2}$ $year^{-1}$. The freestanding solar panels placed horizontally and vertically are used. Also, solar panels have been installed on existing noise barriers. Location for the construction of a solar power plant with marked places for the installation of panels is presented on Fig. 1.

With the construction of solar power plants at this location, it is planned to supply electricity to the next tunnels: November 25 (length: 2,800 m, coordinates: 43°49′00″N 18°05′27″E/43,816693°N 18,090963°E), Grab (length: 400 m, coordinates: 43°48′03″N 18°05′21″E/43,800713°N 18,089161°E) and Ivan (length: 3,192 m, coordinates: 43°45′11″N 18°02′31″E/43,752931°N 18,042083°E) on the A1 highway.

Solar photovoltaic power plant with an installed capacity of 1 MW is built using photovoltaic modules *Canadian Solar CS6U-345M 345W Max Power Solar Module*. The characteristics of the Canadian Solar CS6U-345M 345 W Max Power Solar Module are

Fig. 1. Location for the construction of a solar power plant with marked places for the installation of panels

reflected in the efficiency of the module which is 17.74%. A total of 3,161 Canadian Solar 3S6U 345M modules connected to 185 wires are used, and 37 Sunny Tirpower 24000TL-US inverters are used. Furthermore, the module has a high value of PTC parameter, which is 92.17%, and outstanding performance at low irradiation (200 W/m^2–1,000 W/m^2) with an average relative efficiency of 96.5%. For severe weather conditions, the IP68 standard for long-term durability of the junction box is provided. They can withstand pressures in the amount of up to 5400 Pa under the load of accumulated snowfall, and 2,400 Pa for strong wind conditions. The price of this type of module ranges from € 260 to € 360 (depending on the distributor for sale). The inverters used are type SMA Sunny Tripower STP24000TL-US-10 24 KW Inverter. These inverters provide fully functional network management, as well as superior communication and monitoring. They are suitable for 600 V DC and 1000 V DC applications, which allows flexibility with a wide range of input voltages. The efficiency of the converter used is 98%. The presence of harmonics in the AC output signal is less than 3%. The maximum output current is 29A. Prices for inverters in this series range from € 2,500 to € 8,500, depending on the type selected.

5.2 Calculation of Possible Electricity Production from the Future Solar Power Plant

The total annual consumption for the tunnels; 25 November, Grab and Ivan is 5,845,185.6 kWh, and the amount of electricity produced by the solar power plant is 1,173,500.118 kWh per year. Figures 2 and 3 show the time series of electricity consumption of the tunnels and electricity production by the solar power plant, respectively.

In addition, Figs. 4 and 5 show the differences in production using solar energy and consumption during higher and lower tariff respectively.

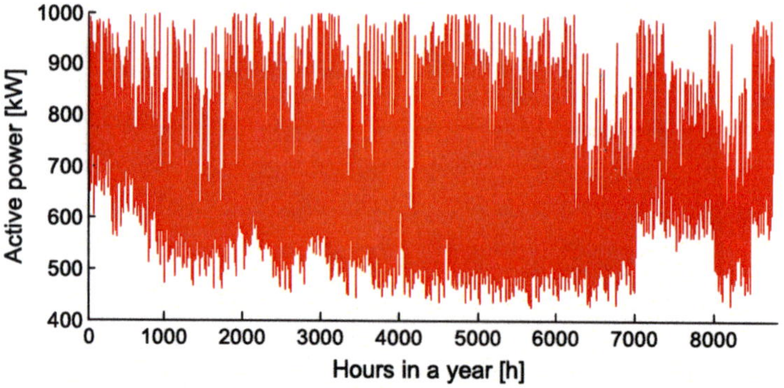

Fig. 2. Total consumption for tunnels: 25. November, Grab, and Ivan for a period of one year

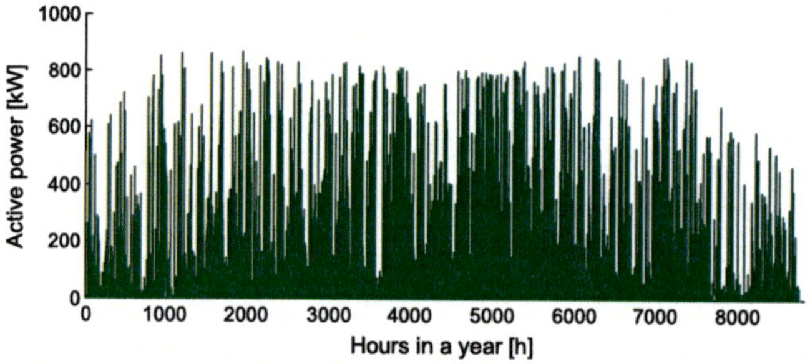

Fig. 3. Total production from the solar power plant for a period of one year

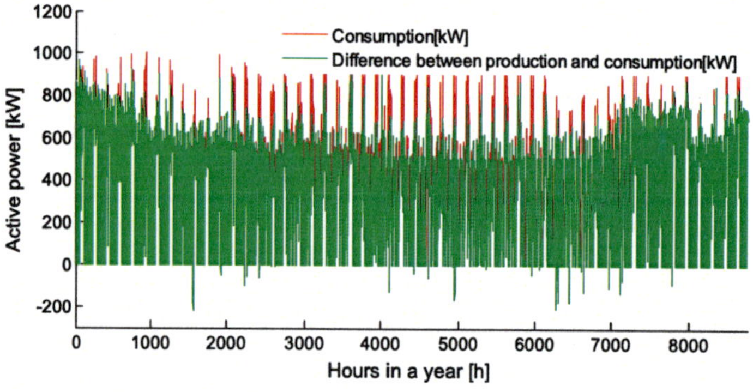

Fig. 4. Difference between production and consumption during higher tariff

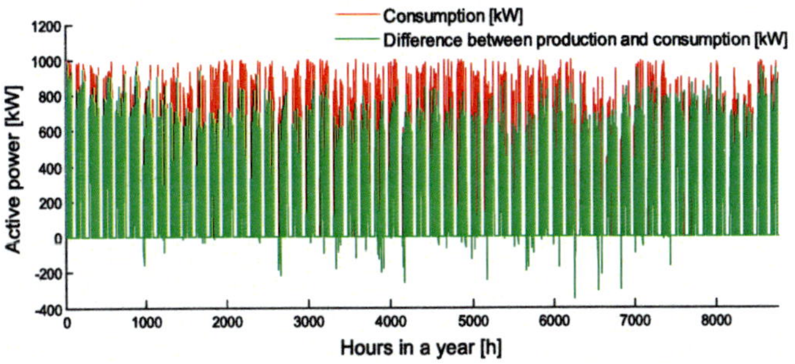

Fig. 5. Difference between production and consumption during lower tariff

5.3 Cost-Effectiveness of the Project

The main indicators of project profitability are cash flow and a report on net financial and economic benefits. They contain projections of annual revenues and expenditures over the life of the project.

The cost-effectiveness of the project can be obtained using several methods, some of which are:

- Return period method
- Net (net) present value method
- Internal rate of return (IRR) method.

During the first year of operation, Canadian Solar guarantees that the output power of the module will not be less than 97.5% of the indicated output power. The rest of the warranty period of 25 years, the annual percentage drop in power output will not exceed 0.7%. Table 1 shows the annual revenues from energy produced, taking into account the stated output power declines.

The value of the investment is estimated at 897,727.5 €. During the years of exploitation, the primary investment cost will be repaid, and in the twelfth year the investment will be repaid, which can be seen in Table 2. At the end of the twenty-fifth year, the total net income will amount to 896,593.26 €. Figure 6 shows the cumulative cash flow by year for the PV power plant.

In this paper is used return on investment method. The payback period is the minimum number of periods (years) in which net cash flows will exceed investment costs. The simplest criterion of financial decision-making. It does not take into account the time value of money. Favoring projects with shorter payback times reduces investment risk. Therefore, the next formula for calculation of payback period is:

$$I \leq \sum_{t=1}^{t_p} F_t$$

Where are:
t_p - payback period,

Table 1. Annual revenues from energy produced

Year of operation	Efficiency	Coefficient	Production [kWh]	Income [€]
1	100%	1	1,173,500	78,402
2	98%	0.993	1,165,286	77,853
3	97%	0.986	1,157,071	77,304
4	96%	0.979	1,132,773	75,680
5	95%	0.972	1,140,642	76,206
6	95%	0.965	1,132,428	75,657
7	94%	0.958	1,124,213	75,109
8	93%	0.951	1,115,999	74,560
9	93%	0.944	1,107,784	74,011
10	92%	0.937	1,099,570	73,462
11	91%	0.93	1,091,355	72,914
12	91%	0.923	1,083,141	72,364
13	90%	0.916	1,074,926	71,816
14	89%	0.909	1,066,712	71,267
15	88%	0.902	1,058,497	70,718
16	88%	0.895	1,050,283	70,169
17	87%	0.888	1,042,068	69,621
18	86%	0.881	1,033,854	69,072
19	86%	0.874	1,025,639	68,523
20	85%	0.867	1,017,425	67,974
21	84%	0.86	1,009,210	67,426
22	84%	0.853	1,000,996	66,876
23	83%	0.846	992,781	66,328
24	82%	0.839	984,567	65,779
25	81%	0.832	976,352	65,230
Sum			**26,857,069**	**1,794,321**

F_t - net cash flow in year t,

I - investment costs.

If the payback period is shorter than the required payback period, the project is eligible.

Cumulative net cash flows become contingent in the 12th period, so the payback period is 12 years.

Table 2. Cumulative net income

Year of operation	Net income (annual amount) [€]	Cumulative net income [€]
0	−897,728	−897,728
1	78,401.79	−819,326
2	77,852.52	−741,473
3	77,303.76	−664,169
4	75,680.43	−588,489
5	76,206.24	−512,283
6	75,657.48	−436,625
7	75,108.72	−361,517
8	74,559.96	−286,957
9	74,011.2	−212,945
10	73,462.44	−139,483
11	72,913.68	−66,569.8
12	72,364.41	5,794.62
13	71,815.65	77,610.78
14	71,266.89	148,877.7
15	70,718.13	219,595.8
16	70,169.37	289,765.2
17	69,620.61	359,385.8
18	69,071.85	428,457.6
19	68,523.09	496,980.7
20	67,974.33	564,954.5
21	67,425.57	632,380.1
22	66,876.3	699,256.4
23	66,327.54	765,584
24	65,778.78	831,363.2
25	65,230.02	896,593.3

5.4 HelioScope Software Package

HelioScope software is web software that can be used to easily design photovoltaic systems in the desired location. Based on the location name or geographic coordinates, the software displays a satellite image of the area via Google Maps or Bing Internet Maps. If neither of these two services provides a good snapshot of the location, it is possible to use your own images as well as the Nearmap application. In this way, it is possible to precisely select the areas in the image on which they want to build photovoltaic power plants. The software provides a choice of different types of panels and inverters, after

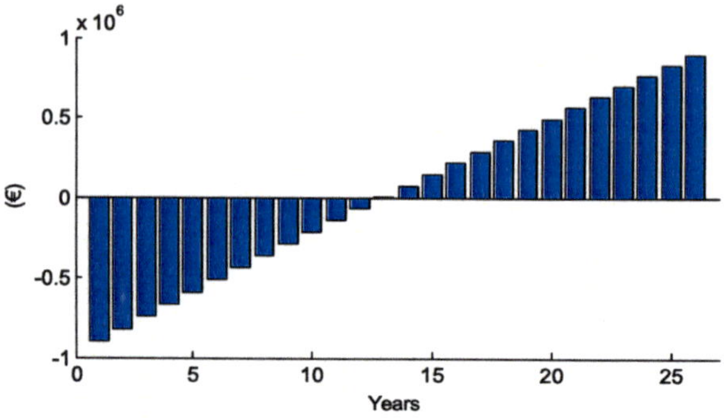

Fig. 6. Cumulative net income

which the power plant display is automatically generated on the selected area together with data on the number of panels and peak power plant production in kW as shown in Fig. 7. By changing parameters such as angle of inclination and azimuth of the panel, the height of the panel installation, the distance between the panels and the number of panels in one row can increase or decrease the power plant production. Finally, the software provides the possibility of a tabular overview of the hourly production of the power plant, losses in the system and other important data at the level of the year. The final appearance of the power plant can be displayed in 2D and 3D display using one of the CAD software.

Fig. 7. Satellite view of a photovoltaic power plant designed in HelioScope software

6 Analysis Results

The results of the analysis are presented through the appropriate tables and figures.

Figure 8 shows the production, consumption and the difference between production and consumption for one week in July where the peak power is in the amount of

1000.8 kW. From the figure, can be seen that with the construction of the solar power plant, the peak power is reduced from 1000.8 kW to 411.76 kW. In addition, with the construction of a solar power plant on this location, the peak power for the month of July is reduced to 845.136 kW. It can be concluded avoiding on-peak uses of energy, the customer can lower their bill. Peak power saving has positive impacts on both customers and the environment. Unlike conventional electricity rates, the time span of its utilization offers different prices for energy at different times of the day and different days of the week, depending on the demand. Demand reduction during peak periods also allows utility company more flexibility to move toward sustainable energy. Having a positive impact on the environment will continue to increase in importance for the utility company. With reducing demand during peak periods, a utility company can expand further into clean energy.

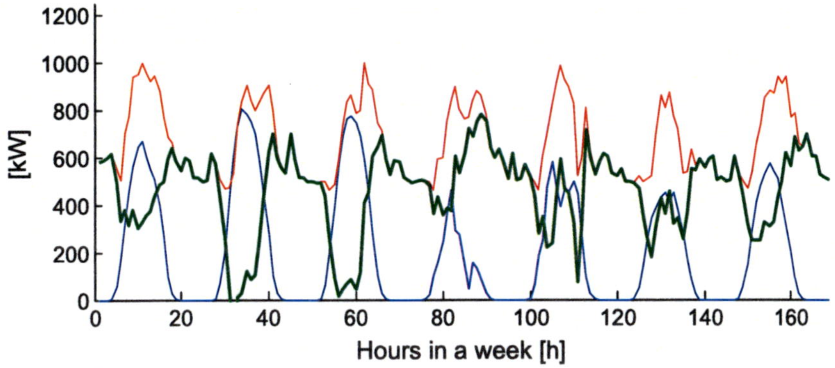

Fig. 8. The difference between production and consumption for one week

Figure 9 shows a comparison of monthly electricity bills. During the summer months, the production of the PV power plant is higher compared to the rest of the year, which is why the costs in the corresponding period are significantly reduced. By analyzing the obtained results, it can be concluded that the given investment represents a cost-effective techno-economic solution for reducing electricity costs. It should be noted that for more detailed analyzes it is necessary to take into account the maintenance costs within the period under review as well as all associated costs during the construction and connection of the PV plant to the transmission network.

Figure 9 presents the total monthly price of electricity, of which the red bar represents the reduction of the price when using electricity obtained by installing the proposed solar power plant. From the figure, it is also noticeable that for each month there are certain savings on the electricity bill presented in green due to the use of a solar power plant to power the mentioned tunnels.

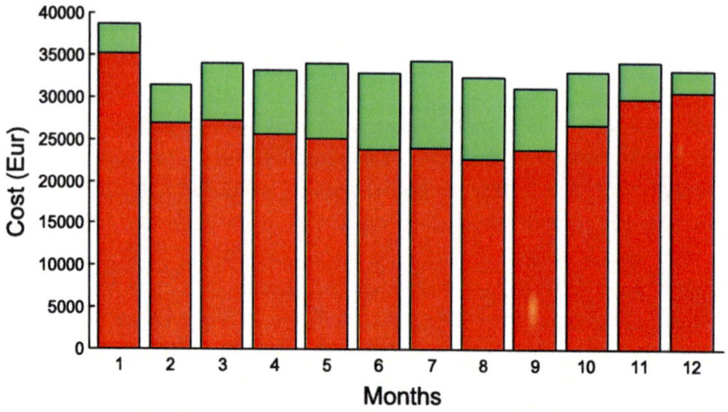

Fig. 9. Comparison of monthly electricity bills

7 Conclusion

There are many advantages of solar energy, but the most important feature is the reduction of electricity or heat costs, and a positive impact on the environment. Inevitably, the installation of solar panels can be considered a large investment, but in each case it is possible to make an accurate calculation and determine the payback period, which is usually several years. In addition, advantages of photovoltaic systems are simple maintenance, energy is clean - photovoltaic systems do not pollute the environment and provide one of the best solutions to reduce harmful emissions of gases and particles. Production is quiet and one of the great advantages of solar energy is the ability to produce energy in remote places where there is no possibility of connection to the grid.

Corridors and highways have large areas, which can be used for the construction of photovoltaic power plants within the corridor highway. The production of electricity from these PV power plants is used to power their own consumption, such as tunnels and lighting today, and tomorrow for mass use to charge electric viechels. In this paper, a techno-economic analysis of a solar photovoltaic power plant with an installed capacity of 1 MW in the village Tarčin, next to the A1 highway, is performed. This power plant would supply lighting on the intersection loop itself and three tunnels near the intersection loop. Also, solar panels have been installed on existing noise barriers. From the analysis, it can be concluded that most of the produced energy is consumed in a higher tariff according to the applicable tariff rates. With the purchase price of electricity during higher tarrife 0.0666 € per produced kWh, and lover tarrife with 0.0447 € per produced kWh, the investment would pay off in the 12 year of operation with cumulative profit of 896,593.26 € after 25 years of operation of the power plant reduction of the cost of photovoltaic applications and increases in investment and technological innovation.

References

1. Wuebbles, D.J., Sanyal, S.: Air quality in a cleaner energy world. Curr. Pollut. Rep. **1**(2), 117–129 (2015). https://doi.org/10.1007/s40726-015-0009-x
2. Jaeger-Waldau, A.: PV Status Report 2019, EUR 29938 EN, Publications Office of the European Union, Luxembourg. ISBN: 978-92-76-12608-9 (2019) (online), JRC118058, 9 May 2020
3. Gabeljić, S.: Renewable Energy Sources in BiH: Question of (UN) Sustainability. Acta Geographica Bosniae et Herzegovinae, Association of Geographers in Bosnia and Herzegovina (2018). ffhal-02121846
4. Pena-Garcia, A., Gomez-Lorente, D.: Installation of solar panels in the surroundings of tunnel portals: A double-targeted strategy to decrease lighting requirements and consumption. https://doi.org/10.1016/j.tust.2019.10325110.1109/ICSCAN.2018.8541253. Accessed 8 Jan 2020
5. Sharma, P.: Solar energy generation potential along national highways. https://doi.org/10.1186/2251-6832-4-16 Accessed 4 Apr 2013
6. Enkhardt, S.: Photovoltaics for highway. https://www.pv-magazine.com/2020/09/01/photovoltaics-for-highways/. Accessed 1 Sept 2020
7. Jia, L., Ma, J., Cheng, P., Liu, Y.: A Perspective on Solar Energy-powered Road and Rail Transportation in China. https://doi.org/10.17775/CSEEJPES.2020.02040. Accessed 4 Dec 2020
8. Schneider, P.: Solardach so groß wie Bremen: Liegt Lösung für Stromproblem auf unseren Autobahnen. https://www.focus.de/perspektiven/nachhaltigkeit/forschungsprojekt-in-oesterreich-koennte-drittel-unseres-bedarfs-decken-liegt-loesung-fuer-stromproblem-auf-den-autobahnen_id_12262256.html. Accessed 19 Oct 2020
9. Qin, X., Zhang, X., Han-Han, S.Q.: Design of Solar Optical Fiber Lighting System for Enhanced Lighting in Highway Tunnel Threshold Zone: A Case Study of Huashuyan Tunnel in China. https://doi.org/10.1155/2015/47136410.1109/REDEC.2016.7577552. Accessed 7 Dec 2015
10. Poe, C., Plovnick, A., Hodges, T., Hastings, A., Dresley, S.: Highway renewable energy: photovoltaic noise barriers (2017). https://rosap.ntl.bts.gov/view/dot/32521. Accessed 7 Jan 2017
11. Mehta, A., Aggrawal, N., Tiwari, A.: Solar Road-ways-The future of roadways. http://large.stanford.edu/courses/2017/ph240/cook2/docs/mehta.pdf. Accessed 1 May 2015
12. Hirsch, A.P.A.: Potential Impacts of Solar Arrays on Highway Environment, Safety and Operations (2015). https://www.codot.gov/programs/research/pdfs/2015-research-reports/solar-arrays. Accessed 01 Oct 2015
13. Parke, G.: The application of Solar-Noise Barriers for UK highways and their combinated benefits for local goverment (2015). https://www.conforg.fr/euronoise2015/proceedings/data/articles/000472.pdf. Accessed 01 June 2015
14. Tokic, A., Akinci, T.C., Zengin, A.: Bosnia and Herzegovina's renewable energy policy and perspective. Int. J. Energy Econ. Policy **10**, 524–530 (2020). https://doi.org/10.32479/ijeep.9777
15. de Oliveira Azevêdo, R., Rotela Junior, P., Rocha, L.C.S., Chicco, G., Aquila, G., Peruchi, R.S.: Identification and analysis of impact factors on the economic feasibility of photovoltaic energy investments. Sustainability **12**, 7173 (2020). https://doi.org/10.3390/su12177173
16. Cengiz, M.S., Mamiş, M.S.: Price-Efficiency relationship for photovoltaic systems on a global basis. Int. J. Photoenergy **2015**, 12. Article ID 256101 (2015). https://doi.org/10.1155/2015/256101
17. Arantegui, R.L., Jäger-Waldau, A.: Photovoltaics and Wind Status in the European Union after the Paris Agreement. Renewable and Sustainable Energy Reviews, vol. 81, Part 2, pp. 2460–2471. ISSN: 1364–0321 (2018). https://doi.org/10.1016/j.rser.2017.06.052

18. MFTERBH: Ministry of Foreign Trade and Economic Relations of Bosnia and Herzegovina. Framework Energy Strategy of Bosnia and Herzegovina until 2035 (2019). http://www.mvteo.gov.ba/data/Home/Dokumenti/Energetika/Framework_Energy_Strategy_of_Bosnia_and_Herzegovina_until_2035_ENG_FINAL.pdf. Accessed 28 Feb 2021
19. Folsom Labs: Helioscope online software tool (2014). https://www.holioscope.com/. Accessed 28 Feb 2021

Economic Analysis of the Application of Photovoltaic Power Plants - Smart City Sarajevo

Mirza Tuco, Maja Muftić Dedović(✉), Nedim Šahovic, Selma Čokljat, Lejla Cogo, Alija Musić, Alma Mašić, and Samir Avdaković

Faculty of Electrical Engineering, University of Sarajevo, Sarajevo, Bosnia and Herzegovina
maja.muftic-dedovic@etf.unsa.ba

Abstract. In the last two decades, a smart city has become increasingly popular in the scientific literature and international politics. To understand this concept, it is important to recognize why cities are considered key elements for the future. Cities around the world have begun to look for solutions that enable high-quality urban services with long-term positive effects on the economy. In times of economic uncertainty, focus should be on identifying renewable sources and enabling sustainable growth of cities. This paper presents the elements for reviewing the achieved energy development in the city of Sarajevo, the construction of existing energy infrastructure, the degree of utilization of energy capacity, consumption, methods of use, on which depends the improvement of economic and environmental conditions (primarily maintaining clean air as a limiting factor). The focus of this paper is the integration of solar panels for 35,000 households that currently use fossil fuels to heat their households. Furthermore, economic indicators of the application of the new approach and new energy policies is given, which would influence the better ranking of the city of Sarajevo through innovative and technologically based solutions. Concluding remarks are given regarding the techno-economic aspects of the impact of solar panel installation on the transmission and distribution network.

Keywords: Solar photovoltaic power · Economic analysis · Smart city · Renewable energy · Payback period

1 Introduction

The realization of the idea of a smart city is by no means simple and requires a comprehensive analysis of the available possibilities regarding the simplification of everyday life of citizens. If we look globally, more than 60% of the world's population lives in cities, more than 70% of people work in cities, while 80% of the world's economic value is created in cities. Precisely because of this rapid growth, scientists and experts around the world are devoting themselves to designing cities as prosperous, smart and liveable places for all residents.

N. Ademović et al. (Eds.): IAT 2021, LNNS 316, pp. 787–797, 2022.
https://doi.org/10.1007/978-3-030-90055-7_63

Smart city means, among other things, digital management of public city transport and communal services such as lighting, heating and water supply, in order to efficiently manage resources, reduce pollution, increase traffic flow and, consequently, improve the quality of life in the urban environment.

In terms of energy, the development trend of smart cities that are fully energy sustainable is significantly strengthening in the world. These are cities that receive a large percentage of electricity and heat from renewable sources, most often hydropower, solar energy, geothermal energy, biomass and wind energy.

The focus of this paper is the integration of solar panels for 35,000 households that currently use fossil fuels to heat their households. The paper will present energy consumption in the city of Sarajevo, then hourly electricity consumption on an annual basis, as well as hourly electricity production by solar panels. Furthermore, an economic analysis of the proposed investment is given, which plans 6kW installations of solar panels for each household. Concluding remarks are also given regarding the technical aspects of the impact of solar panel installation on the transmission and distribution network.

The main goal of this study research is to identify the necessary measures that would enable the accelerated development of a 'smart/modern' city, in this case the city of Sarajevo. Given the very wide area of the strategic framework for the development of a city and which includes a large number of disciplines, a special focus will be on more intensive construction and use of renewable energy sources, primarily low-power photovoltaic power plants. In addition to electricity generation and their integration into modern concepts of 'smart' cities, one of the goals of the paper is to assess the positive effects on reducing air pollution. Considering that gas and electricity are available to citizens in the city of Sarajevo as energy sources, there is a possibility to reduce or completely eliminate the use of solid fuels in the heating function with optimization techniques and stimulative measures, which would result in a significant reduction of pollution.

2 Literature Overview

The idea of smart cities is not new, but the possibilities of its application and development is known for the last 30 years. Some of the good practice examples of smart city development in the world are given below. Systematic planning and the application of appropriate technology have managed to optimize the use of resources, reduce costs and increase revenues, make services better and more efficient, and thus improve the quality of life of its citizens.

REloadIT is a project of the city of Amsterdam in which 16 cars owned by the city are powered based on sustainable energy generated by two solar panels and one wind farm. The goal is to use renewable energy sources and preserve the environment [1].

Barcelona is known for installing solar panels, and the most famous is the Solar Photovoltaic Installation Forum, opened in 2004. Covering an area of 4,500 m^2, it is mounted on four reinforced concrete pillars. The energy produced by this panel is enough to supply 1000 households [2].

Among the most successful examples are often mentioned Copenhagen and Stockholm due to the development of smart technologies and significantly reduced electricity consumption. As leading smart cities, they lead in the use of energy from renewable sources and green IT technology [3].

Vienna is also among the better-ranked smart cities thanks to the establishment of a public-private organization tasked with developing a strategy and solution for Smart City. Among other things, incentives are planned so that 50% of electricity by 2030 would come from solar power plants.It is about optimizing the energy consumption and heating system in the dormitory, private apartment building and school campus, and all three buildings are equipped with solar collectors, solar heaters, hybrid panels, heat pumps and many other technologies [4].

The most common examples applied in practice are those with a combination of both panels and collectors, thus achieving significant energy savings. Home for Life is located in the second largest city in Denmark, Aarhus, where the insolation is in the range of 1,000–1,050 kWh/m^2 per year. The house is designed to produce more energy than it consumes and is the result of research and design in the development and goal of low carbon production and the application of solar energy. Strategic installation of solar cells, solar collectors, facade and roof windows offers a balanced amount of energy flow through the building [5].

The paper [6] highlights the problems that have led to a widely accepted concept of smart city, such as climate change, increasingly aggressive urbanization, excessive use of all resources, etc. The paper explains the role of architectural and urban planning profession in creating a smart city. Modelling of key systems of urban development requires the architects and urban planners to have new knowledge in different fields, as well as an interactive and creative approach to solving the set tasks. The article [7] presents technical description and performance analysis of a 50 kW solar PV plant installed on a business facility of JP Elektroprivreda HZ HB in Tomislavgrad (Bosnia and Herzegovina). The experimental data are based on real time measurements and they can be a good support in the preparation of this paper.

This paper [8] discusses the solar energy potential of built environments and explores how urban planning can contribute to converting existing cities into Solar Smart Cities.

For future urban planning, new buildings can be designed to integrate PV systems in their structure to maximise the installation space. The lack of adequate installation space for PV panels is a key barrier in the context of cities, as PV panels need to be angled in the right direction to maximise solar exposure [9].

3 Energy Consumption in Sarajevo

In this chapter, the energy needs of Sarajevo Canton for one year are presented, which are based on detailed/necessary structural projections of consumption: electricity, natural gas, petroleum products and solid fuels. According to the energy balance of Sarajevo Canton energy needs in 2018 [10] are shown in Table 1.

Table 1. Sarajevo Canton energy needs for 2018 [10]

Energy	Unit	GWh	% in total needs
Electricity	1,417,489 [MWh]	1,417.49	30.59
Natural gas	126,215 [mil.Sm^3]*	1,168.75	25.22
Solid fuels	99,800 [t]	334.66	7.22
Petroleum products	182,807,500 [liters]	1,712.91	36.97
Total		**4,633.81**	**100**

*Oil and gas volumes are often stated in standard cubic meters (Sm^3)

Furthermore, Fig. 1 shows the monthly electricity consumption in the Sarajevo Canton for 2018.

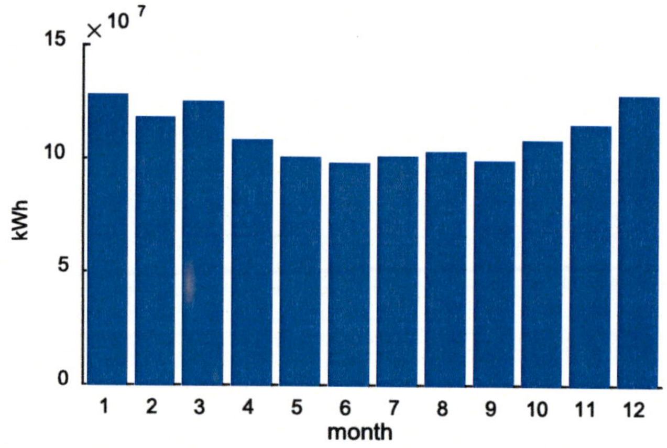

Fig. 1. Monthly electricity consumption in Sarajevo Canton for 2018 [10]

4 Smart City Sarajevo – Case Study

The city of Sarajevo, as the capital of Bosnia and Herzegovina, is the most densely populated and economically most active area of the country. In order to improve life in the city, the installation of a solar panels on 35,000 housholds that currently use fossil fuels to heat their households. Households covered by this study are located on the slopes of the city of Sarajevo where there is less exposure to fog. With solar radiation of about 1,340 kWh/m^2 per year, the conditions for the use of solar energy in the city of Sarajevo are quite favorable. The annual average of daily solar radiation varies between 3.4–4.4 kWh/m^2. According to certain estimates, if solar panels are installed on the roofs of Sarajevo buildings, about 26 MWh of electricity could be produced. Data for the territory of the city of Sarajevo for solar energy show that the average duration of insolation is

2,000 h. About 70% of electricity is generated from April to September, which makes solar a better investment to mitigate peak electricity compared to other renewables [11]. When the average weather conditions, atmospheric pollution and humidity are taken into account, the actual average radiation energy in Sarajevo is around 3.5 kWh/m^2 per day. The project involved the installation of solar panels on the roofs of housholds at an angle of 35° oriented to the west, east and south. The installed power of an individual panel is 6 kW. It is planned to install 6 panels on each household, two facing south, two facing west and two facing east. Accordingly, the hourly production of solar panels for all 35,000 households is given below. Figure 2 present total electricity consumption for Canton Sarajevo on annual basis. On Fig. 3 is plotted production from solar panels for a period of one year.

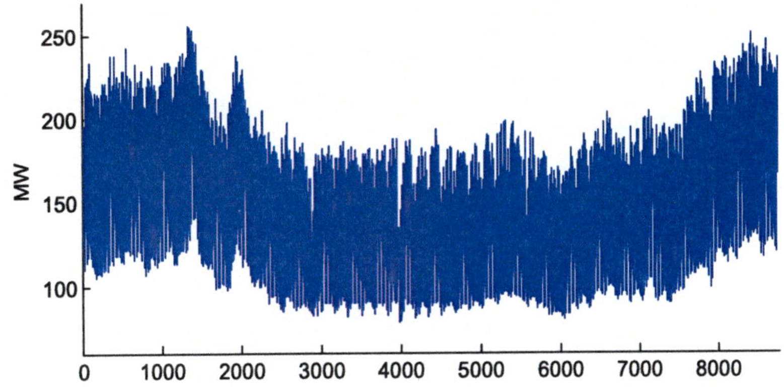

Fig. 2. Total electricity consumption for Canton Sarajevo on annual basis

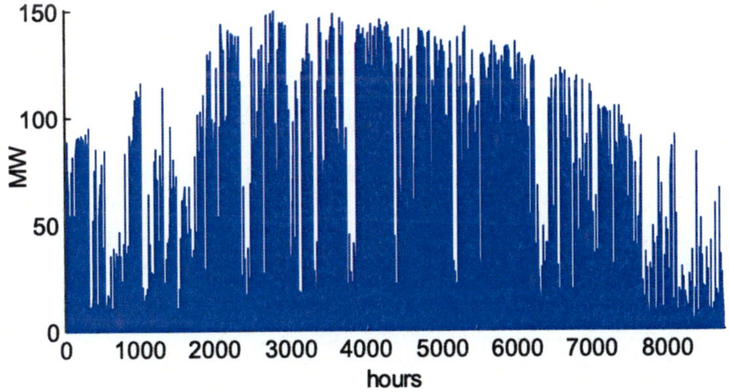

Fig. 3. Total production from the solar panels installed on 35,000 households for a period of one year

From Fig. 2 and Fig. 3 and obtained result on solar panel electricity production it can be seen that the production of electricity by solar panels represents 14.64% of the total electricity consumption at the level of the Sarajevo Canton.

It should be taken into account that electricity produced from the sun should be normalized to take into account annual climate differences. In accordance with this, the figure below shows the normalized load curve on an annual basis for the Canton Sarajevo, as well as the normalized electricity production curve from solar panels. The Fig. 4 in green shows the normalized production of electricity from the sun, while the red color represents the normalized load curve.

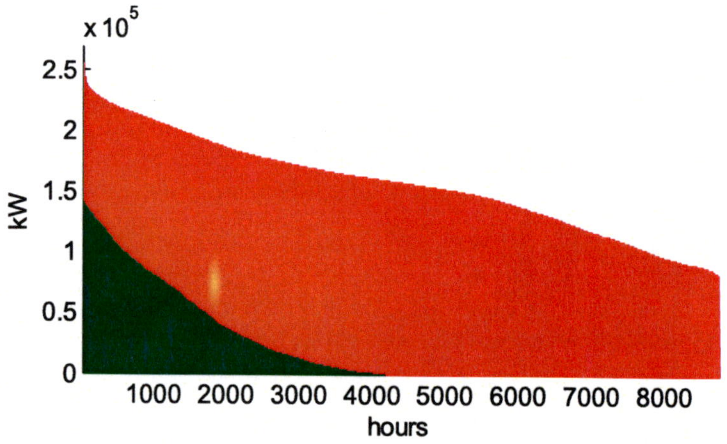

Fig. 4. Normalized load curve (red) and normalized production curve (green) on annual basis

5 Techno-Economic Analysis

In order to elaborate the cost-effectiveness of installing solar panels to produce electricity for to households heating that are currently heated by fossil fuels, the electricity consumption of one household that is currently being heated by electricity is shown. Figure 5 present annual electricity consumption of one household that uses electricity for heating.

The most important inputs for any economic analysis of a photovoltaic system are the initial costs of the system and the amount of energy that is delivered each year.

The decision on whether a system is profitable depends primarily on the price of energy determined by the system, taking into account taxes and duties. A detailed economic analysis includes estimates of operating and maintenance costs and future electricity prices from the grid. Some of the input data are the electricity production of the solar system of one household is 5729 kWh. Then, the total electricity consumption of one household heated by electricity is 11,291 kWh. The average price of produced electricity is 0.08 €/kWh. The investment expressed in monetary units per kWh produced is 765 €/kWh, while the total investment for the installation of solar panels for one household is in the amount of 4,590 €. The cost of the plant includes the network fee, while the income consists of the sold quantities of electricity. Considering that this is a two-tariff system, the calculation of the sold electricity is performed at a higher and lower tariff.

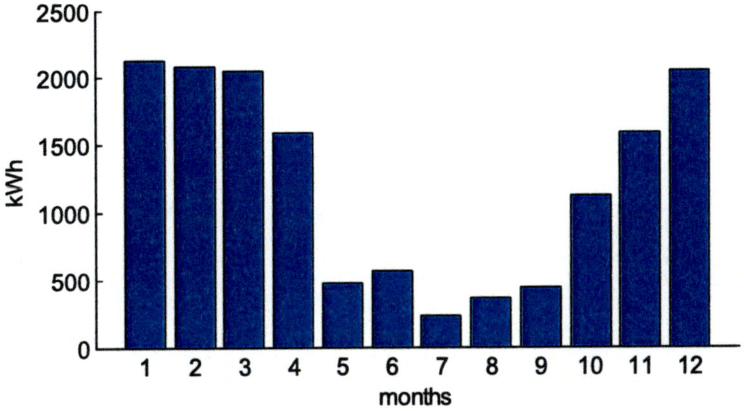

Fig. 5. Annual electricity consumption of one household that uses electricity for heating

In this paper is used return on investment method. The payback period is the minimum number of periods (years) in which net cash flows will exceed investment costs. The simplest criterion of financial decision-making. It does not take into account the time value of money. Favoring projects with shorter payback times reduces investment risk. Therefore, the next formula for calculation of payback period is:

$$I \leq \sum_{t=1}^{t_p} F_t$$

Where are:
t_p - payback period,
F_t - net cash flow in year t,
I - investment costs.

If the payback period is shorter than the required payback period, the project is eligible [12].

The lifespan of solar power plants is 30 years, and the manufacturers guarantee the efficiency of the equipment for up to 25 years. Table 2 present net income and cumulative net income for the period of 25 years. From Table 2 and Fig. 6 is evident that the primary investment cost will be repaid, and that the ten year is a payback period. At the end of the twenty-fifth year, the total net income will amount to € 6,846.75. While at the end of 20 years, a cumulative net income is worth the initial investment.

Furthermore, it is elaborate how much would be reduced fossil fuel usage and how it will contribute in the economical aspect to households' owners. For one household that uses electricity for heating, about 6,300 kWh is needed during the heating season, more precisely in a period of 5 months. Households use fossil fuels, coal or firewood for heating. 7 tons of coal or 16 m^3 of firewood are needed to heat the household during the heating season. The corresponding price for 7 tons of coal is € 536.63, while the price required for the purchase of firewood is around € 674.62. As certain households use coal and some firewood, an average price of € 605.625 is taken. Therefore, the use of photovoltaic panels that would enable the electrification of household heating and eliminate the use of fossil fuels would have annual savings in the amount of € 605.625. Taking into account the cost of installing photovoltaic panels per household and the mentioned savings due to the non-use of fossil fuels for annual heating, the return on investment would take less than 20 years as shown in the Table 2.

Table 2. Cumulative net income

Year of operation	Net income (annual amount) [€]	Cumulative net income [€]
0	−4,590	−4,590
1	457.47	−4,132.53
2	457.47	−3,675.06
3	457.47	−3,217.59
4	457.47	−2,760.12
5	457.47	−2,302.65
6	457.47	−1,845.18
7	457.47	−1,387.71
8	457.47	−930.24
9	457.47	−472.77
10	457.47	−15.3
11	457.47	442.17
12	457.47	899.64
13	457.47	1,357.11
14	457.47	1,814.58
15	457.47	2,272.05
16	457.47	2,729.52
17	457.47	3,186.99
18	457.47	3,644.46
19	457.47	4,101.93
20	457.47	4,559.4
21	457.47	5,016.87
22	457.47	5,474.34
23	457.47	5,931.81
24	457.47	6,389.28
25	457.47	6,846.75

Regarding the technical aspects of the construction of solar panels on households in order to produce electricity for their own needs with the possibility of handing over the surplus produced electricity to the electricity grid, two items are singled out. The

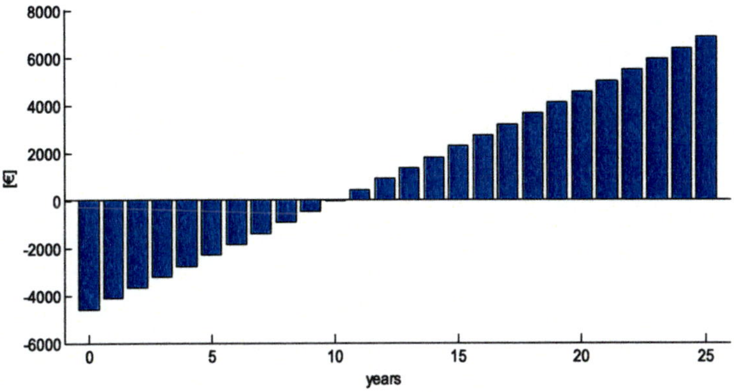

Fig. 6. Cumulative net income

first refers to peak power. Peak electricity can considerably affect the price of electricity, which in turn can affect the economy, as well as quality of life. The second relates to the impact on the transmission and distribution network. The analysis on peak demand includes data for the one week. On Fig. 7 are plotted time series of electricity consumption in blue and electricity production from PV in green. In terms of peak demand reduction, it is interesting to look at the red curve in the figure that represents the difference between production and consumption at the one-week level. From the figure, can be seen that with the construction of the solar panels on households, the peak power is reduced from 177,978.68 kW to 27,491.1 kW. Peak power saving has positive impacts on both customers and the environment. It can be concluded avoiding on-peak uses of energy, the customer can lower their bill. Unlike conventional electricity rates, the time span of its utilization offers different prices for energy at different times of the day and different days of the week, depending on the demand. Demand reduction during peak periods also allows utility company more flexibility to move toward sustainable energy.

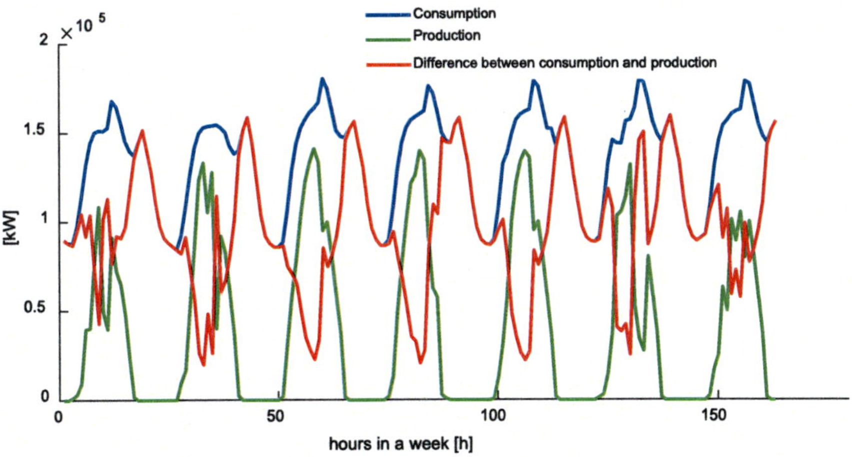

Fig. 7. The difference between production and consumption for one week

According to influence of distributed generation on the transmission network, the model of the electric power system of the city of Sarajevo is considered. Changes in active load power are made for each hour during the day on May 17th. It can be seen that changes in active load power affected the voltage in the power system. 110 kV bus voltage changes are shown in Fig. 8.

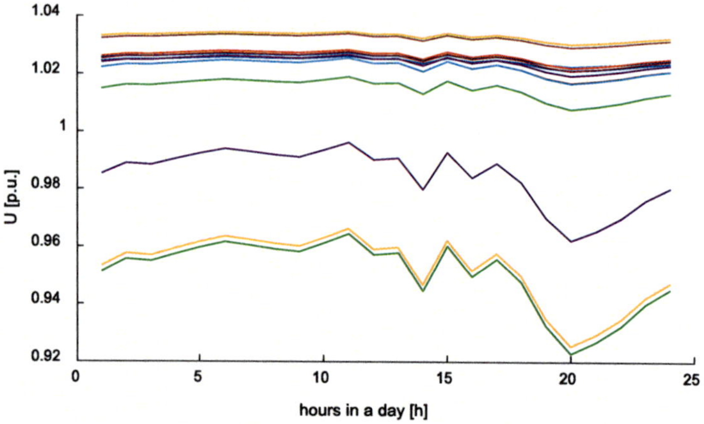

Fig. 8. 110 kV bus voltage change

From Fig. 8, it can be seen that the voltages are within the allowable values, and that the change in the load of the distribution network does not affect the transmission network. It can be conclude that significant changes in the consumption of the city of Sarajevo, caused by the generation from PV, do not have a significant impact on voltage changes in the transmission network. These changes are likely to be significant in the distribution network and will be the subject of further research.

6 Conclusion

The Smart Cities project has become increasingly important in recent years. The need for the development of Smart Cities is a consequence of population growth and the need for more clean energy. This paper presents the integration of renewable energy sources. Special emphasis is given to the energy sense of the smart city. The air in the city of Sarajevo is extremely polluted during the winter months, and many households still use unclean fuels for heating. In terms of promoting the City of Sarajevo towards a smart city, the integration of solar panels on 35,000 households with an installed capacity of 6 kW is analyzed from an economic and technical point of view. In this way, households are heated with electricity, and excess electricity is delivered to the grid. From the economic aspect, this investment pays off in 10 years. From the technical aspect, this integration of electricity obtained from renewable sources has no significant impact on the transmission network, and it is noticeable that the voltage conditions at the busbars remain within acceptable limits. PV systems can often be located near the consumer and thus decrease transmission losses by placing solar panels on any surface that is exposed to sunlight. So, PV systems can be used to mitigate peak demand and hold down costs for current energy suppliers. It is very important to emphasize that the use of solar energy for domestic heating in city of Sarajevo, promotes this city to smart because it reduces pollution due to non-use of fossil fuels, reduces external costs caused by air pollution and there are large savings in terms of no purchase of fossil fuels for household heating.

References

1. Winden, W., et al.: Organising Smart City Projects: Lessons from Amsterdam (2016). https://drive.google.com/file/d/0Bz0U6OArm0dId0labDJRRDhHUjA/view. Accessed 12 Mar 2021
2. Ferrer, J.R.: 'Barcelona's smart city vision: an opportunity for transformation. Field Actions Sci. Rep. **2017**, 70–75 (2017)
3. Taewoo, N., Heungsuk, C.: Platform governance for smart cities: cases of Barcelona, Stockholm, and Copenhagen. J. Korean Urban Manag. Assoc. (2019)
4. Brandl, J., Zielinska, I.: Reviewing the smart city Vienna framework strategy's potential as an eco-social policy in the context of quality of work and socio-ecological transformation. Sustainability **12**, 859 (2020). https://doi.org/10.3390/su12030859
5. Honarvar, A., Sami, A.: Multi-source dataset for urban computing in a smart city. Data Brief **22**, 113 (2018). https://doi.org/10.1016/j.dib.2018.09.113
6. Allam, Z., Newman, P.: Redefining the smart city: culture, metabolism and governance. Smart Cities **1**, 4–25 (2018). https://doi.org/10.3390/smartcities1010002
7. Marin, B., Nikola, K.: Performance analysis of a 50 Kw Pv Plant in Tomislavgrad. In: Annals of DAAAM and Proceedings, p. 868+. Gale Academic OneFile (2020). Accessed 14 Mar 2021
8. Lobaccaro, G., et al.: A cross-country perspective on solar energy in urban planning: lessons learned from international case studies. Renew. Sustain. Energy Rev. **108**, 209–237. ISSN: 1364–0321 (2019). https://doi.org/10.1016/j.rser.2019.03.041
9. Emrah, K., Pranpreya, S.: Barriers to the adoption of photovoltaic systems: the state of the art. Renew. Sustain. Energy Rev. **49**, 58 (2015). https://doi.org/10.1016/j.rser.2015.04.058
10. Dedović, M.M., Avdaković, S., Mujezinović, A., Dautbašić, N.: Integration of PV into the Sarajevo canton energy system-air quality and heating challenges. Energies **14**, 123 (2021). https://doi.org/10.3390/en14010123
11. Djikic, B., Hadziahmetovic, H., Cerimovic, H., Ahmovic, I., Blazevic, R.: Analysis of solar energy potential and optimal panel tilt angle in the area of the city of Sarajevo. In: Katalinic, B. (ed.) Proceedings of the 30th DAAAM International Symposium, pp.0232–0239, Published by DAAAM International, Vienna, Austria. ISBN: 978–3–902734–22–8, ISSN: 1726–9679 (2019). https://doi.org/10.2507/30th.daaam.proceedings.030
12. Usanovic, E., et al.: Techno-economic analysis of solar photovoltaic power plant along national highway corridors. In: Advanced Technologies, Systems, and Applications. LNCS, Springer, New York (2021, in progress)

Author Index

N. Ademović et al. (Eds.): IAT 2021, LNNS 316, pp. 799–801, 2022.
https://doi.org/10.1007/978-3-030-90055-7